Random Graphs and Complex Networks

Volume 2

Complex networks are key to describing the connected nature of the society that we live in. This book, the second of two volumes, describes the local structure of random graph models for real-world networks and determines when these models have a giant component and when they are small-, and ultra-small, worlds.

This is the first book to cover the theory and implications of local convergence, a crucial technique in the analysis of sparse random graphs. Suitable as a resource for researchers and PhD-level courses, it uses examples of real-world networks, such as the Internet and citation networks, as motivation for the models that are discussed, and includes exercises at the end of each chapter to develop intuition. The book closes with an extensive discussion of related models and problems that demonstrate modern approaches to network theory, such as community structure and directed models.

REMCO VAN DER HOFSTAD is Full Professor of Probability at Eindhoven University of Technology. He received the 2003 Prix Henri Poincaré (jointly with Gordon Slade) and the 2007 Rollo Davidson Prize, and he is a laureate of the 2003 Innovative Research VIDI Scheme and the 2008 Innovative Research VICI Scheme. He served as scientific director at Eurandom from 2011 to 2019, and is a member of the Dutch Royal Academy of Science. Van der Hofstad is one of the principal investigators of the prestigious NETWORKS Gravitation Program and is the creator of the interactive website networkpages.nl, aimed at a broad audience.

CAMBRIDGE SERIES IN STATISTICAL AND PROBABILISTIC MATHEMATICS

This series of high-quality upper-division textbooks and expository monographs covers all aspects of stochastic applicable mathematics. The topics range from pure and applied statistics to probability theory, operations research, optimization, and mathematical programming. The books contain clear presentations of new developments in the field and also of the state of the art in classical methods. While emphasizing rigorous treatment of theoretical methods, the books also contain applications and discussions of new techniques made possible by advances in computational practice.

A complete list of books in the series can be found at www.cambridge.org/statistics. Recent titles include the following:

Random Graphs and Complex Networks

Volume 2

Remco van der Hofstad

Technische Universiteit Eindhoven

CAMBRIDGE
UNIVERSITY PRESS

Shaftesbury Road, Cambridge CB2 8EA, United Kingdom

One Liberty Plaza, 20th Floor, New York, NY 10006, USA

477 Williamstown Road, Port Melbourne, VIC 3207, Australia

314–321, 3rd Floor, Plot 3, Splendor Forum, Jasola District Centre, New Delhi – 110025, India

103 Penang Road, #05–06/07, Visioncrest Commercial, Singapore 238467

Cambridge University Press is part of Cambridge University Press & Assessment,
a department of the University of Cambridge.

We share the University's mission to contribute to society through the pursuit of
education, learning and research at the highest international levels of excellence.

www.cambridge.org
Information on this title: www.cambridge.org/9781107174009

DOI: 10.1017/9781316795552

First published 2024

A catalogue record for this publication is available from the British Library

A Cataloging-in-Publication data record for this book is available from the Library of Congress

ISBN 978-1-107-17400-9 Hardback

Aan Mad, Max en Lars
het licht in mijn leven

Ter nagedachtenis aan mijn ouders
die me altijd aangemoedigd hebben

CONTENTS

PREFACE

Targets. In this book, which is Volume 2 of a sequence of two books, we study *local limits*, *connected components*, and *small-world properties* of random graph models for complex networks. Volume 1 describes the preliminaries of *random graphs* as models for *real-world networks*, as investigated since 1999. These networks turned out to be rather different from classical random graph models, for example in the number of connections that the elements make. As a result, a wealth of new models was invented to capture these properties. Volume 1 studies these models as well as their *degree structure*. Volume 2 summarizes the insights developed in this exciting period related to the *local, connectivity*, and *small-world structure* of the proposed random graph models. While Volume 1 is intended to be used for a master level course, where students have a limited prior knowledge of special topics in probability, Volume 2 describes the more involved notions that have been the focus of attention of the research community in the past two decades.

Volume 2 is intended to be used for a PhD level course, a reading seminar, or for researchers wishing to obtain a consistent and extended overview of the results and methodologies developed in this scientific area. Volume 1 includes many of the preliminaries, such as the convergence of random variables, probabilistic bounds, coupling, martingales, and branching processes, and we frequently rely on these results.

The sequence of Volumes 1 and 2 aims to be self-contained. In Volume 2, we briefly repeat some of the preliminaries on random graphs, including an introduction to the key models and their degree distributions, as discussed in detail in Volume 1. In Volume 2, we aim to give detailed and complete proofs. When we do not give proofs, we provide heuristics, as well as extensive pointers to the literature. We further discuss several more recent random graph models that aim to more realistically model real-world networks, as they incorporate their *directed* nature, their *community structure*, and/or their *spatial embedding*.

Developments. The field of random graphs was pioneered in 1959–1960 by Erdős and Rényi (1959; 1960; 1961a; 1961b), in the context of the *probabilistic method*. The initial work by Erdős and Rényi incited a great amount of follow-up in the field, initially mainly in the combinatorics community. See the standard references on the subject by Bollobás (2001) and Janson, Łuczak, and Ruciński (2000) for the state of the art. Erdős and Rényi (1960) gives a rather complete picture of the various phase transitions that occur in the Erdős–Rényi random graph. This initial work did not aim to model real-world networks realistically.

In the period after 1999, owing to the fact that data sets of large real-world networks became abundantly available, their structure has attracted enormous attention in mathematics as well as in various applied domains. This is exemplified by the fact that one of the first articles in the field, by Barabási and Albert (1999), has attracted over 40,000 citations. One of the main conclusions from this overwhelming body of work is that many real-world networks share two fundamental properties. The first is that they are highly *inhomogeneous*, in the sense that different vertices play rather different roles in the networks. This property is exemplified by the degree structure of the real-world networks obeying power laws: these networks are *scale-free*. This scale-free nature of real-world networks has prompted the community to come up with many novel random graph models that, unlike the Erdős–Rényi random graph, do have power-law degree sequences. This was the key focus in Volume 1.

Content. In this book, we pick up on the trail left in Volume 1, where we now focus on the *connectivity structure* between vertices. Connectivity can be summarized in two key aspects of real-world networks: the facts that they are *highly connected*, as exemplified by the fact that they tend to have one giant component containing a large proportion of the vertices (if not all of them), and that they are *small world*, in that most pairs of vertices are separated by short paths. We discuss the available methods for these proofs, including path-counting techniques, branching-process approximations, exchangeable random variables, and de Finetti's theorems. We pay particular attention to a recent technique, called *local convergence*, that makes the statement that random graphs "locally look like trees" precise.

This book consists of four parts. In Part I, consisting of Chapters 1 and 2, we start in Chapter 1 by repeating some definitions from Volume 1, including the random graph models studied in the present book, which are inhomogeneous random graphs, configuration models, and preferential attachment models. We also discuss general topics that are important in random graph theory, such as power-law distributions and their properties. In Chapter 2, we continue by discussing *local convergence*, an extremely powerful technique that plays a central role in the theory of random graphs and in this book. In Part II, consisting of Chapters 3–5, we discuss local limits and large connected components in random graph models. In Chapter 3, we further extend the definition of the generalized random graph to general inhomogeneous random graphs. In Chapter 4, we discuss the local limit and large connected components in the configuration model, and in Chapter 5, we discuss the local structure in, and connectivity of, preferential attachment models. In Part III, consisting of Chapters 6–8, we study the small-world nature of random graphs, starting with inhomogeneous random graphs, continuing with the configuration model, and ending with the preferential attachment model. In Part IV, consisting of Chapter 9, we study related random graph models and their structure.

Along the way, we give many exercises that should help the reader to obtain a deeper understanding of the material by working on the solutions. These exercises appear in the last section of each of the chapters, and, when applicable, we refer to them at the appropriate place in the text. We also provide extensive notes in the penultimate section of each chapter, where we discuss the links to the literature and some extensions.

Literature. We have tried to give as many references to the literature as possible. However, the number of papers on random graphs has exploded. In MathSciNet (see www.ams.org/mathscinet), there were, on December 21, 2006, a total of 1,428 papers that contain the phrase "random graphs" in the review text; on September 29, 2008, this number had increased to 1,614, to 2,346 on April 9, 2013; to 2,986 on April 21, 2016; and to 12,038 on October 5, 2020. These are merely the papers on the topic in the mathematics community. What is special about random graph theory is that it is extremely multidisciplinary, and many papers using random graphs are currently written in economics, biology, theoretical physics, and computer science. For example, in Scopus (see www.scopus.com/scopus/home.url), again on December 21, 2006, there were 5,403 papers that contain the phrase "random graph" in the title, abstract or keywords; on September 29, 2008, this had increased to 7,928; to 13,987 on April 9, 2013; to 19,841 on April 21, 2016; and to 30,251 on October

5, 2020. It can be expected that these numbers will continue to increase, rendering it utterly impossible to review all the literature.

In June 2014, we decided to split the preliminary version of this book up into two books. This has several reasons and advantages, particularly since Volume 2 is more tuned towards a research audience, while Volume 1 is aimed at an audience of master students of varying backgrounds. The pdf-versions of both Volumes 1 and 2 can be obtained from

www.win.tue.nl/~ rhofstad/NotesRGCN.html.

For errata for this book and Volume 1, or possible outlines for courses based on them, readers are encouraged to look at this website or e-mail me. Also, for a more playful approach to networks for a broad audience, including articles, videos, and demos of many of the models treated in this book, we refer all readers to the NetworkPages at www.networkspages.nl. The NetworkPages provide an interactive website developed by and for all those who are interested in networks. Finally, we have relied on various real-world networks data sets provided by the KONECT project; see http://konect.cc as well as Kunegis (2013) for more details.

Thanks. This book, as well as Volume 1, would not have been possible without the help and encouragement of many people. I particularly thank Gerard Hooghiemstra for encouraging me to write it, and for using it at Delft University of Technology almost simultaneously while I was using it at Eindhoven University of Technology in the spring of 2006 and again in the fall of 2008. I thank Gerard for many useful comments, solutions to exercises, and suggestions for improvements of the presentation throughout the book. Together with Piet Van Mieghem, we entered the world of random graphs in 2001, and I have tremendously enjoyed exploring this field together with them, as well as with Henri van den Esker, Dmitri Znamenski, Mia Deijfen, Shankar Bhamidi, Johan van Leeuwaarden, Júlia Komjáthy, Nelly Litvak and many others.

I thank Christian Borgs, Jennifer Chayes, Gordon Slade, and Joel Spencer for joint work on random graphs that are like the Erdős–Rényi random graph but do have geometry. Special thanks go to Gordon Slade, who introduced me to the exciting world of percolation, which is closely linked to the world of random graphs (see the classic text on percolation by Grimmett (1999)). It is striking to see two communities working on two such closely related topics with different methods and even different terminology, and it has taken a long time to build bridges between the two subjects. I am very happy that these bridges are now rapidly appearing, and the level of communication between different communities has increased significantly. I hope that this book helps to further enhance this communication. Frank den Hollander deserves a special mention. Frank, you have been important as a driving force throughout my career, and I am very happy now to be working with you on fascinating random graph problems!

Further, I thank

Marie Albenque, Yeganeh Alimohammadi, Rangel Baldasso, Gianmarco Bet,
Shankar Bhamidi, Finbar Bogerd, Marko Boon, Christian Borgs, Hao Can,
Francesco Caravenna, Rui Castro, Kota Chisaki, Deen Colenbrander, Nicolas Curien,
Umberto De Ambroggio, Mia Deijfen, Michel Dekking, Serte Donderwinkel,
Dylan Dronnier, Henri van den Esker, Lorenzo Federico, Federica Finazzi, Allison Fisher,

Lucas Gerin, Cristian Giardinà, Claudia Giberti, Jesse Goodman, Rowel Gündlach,
Rajat Hazra, Markus Heydenreich, Frank den Hollander, Yusuke Ide, Simon Irons,
Emmanuel Jacob, Svante Janson, Guido Janssen, Lancelot James, Martin van Jole,
Joost Jorritsma, Willemien Kets, Heejune Kim, Bas Kleijn, Júlia Komjáthy, Norio Konno,
Dima Krioukov, John Lapeyre, Lasse Leskelä, Nelly Litvak, Neeladri Maitra,
Abbas Mehrabian, Marta Milewska, Steven Miltenburg, Mislav Mišković, Christian Mönch,
Peter Mörters, Mirko Moscatelli, Jan Nagel, Sidharthan Nair, Alex Olssen,
Mariana Olvera-Cravioto, Helena Peña, Manish Pandey, Rounak Ray, Nathan Ross,
Christoph Schumacher, Matteo Sfragara, Karoly Simon, Lars Smolders, Clara Stegehuis,
Dominik Tomecki, Nicola Turchi, Viktória Vadon, Thomas Vallier, Irène Ayuso Ventura,
Xiaotin Yu, Haodong Zhu and Bert Zwart

for remarks and ideas that have improved the content and presentation of these books substantially. Wouter Kager read the February 2007 version of this book in its entirety, giving many ideas for improvements in the arguments and the methodology. Artëm Sapozhnikov, Maren Eckhoff, and Gerard Hooghiemstra read and commented on the October 2011 version. Haodong Zhu read the December 2023 version completely, and corrected several typos.

Particular thanks go to Dennis Timmers, Eefje van den Dungen, Joop van de Pol, Rowel Gündlach and Lourens Touwen, who, as, my student assistants, have been a great help in the development of this pair of books, in making figures, providing solutions to some of the exercises, checking proofs, and keeping the references up to date. Maren Eckhoff also provided many solutions to the exercises, for which I am grateful! Sándor Kolumbán, Robert Fitzner, and Lourens Touwen helped me to turn all pictures of real-world networks as well as simulations of network models into a unified style, a feat that is beyond my LaTeX skills. A big thanks for that! Also my thanks for suggestions and help with figures to Marko Boon, Alessandro Garavaglia, Dimitri Krioukov, Vincent Kusters, Clara Stegehuis, Piet Van Mieghem, and Yana Volkovich. A special thanks to my running mates Jan and Ruud, whose continuing support has been extremely helpful for me.

Support. This work would not have been possible without the generous support of the Netherlands Organization for Scientific Research (NWO) through VIDI grant 639.032.304, VICI grant 639.033.806, and the Gravitation NETWORKS grant 024.002.003.

POSSIBLE COURSE OUTLINES

The relation between the chapters in Volumes 1 and 2 is as follows:

Here is some more explanation as well as a possible itinerary of a master or PhD course on random graphs, based on Volume 2, in a course outline. For a course outline based on Volume 1, we refer to [V1, Preface] for alternative routes through the material, we refer to the book's website at www.win.tue.nl/~ rhofstad/NotesRGCN.html:

▷ Start with the introduction to real-world networks in [V2, Chapter 1], which forms the inspiration for what follows. For readers wishing for a more substantial introduction, do visit Volume 1 for an extensive introduction to the models discussed here.

▷ Continue with [V2, Chapter 2] on the local convergence of (random and non-random) graphs, as this is a crucial tool in the book and has developed into a key methodology in the field.

The material in this book is rather substantial, and probably too much to be treated in one course. Thus, we give two alternative approaches to teaching coherent parts of this book:

▷ You can either take one of the *models* and discuss the different chapters in Volume 2 that focus on them. [V2, Chapters 3 and 6] discuss inhomogeneous random graphs, [V2, Chapters 4 and 7] discuss configuration models, while [V2, Chapters 5 and 8] focus on preferential attachment models.

▷ The alternative is that you take one of the *topics*, and work through them in detail. [V2, Part II] discusses the local limits and largest connected components or phase transition in our random graph models, while [V2, Part III] treats their small-world nature.

If you have further questions and/or suggestions about course outlines, feel free to contact me. Refer to www.win.tue.nl/~ rhofstad/NotesRGCN.html for further suggestions on how to lecture from Volume 2.

Part I

Preliminaries

INTRODUCTION AND PRELIMINARIES

Abstract

In this chapter, we draw motivation from real-world networks and formulate random graph models for them. We focus on some of the models that have received the most attention in the literature, namely, Erdős–Rényi random graphs, inhomogeneous random graphs, configuration models, and preferential attachment models. We follow van der Hofstad (2017), which we refer to as [V1], both for motivation and for the introduction to the random graph models involved.

Looking Back, and Ahead

In Volume 1 of this pair of books, we discussed various models having flexible degree sequences. The generalized random graph and the configuration model give us *static* flexible models for random graphs with various degree sequences. Because of their *dynamic* nature, preferential attachment models give us a convincing explanation of the abundance of power-law degree sequences in various applications. We will often refer to Volume 1. When we do so, we write [V1, Theorem 2.17] to signify that we refer to Theorem 2.17 in van der Hofstad (2017).

In [V1, Chapters 6–8], we focussed on the properties of the *degrees* of such graphs. However, we noted in [V1, Chapter 1] that not only do many real-world networks have degree sequences that are rather different from the ones of the Erdős–Rényi random graph, also many examples have a *giant connected component* and are *small worlds*.

In Chapters 3–8, we will return to the models discussed in [V1, Chapters 6–8], and focus on their local structure, and their connected components, as well as on their distance structure. Interestingly, a large chunk of the non-rigorous physics literature suggests that the behavior in various *different* random graph models can be described by only a *few* essential parameters. The key parameter of each of these models is the *power-law degree exponent*, and the physics literature predicts the behavior in random graph models with similar degree sequences to be similar. This is an example of the notion of *universality*, a central notion in statistical physics. Despite its importance, there are only a few examples of universality that can be rigorously proved. In Chapters 3–8, we investigate the level of universality present in random graph models.

Organization of this Chapter

This chapter is organized as follows. In Section 1.1 we discuss real-world networks and the inspiration that they provide. In Section 1.2, we then discuss how *graph sequences*, where the size of the involved graphs tends to infinity, aim at describing *large* complex networks. In Section 1.3 we recall the definition of several random graph models, as introduced in Volume 1. In Section 1.4, we discuss *power-law* random variables, as they play an important role in this book. In Section 1.5 we recall some of the standard notation and notions used in this book. We close this chapter with notes and discussion in Section 1.6 and with exercises in Section 1.7. We give few references to the literature within this chapter, but defer a discussion of the history of the various models to the extensive notes in Section 1.6.

1.1 MOTIVATION: REAL-WORLD NETWORKS

In the past two decades, an enormous research effort has been performed with regard to modeling various real-world phenomena using networks. Networks arise in various applications ranging from the connections between friends in friendship networks to the connectivity of neurons in the brain, to the relations between companies and countries in economics, and the hyperlinks between webpages in the World-Wide Web. The advent of the computer era has made many network data sets available. Around 1999–2000, various groups started to investigate network data from an empirical perspective. [V1, Chapter 1] gives many examples of real-world networks and the empirical findings from them. Here we give some basics.

1.1.1 GRAPHS AND NETWORKS

A graph $G = (V, E)$ consists of a collection $V = V(G)$ of vertices, also called a vertex set, and a collection of edges $E = E(G)$, often called an edge set. The vertices correspond to the objects that we model; the edges indicate some relation between pairs of these objects. In our settings, graphs are usually *undirected*. Thus, an edge is an unordered pair $\{u, v\} \in E$ indicating that u and v with $u, v \in V(G)$ are directly connected. When G is undirected, if u is directly connected to v then also v is directly connected to u. Therefore, an edge can be seen as a pair of vertices. When dealing with social networks, the vertices represent the individuals in the population while the edges represent the friendships among them. We sometimes work with *multi-graphs*, which are graphs possibly having *self-loops* or *multiple edges* between vertices, and we will clearly indicate when we do so.

We mainly deal with *finite* graphs and then, for simplicity, we often take $V = [n] := \{1, \ldots, n\}$. The *degree* $d_u^{(G)}$ of a vertex $u \in V(G)$ in the graph G is equal to the number of edges containing u, i.e.,

$$d_u^{(G)} = \#\{v \in V(G) \colon \{u, v\} \in E(G)\}. \tag{1.1.1}$$

Often, we deal with the degree of a *random vertex* in G. Let $o \in V(G)$ be a vertex chosen uniformly at random (uar) in $V(G)$. The *typical degree* is the random variable D_n given by

$$D_n = d_o^{(G)}. \tag{1.1.2}$$

It is not hard to see that the probability mass function of D_n is given by

$$\mathbb{P}(D_n = k) = \frac{1}{|V(G)|} \sum_{v \in V(G)} \mathbb{1}_{\{d_v^{(G)} = k\}}, \tag{1.1.3}$$

where, for a set A, we write $|A|$ for its size. Exercise 1.1 asks you to prove (1.1.3).

The *average degree* in a network is equal to

$$\frac{1}{|V(G)|} \sum_{v \in V(G)} d_v^{(G)} = \frac{2|E(G)|}{|V(G)|}. \tag{1.1.4}$$

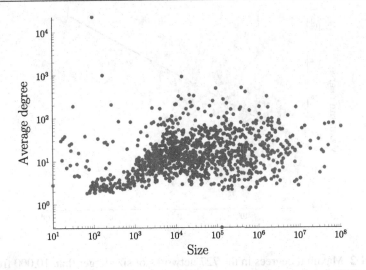

Figure 1.1 Average degrees in the 727 networks of size larger than 10,000 from the KONECT data base.

We can rewrite (1.1.4) as

$$\frac{1}{|V(G)|} \sum_{v \in V(G)} d_v^{(G)} = \mathbb{E}[D_n], \tag{1.1.5}$$

where the expectation is with respect to the random vertex o in $D_n = d_o^{(G)}$ (recall (1.1.2)). The average degree can take any value in between 0 for an empty graph, and $|V(G)| - 1$ for a complete graph. In reality, however, we see that the average degree of many real-world networks is not very large, i.e., these networks tend to be *sparse*. Figure 1.1 shows the average degrees in the KONECT data base, and we see that the average degree does not seem to grow with the network size.

We next discuss some common features that many real-world networks turn out to have.

1.1.2 SCALE-FREE PHENOMENON

The first, maybe quite surprising, fundamental property of many real-world networks is that the number of vertices with degree at least k decays slowly for large k. This implies that degrees are highly variable and that, even though the average degree is not particularly large, there exist vertices with extremely high degree. Often, the tail of the empirical degree distribution seems to fall off as an inverse power of k. This is called a "power-law degree sequence," and the resulting graphs often go under the name "scale-free graphs." This is visualized for the Autonomous Systems (AS) graph from the Internet in Figure 1.5(a), where the degree distribution of the AS graph is plotted on a log–log scale. Thus, we see a plot of $\log k \mapsto \log n_k$, where n_k is the number of vertices with degree k. When n_k is proportional to an inverse power of k, i.e., when, for some normalizing constant c_n and exponent τ,

$$n_k \approx c_n k^{-\tau}, \tag{1.1.6}$$

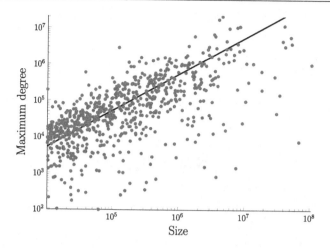

Figure 1.2 Maximal degrees in the 727 networks of size larger than 10,000 from the KONECT data base. Linear regression gives $\log d_{\max} = 0.742 + 0.519 \log n$.

and thus

$$\log n_k \approx \log c_n - \tau \log k, \tag{1.1.7}$$

so that the plot of $\log k \mapsto \log n_k$ is close to a *straight line*. This is the reason why degree sequences in networks are often depicted in a log–log fashion, rather than in the more customary form of $k \mapsto n_k$. Here, and in the remainder of this section, we write $\approx$ to denote an uncontrolled approximation. The power-law exponent τ can be estimated by the absolute value of the slope of the line in the log–log plot. Naturally, we must have that

$$\sum_k n_k = |V(G_n)| < \infty, \tag{1.1.8}$$

so that it is reasonable to assume that $\tau > 1$. In fact, many networks are *sparse*, meaning that their average $\sum_k k n_k / |V(G_n)|$ remains uniformly bounded, which in turn suggests that $\tau > 2$ is to be expected. See Figure 1.2 for the maximal degrees in the KONECT data base in log–log scale, which should be compared with Figure 1.1. While there does not seem to be a trend in Figure 1.1, there does seem to be one in Figure 1.2; this indicates that the log of the maximal degree tends to grow linearly with the log of the network size. The latter is consistent with power-law degrees.

Let us define the *degree distribution* by $p_k^{(G_n)} = n_k / |V(G_n)| = \mathbb{P}(D_n = k)$ (recall (1.1.2) and (1.1.3)), so that $p_k^{(G_n)}$ equals the probability that a *uniformly chosen vertex* in a graph G_n with n vertices has degree k (recall (1.1.3)). Then (1.1.6) can be reexpressed as

$$p_k^{(G_n)} \approx c k^{-\tau}, \tag{1.1.9}$$

where again $\approx$ denotes an uncontrolled approximation.

Vertices with extremely high degrees go under various names, indicating their importance in the field. They are often called *hubs*, like the hubs in airport networks. Another name for them is *super-spreader*, indicating the importance of the high-degree vertices in spreading information or diseases. The hubs quantify the level of inhomogeneity in the real-world

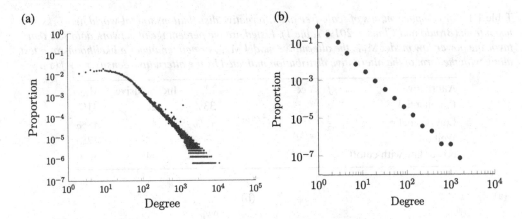

Figure 1.3 (a) Log–log plot of the degree sequence in the 2007 Internet Movie Data base. (b) Log–log plot of the probability mass function of the Autonomous Systems degree sequence on April 2014, on a log–log scale from Krioukov et al. (2012) (data courtesy of Dmitri Krioukov). This degree distribution looks smoother than others (see e.g., Figure 1.3(a) and 1.4), due to binning of the data.

networks, and a large part of this book is centered around rigorously establishing the effect that the high-degree vertices have on various properties of the graphs involved.

Further, a central topic in network science is how the behavior of stochastic processes on networks is affected by degree inhomogeneities. Such effects are especially significant when the networks are "scale-free," meaning that they can be well approximated by power laws with exponents τ satisfying $\tau \in (2,3)$, so that random variables with such degrees have *infinite variance*. Since maximal degrees of networks of size n can be expected to grow as $n^{1/(\tau-1)}$ (see Exercise 1.2 for an illuminating example), Figure 1.2 suggests that, on average, $1/(\tau - 1) \approx 0.519$, so that, again on average, $\tau \approx 2.93$, which is in line with such predictions.

For the Internet, log–log plots of degree sequences first appeared in a paper by the Faloutsos brothers (1999) (see Figure 1.3(b) for the degree sequence in the Autonomous Systems graph, where the degree distribution looks relatively smooth because it is binned). Here, the power-law exponent is estimated as $\tau \approx 2.15$–2.20. Figure 1.3(a) displays the degree distribution in the Internet Movie Data base (IMDb), in which the vertices are actors and two actors are connected when they have acted together in a movie. Figure 1.4 displays the degree-sequence for both the in- as well as the out-degrees in various World-Wide Web data bases.

Recent Discussion on Power-Law Degrees in Real-World Networks

Recently, a vigorous discussion has emerged on how often real-world networks have power-law degree distributions. This discussion was spurred by Broido and Clauset (2019), who claimed (even as the title of their paper) that

Scale-free networks are rare.

What did they do to reach this conclusion? Broido and Clauset (2019) performed the first extensive analysis of a large number of real-world network data sets, and compared degree

Table 1.1 *For comparison, fits of scale-free and alternative distributions to real-world networks taken from (Broido and Clauset, 2019, Table 1). Listed are the percentage of network data sets that favor the power-law model* $\mathrm{M_{PL}}$*, the alternative model* $\mathrm{M_{Alt}}$*, or neither, under a likelihood-ratio test, along with the form of the alternative distribution indicated by the alternative density* $x \mapsto f(x)$*.*

Alternative	$f(x) \propto$	$\mathrm{M_{PL}}$	Inconclusive	$\mathrm{M_{Alt}}$
Exponential	$e^{-\lambda x}$	33%	26%	41%
Log-normal	$\frac{1}{x}e^{-(\log x - \mu)^2/(2\sigma^2)}$	12%	40%	48%
Weibull	$e^{-(x/b)^a}$	33%	20%	47%
Power law with cutoff	$x^{-\tau}e^{-Ax}$	–	44%	56%

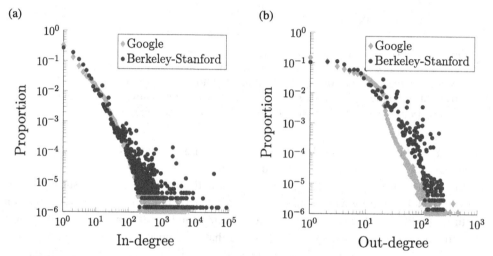

Figure 1.4 The probability mass function of the in- and out-degree sequences in the Berkeley-Stanford and Google competition graph data sets of the World Wide Web in Leskovec et al. (2009). (a) In-degree; (b) out-degree.

sequences of these real-world networks with power-law, as well as with log-normal, exponential and Weibull distributions. They also made comparisons with power-law distributions having exponential truncation. The main conclusion of Broido and Clauset (2019) was that, in many cases, alternative distributions are preferred over power laws (see also Table 1.1).

Clearly this work caused quite a stir, as the conclusion, if correct, would make about 20 years of network science close to redundant from a practical perspective. Barabási (2018) wrote a blog post containing detailed criticism of the methods and results in Broido and Clauset (2019), see also Voitalov et al. (2019). Holme (2019) summarized the status of the arguments in 2019, reaching an almost philosophical conclusion:

Still, it often feels like the topic of scale-free networks transcends science – debating them probably has some dimension of collective soul searching as our field slowly gravitates toward data science, away from complexity science.

So, what did the discussion focus on? Here is a list of questions:

What are power-law data? An important question in the discussion on power-law degree distributions is how to interpret the approximation sign in (1.1.9). Most approaches start by assuming that the data are realizations of *independent and identically distributed* (iid) random variables. This can only be an assumption, as degree distributions are mostly *graphical* (meaning that they can arise as degree sequences of graphs without self-loops and multiple edges), which introduces dependencies between them (if only because the sum of the degrees needs to be even). However, without this assumption, virtually any analysis becomes impossible, so let us assume this as well.

Under the above assumption, one needs to infer the degree distribution from the sample of degrees obtained from a real-world network. We denote the asymptotic degree distribution by p_k, i.e., the proportion of vertices of degree k in the *infinite-graph limit*. Under this assumption, $p_k^{(G_n)}$ in (1.1.9) is the *empirical probability mass function* corresponding to the true underlying degree distribution $(p_k)_{k \geq 0}$. The question is thus what probability mass functions $(p_k)_{k \geq 0}$ correspond to a power law.

Broido and Clauset (2019) interpreted the power-law assumption as

$$p_k = ck^{-\tau} \qquad \text{for all} \quad k \geq k_{\min}, \tag{1.1.10}$$

and p_k arbitrary for $k \in [k_{\min} - 1]$; here $c > 0$ is chosen appropriately. The inclusion of the $k_{\min}$ parameter is based on the observation that small values of k generally do not satisfy the pure power law (see also Clauset et al. (2009), where (1.1.10) first appeared).

Barabási (2018) instead argued from the perspective of *generative models* (such as the preferential attachment models described in Section 1.3.5, as well as in Chapters 5 and 8):

> *In other words, by 2001 it was pretty clear that there is no one-size-fits-all formula for the degree distribution for networks driven by the scale-free mechanism. A pure power law only emerges in simple idealised models, driven by only growth and preferential attachment, and free of any additional effects.*

Bear in mind that this dynamical approach is very different from that of Broido and Clauset (2019), as the degrees in generative models can hardly be expected to be realizations of an iid sample! Barabási (2018) instead advocated a theory that predicts power laws with exponential truncation for many settings, meaning that

$$p_k = ck^{-\tau} e^{-Ak} \qquad \text{for all} \quad k \geq d_{\min}, \tag{1.1.11}$$

where $d_{\min}$ denotes the minimal degree in the graph and $c, A > 0$ are appropriate constants, but the theory also allows for "additional effects," such as vertex fitnesses that describe intrinsic differences in how likely it is to connect to vertices, and that may be realistic in some real-world networks.

Voitalov et al. (2019) took a static approach related to that of Broido and Clauset (2019), but instead assumed more general power laws of the form

$$1 - F(x) = \sum_{k > x} p_k = x^{-(\tau - 1)} L(x) \qquad \text{for all} \quad x \geq 1, \tag{1.1.12}$$

where $x \mapsto L(x)$ is a so-called *slowly varying function*, meaning a function that does not change the power-law exponent, in that it grows or decays more slowly than any power at infinity. See [V1, Definition 1.5], or Definition 1.19 below, for a precise definition. In particular, distributions that satisfy (1.1.10) also satisfy (1.1.12), but not necessarily the other way around.

The advantage of working with (1.1.12) is that this definition is quite general, yet a large body of work within the *extreme-value statistics* community becomes available. These results, as summarized in Voitalov et al. (2019), allow for the "most accurate" ways of estimating the power-law exponent τ, which brings us to the next question.

How to estimate the power-law exponent? Since Broido and Clauset (2019) interpreted the power-law assumption as in (1.1.10), estimating the model parameters then boiled down to estimating $k_{\min}$ and τ. For this, Broido and Clauset (2019) relied on the first paper on estimating power-law exponents in the area of networks, by Clauset et al. (2009), who proposed the *power-law-fit method* (PLFIT). This method chooses the best possible $k_{\min}$ on the basis of the difference between the empirical degree distribution for values above $k_{\min}$ and the power-law distribution function based on (1.1.10) with an appropriately estimated value $\hat{\tau}$ of τ, as proposed by Hill (1975), for realizations above $k_{\min}$.

The estimator $\hat{\tau}_{\mathrm{PLFit}}$ is then the estimator of τ corresponding to the optimal $k_{\min}$. The PLFIT method was recently proved to be a *consistent* method by Bhattacharya et al. (2020), which means that the estimator will, in the limit, converge in probability to the correct value τ, even under the weaker assumption in (1.1.12). Of course, the question remains whether $\hat{\tau}_{\mathrm{PLFit}}$ is a good estimator, for example in the sense that the rate of convergence of $\hat{\tau}_{\mathrm{PLFit}}$ to τ is optimal. The results and simulations in Drees et al. (2020) suggest that, even in the case of a pure power law as in (1.1.10) with $k_{\min} = 1$, $\hat{\tau}_{\mathrm{PLFit}}$ is outperformed by more classical estimators (such as the maximum likelihood estimator for τ). Voitalov et al. (2019) rely on the estimators proposed in the extreme-value literature; see e.g. Danielsson et al. (2001); Draisma et al. (1999); Hall and Welsh (1984) for such methods and Resnick (2007); Beirlant et al. (2006) for extensive overviews of extreme-value statistics.

The dynamical approach by Barabási (2018) instead focusses on estimating the parameters in the proposed dynamical models, a highly interesting topic that is beyond the scope of this book.

How to perform tests? When confronted with a model, or with two competing models such as in Table 1.1, a statistician would often like to compare the *fit* of these models to the data, so as to be able to choose between them. When both models are *parametric*, meaning that they involve a finite number of parameters, like the models in Table 1.1, this can be done using a so-called *likelihood-ratio test*. For this, one computes the likelihood of the data (basically the probability that the model in question gives rise to exactly what was found in the data) for each of the models, and then takes the ratio of the two likelihoods. In the settings in Table 1.1, this means that the likelihood of the data for the power-law model is divided by that for the alternative model. When this exceeds a certain threshold, the test does not reject the possibility that the data comes from a power law, otherwise it rejects the null hypothesis of a power-law degree distribution. This is done for each of the networks in the data base, and Table 1.1 indicates the percentages for which each of the models is deemed the most likely.

Unfortunately, such likelihood ratio tests can be performed only when one compares *parametric* settings. The setting in (1.1.12) is *non-parametric*, as it involves the unknown slowly varying function $x \mapsto L(x)$, and thus, in that setting, no statistical test can be performed unless one makes parametric assumptions on the shape of $x \mapsto L(x)$ (by assuming, for example, that $L(x)$ is a power of $\log x$). Thus, the parametric choice in (1.1.10) is crucial in that it allows for a testing procedure to be performed. Alternatively, if one does not believe in the "pure" power-law form as in (1.1.10), then tests are no longer feasible. What approach should one then follow? See Artico et al. (2020) for a related testing procedure, in which the authors reached a rather different conclusion than that of Broido and Clauset (2019).

How to partition networks? Broido and Clauset (2019) investigated a large body of networks, relying on a data base consisting of 927 real-world networks from the KONECT project; see http://konect.cc as well as Kunegis (2013). We are also relying on this data base for graphs showing network properties, such as average and maximal degrees, etc. These networks vary in size, as well as in their properties (directed versus undirected, static versus temporal, etc.). In their paper, Broido and Clauset (2019) report percentages of networks having certain properties; see for example Table 1.1.

A substantial part of the discussion around Broido and Clauset (2019) focusses on whether these percentages are representative. Take the example of a directed network, which has several degree distributions, namely, in-degree, out-degree, and total degree distributions (in the latter, the directions are simply ignored). This "diversity of degree distributions" becomes even more pronounced when the network is *temporal*, meaning that edges come and go as time progresses. When does one say that a temporal network has a power-law degree distribution? When one of these degree distributions is classified as power-law, when a certain percentage of them is, or when all of them are?

What is our approach in this book? We prefer to avoid the precise debate about whether power laws in degree distributions are omnipresent or rare. We view power laws as a way to *model* settings where there is a large amount of variability in the data, and where the maximum values of the degrees are several orders of magnitude larger than the average values (compare Figures 1.1 and 1.2). Power laws predict such differences in scale.

There is little debate about the fact that degree distributions in networks tend to be highly inhomogeneous. Power laws are the model of choice to model such inhomogeneities, certainly in settings where empirical moments (for example, empirical variances) are very large. Further, inhomogeneities lead to interesting *differences in structure* of the networks in question, which will be a focal point of this book. All the alternative models in Table 1.1 have tails that are too thin for such differences to emerge. Thus, it is natural to focus on models with power-law degrees to highlight the relation between degree structure and network topology. Therefore, we often consider degree distributions that are either *exactly* described by power laws or are bounded above or below by them. The focus then resides in how the degree power-law exponent τ changes the network topology.

After this extensive discussion of degrees in graphs, we continue by discussing *graph distances* and their relation to *small-world phenomena*, a topic that is much less heatedly debated.

1.1.3 SMALL-WORLD PHENOMENON

A second fundamental network property observed in many real-world networks is the fact that typical distances between vertices are small. This is called the "small-world" phenomenon (see, e.g., the book by Watts (1999)). In particular, such networks are highly connected: their largest connected component contains a significant proportion of the vertices. Many networks, such as the Internet, even consist of *one* connected component, since otherwise e-mail messages could not be delivered between pairs of vertices in distinct connected components.

Graph distances between pairs of vertices tend to be quite small in most networks. For example, in the Internet, IP packets cannot use more than a threshold of physical links, and if distances in the Internet were larger than this threshold then the e-mail service would simply break down. Thus, the Internet graph has evolved in such a way that typical distances are relatively small, even though the Internet itself is rather large. As seen in Figure 1.5(a), the number of Autonomous Systems (ASs) traversed by an e-mail data set, sometimes referred to as the AS-count, is typically at most 7. In Figure 1.5(b), the proportion of routers traversed by an e-mail message between two uniformly chosen routers, referred to as the *hopcount*, is shown. It shows that the number of routers traversed is at most 27. Figure 1.6 shows typical distances in the IMDb; the distances are quite small despite the fact that the network contains more than one million vertices.

The small-world nature of real-world networks is highly significant. Indeed, in small worlds, news can spread quickly as relatively few people are needed to spread it between two typical individuals. This is quite helpful in the Internet, where e-mail messages hop along the edges of the network. At the other side of the spectrum, it also implies that infectious diseases can spread quite quickly, as just a few infections can carry the disease to a large part of the population. This implies that diseases have a large potential of becoming pandemic, as the corona pandemic has made painfully clear.

Let us continue this discussion by formally introducing graph distances, as displayed in Figures 1.5 and 1.6. For a graph $G = (V(G), E(G))$ and a pair of vertices $u, v \in V(G)$,

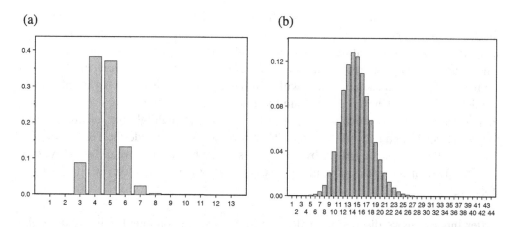

(a)   (b)

Figure 1.5 (a) Number of Autonomous Systems traversed in hopcount data. (b) Internet hopcount data (courtesy of Hongsuda Tangmunarunkit).

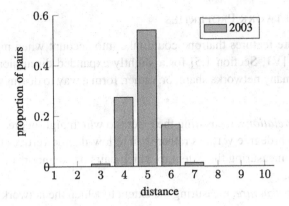

Figure 1.6 Typical distances in the Internet Movie Data base (IMDb) in 2003.

we let the graph distance $\mathrm{dist}_G(u,v)$ between u and v be equal to the minimal number of edges in a path linking u and v. When u and v are not in the same connected component, we set $\mathrm{dist}_G(u,v) = \infty$. We are interested in settings where G has a high amount of connectivity, so that many pairs of vertices are connected to one another by short paths. In order to describe the typical distances between vertices, we draw o_1 and o_2 independently and uar from $V(G)$, and we investigate the random variable

$$\mathrm{dist}_G(o_1, o_2). \tag{1.1.13}$$

The quantity in (1.1.13) is a random variable even for *deterministic* graphs, owing to the presence of the two uar-chosen vertices $o_1, o_2 \in V(G)$. Figures 1.5 and 1.6 display the probability mass functions of this random variable for some real-world networks.

Often, we consider $\mathrm{dist}_G(o_1, o_2)$ conditional on $\mathrm{dist}_G(o_1, o_2) < \infty$. This means that we consider the typical number of edges between a uniformly chosen pair of *connected* vertices. As a result, $\mathrm{dist}_G(o_1, o_2)$ is sometimes referred to as the *typical distance*.

The nice property of $\mathrm{dist}_G(o_1, o_2)$ is that its distribution tells us something about *all possible* distances in the graph. An alternative and frequently used measure of distance in a graph is the *diameter* of the graph G, defined as

$$\mathrm{diam}(G) = \max_{u,v \in V(G)} \mathrm{dist}_G(u,v). \tag{1.1.14}$$

However, the diameter has several disadvantages. First, in many instances, the diameter is algorithmically more difficult to compute than the typical distances (since one has to compute the distances between *all* pairs of vertices and maximize over them). Second, it is a *number* instead of a *distribution of a random variable*, and therefore contains far less information than the distribution $\mathrm{dist}_G(o_1, o_2)$. Finally, the diameter is highly sensitive to relatively small changes in the graph G under consideration. For example, adding a relatively small string of connected vertices to a graph (each of the vertices in the string having degree 2) may drastically change the diameter, while it hardly influences the typical distances.

1.1.4 RELATED NETWORK PROPERTIES

There are many more features that one could take into account when modeling real-world networks. See e.g., [V1, Section 1.5] for a slightly expanded discussion of such features. Other features that many networks share, or, rather, form a way to distinguish between them, are the following:

(a) their *degree correlations*, measuring the extent to which high-degree vertices tend to be connected to high-degree vertices rather than to low-degree vertices (and vice versa);

(b) their *clustering*, measuring the extent to which pairs of neighbors of vertices are neighbors themselves;

(c) their *community structure*, measuring the extent to which the network has more densely-connected subgraphs;

(d) their *spatial structure*, where the spatial component is either describing true vertex locations in real-world networks, or instead some *latent geometry* in them. The spatial structure is such that vertices that are near are more likely to be connected.

See, e.g., the book by Newman (2010) for an extensive discussion of such features, as well as the algorithmic problems that arise from them. We also refer the reader to Chapter 9, where we discuss several related models that focus on these properties.

1.2 RANDOM GRAPHS AND REAL-WORLD NETWORKS

In this section we discuss how *random graph sequences* can be used to model real-world networks. We start by discussing graph sequences.

Graph Sequences

Motivated by the previous section, in which empirical evidence was discussed showing that many real-world networks are *scale free* and *small world*, we set about the question of how to model them. Since many networks are quite large, mathematically, we model real-world networks by *graph sequences* $(G_n)_{n\geq 1}$, where $G_n = (V(G_n), E(G_n))$ has size $|V(G_n)| = n$ and we take the limit $n \to \infty$. Since most real-world networks are such that the average degree remains bounded, we will focus on the *sparse* regime. In the sparse regime (recall (1.1.2) and (1.1.3)), it is assumed that

$$\limsup_{n\to\infty} \mathbb{E}[D_n] = \limsup_{n\to\infty} \frac{1}{|V(G_n)|} \sum_{v\in V(G_n)} d_v^{(G_n)} < \infty. \tag{1.2.1}$$

Furthermore, we aim to study graphs that are asymptotically well behaved. For example, we often either assume, or prove, that the typical degree distribution converges, i.e., there exists a limiting degree random variable D such that

$$D_n \xrightarrow{d} D, \tag{1.2.2}$$

where $\xrightarrow{d}$ denotes weak convergence of random variables. Also, we assume that our graphs are *small worlds*, which is often translated in the asymptotic sense that there exists a constant $K < \infty$ such that

$$\lim_{n\to\infty} \mathbb{P}(\text{dist}_G(o_1, o_2) \leq K \log n) = 1, \tag{1.2.3}$$

where n denotes the network size. Sometimes, we even discuss *ultra-small worlds*, for which

$$\lim_{n\to\infty} \mathbb{P}(\text{dist}_G(o_1, o_2) \leq \varepsilon \log n) = 1 \tag{1.2.4}$$

for every $\varepsilon > 0$. In what follows, we discuss random graph models that share these two features.

Random Graphs as Models for Real-World Networks

Real-world networks tend to be quite complex and unpredictable. This is understandable, since connections often arise rather irregularly. We model such irregular behavior by letting connections arise through a *random process*, thus leading us to study *random graphs*. By the previous discussion, our graphs are large and their sizes n tend to infinity.

In such settings, we can either model the graphs by *fixing* their size to be large, or rather by letting the graphs *grow* to infinite size in a consistent manner. We refer to these two settings as *static* and *dynamic* random graphs. Both are useful viewpoints. Indeed, a static graph is a model for a snapshot of a network at a fixed time, where we do not know how the connections arose over time. Many network data sets are of this form. A dynamic setting, however, may be useful when we know how the network came to be as it is. In the static setting, we can make model assumptions on the degrees such that they are scale free. In the dynamic setting, we can let the evolution of the graphs give rise to power-law degree sequences, so that these settings may provide explanations for the frequent occurrence of power laws in real-world networks.

Most of the random graph models that have been investigated in the (extensive) literature are *caricatures of reality*, in the sense that one cannot confidently argue that they describe any real-world network quantitatively correctly. However, these random graph models *do* provide insight into how any of the above features can influence the global behavior of networks. In this way, they provide possible explanations of the empirical properties of real-world networks that are observed. Also, random graph models can be used as *null models*, where certain aspects of real-world networks are taken into account while others are not. This gives a qualitative way of investigating the importance of such empirical features in the real world. Often, real-world networks are compared with uniform random graphs with certain specified properties, such as their number of edges or even their degree sequence. Below, we will come back to how to generate random graphs uar from the collection of all graphs with these properties.

In the next section we describe four models of random graphs, three of which are static and one is dynamic. Below, we frequently write $f(n) = O(g(n))$ if $|f(n)|/|g(n)|$ is uniformly bounded from above by a positive constant as $n \to \infty$, $f(n) = \Theta(g(n))$ if $f(n) = O(g(n))$ and $g(n) = O(f(n))$, and $f(n) = o(g(n))$ if $f(n)/g(n)$ tends to 0 as $n \to \infty$. We say that $f(n) \gg g(n)$ when $g(n) = o(f(n))$.

1.3 RANDOM GRAPH MODELS

We start with the most basic and simple random graph model, which has proved to be a source of tremendous inspiration, both for its mathematical beauty, as well as for providing a starting point for the analysis of random graphs.

1.3.1 ERDŐS–RÉNYI RANDOM GRAPH

The Erdős–Rényi random graph is the simplest possible random graph. In it, we make every possible edge between a collection of n vertices independently either open or closed with equal probability. This means that the Erdős–Rényi random graph has vertex set $[n] = \{1, \ldots, n\}$, and the edge uv is occupied or present with probability p, and vacant or absent otherwise, independently of all the other edges. Here we denote the edge between vertices $u, v \in [n]$ by uv. The parameter p is called the *edge probability*. The above random graph is denoted by $\mathrm{ER}_n(p)$. The model is named after Erdős and Rényi, since they made profound contributions in the study of this model. Exercise 1.3 investigates the uniform nature of $\mathrm{ER}_n(p)$ with $p = \frac{1}{2}$. Alternatively speaking, $\mathrm{ER}_n(p)$ with $p = \frac{1}{2}$ is the *null model*, where we take no properties of the network into account except for the total number of edges. The vertices in this model have expected degree $(n-1)/2$, which is quite large. As a result, this model is not sparse at all. Thus, we next make this model *sparse* by making p smaller.

Since each edge is occupied with probability p, we obtain that

$$\mathbb{P}(D_n = k) = \binom{n-1}{k} p^k (1-p)^{n-1-k} = \mathbb{P}(\mathrm{Bin}(n-1, p) = k), \qquad (1.3.1)$$

where $\mathrm{Bin}(m, p)$ is a binomial random variable with m trials and success probability p. Note that

$$\mathbb{E}[D_n] = (n-1)p, \qquad (1.3.2)$$

so for this model to be sparse, we need that p becomes small with n. Thus, we take

$$p = \frac{\lambda}{n}, \qquad (1.3.3)$$

and study the graph as λ is held fixed while $n \to \infty$. In this regime, we know that

$$D_n \xrightarrow{d} D, \qquad (1.3.4)$$

with $D \sim \mathrm{Poi}(\lambda)$, where $\mathrm{Poi}(\lambda)$ is a Poisson random variable with mean λ. It turns out that this result can be strengthened to the statement that the proportion of vertices with degree k also converges to the probability mass function of a Poisson random variable (see [V1, Section 5.4], and in particular [V1, Theorem 5.12]), i.e., for every $k \geq 0$,

$$P_k^{(n)} = \frac{1}{n} \sum_{v \in [n]} \mathbb{1}_{\{d_v = k\}} \xrightarrow{\mathbb{P}} p_k \equiv e^{-\lambda} \frac{\lambda^k}{k!}, \qquad (1.3.5)$$

where d_v denotes the degree of $v \in [n]$.

It is well known that the Poisson distribution has very thin tails, even thinner than any exponential, as you are requested to prove in Exercise 1.4. We conclude that the Erdős–Rényi random graph is not a good model for real-world networks with their highly variable degree distributions. In the next subsection, we discuss *inhomogeneous* extensions of Erdős–Rényi random graphs which can have highly variable degrees.

1.3.2 INHOMOGENEOUS RANDOM GRAPHS

In inhomogeneous random graphs, we keep the independence of the edges, but make the edge probabilities different for different edges. We will discuss such general inhomogeneous random graphs in Chapter 3 below. Here, we start with one key example, which has attracted most attention in the literature so far and is also discussed in great detail in [V1, Chapter 6].

Rank-1 Inhomogeneous Random Graphs

The simplest inhomogeneous random graph models are sometimes referred to as *rank*-1 models, since the edge probabilities are (close to) products of vertex weights (see Remark 1.5 below for more details). This means that the expected number of edges between vertices, when viewed as a matrix, is (close to) a rank-1 matrix. We start by discussing one such model, which is the so-called *generalized random graph*.

In the generalized random graph model, the edge probability of the edge between vertices u and v, for $u \neq v$, is equal to

$$p_{uv} = p_{uv}^{(\text{GRG})} = \frac{w_u w_v}{\ell_n + w_u w_v}, \tag{1.3.6}$$

where $\boldsymbol{w} = (w_v)_{v \in [n]}$ are the *vertex weights,* and ℓ_n is the total vertex weight, given by

$$\ell_n = \sum_{v \in [n]} w_v. \tag{1.3.7}$$

We denote the resulting graph by $\text{GRG}_n(\boldsymbol{w})$. In many cases, the vertex weights actually depend on n, and it would be more appropriate (but also more cumbersome), to write the weights as $\boldsymbol{w}^{(n)} = (w_v^{(n)})_{v \in [n]}$. To keep the notation simple, we refrain from making the dependence on n explicit. A special case of the generalized random graph occurs when we take $w_v \equiv \frac{n\lambda}{n-\lambda}$, in which case $p_{uv} = \lambda/n$ for all $u, v \in [n]$ so that we retrieve the Erdős–Rényi random graph $\text{ER}_n(\lambda/n)$.

The generalized random graph $\text{GRG}_n(\boldsymbol{w})$ is close to many other inhomogeneous random graph models, such as the *random graph with prescribed expected degrees* or Chung–Lu model, denoted by $\text{CL}_n(\boldsymbol{w})$, where instead

$$p_{uv} = p_{uv}^{(\text{CL})} = \min(w_u w_v/\ell_n, 1). \tag{1.3.8}$$

A further adaptation is the so-called *Poissonian random graph* or Norros–Reittu model, denoted by $\text{NR}_n(\boldsymbol{w})$, for which

$$p_{uv} = p_{uv}^{(\text{NR})} = 1 - \exp\left(-w_u w_v/\ell_n\right). \tag{1.3.9}$$

See [V1, Sections 6.7 and 6.8] for conditions under which these random graphs are *asymptotically equivalent*, meaning that all events have equal asymptotic probabilities.

Naturally, the topology of the generalized random graph depends sensitively upon the choice of the vertex weights $\boldsymbol{w} = (w_v)_{v \in [n]}$. These vertex weights can be rather general, and we investigate both settings where the weights are *deterministic* as well as settings where they are *random*. In order to describe the empirical proportions of the weights, we define their *empirical distribution function* to be

$$F_n(x) = \frac{1}{n} \sum_{v \in [n]} \mathbb{1}_{\{w_v \leq x\}}, \qquad x \geq 0. \tag{1.3.10}$$

We can interpret F_n as the distribution of the weight of a uniformly chosen vertex in $[n]$ (see Exercise 1.7). We denote the weight of a uniformly chosen vertex o in $[n]$ by $W_n = w_o$, so that, by Exercise 1.7, W_n has distribution function F_n.

The degree distribution can converge only when the vertex weights are sufficiently regular. We often assume that the vertex weights satisfy the following *regularity conditions*, which turn out to imply convergence of the degree distribution in the generalized random graph:

Condition 1.1 (Regularity conditions for vertex weights)   *There exists a distribution function F such that, as $n \to \infty$, the following conditions hold:*

(a) Weak convergence of vertex weights. *As $n \to \infty$,*

$$W_n \xrightarrow{d} W, \tag{1.3.11}$$

where W_n and W have distribution functions F_n and F, respectively. Equivalently, for any x for which $x \mapsto F(x)$ is continuous,

$$\lim_{n \to \infty} F_n(x) = F(x). \tag{1.3.12}$$

(b) Convergence of average vertex weight. *As $n \to \infty$,*

$$\mathbb{E}[W_n] \to \mathbb{E}[W] \in (0, \infty), \tag{1.3.13}$$

where W_n and W have distribution functions F_n and F from part (a) above, respectively.

(c) Convergence of second moment of vertex weights. *As $n \to \infty$,*

$$\mathbb{E}[W_n^2] \to \mathbb{E}[W^2] < \infty, \tag{1.3.14}$$

where W_n and W have distribution functions F_n and F from part (a) above, respectively.

Condition 1.1 is virtually the same as [V1, Condition 6.4]. Condition 1.1(a) guarantees that the weight of a "typical" vertex is close to a random variable W that is independent of n. Condition 1.1(b) implies that the average weight of the vertices in $\mathrm{GRG}_n(\boldsymbol{w})$ converges to the expectation of the limiting weight variable. In turn, this implies that the expectation of the average degree in $\mathrm{GRG}_n(\boldsymbol{w})$ converges to the expectation of this limiting random variable as well. Condition 1.1(c) ensures the convergence of the second moment of the weights to the second moment of the limiting weight variable.

Remark 1.2 (Regularity for random weights)   Sometimes we are interested in cases where the weights of the vertices are *random* themselves. For example, this arises when the weights $\boldsymbol{w} = (w_v)_{v \in [n]}$ are realizations of iid random variables. Then, the function F_n is also a random distribution function. Indeed, in this case F_n is the *empirical distribution function* of the random weights $(w_v)_{v \in [n]}$. We stress that $\mathbb{E}[W_n]$ is then to be interpreted as $\frac{1}{n} \sum_{v \in [n]} w_v$, which is itself random. Therefore, in Condition 1.1 we require random variables to converge, and there are several notions of convergence that may be used. The notion of convergence that we assume is *convergence in probability* (see [V1, Section 6.2]).   ◄

Let us now discuss some canonical examples of weight distributions that satisfy the Regularity Condition 1.1.

Weights Moderated by a Distribution Function

Let F be a distribution function for which $F(0) = 0$ and fix

$$w_v = [1 - F]^{-1}(v/n), \tag{1.3.15}$$

where $[1 - F]^{-1}$ is the generalized inverse function of $1 - F$, defined, for $u \in (0, 1)$, by (recall [V1, (6.2.14) and (6.2.15)])

$$[1 - F]^{-1}(u) = \inf\{x \colon [1 - F](x) \le u\}. \tag{1.3.16}$$

For the choice (1.3.15), we can explicitly compute F_n as (see [V1, (6.2.17)])

$$F_n(x) = \frac{1}{n}\big(\lfloor nF(x) \rfloor + 1\big) \wedge 1, \tag{1.3.17}$$

where $x \wedge y$ denotes the minimum of $x, y \in \mathbb{R}$. It is not hard to see that Condition 1.1(a) holds for $(w_v)_{v \in [n]}$ as in (1.3.15), while Condition 1.1(b) holds when $\mathbb{E}[W] \in (0, \infty)$, and Condition 1.1(c) holds when $\mathbb{E}[W^2] < \infty$, as can be concluded from Exercise 1.9.

Independent and Identically Distributed Weights

We now discuss the setting where the weights are an independent and identically distributed (iid) sequence of random variables, for which Conditions 1.1(b) and (c) follow from the law of large numbers, and Condition 1.1(a) from the Glivenko–Cantelli Theorem. Since we will often deal with ratios of the form $w_u w_v / (\sum_{k \in [n]} w_k)$, we assume that $\mathbb{P}(w = 0) = 0$ to avoid situations where all weights are zero.

Both settings, i.e., with weights $(w_v)_{v \in [n]}$ as in (1.3.15), and with iid weights $(w_v)_{v \in [n]}$, have their own merits. The great advantage of iid weights is that the vertices in the resulting graph are, in distribution, the same. More precisely, the vertices are completely *exchangeable*, as in the Erdős–Rényi random graph $\mathrm{ER}_n(p)$. Unfortunately, when we take the weights to be iid, in the resulting graph the edges are no longer independent (despite the fact that they are *conditionally* independent *given* the weights). In what follows, we focus on the setting where the weights are *prescribed*. When the weights are deterministic, this changes nothing; when the weights are iid, this means that we are *conditioning on the weights*.

Degrees in Generalized Random Graphs

We write d_v for the degree of vertex v in $\mathrm{GRG}_n(\boldsymbol{w})$. Thus, d_v is given by

$$d_v = \sum_{u \in [n]} \mathbb{1}_{\{uv \in E(\mathrm{GRG}_n(\boldsymbol{w}))\}}. \tag{1.3.18}$$

For $k \geq 0$, we let

$$P_k^{(n)} = \frac{1}{n} \sum_{v \in [n]} \mathbb{1}_{\{d_v = k\}} \tag{1.3.19}$$

denote the proportion of vertices with degree k of $\mathrm{GRG}_n(\boldsymbol{w})$. We call $(P_k^{(n)})_{k \geq 0}$ the *degree sequence* of $\mathrm{GRG}_n(\boldsymbol{w})$. We denote the probability mass function of a *mixed-Poisson distribution* by p_k, i.e., for $k \geq 0$,

$$p_k = \mathbb{E}\Big[e^{-W} \frac{W^k}{k!}\Big], \tag{1.3.20}$$

where W is a random variable having distribution function F from Condition 1.1. The main result concerning the vertex degrees is as follows:

Theorem 1.3 (Degree sequence of $\mathrm{GRG}_n(\boldsymbol{w})$) *Assume that Conditions 1.1(a),(b) hold. Then, for every $\varepsilon > 0$,*

$$\mathbb{P}\Big(\sum_{k=0}^{\infty} |P_k^{(n)} - p_k| \geq \varepsilon \Big) \to 0, \tag{1.3.21}$$

where $(p_k)_{k \geq 0}$ is given by (1.3.20).

Proof This is given in [V1, Theorem 6.10]. $\square$

Consequently, with $D_n = d_o$ denoting the degree of a random vertex, we obtain

$$D_n \xrightarrow{d} D, \tag{1.3.22}$$

where $\mathbb{P}(D = k) = p_k$, defined in (1.3.20), as shown in Exercise 1.10.

Recall from Section 1.1.2 that we are often interested in scale-free random graphs, i.e., random graphs for which the degree distribution obeys a power law. We see from Theorem 1.3 that this is true precisely when D obeys a power law. This, in turn, occurs precisely when W obeys a power law, for example, when, for w large,

$$\mathbb{P}(W > w) = \frac{c}{w^{\tau-1}}(1 + o(1)). \tag{1.3.23}$$

Then, for w large,

$$\mathbb{P}(D > w) = \mathbb{P}(W > w)(1 + o(1)). \tag{1.3.24}$$

This follows from Theorem 1.3, in combination with [V1, Exercise 6.12], which shows that the tail behavior of a mixed-Poisson distribution and that of its weight distribution agree for power laws.

Generalized Random Graph Conditioned on its Degrees

The generalized random graph with edge probabilities as in (1.3.6) is rather special. Indeed, when we *condition on its degree sequence*, the graph has a uniform distribution over the set of all graphs with the same degree sequence. For this, note that $\mathrm{GRG}_n(\boldsymbol{w})$ can be equivalently encoded by $(X_{uv})_{1 \leq u \leq v \leq n}$, where X_{uv} is the indicator that the edge uv is occupied. Then, $(X_{uv})_{1 \leq u \leq v \leq n}$ are independent Bernoulli random variables with edge probabilities as in (1.3.6). By convention, let $X_{vv} = 0$ for every $v \in [n]$, and $X_{vu} = X_{uv}$ for

$1 \leq u < v \leq n$. In terms of the variables $X = (X_{uv})_{1 \leq u < v \leq n}$, let $d_v(X) = \sum_{u \in [n]} X_{uv}$ be the degree of vertex v. Then, the uniformity is equivalent to the statement that, for each $x = (x_{uv})_{1 \leq u < v \leq n}$ such that $d_v(x) = d_v$ for every $v \in [n]$,

$$\mathbb{P}(X = x \mid d_v(X) = d_v \ \forall v \in [n]) = \frac{1}{\#\{y \colon d_v(y) = d_v \ \forall v \in [n]\}}, \tag{1.3.25}$$

that is, the distribution is uniform over all graphs with the prescribed degree sequence. This turns out to be rather convenient, and thus we state it formally here:

Theorem 1.4 (GRG conditioned on degrees has a uniform law)   *The generalized random graph* $\mathrm{GRG}_n(\boldsymbol{w})$ *with edge probabilities* $(p_{uv})_{1 \leq u < v \leq n}$ *given by*

$$p_{uv} = \frac{w_u w_v}{\ell_n + w_u w_v}, \tag{1.3.26}$$

conditioned on $\{d_v(X) = d_v \forall v \in [n]\}$, *is uniform over all graphs with degrees* $(d_v)_{v \in [n]}$.

Proof   See [V1, Theorem 6.15].   □

In Chapter 3 below, we discuss a far more general setting of inhomogeneous random graphs. The analysis of such random graphs is substantially more challenging than the rank-1 case. As explained in more detail there, this is due to the fact that these random graphs are no longer locally described by single-type branching processes, but rather by multi-type branching processes.

Remark 1.5 (What's in a name?)   The models discussed here, $\mathrm{GRG}_n(\boldsymbol{w})$ in (1.3.6) as well as $\mathrm{CL}_n(\boldsymbol{w})$ in (1.3.8) and $\mathrm{NR}_n(\boldsymbol{w})$ in (1.3.9), go under various names in the literature. Bollobás et al. (2007) referred to them as a *rank-1 random graph*, because $p_{uv} \approx w_u w_v / \ell_n$ and the matrix $(w_u w_v / \ell_n)_{u,v \in [n]}$ has rank one. In the physics literature, they go under the name of *hidden variable models*, where the weights $(w_v)_{v \in [n]}$ are interpreted as the hidden variables (and they are often assumed to be iid). Owing to the uniformity in the conditional distribution given its degrees, $\mathrm{GRG}_n(\boldsymbol{w})$ is also a *maximal entropy* model, as will be explained in more detail in Section 9.4.4. Finally, some researchers call them *soft configuration models*; see Remark 1.6 for further discussion of this phrase.   ◀

1.3.3 CONFIGURATION MODELS

The configuration model is a model in which the degrees of vertices are fixed beforehand. Such a model is more flexible than the generalized random graph. For example, the generalized random graph always has a positive proportion of vertices of degree 0, 1, 2, etc., as easily follows from Theorem 1.3.

Fix an integer n that denotes the number of vertices in the random graph. Consider a sequence of degrees $\boldsymbol{d} = (d_v)_{v \in [n]}$. Again, it might be more appropriate, but also more cumbersome, to write the degrees as $\boldsymbol{d}^{(n)} = (d_v^{(n)})_{v \in [n]}$, and so we will refrain from this. The aim is to construct an undirected (multi-)graph with n vertices, where vertex v has degree d_v. Here a multi-graph is a graph *possibly* having self-loops and multiple edges between pairs of vertices.

Without loss of generality, we assume throughout this chapter that $d_v \geq 1$ for all $v \in [n]$, since, when $d_v = 0$, vertex v is isolated and can be removed from the graph. One possible random graph model takes the uniform measure over such undirected and simple graphs. Here, we call a multi-graph *simple* when it has no self-loops, and no multiple edges exist between any pair of vertices. However, the set of undirected simple graphs with n vertices where vertex v has degree d_v may be empty. For example, in order for such a graph to exist, we must assume that the total degree

$$\ell_n = \sum_{v \in [n]} d_v \qquad (1.3.27)$$

is even.

We wish to construct a simple graph such that $\boldsymbol{d} = (d_v)_{v \in [n]}$ are the degrees of the n vertices. Even when $\ell_n = \sum_{v \in [n]} d_v$ *is* even, however, this is not always possible. Therefore, instead, we construct a *multi-graph*. One way of obtaining such a multi-graph with the given degree sequence is to pair the half-edges attached to the different vertices in a uniform way. Two half-edges together form an edge, thus creating the edges in the graph. Let us explain this in more detail.

To construct the multi-graph where vertex v has degree d_v for all $v \in [n]$, we have n separate vertices and, incident to vertex v, we have d_v half-edges. Every half-edge needs to be connected to another half-edge to form an edge, and by forming all edges we build the graph. For this, the half-edges are numbered in an arbitrary order from 1 to ℓ_n. We start by randomly connecting the first half-edge with one of the $\ell_n - 1$ remaining half-edges. Once paired, two half-edges form a single edge of the multi-graph, and these half-edges are removed from the list of half-edges that need to be paired. Hence, a half-edge can be seen as the left or the right half of an edge. We continue the procedure of randomly choosing and pairing the half-edges until all half-edges are connected, and we call the resulting graph the *configuration model with degree sequence* $\boldsymbol{d}$, abbreviated as $\mathrm{CM}_n(\boldsymbol{d})$. The pairing of the half-edges that induces the configuration model graph is sometimes called a *configuration*l.

A careful reader may worry about the order in which the half-edges are being paired. In fact, this ordering turns out to be irrelevant since the random pairing of half-edges is completely *exchangeable*. It can even be done in a *random* fashion, which will be useful when investigating neighborhoods in the configuration model. See e.g., [V1, Definition 7.5 and Lemma 7.6] for more details on this exchangeability.

Interestingly, one can rather explicitly compute the distribution of $\mathrm{CM}_n(\boldsymbol{d})$. To do so, note that $\mathrm{CM}_n(\boldsymbol{d})$ is characterized by the random vector $(X_{uv})_{1 \leq u \leq v \leq n}$. Here X_{uv} is the number of edges between vertex u and v, and X_{vv} is the number of self-loops incident to vertex v, so that

$$d_v = X_{vv} + \sum_{u \in [n]} X_{uv}. \qquad (1.3.28)$$

Note furthermore that X_{vv} appears *twice* in (1.3.28), which is natural, since a self-loop consists of *two* half-edges. This does not conflict with the definition of d_v for $\mathrm{GRG}_n(\boldsymbol{w})$, since $X_{uu} = 0$ and $X_{u,v} \in \{0,1\}$ for $\mathrm{GRG}_n(\boldsymbol{w})$.

In terms of this notation, and writing $G = (x_{uv})_{u,v \in [n]}$ to denote a multi-graph on $[n]$,

$$\mathbb{P}(\mathrm{CM}_n(\boldsymbol{d}) = G) = \frac{1}{(\ell_n - 1)!!} \frac{\prod_{v \in [n]} d_v!}{\prod_{v \in [n]} 2^{x_{vv}} \prod_{1 \le u \le v \le n} x_{uv}!}. \qquad (1.3.29)$$

See, e.g., [V1, Proposition 7.7] for this result. In particular, $\mathbb{P}(\mathrm{CM}_n(\boldsymbol{d}) = G)$ is the *same* for each *simple* G, where G is simple when $x_{vv} = 0$ for every $v \in [n]$ and $x_{uv} \in \{0, 1\}$ for every $1 \le u < v \le n$. Thus, the configuration model conditioned on simplicity is a *uniform* random graph with the prescribed degree distribution. This is quite relevant, as it gives a convenient way to *obtain* such a uniform graph, which is a highly non-trivial fact.

Remark 1.6 (What's in a name continued?)   The name *configuration model* was invented by Bollobás (1980), who considered the matching of half-edges to be the *configuration* on which the model is based. The model of study for Bollobás (1980) was the uniform simple random regular graph, where all degrees are the same, as we discuss further below. Molloy and Reed (1995, 1998) extended it to general degrees. As a result, it is sometimes also called the *Molloy–Reed model*. With X_{uv} equal to the number of edges between vertices u and v,

$$\mathbb{E}[X_{uv}] = \frac{d_u d_v}{\ell_n - 1}, \qquad (1.3.30)$$

since each of the d_v half-edges incident to vertex v has probability $d_u/(\ell_n - 1)$ to be connected to vertex u. Since (1.3.30) is close to the edge probability p_{uv} in rank-1 random graphs (recall Remark 1.5), rank-1 random graphs are sometimes called *soft configuration models*. The configuration-model degree constraint is instead viewed as a *hard* constraint. ◄

The uniform nature of the configuration model conditioned on simplicity partly explains its popularity, and it has become one of the most highly studied random graph models. It also implies that, conditioned on simplicity, the configuration model is the *null model* for a real-world network where all the degrees are fixed. This allows one to distinguish the relevance of the *degree inhomogeneity* from other features of the network, such as its community structure, clustering, etc.

As for $\mathrm{GRG}_n(\boldsymbol{w})$, we again impose *regularity conditions* on the degree sequence $\boldsymbol{d}$. In order to state these assumptions, we introduce some notation. We denote the degree of a uniformly chosen vertex o in $[n]$ by $D_n = d_o$. The random variable D_n has distribution function F_n given by

$$F_n(x) = \frac{1}{n} \sum_{v \in [n]} \mathbb{1}_{\{d_v \le x\}}, \qquad (1.3.31)$$

which is the *empirical distribution of the degrees*. We assume that the vertex degrees satisfy the following *regularity conditions*:

Condition 1.7 (Regularity conditions for vertex degrees)

(a) **Weak convergence of vertex degrees.** *There exists a distribution function F such that, as $n \to \infty$,*

$$D_n \xrightarrow{d} D, \qquad (1.3.32)$$

where D_n and D have distribution functions F_n and F, respectively.

Equivalently, for any $x \in \mathbb{R}$,

$$\lim_{n \to \infty} F_n(x) = F(x). \tag{1.3.33}$$

Further, we assume that $F(0) = 0$, i.e., $\mathbb{P}(D \geq 1) = 1$.

(b) Convergence of average vertex degree. *As $n \to \infty$,*

$$\mathbb{E}[D_n] \to \mathbb{E}[D] < \infty, \tag{1.3.34}$$

where D_n and D have the distribution functions F_n and F from part (a) above, respectively.

(c) Convergence of second moment of vertex degrees. *As $n \to \infty$,*

$$\mathbb{E}[D_n^2] \to \mathbb{E}[D^2] \in (0, \infty), \tag{1.3.35}$$

where D_n and D have distribution functions F_n and F from part (a) above, respectively.

The possibility that one will obtain a non-simple graph is a major disadvantage of the configuration model. There are two ways of dealing with this complication, as follows:

Erased Configuration Model

The first way of dealing with self-loops and multi-edges is to *erase* the problems. This means that we replace $\mathrm{CM}_n(\boldsymbol{d}) = (X_{uv})_{1 \leq u \leq v \leq n}$ by its erased version $\mathrm{ECM}_n(\boldsymbol{d}) = (X_{uv}^{(\mathrm{er})})_{1 \leq u \leq v \leq n}$, where $X_{vv}^{(\mathrm{er})} \equiv 0$, while $X_{uv}^{(\mathrm{er})} = 1$ precisely when $X_{uv} \geq 1$. In words, we remove the self-loops and merge all multiple edges to a single edge. Of course, this changes the precise degree distribution. However, [V1, Theorem 7.10] (see also Theorem 1.8 below) shows that only a small proportion of the edges is erased, so that the erasing does not change the asymptotic degree distribution. See [V1, Section 7.3] for more details. Of course, the downside of this approach is that the degrees are changed by the procedure, while we would like to keep the degrees *precisely* as specified.

Let us describe the degree distribution in the erased configuration model in more detail, to study the effect of the erasure of self-loops and multiple edges. We denote the degrees in the erased configuration model by $\boldsymbol{D}^{(\mathrm{er})} = (D_v^{(\mathrm{er})})_{v \in [n]}$, so that

$$D_v^{(\mathrm{er})} = d_v - 2s_v - m_v, \tag{1.3.36}$$

where $(d_v)_{v \in [n]}$ are the degrees in $\mathrm{CM}_n(\boldsymbol{d})$, $s_v = x_{vv}$ is the number of self-loops of vertex v in $\mathrm{CM}_n(\boldsymbol{d})$, and

$$m_v = \sum_{u \neq v} (x_{uv} - 1) \mathbb{1}_{\{x_{uv} \geq 2\}} \tag{1.3.37}$$

is the number of multiple edges removed from v.

Denote the empirical degree sequence $(p_k^{(n)})_{k \geq 1}$ in $\mathrm{CM}_n(\boldsymbol{d})$ by

$$p_k^{(n)} = \mathbb{P}(D_n = k) = \frac{1}{n} \sum_{v \in [n]} \mathbb{1}_{\{d_v = k\}}, \tag{1.3.38}$$

and denote the related degree sequence in the erased configuration model $(P_k^{(\mathrm{er})})_{k \geq 1}$ by

$$P_k^{(\mathrm{er})} = \frac{1}{n} \sum_{v \in [n]} \mathbb{1}_{\{D_v^{(\mathrm{er})} = k\}}. \tag{1.3.39}$$

From the notation it should be clear that $(p_k^{(n)})_{k \geq 1}$ is a *deterministic* sequence when $d = (d_v)_{v \in [n]}$ is deterministic, while $(P_k^{(\mathrm{er})})_{k \geq 1}$ is a *random* sequence, since the erased degrees $(D_v^{(\mathrm{er})})_{v \in [n]}$ form a random vector even when $d = (d_v)_{v \in [n]}$ is deterministic.

Now we are ready to state the main result concerning the degree sequence of the erased configuration model:

Theorem 1.8 (Degree sequence of erased configuration model with fixed degrees) *For fixed degrees d satisfying Conditions 1.7(a),(b), the degree sequence of the erased configuration model $(P_k^{(\mathrm{er})})_{k \geq 1}$ converges in probability to $(p_k)_{k \geq 1}$. More precisely, for every $\varepsilon > 0$,*

$$\mathbb{P}\Big(\sum_{k=1}^{\infty} |P_k^{(\mathrm{er})} - p_k| \geq \varepsilon \Big) \to 0, \tag{1.3.40}$$

where $p_k = \mathbb{P}(D = k)$ as in Condition 1.7(a).

Proof See [V1, Theorem 7.10]. □

Theorem 1.8 indeed shows that most of the edges are kept in the erasure procedure; see Exercise 1.17.

Configuration Model Conditioned on Simplicity

The second solution to the multi-graph problem of the configuration model is to throw away the result when it is not simple, and try again. Therefore, this construction is sometimes called the *repeated configuration model*. It turns out that, when Conditions 1.7(a)–(c) hold (see [V1, Theorem 7.12]),

$$\lim_{n \to \infty} \mathbb{P}(\mathrm{CM}_n(d) \text{ is a simple graph}) = e^{-\nu/2 - \nu^2/4}, \tag{1.3.41}$$

where

$$\nu = \frac{\mathbb{E}[D(D-1)]}{\mathbb{E}[D]} \tag{1.3.42}$$

is the expected forward degree. This is a realistic option when $\mathbb{E}[D^2] < \infty$. Unfortunately, this is not an option when the asymptotic degrees obey an asymptotic power law with $\tau \in (2, 3)$ (as, e.g., in (1.1.12)), since then $\mathbb{E}[D^2] = \infty$. Note that, by (1.3.29), $\mathrm{CM}_n(d)$ conditioned on simplicity is a *uniform random graph* with the prescribed degree sequence. We denote this random graph by $\mathrm{UG}_n(d)$. We return to the difficulty of generating simple graphs with infinite-variance degrees in Section 1.3.4 below.

Relation between Generalized Random Graph and Configuration Model

Since $\mathrm{CM}_n(d)$ conditioned on simplicity yields a uniform (simple) random graph with these degrees, and, also, by (1.3.25), $\mathrm{GRG}_n(w)$ conditioned on its degrees is a uniform (simple) random graph with the given degree distribution, the laws of these (conditioned) random graph models are the same. As a result, one can prove results for $\mathrm{GRG}_n(w)$ by proving them for $\mathrm{CM}_n(d)$ under the appropriate degree conditions, and then proving that $\mathrm{GRG}_n(w)$ satisfies these conditions in probability.

A further useful result in this direction is that the weight regularity conditions in Conditions 1.1(a),(b) imply the degree regularity conditions in Conditions 1.7(a),(b):

Theorem 1.9 (Regularity conditions for weights and degrees) *Let d_v be the degree of vertex v in $\mathrm{GRG}_n(\boldsymbol{w})$, and let $\boldsymbol{d} = (d_v)_{v \in [n]}$. Then, $\boldsymbol{d}$ satisfies Conditions 1.7(a),(b) in probability when $\boldsymbol{w}$ satisfies Conditions 1.1(a),(b), where*

$$\mathbb{P}(D = k) = \mathbb{E}\left[\frac{W^k}{k!}e^{-W}\right] \tag{1.3.43}$$

denotes the mixed-Poisson distribution with mixing distribution W having distribution function F in Condition 1.1(a). Further, $\boldsymbol{d}$ satisfies Conditions 1.7(a)–(c) in probability when $\boldsymbol{w}$ satisfies Conditions 1.1(a)–(c).

Proof See [V1, Theorem 7.19]. The weak convergence in Condition 1.7(a) follows from Theorem 1.3. $\qquad\square$

Remark 1.10 (Proving results for $\mathrm{GRG}_n(\boldsymbol{w})$ through $\mathrm{CM}_n(\boldsymbol{d})$) Combined with Theorem 1.4, Theorem 1.9 allows us to prove many results for the generalized random graph by first proving them for the configuration model under appropriate conditions on its degrees, and then extending them to the generalized random graph by proving that its degrees satisfy the assumptions made. In particular, any property that holds *in probability* for $\mathrm{CM}_n(\boldsymbol{d})$ can be extended to $\mathrm{GRG}_n(\boldsymbol{w})$ in this way. See [V1, Sections 6.6 and 7.5] for more details. This strategy is also frequently used in the present volume. $\qquad\blacktriangleleft$

A Useful Degree-Truncation Argument for Heavy-Tailed Degrees

Recall from Section 1.1.2 that many real-world networks have substantial inhomogeneities in their degrees. As a result, we frequently discuss configuration models with power-law degrees, giving rise to degree distributions with maxima that grow as a positive power of n. Such large degrees can be inconvenient in technical estimates. We next present a useful *degree-truncation argument* for the configuration model, which allows us to compare such a model with an alternative configuration model with *bounded* degrees. In its statement, we write $x \wedge y$ for the minimum of $x, y \in \mathbb{R}$:

Theorem 1.11 (Degree truncation for configuration models) *Consider $\mathrm{CM}_n(\boldsymbol{d})$ with general degrees. Fix $b \geq 1$. There exists a related configuration model $\mathrm{CM}_{n'}(\boldsymbol{d}')$ with $n' \geq n$ that is coupled to $\mathrm{CM}_n(\boldsymbol{d})$ and satisfies the following:*

(a) *the degrees in $\mathrm{CM}_{n'}(\boldsymbol{d}')$ are a truncated version of those in $\mathrm{CM}_n(\boldsymbol{d})$, i.e., $d'_v = (d_v \wedge b)$ for $v \in [n]$, and $d'_v = 1$ for $v \in [n'] \setminus [n]$;*
(b) *the total degree in $\mathrm{CM}_{n'}(\boldsymbol{d}')$ is the same as that in $\mathrm{CM}_n(\boldsymbol{d})$, i.e., $\sum_{v \in [n']} d'_v = \sum_{v \in [n]} d_v$;*
(c) *for all $u, v \in [n]$, if u and v are connected in $\mathrm{CM}_{n'}(\boldsymbol{d}')$, then so are u and v in $\mathrm{CM}_n(\boldsymbol{d})$, i.e., $\mathrm{dist}_{\mathrm{CM}_n(\boldsymbol{d})}(u, v) \leq \mathrm{dist}_{\mathrm{CM}_{n'}(\boldsymbol{d}')}(u, v)$ almost surely.*

Remark 1.12 (Truncation of degrees in range) The construction that proves Theorem 1.11 is highly flexible, and also allows for a degree truncation that maintains restrictions on the minimal degree $d_{\min} = \min_{v \in [n]} d_v$. Indeed, fix $b \geq 2$. There exists a related configuration model $\mathrm{CM}_{n'}(\boldsymbol{d}')$ satisfying (b) and (c) in Theorem 1.11, while (a) is replaced by $d'_v = d_v$ when $d_v < 2b$, by $d'_v = b$ when $d_v \geq 2b$ for $v \in [n]$, and by $b \leq d'_v < 2b$ for $v \in [n'] \setminus [n]$, so that $d'_{\min} = \min_{v \in [n']} d'_v \geq d_{\min} \wedge b$. $\qquad\blacktriangleleft$

Proof The proof relies on an "explosion" or "fragmentation" of the vertices $[n]$ in $\mathrm{CM}_n(\boldsymbol{d})$. Label the half-edges from 1 to ℓ_n. We go through the vertices $v \in [n]$ one by one. When $d_v \leq b$, we do nothing. When $d_v > b$, we let $d'_v = b$ and keep the b half-edges with the lowest labels. The remaining $d_v - b$ half-edges are exploded from vertex v, in that they are incident to vertices of degree 1 in $\mathrm{CM}_{n'}(\boldsymbol{d}')$, and are given vertex labels above n. We give the exploded half-edges the remaining labels of the half-edges incident to v. Thus, the half-edges receive labels both in $\mathrm{CM}_n(\boldsymbol{d})$ as well as in $\mathrm{CM}_{n'}(\boldsymbol{d}')$, and the labels of the half-edges incident to $v \in [n]$ in $\mathrm{CM}_{n'}(\boldsymbol{d}')$ are a subset of those in $\mathrm{CM}_n(\boldsymbol{d})$. In total, we thus create an extra $n^+ = \sum_{v \in [n]} (d_v - b) \vee 0$ "exploded" vertices of degree 1, and $n' = n + n^+$, where $x \vee y$ denotes the maximum of $x, y \in \mathbb{R}$.

We then pair the half-edges randomly, in the same way in $\mathrm{CM}_n(\boldsymbol{d})$ as in $\mathrm{CM}_{n'}(\boldsymbol{d}')$. This means that when the half-edge with label x is paired with the half-edge with label y in $\mathrm{CM}_n(\boldsymbol{d})$, then also the half-edge with label x is paired with the half-edge with label y in $\mathrm{CM}_{n'}(\boldsymbol{d}')$, for all $x, y \in [\ell_n]$.

We now check parts (a)–(c). Obviously parts (a) and (b) follow from the construction. For part (c), we note that all exploded vertices in $[n^+] \setminus [n]$ have degree 1. Further, for vertices $u, v \in [n]$, if there exists a path in $\mathrm{CM}_{n'}(\boldsymbol{d}')$ connecting them then the intermediate vertices have degree at least 2, so that they cannot correspond to exploded vertices and must therefore in $\mathrm{CM}_{n'}(\boldsymbol{d}')$ have labels in $[n]$. Thus, the same path of paired half-edges also exists in $\mathrm{CM}_n(\boldsymbol{d})$, so that u and v are also connected in $\mathrm{CM}_n(\boldsymbol{d})$.

We conclude by adapting the construction to prove the statement in Remark 1.12. We again go through the vertices $v \in [n]$ one by one. When $d_v < 2b$, we do nothing. When $d_v \geq 2b$, we let $d'_v = b$ and keep the b half-edges with the lowest labels. The remaining $d_v - b$ half-edges are exploded from vertex v, in that they are incident to "exploded" vertices that all have degree b in $\mathrm{CM}_{n'}(\boldsymbol{d}')$ possibly except for one vertex that has degree in $[b, 2b)$, and are given vertex labels above n. This means that a vertex of degree $d_v \geq 2b$ is replaced by one vertex in $[n]$ and $\lfloor d_v/b \rfloor - 1$ vertices in $[n'] \setminus [n]$, of which all, possibly except for the last vertex, have degree b, and the degree of the last vertex equals $d_v - b(\lfloor d_v/b \rfloor - 1) \in [b, 2b)$. We again give the exploded half-edges the remaining labels of the half-edges incident to v. This identifies the desired construction for Remark 1.12. For part (c), we note that the half-edges incident to exploded vertices arise from the same vertex in $[n]$ as before explosion, so a path between vertices $u', v' \in [n']$ in $\mathrm{CM}_{n'}(\boldsymbol{d}')$ implies that a path between the vertices $u, v \in [n]$ that correspond to u', v' exists. This implies that part (c) holds. □

1.3.4 Uniform Random Graphs and Switching Algorithms for them

So far, we have focussed on obtaining a uniform random graph with a prescribed degree sequence by conditioning the configuration model on being simple. As explained above, this does not work so well when the degrees have infinite variance. Another setting where this method fails to deliver occurs when the average degree is *large* rather than bounded, so that the graph is no longer *sparse* in the strict sense (recall Section 1.1.1).

An alternative method for producing a sample from the uniform distribution on simple graphs uses a *switching algorithm*. A switching algorithm is a Markov chain on the space of simple graphs, where, in each step, some edges in the graph are rewired while keeping the graph simple. Under mild conditions on the precise switching dynamics, the uniform distribution is the stationary distribution of this Markov chain, so letting the switching algorithm run for an infinitely long time, we obtain a perfect sample from the uniform distribution. The "mild" conditions follow, for example, when the switch chain is *doubly stochastic*.

Switching algorithms can also be used rather effectively to compute probabilities of certain events for uniform random graphs with specified degrees, as we explain later. As such, switching methods form an indispensable tool in studying uniform random graphs with prescribed degrees. We start by explaining the basic switching algorithms and their relation to uniform sampling.

Switch Markov Chain

The switch Markov chain is a Markov chain on the space of simple graphs with prescribed degrees given by d. Fix a simple graph $G = ([n], E(G))$ for which the degree of vertex v equals d_v for all $v \in [n]$. We assume that such a simple graph exists, i.e., we assume that $d = (d_v)_{v \in [n]}$ is *graphical*.

In order to describe the dynamics of the switch chain, choose two edges $\{u, v\}$ and $\{x, y\}$ uar from the edge set $E(G)$, where G is the current simple graph. The possible switches of these two edges are (1) $\{u, x\}$ and $\{v, y\}$; (2) $\{v, x\}$ and $\{u, y\}$; and (3) $\{u, v\}$ and $\{x, y\}$ (so that no change is made). Choose each of these three options with probability equal to $\frac{1}{3}$, and write the chosen edges as e_1, e_2. *Accept* the switch when the resulting graph with edges $\{e_1, e_2\} \cup (E(G) \setminus \{\{u, v\}, \{x, y\}\})$ is simple, and *reject* the switch otherwise (so that the graph remains unchanged under the dynamics).

It is not hard to see that the resulting Markov chain is aperiodic and irreducible. Further, the switch chain is doubly stochastic since it is reversible. As a result, its stationary distribution is the uniform random graph with prescribed degree sequence d, which we have denoted by $\mathrm{UG}_n(d)$, as required.

The above method works rather generally, and, in the limit of infinitely many switches, produces a sample from $\mathrm{UG}_n(d)$ for *every* graphical degree sequence, even when the degrees are large. As a result, this chain is the method of choice to produce a sample of $\mathrm{UG}_n(d)$ when the probability of simplicity of the configuration model vanishes. However, it is unclear *precisely* how often one needs to switch in order for the Markov chain to be sufficiently close to the uniform (and thus stationary) distribution. See the notes in Section 1.6 for a discussion of the history of the switch chain, as well as the available results about its convergence.

Switching Methods for Random Graphs with Prescribed Degrees

Switching algorithms can also be used to prove properties about uniform random graphs with prescribed degrees. Here, we explain how switching can be used to estimate the connection probability between vertices of specific degrees in a uniform random graph. Recall that $\ell_n = \sum_{v \in [n]} d_v$. Then, the asymptotics for the edge probabilities for $\mathrm{UG}_n(d)$ are given in the following theorem, where $E(\mathrm{UG}_n(d))$ denotes the edge set of $\mathrm{UG}_n(d)$:

Theorem 1.13 (Edge probabilities for uniform random graphs with prescribed degrees) *Assume that the empirical distribution F_n of d satisfies, for all $x \geq 1$,*

$$[1 - F_n](x) \leq c_F x^{-(\tau-1)}, \qquad (1.3.44)$$

for some $c_F > 0$ and $\tau \in (2,3)$. Let U denote a set of unordered pairs of vertices and let $\mathcal{E}_U = \{\{s,t\} \in E(\mathrm{UG}_n(\boldsymbol{d})) \, \forall \{s,t\} \in U\}$ denote the event that $\{s,t\}$ is an edge for every $\{s,t\} \in U$. Then, assuming that $|U| = O(1)$, for every $\{u,v\} \notin U$,

$$\mathbb{P}(\{u,v\} \in E(\mathrm{UG}_n(\boldsymbol{d})) \mid \mathcal{E}_U) = (1 + o(1))\frac{(d_u - |U_u|)(d_v - |U_v|)}{\ell_n + (d_u - |U_u|)(d_v - |U_v|)}, \qquad (1.3.45)$$

where U_v denotes the set of pairs in U that contain $v \in [n]$.

Remark 1.14 (Relation to $\mathrm{ECM}_n(\boldsymbol{d})$ and $\mathrm{GRG}_n(\boldsymbol{w})$)   Theorem 1.13 shows that, when $d_u d_v \gg \ell_n$,

$$1 - \mathbb{P}(\{u,v\} \in E(\mathrm{UG}_n(\boldsymbol{d}))) = (1 + o(1))\frac{\ell_n}{d_u d_v}. \qquad (1.3.46)$$

In the erased configuration model, on the other hand,

$$1 - \mathbb{P}(\{u,v\} \in E(\mathrm{ECM}_n(\boldsymbol{d}))) \leq \mathrm{e}^{-d_u d_v / (2\ell_n)}, \qquad (1.3.47)$$

as will be crucially used in Chapter 7 below (see Lemma 7.12 for a proof of (1.3.47)). Thus, the probability that two high-degree vertices are not connected is much smaller for $\mathrm{ECM}_n(\boldsymbol{d})$ than for $\mathrm{UG}_n(\boldsymbol{d})$. On a related note, the fact that

$$\mathbb{P}(\{u,v\} \in E(\mathrm{UG}_n(\boldsymbol{d}))) \approx \frac{d_u d_v}{\ell_n + d_u d_v},$$

as in $\mathrm{GRG}_n(\boldsymbol{w})$ when $\boldsymbol{w} = \boldsymbol{d}$, indicates once more that $\mathrm{GRG}_n(\boldsymbol{w})$ and $\mathrm{UG}_n(\boldsymbol{d})$ are closely related.   ◄

We now proceed with the proof of Theorem 1.13. We first prove a useful lemma about the number of 2-paths starting from a specified vertex, where a 2-path is a path consisting of two edges:

Lemma 1.15 (The number of 2-paths)   *Assume that $\boldsymbol{d}$ satisfies (1.3.44) for some $c_F > 0$ and $\tau \in (2,3)$. For any graph G whose degree sequence is $\boldsymbol{d}$, the number of 2-paths starting from any specified vertex is $O(n^{(2\tau-3)/(\tau-1)^2}) = o(n)$.*

Proof   Without loss of generality we may assume that the degrees are ordered from large to small as $d_1 \geq d_2 \geq \cdots \geq d_n$. Then, for every $v \in [n]$, the number of vertices with degree at least d_v is at least v. By (1.3.44), for every $v \in [n]$,

$$c_F n(d_v - 1)^{1-\tau} \geq n[1 - F_n](d_v - 1) \geq v. \qquad (1.3.48)$$

Thus, $d_v \leq (c_F n / v)^{1/(\tau-1)} + 1$. The number of 2-paths from any vertex is bounded by $\sum_{v=1}^{d_1} d_v$, which is at most

$$\sum_{v=1}^{d_1} \left(\left(\frac{c_F n}{v} \right)^{1/(\tau-1)} + 1 \right) = (c_F n)^{1/(\tau-1)} \sum_{v=1}^{d_1} v^{-1/(\tau-1)} + d_1 \tag{1.3.49}$$

$$= O\left(n^{1/(\tau-1)} \right) d_1^{(\tau-2)/(\tau-1)} = O\left(n^{(2\tau-3)/(\tau-1)^2} \right),$$

since $d_1 \leq (c_F n)^{1/(\tau-1)} + 1$. Since $\tau \in (2,3)$, the above is $o(n)$. $\qquad\square$

Proof of Theorem 1.13. To compute the asymptotics of $\mathbb{P}(\{u,v\} \in E(\mathrm{UG}_n(\boldsymbol{d})) \mid \mathcal{E}_U)$, we switch between two classes of graphs, $\mathcal{S}$ and $\bar{\mathcal{S}}$. Class $\mathcal{S}$ consists of graphs where all edges in $\{u,v\} \cup U$ are present, whereas $\bar{\mathcal{S}}$ consists of all graphs where every $\{s,t\} \in U$ is present, but $\{u,v\}$ is not. Recall that $\mathcal{E}_U = \{\{s,t\} \in E(\mathrm{UG}_n(\boldsymbol{d})) \, \forall \{s,t\} \in U\}$ denotes the event that $\{s,t\}$ is an edge for every $\{s,t\} \in U$. Then, since the law on simple graphs is *uniform* (see also Exercise 1.18),

$$\mathbb{P}(\{u,v\} \in E(\mathrm{UG}_n(\boldsymbol{d})) \mid \mathcal{E}_U) = \frac{|\mathcal{S}|}{|\mathcal{S}| + |\bar{\mathcal{S}}|} = \frac{1}{1 + |\bar{\mathcal{S}}|/|\mathcal{S}|}, \tag{1.3.50}$$

and we are left to compute the asymptotics of $|\bar{\mathcal{S}}|/|\mathcal{S}|$.

For this, we define an operation called a *forward switching* that converts a graph in $G \in \mathcal{S}$ to a graph $G' \in \bar{\mathcal{S}}$. The reverse operation, converting G' to G, is called a *backward switching*. Then we estimate $|\bar{\mathcal{S}}|/|\mathcal{S}|$ by counting the number of forward switchings that can be applied to the graph $G \in \mathcal{S}$, and the number of backward switchings that can be applied to the graph $G' \in \bar{\mathcal{S}}$. In our switching, we wish to have control on whether $\{u,v\}$ is present or not, so we tune it to take this restriction into account.

The forward switching on $G \in \mathcal{S}$ is defined by choosing two edges and specifying their ends as $\{x,a\}$ and $\{y,b\}$. We write this as *directed* edges (x,a) since the roles of x and a are different, as indicated in Figure 1.7. We assume that $\mathcal{E}_U$ occurs. The choice must satisfy the following constraints:

(1) none of $\{u,x\}$, $\{v,y\}$, or $\{a,b\}$ is an edge in G;
(2) $\{x,a\}, \{y,b\} \notin U$;
(3) all of $u, v, x, y, a,$ and b must be distinct except that $x = y$ is permitted.

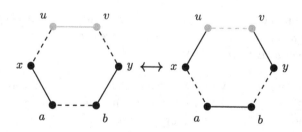

Figure 1.7 Forward and backward switchings. The edge $\{u,v\}$ is present on the left, but not on the right.

Given a valid choice, forward switching replaces the three edges $\{u, v\}$, $\{x, a\}$, and $\{y, b\}$ by $\{u, x\}$, $\{v, y\}$, and $\{a, b\}$, while ensuring that the graph after switching is simple. Note that forward switching preserves the degree sequence, and converts a graph in $\mathcal{S}$ to a graph in $\bar{\mathcal{S}}$. See Figure 1.7 for an illustration of both the forward and backward switchings.

Next, we estimate the number of ways to perform a forward switching to a graph G in $\mathcal{S}$, denoted by $f(G)$, and the number of ways to perform a backward switching to a graph G' in $\bar{\mathcal{S}}$, denoted by $b(G)$. The number of total switchings between $\mathcal{S}$ and $\bar{\mathcal{S}}$ is equal to (see Exercise 1.19)

$$|\mathcal{S}|\mathbb{E}[f(G)] = |\bar{\mathcal{S}}|\mathbb{E}[b(G')], \tag{1.3.51}$$

where the expectation is over a uniformly random $G \in \mathcal{S}$ on the left-hand side, and over a uniformly random $G' \in \bar{\mathcal{S}}$ on the right-hand side, respectively. Consequently,

$$\frac{|\bar{\mathcal{S}}|}{|\mathcal{S}|} = \frac{\mathbb{E}[f(G)]}{\mathbb{E}[b(G')]}. \tag{1.3.52}$$

We next compute each of these factors.

The Number of Forward Switchings: Computing $\mathbb{E}[f(G)]$

Given an arbitrary graph $G \in \mathcal{S}$, the number of ways to carry out a forward switching is at most ℓ_n^2, since there are at most ℓ_n ways to choose (x, a), and at most ℓ_n ways to choose (y, b). Note that choosing (x, a) for the first directed edge and (y, b) for the second directed edge results in a different switch from vice versa.

To find a lower bound on the number of ways of performing a forward switching, we subtract from ℓ_n^2 an upper bound on the number of invalid choices for (x, a) and (y, b). Such invalid choices can be categorized as follows:

(a) at least one of $\{u, x\}$, $\{a, b\}$, $\{v, y\}$ is an edge in G;
(b) at least one of $\{x, a\}$ or $\{y, b\}$ is in U;
(c) any vertex overlap other than $x = y$ (i.e., if one of a or b is equal to one of x or y, or if $a = b$, or if one of u or v is one of $\{a, b, x, y\}$).

We now bound all these different categories of invalid choices. To find an upper bound for (a), note that any choice in case (a) must involve a single edge, and a 2-path starting from a specified vertex. By Lemma 1.15, the number of choices for (a) is then upper bounded by $3 \times o(\ell_n) \times \ell_n = o(\ell_n^2)$ (noting that $n = \Theta(\ell_n)$). The number of choices for case (b) is $O(\ell_n)$, as $|U| = O(1)$, and there are at most ℓ_n ways to choose the other directed edge, which is not restricted to be in U.

To bound the number of choices for (c), we investigate each case:

(c1) Either a or b is equal to x or y, or $a = b$. In this case, x, y, a, b forms a 2-path in G. Thus, there are at most $5 \times n \times o(\ell_n) = o(\ell_n^2)$ choices (noting that $n = \Theta(\ell_n)$), where n is the number of ways to choose a vertex, and $o(\ell_n)$ bounds the number of 2-paths starting from this specified vertex.

(c2) One of u and v is one of $\{a, b, x, y\}$. In this case, there is one 2-path starting from u or v, and a single edge. Thus, there are at most $8 \times \ell_n \times d_{\max} = o(\ell_n^2)$ choices, where $d_{\max} = \max_{v \in [n]} d_v$ bounds the number of ways to choose a vertex adjacent to u or v and ℓ_n bounds the number of ways to choose a single edge, by Lemma 1.15.

Thus, the number of invalid choices for (x, a) and (y, b) is $o(\ell_n^2)$, so that the number of forward switchings which can be applied to any $G \in \mathcal{S}$ is $(1 + o(1))\ell_n^2$. We conclude that

$$\mathbb{E}[f(G)] = (1 + o(1))\ell_n^2. \tag{1.3.53}$$

The Number of Backward Switchings: Computing $\mathbb{E}[b(G')]$

Given a graph $G' \in \bar{\mathcal{S}}$, consider the backward switchings that can be applied to G'. There are at most $\ell_n(d_u - |U_u|)(d_v - |U_v|)$ ways to do the backward switching, since we are choosing an edge that is adjacent to u but not in U, an edge that is adjacent to v but not in U, and another directed edge (a, b). For a lower bound, we consider the following forbidden choices:

(a') at least one of $\{x, a\}$ or $\{y, b\}$ is an edge;

(b') $\{a, b\} \in U$;

(c') any vertices overlapping other than $x = y$ (i.e., when $\{a, b\} \cap \{u, v, x, y\} \neq \varnothing$).

We now go through each of these forbidden cases.

For (a'), suppose that $\{x, a\}$ is present, giving the 2-path $\{x, a\}, \{a, b\}$ in G'. There are at most $(d_u - |U_u|)(d_v - |U_v|)$ ways to choose x and y. Given any choice for x and y, by Lemma 1.15, there are at most $o(\ell_n)$ ways to choose a 2-path starting from x in G', and hence $o(\ell_n)$ ways to choose a, b. Thus, the total number of choices is at most $o((d_u - |U_u|)(d_v - |U_v|)\ell_n)$. The case where $\{y, b\}$ is an edge is symmetric.

For (b'), there are $O(1)$ choices for choosing $\{a, b\}$ since $|U| = O(1)$, and at most $(d_u - |U_u|)(d_v - |U_v|)$ choices for x and y. Thus, the number of choices for case (b') is $O((d_u - |U_u|)(d_v - |U_v|)) = o((d_u - |U_u|)(d_v - |U_v|)\ell_n)$.

For (c'), the case where a or b is equal to x or y corresponds to a 2-path starting from u or v together with a single edge from u or v. Since $o(\ell_n)$ bounds the number of 2-paths starting from u or v and $d_u - |U_u| + d_v - |U_v|$ bounds the number of ways to choose the single edge, there are $o(\ell_n(d_v - |U_v|)) + o(\ell_n(d_u - |U_u|))$ total choices. If a or b is equal to u or v, there are $(d_u - |U_u|)(d_v - |U_v|)$ ways to choose x and y, and at most $d_u + d_v$ ways to choose the last vertex as a neighbor of u or v. Thus, there are $O((d_u - |U_u|)(d_v - |U_v|)d_{\max}) = o((d_u - |U_u|)(d_v - |U_v|)\ell_n)$ total choices, since $d_{\max} = O(n^{1/(\tau - 1)}) = o(n) = o(\ell_n)$.

We conclude that the number of backward switchings that can be applied to any graph $G' \in \mathcal{S}'$ is $(d_u - |U_u|)(d_v - |U_v|)\ell_n(1 + o(1))$, so that

$$\mathbb{E}[b(G')] = (d_u - |U_u|)(d_v - |U_v|)\ell_n(1 + o(1)). \tag{1.3.54}$$

Conclusion

Combining (1.3.52), (1.3.53), and (1.3.54) results in

$$|\bar{\mathcal{S}}|/|\mathcal{S}| = (1 + o(1))\frac{\ell_n^2}{(d_u - |U_u|)(d_v - |U_v|)\ell_n}, \tag{1.3.55}$$

and thus (1.3.50) yields

$$\mathbb{P}(\{u,v\} \in E(\mathrm{UG}_n(\boldsymbol{d})) \mid \mathcal{E}_U) = \frac{1}{1 + |\bar{S}|/|S|}$$

$$= (1 + o(1)) \frac{(d_u - |U_u|)(d_v - |U_v|)}{\ell_n + (d_u - |U_u|)(d_v - |U_v|)}. \quad (1.3.56)$$

$\square$

Remark 1.16 (Uniform random graphs and configuration models) Owing to the close links between uniform random graphs with prescribed degrees and configuration models, we treat the two models together, in Chapters 4 and 7. ◄

1.3.5 PREFERENTIAL ATTACHMENT MODELS

Most networks grow in time. Preferential attachment models describe growing networks, where the numbers of edges and vertices grow with time. Here we give a brief introduction. The model that we investigate produces a *graph sequence* denoted by $(\mathrm{PA}_n^{(m,\delta)}(a))_{n \geq 1}$ and which, for every time n, yields a graph of n vertices and mn edges for some $m = 1, 2, \ldots$ This model is denoted by $(\mathrm{PA}_n^{(m,\delta)}(a))_{n \geq 1}$ in [V1, Chapter 8]. Below, we define $(\mathrm{PA}_n^{(m,\delta)}(b))_{n \geq 1}$ and $(\mathrm{PA}_n^{(m,\delta)}(d))_{n \geq 1}$, which are variations of this model.

We start by defining the model for $m = 1$, for which the graph consists of a collection of trees. In this case, $\mathrm{PA}_1^{(1,\delta)}(a)$ consists of a single vertex with a single self-loop. We denote the vertices of $\mathrm{PA}_n^{(1,\delta)}(a)$ by $v_1^{(1)}, \ldots, v_n^{(1)}$. We denote the degree of vertex $v_i^{(1)}$ in $\mathrm{PA}_n^{(1,\delta)}(a)$ by $D_i(n)$, where, by convention, a self-loop increases the degree by 2.

We next describe the evolution of the graph. Conditional on $\mathrm{PA}_n^{(1,\delta)}(a)$, the growth rule to obtain $\mathrm{PA}_{n+1}^{(1,\delta)}(a)$ is as follows. Add a single vertex $v_{n+1}^{(1)}$ having a single edge. This edge is connected to a second vertex (including itself), according to the probabilities

$$\mathbb{P}\big(v_{n+1}^{(1)} \to v_i^{(1)} \mid \mathrm{PA}_n^{(1,\delta)}(a)\big) = \begin{cases} \dfrac{1+\delta}{n(2+\delta)+(1+\delta)} & \text{for } i = n+1, \\[2ex] \dfrac{D_i(n)+\delta}{n(2+\delta)+(1+\delta)} & \text{for } i \in [n]. \end{cases} \quad (1.3.57)$$

This preferential attachment mechanism is called *affine*, since the attachment probabilities in (1.3.57) depend in an affine way on the degrees of the random graph $\mathrm{PA}_n^{(1,\delta)}(a)$.

The model with $m > 1$ is defined in terms of the model for $m = 1$ as follows. Fix $\delta \geq -m$. We start with $\mathrm{PA}_{mn}^{(1,\delta/m)}(a)$, and denote the vertices in $\mathrm{PA}_{mn}^{(1,\delta/m)}(a)$ by $v_1^{(1)}, \ldots, v_{mn}^{(1)}$. Then we identify or collapse the m vertices $v_1^{(1)}, \ldots, v_m^{(1)}$ in $\mathrm{PA}_{mn}^{(1,\delta/m)}(a)$ to become vertex $v_1^{(m)}$ in $\mathrm{PA}_n^{(m,\delta)}(a)$. In doing so, we let all the edges that are incident to any of the vertices in $v_1^{(1)}, \ldots, v_m^{(1)}$ be incident to the new vertex $v_1^{(m)}$ in $\mathrm{PA}_n^{(m,\delta)}(a)$. Then, we collapse the m vertices $v_{m+1}^{(1)}, \ldots, v_{2m}^{(1)}$ in $\mathrm{PA}_{mn}^{(1,\delta/m)}(a)$ to become vertex $v_2^{(m)}$ in $\mathrm{PA}_n^{(m,\delta)}(a)$, etc. More generally, we collapse the m vertices $v_{(j-1)m+1}^{(1)}, \ldots, v_{jm}^{(1)}$ in $\mathrm{PA}_{mn}^{(1,\delta/m)}(a)$ to become vertex $v_j^{(m)}$ in $\mathrm{PA}_n^{(m,\delta)}(a)$. This defines the model for general $m \geq 1$.

The resulting graph $\mathrm{PA}_n^{(m,\delta)}(a)$ is a multi-graph with precisely n vertices and mn edges, so that the total degree is equal to $2mn$. The model with $\delta = 0$ is sometimes called the *proportional* model. The inclusion of the extra parameter $\delta > -m$ is relevant, though, as we will see later. It can be useful to think of edges and vertices as carrying *weights*, where a vertex has weight δ and an edge has weight 1. Then, the vertex $v_{n+1}^{(1)}$ attaches its edges with a probability proportional to the weight of the vertex plus the edges to which it is incident. This, for example, explains why $\mathrm{PA}_{mn}^{(1,\delta/m)}(a)$ needs to be used in the collapsing procedure, rather than $\mathrm{PA}_{mn}^{(1,\delta)}(a)$.

The preferential attachment model $(\mathrm{PA}_n^{(m,\delta)}(a))_{n \geq 1}$ is increasing in time, in the sense that vertices and edges, once they have appeared, remain there forever. Thus, the degrees are monotonically increasing in time. Moreover, vertices with a high degree have a higher chance of attracting further edges of later vertices. Therefore, the model is sometimes called the *rich-get-richer model*. It is not hard to see that $D_i(n) \overset{a.s.}{\longrightarrow} \infty$ for each fixed $i \geq 1$, as $n \to \infty$ (see Exercise 1.20). As a result, one could also call the preferential attachment model the *old-get-richer model*.

Degrees of Fixed Vertices

We start by investigating the degrees of fixed vertices as $n \to \infty$, i.e., we will study $D_i(n)$ for fixed i as $n \to \infty$. To formulate our results, we define the *Gamma function* $t \mapsto \Gamma(t)$ for $t > 0$ by

$$\Gamma(t) = \int_0^\infty x^{t-1} e^{-x} \mathrm{d}x. \tag{1.3.58}$$

The following theorem describes the evolution of the degree of fixed vertices:

Theorem 1.17 (Degrees of fixed vertices)   *Consider* $\mathrm{PA}_n^{(m,\delta)}(a)$ *with* $m \geq 1$ *and* $\delta > -m$. *Then,* $D_i(n)/n^{1/(2+\delta/m)}$ *converges almost surely to a random variable* ξ_i *as* $n \to \infty$.

Proof   This is to be found in [V1, Theorem 8.2 and (8.3.11)].   □

It turns out that also $n^{-1/(2+\delta/m)} \max_{v \in [n]} D_v(n) \overset{a.s.}{\longrightarrow} M$ for some limiting positive and finite random variable M (see [V1, Section 8.7]). In analogy to iid random variables, this fact suggests that the degree of a random vertex satisfies a power law with power-law exponent $\tau = 3 + \delta/m$, and this is our next item on the agenda.

Degree Sequence of the Preferential Attachment Model

We write

$$P_k(n) = \frac{1}{n} \sum_{i \in [n]} \mathbb{1}_{\{D_i(n) = k\}} \tag{1.3.59}$$

for the (random) proportion of vertices with degree k at time n. For $m \geq 1$ and $\delta > -m$, we define $(p_k)_{k \geq 0}$ by $p_k = 0$ for $k = 0, \ldots, m-1$ and, for $k \geq m$,

$$p_k = (2 + \delta/m) \frac{\Gamma(k+\delta)\Gamma(m+2+\delta+\delta/m)}{\Gamma(m+\delta)\Gamma(k+3+\delta+\delta/m)}. \tag{1.3.60}$$

It turns out that $(p_k)_{k \geq 0}$ is a probability mass function (see [V1, Section 8.4]). It arises as the limiting degree distribution for $\mathrm{PA}_n^{(m,\delta)}(a)$, as shown in the following theorem:

Theorem 1.18 (Degree sequence in preferential attachment model)   *Consider* $\mathrm{PA}_n^{(m,\delta)}(a)$ *with* $m \geq 1$ *and* $\delta > -m$. *There exists a constant* $C = C(m,\delta) > 0$ *such that, as* $n \to \infty$,

$$\mathbb{P}\left(\max_k |P_k(n) - p_k| \geq C\sqrt{\frac{\log n}{n}} \right) = o(1). \tag{1.3.61}$$

Proof   See [V1, Theorem 8.3].   □

We next investigate the scale-free properties of $(p_k)_{k \geq 0}$ by investigating the asymptotics of p_k for k large. By (1.3.60) and Stirling's formula, as $k \to \infty$ we have

$$p_k = c_{m,\delta} k^{-\tau}(1 + O(1/k)), \tag{1.3.62}$$

where

$$\tau = 3 + \delta/m > 2, \qquad \text{and} \qquad c_{m,\delta} = (2 + \delta/m)\frac{\Gamma(m + 2 + \delta + \delta/m)}{\Gamma(m + \delta)}. \tag{1.3.63}$$

Therefore, by Theorem 1.18 and (1.3.62), the asymptotic degree sequence of $\mathrm{PA}_n^{(m,\delta)}(a)$ is close to a power law with exponent $\tau = 3 + \delta/m$. We note that any exponent $\tau > 2$ can be obtained by choosing $\delta > -m$ and $m \geq 1$ appropriately.

Related Preferential Attachment Rules

In this book, we also sometimes investigate the related $(\mathrm{PA}_n^{(m,\delta)}(b))_{n \geq 1}$ model, in which self-loops for $m = 1$ in (1.3.57) are not allowed, so that

$$\mathbb{P}\left(v_{n+1}^{(1)} \to v_i^{(1)} \mid \mathrm{PA}_n^{(1,\delta)}(b)\right) = \frac{D_i(n) + \delta}{n(2 + \delta)} \qquad \text{for } i \in [n]. \tag{1.3.64}$$

For $m = 1$, this model starts with two vertices and two edges in between, so that at time n, there are precisely n edges. The model for $m \geq 2$ is again defined in terms of the model $(\mathrm{PA}_{nm}^{(1,\delta/m)}(b))_{n \geq 1}$ for $m = 1$ by collapsing blocks of m vertices, so that $\mathrm{PA}_n^{(m,\delta)}(b)$ has n vertices and mn edges. The advantage of $(\mathrm{PA}_n^{(m,\delta)}(b))_{n \geq 1}$ compared to $(\mathrm{PA}_n^{(m,\delta)}(a))_{n \geq 1}$ is that $(\mathrm{PA}_n^{(m,\delta)}(b))_{n \geq 1}$ is naturally *connected*, while $(\mathrm{PA}_n^{(m,\delta)}(a))_{n \geq 1}$ may not be. Note that $\mathrm{PA}_n^{(m,\delta)}(b)$ can contain self-loops when $m \geq 2$, due to the collapsing procedure.

(Model $(\mathrm{PA}_n^{(m,\delta)}(c))_{n \geq 1}$, as formulated in [V1, Section 8.3], is defined by connecting edges with probability α to a uniformly chosen vertex and with probability $1 - \alpha$ to a vertex chosen proportionally to its degrees. It turns out to be equivalent to $(\mathrm{PA}_n^{(m,\delta)}(a))_{n \geq 1}$. We will not discuss this model further here.)

Another adaptation of the preferential attachment rule arises when no self-loops are ever allowed, while the degrees are updated when the m edges incident to the new vertex are being attached. We denote this model by $(\mathrm{PA}_n^{(m,\delta)}(d))_{n \geq 1}$. In this case, the model starts at time $n = 2$ with two vertices labeled 1 and 2, and m edges between them. $\mathrm{PA}_n^{(m,\delta)}(d)$ has vertex set $[n]$, and $m(n-1)$ edges. At time $n + 1$, for $m \geq 1$ and $j \in \{0, \ldots, m-1\}$, we attach the $(j+1)$th edge of vertex $v_{n+1}^{(m)}$ to vertex $v_i^{(m)}$ with probability

$$\mathbb{P}\left(v_{n+1,j+1}^{(m)} \to v_i^{(m)} \mid \mathrm{PA}_{n,j}^{(m,\delta)}(d)\right) = \frac{D_i(n,j) + \delta}{n(2m + \delta)} \qquad \text{for } i \in [n]. \tag{1.3.65}$$

Here, $D_i(n, j)$ is the degree of vertex $v_i^{(m)}$ after the connection of the edges incident to the first $n + 1$ vertices, as well as the first j edges incident to vertex $v_{n+1}^{(m)}$, and $\text{PA}_{n,j}^{(m,\delta)}(d)$ is the graph of the first n vertices, as well as the first j edges incident to vertex $v_{n+1}^{(m)}$. The model is by default connected, and at time n consists of $n + 1$ vertices and mn edges. For $m = 1$, apart from the different starting graphs, models (b) and (d) are identical. Indeed, $\text{PA}_n^{(1,\delta)}(b)$ for $n = 2$ consist of two vertices with two edges between them, while $\text{PA}_n^{(1,\delta)}(d)$ for $n = 2$ consists of two vertices with one edge between them for $\text{PA}_n^{(1,\delta)}(d)$.

Many other adaptations are possible and have been investigated in the literature, such as settings where the m edges incident to $v_{n+1}^{(m)}$ are *independently* connected as in (1.3.65) when $j = 0$. We refrain from discussing these. It is not hard to verify that Theorem 1.18 holds for all these adaptations, which explains why authors have often opted for the version of the model that is most convenient for them. From the perspective of local convergence, it turns out that $(\text{PA}_n^{(m,\delta)}(d))_{n \geq 1}$ is the most convenient, as we will see in Chapter 5. On the other hand, Theorem 1.17 contains minor adaptations between models, particularly since the limiting random variables $(\xi_i)_{i \geq 1}$ *do* depend on the precise model.

Bernoulli Preferential Attachment Model

We finally discuss a model that is quite a bit different from the other preferential attachment models discussed above. The main difference is that in this model, the number of edges is *not* fixed, but instead there is *conditional independence* in the edge attachments. We call this model the *Bernoulli preferential attachment model*, as the attachment indicators are all conditionally independent Bernoulli variables. Let us now give the details.

Fix a preferential attachment function $f\colon \mathbb{N}_0 \mapsto (0, \infty)$. Then, the graph evolves as follows. We start with a graph $\text{BPA}_1^{(f)}$ containing one vertex v_1 and no edges. At each time $n \geq 2$, we add a vertex v_n. Conditional on $\text{BPA}_{n-1}^{(f)}$, and independently for every $v \in [n - 1]$, we connect this vertex to v by a directed edge with probability

$$\frac{f(D_v^{(\text{in})}(n - 1))}{n - 1}, \tag{1.3.66}$$

where $D_v^{(\text{in})}(n - 1)$ is the in-degree of vertex v at time $n - 1$. This creates the random graph $\text{BPA}_n^{(f)}$. Note that the number of edges in the random graph process $(\text{BPA}_n^{(f)})_{n \geq 1}$ is a random variable, and thus *not* fixed. In particular, it makes a difference whether we use the in-degree in (1.3.66) or the total degree.

We consider functions $f\colon \mathbb{N} \mapsto (0, \infty)$ that satisfy that $f(k + 1) - f(k) < 1$ for every $k \geq 0$. Under this assumption and when $f(0) \leq 1$, the empirical degree sequence converges as $n \to \infty$, i.e.,

$$P_k(n) \equiv \frac{1}{n} \sum_{i \in [n]} \mathbb{1}_{\{D_i(n)=k\}} \xrightarrow{\mathbb{P}} p_k, \quad \text{where} \quad p_k = \frac{1}{1 + f(k)} \prod_{l=0}^{k-1} \frac{f(l)}{1 + f(l)}. \tag{1.3.67}$$

In particular, $\log(1/p_k)/\log k \to 1 + 1/\gamma$ when $f(k)/k \to \gamma \in (0, 1)$ (see Exercise 1.23). Remarkably, when $f(k) = \gamma k + \beta$, the power-law exponent of the degree distribution does not depend on β. The restriction that $f(k + 1) - f(k) < 1$ is needed to prevent the degrees from exploding. Further, $\log(1/p_k) \sim k^{1-\alpha}/(\gamma(1-\alpha))$ when $f(k) \sim \gamma k^\alpha$ for some $\alpha \in (0,1)$ (see Exercise 1.24). Interestingly, there exists a persistent hub, i.e., a vertex that has maximal degree for all but finitely many times, when $\sum_{k\geq 1} 1/f(k)^2 < \infty$. When $\sum_{k\geq 1} 1/f(k)^2 = \infty$, this does not happen.

1.3.6 UNIVERSALITY OF RANDOM GRAPHS

There are many other graph topologies where one can expect results similar to those in the random graphs discussed above. We will discuss many related models in Chapter 9, where we include several settings that are relevant in practice, such as *directed graphs* as well as graphs with *community structure* and *geometry*. The random graph models that we investigate are *inhomogeneous*, and one can expect that the results depend sensitively on the amount of inhomogeneity present. This is reflected in the results that we prove, where the precise asymptotics is different when the vertices have heavy-tailed rather than light-tailed degrees. However, interestingly, what is "heavy tailed" and what is "light tailed" depends on the precise model and setting at hand. Often, as we will see, the distinction depends on how many moments the degree distribution has.

We have proposed many random graph models for real-world networks. Since these models aim at describing similar real-world networks, one would hope that they also give similar answers. Indeed, for a real-world network with a power-law degree sequence, we could model its static structure by the configuration model with the same degree sequence, and its dynamical properties by the preferential attachment model with similar scale-free degrees. How do we interpret implications to the real world when these attempts give completely different predictions?

Universality is the phrase physicists use when different models display similar behavior. Models that show similar behavior are then in the same *universality class*. Enormous effort has gone, and is currently going, into deciding whether various random graph models are in the same universality class, or rather in different ones, and why. We will see that the *degree distribution* decides the universality class for a wide range of models, as one might possibly hope. This also explains why the degree distribution plays such a dominant role in the investigation of random graphs. See Chapter 9 for more details.

1.4 POWER LAWS AND THEIR PROPERTIES

In this book, we frequently deal with random variables having an (asymptotic) power-law distribution. For such random variables, we often need to investigate *truncated moments*, and we also often deal with their *sized-biased distribution*. In this section, we collect some results concerning power-law random variables. We start by recalling the definition of a power-law distribution:

Definition 1.19 (Power-law distributions)   We say that X has a *power-law distribution* with exponent τ when there exists a function $x \mapsto L(x)$ that is slowly varying at infinity such that

$$1 - F_X(x) = \mathbb{P}(X > x) = L(x)x^{-(\tau-1)}. \tag{1.4.1}$$

Here, we recall that a function $x \mapsto L(x)$ is *slowly varying at infinity* when, for every $t > 0$,

$$\lim_{x \to \infty} \frac{L(xt)}{L(x)} = 1. \tag{1.4.2}$$

◄

A crucial result about slowly varying functions is *Potter's Theorem*, which we next recall:

Theorem 1.20 (Potter's Theorem)   *Let $x \mapsto L(x)$ be slowly varying at infinity. For every δ, there exists a constant $C_\delta \geq 1$ such that, for all $x \geq 1$,*

$$x^{-\delta}/C_\delta \leq L(x) \leq C_\delta x^\delta. \tag{1.4.3}$$

Theorem 1.20 implies that the tail of any general power-law distribution, as in Definition 1.19, can be bounded above and below by that of a *pure* power-law distribution (i.e., one without a slowly varying function) with a slightly adapted power-law exponent. As a result, we can often deal with pure power laws instead.

We continue by studying the relation between power-law tails of the empirical degree distribution and bounds on the degrees themselves:

Lemma 1.21 (Tail and degree bounds)   *Let $\boldsymbol{d} = (d_v)_{v \in [n]}$ be a degree distribution, $d_{(1)} \geq d_{(2)} \geq \cdots \geq d_{(n-1)} \geq d_{(n)}$ its non-increasing ordered version, and*

$$F_n(x) = \frac{1}{n} \sum_{v \in [n]} \mathbb{1}_{\{d_v \leq x\}} \tag{1.4.4}$$

its empirical distribution. Then

$$[1 - F_n](x) \leq c_F x^{-(\tau-1)} \qquad \forall x \geq 1 \tag{1.4.5}$$

implies that

$$d_{(v)} \leq (c_F n/v)^{1/(\tau-1)} + 1 \qquad \forall v \in [n], \tag{1.4.6}$$

while

$$d_{(v)} \leq (c_F n/v)^{1/(\tau-1)} \qquad \forall v \in [n] \tag{1.4.7}$$

implies that

$$[1 - F_n](x) \leq c_F x^{-(\tau-1)} \qquad \forall x \geq 1. \tag{1.4.8}$$

Proof   Assume first that (1.4.5) holds. For every $v \in [n]$, the number of vertices with degree at least $d_{(v)}$ is at least v. By (1.4.5), for every $v \in [n]$,

$$c_F n(d_{(v)} - 1)^{1-\tau} \geq n[1 - F_n](d_{(v)} - 1) \geq v. \tag{1.4.9}$$

Thus, $d_{(v)} \leq (c_F n/v)^{1/(\tau-1)} + 1$, as required.

Next, assume that (1.4.7) holds. Then

$$[1 - F_n](x) = \frac{1}{n} \sum_{v \in [n]} \mathbb{1}_{\{d_v > x\}} = \frac{1}{n} \sum_{v \in [n]} \mathbb{1}_{\{d_{(v)} > x\}}$$

$$\leq \frac{1}{n} \sum_{v \in [n]} \mathbb{1}_{\{(c_F n/v)^{1/(\tau-1)} > x\}}$$

$$= \frac{1}{n} \sum_{v \in [n]} \mathbb{1}_{\{v < n c_F x^{-(\tau-1)}\}} \leq c_F x^{-(\tau-1)}, \qquad (1.4.10)$$

as required. $\qquad\qquad\qquad\qquad\qquad\qquad\qquad\qquad\qquad\qquad\qquad\qquad\qquad\qquad$ $\square$

We next study truncated moments of random variables whose tail is bounded by that of a power law:

Lemma 1.22 (Truncated moments) *Let X be a non-negative random variable whose distribution function $F_X(x) = \mathbb{P}(X \leq x)$ satisfies, for every $x \geq 1$,*

$$1 - F_X(x) \leq C_X x^{-(\tau-1)}. \qquad (1.4.11)$$

Then, for all $a < \tau - 1$, there exists a constant $C_X(a)$ such that, for all $\ell \geq 1$,

$$\mathbb{E}[X^a \mathbb{1}_{\{X > \ell\}}] \leq C_X(a) \ell^{a-(\tau-1)}, \qquad (1.4.12)$$

while, for $a > \tau - 1$ and all $\ell \geq 1$,

$$\mathbb{E}[X^a \mathbb{1}_{\{X \leq \ell\}}] \leq C_X(a) \ell^{a-(\tau-1)}. \qquad (1.4.13)$$

Proof We note that for any cumulative distribution function $x \mapsto F_X(x)$ on the non-negative reals, we have a *partial integration identity*, stating that, for every $f \colon \mathbb{R} \to \mathbb{R}$,

$$\int_u^\infty f(x) F_X(\mathrm{d}x) = f(u)[1 - F_X(u)] + \int_u^\infty [f(x) - f(u)] F_X(\mathrm{d}x)$$

$$= f(u)[1 - F_X(u)] + \int_u^\infty \int_u^x f'(y)\mathrm{d}y F_X(\mathrm{d}x)$$

$$= f(u)[1 - F_X(u)] + \int_u^\infty f'(y) \int_y^\infty F_X(\mathrm{d}x)\mathrm{d}y$$

$$= f(u)[1 - F_X(u)] + \int_u^\infty f'(y)[1 - F_X(y)]\mathrm{d}y, \qquad (1.4.14)$$

provided that either (a) $y \mapsto f'(y)[1 - F_X(y)]$ is absolutely integrable, or (b) $x \mapsto f(x)$ is either non-decreasing or non-increasing, so that $f'(y)[1 - F_X(y)]$ has a fixed sign. Here, the interchange of the integration order is allowed by Fubini's Theorem for non-negative functions (Halmos, 1950, Section 36, Theorem B) when $x \mapsto f(x)$ is non-decreasing, and by Fubini's Theorem for absolutely–integrable functions (Halmos, 1950, Section 36, Theorem C) when $y \mapsto f'(y)[1 - F_X(y)]$ is absolutely integrable. Similarly, for f with $f(0) = 0$,

$$\int_0^u f(x) F_X(\mathrm{d}x) = \int_0^u \int_0^x f'(y)\mathrm{d}y F_X(\mathrm{d}x) = \int_0^u f'(y) \int_y^u F_X(\mathrm{d}x)\mathrm{d}y$$

$$= \int_0^u f'(y)[F_X(u) - F_X(y)]\mathrm{d}y. \qquad (1.4.15)$$

When $X \geq 0$, using (1.4.11) and (1.4.14), for $a < \tau - 1$ and $\ell > 0$,

$$\mathbb{E}\left[X^a \mathbf{1}_{\{X > \ell\}}\right] = \ell^a \mathbb{P}(X > \ell) + \int_\ell^\infty a x^{a-1} \mathbb{P}(X > x) \mathrm{d}x$$

$$\leq C_X \ell^{a-(\tau-1)} + a C_X \int_\ell^\infty x^{a-1} x^{-(\tau-1)} \mathrm{d}x \leq C_X(a) \ell^{a-(\tau-1)}. \quad (1.4.16)$$

as required. Further, for $a > \tau - 1$ and $\ell > 0$, now using (1.4.15),

$$\mathbb{E}\left[X^a \mathbf{1}_{\{X \leq \ell\}}\right] \leq a C_X \int_0^\ell x^{a-1} x^{-(\tau-1)} \mathrm{d}x \leq C_X(a) \ell^{a-(\tau-1)}. \quad (1.4.17)$$

$\square$

An important notion in many graphs is the *size-biased* version $X^\star$ of a non-negative random variable X, which is given by

$$\mathbb{P}(X^\star \leq x) = \frac{\mathbb{E}[X \mathbf{1}_{\{X \leq x\}}]}{\mathbb{E}[X]}. \quad (1.4.18)$$

Exercise 1.25 shows that the size-biased distribution of the degree of a random vertex is the degree of a random vertex in a random edge. Let $F_X^\star$ denote the distribution function of $X^\star$. The following lemma gives bounds on the tail of the distribution function $F_X^\star$:

Lemma 1.23 (Size-biased tail distribution) *Let X be a non-negative random variable whose distribution function $F_X(x) = \mathbb{P}(X \leq x)$ satisfies that there exists a C_X such that, for every $x \geq 1$,*

$$1 - F_X(x) \leq C_X x^{-(\tau-1)}. \quad (1.4.19)$$

Assume that $\tau > 2$, so that $\mathbb{E}[X] < \infty$. Further, assume that $\mathbb{E}[X] > 0$. Then, there exists a constant $C_X^\star$ such that

$$1 - F_X^\star(x) \leq C_X^\star x^{-(\tau-2)}. \quad (1.4.20)$$

Proof This follows immediately from (1.4.18), by using (1.4.12) with $a = 1$. $\square$

1.5 Notation and Preliminaries

Let us introduce some standard notation used throughout this book, and recall some properties of trees and Poisson processes.

Abbreviations

We write rhs for *right-hand side*, and lhs for *left-hand side*. Further, we abbreviate *with respect to* by wrt.

Random variables

We write $X \overset{d}{=} Y$ to denote that X and Y have the same distribution. We write $X \sim \mathrm{Be}(p)$ when X has a Bernoulli distribution with success probability p, i.e., $\mathbb{P}(X = 1) = 1 - \mathbb{P}(X = 0) = p$. We write $X \sim \mathrm{Bin}(n, p)$ when the random variable X has a binomial distribution with parameters n and p, and we write $X \sim \mathsf{Poi}(\lambda)$ when X has a Poisson distribution with parameter λ.

We write $X \sim \mathsf{Exp}(\lambda)$ when X has an exponential distribution with mean $1/\lambda$. We write $X \sim \mathsf{Gam}(r, \lambda)$ when X has a gamma distribution with scale parameter λ and shape parameter r, for which the density, for $x \geq 0$, is given by

$$f_X(x) = \lambda^r x^{r-1} \mathrm{e}^{-\lambda x}/\Gamma(r), \tag{1.5.1}$$

where $r, \lambda > 0$, and we recall (1.3.58), while $f_X(x) = 0$ for $x < 0$. The random variable $\mathsf{Gam}(r, \lambda)$ has mean r/λ and variance r/λ^2. Finally, we write $X \sim \mathsf{Beta}(\alpha, \beta)$ when X has a beta distribution with parameters $\alpha, \beta > 0$, so that X has density, for $x \in [0, 1]$,

$$f_X(x) = x^{\alpha-1}(1-x)^{\beta-1}/B(\alpha, \beta), \tag{1.5.2}$$

where

$$B(\alpha, \beta) = \frac{\Gamma(\alpha)\Gamma(\beta)}{\Gamma(\alpha + \beta)} \tag{1.5.3}$$

is the Beta-function, while $f_X(x) = 0$ for $x \notin [0, 1]$. We sometimes abuse notation, and write e.g., $\mathbb{P}(\mathsf{Bin}(n, p) = k)$ to denote $\mathbb{P}(X = k)$ when $X \sim \mathsf{Bin}(n, p)$.

We call a sequence of random variables $(X_i)_{i \geq 1}$ *independent and identically distributed* (iid) when they are independent, and X_i has the same distribution as X_1 for every $i \geq 1$. For a finite set $\mathcal{X}$, we say that $X \in \mathcal{X}$ is drawn *uniformly at random* (uar) when X has the uniform distribution on $\mathcal{X}$.

Convergence of Random Variables

We say that a sequence of events $(\mathcal{E}_n)_{n \geq 1}$ occurs *with high probability* (whp) when $\lim_{n \to \infty} \mathbb{P}(\mathcal{E}_n) = 1$.

For sequences of random variables $(X_n)_{n \geq 1}$, $X_n \xrightarrow{d} X$ denotes that X_n converges in distribution to X, while $X_n \xrightarrow{\mathbb{P}} X$ denotes that X_n converges in probability to X and $X_n \xrightarrow{a.s.} X$ denotes that X_n converges almost surely to X. We write that $X_n = O_{\mathbb{P}}(Y_n)$ when $|X_n|/Y_n$ is a tight sequence of random variables and $X_n = \Theta_{\mathbb{P}}(Y_n)$ when $X_n = O_{\mathbb{P}}(Y_n)$ and $Y_n = O_{\mathbb{P}}(X_n)$. Finally, we write that $X_n = o_{\mathbb{P}}(Y_n)$ when $X_n/Y_n \xrightarrow{\mathbb{P}} 0$.

Stochastic Domination

We recall that a random variable X is *stochastically dominated* by a random variable Y when $F_X(x) = \mathbb{P}(X \leq x) \geq F_Y(x) = \mathbb{P}(Y \leq x)$ for every $x \in \mathbb{R}$. We write this as $X \preceq Y$. See [V1, Section 2.3] for more details on stochastic ordering.

Two Useful Martingale Inequalities

Recall [V1, Section 2.5] for the definition of a martingale $(M_n)_{n \geq 0}$. We rely on Doob's martingale inequality, which for a martingale $(M_n)_{n \geq 0}$ states that

$$\mathbb{E}\left[\sup_{m \leq n} \left|M_m - \mathbb{E}[M_m]\right|^2\right] \leq \mathrm{Var}(M_n). \tag{1.5.4}$$

An immediate consequence is Kolmogorov's martingale inequality, which states that

$$\mathbb{P}(\sup_{m \leq n} |M_m - \mathbb{E}[M_m]| \geq \varepsilon) \leq \varepsilon^{-2} \operatorname{Var}(M_n). \tag{1.5.5}$$

Densities in Inhomogeneous Poisson Processes

Let Π be an inhomogeneous Poisson process with intensity measure $\Lambda \colon \mathcal{X} \to \mathbb{N}$, where $\mathcal{X} \subseteq [0, \infty)$. This means that the number of points $\Pi(\mathcal{A})$ in $\mathcal{A} \subseteq \mathcal{X}$ has a Poisson distribution with parameter $\int_{\mathcal{A}} \Lambda(\mathrm{d}x)$, and the number of points in disjoint sets are independent (see Last and Penrose (2018) for details on Poisson processes).

Let $\mathcal{A}$ be a bounded set. We wish to give a formula for the probability of the event that the points of Π in $\mathcal{A}$ (of which there are $\Pi(\mathcal{A})$) are in $(a_1 + \mathrm{d}a_1, \ldots, a_k + \mathrm{d}a_k)$, where we assume that $a_1 < a_2 < \cdots < a_k$. Thus, there is one point in $a_1 + \mathrm{d}a_1$, one in $a_2 + \mathrm{d}a_2$, etc. Note that this event is a subset of the event $\Pi(\mathcal{A}) = k$. Denote this event by $\mathcal{P}(a_1 + \mathrm{d}a_1, \ldots, a_k + \mathrm{d}a_k)$. Assume that $x \mapsto \Lambda(x)$ is continuous almost everywhere. Then, for all measurable $\mathcal{A} \subseteq \mathcal{X}$ and all ordered elements $a_1, \ldots, a_k \in \mathcal{A}$,

$$\mathbb{P}(\mathcal{P}(a_1 + \mathrm{d}a_1, \ldots, a_k + \mathrm{d}a_k)) = \mathrm{e}^{- \int_{\mathcal{A}} \Lambda(\mathrm{d}x)} \prod_{i=1}^{k} \Lambda(a_i) \mathrm{d}a_i. \tag{1.5.6}$$

We refer to $\mathrm{e}^{- \int_{\mathcal{A}} \Lambda(\mathrm{d}x)}$ as the *no-further-point probability*, in that it ensures that Π has precisely k points in $\mathcal{A}$. We refer to Exercise 1.26 for a proof that (1.5.6) implies that $\Pi(\mathcal{A}) \sim \operatorname{Poi}(\int_{\mathcal{A}} \Lambda(\mathrm{d}x))$, and Exercise 1.27 for a proof that (1.5.6) implies that $\Pi(\mathcal{A})$ and $\Pi(\mathcal{B})$ are independent when $\mathcal{A}$ and $\mathcal{B}$ are disjoint.

Ordered Trees and Their Exploration

In this book, *trees* play a central role, and it is important to be clear about exactly what we mean by a tree. Trees are rooted and ordered. A tree $\mathbf{t}$ has root $\varnothing$, vertex set $V(\mathbf{t})$ and edge set $E(\mathbf{t})$, and the vertex set will be given an ordering below.

It is convenient to think of a tree $\mathbf{t}$ with root $\varnothing$ as being labeled in the Ulam–Harris way, so that a vertex v in generation k has a label $\varnothing v_1 \cdots v_k$, where $v_i \in \mathbb{N}$. Naturally, there are some restrictions, in that if $\varnothing v_1 \cdots v_k \in V(\mathbf{t})$, then also $\varnothing v_1 \cdots v_{k-1} \in V(\mathbf{t})$, and $\varnothing v_1 \cdots (v_k - 1) \in V(\mathbf{t})$ when $v_k \geq 2$. We refer to [V1, Chapter 3] for details.

It will sometimes also be useful to explore trees in a breadth-first manner. This corresponds to the lexicographical ordering in the Ulam–Harris encoding of the tree. Ulam–Harris trees are also sometimes called *plane trees* (see, e.g., (Drmota, 2009, Chapter 1)). Let us now make the breadth-first ordering of the tree precise:

Definition 1.24 (Breadth-first order on a tree) For $v \in V(\mathbf{t})$, let $|v|$ be its height. Thus $|v| = k$ when $v = \varnothing v_1 \cdots v_k$ and $|\varnothing| = 0$. Let $u, v \in V(\mathbf{t})$. Then $u < v$ when either $|u| < |v|$ or $|u| = |v|$ and $u = \varnothing u_1 \cdots u_k$ and $v = \varnothing v_1 \cdots v_k$ are such that $(u_1, \ldots, u_k) < (v_1, \ldots, v_k)$ in the lexicographic sense. ◀

We next explain the breadth-first exploration of $\mathbf{t}$:

Definition 1.25 (Breadth-first exploration of a tree) For a tree $\mathbf{t}$ of size $|V(\mathbf{t})| = t$, we let $(a_k)_{k=0}^{t}$ be the elements of $V(\mathbf{t})$ ordered according to the breadth-first ordering of $\mathbf{t}$ (recall

Definition 1.24). For $i \geq 1$, let x_i denote the number of children of vertex a_i. Thus, if d_v denotes the degree of $v \in V(\mathbf{t})$ in the tree $\mathbf{t}$, we have $x_1 = d_{a_0} = d_\varnothing$ and $x_i = d_{a_i} - 1$ for $i \geq 2$. The recursion

$$s_i = s_{i-1} + x_i - 1 \quad \text{for } i \geq 1, \qquad \text{with} \qquad s_0 = 1, \qquad (1.5.7)$$

describes the evolution of the number of unexplored vertices in the breadth-first exploration. For a finite tree $\mathbf{t}$ of size $|V(\mathbf{t})| = t$, thus $s_i > 0$ for $i \in \{0, \dots, t-1\}$ while $s_t = 0$. ◄

The sequence $(x_i)_{i=1}^t$ gives an alternative encoding of the tree $\mathbf{t}$ that is often convenient. Indeed, by Exercise 1.28, the sequence $(x_i)_{i=1}^t$ is in one-to-one correspondence to $\mathbf{t}$.

Unimodular Branching Process Trees

We next describe one type of random tree that occurs frequently in our analyses, the so-called *unimodular branching process tree*:

Definition 1.26 (Unimodular branching process tree) Fix a probability distribution $(p_k)_{k \geq 1}$, where $p_k = \mathbb{P}(D = k)$ for some integer-valued random variable D. The *unimodular branching process tree* with root-offspring distribution $(p_k)_{k \geq 1}$ is the branching process where the root has offspring distribution $(p_k)_{k \geq 1}$, while all vertices in other generations have offspring distribution $p_k^\star$ given by

$$p_k^\star = \mathbb{P}(D^\star - 1 = k) = \frac{(k+1)}{\mathbb{E}[D]} \mathbb{P}(D = k+1), \qquad (1.5.8)$$

where we recall that $D^\star$ denotes the size-biased version of D in (1.4.18). ◄

It turns out that unimodular branching process trees arise as local limits of random graphs, seen from a uniform vertex. The distribution $(p_k)_{k \geq 1}$, where $p_k = \mathbb{P}(D = k)$, describes the *degree distribution* in the graph, while the law (1.5.8) is related to the degree distribution of other vertices that are close to a uniform vertex.

We let $\mathsf{BP}_{\leq r}$ denote the branching process up to and including generation r, and write $\mathsf{BP}_r = \mathsf{BP}_{\leq r} \setminus \mathsf{BP}_{\leq r-1}$ for the rth generation. It is convenient to think of the branching process tree, denoted as BP, as being labeled in the Ulam–Harris way (recall Definitions 1.24 and 1.25), so that a vertex v in generation r has a label $\varnothing a_1 \cdots a_r$, where $a_i \in \mathbb{N}$. When applied to BP, we denote this process by $(\mathsf{BP}(t))_{t \geq 1}$, where $\mathsf{BP}(t)$ consists of precisely $t+1$ vertices (with $\mathsf{BP}(0)$ equal to the root $\varnothing$).

1.6 Notes and Discussion for Chapter 1

Sections 1.1–1.3 are in the majority summaries of chapters in Volume 1, to which we refer for notes and discussion, so we restrict ourselves here to the exceptions.

Notes on Sections 1.1 and 1.2

See Barabási (2002) and Watts (2003) for expository accounts of the discovery of network properties by Barabási, Watts, and co-authors. Newman et al. (2006) bundles together some of the original papers detailing the empirical findings of real-world networks and the network models invented for them. The introductory book by Newman (2010) lists many of the empirical properties of, and scientific methods for, networks.

See also Barabási (2016) for an online book giving an extensive background to the science of networks, and Coscia (2021) for an "atlas to the aspiring network scientist."

The discussion of the scale-free phenomenon in Section 1.1.2 has been substantially extended compared with [V1, Section 1.4.1]. Artico et al. (2020) considered another static definition of the degree distribution, based on that in preferential attachment models (which they call the degree distribution of the *de Solla Price model* in honor of Price (1965), who invented the model for citation networks; see also Section 9.1.1). This can be seen as an interpolation between the static approach of Broido and Clauset (2019) and the dynamic approach advocated by Barabási (2018). Artico et al. (2020) used maximum-likelihood techniques to argue that power-law network degree distributions are not rare, classifying almost 65% of the tested networks as having a power-law tail with at least 80% power. We further refer to Nair et al. (2022) for an extensive discussion on heavy-tailed phenomena.

Notes on Section 1.3

Erdős–Rényi random graph. The seminal papers Erdős and Rényi (1959, 1960, 1961a,b) investigated a related model in which a collection of m edges is chosen uar from the collection of $\binom{n}{2}$ possible edges. The model described here as the Erdős–Rényi random graph was actually *not* invented by Erdős and Rényi, but rather by Gilbert (1959). When adding a fixed number of edges, the proportion of edges is $2m/n(n-1) \approx 2m/n^2$, so we should think of $m \approx 2\lambda n$ for a fair comparison. Note that when we condition the total number of edges in the independent-edge model to be equal to m, the law of the Erdős–Rényi random graph is equal to the model where a collection of m uniformly chosen edges is added, which explains the close relation between the two models. Owing to the concentration of the total number of edges, we can indeed roughly exchange the binomial model with $p = \lambda/n$ with the combinatorial model with $m = 2\lambda n$. The combinatorial model has the nice feature that it produces a uniform graph from the collection of all graphs with m edges, and thus could serve as a *null model* for a real-world network in which only the number of edges is fixed. We can also view $\mathrm{ER}_n(\lambda/n)$ as percolation on the complete graph. Percolation is a paradigmatic model in statistical physics describing random failures in networks (see Grimmett (1999) for an extensive overview of percolation theory focussing on $\mathbb{Z}^d$).

Inhomogeneous random graphs were first proposed by Söderberg (2002, 2003a,c,b). A general formalism for inhomogeneous random graphs is described in the seminal work of Bollobás et al. (2007). The generalized random graph was first introduced by Britton et al. (2006). The random graph with prescribed expected degrees, or Chung–Lu model, was introduced, and studied intensively, by Chung and Lu (2002a,b, 2003, 2006a,b). The Poissonian random graph or Norros–Reittu model was introduced by Norros and Reittu (2006). The conditions under which these random graphs are asymptotically equivalent [V1, Sections 6.7 and 6.8] were derived by Janson (2010a). Condition 1.1 has been slightly modified compared with [V1, Condition 6.4], in that we now assume that $\mathbb{E}[W] \in (0, \infty)$, which excludes trivial cases where the graph is almost empty.

Configuration model and uniform random graphs with prescribed degrees. The configuration model was invented by Bollobás (1980) to study uniform random regular graphs (see also (Bollobás, 2001, Section 2.4)). The introduction was inspired by, and generalized the results in, the work of Bender and Canfield (1978). The original work allowed for a careful computation of the number of regular graphs, using a probabilistic argument. This is the *probabilistic method* at its best, and also explains the emphasis on the study of the probability for the graph to be simple, as we will see below. The configuration model, as well as uniform random graphs with a prescribed degree sequence, were further studied in greater generality by Molloy and Reed (1995, 1998). This extension is quite relevant to us, as the scale-free nature of many real-world applications encourages us to investigate configuration models with power-law degree sequences. Condition 1.7 is a minor modification of [V1, Condition 7.8]. The terms "erased" and "repeated configuration model" were coined by Britton et al. (2006).

The degree-truncation argument for the configuration model is, to the best of our knowledge, novel. Switching algorithms, as discussed in Section 1.3.4 have a long history, dating back at least to McKay (1981), see also Erdős et al. (2022); Gao and Greenhill (2021); Gao and Wormald (2016); McKay and Wormald (1990), as well as McKay (2011) and the references therein for overviews. The literature on switch chains focusses on two key aspects: first, their rapid mixing (Erdős et al. (2022); Gao and Greenhill (2021), and various related papers, for which we refer to Erdős et al. (2022)), and, second, counting the number of simple graphs using switch chain arguments (as in Gao and Wormald (2016)), which is the approach that we take in this section. Rapid mixing means that the mixing time of the switch chain is bounded by an explicit power of the number of vertices (or number of edges, or both combined). The powers, however, tend to be large, and thus "rapid mixing" may not be rapid enough to give good guarantees when one is trying to sample a uniform random graph of the degree distribution of some real-world network. Theorem 1.13 is adapted from Gao et al. (2020), where it was used to compute the number of triangles in uniform random graphs with power-law degree distributions having infinite variance. See also Janson (2020b) for a relation between the configuration model and uniform random graphs using switchings.

Preferential attachment models were first introduced in the context of complex networks by Barabási and Albert (1999). Bollobás et al. (2001) studied the model by Barabási and Albert (1999), and later many other papers followed on this, and related, models. Barabási and Albert (1999) and Bollobás et al. (2001) focussed on the proportional model, for which $\delta = 0$. The affine model was proposed by Dorogovtsev et al. (2000). All these works were pre-dated by Price (1965); Simon (1955); Yule (1925); see [V1, Chapter 8] for more details on the literature. The Bernoulli preferential attachment model was introduced and investigated by Dereich and Mörters (2009, 2011, 2013).

Notes on Section 1.4

This material is folklore. A lot is known about slowly varying functions; we refer to the classic book on the topic by Bingham et al. (1989) for details.

Notes on Section 1.5

Our choice of notation was heavily influenced by Janson (2011), to which we refer for further background and equivalent notation.

1.7 Exercises for Chapter 1

Exercise 1.1 (Probability mass function typical degree)   *Prove that the probability mass function of the degree of a uniform vertex is given by* (1.1.3).

Exercise 1.2 (Growth maxima power-law random variables)   *Suppose that the non-negative random variable X satisfies that there exists $\tau > 1$,*

$$\mathbb{P}(X > x) = c_X x^{-(\tau-1)}. \tag{1.7.1}$$

Let $(X_v)_{v \in [n]}$ be a sequence of iid copies of X. Show that $M_n = \max_{v \in [n]} X_v$ satisfies $n^{-1/(\tau-1)} M_n \xrightarrow{d} M$ for some limiting random variable M.

Exercise 1.3 (Uniform random graph)   *Consider $\mathrm{ER}_n(p)$ with $p = \frac{1}{2}$. Show that the result is a uniform graph, i.e., it has the same distribution as a uniform choice from all the graphs on n vertices.*

Exercise 1.4 (Thin-tailed Poisson)   *Show that, $\lim_{k \to \infty} e^{\alpha k} p_k = 0$ for every $\alpha > 0$, where $p_k = e^{-\lambda} \lambda^k / k!$ denotes the Poisson probability mass function.*

Exercise 1.5 (A nice power-law distribution)   *Let the random variable X have generating function*

$$G_X(s) = \mathbb{E}[s^X] = 1 - (1 - s)^\alpha. \tag{1.7.2}$$

Fix $\alpha \in (0, 1)$. Identify the probability mass function $\mathbb{P}(X = k)$ of X.

Exercise 1.6 (A power-law distribution?)   *Consider $G(s) = 1 - (1 - s)^\alpha$ as in Exercise 1.5, now for $\alpha > 1$. Is $G(s)$ the generating function of a random variable?*

Exercise 1.7 (Weight of uniformly chosen vertex)   *Let o be a vertex chosen uar from $[n]$. Show that the weight w_o of o has the distribution function F_n given in* (1.3.10).

Exercise 1.8 (Maximal weight bound)   *Assume that Conditions 1.1(a) and (b) hold. Show that $\max_{v \in [n]} w_v = o(n)$. Further, show that $\max_{v \in [n]} w_v = o(\sqrt{n})$ when Conditions 1.1(a)–(c) hold.*

Exercise 1.9 (Domination weights)   *Let W_n have the distribution function F_n from* (1.3.17). *Show that W_n is stochastically dominated by the random variable W having distribution function F. Here we recall that W_n is stochastically dominated by W when $\mathbb{P}(W_n \leq w) \geq \mathbb{P}(W \leq w)$ for all $w \in \mathbb{R}$.*

Exercise 1.10 (Degree of uniformly chosen vertex in $\mathrm{GRG}_n(\boldsymbol{w})$)   *Prove that the asymptotic degree in $\mathrm{GRG}_n(\boldsymbol{w})$ satisfies* (1.3.22) *under the conditions of Theorem 1.3.*

Exercise 1.11 (Power-law degrees in generalized random graphs) *Prove that, under the conditions of Theorem 1.3, the degree power-law tail in (1.3.24) for $\mathrm{GRG}_n(\boldsymbol{w})$ follows from the weight power-law tail in (1.3.23). Does the converse also hold?*

Exercise 1.12 (Degree example) *Let the degree sequence $\boldsymbol{d} = (d_v)_{v \in [n]}$ be given by*

$$d_v = 1 + (v \bmod 3). \tag{1.7.3}$$

Show that Conditions 1.7(a)–(c) hold. What is the limiting degree variable D?

Exercise 1.13 (Poisson degree example) *Let the degree sequence $\boldsymbol{d} = (d_v)_{v \in [n]}$ satisfy*

$$n_k / n \to e^{-\lambda} \frac{\lambda^k}{k} \tag{1.7.4}$$

and

$$\sum_{k \geq 0} k n_k / n \to \lambda, \qquad \sum_{k \geq 0} k^2 n_k / n \to \lambda(\lambda + 1). \tag{1.7.5}$$

Show that Conditions 1.7(a)–(c) hold. What is the limiting degree variable D?

Exercise 1.14 (Power-law degree example) *Consider the random variable D having generating function, for $\alpha \in (0, 1)$,*

$$G_D(s) = s - (1 - s)^{\alpha+1} / (\alpha + 1). \tag{1.7.6}$$

What is the probability mass function of D?

Exercise 1.15 (Power-law degree example) *Consider the random variable D having generating function (1.7.6) with $\alpha \in (0, 1)$. Show that D has an asymptotic power-law distribution and compute its power-law exponent.*

Exercise 1.16 (Power-law degree example (cont.)) *Consider the degree sequence $\boldsymbol{d} = (d_v)_{v \in [n]}$ with $d_v = [1 - F]^{-1}(v/n)$, where F is the distribution of a random variable D having generating function (1.7.6) with $\alpha \in (0, 1)$. Show that Conditions 1.7(a) and (b) hold, but Condition 1.7(c) does not.*

Exercise 1.17 (Number of erased edges) *Assume that Conditions 1.7(a) and (b) hold. Show that Theorem 1.8 implies that the number of erased edges in $\mathrm{ECM}_n(\boldsymbol{d})$ is $o_{\mathbb{P}}(n)$.*

Exercise 1.18 (Edge probability of uniform random graphs with prescribed degrees) *Prove the formula for the (conditional) edge probabilities in uniform random graphs with prescribed degrees in (1.3.50).*

Exercise 1.19 (Edge probability of uniform random graphs with prescribed degrees (cont.)) *Prove the formula for the number of switches with and without a specific edge in uniform random graphs with prescribed degrees in (1.3.51). Hint: Use an "out-is-in" argument that the number of switches from S to $\bar{S}$ is the same as the number of switches that enter $\bar{S}$ from S.*

Exercise 1.20 (Degrees grow to infinity almost surely) *Consider the preferential attachment model $\mathrm{PA}_n^{(m,\delta)}(a)$. Fix $m = 1$ and $i \geq 1$. Prove that $D_i(n) \xrightarrow{a.s.} \infty$ as $n \to \infty$, by using $\sum_{s=i}^n I_s \preceq D_i(n)$, where $(I_n)_{n \geq i}$ is a sequence of independent Bernoulli random variables with $\mathbb{P}(I_n = 1) = (1+\delta)/(n(2+\delta) + 1 + \delta)$. What does this imply for $m > 1$?*

Exercise 1.21 (Degrees of fixed vertices) *Consider the preferential attachment model $\mathrm{PA}_n^{(m,\delta)}(a)$. Prove Theorem 1.17 for $m = 1$ and $\delta > -1$ using the martingale convergence theorem and the fact that*

$$M_i(n) = \frac{D_i(n) + \delta}{1 + \delta} \prod_{s=i-1}^{n-1} \frac{(2+\delta)s + 1 + \delta}{(2+\delta)(s+1)} \tag{1.7.7}$$

is a martingale for every $i \geq 1$ and for $n \geq i$.

Exercise 1.22 (Power-law degree sequence) *Prove that the limiting degree distribution of preferential attachment models in (1.3.60) satisfies the power-law asymptotics in (1.3.63) by using Stirling's formula.*

Exercise 1.23 (Degrees distribution of affine Bernoulli preferential attachment model) *Recall the limiting degree distribution* $(p_k)_{k \geq 0}$ *in (1.3.67). Show that* $p_k \sim c_{\gamma, \beta} k^{-(1+1/\gamma)}$ *when* $f(k) = \gamma k + \beta$. *What is* $c_{\gamma, \beta}$?

Exercise 1.24 (Degrees distribution of sublinear Bernoulli preferential attachment model) *Recall the limiting degree distribution* $(p_k)_{k \geq 0}$ *in (1.3.67). Show that* $\log(1/p_k) \sim k^{1-\alpha}/(\gamma(1-\alpha))$ *when* $f(k) \sim \gamma k^\alpha$ *for some* $\alpha \in (0, 1)$.

Exercise 1.25 (Size-biased degree distribution and random edges) *Let* D_n *be the degree of a random vertex in a graph* $G_n = (V(G_n), E(G_n))$ *of size* $|V(G_n)| = n$. *Let* $D_n^\star$ *be the degree of a random vertex in an edge drawn uar from* $E(G_n)$. *Show that* $D_n^\star$ *has the size-biased distribution of* D_n, *where we recall the definition of the size-biased distribution of a random variable from (1.4.18).*

Exercise 1.26 (Number of points in an inhomogeneous Poisson process) *Prove that the Poisson density formula in (1.5.6) implies that the number of points of the Poisson process in* $\mathcal{A}$ *has the appropriate Poisson distribution, i.e.,* $\Pi(\mathcal{A}) \sim \text{Poi}(\int_{\mathcal{A}} \Lambda(dx))$.

Exercise 1.27 (Number of points in an inhomogeneous Poisson process (cont.)) *In the setting of Exercise 1.26, show that (1.5.6) implies that* $\Pi(\mathcal{A})$ *and* $\Pi(\mathcal{B})$ *are independent when* $\mathcal{A}$ *and* $\mathcal{B}$ *are disjoint.*

Exercise 1.28 (Breadth-first encoding ordered rooted tree) *Recall Definitions 1.24 and 1.25 for the breadth-first order on, and exploration of, a rooted ordered tree. Show that the sequence* $(x_i)_{i=1}^t$ *is in one-to-one correspondence to the rooted ordered tree* t.

LOCAL CONVERGENCE
OF RANDOM GRAPHS

Abstract

In this chapter we discuss local convergence, which describes the intuitive notion that a finite graph, seen from the perspective of a typical vertex, looks like a certain limiting graph. Local convergence plays a profound role in random graph theory.

We give general definitions of local convergence in several probabilistic senses. We then show that local convergence in its various forms is equivalent to the appropriate convergence of subgraph counts. We continue by discussing several implications of local convergence, concerning local neighborhoods, clustering, assortativity, and PageRank. We further investigate the relation between local convergence and the size of the giant component, making the statement that the giant is "almost local" precise.

2.1 MOTIVATION: LOCAL LIMITS

The *local convergence* of finite graphs was first introduced by Benjamini and Schramm (2001) and, in a different context, independently by Aldous and Steele (2004). It describes the intuitive notion that a finite graph, seen from the perspective of a vertex that is chosen uar from the vertex set, looks like a certain limiting graph. This is already useful to in that it makes precise the notion that a finite cube in $\mathbb{Z}^d$ with large side length is locally much like $\mathbb{Z}^d$ itself. However, it plays an even more profound role in random graph theory. For example, local convergence to some limiting tree, which often occurs in random graphs, as we will see throughout this book, is referred to as *locally tree-like* behavior. Such trees are often branching processes; see for example [V1, Section 4.1] where this is worked out for the Erdős–Rényi random graph. Since trees are generally simpler objects than graphs, this means that, to understand a random graph, it often suffices to understand a branching process tree instead.

Local convergence is a central technique in random graph theory, since many properties of random graphs are in fact determined by a local limit. For example, the asymptotic number of spanning trees, the partition function of the Ising model, and the spectral distribution of the adjacency matrix of the graph all turn out to be computable in terms of the local limit. We refer to Section 2.7 for an extensive discussion of the highly non-trivial consequences of local convergence. Owing to its enormous power, local convergence has become an indispensable tool in the random graph theory of sparse graphs. In this book we will see several examples of quantities whose convergence and limit are determined by the local limit, including clustering, the size of the giant in most cases, and the PageRank distribution of sparse random graphs. In this chapter, we lay the general foundations of local convergence.

Organization of the Chapter

This chapter is organized as follows. In Section 2.2 we start by discussing the metric space of rooted graphs, which plays a crucial role in local convergence. In Section 2.3 we give the formal definition of local weak convergence for deterministic graphs. There, we discuss their convergence to some, surprisingly possibly random, limit. The randomness of the local weak limit originates from the fact that we consider the graph as rooted at a *random* vertex, and this randomness may persist in the limit. In Section 2.4 we then extend the notion of local convergence to *random* graphs, for which there are several notions of convergence, such as local weak convergence and local convergence in probability. In Section 2.5 we discuss the consequences of local convergence to local functionals, such as clustering and assortativity. For the latter, we extend the convergence of the neighborhood of a uniformly chosen vertex to that of a uniformly chosen *edge*. In Section 2.6 we discuss the consequences of local convergence on the giant component. While the proportion of vertices in the giant is not a continuous functional in the local topology, one could argue that it is "almost local", in a way that can be made precise. We close the chapter with notes and discussion in Section 2.7 and with exercises in Section 2.8. Some of the technicalities are deferred to the Appendix.

2.2 METRIC SPACE OF ROOTED GRAPHS

Local weak convergence is a notion of the weak convergence of finite rooted graphs. In general, weak convergence is equivalent to the convergence of expectations of continuous functions. For continuity, one needs a topology. Therefore, we start by discussing the topology of rooted graphs that is at the center of local weak convergence. We start with some definitions:

Definition 2.1 (Locally finite and rooted graphs) A *rooted graph* is a pair (G, o), where $G = (V(G), E(G))$ is a graph with vertex set $V(G)$, edge set $E(G)$, and root vertex $o \in V(G)$. Further, a rooted or non-rooted graph is called *locally finite* when each of its vertices has finite degree (though not necessarily uniformly bounded). ◀

In Definition 2.1, graphs can have finitely or infinitely many vertices, but we always have graphs in mind that are locally finite. Also, in the definitions below, the graphs are *deterministic* and we clearly indicate when we move to *random graphs* instead. We next define *neighborhoods* as rooted subgraphs of a rooted graph, for which we recall that dist_G denotes the graph distance in the graph G:

Definition 2.2 (Neighborhoods as rooted graphs) For a rooted graph (G, o), we let $B_r^{(G)}(o)$ denote the (rooted) subgraph of (G, o) of all vertices at graph distance at most r away from o. Formally, this means that $B_r^{(G)}(o) = ((V(B_r^{(G)}(o)), E(B_r^{(G)}(o))), o)$, where

$$V(B_r^{(G)}(o)) = \{u \colon \mathrm{dist}_G(o, u) \le r\}, \tag{2.2.1}$$
$$E(B_r^{(G)}(o)) = \{\{u, v\} \in E(G) \colon \mathrm{dist}_G(o, u), \mathrm{dist}_G(o, v) \le r\}.$$

Also, let $\partial B_r^{(G)}(o)$ denote the (unrooted) graph with vertex set $V(\partial B_r^{(G)}(o)) = V(B_r^{(G)}(o)) \setminus V(B_{r-1}^{(G)}(o))$ and edge set $E(\partial B_r^{(G)}(o)) = E(B_r^{(G)}(o)) \setminus E(B_{r-1}^{(G)}(o))$. ◀

We continue by introducing the notion of *isomorphisms* between graphs, which basically describes graphs that "look the same." Here is the formal definition:

Definition 2.3 (Graph isomorphism)

(a) Two (finite or infinite) graphs $G_1 = (V(G_1), E(G_1))$ and $G_2 = (V(G_2), E(G_2))$ are called *isomorphic*, which we write as $G_1 \simeq G_2$, when there exists a bijection ϕ mapping $V(G_1)$ to $V(G_2)$ such that $\{u, v\} \in E(G_1)$ precisely when $\{\phi(u), \phi(v)\} \in E(G_2)$.

(b) Similarly, two rooted (finite or infinite) graphs (G_1, o_1) and (G_2, o_2) are called *isomorphic*, abbreviated as $(G_1, o_1) \simeq (G_2, o_2)$, when there exists a bijection ϕ mapping $V(G_1)$ to $V(G_2)$ such that $\phi(o_1) = o_2$ and $\{u, v\} \in E(G_1)$ precisely when $\{\phi(u), \phi(v)\} \in E(G_2)$. ◀

Exercises 2.1 and 2.2 investigate the notion of graph isomorphisms. We let $\mathscr{G}_\star$ denote the space of rooted graphs modulo isomorphisms. We often omit the equivalence classes, and write $(G, o) \in \mathscr{G}_\star$, bearing in mind that all (G', o') such that $(G', o') \simeq (G, o)$ are considered to be the same. Thus, formally we deal with the *equivalence classes* of rooted graphs. In the literature, the equivalence class containing (G, o) is often denoted as $[G, o]$.

Remark 2.4 (Multi-graphs)   Sometimes, we consider multi-graphs, i.e., graphs that contain self-loops or multi-edges, and that can be characterized as $G = (V(G), (x_{u,v})_{u,v \in V(G)})$, where $x_{u,v}$ denotes the number of edges between the vertices u and v, and $x_{u,u}$ denotes the number of self-loops at u. The above notions easily extend to such a setting. Indeed, let $G = (V(G), (x_{u,v})_{u,v \in V(G)})$ and $G' = (V(G'), (x'_{u,v})_{u,v \in V(G')})$ be two multi-graphs. An isomorphism $\phi \colon V(G) \mapsto V(G')$ is then instead required to be a bijection, satisfying $x_{u,v} = x'_{\phi(u),\phi(v)}$ for every $\{u, v\} \in E(G)$. We will not place much emphasis on multi-graphs in the main text. ◀

These notions allow us to turn the space of connected rooted graphs into a metric space:

Definition 2.5 (Metric on rooted graphs)   Let (G_1, o_1) and (G_2, o_2) be two rooted *connected* graphs, and write $B_r^{(G_i)}(o_i)$ for the neighborhood of vertex $o_i \in V(G_i)$. Let

$$R^\star = \sup\{r \colon B_r^{(G_1)}(o_1) \simeq B_r^{(G_2)}(o_2)\}, \tag{2.2.2}$$

and define the metric

$$\mathrm{d}_{\mathscr{G}_\star}((G_1, o_1), (G_2, o_2)) = \frac{1}{R^\star + 1}. \tag{2.2.3}$$

◀

The value $R^\star$ is the largest value of r for which $B_r^{(G_1)}(o_1)$ is isomorphic to $B_r^{(G_2)}(o_2)$. When $R^\star = \infty$, $B_r^{(G_1)}(o_1)$ is isomorphic to $B_r^{(G_2)}(o_2)$ for *every* $r \geq 1$, and then the rooted graphs (G_1, o_1) and (G_2, o_2) are the same apart from an isomorphism; see Lemma A.11 in the Appendix where this is worked out in detail.

The space $\mathscr{G}_\star$ of rooted graphs is a nice metric space under the metric $\mathrm{d}_{\mathscr{G}_\star}$ in (2.2.3), in that $(\mathscr{G}_\star, \mathrm{d}_{\mathscr{G}_\star})$ is separable and thus Polish. Here we recall that a metric space is called *separable* when there exists a countable dense subset of elements. Later on we will see that such a countable dense set can be created by looking at *finite* rooted graphs. Since graphs that agree up to distance r are at distance at most $1/(r + 1)$ from each other (see Exercise

2.3), this is indeed a dense countable subset. We discuss the metric structure of the space of rooted graphs in more detail in Appendix A.3. Exercises 2.4 and 2.5 study such aspects.

2.3 LOCAL WEAK CONVERGENCE OF DETERMINISTIC GRAPHS

In this section, we discuss the local weak convergence of deterministic graphs (G_n, o_n), rooted at a uniform vertex $o_n \in V(G_n)$, whose size tends to infinity as $n \to \infty$. This section is organized as follows. In Section 2.3.1, we give the definitions of local weak convergence of (possibly disconnected) finite graphs. In Section 2.3.2, we provide a convenient criterion to prove local weak convergence and discuss tightness. In Section 2.3.3, we show that when the limit has full support on some subset of rooted graphs, convergence can be restricted to that set. In Section 2.3.4, we discuss two examples of graphs that converge locally weakly. We close in Section 2.3.5 by discussing the local weak convergence of *marked* graphs, which turns out to be useful in many applications of local weak convergence.

2.3.1 DEFINITION OF LOCAL WEAK CONVERGENCE

Above, we worked with *connected* graphs (see, e.g., Definition 2.5). We often wish to apply local weak convergence arguments to *disconnected* graphs. For such examples, we think of the rooted connected graph (G_n, o_n) as corresponding to the *connected component* $\mathscr{C}(o_n)$ of o_n in G_n. Here, for $v \in V(G_n)$, we let $\mathscr{C}(v)$ denote its connected component. Then, we define, similarly to (2.2.1), the rooted graph $\mathscr{C}(o_n) = ((V(\mathscr{C}(o_n)), E(\mathscr{C}(o_n))), o_n)$ as

$$V(\mathscr{C}(o_n)) = \{u : \mathrm{dist}_{G_n}(o, u) < \infty\}, \tag{2.3.1}$$
$$E(\mathscr{C}(o_n)) = \{\{u, v\} \in E(G_n) : \mathrm{dist}_{G_n}(o, u), \mathrm{dist}_{G_n}(o, v) < \infty\}.$$

For $h : \mathscr{G}_\star \to \mathbb{R}$, by convention, we extend the definition to all (not necessarily connected) graphs by letting

$$h(G_n, o_n) \equiv h(\mathscr{C}(o_n)). \tag{2.3.2}$$

We next use these conventions to define the local weak convergence of finite graphs:

Definition 2.6 (Local weak convergence) Let $G_n = (V(G_n), E(G_n))$ denote a finite (possibly disconnected) graph. Let (G_n, o_n) be the rooted graph obtained by letting $o_n \in V(G_n)$ be chosen uar, and restricting G_n to the connected component $\mathscr{C}(o_n)$ of o_n in G_n. We say that (G_n, o_n) *converges locally weakly* to the connected rooted graph (G, o), which is a (possibly random) element of $\mathscr{G}_\star$ having law μ, when, for every bounded and continuous function $h : \mathscr{G}_\star \mapsto \mathbb{R}$,

$$\mathbb{E}[h(G_n, o_n)] \to \mathbb{E}_\mu[h(G, o)], \tag{2.3.3}$$

where the expectation on the rhs of (2.3.3) is wrt (G, o) having law μ, while the expectation on the lhs is wrt the random vertex o_n. We denote the above convergence by $(G_n, o_n) \xrightarrow{d} (G, o)$. ◀

Of course, by (2.3.2), the values $h(G_n, o_n)$ give you information only about $\mathscr{C}(o_n)$, which may be only a small portion of the graph when G_n is disconnected. However, since we are sampling $o_n \in V(G_n)$ uar, actually we *may* "see" every connected component, so in distribution we *do* observe the graph as a whole.

Since later we apply local weak convergence ideas to random graphs, we need to be absolutely clear about with respect to what we take the expectation. Indeed, the expectation in (2.3.3) is wrt the random root $o_n \in V(G_n)$, and is thus equal to

$$\mathbb{E}[h(G_n, o_n)] = \frac{1}{|V(G_n)|} \sum_{v \in V(G_n)} h(G_n, v). \tag{2.3.4}$$

The notion of local weak convergence plays a central role in this book. It may be hard to grasp, and it also may appear to be rather weak. In what follows, we discuss examples of graphs that converge locally weakly. Further, in Section 2.5 we discuss examples of how local weak convergence may be used to obtain interesting consequences for graphs, such as their clustering and degree–degree dependencies, measured through the assortativity coefficient. We continue by discussing a convenient criterion for proving local weak convergence.

2.3.2 LOCAL WEAK CONVERGENCE: CRITERION

We next provide a convenient criterion for local weak convergence:

Theorem 2.7 (Criterion for local weak convergence) *The sequence of finite rooted graphs* $((G_n, o_n))_{n \geq 1}$ *converges locally weakly to* $(G, o) \sim \mu$ *precisely when, for every rooted graph* $H_\star \in \mathscr{G}_\star$ *and all integers* $r \geq 0$,

$$p^{(G_n)}(H_\star) = \frac{1}{|V(G_n)|} \sum_{v \in V(G_n)} \mathbb{1}_{\{B_r^{(G_n)}(v) \simeq H_\star\}} \to \mu(B_r^{(G)}(o) \simeq H_\star), \tag{2.3.5}$$

where $B_r^{(G_n)}(v)$ *is the rooted* r-*neighborhood of* u *in* G_n, *and* $B_r^{(G)}(o)$ *is the rooted* r-*neighborhood of* v *in the limiting graph* (G, o).

Proof This is a standard weak convergence argument. First, the local weak convergence in Definition 2.6 implies that (2.3.5) holds, since we can take $p^{(G_n)}(H_\star) = \mathbb{E}[h(G, o_n)]$ and $h(G, o) = \mathbb{1}_{\{B_r^{(G)}(o) \simeq H_\star\}}$, and $h \colon \mathscr{G}_\star \mapsto \{0, 1\}$ is bounded and continuous (see Exercise 2.6). For the other direction, since μ is a *probability measure* on $\mathscr{G}_\star$, the sequence $((G_n, o_n))_{n \geq 1}$ is tight; see Theorem A.7 in Appendix A.2. By tightness, every subsequence of $((G_n, o_n))_{n \geq 1}$ has a further subsequence that converges in distribution. We work along that subsequence, and note that the limiting law is that of (G, o), since the laws of $B_r^{(G)}(o)$ for all $r \geq 1$ uniquely identify the law of (G, o) (see Proposition A.15 in Appendix A.3.5). Since this is true for every subsequence, the local weak limit is (G, o). $\qquad\square$

Theorem 2.7 shows that the proportion of vertices in G_n whose neighborhoods look like $H_\star$ converges to a (possibly random) limit. See Exercise 2.8, where you are asked to construct an example where the local weak limit of a sequence of deterministic graphs actually is *random*. You are asked to prove local weak convergence for some examples in Exercises 2.9 and 2.10. Appendix A.3.6 discusses tightness in $\mathscr{G}_\star$ in more detail.

2.3.3 LOCAL WEAK CONVERGENCE AND COMPLETENESS OF THE LIMIT

In many settings, the local weak limit is almost surely contained in a smaller set of rooted graphs, for example rooted trees. In this case, it turns out to be enough to prove the convergence in Definition 2.11 below only for those elements in which the limit (G, o) takes values. Let us explain this in more detail.

Let $\mathcal{T}_\star \subset \mathcal{G}_\star$ be a subset of the space of rooted graphs. Let $\mathcal{T}_\star(r) \subseteq \mathcal{T}_\star$ be the subset of $\mathcal{T}_\star$ of graphs for which the distance between any vertex and the root is at most r. Then, we have the following result:

Theorem 2.8 (Local weak convergence and subsets) *Let $(G_n)_{n \geq 1}$ be a sequence of rooted graphs, and $o_n \in V(G_n)$ be a vertex chosen uar. Let (G, o) be a random variable on $\mathcal{G}_\star$ having law μ. Let $\mathcal{T}_\star \subset \mathcal{G}_\star$ be a subset of the space of rooted graphs. Assume that $\mu((G, o) \in \mathcal{T}_\star) = 1$. Then, $(G_n, o_n) \xrightarrow{d} (G, o)$ when (2.3.5) holds for all $H_\star \in \mathcal{T}_\star(r)$ and all $r \geq 1$.*

We will apply Theorem 2.8 in particular when the limit is almost surely a tree. Then Theorem 2.8 implies that we have to investigate only rooted graphs $H_\star$ that are finite trees of height at most r themselves.

Proof The set $\mathcal{T}_\star(r)$ is countable. Therefore, since $\mu((G, o) \in \mathcal{T}_\star) = 1$, for every $\varepsilon > 0$, there exists an $m = m(\varepsilon)$ and a subset $\mathcal{T}_\star(r, m)$ of size at most m such that $\mu(B_r^{(G)}(o) \in \mathcal{T}_\star(r, m)) \geq 1 - \varepsilon$. Fix this set. Then we bound

$$\mathbb{P}(B_r^{(G_n)}(o_n) \notin \mathcal{T}_\star(r)) = 1 - \mathbb{P}(B_r^{(G_n)}(o_n) \in \mathcal{T}_\star(r))$$
$$\leq 1 - \mathbb{P}(B_r^{(G_n)}(o_n) \in \mathcal{T}_\star(r, m)). \qquad (2.3.6)$$

Therefore,

$$\limsup_{n \to \infty} \mathbb{P}(B_r^{(G_n)}(o_n) \notin \mathcal{T}_\star(r)) \leq 1 - \liminf_{n \to \infty} \mathbb{P}(B_r^{(G_n)}(o_n) \in \mathcal{T}_\star(r, m)) \qquad (2.3.7)$$

$$= 1 - \mu(B_r^{(G)}(o) \in \mathcal{T}_\star(r, m)) \leq 1 - (1 - \varepsilon) = \varepsilon.$$

Since $\varepsilon > 0$ is arbitrary, we conclude that $\mathbb{P}(B_r^{(G_n)}(o_n) \notin \mathcal{T}_\star(r)) \to 0$. In particular, this means that, for any $H_\star \notin \mathcal{T}_\star(r)$,

$$\mathbb{P}(B_r^{(G_n)}(o_n) \simeq H_\star) \to 0 = \mu(B_r^{(G)}(o) \simeq H_\star). \qquad (2.3.8)$$

Thus, when the required convergence holds for every $H_\star \in \mathcal{T}_\star(r)$, it follows for every $H_\star \notin \mathcal{T}_\star(r)$ with limit zero when $\mu((G, o) \in \mathcal{T}_\star) = 1$. $\qquad \square$

2.3.4 EXAMPLES OF LOCAL WEAK CONVERGENCE

We close this section by discussing two important examples of local weak convergence. We start with large boxes in $\mathbb{Z}^d$, and then discuss the local weak limit of finite trees.

Local Weak Convergence of Boxes in $\mathbb{Z}^d$

Consider the nearest-neighbor box $[n]^d$, where $x = (x_1, \ldots, x_d) \in \mathbb{Z}^d$ is a neighbor of $y \in \mathbb{Z}^d$ precisely when there is a unique $i \in [d]$ such that $|x_i - y_i| = 1$. Take a root uar, and denote the resulting graph by (G_n, o_n). We claim that $(G_n, o_n) \xrightarrow{d} (\mathbb{Z}^d, o)$, which we now prove. We rely on Theorem 2.7, which shows that we need to prove the convergence of subgraph proportions.

Let μ be the point measure on $(\mathbb{Z}^d, o)$, so that $\mu(B_r^{(G)}(o) \simeq B_r^{(\mathbb{Z}^d)}(o)) = 1$. Thus, by Theorem 2.8, it remains to show that $p^{(G_n)}(B_r^{(\mathbb{Z}^d)}(o)) \to 1$ (recall (2.3.5)). For this, we note that $B_r^{(G_n)}(o_n) \simeq B_r^{(\mathbb{Z}^d)}(o)$ *unless* o_n happens to lie within a distance strictly smaller than r from one of the boundaries of $[n]^d$. This means that one of the coordinates of o_n is either in $[r-1]$, or in $[n] \setminus [n-r+1]$. Since the latter occurs with vanishing probability, the claim follows.

In the above case, we see that the local weak limit is deterministic, as one would have expected. One can generalize the above to the local weak convergence of tori as well.

Local Weak Convergence of Truncated Trees

Recall the notion of a tree in Section 1.5. Fix a degree d, and a height n that we take to infinity. We now define the regular tree $\mathbb{T}_{d,n}$ truncated at height n. The graph $\mathbb{T}_{d,n}$ has vertex set

$$V(\mathbb{T}_{d,n}) = \{\varnothing\} \cup \bigcup_{k=1}^{n} \{\varnothing\} \times [d] \times [d-1]^{k-1}, \qquad (2.3.9)$$

and edge set as follows. Let $v = \varnothing v_1 \cdots v_k$ and $u = \varnothing u_1 \cdots u_\ell$ be two vertices. We say that u is the *parent* of v when $\ell = k - 1$ and $u_i = v_i$ for all $i \in [k-1]$. Then we say that two vertices u and v are neighbors when u is the parent of v or vice versa. We obtain a graph with

$$|V(\mathbb{T}_{d,n})| = 1 + d + \cdots + d(d-1)^{n-1} \qquad (2.3.10)$$

vertices.

Let o_n denote a vertex chosen uar from $V(\mathbb{T}_{d,n})$. To study the local weak limit of $(\mathbb{T}_{d,n}, o_n)$, we first consider the so-called *canopy tree*. For this, we take the graph $\mathbb{T}_{d,n}$, root it at any leaf, which we will call the root-leaf, and take the limit of $n \to \infty$. Denote this graph by $\mathbb{T}_d^c$, which we consider to be an unrooted graph, but we keep the root-leaf for reference purposes. This graph has a unique infinite path from the root-leaf. Let o_ℓ be the ℓth vertex on this infinite path (the root-leaf being o_0), and consider $(\mathbb{T}_d^c, o_\ell)$. Define the limiting measure μ by

$$\mu((\mathbb{T}_d^c, o_\ell)) \equiv \mu_\ell = (d-2)(d-1)^{-(\ell+1)}, \qquad \ell \geq 0. \qquad (2.3.11)$$

Fix $G_n = \mathbb{T}_{d,n}$. We claim that $(G_n, o_n) \equiv (\mathbb{T}_{d,n}, o_n) \xrightarrow{d} (G, o)$ with law μ in (2.3.11). We again rely on Theorem 2.7, which shows that we need to prove only the convergence of the subgraph proportions.

By Theorem 2.8, it remains to show that $p^{(G_n)}(B_r^{(\mathbb{T}_d^c)}(o_\ell)) \to \mu_\ell$ (recall (2.3.5)). When n is larger than r (which we now assume), $B_r^{(G_n)}(o_n) \simeq B_r^{(\mathbb{T}_d^c)}(o_\ell)$ precisely when o_n has distance ℓ from the closest leaf. There are

$$d(d-1)^{k-1} \qquad (2.3.12)$$

vertices at distance k from the root, out of a total of $|V(\mathbb{T}_{d,n})| = d(d-1)^n/(d-2)(1+o(1))$. Having distance ℓ to the closest leaf in $\mathbb{T}_{d,n}$ is the same as having distance $k = n - \ell$ from the root. Thus,

$$p^{(G_n)}(B_r^{(\mathbb{T}_d^G)}(o_\ell)) = \frac{d(d-1)^{k-1}}{|V(\mathbb{T}_{d,n})|} \to (d-2)(d-1)^{-(\ell+1)} = \mu_\ell, \qquad (2.3.13)$$

as required.

We see that, for truncated regular trees, the local weak limit is *random*, where the randomness originates from the choice of the random root of the graph. More precisely, this is due to the choice how far away the chosen root is from the leaves of the finite tree. Perhaps surprisingly, the local limit of a truncated regular tree is *not* the infinite regular tree.

2.3.5 LOCAL WEAK CONVERGENCE OF MARKED ROOTED GRAPHS

We next extend the notion of local weak convergence to include *marks*. Such marks are associated with the vertices as well as the edges, and can take values in a general complete separable metric space. Rooted graphs with marks are called *marked graphs*. Such more general set-ups are highly relevant in many applications, for example when dealing with general inhomogeneous graphs. Let us explain this in some more detail:

Definition 2.9 (Marked rooted graphs)   A *marked (multi-)graph* is a (multi-)graph $G = (V(G), E(G))$ together with a set $M(G)$ of marks taking values in a complete separable metric space Ξ, called the mark space. Here M maps from $V(G)$ and $E(G)$ to Ξ. Images in Ξ are called marks. Each edge is given two marks, one associated with ("at") each of its endpoints. Thus, we could think of the marks as located at the "half-edges" incident to a vertex. Such a half-edge can be formalized as $(u, \{u, v\})$ for $u \in V(G)$ and $\{u, v\} \in E(G)$. Alternatively, we can think of the mark as located on a directed edge (u, v) where $\{u, v\} \in E(G)$; we will use both perspectives. The only assumption on the degrees is that they are locally finite. We denote the marked rooted graph by $(G, o, M(G)) = ((V(G), E(G)), o, M(G))$.   ◄

We next extend the metric to the above setting of marked rooted graphs. Let the distance between $(G_1, o_1, M(G_1))$ and $(G_2, o_2, M(G_2))$ be $1/(1+R^\star)$, where $R^\star$ is the supremum of those $r > 0$ such that there is some rooted isomorphism of the balls of (graph distance) radius $\lfloor r \rfloor$ around the roots of G_i, such that each pair of corresponding marks has distance in Ξ less than $1/r$. This is formalized as follows:

Definition 2.10 (Metric on marked rooted graphs)   Let d_Ξ be a metric on the space of marks Ξ. Then, we let

$$d_{\mathcal{G}_\star}((G_1, o_1, M_1(G_1)), (G_2, o_2, M_2(G_2))) = \frac{1}{1+R^\star}, \qquad (2.3.14)$$

where

$$R^\star = \sup \Big\{ r \colon B_r^{(G_1)}(o_1) \simeq B_r^{(G_2)}(o_2), \text{ and there exists } \phi \text{ such that}$$

$$d_\Xi(m_1(u), m_2(\phi(u))) \leq 1/r \ \forall u \in V(B_r^{(G_1)}(o_1)), \text{ and}$$

$$d_\Xi(m_1(u,v), m_2(\phi(u,v))) \leq 1/r \ \forall \{u,v\} \in E(B_r^{(G_1)}(o_1)) \Big\}, \quad (2.3.15)$$

with $\phi \colon V(B_r^{(G_1)}(o_1)) \to V(B_r^{(G_2)}(o_2))$ running over all isomorphisms between $B_r^{(G_1)}(o_1)$ and $B_r^{(G_2)}(o_2)$ satisfying $\phi(o_1) = o_2$. ◀

When Ξ is a finite set, we can simply let $d_\Xi(a,b) = \mathbb{1}_{\{a \neq b\}}$, so that (2.3.14)–(2.3.15) state that not only should the neighborhoods $B_r^{(G_1)}(o_1)$ and $B_r^{(G_2)}(o_2)$ be isomorphic but also the corresponding marks on the vertices and half-edges in $B_r^{(G_1)}(o_1)$ and $B_r^{(G_2)}(o_2)$ should all be the same.

Definition 2.10 puts a metric structure on marked rooted graphs. With this metric topology in hand, we can simply adapt all convergence statements to this setting. We refrain from stating all these extensions explicitly. See Exercise 2.17 for an application of marked graphs to *directed* graphs. For example, the marked rooted graph setting is a way to formalize the setting of multi-graphs in Remark 2.4 (see Exercise 2.18).

Having discussed the notion of local weak convergence for *deterministic* graphs, we now move on to *random* graphs. Here the situation becomes more delicate, as now we have *double* randomness, both in the *random root* as well as the *random graph*. This gives rise to surprising subtleties.

2.4 LOCAL CONVERGENCE OF RANDOM GRAPHS

We next discuss the local convergence of *random* graphs. This section is organized as follows. In Section 2.4.1 we define what it means for a sequence of random graphs to converge locally, as well as which different versions thereof exist. In Section 2.4.2 we then give a useful criterion to verify the local convergence of random graphs. In Section 2.4.3 we prove the *completeness of the limit* by showing that, when the limit is supported on a subset of rooted graphs, then one needs only to verify the convergence for that subset. In many examples that we encounter in this book, this subset is the collection of *trees*. We close with two examples: that of random regular graphs in Section 2.4.4 and that of the Erdős–Rényi random graph $ER_n(\lambda/n)$ in Section 2.4.5.

2.4.1 DEFINITION OF LOCAL CONVERGENCE OF RANDOM GRAPHS

Even for random variables, there are different notions of convergence that are relevant, such as convergence in distribution and in probability. Also for local convergence, there are several related notions of convergence that we may consider:

Definition 2.11 (Local weak convergence of random graphs) Let $(G_n)_{n \geq 1}$ with $G_n = (V(G_n), E(G_n))$ denote a finite sequence of (possibly disconnected) random graphs. Then,

(a) G_n converges *locally weakly* to $(\bar{G}, \bar{o})$ having law $\bar{\mu}$ when

$$\mathbb{E}\big[h(G_n, o_n)\big] \to \mathbb{E}_{\bar{\mu}}\big[h(\bar{G}, \bar{o})\big], \tag{2.4.1}$$

for every bounded and continuous function $h\colon \mathscr{G}_\star \mapsto \mathbb{R}$, where the expectation $\mathbb{E}$ on the lhs of (2.4.1) is wrt the random vertex o_n *and* the random graph G_n. This is equivalent to $(G_n, o_n) \xrightarrow{d} (G, o)$.

(b) G_n converges *locally in probability* to (G, o) having law μ when

$$\mathbb{E}\big[h(G_n, o_n) \mid G_n\big] \xrightarrow{\mathbb{P}} \mathbb{E}_\mu\big[h(G, o)\big], \tag{2.4.2}$$

for every bounded and continuous function $h\colon \mathscr{G}_\star \mapsto \mathbb{R}$.

(c) G_n converges *locally almost surely* to (G, o) having law μ when

$$\mathbb{E}\big[h(G_n, o_n) \mid G_n\big] \xrightarrow{a.s.} \mathbb{E}_\mu\big[h(G, o)\big], \tag{2.4.3}$$

for every bounded and continuous function $h\colon \mathscr{G}_\star \mapsto \mathbb{R}$. ◄

When we have local convergence in probability, $\mathbb{E}\big[h(G_n, o_n) \mid G_n\big]$, which is a random variable due to its dependence on the random graph G_n, converges in probability to $\mathbb{E}_\mu\big[h(G, o)\big]$, which is possibly also a random variable in that μ might be a *random* probability distribution on $\mathscr{G}_\star$. When, instead, we have local weak convergence, only *expectations* wrt the random graph of the form $\mathbb{E}\big[h(G_n, o_n)\big]$ converge, and the limiting measure $\bar{\mu}$ is *deterministic*.

Remark 2.12 (Local convergence in probability and rooted versus unrooted graphs) Usually, if we have a sequence of objects x_n living in some space $\mathcal{X}$, and x_n converges to x, then x also lives in $\mathcal{X}$. In the above definitions of local convergence in probability and almost surely, respectively, we take a graph sequence $(G_n)_{n \geq 1}$ that converges locally in probability and almost surely, respectively, to a *rooted* graph $(G, o) \sim \mu$. One might have guessed that this is related to $(G_n, o_n) \xrightarrow{\mathbb{P}} (G, o)$ and $(G_n, o_n) \xrightarrow{a.s.} (G, o)$, but in fact it is quite different. Let us restrict attention to local convergence in probability. Indeed, $(G_n, o_n) \xrightarrow{\mathbb{P}} (G, o)$ is a very strong and arguably not so useful statement. For one, it requires that $((G_n, o_n))_{n \geq 1}$ and (G, o) live on the same probability space, which is not often evidently the case. Further, sampling o_n gives rise to a variability in (G_n, o_n) that is hard to capture by the limit (G, o). Indeed, when n varies, the root o_n in (G_n, o_n) will have to be redrawn every once in a while, and it seems difficult to do this in such a way that (G_n, o_n) is consistently close to (G, o). ◄

Remark 2.13 (Random measure interpretation of local convergence in probability) The following observations turn local convergence in probability into the convergence of objects living on the same space, namely, the space of *probability measures* on rooted graphs. Denote the empirical neighborhood measure μ_n on $\mathscr{G}_\star$ by

$$\mu_n(\mathscr{H}_\star) = \frac{1}{|V(G_n)|} \sum_{v \in V(G_n)} \mathbf{1}_{\{(G_n, v) \in \mathscr{H}_\star\}}, \tag{2.4.4}$$

for every measurable subset $\mathscr{H}_\star$ of $\mathscr{G}_\star$. Then, $(G_n)_{n \geq 1}$ converges locally in probability to the random rooted graph $(G, o) \sim \mu$ when

$$\mu_n(\mathscr{H}_\star) \xrightarrow{\mathbb{P}} \mu(\mathscr{H}_\star) \tag{2.4.5}$$

for every measurable subset $\mathcal{H}_\star$ of $\mathcal{G}_\star$. This is equivalent to Definition 2.11(b), since, for every bounded and continuous $h\colon \mathcal{G}_\star \mapsto \mathbb{R}$, and denoting the conditional expected value of $h(G_n, o_n)$ when $(G_n, o_n) \sim \mu_n$ by $\mathbb{E}_{\mu_n}[h(G_n, o_n) \mid G_n]$, we have

$$\mathbb{E}_{\mu_n}[h(G_n, o_n) \mid G_n] = \frac{1}{|V(G_n)|} \sum_{v \in V(G_n)} h(G_n, v) = \mathbb{E}[h(G_n, o_n) \mid G_n], \quad (2.4.6)$$

which converges to $\mathbb{E}_\mu[h(G, o)]$ by (2.4.2). Thus, (2.4.5) is equivalent to the convergence in probability of the empirical neighborhood measure in Definition 2.11(b). This explains why we view local convergence in probability as a property of the *graph* G_n rather than of the *rooted graph* (G_n, o_n). In turn, it can be shown that convergence in probability of the empirical neighborhood measure in Definition 2.11(b) is equivalent to convergence in distribution of the two-vertex measure

$$\bar{\mu}_n^{(2)}(\mathcal{H}_\star^{(1)}, \mathcal{H}_\star^{(2)}) = \frac{1}{|V(G_n)|^2} \sum_{v_1, v_2 \in V(G_n)} \mathbb{P}((G_n, v_1) \in \mathcal{H}_\star^{(1)}, (G_n, v_2) \in \mathcal{H}_\star^{(2)})$$

$$(2.4.7)$$

to $\bar{\mu}^{(2)}(\mathcal{H}_\star^{(1)}, \mathcal{H}_\star^{(2)}) = \mathbb{E}[\mu(\mathcal{H}_\star^{(1)})\mu(\mathcal{H}_\star^{(2)})]$, for every pair of measurable subsets $\mathcal{H}_\star^{(1)}$ and $\mathcal{H}_\star^{(2)}$ of $\mathcal{G}_\star$ (see e.g., Exercise 2.16). A similar comment applies to local convergence almost surely. ◀

The limiting probability measures $\bar{\mu}$ for local convergence in distribution and μ for local convergence in probability and local convergence almost surely are closely related:

Corollary 2.14 (Relation between local limits)   *Suppose that $(G_n)_{n \geq 1}$ converges locally in probability to (G, o) having law μ. Then $(G_n, o_n) \overset{d}{\longrightarrow} (\bar{G}, \bar{o})$ having law $\bar{\mu}$, and $\bar{\mu}(\cdot) = \mathbb{E}[\mu(\cdot)]$ is defined, for every bounded and continuous $h\colon \mathcal{G}_\star \to \mathbb{R}$, by*

$$\mathbb{E}_{\bar{\mu}}[h(\bar{G}, \bar{o})] = \mathbb{E}\Big[\mathbb{E}_\mu[h(G, o)]\Big]. \quad (2.4.8)$$

Further, if $(G_n)_{n \geq 1}$ converges almost surely to (G, o), then $(G_n)_{n \geq 1}$ also converges locally in probability to (G, o).

Proof   Note that $\mathbb{E}_\mu[h(G, o)]$ is a bounded random variable, and so is $\mathbb{E}[h(G_n, o_n) \mid G_n]$. Therefore, by the Dominated Convergence Theorem [V1, Theorem A.1], the expectations also converge. We conclude that

$$\mathbb{E}[h(G_n, o_n)] = \mathbb{E}\Big[\mathbb{E}[h(G_n, o_n) \mid G_n]\Big] \to \mathbb{E}\Big[\mathbb{E}_\mu[h(G, o)]\Big], \quad (2.4.9)$$

which proves that the claim with the limit identified in (2.4.8) holds. The relation between local convergence almost surely and in probability follows from that for random variables and Definition 2.11. □

In most of our examples, the law μ of the local limit in probability is actually *deterministic*, in which case $\bar{\mu} = \mu$. However, there are some cases where this is not true. A simple example arises as follows. For $\mathrm{ER}_n(\lambda/n)$, the local limit in probability turns out to be a $\mathrm{Poi}(\lambda)$ branching process (see Section 2.4.5). Therefore, when considering $\mathrm{ER}_n(X/n)$, where X is uniform on $[0, 2]$, the local limit in probability will be a $\mathrm{Poi}(X)$ branching process. Here, the expected offsprings conditioned on the random variable X are *random* and related, as they are all equal to X. This is *not* the same as a mixed-Poisson branching process with offspring distribution $\mathrm{Poi}(X)$, since, for the local limit in probability of $\mathrm{ER}_n(X/n)$, we draw X *only once*. We refer to Section 2.4.5 for more details on local convergence for $\mathrm{ER}_n(\lambda/n)$.

We have added the notion of local convergence in the almost sure sense, even though for random graphs this notion is often not highly useful. Indeed, almost sure convergence for random graphs can already be tricky, since for static models such as the Erdős–Rényi random graph and the configuration model, there is no obvious relation between the graphs of size n and those of size $n + 1$. This of course is different for the preferential attachment model, which forms a (consistent) random graph process.

2.4.2 CRITERION FOR LOCAL CONVERGENCE OF RANDOM GRAPHS

We next discuss a convenient criterion for local convergence, inspired by Theorem 2.7:

Theorem 2.15 (Criterion for local convergence of random graphs) *Let $(G_n)_{n \geq 1}$ be a sequence of graphs. Then,*

(a) $(G_n, o_n) \xrightarrow{d} (\bar{G}, \bar{o}) \sim \bar{\mu}$ *precisely when, for every rooted graph $H_\star \in \mathscr{G}_\star$ and all integers $r \geq 0$,*

$$\mathbb{E}[p^{(G_n)}(H_\star)] = \frac{1}{|V(G_n)|} \sum_{v \in V(G_n)} \mathbb{P}(B_r^{(G_n)}(v) \simeq H_\star) \to \bar{\mu}(B_r^{(G)}(\bar{o}) \simeq H_\star).$$

(2.4.10)

(b) G_n *converges locally in probability to $(G, o) \sim \mu$ precisely when, for every rooted graph $H_\star \in \mathscr{G}_\star$ and all integers $r \geq 0$,*

$$p^{(G_n)}(H_\star) = \frac{1}{|V(G_n)|} \sum_{v \in V(G_n)} \mathbb{1}_{\{B_r^{(G_n)}(v) \simeq H_\star\}} \xrightarrow{\mathbb{P}} \mu(B_r^{(G)}(o) \simeq H_\star). \quad (2.4.11)$$

(c) G_n *converges locally almost surely to $(G, o) \sim \mu$ precisely when, for every rooted graph $H_\star \in \mathscr{G}_\star$ and all integers $r \geq 0$,*

$$p^{(G_n)}(H_\star) = \frac{1}{|V(G_n)|} \sum_{v \in V(G_n)} \mathbb{1}_{\{B_r^{(G_n)}(v) \simeq H_\star\}} \xrightarrow{a.s.} \mu(B_r^{(G)}(o) \simeq H_\star). \quad (2.4.12)$$

Proof This follows from Theorem 2.7. Indeed, for part (a), it follows directly, as part (a) deals with local weak convergence as in Theorem 2.7. For convergence almost surely as in part (c), this also follows directly. For part (b), we need an extra argument. By the Skorokhod Embedding Theorem, for each $H_\star$, there exists a probability space for which the convergence in (2.4.12) occurs almost surely. The same holds for any finite subcollection of $H_\star \in \mathscr{G}_\star$, and since the set of graphs $\mathscr{H}_\star$ that can occur as r-neighborhoods is countable, it can even be extended to all such $H_\star \in \mathscr{H}_\star$. Thus, the statement again follows from Theorem 2.7. □

In what follows, we are mainly interested in local convergence in probability, since this is the notion that is the most powerful and useful in the setting of random graphs.

2.4.3 LOCAL CONVERGENCE AND COMPLETENESS OF THE LIMIT

For many random graph models, the local limit is almost surely contained in a smaller set of rooted graphs, as already mentioned in Section 2.3.3 and as extensively used in Section 2.3.4. The most common example occurs when the random graph converges locally to a *tree* (recall that trees are rooted, see Section 1.5), but it can apply more generally. In this case, and similarly to Theorem 2.8, it turns out to be enough to prove the convergence in Definition 2.11 only for elements of this subset. Let us explain this in more detail.

Recall that $\mathscr{T}_\star \subset \mathscr{G}_\star$ is a subset of the space of rooted graphs, and that $\mathscr{T}_\star(r) \subseteq \mathscr{T}_\star$ is the subset of $\mathscr{T}_\star$ of graphs for which the distance between any vertex and the root is at most r. Then, we have the following result:

Theorem 2.16 (Local convergence and subsets) *Let $(G_n)_{n \geq 1}$ be a sequence of rooted graphs. Let $(\bar{G}, \bar{o})$ be a random variable on $\mathscr{G}_\star$ having law $\bar{\mu}$. Let $\mathscr{T}_\star \subset \mathscr{G}_\star$ be a subset of the space of rooted graphs. Assume that $\bar{\mu}\big((\bar{G}, \bar{o}) \in \mathscr{T}_\star\big) = 1$. Then, $(G_n, o_n) \xrightarrow{d} (\bar{G}, \bar{o})$ when (2.4.10) holds for all $H_\star \in \mathscr{T}_\star(r)$ and all $r \geq 1$.*

Similar extensions hold for the local convergence in probability in (2.4.11) and local convergence almost surely in (2.4.12), with $\bar{\mu}$ replaced by μ and $(\bar{G}, \bar{o})$ by (G, o).

Proof The proof for local weak convergence is identical to that of Theorem 2.8. The extensions to local convergence in probability and almost surely follow similarly. See Exercises 2.20 and 2.21. □

2.4.4 LOCAL CONVERGENCE OF RANDOM REGULAR GRAPHS

In this subsection we give the very first example of a random graph that converges locally in probability. This is the random regular graph, which is obtained by taking the configuration model $\mathrm{CM}_n(\boldsymbol{d})$, letting $d_v = d$ for all $v \in [n]$, and conditioning it on simplicity (recall the discussion below (1.3.29)). Here, we assume that nd is even. The main result is the following:

Theorem 2.17 (Local convergence of random regular graphs) *Fix $d \geq 1$ and assume that nd is even. The random regular graph of degree d and size n converges locally in probability to the rooted d-regular tree.*

Looking back at the local weak limit of truncated trees discussed in Section 2.3.4, we see that a random regular graph is a *much* better approximation to a d-regular tree than a truncation of the latter. One could see this as another example of the *probabilistic method*, where a certain amount of randomization is useful to construct deterministic objects.

Proof Let G_n be the random regular graph, and let $(\mathbb{T}_d, o)$ be the rooted d-regular tree. We view G_n as a configuration model $\mathrm{CM}_n(\boldsymbol{d})$ with $d_v = d$ for all $v \in [n]$, conditioned on simplicity. We first prove Theorem 2.17 for this configuration model instead.

We see that the only requirement is to show that $p^{(G_n)}(B_r^{(\mathbb{T}_d)}(o)) \overset{\mathbb{P}}{\longrightarrow} 1$, for which we use a second-moment method. We start by showing that $\mathbb{E}[p^{(G_n)}(B_r^{(\mathbb{T}_d)}(o))] \to 1$. We write

$$1 - \mathbb{E}[p^{(G_n)}(B_r^{(\mathbb{T}_d)}(o))] = \mathbb{P}(B_r^{(G_n)}(o_n) \text{ is not a tree}). \qquad (2.4.13)$$

For $B_r^{(G_n)}(o_n)$ not to be a tree, a cycle needs to occur within a distance r. We grow the neighborhood $B_r^{(G_n)}(o_n)$ by pairing the half-edges incident to discovered vertices one by one in a breadth-first way (recall Definitions 1.24 and 1.25). Since there is just a bounded number of unpaired half-edges incident to the vertices found at any moment in the exploration, and since we need to pair at most

$$d + d(d-1) + \cdots + d(d-1)^{r-1} \qquad (2.4.14)$$

half-edges, the probability that any one of them creates a cycle vanishes. We conclude that $1 - \mathbb{E}[p^{(G_n)}(B_r^{(\mathbb{T}_d)}(o))] \to 0$. Next, we show that $\mathbb{E}[p^{(G_n)}(B_r^{(\mathbb{T}_d)}(o))^2] \to 1$, which shows that $\mathrm{Var}(p^{(G_n)}(B_r^{(\mathbb{T}_d)}(o))) \to 0$, and thus $p^{(G_n)}(B_r^{(\mathbb{T}_d)}(o)) \overset{\mathbb{P}}{\longrightarrow} 1$. Now,

$$1 - \mathbb{E}[p^{(G_n)}(B_r^{(\mathbb{T}_d)}(o))^2] = \mathbb{P}(B_r^{(G_n)}(o_n^{(1)}) \text{ or } B_r^{(G_n)}(o_n^{(2)}) \text{ is not a tree}) \to 0, \qquad (2.4.15)$$

as before, where $o_n^{(1)}$ and $o_n^{(2)}$ are two independent and uniformly chosen vertices in $[n]$. This completes the proof for the configuration model. Since we have proved the convergence in probability of the subgraph proportions, convergence in probability follows when we condition on simplicity (recall [V1, Corollary 7.17]), and thus the proof also follows for random regular graphs. We leave the details of this argument as Exercise 2.19. $\qquad \square$

2.4.5 LOCAL CONVERGENCE OF ERDŐS–RÉNYI RANDOM GRAPHS

In this subsection, we work out one more example and show that the Erdős–Rényi random graph $\mathrm{ER}_n(\lambda/n)$ converges locally in probability to a $\mathrm{Poi}(\lambda)$ branching process:

Theorem 2.18 (Local convergence of Erdős–Rényi random graph) *Fix $\lambda > 0$. $\mathrm{ER}_n(\lambda/n)$ converges locally in probability to a Poisson branching process with mean offspring λ.*

Proof We start by reducing the proof to a convergence in probability statement.

Setting the Stage for the Proof

We start by using the convenient criterion in Theorem 2.15, so that we are left to prove (2.4.11) in Theorem 2.15.

We then rely on Theorem 2.16, and prove the convergence of subgraph proportions as in (2.4.12), but only for *trees*. Recall the discussion of rooted and ordered trees in Section 1.5. There, we considered trees to be *ordered* as described in Definition 1.24, so that, in particular, every vertex has an ordered set of forward neighbors.

Fix a tree $\mathbf{t}$. Then, in order to prove that $\mathrm{ER}_n(\lambda/n)$ converges locally in probability to a $\mathrm{Poi}(\lambda)$ branching process, we need to show that, for every tree $\mathbf{t}$ and all integers $r \geq 0$,

$$p^{(G_n)}(\mathbf{t}) = \frac{1}{n} \sum_{u \in [n]} \mathbb{1}_{\{B_r^{(G_n)}(u) \simeq \mathbf{t}\}} \overset{\mathbb{P}}{\longrightarrow} \mu(B_r^{(G)}(o) \simeq \mathbf{t}), \qquad (2.4.16)$$

where $G_n = \mathrm{ER}_n(\lambda/n)$, and the law μ of (G, o) is that of a $\mathrm{Poi}(\lambda)$ branching process. We see that, in this case, μ is *deterministic*, as it will be in most examples encountered in this book. In (2.4.20), we may without loss of generality assume that $\mathbf{t}$ is a finite tree of depth at most r, since otherwise both sides are zero.

Ordering Trees and Subgraphs

Of course, for the event $\{B_r^{(G)}(o) \simeq \mathbf{t}\}$ to occur, the order of the tree $\mathbf{t}$ is irrelevant. Recall the breadth-first exploration of the tree $\mathbf{t}$ in Definition 1.25, which is described in terms of $(x_i)_{i=0}^t$ as in (1.5.7) and the corresponding vertices $(a_i)_{i=0}^t$, where $t = |V(\mathbf{t})|$ denotes the number of vertices in $\mathbf{t}$. Further, note that (G, o) is, by construction, an ordered tree, and therefore $B_r^{(G)}(o)$ inherits this ordering. We make this explicit by writing $\bar{B}_r^{(G)}(o)$ for the ordered version of $B_r^{(G)}(o)$. Therefore, we can write $\bar{B}_r^{(G)}(o) = \mathbf{t}$ to indicate that the two ordered trees $\bar{B}_r^{(G)}(o)$ and $\mathbf{t}$ agree. In terms of this notation, one can compute

$$\mu(\bar{B}_r^{(G)}(o) = \mathbf{t}) = \prod_{i \in [t]:\, \mathrm{dist}(\varnothing, a_i) < r} \mathrm{e}^{-\lambda} \frac{\lambda^{x_i}}{x_i!}, \tag{2.4.17}$$

where $\mathrm{dist}(\varnothing, v)$ is the tree distance between $v \in V(\mathbf{t})$ and the root $\varnothing \in V(\mathbf{t})$. We note that $\bar{B}_r^{(G)}(o) = \mathbf{t}$ says nothing about the degrees of the vertices that are at distance exactly r away from the root $\varnothing$, which is why we restrict to vertices v with $\mathrm{dist}(\varnothing, a_i) < r$ in (2.4.18). Further, $\mu(\bar{B}_r(o) = \mathbf{t}') = \mu(\bar{B}_r(o) = \mathbf{t})$ for each ordered tree $\mathbf{t}'$ that is isomorphic to the tree $\mathbf{t}$. This is because the root degrees and degree sequences of the non-root vertices are the same for all trees that are isomorphic to $\mathbf{t}$, and the right-hand side of (2.4.17) depends only on the degree of the root, and the degrees of all other non-root vertices (recall also Definition 1.25 and Exercise 1.28). Therefore,

$$\mu(B_r^{(G)}(o) \simeq \mathbf{t}) = \#(\mathbf{t}) \prod_{i \in [t]:\, \mathrm{dist}(\varnothing, a_i) < r} \mathrm{e}^{-\lambda} \frac{\lambda^{x_i}}{x_i!}, \tag{2.4.18}$$

where $\#(\mathbf{t})$ is the number of ordered trees that are isomorphic to $\mathbf{t}$. This identifies the right-hand side of (2.4.20).

We note further that by permuting the labels of all the children of any vertex in $\mathbf{t}$, we obtain a rooted tree that is isomorphic to $\mathbf{t}$, and there are $\prod_{i \in [t]} x_i!$ such permutations. However, not all of them may lead to distinct ordered trees. In our analysis, the precise value of $\#(\mathbf{t})$ will be irrelevant.

It is convenient to order also the vertices in $B_r^{(G_n)}(o_n)$, where $G_n = \mathrm{ER}_n(\lambda/n)$. This can be achieved by ordering the forward children of a vertex in $B_r^{(G_n)}(o_n)$ according to their vertex labels. We denote the result as $\bar{B}_r^{(G_n)}(o)$, which is an ordered graph. Then, we can again write $\bar{B}_r^{(G_n)}(o) = \mathbf{t}$ to indicate that the two *ordered* graphs $B_r^{(G_n)}(o)$ and $\mathbf{t}$ agree. This implies that $B_r^{(G_n)}(o)$ is a tree (so there are no cycles within depth r), and that its ordered version is equal to the ordered tree $\mathbf{t}$. Then, as in (2.4.18),

$$p^{(G_n)}(\mathbf{t}) = \frac{1}{n} \sum_{v \in [n]} \mathbb{1}_{\{B_r^{(G_n)}(v) \simeq \mathbf{t}\}} = \#(\mathbf{t}) \frac{1}{n} \sum_{v \in [n]} \mathbb{1}_{\{\bar{B}_r^{(G_n)}(v) = \mathbf{t}\}}. \tag{2.4.19}$$

We will prove below that

$$\frac{1}{n} \sum_{v \in [n]} \mathbb{1}_{\{\bar{B}_r^{(G_n)}(v)=\mathbf{t}\}} \xrightarrow{\mathbb{P}} \mu(\bar{B}_r^{(G)}(o) = \mathbf{t}). \tag{2.4.20}$$

Second-Moment Method: First Moment

To prove (2.4.20), we use a second-moment method. Denote

$$N_{n,r}(\mathbf{t}) = \sum_{v \in [n]} \mathbb{1}_{\{\bar{B}_r^{(G_n)}(v)=\mathbf{t}\}}. \tag{2.4.21}$$

The second-moment method shows that $N_{n,r}(\mathbf{t})$ is well concentrated around $n\mu(B_r^{(G)}(o)=\mathbf{t})$. We start by investigating the first moment of $N_{n,r}(\mathbf{t})$, which equals

$$\mathbb{E}[N_{n,r}(\mathbf{t})] = \sum_{v \in [n]} \mathbb{P}(\bar{B}_r^{(G_n)}(v) = \mathbf{t}) = n\mathbb{P}(\bar{B}_r^{(G_n)}(1) = \mathbf{t}), \tag{2.4.22}$$

where the latter step uses the fact that the distributions of the neighborhoods of all vertices in $\mathrm{ER}_n(\lambda/n)$ are the same.

We recall the breadth-first description of an ordered tree in Definitions 1.24 and 1.25 in Section 1.5. Let $v_i \in [n]$ denote the vertex label of the ith vertex that is explored in the breadth-first exploration. Let X_i denote the number of forward neighbors of v_i, except when v_i is at graph distance r from vertex 1, in which case we set $X_i = 0$ by convention. Further, let Y_i denote the number of edges leading to already found, but not yet explored, vertices. Then, $\bar{B}_r^{(G_n)}(1) = \mathbf{t}$ occurs precisely when $(X_i, Y_i) = (x_i, 0)$ for all $i \in [t]$. Therefore,

$$\mathbb{P}(\bar{B}_r^{(G_n)}(1) = \mathbf{t}) = \mathbb{P}((X_i, Y_i) = (x_i, 0) \, \forall i \in [t])$$
$$= \mathbb{P}((X_{[t]}, Y_{[t]}) = (x_{[t]}, 0_{[t]}) \, \forall i \in [t]), \tag{2.4.23}$$

where we use the abbreviation $x_{[t]} = (x_1, \ldots, x_t)$. Conditioning, we write this as

$$\prod_{i=1}^{t} \mathbb{P}\left((X_i, Y_i) = (x_{[i]}, 0_{[i]}) \mid (X_{[i-1]}, Y_{[i-1]}) = (x_{[i-1]}, 0_{[i-1]})\right). \tag{2.4.24}$$

Conditional on $(X_{[i-1]}, Y_{[i-1]}) = (x_{[i-1]}, 0_{[i-1]})$, for all i for which v_i is at a distance at most $r - 1$ from vertex 1, we have

$$X_i \sim \mathrm{Bin}(n_i, \lambda/n), \tag{2.4.25}$$

when $n_i = n - s_{i-1} - i + 1$, and $X_i = 0$ otherwise. Here, we recall from (1.5.7) in Definition 1.25 that $(s_i)_{i \geq 0}$ satisfies $s_0 = 1$ and $s_i = s_{i-1} + x_i - 1$ for $i \geq 1$. Further, for all $i \in [t]$,

$$Y_i \sim \mathrm{Bin}(s_{i-1} - 1, \lambda/n), \tag{2.4.26}$$

since there are s_{i-1} active vertices, and Y_i counts the number of edges between v_i and any other vertex. Finally, X_i and Y_i are conditionally independent given $(X_{[i-1]}, Y_{[i-1]}) = (x_{[i-1]}, 0_{[i-1]})$, owing to the independence of the edges in $\mathrm{ER}_n(\lambda/n)$. Note that the distance of v_i from vertex 1 is exactly equal to the distance of the corresponding vertex $a_i \in V(\mathbf{t})$ to the root $\varnothing \in V(\mathbf{t})$. Therefore, since $\mathbb{P}(\mathrm{Bin}(n_i, \lambda/n) = x_i) \to e^{-\lambda}\lambda^{x_i}/x_i!$,

$$\mathbb{P}(\bar{B}_r^{(G_n)}(1) = \mathbf{t}) = \prod_{i \in [t]\,:\, \mathrm{dist}(\varnothing, a_i) < r} \mathbb{P}(\mathrm{Bin}(n_i, \lambda/n) = x_i) \times \prod_{i \in [t]} \left(1 - \frac{\lambda}{n}\right)^{s_i - 1}$$

$$\to \prod_{i \in [t]\,:\, \mathrm{dist}(\varnothing, a_i) < r} e^{-\lambda}\frac{\lambda^{x_i}}{x_i!} = \mu(\bar{B}_r^{(G)}(o) = \mathbf{t}), \tag{2.4.27}$$

where the last equality follows from (2.4.17).

We conclude that

$$\mathbb{P}(\bar{B}_r^{(G_n)}(1) = \mathbf{t}) = \frac{1}{n}\mathbb{E}[N_{n,r}(\mathbf{t})] \to \mu(\bar{B}_r^{(G)}(o) = \mathbf{t}). \tag{2.4.28}$$

Second-Moment Method: Second Moment

For the second moment of $N_{n,r}(\mathbf{t})$, we compute

$$\mathbb{E}[N_{n,r}(\mathbf{t})^2] = \sum_{u_1, u_2 \in [n]} \mathbb{P}(\bar{B}_r^{(G_n)}(u_1) = \mathbf{t}, \bar{B}_r^{(G_n)}(u_2) = \mathbf{t}) \tag{2.4.29}$$

$$= n\mathbb{P}(\bar{B}_r^{(G_n)}(1) = \mathbf{t}) + n(n-1)\mathbb{P}(\bar{B}_r^{(G_n)}(1) = \mathbf{t}, \bar{B}_r^{(G_n)}(2) = \mathbf{t}).$$

We have already computed the asymptotics of the first term, so we are left with the second term. We claim that

$$\mathbb{P}(\bar{B}_r^{(G_n)}(1) = \mathbf{t}, \bar{B}_r^{(G_n)}(2) = \mathbf{t}) \to \mu(\bar{B}_r(o) = \mathbf{t})^2. \tag{2.4.30}$$

We are left with showing (2.4.30). Let $\mathrm{dist}_{G_n}(1, 2)$ denote the graph distance between vertices 1 and 2 in $G_n = \mathrm{ER}_n(\lambda/n)$. We make the following split:

$$\mathbb{P}(\bar{B}_r^{(G_n)}(1) = \mathbf{t}, \bar{B}_r^{(G_n)}(2) = \mathbf{t})$$
$$= \mathbb{P}(\bar{B}_r^{(G_n)}(1) = \mathbf{t}, \bar{B}_r^{(G_n)}(2) = \mathbf{t}, \mathrm{dist}_{G_n}(1, 2) > 2r)$$
$$+ \mathbb{P}(\bar{B}_r^{(G_n)}(1) = \mathbf{t}, \bar{B}_r^{(G_n)}(2) = \mathbf{t}, \mathrm{dist}_{G_n}(1, 2) \le 2r). \tag{2.4.31}$$

We bound these terms one by one. Using the fact that all vertices are exchangeable, so that vertex 2 has the same distribution as a uniform vertex different from vertex 1, the second term in (2.4.31) can be bounded as follows:

$$\mathbb{P}(\bar{B}_r^{(G_n)}(1) = \mathbf{t}, \bar{B}_r^{(G_n)}(2) = \mathbf{t}, \mathrm{dist}_{G_n}(1, 2) \le 2r)$$

$$\le \mathbb{P}(\mathrm{dist}_{G_n}(1, 2) \le 2r) \le \mathbb{E}\left[\frac{|B_{2r}^{(G_n)}(1)| - 1}{n - 1}\right] \le \frac{1}{n}\mathbb{E}[|B_{2r}^{(G_n)}(1)|]. \tag{2.4.32}$$

Recall the definition of $\partial B_k^{(G_n)}(v)$ in Definition 2.2, and write

$$\mathbb{E}[|B_{2r}^{(G_n)}(1)|] = \sum_{k=0}^{2r} \mathbb{E}[|\partial B_k^{(G_n)}(1)|]. \tag{2.4.33}$$

It is not hard to show by induction on k that (see Exercise 2.25)

$$\mathbb{E}[|\partial B_k^{(G_n)}(1)|] \le \lambda^k. \tag{2.4.34}$$

This implies that

$$\mathbb{P}(\text{dist}_{G_n}(1,2) \leq 2r) = o(1). \tag{2.4.35}$$

We rewrite the first term in (2.4.31) as

$$\mathbb{P}(\bar{B}_r^{(G_n)}(2) = \mathbf{t}, \text{dist}_{G_n}(1,2) > 2r \mid \bar{B}_r^{(G_n)}(1) = \mathbf{t})\mathbb{P}(\bar{B}_r^{(G_n)}(1) = \mathbf{t}). \tag{2.4.36}$$

The second probability converges by (2.4.28). For the first probability, we note that, conditional on $\bar{B}_r^{(G_n)}(1) = \mathbf{t}$, $\text{dist}_{G_n}(1,2) > 2r$ occurs when vertex 2 is not in $\bar{B}_r^{(G_n)}(1)$ and no vertex appearing in the exploration process of $\bar{B}_r^{(G_n)}(2)$ connects to $\partial \bar{B}_r^{(G_n)}(1)$. On the realizations where the event $\bar{B}_r^{(G_n)}(1) = \mathbf{t}$ holds, $|\partial \bar{B}_r^{(G_n)}(1)|$ equals the number of (non-root) leafs of $\mathbf{t}$, which is bounded, and thus $\text{dist}_{G_n}(1,2) > 2r$ occurs whp. Further, for $\{\bar{B}_r^{(G_n)}(2) = \mathbf{t}\} \cap \{\text{dist}_{G_n}(1,2) > 2r\}$ to occur, the exploration needs to use only vertices in $[n] \setminus \bar{B}_r^{(G_n)}(1)$, of which there are $n - |\bar{B}_r^{(G_n)}(1)| = n - |V(\mathbf{t})|$.

Since $|V(\mathbf{t})|$ is bounded, as in (2.4.28) we conclude that

$$\mathbb{P}(\bar{B}_r^{(G_n)}(2) = \mathbf{t}, \text{dist}_{G_n}(1,2) > 2r \mid \bar{B}_r^{(G_n)}(1) = \mathbf{t}) \to \mu(\bar{B}_r^{(G)}(o) = \mathbf{t}), \tag{2.4.37}$$

which completes the proof of (2.4.30).

Completion of the Proof

The convergence in (2.4.30) implies that $\text{Var}(N_{n,r}(\mathbf{t}))/\mathbb{E}[N_{n,r}(\mathbf{t})]^2 \to 0$, so that, by the Chebychev inequality [V1, Theorem 2.18],

$$\frac{N_{n,r}(\mathbf{t})}{\mathbb{E}[N_{n,r}(\mathbf{t})]} \xrightarrow{\mathbb{P}} 1. \tag{2.4.38}$$

In turn, by (2.4.28), this implies that

$$\frac{1}{n} N_{n,r}(\mathbf{t}) \xrightarrow{\mathbb{P}} \mu(\bar{B}_r^{(G)}(o) = \mathbf{t}), \tag{2.4.39}$$

which, by (2.4.19), implies that

$$p^{(G_n)}(\mathbf{t}) = \#(\mathbf{t})N_{n,r}(\mathbf{t})/n \xrightarrow{\mathbb{P}} \#(\mathbf{t})\mu(\bar{B}_r^{(G)}(o) = \mathbf{t}) = \mu(B_r^{(G)}(o) \simeq \mathbf{t}), \tag{2.4.40}$$

as required. $\qquad \square$

Local Convergence Proofs in the Remainder of the Book

In Chapters 3, 4, and 5 below we extend the above analysis to inhomogeneous random graphs, the configuration model (as well as uniform random graphs with prescribed degrees), and preferential attachment models, respectively. In many cases, the steps taken are like the above. We always combine a second-moment method for $N_{n,r}(\mathbf{t})$ with explicit computations that allow us to show the appropriate adaptation of (2.4.39).

2.5 CONSEQUENCES OF LOCAL CONVERGENCE: LOCAL FUNCTIONALS

In this section we discuss some consequences of local convergence that will either prove to be useful in what follows or describe how network statistics are determined by the local limit.

2.5.1 Local Convergence and Convergence of Neighborhoods

We start by showing that the number of vertices at distance up to m from a uniform vertex weakly converges to the neighborhood sizes of the limiting rooted graph:

Corollary 2.19 (Weak convergence of neighborhood sizes)   *Let $(G_n)_{n \geq 1}$ be a sequence of graphs whose sizes $|V(G_n)|$ tend to infinity.*

(a) *Assume that $(G_n, o_n) \xrightarrow{d} (\bar{G}, \bar{o}) \sim \bar{\mu}$ on $\mathscr{G}_\star$. Then, for every $m \geq 1$,*

$$\left(|\partial B_r^{(G_n)}(o_n)| \right)_{r=0}^{m} \xrightarrow{d} \left(|\partial B_r^{(\bar{G})}(\bar{o})| \right)_{r=1}^{m}. \tag{2.5.1}$$

(b) *Assume that G_n converges locally in probability to $(G, o) \sim \mu$ on $\mathscr{G}_\star$. Then, for every $m \geq 1$, with $o_n^{(1)}, o_n^{(2)}$ two independent uniformly chosen vertices in $V(G_n)$,*

$$\left(\left(|\partial B_r^{(G_n)}(o_n^{(1)})|, |\partial B_r^{(G_n)}(o_n^{(2)})| \right) \right)_{r=1}^{m}$$
$$\xrightarrow{d} \left(\left(|\partial B_r^{(G)}(o^{(1)})|, |\partial B_r^{(G)}(o^{(2)})| \right) \right)_{r=1}^{m}, \tag{2.5.2}$$

where the two limiting neighborhood sizes are independent *given μ.*

Proof   Part (a) follows immediately, since

$$h(G, o) = \mathbb{1}_{\{|\partial B_r^{(G)}(o)| = \ell_r \forall r \in [m]\}} \tag{2.5.3}$$

is a bounded continuous function for every m and $\ell_1, \dots, \ell_m$ (see Exercise 2.29). The proof of (2.5.2) in part (b) follows by noting that

$$\mathbb{P}\left(B_m^{(G_n)}(o_n^{(1)}) \simeq t_1, B_m^{(G_n)}(o_n^{(2)}) \simeq t_2 \mid G_n \right) = p^{(G_n)}(t_1) p^{(G_n)}(t_2), \tag{2.5.4}$$

by the independence and uniformity of $o_n^{(1)}, o_n^{(2)}$. Therefore,

$$\mathbb{P}\left(B_m^{(G_n)}(o_n^{(1)}) \simeq t_1, B_m^{(G_n)}(o_n^{(2)}) \simeq t_2 \mid G_n \right)$$
$$\xrightarrow{\mathbb{P}} \mu(B_m^{(G)}(o^{(1)}) \simeq t_1) \mu(B_m^{(G)}(o^{(2)}) \simeq t_2). \tag{2.5.5}$$

Taking the expectation proves the claim (the reader is invited to provide the fine details of this argument in Exercise 2.30).

In the above discussion, it is crucial to note that the limits in (2.5.2) correspond to two independent copies of (G, o) having law μ, but with the *same* μ, where μ is a *random* probability measure on $\mathscr{G}_\star$. It is here that the possible randomness of μ manifests itself. Recall also the example below Corollary 2.14.   □

We continue by showing that local convergence implies that the graph distance between two uniform vertices tends to infinity:

Corollary 2.20 (Large distances)   *Let $(G_n)_{n \geq 1}$ be a graph sequence whose sizes $|V(G_n)|$ tend to infinity. Let $o_n^{(1)}, o_n^{(2)}$ be two vertices chosen independently and uar from $V(G_n)$. Assume that $(G_n, o_n) \xrightarrow{d} (\bar{G}, \bar{o}) \sim \bar{\mu}$. Then*

$$\mathrm{dist}_{G_n}(o_n^{(1)}, o_n^{(2)}) \xrightarrow{\mathbb{P}} \infty. \tag{2.5.6}$$

Proof It suffices to prove that, for every $r \geq 1$,

$$\mathbb{P}(\text{dist}_{G_n}(o_n^{(1)}, o_n^{(2)}) \leq r) = o(1). \tag{2.5.7}$$

For this, we use that $o_n^{(2)}$ is chosen uar from $V(G_n)$ independently of $o_n^{(1)}$, so that

$$\mathbb{P}(\text{dist}_{G_n}(o_n^{(1)}, o_n^{(2)}) \leq r) = \mathbb{E}\big[|B_r^{(G_n)}(o_n^{(1)})|/|V(G_n)|\big]$$
$$= \mathbb{E}\big[|B_r^{(G_n)}(o_n)|/|V(G_n)|\big]. \tag{2.5.8}$$

By Corollary 2.19(a), $|B_r^{(G_n)}(o_n)|$ is a tight random variable, so that

$$|B_r^{(G_n)}(o_n)|/|V(G_n)| \overset{\mathbb{P}}{\longrightarrow} 0. \tag{2.5.9}$$

Further, $|B_r^{(G_n)}(o_n)|/|V(G_n)| \leq 1$ almost surely. Thus, by the Dominated Convergence Theorem ([V1, Theorem A.1]), $\mathbb{E}\big[|B_r^{(G_n)}(o_n)|/n\big] = o(1)$ for every $r \geq 1$, so that the claim follows. $\square$

We close this section by showing that local convergence implies that the number of connected components converges:

Corollary 2.21 (Number of connected components)   *Let $(G_n)_{n \geq 1}$ be a sequence of graphs whose sizes $|V(G_n)|$ tend to infinity, and let Q_n denote the number of connected components in G_n.*

(a) Assume that $(G_n, o_n) \overset{d}{\longrightarrow} (\bar{G}, \bar{o}) \sim \bar{\mu}$. Then

$$\mathbb{E}[Q_n/|V(G_n)|] \to \mathbb{E}_{\bar{\mu}}[1/|\mathscr{C}(\bar{o})|], \tag{2.5.10}$$

where $|\mathscr{C}(\bar{o})|$ is the size of the connected component of $\bar{o}$ in $\bar{G}$.
(b) Assume that G_n converges locally in probability to $(G, o) \sim \mu$. Then

$$Q_n/|V(G_n)| \overset{\mathbb{P}}{\longrightarrow} \mathbb{E}_{\mu}[1/|\mathscr{C}(o)|]. \tag{2.5.11}$$

Proof   Note that

$$Q_n = \sum_{v \in V(G_n)} \frac{1}{|\mathscr{C}(v)|}. \tag{2.5.12}$$

Thus, for part (a), we write

$$\mathbb{E}[Q_n/|V(G_n)|] = \mathbb{E}\left[\frac{1}{|V(G_n)|} \sum_{v \in V(G_n)} \frac{1}{|\mathscr{C}(v)|}\right]$$
$$= \mathbb{E}\left[\frac{1}{|\mathscr{C}(o_n)|}\right], \tag{2.5.13}$$

where $o_n \in V(G_n)$ is chosen uar. Since $h(G, o) = 1/|\mathscr{C}(o)|$ is a bounded and continuous function (where, by convention, $h(G, o) = 0$ when $|\mathscr{C}(o)| = \infty$; see Exercise 2.22), the claim follows.

For part (b), instead, we have

$$Q_n/|V(G_n)| = \mathbb{E}\left[\frac{1}{|\mathscr{C}(o_n)|} \;\Big|\; G_n\right] \overset{\mathbb{P}}{\longrightarrow} \mathbb{E}_\mu\left[\frac{1}{|\mathscr{C}(o)|}\right], \tag{2.5.14}$$

as required.   □

2.5.2 LOCAL CONVERGENCE AND CLUSTERING COEFFICIENTS

In this subsection we discuss the convergence of various local and global clustering coefficients when a random graph converges locally. We start by recalling the global clustering coefficient, following [V1, Section 1.5]. For a graph $G_n = (V(G_n), E(G_n))$, we let

$$W_{G_n} = \sum_{i,j,k \in V(G_n)} \mathbb{1}_{\{ij,jk \in E(G_n)\}} = \sum_{v \in V(G_n)} d_v(d_v - 1) \tag{2.5.15}$$

denote twice the number of wedges in the graph G_n. The factor two arises because the wedge ij, jk is the same as the wedge kj, ji, but it is counted twice in (2.5.15). We further let

$$\Delta_{G_n} = \sum_{i,j,k \in V(G_n)} \mathbb{1}_{\{ij,jk,ik \in E(G_n)\}} \tag{2.5.16}$$

denote six times the number of triangles in G_n. The global clustering coefficient CC_{G_n} in G_n is defined as

$$\mathsf{CC}_{G_n} = \frac{\Delta_{G_n}}{W_{G_n}}. \tag{2.5.17}$$

The global clustering coefficient measures the proportion of wedges for which the closing edge is also present. As such, it can be thought of as the probability that two random friends of a random individual are friends themselves.

The following theorem describes the conditions for the clustering coefficient to converge. In its statement, we recall that a sequence $(X_n)_{n \geq 1}$ of random variables is *uniformly integrable* when

$$\lim_{K \to \infty} \limsup_{n \to \infty} \mathbb{E}[|X_n| \mathbb{1}_{\{|X_n| > K\}}] = 0. \tag{2.5.18}$$

Theorem 2.22 (Convergence of global clustering coefficient) *Let $(G_n)_{n \geq 1}$ be a sequence of graphs whose sizes $|V(G_n)|$ tend to infinity. Assume that G_n converges locally in probability to $(G, o) \sim \mu$. Further, assume that $D_n = d_{o_n}^{(G_n)}$ is such that $(D_n^2)_{n \geq 1}$ is uniformly integrable, and that $\mu(d_o > 1) > 0$. Then*

$$\mathsf{CC}_{G_n} \overset{\mathbb{P}}{\longrightarrow} \frac{\mathbb{E}_\mu[\Delta_G(o)]}{\mathbb{E}_\mu[d_o(d_o - 1)]}, \tag{2.5.19}$$

where $\Delta_G(o) = \sum_{u,v \in \partial B_1(o)} \mathbb{1}_{\{\{u,v\} \in E(G)\}}$ denotes twice the number of triangles in G that contain o as a vertex.

Proof We write

$$\mathsf{CC}_{G_n} = \frac{\mathbb{E}[\Delta_{G_n}(o_n) \mid G_n]}{\mathbb{E}[d_{o_n}^{(G_n)}(d_{o_n}^{(G_n)} - 1) \mid G_n]}, \tag{2.5.20}$$

where the expectations are wrt the uniform choice of $o_n \in V(G_n)$, and $d_{o_n}^{(G_n)}$ denotes the degree of o_n in G_n while $\Delta_{G_n}(o_n) = \sum_{u,v \in \partial B_1^{(G_n)}(o_n)} \mathbb{1}_{\{\{u,v\} \in E(G_n)\}}$ denotes twice the number of triangles of which o_n is part.

By local convergence in probability, which implies local weak convergence, $\Delta_{G_n}(o_n) \xrightarrow{d} \Delta_G(o)$ and $d_{o_n}^{(G_n)}(d_{o_n}^{(G_n)} - 1) \xrightarrow{d} d_o(d_o - 1)$. However, both are *unbounded* functionals, so that the convergence of their expectations over o_n does not follow immediately from local convergence in probability. It is here that we need to make use of the uniform integrability of $(D_n^2)_{n \geq 1}$, where $D_n = d_{o_n}^{(G_n)}$. We make the split

$$\mathbb{E}[d_{o_n}^{(G_n)}(d_{o_n}^{(G_n)} - 1) \mid G_n] \qquad (2.5.21)$$
$$= \mathbb{E}[d_{o_n}^{(G_n)}(d_{o_n}^{(G_n)} - 1)\mathbb{1}_{\{(d_{o_n}^{(G_n)})^2 \leq K\}} \mid G_n] + \mathbb{E}[d_{o_n}^{(G_n)}(d_{o_n}^{(G_n)} - 1)\mathbb{1}_{\{(d_{o_n}^{(G_n)})^2 > K\}} \mid G_n].$$

Again by local convergence in probability (recall Corollary 2.19),

$$\mathbb{E}[d_{o_n}^{(G_n)}(d_{o_n}^{(G_n)} - 1)\mathbb{1}_{\{(d_{o_n}^{(G_n)})^2 \leq K\}} \mid G_n] \xrightarrow{\mathbb{P}} \mathbb{E}_\mu[d_o(d_o - 1)\mathbb{1}_{\{d_o^2 \leq K\}}], \qquad (2.5.22)$$

since $h(G, o) = d_o(d_o - 1)\mathbb{1}_{\{d_o^2 \leq K\}}$ is a bounded continuous function. Further, by the uniform integrability of $(D_n^2)_{n \geq 1} = ((d_{o_n}^{(G_n)})^2)_{n \geq 1}$ and with $\mathbb{E}$ denoting the expectation wrt o_n as well as wrt the random graph, for every $\varepsilon > 0$ there exists an $N = N(\varepsilon)$ sufficiently large such that, uniformly in $n \geq N(\varepsilon)$,

$$\mathbb{E}[d_{o_n}^{(G_n)}(d_{o_n}^{(G_n)} - 1)\mathbb{1}_{\{(d_{o_n}^{(G_n)})^2 > K\}}] \leq \mathbb{E}[(d_{o_n}^{(G_n)})^2\mathbb{1}_{\{(d_{o_n}^{(G_n)})^2 > K\}}] \leq \varepsilon^2. \qquad (2.5.23)$$

By the Markov inequality,

$$\mathbb{P}\left(\mathbb{E}[d_{o_n}^{(G_n)}(d_{o_n}^{(G_n)} - 1)\mathbb{1}_{\{(d_{o_n}^{(G_n)})^2 > K\}} \mid G_n] \geq \varepsilon\right)$$
$$\leq \frac{1}{\varepsilon}\mathbb{E}[(d_{o_n}^{(G_n)})^2\mathbb{1}_{\{(d_{o_n}^{(G_n)})^2 > K\}}] \leq \varepsilon. \qquad (2.5.24)$$

It follows that $\mathbb{E}[d_{o_n}^{(G_n)}(d_{o_n}^{(G_n)} - 1) \mid G_n] \xrightarrow{\mathbb{P}} \mathbb{E}_\mu[d_o(d_o - 1)]$, as required. Since $\mu(d_o > 1) > 0$, we have $\mathbb{E}_\mu[d_o(d_o - 1)] > 0$, so that also

$$1/\mathbb{E}[d_{o_n}^{(G_n)}(d_{o_n}^{(G_n)} - 1) \mid G_n] \xrightarrow{\mathbb{P}} 1/\mathbb{E}_\mu[d_o(d_o - 1)].$$

The proof that $\mathbb{E}[\Delta_{G_n}(o_n) \mid G_n] \xrightarrow{\mathbb{P}} \mathbb{E}_\mu[\Delta_G(o)]$ is similar, where now we make the split

$$\mathbb{E}[\Delta_{G_n}(o_n) \mid G_n] = \mathbb{E}[\Delta_{G_n}(o_n)\mathbb{1}_{\{(d_{o_n}^{(G_n)})^2 \leq K\}} \mid G_n]$$
$$+ \mathbb{E}[\Delta_{G_n}(o_n)\mathbb{1}_{\{(d_{o_n}^{(G_n)})^2 > K\}} \mid G_n]. \qquad (2.5.25)$$

Since $\Delta_{G_n}(o_n) \leq d_{o_n}^{(G_n)}(d_{o_n}^{(G_n)} - 1)$, the first term is again the expectation of a bounded and continuous functional, and therefore converges in probability. The second term, on the other hand, satisfies

$$\mathbb{E}[\Delta_{G_n}(o_n)\mathbb{1}_{\{(d_{o_n}^{(G_n)})^2 > K\}} \mid G_n] \leq \mathbb{E}[(d_{o_n}^{(G_n)})^2\mathbb{1}_{\{(d_{o_n}^{(G_n)})^2 > K\}} \mid G_n], \qquad (2.5.26)$$

which can be treated as in the analysis of $\mathbb{E}[d_{o_n}^{(G_n)}(d_{o_n}^{(G_n)} - 1) \mid G_n]$ above, as required. $\qquad \square$

We see that in order to obtain convergence of the clustering coefficient, we need an additional uniform integrability condition on the degree distribution. Indeed, more precisely, we need that $(D_n^2)_{n\geq 1}$ is uniformly integrable, with $D_n = d_{o_n}^{(G_n)}$ the degree of a uniform vertex. This is a recurring theme below.

We next discuss a related clustering coefficient, where such additional assumptions are not needed. For this, we define the *local* clustering coefficient for vertex $v \in V(G_n)$ to be

$$\mathsf{CC}_{G_n}(v) = \frac{\Delta_{G_n}(v)}{d_v(d_v - 1)}, \tag{2.5.27}$$

where $\Delta_{G_n}(v) = \sum_{s,t \in \partial B_1^{(G_n)}(v)} \mathbb{1}_{\{\{s,t\}\in E(G_n)\}}$ is again twice the number of triangles of which v is part. Then, we let the local clustering coefficient be

$$\overline{\mathsf{CC}}_{G_n} = \frac{1}{|V(G_n)|} \sum_{v \in V(G_n)} \mathsf{CC}_{G_n}(v). \tag{2.5.28}$$

Here, we can think of $\Delta_{G_n}(v)/[d_v(d_v - 1)]$ as the proportion of edges present between neighbors of v, and then (2.5.28) takes the average of this. The following theorem implies its convergence without any further uniform integrability conditions, and thus justifies the name *local* clustering coefficient:

Theorem 2.23 (Convergence of local clustering coefficient) *Let $(G_n)_{n\geq 1}$ be a sequence of graphs whose sizes $|V(G_n)|$ tend to infinity. Assume that G_n converges locally in probability to $(G, o) \sim \mu$. Then*

$$\overline{\mathsf{CC}}_{G_n} \xrightarrow{\mathbb{P}} \mathbb{E}_\mu\left[\frac{\Delta_G(o)}{d_o(d_o - 1)}\right]. \tag{2.5.29}$$

Proof We now write

$$\overline{\mathsf{CC}}_{G_n} = \mathbb{E}\left[\frac{\Delta_{G_n}(o_n)}{d_{o_n}^{(G_n)}(d_{o_n}^{(G_n)} - 1)} \mid G_n\right], \tag{2.5.30}$$

and note that $h(G, o) = \Delta_G(o)/[d_o(d_o-1)]$ is a bounded continuous functional. Therefore,

$$\mathbb{E}\left[\frac{\Delta_{G_n}(o_n)}{d_{o_n}^{(G_n)}(d_{o_n}^{(G_n)} - 1)} \mid G_n\right] \xrightarrow{\mathbb{P}} \mathbb{E}_\mu\left[\frac{\Delta_G(o)}{d_o(d_o - 1)}\right], \tag{2.5.31}$$

as required. $\square$

There are more versions of clustering coefficients. Convergence of the so-called *clustering spectrum* is discussed in the notes in Section 2.7.

2.5.3 NEIGHBORHOODS OF EDGES AND DEGREE–DEGREE DEPENDENCIES

In this subsection we discuss the convergence of *degree–degree dependencies* when a random graph converges locally. By degree–degree dependencies, we mean the dependencies between the degrees of the vertices at the two ends of edges in the graph. Often, this is described in terms of the degree–degree dependencies in an edge drawn uar from the collection of all edges in the graph. Such dependencies are often described in terms of the so-called *assortativity coefficient*. For this, it is also crucial to discuss the convergence of the local neighborhood of a *uniformly chosen edge*. We will again see that an extra uniform integrability condition is needed for the assortativity coefficient to converge.

We start by defining the neighborhood structure of edges. It will be convenient to consider *directed* edges. We let $e = (u, v)$ be an edge directed from u to v, and let $\vec{E}(G_n) = \{(u, v) : \{u, v\} \in E(G_n)\}$ denote the collection of directed edges, so that $|\vec{E}(G_n)| = 2|E(G_n)|$. Directed edges are convenient, as they assign a (root-)vertex to an edge. For $e = (u, v)$, we often write $\underline{e} = u$ and $\bar{e} = v$ for its start and end vertices.

For $H_\star \in \mathscr{G}_\star$, let

$$p_e^{(G_n)}(H_\star) = \frac{1}{2|E(G_n)|} \sum_{(u,v) \in \vec{E}(G_n)} \mathbb{1}_{\{B_r^{(G_n)}(u) \simeq H_\star\}}. \tag{2.5.32}$$

Note that

$$p_e^{(G_n)}(H_\star) = \mathbb{P}(B_r^{(G_n)}(\underline{e}) \simeq H_\star \mid G_n), \tag{2.5.33}$$

where $e = (\underline{e}, \bar{e})$ is a uniformly chosen directed edge from $\vec{E}(G_n)$. Thus, $p_e^{(G_n)}(H_\star)$ is the edge-equivalent of $p^{(G_n)}(H_\star)$ in (2.3.5). We next study its asymptotics:

Theorem 2.24 (Convergence neighborhoods of edges) *Let $(G_n)_{n \geq 1}$ be a sequence of graphs whose sizes $|V(G_n)|$ tend to infinity. Assume that G_n converges locally in probability to $(G, o) \sim \mu$. Assume further that $(d_{o_n}^{(G_n)})_{n \geq 1}$ is a uniformly integrable sequence of random variables, and that $\mu(d_o \geq 1) > 0$. Then, for every $H_\star \in \mathscr{G}_\star$,*

$$p_e^{(G_n)}(H_\star) \overset{\mathbb{P}}{\longrightarrow} \frac{\mathbb{E}_\mu[d_o \mathbb{1}_{\{B_r^{(G)}(o) \simeq H_\star\}}]}{\mathbb{E}_\mu[d_o]}. \tag{2.5.34}$$

Proof We recall (2.5.32), and note that

$$\frac{2}{|V(G_n)|}|E(G_n)| = \frac{1}{|V(G_n)|} \sum_{v \in V(G_n)} d_v^{(G_n)} = \mathbb{E}[d_{o_n}^{(G_n)} \mid G_n]. \tag{2.5.35}$$

Therefore, since $(d_{o_n}^{(G_n)})_{n \geq 1}$ is uniformly integrable, local convergence in probability implies that

$$\frac{2}{|V(G_n)|}|E(G_n)| \overset{\mathbb{P}}{\longrightarrow} \mathbb{E}_\mu[d_o]. \tag{2.5.36}$$

Since $\mu(d_o \geq 1) > 0$, it follows that $\mathbb{E}_\mu[d_o] > 0$.

Further, we rewrite

$$\frac{1}{|V(G_n)|} \sum_{(u,v) \in \vec{E}(G_n)} \mathbb{1}_{\{B_r^{(G_n)}(u) \simeq H_\star\}} = \frac{1}{|V(G_n)|} \sum_{u \in V(G_n)} d_u^{(G_n)} \mathbb{1}_{\{B_r^{(G_n)}(u) \simeq H_\star\}}$$

$$= \mathbb{E}\left[d_{o_n}^{(G_n)} \mathbb{1}_{\{B_r^{(G_n)}(o_n) \simeq H_\star\}} \mid G_n\right], \tag{2.5.37}$$

where o_n is a uniformly chosen vertex in $V(G_n)$. Again, since $(d_{o_n}^{(G_n)})_{n \geq 1}$ is uniformly integrable and by local convergence in probability,

$$\mathbb{E}\left[d_{o_n}^{(G_n)}\mathbb{1}_{\{B_r^{(G_n)}(o_n)\simeq H_\star\}} \mid G_n\right] \overset{\mathbb{P}}{\longrightarrow} \mathbb{E}_\mu[d_o\mathbb{1}_{\{B_r^{(G)}(o)\simeq H_\star\}}]. \tag{2.5.38}$$

Therefore, by (2.5.32), taking the ratio of the terms in (2.5.36) and (2.5.38) proves the claim.
□

We continue by considering the *degree–degree distribution* given by

$$p_{k,l}^{(G_n)} = \frac{1}{2|E(G_n)|} \sum_{e\in\vec{E}(G_n)} \mathbb{1}_{\{d_{\underline{e}}^{(G_n)}=k,d_{\bar{e}}^{(G_n)}=l\}}. \tag{2.5.39}$$

Thus, $p_{k,l}^{(G_n)}$ is the probability that a random directed edge connects a vertex of degree k with one of degree l. By convention, we define $p_{k,l}^{(G_n)} = 0$ when $k = 0$. The following theorem proves that the degree–degree distribution converges when the graph locally converges in probability:

Theorem 2.25 (Degree–degree convergence)   *Let $(G_n)_{n\geq 1}$ be a sequence of graphs whose sizes $|V(G_n)|$ tend to infinity. Assume that G_n converges locally in probability to $(G, o) \sim \mu$. Assume further that $(d_{o_n}^{(G_n)})_{n\geq 1}$ is a uniformly integrable sequence of random variables, and that $\mu(d_o \geq 1) > 0$. Then, for every k, l with $k \geq 1$,*

$$p_{k,l}^{(G_n)} \overset{\mathbb{P}}{\longrightarrow} k\mu(d_o = k, d_V = l), \tag{2.5.40}$$

where V is a neighbor of o chosen uar.

Proof   Recall (2.5.36). We rewrite

$$\frac{1}{|V(G_n)|} \sum_{e\in\vec{E}(G_n)} \mathbb{1}_{\{d_{\underline{e}}^{(G_n)}=k,d_{\bar{e}}^{(G_n)}=l\}}$$

$$= k\frac{1}{|V(G_n)|} \sum_{u\in V(G_n)} \mathbb{1}_{\{d_u^{(G_n)}=k\}}\left(\frac{1}{k}\sum_{v:\,v\sim u}\mathbb{1}_{\{d_u^{(G_n)}=l\}}\right)$$

$$= k\mathbb{E}\left[\mathbb{1}_{\{d_{o_n}^{(G_n)}=k,d_V^{(G_n)}=l\}} \mid G_n\right], \tag{2.5.41}$$

where V is a uniform neighbor of o_n, which itself is uniform in $V(G_n)$. Again, by local convergence in probability and the fact that the distribution of (d_o, d_V) conditional on (G, o) is a deterministic function of $B_2^{(G_n)}(o_n)$,

$$\frac{1}{|V(G_n)|} \sum_{e\in\vec{E}(G_n)} \mathbb{1}_{\{d_{\underline{e}}^{(G_n)}=k,d_{\bar{e}}^{(G_n)}=l\}} \overset{\mathbb{P}}{\longrightarrow} k\mu(d_o = k, d_V = l). \tag{2.5.42}$$

Thus, by (2.5.39), taking the ratio of the terms in (2.5.36) and (2.5.42) proves the claim.   □

We finally discuss the consequences for the assortativity coefficient (recall [V1, Section 1.5]). We now write the degrees in G_n as $(d_v)_{v\in V(G_n)}$ to avoid notational clutter. Define the *assortativity coefficient* as

$$\rho_{G_n} = \frac{\sum_{i,j\in V(G_n)} d_id_j\big(\mathbb{1}_{\{(i,j)\in\vec{E}(G_n)\}} - d_id_j/|\vec{E}(G_n)|\big)}{\sum_{i,j\in V(G_n)} \big(d_i\mathbb{1}_{\{i=j\}} - d_id_j/|\vec{E}(G_n)|\big)d_id_j}, \tag{2.5.43}$$

where we recall that $\vec{E}(G_n)$ is the collection of directed edges and we make the abbreviation $d_i = d_i^{(G_n)}$ for $i \in V(G_n)$. We can recognize ρ_{G_n} in (2.5.43) as the *empirical correlation coefficient* of the two-dimensional sequence of variables $(d_{\underline{e}}, d_{\bar{e}})_{e \in \vec{E}(G_n)}$. As a result, it is the correlation between the coordinates of the two-dimensional random variable of which $(p_{k,l}^{(G_n)})_{k,l \geq 1}$ is the joint probability mass function. We can rewrite the assortativity coefficient ρ_{G_n} more conveniently as

$$\rho_{G_n} = \frac{\sum_{(i,j) \in \vec{E}(G_n)} d_i d_j - (\sum_{i \in V(G_n)} d_i^2)^2 / |\vec{E}(G_n)|}{\sum_{i \in V(G_n)} d_i^3 - (\sum_{i \in V(G_n)} d_i^2)^2 / |\vec{E}(G_n)|}. \tag{2.5.44}$$

The following theorem gives conditions for the convergence of ρ_{G_n} when G_n converges locally in probability:

Theorem 2.26 (Assortativity convergence) *Let $(G_n)_{n \geq 1}$ be a sequence of graphs whose sizes $|V(G_n)|$ tend to infinity. Assume that G_n converges locally in probability to $(G, o) \sim \mu$. Assume further that $D_n = d_{o_n}^{(G_n)}$ is such that $(D_n^3)_{n \geq 1}$ is uniformly integrable, and that $\mu(d_o = r) < 1$ for every $r \geq 0$. Then,*

$$\rho_{G_n} \overset{\mathbb{P}}{\longrightarrow} \frac{\mathbb{E}_\mu[d_o^2 d_V] - \mathbb{E}_\mu[d_o^2]^2 / \mathbb{E}_\mu[d_o]}{\mathbb{E}_\mu[d_o^3] - \mathbb{E}_\mu[d_o^2]^2 / \mathbb{E}_\mu[d_o]}, \tag{2.5.45}$$

where V is a neighbor of o chosen uar.

Proof We start with (2.5.44), and consider the various terms. We divide all the sums by n. Then, by local convergence in probability and the uniform integrability of $((d_{o_n}^{(G_n)})^3)_{n \geq 1}$, which implies that $(d_{o_n}^{(G_n)})_{n \geq 1}$ is also uniformly integrable,

$$\frac{1}{n} |\vec{E}(G_n)| = \mathbb{E}[d_{o_n}^{(G_n)} \mid G_n] \overset{\mathbb{P}}{\longrightarrow} \mathbb{E}_\mu[d_o]. \tag{2.5.46}$$

Again by local convergence and the uniform integrability of $((d_{o_n}^{(G_n)})^3)_{n \geq 1}$, which implies that $((d_{o_n}^{(G_n)})^2)_{n \geq 1}$ is also uniformly integrable,

$$\frac{1}{|V(G_n)|} \sum_{i \in V(G_n)} d_i^2 = \mathbb{E}[(d_{o_n}^{(G_n)})^2 \mid G_n] \overset{\mathbb{P}}{\longrightarrow} \mathbb{E}_\mu[d_o^2]. \tag{2.5.47}$$

Further, again by local convergence in probability and the assumed uniform integrability of $((d_{o_n}^{(G_n)})^3)_{n \geq 1}$,

$$\frac{1}{|V(G_n)|} \sum_{i \in V(G_n)} d_i^3 = \mathbb{E}[(d_{o_n}^{(G_n)})^3 \mid G_n] \overset{\mathbb{P}}{\longrightarrow} \mathbb{E}_\mu[d_o^3]. \tag{2.5.48}$$

This identifies the limits of all but one of the sums appearing in (2.5.44). Details are left to the reader in Exercise 2.26. Further, $\mathbb{E}_\mu[d_o^3] - \mathbb{E}_\mu[d_o^2]^2 / \mathbb{E}_\mu[d_o] > 0$ since $\mu(d_o = r) < 1$ for every $r \geq 0$ (see Exercise 2.27).

We finally consider the last term, involving the product of the degrees across edges, i.e.,

$$\frac{1}{|V(G_n)|} \sum_{(i,j) \in \vec{E}(G_n)} d_i d_j = \frac{1}{|V(G_n)|} \sum_{u \in V(G_n)} d_u^2 \left(\frac{1}{d_u} \sum_{v: v \sim u} d_v \right) \tag{2.5.49}$$

$$= \mathbb{E}[d_{o_n}^2 d_V \mid G_n],$$

where V is a uniform neighbor of o_n. When the degrees are uniformly bounded, the functional $h(G, o) = d_o^2 \mathbb{E}[d_V \mid G]$ is bounded and continuous, so that it will converge. However, the degrees are not necessarily bounded, so a truncation argument is needed.

We make the split

$$\frac{1}{|V(G_n)|} \sum_{(i,j)\in\vec{E}(G_n)} d_i d_j = \frac{1}{|V(G_n)|} \sum_{(i,j)\in\vec{E}(G_n)} d_i d_j \mathbb{1}_{\{d_i \leq K, d_j \leq K\}}$$

$$+ \frac{1}{|V(G_n)|} \sum_{(i,j)\in\vec{E}(G_n)} d_i d_j (1 - \mathbb{1}_{\{d_i \leq K, d_j \leq K\}}). \quad (2.5.50)$$

We now rewrite, in the same way as above,

$$\frac{1}{|V(G_n)|} \sum_{(i,j)\in\vec{E}(G_n)} d_i d_j \mathbb{1}_{\{d_i \leq K, d_j \leq K\}} = \mathbb{E}[d_{o_n}^2 d_V \mathbb{1}_{\{d_{o_n} \leq K, d_V \leq K\}} \mid G_n]. \quad (2.5.51)$$

By local convergence in probability (or by Theorem 2.25), since the functional is now bounded and continuous,

$$\frac{1}{|V(G_n)|} \sum_{(i,j)\in\vec{E}(G_n)} d_i d_j \mathbb{1}_{\{d_i \leq K, d_j \leq K\}} \overset{\mathbb{P}}{\longrightarrow} \mathbb{E}_\mu[d_o^2 d_V \mathbb{1}_{\{d_{o_n} \leq K, d_V \leq K\}}]. \quad (2.5.52)$$

We are left with showing that the second contribution in (2.5.50) is small. We bound this contribution as follows:

$$\frac{1}{|V(G_n)|} \sum_{(i,j)\in\vec{E}(G_n)} d_i d_j (\mathbb{1}_{\{d_i > K\}} + \mathbb{1}_{\{d_j > K\}})$$

$$= \frac{2}{|V(G_n)|} \sum_{(i,j)\in\vec{E}(G_n)} d_i d_j \mathbb{1}_{\{d_i > K\}}. \quad (2.5.53)$$

We now use the Cauchy–Schwarz inequality to bound this:

$$\frac{1}{|V(G_n)|} \sum_{(i,j)\in\vec{E}(G_n)} d_i d_j (\mathbb{1}_{\{d_i > K\}} + \mathbb{1}_{\{d_j > K\}})$$

$$\leq \frac{2}{|V(G_n)|} \sqrt{\sum_{(i,j)\in\vec{E}(G_n)} d_i^2 \mathbb{1}_{\{d_i > K\}}} \sqrt{\sum_{(i,j)\in\vec{E}(G_n)} d_j^2}$$

$$= 2\mathbb{E}[(d_{o_n}^{(G_n)})^3 \mathbb{1}_{\{d_{o_n}^{(G_n)} > K\}} \mid G_n]^{1/2} \mathbb{E}[(d_{o_n}^{(G_n)})^3 \mid G_n]^{1/2}. \quad (2.5.54)$$

By the uniform integrability of $((d_{o_n}^{(G_n)})^3)_{n \geq 1}$, there exists $K = K(\varepsilon)$ and $N = N(\varepsilon)$ such that, for all $n \geq N$,

$$\mathbb{E}[(d_{o_n}^{(G_n)})^3 \mathbb{1}_{\{d_{o_n}^{(G_n)} > K\}}] \leq \varepsilon^4/4. \quad (2.5.55)$$

In turn, by the Markov inequality, this implies that

$$\mathbb{P}\Big(\mathbb{E}[(d_{o_n}^{(G_n)})^3 \mathbb{1}_{\{d_{o_n}^{(G_n)} > K\}} \mid G_n] \geq \frac{\varepsilon^3}{4}\Big) \leq \frac{4}{\varepsilon^3} \mathbb{E}[(d_{o_n}^{(G_n)})^3 \mathbb{1}_{\{d_{o_n}^{(G_n)} > K\}}] \leq \varepsilon. \quad (2.5.56)$$

As a result, with probability at least $1 - \varepsilon$ and for $\varepsilon > 0$ sufficiently small to accommodate the factor $\mathbb{E}_n[(d_{o_n}^{(G_n)})^3]^{1/2}$ (which is uniformly bounded by the uniform integrability of $((d_{o_n}^{(G_n)})^3)_{n\geq 1}$),

$$\frac{1}{|V(G_n)|} \sum_{(i,j)\in\vec{E}(G_n)} d_i d_j (\mathbb{1}_{\{d_i>K\}} + \mathbb{1}_{\{d_j>K\}}) \leq \varepsilon^{3/2} \mathbb{E}[(d_{o_n}^{(G_n)})^3 \mid G_n]^{1/2} \leq \varepsilon. \quad (2.5.57)$$

This completes the proof. $\qquad\square$

2.6 GIANT COMPONENT IS ALMOST LOCAL

We continue by investigating the size of the giant component when the graph converges locally. Here, we simplify the notation by assuming that $G_n = (V(G_n), E(G_n))$ is such that $|V(G_n)| = n$, and we recall that

$$|\mathscr{C}_{\max}| = \max_{v\in V(G_n)} |\mathscr{C}(v)| \quad (2.6.1)$$

denotes the maximal connected component size. While Corollary 2.21 shows that the number of connected components is well behaved in the local topology, the proportion of vertices in the giant is not so nicely behaved.

2.6.1 ASYMPTOTICS OF THE GIANT

Clearly, the proportion of vertices in the largest connected component $|\mathscr{C}_{\max}|/n$ is *not* continuous in the local convergence topology (see Exercise 2.31), as it is a *global* object. In fact, $|\mathscr{C}(o_n)|/n$ also does not converge in distribution when $(G_n, o_n) \xrightarrow{d} (G, o)$. However, local convergence still tells us a useful story about the existence of a giant, as well as its size:

Corollary 2.27 (Upper bound on the giant) *Let $(G_n)_{n\geq 1}$ be a sequence of graphs whose sizes $|V(G_n)| = n$ tend to infinity. Assume that G_n converges locally in probability to $(G, o) \sim \mu$. Write $\zeta = \mu(|\mathscr{C}(o)| = \infty)$ for the survival probability of the limiting graph (G, o). Then, for every $\varepsilon > 0$ fixed,*

$$\mathbb{P}(|\mathscr{C}_{\max}| \leq n(\zeta + \varepsilon)) \to 1. \quad (2.6.2)$$

In particular, Corollary 2.27 implies that $|\mathscr{C}_{\max}|/n \xrightarrow{\mathbb{P}} 0$ when $\zeta = 0$ (see Exercise 2.33), so that there can only be a giant when the local limit has a positive survival probability.

Proof Define

$$Z_{\geq k} = \sum_{v\in V(G_n)} \mathbb{1}_{\{|\mathscr{C}(v)|\geq k\}}. \quad (2.6.3)$$

Assume that G_n converges locally in probability to (G, o). Then, we conclude that, with $\zeta_{\geq k} = \mu(|\mathscr{C}(o)| \geq k)$ (see Exercise 2.32),

$$\frac{Z_{\geq k}}{n} = \mathbb{E}[\mathbb{1}_{\{|\mathscr{C}(o_n)|\geq k\}} \mid G_n] \xrightarrow{\mathbb{P}} \zeta_{\geq k}. \quad (2.6.4)$$

For every $k \geq 1$,

$$\{|\mathscr{C}_{\max}| \geq k\} = \{Z_{\geq k} \geq k\}, \tag{2.6.5}$$

and $|\mathscr{C}_{\max}| \leq Z_{\geq k}$ on those realizations where the event that $Z_{\geq k} \geq 1$ holds. Note that $\zeta = \lim_{k \to \infty} \zeta_{\geq k} = \mu(|\mathscr{C}(o)| = \infty)$. We take k large enough that $\zeta \geq \zeta_{\geq k} - \varepsilon/2$. Then, for every $k \geq 1$, $\varepsilon > 0$, and all n large enough that $n(\zeta + \varepsilon) \geq k$,

$$\begin{aligned}
\mathbb{P}(|\mathscr{C}_{\max}| \geq n(\zeta + \varepsilon)) &\leq \mathbb{P}(Z_{\geq k} \geq n(\zeta + \varepsilon)) \\
&\leq \mathbb{P}(Z_{\geq k} \geq n(\zeta_{\geq k} + \varepsilon/2)) = o(1). \tag{2.6.6}
\end{aligned}$$

$\square$

We conclude that while local convergence cannot determine the size of the largest connected component, it *can* prove an upper bound on $|\mathscr{C}_{\max}|$. In this book, we often extend this to $|\mathscr{C}_{\max}|/n \overset{\mathbb{P}}{\longrightarrow} \zeta = \mu(|\mathscr{C}(o)| = \infty)$, but this is no longer a consequence of local convergence alone. In Exercise 2.31, you are asked to give an example where $|\mathscr{C}_{\max}|/n \overset{\mathbb{P}}{\longrightarrow} \eta < \zeta$, even though G_n does converge locally in probability to $(G, o) \sim \mu$. Therefore, in general, more involved arguments must be used. The next theorem shows that one, relatively simple, condition suffices. In its statement, we write $x \longleftrightarrow\!\!\!\!/\; y$ for the statement that $\mathscr{C}(x)$ and $\mathscr{C}(y)$ are disjoint:

Theorem 2.28 (The giant is almost local) *Let $G_n = (V(G_n), E(G_n))$ denote a random graph of size $|V(G_n)| = n$. Assume that G_n converges locally in probability to $(G, o) \sim \mu$. Assume further that*

$$\lim_{k \to \infty} \limsup_{n \to \infty} \frac{1}{n^2} \mathbb{E}\Big[\#\{(x, y) \in V(G_n) \times V(G_n) \colon |\mathscr{C}(x)|, |\mathscr{C}(y)| \geq k, x \longleftrightarrow\!\!\!\!/\; y\}\Big] = 0. \tag{2.6.7}$$

Then, if $\mathscr{C}_{\max}$ and $\mathscr{C}_{(2)}$ denote the largest and second largest connected components (with ties broken arbitrarily),

$$\frac{|\mathscr{C}_{\max}|}{n} \overset{\mathbb{P}}{\longrightarrow} \zeta = \mu(|\mathscr{C}(o)| = \infty), \qquad \frac{|\mathscr{C}_{(2)}|}{n} \overset{\mathbb{P}}{\longrightarrow} 0. \tag{2.6.8}$$

Remark 2.29 ("Giant is almost local" proofs) Theorem 2.28 shows that the relatively mild condition in (2.6.7) suffices for the giant to have the expected limit. In fact, it is necessary *and sufficient*; see Exercise 2.34. It is most useful when we can easily show that vertices with large clusters are likely to be connected, and it will be applied to the Erdős–Rényi random graph below, to configuration models in Section 4.3, and to inhomogeneous random graphs with finitely many types in Section 6.5.3. ◀

We now start with the proof of Theorem 2.28. Recall that $\zeta = \mu(|\mathscr{C}(o) = \infty)$ might be a random variable when μ is a random probability measure on rooted graphs. We first note that, by Corollary 2.27, the statement follows on the event that $\zeta = \mu(|\mathscr{C}(o) = \infty) = 0$, so that it suffices to prove Theorem 2.28 on the event that $\zeta > 0$. By conditioning on this event, we may assume that $\zeta > 0$ almost surely.

We recall that the vector $(|\mathscr{C}_{(i)}|)_{i \geq 1}$ denotes the cluster sizes ordered in size, from large to small with ties broken arbitrarily, so that $|\mathscr{C}_{(1)}| = |\mathscr{C}_{\max}|$. The following lemma gives a useful estimate of the sum of squares of these ordered cluster sizes. In its statement, we write $X_{n,k} = o_{k,\mathbb{P}}(1)$ when

$$\lim_{k\to\infty} \limsup_{n\to\infty} \mathbb{P}(|X_{n,k}| > \varepsilon) = 0. \tag{2.6.9}$$

Exercise 2.23 shows that $Z_{\geq k}/n = \zeta + o_{k,\mathbb{P}}(1)$.

Lemma 2.30 (Convergence of sum of squares of cluster sizes) *Under the conditions of Theorem 2.28,*

$$\frac{1}{n^2} \sum_{i\geq 1} |\mathscr{C}_{(i)}|^2 \mathbb{1}_{\{|\mathscr{C}_{(i)}|\geq k\}} = \zeta^2 + o_{k,\mathbb{P}}(1). \tag{2.6.10}$$

Proof We use that, by local convergence in probability and for any $k \geq 1$ fixed (recall (2.6.4))

$$\frac{1}{n} Z_{\geq k} = \frac{1}{n} \sum_{i\geq 1} |\mathscr{C}_{(i)}| \mathbb{1}_{\{|\mathscr{C}_{(i)}|\geq k\}} = \zeta + o_{k,\mathbb{P}}(1), \tag{2.6.11}$$

by Exercise 2.23. Further,

$$\zeta^2 + o_{k,\mathbb{P}}(1) = \frac{Z_{\geq k}^2}{n^2} = \frac{1}{n} \sum_{i\geq 1} |\mathscr{C}_{(i)}|^2 \mathbb{1}_{\{|\mathscr{C}_{(i)}|\geq k\}} + o_{k,\mathbb{P}}(1). \tag{2.6.12}$$

Indeed,

$$\frac{1}{n^2} \sum_{\substack{i,j\geq 1 \\ i\neq j}} |\mathscr{C}_{(i)}||\mathscr{C}_{(j)}| \mathbb{1}_{\{|\mathscr{C}_{(i)}|,|\mathscr{C}_{(j)}|\geq k\}}$$

$$= \frac{1}{n^2} \#\{(x,y) \in V(G_n) \times V(G_n) : |\mathscr{C}(x)|, |\mathscr{C}(y)| \geq k, x \not\longleftrightarrow y\}, \tag{2.6.13}$$

which, by the Markov inequality, and abbreviating $(x,y) \in V(G_n) \times V(G_n)$ to (x,y), satisfies

$$\lim_{k\to\infty} \limsup_{n\to\infty} \mathbb{P}\Big(\frac{1}{n^2} \#\{(x,y) : |\mathscr{C}(x)|, |\mathscr{C}(y)| \geq k, x \not\longleftrightarrow y\} \geq \varepsilon\Big)$$

$$\leq \lim_{k\to\infty} \limsup_{n\to\infty} \frac{1}{\varepsilon n^2} \mathbb{E}\Big[\#\{(x,y) : |\mathscr{C}(x)|, |\mathscr{C}(y)| \geq k, x \not\longleftrightarrow y\}\Big] = 0, \tag{2.6.14}$$

by our main assumption in (2.6.7). We conclude that

$$\frac{1}{n^2} \sum_{i\geq 1} |\mathscr{C}_{(i)}|^2 \mathbb{1}_{\{|\mathscr{C}_{(i)}|\geq k\}} = \frac{Z_{\geq k}^2}{n^2} = \zeta^2 + o_{k,\mathbb{P}}(1), \tag{2.6.15}$$

by (2.6.11). This proves (2.6.10). Exercise 2.24 proves that $\frac{1}{n^2} \sum_{i\geq 1} |\mathscr{C}_{(i)}|^2 \xrightarrow{\mathbb{P}} \zeta^2$. $\qquad\square$

We are now ready to complete the proof of Theorem 2.28:

Proof of Theorem 2.28. By Lemma 2.30,

$$\zeta^2 + o_{k,\mathbb{P}}(1) = \frac{1}{n^2}\sum_{i\geq 1}|\mathscr{C}_{(i)}|^2 \mathbb{1}_{\{|\mathscr{C}_{(i)}|\geq k\}} \leq \frac{|\mathscr{C}_{\max}|}{n}\frac{1}{n}\sum_{i\geq 1}|\mathscr{C}_{(i)}|\mathbb{1}_{\{|\mathscr{C}_{(i)}|\geq k\}}$$

$$= \frac{|\mathscr{C}_{\max}|}{n}(\zeta + o_{k,\mathbb{P}}(1)), \tag{2.6.16}$$

where the last step follows by (2.6.11). Thus, $|\mathscr{C}_{\max}|/n \geq \zeta + o_{k,\mathbb{P}}(1)$. Since k is arbitrary, this proves that $|\mathscr{C}_{\max}| \geq n\zeta(1 + o_{\mathbb{P}}(1))$. By Corollary 2.27, also $|\mathscr{C}_{\max}| \leq n\zeta(1 + o_{\mathbb{P}}(1))$. Therefore, $|\mathscr{C}_{\max}|/n \xrightarrow{\mathbb{P}} \zeta$. In turn, since $|\mathscr{C}_{\max}|/n \xrightarrow{\mathbb{P}} \zeta$ and by (2.6.11), on the event that $|\mathscr{C}_{(2)}| \geq k$,

$$\frac{1}{n}|\mathscr{C}_{(2)}| \leq \frac{1}{n}\sum_{i\geq 2}|\mathscr{C}_{(i)}|\mathbb{1}_{\{|\mathscr{C}_{(i)}|\geq k\}} = \frac{1}{n}\sum_{i\geq 1}|\mathscr{C}_{(i)}|\mathbb{1}_{\{|\mathscr{C}_{(i)}|\geq k\}} - \frac{|\mathscr{C}_{\max}|}{n} = o_{k,\mathbb{P}}(1), \tag{2.6.17}$$

which, again since k is arbitrary, implies that $|\mathscr{C}_{(2)}|/n \xrightarrow{\mathbb{P}} 0$.   □

2.6.2 PROPERTIES OF THE GIANT

We next extend Theorem 2.28 somewhat, and investigate the structure of the giant. For this, we first let $v_\ell(\mathscr{C}_{\max})$ denote the number of vertices with degree ℓ in the giant component, and we recall that $|E(\mathscr{C}_{\max})|$ denotes the number of edges in the giant component:

Theorem 2.31 (Properties of the giant)   *Under the assumptions of Theorem 2.28, when* $\zeta = \mu(|\mathscr{C}(o)| = \infty) > 0$,

$$\frac{v_\ell(\mathscr{C}_{\max})}{n} \xrightarrow{\mathbb{P}} \mu(|\mathscr{C}(o)| = \infty, d_o = \ell). \tag{2.6.18}$$

Further, assume that $D_n = d_{o_n}^{(G_n)}$ *is such that* $(D_n)_{n\geq 1}$ *is uniformly integrable. Then,*

$$\frac{|E(\mathscr{C}_{\max})|}{n} \xrightarrow{\mathbb{P}} \frac{1}{2}\mathbb{E}_\mu\left[d_o\mathbb{1}_{\{|\mathscr{C}(o)|=\infty\}}\right]. \tag{2.6.19}$$

Proof   The proof follows that of Theorem 2.28. We now define, for $k \geq 1$, $A \subseteq \mathbb{N}$, and with d_v the degree of v in G_n,

$$Z_{A,\geq k} = \sum_{v\in V(G_n)}\mathbb{1}_{\{|\mathscr{C}(v)|\geq k, d_v\in A\}}. \tag{2.6.20}$$

Assume that G_n converges locally in probability to (G, o). Then we conclude that

$$\frac{Z_{A,\geq k}}{n} \xrightarrow{\mathbb{P}} \mu(|\mathscr{C}(o)| \geq k, d_o \in A). \tag{2.6.21}$$

Since $|\mathscr{C}_{\max}| \geq k$ whp by Theorem 2.28, we thus obtain, for every $A \subseteq \mathbb{N}$,

$$\frac{1}{n}\sum_{a\in A}v_a(\mathscr{C}_{\max}) \leq \frac{Z_{A,\geq k}}{n} \xrightarrow{\mathbb{P}} \mu(|\mathscr{C}(o)| \geq k, d_o \in A), \tag{2.6.22}$$

so that, with $A = \{\ell\}$,

$$\frac{1}{n}v_\ell(\mathscr{C}_{\max}) \leq \mu(|\mathscr{C}(o)| = \infty, d_o = \ell) + o_{k,\mathbb{P}}(1), \qquad (2.6.23)$$

which establishes the required upper bound. We are left to prove the corresponding lower bound.

Applying (2.6.22) to $A = \{\ell\}^c$, we obtain that, for all $\varepsilon > 0$,

$$\lim_{n\to\infty} \mathbb{P}\left(\frac{1}{n}\left[|\mathscr{C}_{\max}| - v_\ell(\mathscr{C}_{\max})\right] \leq \mu(|\mathscr{C}(o)| \geq k, d_o \neq \ell) + \varepsilon/2\right) = 1. \qquad (2.6.24)$$

We argue by contradiction. Suppose that, for some ℓ,

$$\liminf_{n\to\infty} \mathbb{P}\left(\frac{v_\ell(\mathscr{C}_{\max})}{n} \leq \mu(|\mathscr{C}(o)| = \infty, d_o = \ell) - \varepsilon\right) = \kappa > 0. \qquad (2.6.25)$$

Then, by (2.6.24), we have that also the lim inf of the probability of the intersection of $\{[v_\ell(\mathscr{C}_{\max})/n \leq \mu(|\mathscr{C}(o)| = \infty, d_o = \ell) - \varepsilon\}$ and $\{[|\mathscr{C}_{\max}| - v_\ell(\mathscr{C}_{\max})]/n \leq \mu(|\mathscr{C}(o)| \geq k, d_o \neq \ell) + \varepsilon/2\}$ is at least $\kappa > 0$.

Therefore, along the subsequence $(n_l)_{l\geq 1}$ that attains the lim inf in (2.6.25), with asymptotic probability $\kappa > 0$, and using (2.6.24),

$$\begin{aligned}
\frac{|\mathscr{C}_{\max}|}{n} &= \frac{1}{n}[|\mathscr{C}_{\max}| - v_\ell(\mathscr{C}_{\max})] + \frac{v_\ell(\mathscr{C}_{\max})}{n} \\
&\leq \mu(|\mathscr{C}(o)| \in \{\ell\}^c) + \varepsilon/2 + \mu(|\mathscr{C}(o)| = \infty, d_o = \ell) - \varepsilon \\
&\leq \mu(|\mathscr{C}(o)| = \infty) - \varepsilon/2, \qquad (2.6.26)
\end{aligned}$$

which contradicts Theorem 2.28. We conclude that (2.6.25) cannot hold, so that (2.6.18) follows.

For (2.6.19), we note that

$$|E(\mathscr{C}_{\max})| = \frac{1}{2}\sum_{\ell \geq 1} \ell v_\ell(\mathscr{C}_{\max}). \qquad (2.6.27)$$

We divide by n and split the sum over ℓ into small and large ℓ:

$$\frac{|E(\mathscr{C}_{\max})|}{n} = \frac{1}{2n}\sum_{\ell \in [K]} \ell v_\ell(\mathscr{C}_{\max}) + \frac{1}{2n}\sum_{\ell > K} \ell v_\ell(\mathscr{C}_{\max}). \qquad (2.6.28)$$

For the first term in (2.6.28), by (2.6.18), we have

$$\frac{1}{2n}\sum_{\ell \in [K]} \ell v_\ell(\mathscr{C}_{\max}) \xrightarrow{\mathbb{P}} \frac{1}{2}\sum_{\ell \in [K]} \mu(|\mathscr{C}(o)| = \infty, d_o = \ell)$$

$$= \frac{1}{2}\mathbb{E}_\mu\left[d_o \mathbf{1}_{\{|\mathscr{C}(o)|=\infty, d_o \in [K]\}}\right]. \qquad (2.6.29)$$

For the second term in (2.6.28), we obtain the bound, with n_ℓ the number of degree ℓ vertices in G_n,

$$\frac{1}{2n}\sum_{\ell > K} \ell v_\ell(\mathscr{C}_{\max}) \leq \frac{1}{2}\sum_{\ell > K} \ell \frac{n_\ell}{n} = \frac{1}{2}\mathbb{E}\left[d_{o_n}^{(G_n)} \mathbf{1}_{\{d_{o_n}^{(G_n)} > K\}} \mid G_n\right]. \qquad (2.6.30)$$

By uniform integrability,

$$\lim_{K \to \infty} \limsup_{n \to \infty} \mathbb{E}\big[d_{o_n}^{(n)} \mathbb{1}_{\{d_{o_n}^{(G_n)} > K\}}\big] = 0. \tag{2.6.31}$$

As a result, by the Markov inequality and for every $\varepsilon > 0$,

$$\mathbb{P}\Big(\mathbb{E}\big[d_{o_n}^{(G_n)} \mathbb{1}_{\{d_{o_n}^{(G_n)} > K\}} \mid G_n\big] > \varepsilon\Big) \leq \mathbb{E}\big[d_{o_n}^{(n)} \mathbb{1}_{\{d_{o_n}^{(G_n)} > K\}}\big] / \varepsilon \to 0, \tag{2.6.32}$$

when first $n \to \infty$ followed by $K \to \infty$. This completes the proof of (2.6.19). $\qquad\square$

It is not hard to extend the above analysis to the local convergence in probability of the giant, as well as its complement, as formulated in the following theorem:

Theorem 2.32 (Local limit of the giant)   *Under the assumptions of Theorem 2.28, when* $\zeta = \mu(|\mathscr{C}(o)| = \infty) > 0$,

$$\frac{1}{n} \sum_{v \in \mathscr{C}_{\max}} \mathbb{1}_{\{B_r^{(G_n)}(v) \simeq H_\star\}} \overset{\mathbb{P}}{\longrightarrow} \mu(|\mathscr{C}(o)| = \infty, B_r^{(G)}(o) \simeq H_\star), \tag{2.6.33}$$

and

$$\frac{1}{n} \sum_{v \notin \mathscr{C}_{\max}} \mathbb{1}_{\{B_r^{(G_n)}(v) \simeq H_\star\}} \overset{\mathbb{P}}{\longrightarrow} \mu(|\mathscr{C}(o)| < \infty, B_r^{(G)}(o) \simeq H_\star). \tag{2.6.34}$$

Proof   The convergence in (2.6.34) follows from that in (2.6.33) combined with the fact that, by assumption,

$$\frac{1}{n} \sum_{v \in V(G_n)} \mathbb{1}_{\{B_r^{(G_n)}(v) \simeq H_\star\}} \overset{\mathbb{P}}{\longrightarrow} \mu(B_r^{(G)}(o) \simeq H_\star). \tag{2.6.35}$$

The convergence in (2.6.34) can be proved as for Theorem 2.31, now using that, for every $\mathscr{H}_\star \subseteq \mathscr{G}_\star$,

$$\frac{1}{n} Z_{\mathscr{H}_\star, \geq k} \equiv \frac{1}{n} \sum_{v \in V(G_n)} \mathbb{1}_{\{|\mathscr{C}(v)| \geq k, B_r^{(G_n)}(v) \in \mathscr{H}_\star\}}$$

$$\overset{\mathbb{P}}{\longrightarrow} \mu(|\mathscr{C}(o)| \geq k, B_r^{(G)}(o) \in \mathscr{H}_\star), \tag{2.6.36}$$

and, since $|\mathscr{C}_{\max}|/n \overset{\mathbb{P}}{\longrightarrow} \zeta > 0$ by Theorem 2.28,

$$\frac{1}{n} \sum_{v \in \mathscr{C}_{\max}} \mathbb{1}_{\{B_r^{(G_n)}(v) \in \mathscr{H}_\star\}} \leq \frac{Z_{\mathscr{H}_\star, \geq k}}{n}. \tag{2.6.37}$$

We then argue by contradiction again as in (2.6.25) and (2.6.26). We leave the details to the reader. $\qquad\square$

2.6.3 "Giant is Almost Local" Condition Revisited

The "giant is almost local" condition (2.6.7) is sometimes inconvenient to verify, and we now give an alternative form that is often easier to work with. Recall that when we count vertex pairs (x, y) with certain properties, we mean that $(x, y) \in V(G_n) \times V(G_n)$:

Lemma 2.33 (Condition (2.6.7) revisited) *Under the assumptions of Theorem 2.28, the condition in (2.6.7) holds when*

$$\lim_{r\to\infty} \limsup_{n\to\infty} \frac{1}{n^2} \mathbb{E}\Big[\#\big\{(x,y)\colon |\partial B_r^{(G_n)}(x)|, |\partial B_r^{(G_n)}(y)| \geq r, x \not\longleftrightarrow y\big\}\Big] = 0, \quad (2.6.38)$$

provided that there exists $r = r_k \to \infty$ such that, as $k \to \infty$,

$$\mu(|\mathscr{C}(o)| \geq k, |\partial B_r^{(G)}(o)| < r) \to 0, \qquad \mu(|\mathscr{C}(o)| < k, |\partial B_r^{(G)}(o)| \geq r) \to 0. \quad (2.6.39)$$

Proof Denote

$$P_k = \#\big\{(x,y)\colon |\mathscr{C}(x)|, |\mathscr{C}(y)| \geq k, x \not\longleftrightarrow y\big\}, \quad (2.6.40)$$

$$P_r' = \#\big\{(x,y)\colon |\partial B_r^{(G_n)}(x)|, |\partial B_r^{(G_n)}(y)| \geq r, x \not\longleftrightarrow y\big\}. \quad (2.6.41)$$

Then,

$$|P_k - P_r'| \leq 2n\big[Z_{<r,\geq k} + Z_{\geq r,<k}\big], \quad (2.6.42)$$

where

$$Z_{<r,\geq k} = \sum_{v\in V(G_n)} \mathbb{1}_{\{|\partial B_r^{(G_n)}(v)| < r, |\mathscr{C}(v)| \geq k\}}, \quad (2.6.43)$$

$$Z_{\geq r,<k} = \sum_{v\in V(G_n)} \mathbb{1}_{\{|\partial B_r^{(G_n)}(v)| \geq r, |\mathscr{C}(v)| < k\}}. \quad (2.6.44)$$

Therefore, by local convergence in probability,

$$\frac{1}{n^2}|P_k - P_r'| \leq \frac{2}{n}\big[Z_{<r,\geq k} + Z_{\geq r,<k}\big] \quad (2.6.45)$$

$$\overset{\mathbb{P}}{\longrightarrow} 2\mu(|\mathscr{C}(o)| \geq k, |\partial B_r^{(G)}(o)| < r) + 2\mu(|\mathscr{C}(o)| < k, |\partial B_r^{(G)}(o)| \geq r).$$

Take $r = r_k$ as in (2.6.39). Then, the rhs of (2.6.45) vanishes, so that by the Dominated Convergence Theorem [V1, Theorem A.1], we also have

$$\lim_{k\to\infty} \limsup_{n\to\infty} \frac{1}{n^2} \mathbb{E}\big[|P_k - P_{r_k}'|\big] = 0. \quad (2.6.46)$$

We arrive at

$$\lim_{k\to\infty} \limsup_{n\to\infty} \frac{1}{n^2} \mathbb{E}[P_k] \leq \lim_{k\to\infty} \limsup_{n\to\infty} \frac{1}{n^2} \mathbb{E}\big[|P_k - P_{r_k}'|\big] + \frac{1}{n^2} \mathbb{E}[P_{r_k}'] = 0, \quad (2.6.47)$$

by (2.6.38), since $r_k \to \infty$ when $k \to \infty$. $\qquad\square$

The assumption in (2.6.39) on the local limit is often easily verified. For example, for the Erdős–Rényi random graph $ER_n(\lambda/n)$ with $\lambda > 1$, to which we apply it below, we can take $r = k$ and use that, on the event of survival (recall [V1, Theorem 3.9]),

$$\lambda^{-r}|\partial B_r^{(G)}(o)| \overset{a.s.}{\longrightarrow} M. \quad (2.6.48)$$

Here $(G, o) \sim \mu$ denotes a Poisson branching process with mean λ offspring, for which we know that $M > 0$ on the event of survival by [V1, Theorem 3.10]. Therefore, $\mu(|\mathscr{C}(o)| \geq k, |\partial B_k^{(G)}(o)| < k) \to 0$ as $k \to \infty$. Further, $\mu(|\mathscr{C}(o)| < k, |\partial B_k^{(G)}(o)| \geq k) = 0$

trivially. However, there are examples where (2.6.39) fails, and then also the equivalence of (2.6.7) and (2.6.38) may also be false.

2.6.4 GIANT IN ERDŐS–RÉNYI RANDOM GRAPHS

In this subsection we use the local convergence in probability of $ER_n(\lambda/n)$ in Theorem 2.18, combined with the fact that the "giant is almost local" in Theorem 2.28, to identify the phase transition and size of the giant in the Erdős–Rényi random graph $ER_n(\lambda/n)$:

Theorem 2.34 (Phase transition in Erdős–Rényi random graph) *Fix $\lambda > 0$, and let $\mathscr{C}_{\max}$ be the largest connected component of the Erdős–Rényi random graph $ER_n(\lambda/n)$ and $\mathscr{C}_{(2)}$ the second largest connected component (breaking ties arbitrarily). Then,*

$$\frac{|\mathscr{C}_{\max}|}{n} \xrightarrow{\mathbb{P}} \zeta_\lambda, \qquad \frac{|\mathscr{C}_{(2)}|}{n} \xrightarrow{\mathbb{P}} 0, \tag{2.6.49}$$

where ζ_λ is the survival probability of a Poisson branching process with mean offspring λ. In particular, $\zeta_\lambda > 0$ precisely when $\lambda > 1$.

Further, for $\lambda > 0$, with $\eta_\lambda = 1 - \zeta_\lambda$ and for all $\ell \geq 0$,

$$\frac{v_\ell(\mathscr{C}_{\max})}{n} \xrightarrow{\mathbb{P}} e^{-\lambda}\frac{\lambda^\ell}{\ell!}[1 - \eta_\lambda^\ell], \tag{2.6.50}$$

and

$$\frac{|E(\mathscr{C}_{\max})|}{n} \xrightarrow{\mathbb{P}} \frac{1}{2}\lambda[1 - \eta_\lambda^2]. \tag{2.6.51}$$

The law of large numbers for the giant in Theorem 2.34 for $\lambda > 1$ was also proved in [V1, Theorem 4.8], where a more precise bound was given on the convergence rate. There, the proof was given using explicit computations. Here we show that it follows rather directly from local convergence considerations. We refer to [V1, Section 4.6] for a discussion of the history of the phase transition for $ER_n(\lambda/n)$.

Proof The main work resides in showing that the condition (2.6.7) in Theorem 2.28 holds. In turn, (2.6.7) in Theorem 2.28 can be replaced by (2.6.38) in Lemma 2.33.

Indeed, local convergence in probability follows from Theorem 2.18. Then, the claim in (2.6.49) follows directly from Theorem 2.28, while (2.6.50) and (2.6.51) follow from Theorem 2.31, and the observations that, for a Poisson branching process with mean offspring $\lambda > 1$,

$$\mu(|\mathscr{C}(o)| = \infty, d_o = \ell) = e^{-\lambda}\frac{\lambda^\ell}{\ell!}[1 - \eta_\lambda^\ell], \tag{2.6.52}$$

and thus

$$\begin{aligned}
\mathbb{E}_\mu\Big[d_o\mathbb{1}_{\{|\mathscr{C}(o)|=\infty\}}\Big] &= \sum_\ell \ell\mu(|\mathscr{C}(o)| = \infty, d_o = \ell) \\
&= \sum_\ell \ell e^{-\lambda}\frac{\lambda^\ell}{\ell!}[1 - \eta_\lambda^\ell] = \lambda[1 - \eta_\lambda e^{-\lambda(1-\eta_\lambda)}] \\
&= \lambda[1 - \eta_\lambda^2],
\end{aligned} \tag{2.6.53}$$

since η_λ satisfies $\eta_\lambda = e^{-\lambda(1-\eta_\lambda)}$. Therefore, we are left to show that (2.6.38) holds, which is the key to the proof. The proof proceeds in several steps.

Step 1: Relation to, and Concentration of, Binomials

Fix $G_n = \mathrm{ER}_n(\lambda/n)$ with $\lambda > 1$. Note that, with $o_1, o_2 \in V(G_n) = [n]$ chosen independently and uar,

$$\frac{1}{n^2}\mathbb{E}\Big[\#\{(x,y) \in [n]: |\partial B_r^{(G_n)}(x)|, |\partial B_r^{(G_n)}(x)| \geq r, x \not\leftrightarrow y\}\Big] \qquad (2.6.54)$$

$$= \mathbb{P}(|\partial B_r^{(G_n)}(o_1)|, |\partial B_r^{(G_n)}(o_2)| \geq r, o_1 \not\leftrightarrow o_2).$$

Throughout this proof, we use the abbreviations $B_r(o_i) = B_r^{(G_n)}(o_i)$ and $\partial B_r(o_i) = \partial B_r^{(G_n)}(o_i)$ to simplify notation. We let $\mathbb{P}_r$ denote the conditional distribution of $\mathrm{ER}_n(\lambda/n)$ given $|\partial B_r(o_1)| = b_0^{(i)}$ with $b_0^{(i)} \geq r$, and $|B_r(o_i)| = s_0^{(i)}$, so that

$$\mathbb{P}(|\partial B_r(o_1)|, |\partial B_r(o_2)| \geq r, o_1 \not\leftrightarrow o_2) \qquad (2.6.55)$$

$$= \sum_{b_0^{(1)}, b_0^{(2)}, s_0^{(1)}, s_0^{(2)}} \mathbb{P}_r(o_1 \not\leftrightarrow o_2)\mathbb{P}(|\partial B_r(o_i)| = b_0^{(i)}, |B_{r-1}(o_i)| = s_0^{(i)}, i \in \{1,2\}).$$

Our aim is to show that, for every $\varepsilon > 0$, we can find $r = r_\varepsilon$ such that, for every $b_0^{(1)}, b_0^{(2)} \geq r$ and $s_0^{(1)}, s_0^{(2)}$ fixed,

$$\limsup_{n\to\infty} \mathbb{P}_r(o_1 \not\leftrightarrow o_2) \leq \varepsilon. \qquad (2.6.56)$$

Under $\mathbb{P}_r$,

$$|\partial B_{r+1}(o_1) \setminus B_r(o_2)| \sim \mathrm{Bin}(n_1^{(1)}, p_1^{(1)}), \qquad (2.6.57)$$

where

$$n_1^{(1)} = n - s_0^{(1)} - s_0^{(2)}, \qquad p_1^{(1)} = 1 - \Big(1 - \frac{\lambda}{n}\Big)^{b_0^{(1)}}. \qquad (2.6.58)$$

Here, we note that the vertices in $\partial B_r(o_2)$ play a different role from those in $\partial B_r(o_1)$, as they can be in $\partial B_{r+1}(o_1)$, but those in $\partial B_r(o_1)$ cannot. This explains the slightly asymmetric form with respect to vertices 1 and 2 in (2.6.58).

We are led to studying concentration properties of binomial random variables. For this, we rely on the following lemma:

Lemma 2.35 (Concentration binomials) *Let $X \sim \mathrm{Bin}(m, p)$. Then, for every $\delta > 0$,*

$$\mathbb{P}(|X - \mathbb{E}[X]| \geq \delta\mathbb{E}[X]) \leq 2\exp\Big(-\frac{\delta^2\mathbb{E}[X]}{2(1 + \delta/3)}\Big). \qquad (2.6.59)$$

Proof This is a direct consequence of [V1, Theorem 2.21]. $\qquad\square$

Lemma 2.35 ensures that whp the boundary of $|\partial B_{r+1}(o_1)|$ is close to $\lambda|\partial B_r(o_1)|$ for r and n large, so that the boundary grows by a factor $\lambda > 1$. Further applications lead to the statement that $|\partial B_{r+k}(o_1)| \approx \lambda^k|\partial B_r(o_1)|$. Thus, in roughly $a\log_\lambda n$ steps, the boundary will have expanded to n^a vertices. However, in order to make this precise, we need that (1) the *sum* of complementary probabilities in Lemma 2.35 is still quite small uniformly in k and for r large; and (2) we have good control over the number of vertices in the boundaries, not just in terms of lower bounds, but also in terms of upper bounds, as that gives control over the number of vertices that have not yet been used. For the latter, we also need to deal with the δ-dependence in (2.6.59).

We prove (2.6.56) by first growing $|\partial B_{r+k}(o_1)|$ for $k \geq 1$ until $|\partial B_{r+k}(o_1)|$ is very large (much larger than $\sqrt{n}$ will suffice), and then, outside $B_{r+k}(o_1)$ for the appropriate k, growing $|\partial B_{r+k}(o_2)|$ for $k \geq r$ until $|\partial B_{r+k}(o_2)|$ is also very large (now $\sqrt{n}$ will suffice). Then, it is very likely that there is a direct edge between the resulting boundaries. We next provide the details.

Step 2: Erdős–Rényi Neighborhood Growth of First Vertex

To make the above analysis precise, we start by introducing some notation. We grow the sets $\partial B_{r+l}(o_1)$ *outside* $B_r(o_2)$, and denote $\partial B'_{r+l-1}(o_1)$ to indicate this (minor) change. We let $\overline{b}_k^{(1)} = b_0^{(1)}[\lambda(1+\varepsilon)]^k$ and $\underline{b}_k^{(1)} = b_0^{(1)}[\lambda(1-\varepsilon)]^k$ denote upper and lower bounds on $|\partial B'_{r+k}(o_1)|$ that we will prove to hold whp, where we choose $\varepsilon > 0$ small enough that $\lambda(1-\varepsilon) > 1$. We let

$$\overline{s}_{k-1}^{(1)} = s_0^{(1)} + \sum_{l=0}^{k-1} \overline{b}_l^{(1)} \tag{2.6.60}$$

denote the resulting upper bound on $|B'_{r+k-1}(o_1)|$. We fix $a \in (\frac{1}{2}, 1)$, and let

$$k \leq k_n^\star = k_n^\star(\varepsilon) = \lceil a \log_{\lambda(1-\varepsilon)} n \rceil, \tag{2.6.61}$$

and note that there exists $C > 1$ such that

$$\overline{s}_{k-1}^{(1)} \leq s_0^{(1)} + \sum_{l=0}^{k-1} [\lambda(1+\varepsilon)]^l \leq s_0^{(1)} + \frac{b_0^{(1)}}{\lambda(1+\varepsilon) - 1}[\lambda(1+\varepsilon)]^l$$
$$\leq C n^{a \log \lambda(1+\varepsilon)/\log \lambda(1-\varepsilon)}, \tag{2.6.62}$$

uniformly in $k \leq k_n^\star$. We choose $a \in (\frac{1}{2}, 1)$ so that $a \log \lambda(1+\varepsilon)/\log \lambda(1-\varepsilon) \in (\frac{1}{2}, 1)$. Define the *good event* by

$$\mathcal{E}_{r,[k]}^{(1)} = \bigcap_{l \in [k]} \mathcal{E}_{r,l}^{(1)}, \quad \text{where} \quad \mathcal{E}_{r,k}^{(1)} = \{\underline{b}_k^{(1)} \leq |\partial B'_{r+k}(o_1)| \leq \overline{b}_k^{(1)}\}. \tag{2.6.63}$$

We write

$$\mathbb{P}_r\left(\mathcal{E}_{r,[k]}^{(1)}\right) = \prod_{l \in [k]} \mathbb{P}_r\left(\mathcal{E}_{r,l}^{(1)} \mid \mathcal{E}_{r,[l-1]}^{(1)}\right), \tag{2.6.64}$$

so that

$$\mathbb{P}_r\left(\mathcal{E}_{r,[k]}^{(1)}\right) \geq 1 - \sum_{l \in [k]} \mathbb{P}_r\left((\mathcal{E}_{r,l}^{(1)})^c \mid \mathcal{E}_{r,[l-1]}^{(1)}\right). \tag{2.6.65}$$

With the above choices, we have, conditional on $|\partial B'_{r+l-1}(o_1)| = b_{l-1}^{(1)} \in [\underline{b}_{l-1}^{(1)}, \overline{b}_{l-1}^{(1)}]$ and $|B'_{r+l-1}(o_1)| = s_{l-1}^{(1)} \leq \overline{s}_{l-1}^{(1)}$,

$$|\partial B'_{r+l}(o_1)| \sim \mathrm{Bin}(n_l^{(1)}, p_l^{(1)}), \tag{2.6.66}$$

where

$$n_l^{(1)} = n - s_{l-1}^{(1)} - s_0^{(2)}, \qquad p_l^{(1)} = 1 - \left(1 - \frac{\lambda}{n}\right)^{b_{l-1}^{(1)}}. \tag{2.6.67}$$

The fact that we grow $\partial B'_{r+l-1}(o_1)$ *outside of* $B_r(o_2)$ is reflected in the subtraction of $s_0^{(2)}$ in $n_l^{(1)}$. We aim to apply Lemma 2.35 to $|\partial B'_{r+l}(o_1)|$, with $\delta = \varepsilon/2$, for which it suffices to prove bounds on the (conditional) expectation $n_l^{(1)} p_l^{(1)}$. We use that

$$\frac{b_{l-1}^{(1)}\lambda}{n} - \frac{(b_{l-1}^{(1)}\lambda)^2}{2n^2} \le p_l^{(1)} \le \frac{b_{l-1}^{(1)}\lambda}{n}. \tag{2.6.68}$$

Therefore, with $\mathbb{E}_r$ denoting expectation wrt $\mathbb{P}_r$,

$$\mathbb{E}_r[|\partial B'_{r+l}(o_1)| \mid \mathcal{E}_{r,[l-1]}^{(1)}] = n_l^{(1)} p_l^{(1)} \le n \frac{b_{l-1}^{(1)}\lambda}{n} = \lambda b_{l-1}^{(1)}, \tag{2.6.69}$$

which provides the upper bound on $n_l^{(1)} p_l^{(1)}$. For the lower bound, we use the lower bound in (2.6.68) to note that $p_l^{(1)} \ge (1 - \varepsilon/4)\lambda b_{l-1}^{(1)}/n$ for n sufficiently large, since we are on $\mathcal{E}_{r,[l-1]}^{(1)}$. Further, $n_l^{(1)} \ge (1 - \varepsilon/4)n$ on $\mathcal{E}_{r,[l-1]}^{(1)}$, uniformly in $l \le k_n^\star$. We conclude that, for n sufficiently large,

$$n_l^{(1)} p_l^{(1)} \ge (1 - \varepsilon/4)^2 n \frac{b_{l-1}^{(1)}\lambda}{n} \ge (1 - \varepsilon/2)\lambda b_{l-1}^{(1)}. \tag{2.6.70}$$

Recall the definition of $\mathcal{E}_{r,l}^{(1)}$ in (2.6.63). As a result, $\underline{b}_k^{(1)} \le |\partial B'_{r+k}(o_1)| \le \overline{b}_k^{(1)}$ implies that $||\partial B'_{r+k}(o_1)| - n_l^{(1)} p_l^{(1)}| \le (\varepsilon/2)n_l^{(1)} p_l^{(1)}$. Thus, by Lemma 2.35 with $\delta = \varepsilon/2$,

$$\mathbb{P}_r\left((\mathcal{E}_{r,l}^{(1)})^c \mid \mathcal{E}_{r,[l-1]}^{(1)}\right) \le \mathbb{P}_r\left(||\partial B'_{r+l}(o_1)| - n_l^{(1)} p_l^{(1)}| \ge (\varepsilon/2)n_l^{(1)} p_l^{(1)} \mid \mathcal{E}_{r,[l-1]}^{(1)}\right)$$

$$\le 2\exp\left(-\frac{\varepsilon^2(1 - \varepsilon/2)\lambda b_{l-1}^{(1)}}{8(1 + \varepsilon/6)}\right) = 2\exp\left(-q\lambda b_{l-1}^{(1)}\right), \tag{2.6.71}$$

where $q = \varepsilon^2(1 - \varepsilon/2)/[8(1 + \varepsilon/6)] > 0$.

We conclude that, for n sufficiently large,

$$\mathbb{P}_r\left(\mathcal{E}_{r,[k]}^{(1)}\right) \ge 1 - 2\sum_{l=1}^{k} e^{-q\lambda \underline{b}_{l-1}^{(1)}}, \tag{2.6.72}$$

which is our key estimate for the neighborhood growth in $\mathrm{ER}_n(\lambda/n)$.

Step 3: Erdős–Rényi Neighborhood Growth of Second Vertex

We next grow the neighborhoods from vertex 2 in a similar way, and we focus on the differences only. In the whole argument below, we are conditioning on $|\partial B_{r+l}(o_1)| = b_l^{(1)} \in [\underline{b}_l^{(1)}, \overline{b}_l^{(1)}]$ and $|B_{r+l}(o_1)| = s_l^{(1)} \le \overline{s}_l^{(1)}$ for all $l \in [k_n^\star]$.

Further, rather than exploring $(\partial B_{r+k}(o_2))_{k\ge0}$, we again explore these neighborhoods *outside* $B'_{r+k_n^\star}(o_1)$, and denote them by $(\partial B'_{r+k}(o_2))_{k\ge0}$.

We define $\overline{b}_k^{(2)} = b_0^{(2)}[\lambda(1 + \varepsilon)]^k$ and $\underline{b}_k^{(2)} = b_0^{(2)}[\lambda(1 - \varepsilon)]^k$ in the same way as for o_1, and, as in (2.6.60),

$$\overline{s}_{k-1}^{(2)} = s_0^{(2)} + s_k^{(2)} + \sum_{l=0}^{k-1} \overline{b}_l^{(2)}, \tag{2.6.73}$$

so that also, uniformly in $k \leq k_n^\star$,

$$\overline{s}_{k-1}^{(2)} \leq C n^{a \log \lambda (1+\varepsilon) / \log \lambda (1-\varepsilon)}. \tag{2.6.74}$$

We further define, as in (2.6.63) and for some $C > 1$,

$$\mathcal{E}_{r,[k]}^{(2)} = \bigcap_{l \in [k]} \mathcal{E}_{r,l}^{(2)}, \qquad \text{where} \qquad \mathcal{E}_{r,k}^{(2)} = \{\underline{b}_k^{(2)} \leq |\partial B'_{r+k}(o_2)| \leq \overline{b}_k^{(2)}\}. \tag{2.6.75}$$

Then, conditional on the above, as well as on $|\partial B'_{r+k-1}(o_2)| = b_{k-1}^{(2)} \in [\underline{b}_{k-1}^{(2)}, \overline{b}_{k-1}^{(2)}]$ and $|B'_{r+k-1}(o_2)| = s_{k-1}^{(2)} \leq \overline{s}_{k-1}^{(2)}$, we have

$$|\partial B'_{r+k}(o_2)| \sim \mathsf{Bin}(n_k^{(2)}, p_k^{(2)}), \tag{2.6.76}$$

where now

$$n_k^{(2)} = n - s_{k_n^\star}^{(1)} - s_{k-1}^{(2)}, \qquad p_k^{(2)} = 1 - \left(1 - \frac{\lambda}{n}\right)^{b_{k-1}^{(2)}}. \tag{2.6.77}$$

Let $\mathbb{P}_{r,k_n^\star}$ denote the conditional probability given $|\partial B_r(o_1)| = b_0^{(i)}$ with $b_0^{(i)} \geq r$, $|B_{r-1}(o_i)| = s_0^{(i)}$, and $\mathcal{E}_{r,[k_n^\star]}^{(1)}$. Following the argument in the previous paragraph, we are led to the conclusion that

$$\mathbb{P}_{r,k_n^\star}\left(\mathcal{E}_{r,[k_n^\star]}^{(2)}\right) \geq 1 - 2 \sum_{l=1}^{k_n^\star} e^{-q\lambda \underline{b}_{l-1}^{(2)}}, \tag{2.6.78}$$

where again $q = \varepsilon^2 (1 - \varepsilon/2)/[8(1 + \varepsilon/6)] > 0$.

Completion of the Proof of Theorem 2.34

We will use Lemma 2.33, and conclude that we need to show (2.6.56). Recall (2.6.78), and that $\underline{b}_k^{(i)} = b_0^{(i)}[\lambda(1-\varepsilon)]^k$, where $b_0^{(i)} \geq r = r_\varepsilon$. Fix $k = k_n^\star = k_n^\star(\varepsilon) = \lceil a \log_{\lambda(1-\varepsilon)} n \rceil$ as in (2.6.61). For this $k = k_n^\star$, take $r = r_\varepsilon$ sufficiently large that

$$\sum_{l=1}^{k} [e^{-q\lambda \underline{b}_{l-1}^{(1)}} + e^{-q\lambda \underline{b}_{l-1}^{(2)}}] \leq \varepsilon/2. \tag{2.6.79}$$

Denote the *good event* by

$$\mathcal{E}_{r_\varepsilon,[k_n^\star]} = \mathcal{E}_{r_\varepsilon,[k_n^\star]}^{(1)} \cap \mathcal{E}_{r_\varepsilon,[k_n^\star]}^{(2)}, \tag{2.6.80}$$

where we recall (2.6.63) and (2.6.75). Then,

$$\liminf_{n \to \infty} \mathbb{P}_r\left(\mathcal{E}_{r_\varepsilon,[k_n^\star]}\right) \geq 1 - \varepsilon. \tag{2.6.81}$$

On $\mathcal{E}_{r_\varepsilon,[k_n^\star]}$,

$$|\partial B_{r+k_n^\star}(o_1)| \geq \underline{b}_{k_n^\star}^{(1)} = b_0^{(1)}[\lambda(1-\varepsilon)]^k \geq r_\varepsilon n^a, \tag{2.6.82}$$

where we choose a such that $a \in (\frac{1}{2}, 1)$. An identical bound holds for $|\partial B'_{r+k_n^\star}(o_2)|$. Therefore, the total number of direct edges between $\partial B_{r_\varepsilon+k_n^\star}(o_1)$ and $\partial B'_{r_\varepsilon+k_n^\star}(o_2)$ is at least $(r_\varepsilon n^a)^2 \gg n$ when $a > \frac{1}{2}$. Each of these potential edges is present independently with probability λ/n. Therefore,

$$\mathbb{P}_r\Big(\mathrm{dist}_{\mathrm{ER}_n(\lambda/n)}(o_1, o_2) > 2(k_n^\star + r_\varepsilon) + 1 \mid \mathcal{E}_{r_\varepsilon,[k_n^\star]}\Big) \leq \Big(1 - \frac{\lambda}{n}\Big)^{(r_\varepsilon n^a)^2} = o(1). \quad (2.6.83)$$

We conclude that, for n sufficiently large,

$$\mathbb{P}_r\Big(\mathrm{dist}_{\mathrm{ER}_n(\lambda/n)}(o_1, o_2) \leq 2(k_n^\star + r_\varepsilon) + 1 \mid \mathcal{E}_{r_\varepsilon,[k_n^\star]}\Big) = 1 - o(1). \quad (2.6.84)$$

Thus, by (2.6.81), for n sufficiently large,

$$\mathbb{P}_r(o_1 \longleftrightarrow\hspace{-1.4em}/\hspace{0.6em} o_2) \leq \mathbb{P}_r\big(\mathcal{E}_{r_\varepsilon,[k_n^\star]}^c\big) + \mathbb{P}_r\big(o_1 \longleftrightarrow\hspace{-1.4em}/\hspace{0.6em} o_2 \mid \mathcal{E}_{r_\varepsilon,[k_n^\star]}\big) \leq \varepsilon/2 + \varepsilon/2 \leq \varepsilon. \quad (2.6.85)$$

Since $\varepsilon > 0$ is arbitrary, the claim in (2.6.56) follows. $\qquad\square$

Small-World Nature of $\mathrm{ER}_n(\lambda/n)$

In the above proof, we also identified an upper bound on the typical distances in $\mathrm{ER}_n(\lambda/n)$, that is, the graph distance in $\mathrm{ER}_n(\lambda/n)$ between o_1 and o_2, as formulated in the following theorem:

Theorem 2.36 (Small-world nature Erdős–Rényi random graph) *Consider $\mathrm{ER}_n(\lambda/n)$ with $\lambda > 1$. Then, conditional on $o_1 \longleftrightarrow o_2$,*

$$\frac{\mathrm{dist}_{\mathrm{ER}_n(\lambda/n)}(o_1, o_2)}{\log n} \xrightarrow{\mathbb{P}} \frac{1}{\log \lambda}. \quad (2.6.86)$$

Proof   The lower bound follows directly from (2.4.34), which implies that

$$\mathbb{P}\big(\mathrm{dist}_{\mathrm{ER}_n(\lambda/n)}(o_1, o_2) \leq k\big) = \mathbb{E}\Big[|B_k(o_1)|/n\Big] \leq \sum_{l=0}^{k} \frac{\lambda^l}{n} = \frac{\lambda^{k+1} - 1}{n(\lambda - 1)}. \quad (2.6.87)$$

Applying this to $k = \lceil (1 - \eta) \log_\lambda n \rceil$ shows that, for any $\eta > 0$,

$$\mathbb{P}\big(\mathrm{dist}_{\mathrm{ER}_n(\lambda/n)}(o_1, o_2) \leq (1 - \eta) \log_\lambda n\big) \to 0. \quad (2.6.88)$$

For the upper bound, we start by noting that

$$\mathbb{P}\big(o_1 \longleftrightarrow o_2 \mid \mathrm{ER}_n(\lambda/n)\big) = \frac{1}{n^2} \sum_{i \geq 1} |\mathscr{C}_{(i)}|^2 \xrightarrow{\mathbb{P}} \zeta_\lambda^2, \quad (2.6.89)$$

by Exercise 2.24 below. Thus, conditioning on $o_1 \longleftrightarrow o_2$ is asymptotically the same as $o_1, o_2 \in \mathscr{C}_{\max}$. The upper bound then follows from the fact that the event $\mathcal{E}_{r_\varepsilon,[k_n^\star]}$ in (2.6.80) holds with probability at least $1 - \varepsilon$, and, on the event $\mathcal{E}_{r_\varepsilon,[k_n^\star]}$ and recalling (2.6.61),

$$\mathrm{dist}_{\mathrm{ER}_n(\lambda/n)}(o_1, o_2) \leq 2(k_n^\star + r_\varepsilon) + 1 \leq 2(a \log_{(1-\varepsilon)\lambda}(n) + r_\varepsilon) + 1, \quad (2.6.90)$$

with probability at least $1 - \varepsilon$, by (2.6.84). For any $\eta > 0$, we can take $\varepsilon > 0$ so small and $a > \frac{1}{2}$ so close to $\frac{1}{2}$ that the rhs can be bounded from above by $(1 + \eta) \log_\lambda n$ We conclude that $\mathrm{dist}_{\mathrm{ER}_n(\lambda/n)}(o_1, o_2) \leq (1 + \eta) \log_\lambda n$ whp on the event that $o_1 \longleftrightarrow o_2$. $\qquad\square$

2.6.5 OUTLINE OF THE REMAINDER OF THIS BOOK

The results proved for the Erdős–Rényi random graph complete the preliminaries for this book given in Part I. They further allow us to provide a brief glimpse into the content of the remainder of this book. We will prove results about local convergence, the existence of the giant component, and the small-world nature of various random graph models. We focus on random graphs with substantial inhomogeneities, such as the generalized random graph, the configuration model and its related uniform random graph with prescribed degrees, and various preferential attachment models. The remainder of this book consists of three parts.

In **Part II**, consisting of Chapters 3, 4 and 5, we focus on local convergence as in Theorem 2.18, but then applied to these models. Further, we investigate the size of the giant component as in Theorems 2.28 and 2.34. The proofs are often more involved than the corresponding results for the Erdős–Rényi random graph, since these random-graph models are inhomogeneous and often also lack independence of the edge statuses. Therefore, the proofs require detailed knowledge of the structure of the models in question.

In **Part III**, consisting of Chapters 6, 7, and 8, we focus on the small-world nature of these random graph models, as in Theorem 2.36. It turns out that the exact *scaling* of the graph distances depends on the level of inhomogeneity present in the random graph model. In particular, we will see that when the second moment of the degrees remains bounded, then graph distances grow logarithmically as in Theorem 2.36. If, on the other hand, the second moment blows up with the graph size, then distances are smaller. In particular, often these typical distances are *doubly logarithmic* when the degrees obey a power law with exponent τ that satisfies $\tau \in (2, 3)$, so that even a moment of order $2 - \varepsilon$ is infinite for some $\varepsilon > 0$. Anyone who has done some numerical work will realize that in practice there is little difference between $\log \log n$ and a constant, even when n is quite large.

One of the main conclusions of the local convergence results in Part II is that the most popular random graph models for inhomogeneous real-world networks are *locally tree-like*, in that the majority of neighborhoods of vertices have no cycles. This is for example true for the Erdős–Rényi random graph, see Theorem 2.18, since the local limit is a branching process tree. In many real-world settings, however, this is not realistic. Certainly in social networks, many triangles and even cliques of larger size exist. Therefore, in **Part IV**, consisting of Chapter 9, we investigate some adaptations of the models discussed in Parts II and III. These models may incorporate clustering or community structure; they may be directed or living in a geometric space. All these aspects have received tremendous attention in the literature. Therefore, with Part IV in hand, the reader will be able to access the literature more easily.

2.7 NOTES AND DISCUSSION FOR CHAPTER 2

There is an extensive body of work studying dense graph limits, using the theory of graphons; see Lovász (2012) and the references therein. Links have been built between this theory and the so-called "local–global" limits of sparse graphs called *graphings*. Elek (2007) proved that local limits of bounded-degree graphs are graphings; see also Hatami et al. (2014) for an extension. The related notion of *graphops* was defined in Backhausz and Szegedy (2022).

Notes on Section 2.1

There is no definitive source on local convergence techniques. Aside from the classics Aldous and Steele (2004) and Benjamini and Schramm (2001), I have been inspired by the introductions in Bordenave (2016), Curien (2018), and Leskelä (2019). Further, the discussions on local convergence by Bordenave and Caputo (2015) and Anantharam and Salez (2016) are clean and clear, and have been very helpful. In the literature, often the notation $[G, o]_r$ is used for the isomorphism classes of $B_r^{(G)}(o)$.

Notes on Section 2.2

The discussion in this and the following section follows Aldous and Steele (2004) and Benjamini and Schramm (2001). We refer to Appendix A.3 for proofs of various properties of the metric $d_{\mathscr{G}_\star}$ on rooted graphs, including the fact that it turns $\mathscr{G}_\star$ into a Polish space.

Notes on Section 2.3

Various generalizations of local convergence are possible. For example, Aldous and Steele (2004) introduced the notion of *geometric* rooted graphs, which are rooted graphs where each edge e receives a weight $\ell(e)$, turning the rooted graph into a metric space itself. Benjamini, Lyons, and Schramm (2015) allowed for the more general marks as discussed in Section 2.3.5. Dembo and Montanari (2010b) defined a version of local weak convergence in terms of convergence of subgraph counts (see (Dembo and Montanari, 2010b, Definition 2.1)). Aldous and Lyons (2007) also studied the implications for stochastic processes, such as percolation and random walks, on unimodular graphs.

Notes on Section 2.4

Dembo and Montanari (2010b) proved that their local weak convergence in terms of convergence of subgraph counts defined in (Dembo and Montanari, 2010b, Definition 2.1) holds for several models, including $\mathrm{ER}_n(\lambda/n)$ and $\mathrm{CM}_n(\boldsymbol{d})$ under appropriate conditions, while Dembo and Montanari (2010a) provides more details. See e.g., (Dembo and Montanari, 2010a, Lemma 2.4) for a proof for the configuration model, and (Dembo and Montanari, 2010a, Proposition 2.6) for a proof for the Erdős–Rényi random graph.

The local limit of the random regular graph in Theorem 2.17 was first identified by McKay (1981). An intuitive analysis of $\mathbb{P}(\bar{B}_r^{(G_n)}(1) = \boldsymbol{t})$ for $G_n = \mathrm{ER}_n(\lambda/n)$ in (2.4.28) has already appeared in [V1, Section 4.1.2]; see in particular [V1, (4.1.12)]. The local limit was identified by Dembo and Montanari (2010b,a), and can be obtained from the breadth-first exploration first analyzed by Karp (1990) in the context of random digraphs.

We thank Shankar Bhamidi for useful discussions about possibly random limits in Definition 2.11. For background on the convergence of measures on rooted graphs in Remark 2.13, we refer to (Kallenberg, 2017, Chapter 4, and in particular Lemma 4.8). Local weak convergence, i.e., $(G_n, o_n) \xrightarrow{d} (G, o)$ is, interestingly enough, not the same as convergence in distribution of the measure μ_n defined in Remark 2.13 (see (Kallenberg, 2017, Theorem 4.11)).

The local convergence of dynamic random graphs has been studied in Dort and Jacob (2023) and Milewska et al. (2023).

Notes on Section 2.5

While many of the results in this section are "folklore," appropriate references are not obvious.

Theorems 2.22 and 2.23 discuss the convergence of two clustering coefficients. In the literature, the *clustering spectrum* has also attracted attention. For this, we recall that n_k denotes the number of vertices with degree k in G_n, and define the clustering coefficient of vertices of degree k to be

$$c_{G_n}(k) = \frac{1}{n_k} \sum_{v \in V(G_n) \,:\, d_v^{(n)} = k} \frac{\Delta_{G_n}(v)}{k(k-1)}, \tag{2.7.1}$$

where n_k is the number of vertices of degree k in G_n. It is not hard to adapt the proof of Theorem 2.23 to show that, under its assumptions, $c_{G_n}(k) \xrightarrow{\mathbb{P}} c_G(k)$, where

$$c_G(k) = \mathbb{E}_\mu\left[\frac{\Delta_G(o)}{k(k-1)} \mid d_o = k \right], \tag{2.7.2}$$

and the convergence holds for all k for which $p_k^{(G)} = \mu(d_o^{(G)} = k) > 0$. See Exercise 2.37.

The convergence of the assortativity coefficient in Theorem 2.26 is restricted to degree distributions that have uniformly integrable third moments. In general, an empirical correlation coefficient needs a finite variance of the random variables to converge to the correlation coefficient. Nelly Litvak and the author (see van der Hofstad and Litvak (2014) and Litvak and van der Hofstad (2013)) proved that when the random variables do not have finite variance, such convergence (even for an iid sample) can be to a proper random variable that has support containing a subinterval of $[-1, 0]$ *and* a subinterval of $[0, 1]$, giving problems in interpretation.

For networks, ρ_{G_n} in (2.5.44) is always well defined, and gives a value in $[-1, 1]$. However, also for networks there is a problem with this definition. Indeed, van der Hofstad and Litvak (2014) and Litvak and van der Hofstad (2013) proved that if a limiting value of ρ_{G_n} exists for a sequence of networks and the third moment of the degree of a random vertex is not uniformly integrable, then $\liminf_{n\to\infty} \rho_{G_n} \geq 0$, so no asymptotically disassortative graph sequences exist for power-law networks with infinite third-moment degrees. Naturally, other ways of classifying the degree–degree dependence can be proposed, such as the correlation of their *ranks*. Here, a sequence of numbers $x_1, \ldots, x_n$ has ranks $r_1, \ldots, r_n$ when x_i is the r_ith largest of $x_1, \ldots, x_n$. Ties tend to be broken by giving random ranks for the equal values. For practical purposes, a scatter plot of the values might be the most useful way to gain insight into degree–degree dependencies.

Several related graph properties or parameters have been investigated using local convergence. Lyons (2005) showed that the exponential growth rate of the number of spanning trees of a finite connected graph can be computed through the local limit. See also Salez (2013) for weighted spanning subgraphs, and Gamarnik et al. (2006) for maximum-weight independent sets. Bhamidi et al. (2012) identified the limiting spectral distribution of the graph adjacency matrix of finite random trees using local convergence, and Bordenave et al. (2011) proved the convergence of the spectral measure of sparse random graphs (see also Bordenave and Lelarge (2010) and Bordenave et al. (2013) for related results). A property that is almost local is the density of the densest subgraph in a random graph, as shown by Anantharam and Salez (2016) and studied in more detail in Section 4.5.

Notes on Section 2.6

The results in this section appeared in van der van der Hofstad (2021), and were developed for this book.

2.8 EXERCISES FOR CHAPTER 2

Exercise 2.1 (Graph isomorphisms fix vertex and edge numbers)   *Assume that $G_1 \simeq G_2$. Show that G_1 and G_2 have the same number of vertices and edges.*

Exercise 2.2 (Graph isomorphisms fix degree sequence)   *Let G_1 and G_2 be two finite graphs. Assume that $G_1 \simeq G_2$. Show that G_1 and G_2 have the same degree sequences. Here, for a graph G, we let the degree sequence be $(p_k^{(G)})_{k \geq 0}$, where*

$$p_k^{(G)} = \frac{1}{|V(G)|} \sum_{v \in V(G)} \mathbb{1}_{\{d_v^{(G)} = k\}}, \tag{2.8.1}$$

and $d_v^{(G)}$ is the degree of v in G.

Exercise 2.3 (Distance to rooted graph ball)   *Recall the definition of the ball $B_r^{(G)}(o)$ around o in the graph G in (2.2.1). Show that $d_{\mathscr{G}_\star}\left(B_r^{(G)}(o), (G, o)\right) \leq 1/(r+1)$. When does equality hold?*

Exercise 2.4 (Countable number of graphs with bounded radius)   *Fix r. Show that there is a countable number of isomorphism classes of rooted graphs (G, o) with radius at most r. Here, we let the radius $\mathrm{rad}(G, o)$ of a rooted graph (G, o) be equal to $\mathrm{rad}(G, o) = \max_{v \in V(G)} \mathrm{dist}_G(o, v)$ where dist_G denotes the graph distance in G.*

Exercise 2.5 ($\mathscr{G}_\star$ is separable)   *Use Exercise 2.4 above to show that the set of rooted graphs $\mathscr{G}_\star$ has a countable dense set, and is thus separable. (See also Proposition A.12 in Appendix A.3.2.)*

Exercise 2.6 (Continuity of local neighborhood functions)   *Fix* $H_\star \in \mathscr{G}_\star$. *Show that* $h \colon \mathscr{G}_\star \mapsto \{0, 1\}$ *given by* $h(G, o) = \mathbb{1}_{\{B_r^{(G)}(o) \simeq H_\star\}}$ *is continuous.*

Exercise 2.7 (Bounded number of graphs with bounded radius and degrees)   *Show that there are only a bounded number of isomorphism classes of rooted graphs* (G, o) *with radius at most* r *for which the degree of every vertex is at most* k.

Exercise 2.8 (Random local weak limit)   *Construct the simplest (in your opinion) possible example where the local weak limit of a sequence of deterministic graphs is random.*

Exercise 2.9 (Local weak limit of line and cycle)   *Let* G_n *be given by* $V(G_n) = [n]$, $E(G_n) = \{\{i, i + 1\} \colon i \in [n - 1]\}$ *be a line. Show that* (G_n, o_n) *converges to* $(\mathbb{Z}, 0)$. *Show that the same is true for the cycle, for which* $E(G_n) = \{\{i, i + 1\} \colon i \in [n - 1]\} \cup \{\{1, n\}\}$.

Exercise 2.10 (Local weak limit of finite tree)   *Let* G_n *be the tree of depth* k, *in which every vertex except the* $3 \times 2^{k-1}$ *leaves have degree 3. Here* $n = 3(2^k - 1)$. *What is the local weak limit of* G_n?

Exercise 2.11 (Uniform integrability and convergence of size-biased degrees)   *Show that when* $(d_{o_n}^{(G_n)})_{n \geq 1}$ *forms a uniformly integrable sequence of random variables, there exists a subsequence along which* $D_n^\star$, *the size-biased version of* $D_n = d_{o_n}^{(G_n)}$, *converges in distribution.*

Exercise 2.12 (Uniform integrability and degree regularity condition)   *For* $G_n = \mathrm{CM}_n(\boldsymbol{d})$, *show that Conditions 1.7(a),(b) imply that* $(d_{o_n}^{(G_n)})_{n \geq 1}$ *is a uniformly integrable sequence of random variables.*

Exercise 2.13 (Adding a small disjoint graph does not change local weak limit)   *Let* G_n *be a graph that converges in the local weak sense. Let* $a_n \in \mathbb{N}$ *be such that* $a_n = o(n)$, *and add a disjoint copy of an arbitrary graph of size* a_n *to* G_n. *Denote the resulting graph by* G_n'. *Show that* G_n' *has the same local weak limit as* G_n.

Exercise 2.14 (Local weak convergence does not imply uniform integrability of the degree of a random vertex)   *In the setting of Exercise 2.13, add a complete graph of size* a_n *to* G_n. *Let* $a_n^2 \gg n$. *Show that the degree of a vertex chosen uar in* G_n' *is not uniformly integrable.*

Exercise 2.15 (Local limit of random 2-regular graph)   *Show that the configuration model* $\mathrm{CM}_n(\boldsymbol{d})$ *with* $d_v = 2$ *for all* $v \in [n]$ *converges locally in probability to* $(\mathbb{Z}, 0)$. *Conclude that the same applies to the random 2-regular graph.*

Exercise 2.16 (Independent neighborhoods of different vertices)   *Let* G_n *converge locally in probability to* (G, o). *Let* $(o_n^{(1)}, o_n^{(2)})$ *be two independent uniformly chosen vertices in* $V(G_n)$. *Show that* $(G_n, o_n^{(1)})$ *and* $(G_n, o_n^{(2)})$ *jointly converge to two* conditionally independent *copies of* (G, o) *given* μ.

Exercise 2.17 (Directed graphs as marked graphs)   *There are several ways to describe directed graphs as marked graphs. Give one.*

Exercise 2.18 (Multi-graphs as marked graphs)   *Use the formalism of marked rooted graphs in Definition 2.10 to cast the setting of multi-graphs discussed in Remark 2.4 into this framework.*

Exercise 2.19 (Uniform d-regular simple graph)   *Use Theorem 2.17 and (1.3.41) to show that the uniform random d-regular graph (which is the same as the d-regular configuration model conditioned on simplicity) also converges locally in probability to the infinite d-regular tree.*

Exercise 2.20 (Local weak convergence and subsets)   *Recall the statement of Theorem 2.16. Prove that local weak convergence* $(G_n, o_n) \xrightarrow{d} (\bar{G}, \bar{o})$ *when (2.4.10) holds for all* $H_\star \in \mathscr{T}_\star(r)$ *and all* $r \geq 1$.

Exercise 2.21 (Local convergence in probability and subsets)   *Recall the statement of Theorem 2.16. Prove that* G_n *converges locally in probability to* (G, o) *when (2.4.11) holds for all* $H_\star \in \mathscr{T}_\star(r)$ *and all* $r \geq 1$. *Extend this to almost sure local convergence and (2.4.12).*

Exercise 2.22 (Functional for number of connected components is continuous)   *Prove that* $h(G, o) = 1/|\mathscr{C}(o)|$ *is a bounded and continuous function, where, by convention,* $h(G, o) = 0$ *when* $|\mathscr{C}(o)| = \infty$.

Exercise 2.23 (Functional for number of connected components is continuous)   *Recall the notion of a random variable* $X_{n,k}$ *being* $o_{k,\mathbb{P}}(1)$ *in (2.6.9). Recall the definition of* $Z_{\geq k}$ *in (2.6.3). Show that* $Z_{\geq k}/n = \zeta + o_{k,\mathbb{P}}(1)$.

Exercise 2.24 (Convergence of sum of squares of cluster sizes) *Show that, under the conditions of Theorem 2.28 and with $\zeta = \mu(|\mathscr{C}(o)| = \infty)$,*

$$\frac{1}{n^2} \sum_{i \geq 1} |\mathscr{C}_{(i)}|^2 \overset{\mathbb{P}}{\longrightarrow} \zeta^2. \tag{2.8.2}$$

Exercise 2.25 (Expected boundary of balls in Erdős–Rényi random graphs) *Prove that $\mathbb{E}\big[|\partial B_r^{(G_n)}(1)|\big] \leq \lambda^r$ for $G_n = \mathrm{ER}_n(\lambda/n)$ and every $r \geq 0$. This can be done, for example, by using induction and showing that, for every $r \geq 1$,*

$$\mathbb{E}\big[|\partial B_r^{(G_n)}(1)| \mid B_{r-1}^{(G_n)}(1)\big] \leq \lambda |\partial B_{r-1}^{(G_n)}(1)|. \tag{2.8.3}$$

Exercise 2.26 (Uniform integrability and moment convergence) *Assume that $D_n = d_{o_n}^{(G_n)}$ is such that $(D_n^3)_{n \geq 1}$ is uniformly integrable. Assume further that G_n converges locally in probability to (G, o). Prove that $\mathbb{E}[(d_{o_n}^{(G_n)})^3 \mid G_n] \overset{\mathbb{P}}{\longrightarrow} \mathbb{E}_\mu[d_o^3]$. Conclude that (2.5.47) and (2.5.48) hold. Hint: You need to be very careful, as $\mathbb{E}_\mu[d_o^3]$ may be a random variable when μ is a random measure.*

Exercise 2.27 (Uniform integrability and moment convergence) *Use Cauchy–Schwarz to show that $\mathbb{E}_\mu[d_o^3] - \mathbb{E}_\mu[d_o^2]^2/\mathbb{E}_\mu[d_o] > 0$ when $\mu(d_o = r) < 1$ for every $r \geq 0$.*

Exercise 2.28 (Example of weak convergence where convergence in probability fails) *Construct an example where G_n converges locally weakly to (G, o), but not locally in probability.*

Exercise 2.29 (Continuity of neighborhood functions) *Fix $m \geq 1$ and $\ell_1, \dots, \ell_m$. Show that*

$$h(G, o) = \mathbb{1}_{\{|\partial B_r^{(G)}(o)| = \ell_k \, \forall k \leq m\}} \tag{2.8.4}$$

is a bounded continuous function in $(\mathscr{G}_\star, d_{\mathscr{G}_\star})$.

Exercise 2.30 (Proof of (2.5.2)) *Let G_n converge locally in probability to (G, o). Prove the joint convergence in distribution of the neighborhood sizes in (2.5.2) using Exercise 2.16.*

Exercise 2.31 (Example where the proportion in the giant is smaller than the survival probability) *Construct an example where G_n converges locally in probability to $(G, o) \sim \mu$, while $|\mathscr{C}_{\max}|/n \overset{\mathbb{P}}{\longrightarrow} \eta < \zeta = \mu(|\mathscr{C}(o)| = \infty)$.*

Exercise 2.32 (Convergence of the proportion of vertices in clusters of size at least k) *Let G_n converge locally in probability to (G, o) as $n \to \infty$. Show that $Z_{\geq k}$ in (2.6.3) satisfies that $Z_{\geq k}/n \overset{\mathbb{P}}{\longrightarrow} \zeta_{\geq k} = \mu(|\mathscr{C}(o)| \geq k)$ for every $k \geq 1$.*

Exercise 2.33 (Upper bound on $|\mathscr{C}_{\max}|$ using local convergence) *Let $G_n = (V(G_n), E(G_n))$ denote a random graph of size $|V(G_n)| = n$. Assume that G_n converges locally in probability to $(G, o) \sim \mu$ as $n \to \infty$, and assume that the survival probability of the limiting graph (G, o) satisfies $\zeta = \mu(|\mathscr{C}(o)| = \infty) = 0$. Show that $|\mathscr{C}_{\max}|/n \overset{\mathbb{P}}{\longrightarrow} 0$.*

Exercise 2.34 (Sufficiency of (2.6.7) for almost locality of the giant) *Let $G_n = (V(G_n), E(G_n))$ denote a random graph of size $|V(G_n)| = n$. Assume that G_n converges locally in probability to $(G, o) \sim \mu$ and write $\zeta = \mu(|\mathscr{C}(o)| = \infty)$ for the survival probability of the limiting graph (G, o). Assume that*

$$\limsup_{k \to \infty} \frac{1}{n^2} \limsup_{n \to \infty} \mathbb{E}\Big[\#\big\{ (x, y) \in V(G_n) \times V(G_n) : |\mathscr{C}(x)|, |\mathscr{C}(y)| \geq k, x \overset{}{\longleftrightarrow\!\!\!\!/} \, y \big\} \Big] > 0. \tag{2.8.5}$$

Then, prove that for some $\varepsilon > 0$,

$$\limsup_{n \to \infty} \mathbb{P}(|\mathscr{C}_{\max}| \leq n(\zeta - \varepsilon)) > 0. \tag{2.8.6}$$

Exercise 2.35 (Lower bound on graph distances in Erdős–Rényi random graphs) *Use Exercise 2.25 to show that, for every $\varepsilon > 0$,*

$$\lim_{n \to \infty} \mathbb{P}\Big(\frac{\mathrm{dist}_{\mathrm{ER}_n(\lambda/n)}(o_1, o_2)}{\log_\lambda n} \leq 1 - \varepsilon \Big) = 0. \tag{2.8.7}$$

Exercise 2.36 (Lower bound on graph distances in Erdős–Rényi random graphs)   *Use Exercise 2.25 to show that*

$$\lim_{K \to \infty} \limsup_{n \to \infty} \mathbb{P}\big(\mathrm{dist}_{\mathrm{ER}_n(\lambda/n)}(o_1, o_2) \leq \log_\lambda n - K\big) = 0, \tag{2.8.8}$$

which is a significant extension of Exercise 2.35.

Exercise 2.37 (Convergence of the clustering spectrum)   *Prove that, under the conditions of Theorem 2.23, the convergence of the clustering spectrum in* (2.7.2) *holds for all k such that* $p_k^{(G)} = \mu(d_o^{(G)} = k) > 0$.

Part II

Connected Components in Random Graphs

Overview of Part II

In this Part, we study *local limits* and *connected components* in random graphs, and the relation between them. In more detail, we investigate the connected components of uniform vertices, thus also describing the *local limits* of these random graphs. Further, we study the existence and structure of the *largest connected component*, sometimes also called the *giant component* when it contains a positive (as opposed to zero) proportion of the vertices in the graph.

In many random graphs, such a giant component exists when there are sufficiently many connections, while the largest connected component is much smaller than the number of vertices when there are few connections. Thus, these random graphs undergo a *phase transition*. We identify the size of the giant component, as well as its structure in terms of the degrees of its vertices. We also investigate whether the graph is *fully* connected. General inhomogeneous random graphs are studied in Chapter 3, and the configuration model, as well the closely related uniform random graph with prescribed degrees, in Chapter 4. In the last chapter of this part, Chapter 5, we study the connected components and local limits of preferential attachment models.

CONNECTED COMPONENTS IN GENERAL INHOMOGENEOUS RANDOM GRAPHS

Abstract

In this chapter, we introduce the general setting of inhomogeneous random graphs that are generalizations of the Erdős–Rényi and generalized random graphs. In inhomogeneous random graphs the status of edges is independent, with unequal edge-occupation probabilities. While these edge probabilities are moderated by vertex weights in generalized random graphs, in the general setting they are described in terms of a *kernel*.

The main results in this chapter concern the degree structure, the multi-type branching process local limits, and the phase transition in these inhomogeneous random graphs. We also discuss various examples, and indicate that they can have rather different structures.

3.1 MOTIVATION: EXISTENCE AND SIZE OF THE GIANT

In this chapter we discuss general inhomogeneous random graphs, which are sparse random graphs in which the edge statuses are independent. We investigate their local limits, as well as their connectivity structure, and their giant component. This is inspired by the fact that many real-world networks are highly connected, in the sense that their largest connected component contains a large proportion of the total vertices of the graph. See Table 3.1 for many examples and Figure 3.1 for the proportion of vertices in the maximal connected components in the KONECT data base.

Table 3.1 and Figure 3.1 raise the question of how one can view settings where giant components exist. We know that there is a *phase transition* in the size of the giant component in $\mathrm{ER}_n(\lambda/n)$; recall [V1, Chapter 4]. A main topic in the present chapter is to investigate the conditions for a giant component to be present in general inhomogeneous random graphs; this occurs precisely when the local limit has a positive survival probability (recall Section 2.6). Therefore, we also investigate the local convergence of inhomogeneous random graphs in this chapter.

We will study much more general models, where edges are present independently, than in the generalized random graph in [V1, Chapter 6]; see also Section 1.3.2. There, vertices have weights associated to them, and the edge-occupation probabilities are approximately proportional to the product of the weights of the vertices that the edge connects. This means that vertices with high weights have relatively large probabilities of connections to *all* other vertices, a property that may not always be appropriate. Let us illustrate this by an example, which is a continuation of [V1, Example 6.1]:

Example 3.1 (Population of two types: general setting) Suppose that we have a complex network in which there are n_1 vertices of type-1 and n_2 of type-2. Type-1 individuals have on average m_1 neighbors, type-2 individuals m_2, where $m_1 \neq m_2$. Further, suppose that the probability that a type-1 individual is a friend of a type-2 individual is quite different from the probability that a type-1 individual is a friend of another type-1 individual.

Table 3.1 *The rows in this table correspond to the following real-world networks:*
Protein–protein interactions in the blood of people with Alzheimer's disease.
Protein–protein interactions in the blood of people with multiple sclerosis.
IMDb collaboration network, where actors are connected when they have co-acted in a movie.
DBLP collaboration network, where scientist are connected when they have co-authored a paper.
Interactions between zebras, where zebras are connected when they have interacted during the
observation phase.

Subject	% in giant	Total	Original source	Data
AD-blood	0.8542	96	Goñi et al. (2008)	Goñi et al. (2008)
MS-blood	0.8780	205	Goñi et al. (2008)	Goñi et al. (2008)
actors	0.8791	2,180,759	Boldi and Vigna (2004)	Boldi et al. (2011)
DBLP	0.8162	986,324	Boldi and Vigna (2004)	Boldi et al. (2011)
Zebras	0.8214	28	Sundaresan et al. (2007)	Sundaresan et al. (2007)

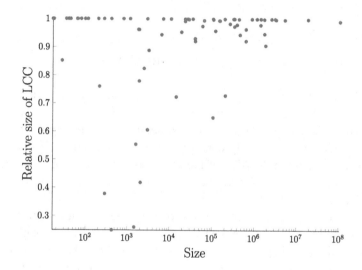

Figure 3.1 Proportion of vertices in the maximal connected component in the
1,203 networks from the KONECT data base.

In the generalized random graph model proposed in [V1, Example 6.3], the probability that a type-s individual is a friend of a type-r individual (where $s, r \in [2]$) equals $m_s m_r / (\ell_n + m_s m_r)$, where $\ell_n = n_1 m_1 + n_2 m_2$. Approximating this probability by $m_s m_r / \ell_n$, we see that the probability that a type-1 individual is a friend of a type-2 individual is highly related to the probability that a type-1 individual is friend of a type-1 individual. Indeed, take two type-1 and two type-2 individuals. Then, the probability that the type-1 individuals are friends and the type-2 individuals are friends is *almost* the same as the probability that the first type-1 individual is friend of the first type-2 individual, and that the second type-1 individual is a friend of the second type-2 individual. Thus, there is some, possibly unwanted and artificial, *symmetry* in the model.

How can one create instances where the edge probabilities between vertices of the same type are much larger, or alternatively much smaller, than they would be for the generalized random graph? In sexual networks, there are likely to be more edges between the different sexes than amongst them, while in highly polarized societies most connections are within the groups. In the two extremes, we either have a bipartite graph, where vertices are connected only to vertices of the other type, or a disjoint union of two Erdős–Rényi random graphs, consisting of the vertices of the two types and no edges between them. We aim to be able to obtain anything in between. In particular, the problem with the generalized random graph originates in the approximate *product structure* of the edge probabilities. In this chapter, we deviate from such a product structure. ◄

As explained above, we wish to be quite flexible in our choices of edge probabilities. However, we also aim for settings where the random graph is sufficiently "regular," as for example exemplified by its degree sequences converging to some deterministic distribution, or even by a local limit existing. In particular, we aim for settings where the random graphs are *sparse*. As a result, we need to build this regularity into the precise structure of the edge probabilities. This will be achieved by introducing a sufficiently regular *kernel* that moderates the edge probabilities.

Organization of this Chapter

This chapter is organized as follows. In Section 3.2, we introduce general inhomogeneous random graphs. In Section 3.3 we study the degree distribution in such random graphs. In Section 3.4 we treat multi-type branching processes, the natural generalization of branching processes for inhomogeneous random graphs. In Section 3.5 we use these multi-type branching processes to identify the local limit of inhomogeneous random graphs. In Section 3.6 we study the phase transitions of inhomogeneous random graphs. In Section 3.7 we state some related results. We close this chapter with notes and discussion in Section 3.8 and exercises in Section 3.9.

3.2 DEFINITION OF THE MODEL

We assume that our individuals (vertices) have types which are in a certain type space $\mathcal{S}$. When there are individuals of just two types, as in Example 3.1, then it suffices to take $\mathcal{S} = \{1, 2\}$. However, the model allows for rather general sets of types of the individuals, both finite as well as (countably or even uncountably) infinite type spaces. An example of an uncountably infinite type space arises when the types are related to the *ages* of the individuals in the population. Also the setting of the generalized random graph with w_i satisfying (1.3.15) correponds to the uncountable type-space setting when the distribution function F is that of a continuous random variable W. We therefore also need to know how many individuals there are of a given type. This is described in terms of a *measure* μ_n, where, for $\mathcal{A} \subseteq \mathcal{S}$, $\mu_n(\mathcal{A})$ denotes the proportion of individuals having a type in $\mathcal{A}$.

In our general model, instead of vertex weights, the edge probabilities are moderated by a *kernel* $\kappa \colon \mathcal{S}^2 \to [0, \infty)$. The probability that two vertices of types x_1 and x_2 are connected is approximately $\kappa(x_1, x_2)/n$, and different edges are present independently. Since there are many choices for κ, we arrive at a rather flexible model.

3.2.1 INHOMOGENEOUS RANDOM GRAPHS AND THEIR KERNELS

We start by making the above definitions formal, by defining the ground space and kernel:

Definition 3.2 (Setting: ground space and kernel)

(a) A *ground space* is a pair $(\mathcal{S}, \mu)$, where $\mathcal{S}$ is a separable metric space and μ is a Borel probability measure on $\mathcal{S}$.

(b) A *vertex space* is a triple $(\mathcal{S}, \mu, (\boldsymbol{x}_n)_{n \geq 1})$, where $(\mathcal{S}, \mu)$ is a ground space and, for each $n \geq 1$, $\boldsymbol{x}_n$ is a (possibly random) sequence $(x_1, x_2, \ldots, x_n)$ of n points of $\mathcal{S}$, such that

$$\mu_n(\mathcal{A}) = \#\{v \in [n] : x_v \in \mathcal{A}\}/n \to \mu(\mathcal{A}) \tag{3.2.1}$$

for every μ-continuity set $\mathcal{A} \subseteq \mathcal{S}$.

(c) A *kernel* $\kappa \colon \mathcal{S}^2 \to [0, \infty)$ is a symmetric non-negative (Borel) measurable function. By a kernel on a vertex space $(\mathcal{S}, \mu, (\boldsymbol{x}_n)_{n \geq 1})$, we mean a kernel on $(\mathcal{S}, \mu)$. ◀

Before defining the random graph model, we state the necessary conditions on our kernels:

Definition 3.3 (Setting: graphical and irreducible kernels)

(a) A kernel κ is *graphical* if the following conditions hold:

   (i) κ is continuous a.e. on $\mathcal{S}^2$;

   (ii)

$$\iint_{\mathcal{S}^2} \kappa(x, y)\mu(\mathrm{d}x)\mu(\mathrm{d}y) < \infty; \tag{3.2.2}$$

   (iii)

$$\frac{1}{n^2} \sum_{1 \leq u < v \leq n} [\kappa(x_u, x_v) \wedge n] \to \frac{1}{2} \iint_{\mathcal{S}^2} \kappa(x, y)\mu(\mathrm{d}x)\mu(\mathrm{d}y). \tag{3.2.3}$$

Similarly, a sequence $(\kappa_n)_{n \geq 1}$ of kernels is called *graphical with limit* κ when, for μ-almost every y, z,

$$y_n \to y \quad \text{and} \quad z_n \to z \quad \text{imply that} \quad \kappa_n(y_n, z_n) \to \kappa(y, z), \tag{3.2.4}$$

where κ satisfies conditions (a) and (b) above, and

$$\frac{1}{n} \sum_{1 \leq u < v \leq n} [\kappa_n(x_u, x_v) \wedge n] \to \frac{1}{2} \iint_{\mathcal{S}^2} \kappa(x, y)\mu(\mathrm{d}x)\mu(\mathrm{d}y). \tag{3.2.5}$$

(b) A kernel κ is called *reducible* if

$$\exists \mathcal{A} \subseteq \mathcal{S} \quad \text{with} \quad 0 < \mu(\mathcal{A}) < 1 \quad \text{such that} \quad \kappa = 0 \quad \text{a.e. on} \quad \mathcal{A} \times (\mathcal{S} \backslash \mathcal{A});$$

otherwise κ is *irreducible*. ◀

We now discuss the above definitions. Below, we will take $p_{uv} = [\kappa_n(x_u, x_v) \wedge n]/n$. Then the assumptions in (3.2.2), (3.2.3), (3.2.5) imply that the expected number of edges $\mathbb{E}[|E(\mathrm{IRG}_n(\kappa_n))|]$ is proportional to n, and that the proportionality constant is precisely $\frac{1}{2} \iint_{\mathcal{S}^2} \kappa(x, y)\mu(\mathrm{d}x)\mu(\mathrm{d}y)$. Thus, in the terminology of [V1, Chapter 1], the model is *sparse* (recall Section 1.1.1). This sparsity allows us to *approximate* graphical kernels by *bounded* ones in such a way that the number of removed edges is $o_{\mathbb{P}}(n)$, a fact that will be crucially used in what follows. Indeed, bounded graphical kernels can be well approximated by step functions similarly to the way in which continuous functions on $\mathbb{R}$ can be well approximated by step functions. In turn, such step functions on $\mathcal{S} \times \mathcal{S}$ correspond to random graphs with vertices having only *finitely* many different types.

We extend the setting to n-dependent sequences $(\kappa_n)_{n \geq 1}$ of kernels in (3.2.4), as in many natural cases the kernels do depend on n. In particular, this allows us to deal with several closely related and natural notions of the edge probabilities, all at the same time (see, e.g., (3.2.6) and (3.2.7) below), showing that identical results hold in each of these cases.

Roughly speaking, κ is *reducible* if the vertex set $[n]$ of $\mathrm{IRG}_n(\kappa)$ can be split into two parts in such a way that the probability of an edge from one part to the other is zero, and κ is irreducible otherwise. For reducible kernels, we could equally well have started with each of these parts separately, explaining why the notion of irreducibility is quite natural.

In many cases, we take $\mathcal{S} = [0, 1]$, $x_i = i/n$, and μ the Lebesgue measure on $[0, 1]$. Then, clearly, (3.2.1) is satisfied. In fact, Janson (2009) shows that we can always restrict to $\mathcal{S} = [0, 1]$ by suitably adapting the other choices of our model. However, for notational purposes, it is more convenient to work with general $\mathcal{S}$. For example, when $\mathcal{S} = \{1\}$ is just a single type, the model reduces to the Erdős–Rényi random graph, and, in the setting where $\mathcal{S} = [0, 1]$, this is slightly more cumbersome, as can be worked out in detail in Exercise 3.1.

3.2.2 INHOMOGENEOUS RANDOM GRAPHS AND THEIR EDGE PROBABILITIES

Now we come to the definition of our random graph. Given a sequence of kernels $(\kappa_n)_{n \geq 1}$, for $n \in \mathbb{N}$, we let $\mathrm{IRG}_n(\kappa_n)$ be the random graph on $[n]$ in which each possible edge $uv = \{u, v\}$, where $u, v \in [n]$ with $u \neq v$, is present with probability

$$p_{uv}(\kappa_n) = p_{uv} = \frac{1}{n}[\kappa_n(x_u, x_v) \wedge n], \qquad (3.2.6)$$

and the events that different edges are present are independent. Exercise 3.2 shows that the lower bound in (3.2.3) always holds for $\mathrm{IRG}_n(\kappa)$ when κ is independent of n and continuous a.e. Further, Exercise 3.3 shows that (3.2.3) holds for $\mathrm{IRG}_n(\kappa)$ when κ is bounded and continuous.

We also allow for the choices, inspired by the Norros–Reittu and generalized random graphs,

$$p_{uv}^{(\mathrm{NR})}(\kappa_n) = 1 - \mathrm{e}^{-\kappa_n(x_u, x_v)/n}, \quad \text{and} \quad p_{uv}^{(\mathrm{GRG})}(\kappa_n) = \frac{\kappa_n(x_u, x_v)}{n + \kappa_n(x_u, x_v)}. \qquad (3.2.7)$$

All the results presented here remain valid for the choices in (3.2.7). When

$$\sum_{u, v \in [n]} \kappa_n(x_u, x_v)^3 = o(n^3), \qquad (3.2.8)$$

this follows immediately from [V1, Theorem 6.18] (see Exercise 3.4). In the next section, we discuss some examples of inhomogeneous random graphs.

3.2.3 EXAMPLES OF INHOMOGENEOUS RANDOM GRAPHS

Erdős–Rényi Random Graph

If $\mathcal{S}$ is general and $\kappa(x, y) = \lambda$ for every $x, y \in \mathcal{S}$ then the edge probabilities p_{uv} given by (3.2.6) are all equal to λ/n (for $n > \lambda$). It follows that $\mathrm{IRG}_n(\kappa) = \mathrm{ER}_n(\lambda/n)$. The simplest choice here is to take $\mathcal{S} = \{1\}$.

Chung–Lu Model

For $\mathrm{CL}_n(\boldsymbol{w})$ with $\boldsymbol{w} = (w_v)_{v \in [n]}$, where $w_v = [1 - F]^{-1}(v/n)$ as in (1.3.15), we take $\mathcal{S} = [0, 1]$, $x_v = v/n$ and, with $\psi(x) = [1 - F]^{-1}(x)$,

$$\kappa_n(x, y) = \psi(x)\psi(y)n/\ell_n. \tag{3.2.9}$$

For $\mathrm{CL}_n(\boldsymbol{w})$ with $\boldsymbol{w} = (w_v)_{v \in [n]}$ satisfying Condition 1.1 in Section 1.3.2, instead, we take $\mathcal{S} = [0, 1]$, $x_v = v/n$, and

$$\kappa_n(u/n, v/n) = w_u w_v / \mathbb{E}[W_n]. \tag{3.2.10}$$

Exercises 3.5 and 3.6 study the Chung–Lu random graph in the present framework.

Homogeneous Bipartite Random Graph

Let n be even, let $\mathcal{S} = \{1, 2\}$, and let $x_v = 1$ for $v \in [n/2]$ and $x_v = 2$ for $v \in [n] \setminus [n/2]$. Further, let κ be defined by $\kappa(x, y) = \lambda$ when $x \neq y$ and $\kappa(x, y) = 0$ when $x = y$. Then $\mathrm{IRG}_n(\kappa)$ is the random bipartite graph with $n/2$ vertices in each class, where each possible edge between classes is present with probability λ/n, independently of the other edges, while the edges within each of the two classes are all absent. Exercise 3.7 investigates the validity of Definitions 3.2 and 3.3 for homogeneous bipartite graphs.

Stochastic Block Model

The stochastic block model generalizes the above setting. Again, let n be even, let $\mathcal{S} = \{1, 2\}$, and let $x_v = 1$ for $v \in [n/2]$ and $x_v = 2$ for $v \in [n] \setminus [n/2]$. Further, let κ be defined by $\kappa(x, y) = b$ when $x \neq y$ and $\kappa(x, y) = a$ when $x = y$. This means that vertices of the same type are connected with probability a/n, while vertices with different types are connected with probability b/n. A major research effort has been devoted to studying when it can be *statistically detected* that $a > b$. Below, we also investigate more general stochastic block models.

Homogeneous Random Graphs

We call an inhomogeneous random graph *homogeneous* when, for almost every $x \in \mathcal{S}$,

$$\lambda(x) = \int_{\mathcal{S}} \kappa(x, y)\mu(\mathrm{d}y) \equiv \lambda. \tag{3.2.11}$$

Thus, despite the inhomogeneity that is present, every vertex in the graph has (asymptotically) the same number of expected offspring. Exercise 3.8 shows that the Erdős–Rényi random graph, the homogeneous bipartite random graph, and the stochastic block model are all homogeneous random graphs. In such settings, the level of inhomogeneity is limited.

Inhomogeneous Random Graphs with Finitely Many Types

Fix $t \geq 2$ and suppose we have a graph with t different types of vertices. Let $\mathcal{S} = [t]$. Let n_s denote the number of vertices of type s, and let $\mu_n(s) = n_s/n$. Let $\mathrm{IRG}_n(\kappa_n)$ be the random graph where two vertices, of types s and r, respectively, are independently joined by an edge with probability $\kappa_n(s, r)/n \wedge 1$. Then κ_n is equivalent to a $t \times t$ matrix, and the random graph $\mathrm{IRG}_n(\kappa_n)$ has vertices of t different types (or colors). We conclude that our general inhomogeneous random graph covers the cases of a finite (or even countably infinite) number of types. Exercises 3.9–3.11 study the setting of inhomogeneous random graphs with finitely many types. It will turn out that this case is particularly important, as many other settings can be arbitrarily well approximated by inhomogeneous random graphs with finitely many types. As such, this model will be the building block upon which most of our results are built.

Uniformly Grown Random Graph

The uniformly grown random graph model, or Dubins model, is an example of the general inhomogeneous random graphs as discussed in the previous section. We take a vertex space $[n]$, and assume that each edge uv is present with probability

$$p_{uv} = \frac{\lambda}{\max\{u, v\}}, \tag{3.2.12}$$

all edge statuses being independent random variables. Equivalently, we can view this random graph as arising *dynamically*, where vertex n connects to a vertex $m < n$ with probability λ/n independently for all $m \in [n-1]$. In particular, the model with $n = \infty$ is well defined.

Sum Kernels

We have already seen that *product kernels* are special, as they give rise to the Chung–Lu model or its close relatives, the generalized random graph and the Norros–Reittu model. For sum kernels, instead, we take $\kappa(x, y) = \psi(x) + \psi(y)$, so that

$$p_{uv} = \min\{(\psi(u/n) + \psi(v/n))/n, 1\}. \tag{3.2.13}$$

Conclusion on Inhomogeneous Random Graph Models

We conclude that there are many examples of random graphs with independent edges that fall into the general class of inhomogeneous random graphs, some of them leading to rather interesting behavior. In what follows, we investigate them in general. We start by investigating their degree structure. Let us make some notational conventions. We will use $\mathcal{S} = [t]$ for the type space of finite-type inhomogeneous random graphs, and write s, r for elements in it. However, we write $x, y \in \mathcal{S}$ for pairs of types when $\mathcal{S}$ is continuous.

3.3 DEGREE SEQUENCE OF INHOMOGENEOUS RANDOM GRAPHS

We start by investigating the degrees of the vertices of $\mathrm{IRG}_n(\kappa_n)$. As we shall see, the degree of a vertex of a given type x is asymptotically Poisson with mean

$$\lambda(x) = \int_{\mathcal{S}} \kappa(x, y)\mu(\mathrm{d}y) \tag{3.3.1}$$

that (possibly) depends on the type $x \in \mathcal{S}$. This leads to a mixed-Poisson distribution for the degree D of a (uniformly chosen) random vertex of $\mathrm{IRG}_n(\kappa_n)$. We recall that $N_k(n)$ denotes the number of vertices of $\mathrm{IRG}_n(\kappa_n)$ with degree k, i.e.,

$$N_k(n) = \sum_{v \in [n]} \mathbb{1}_{\{d_v = k\}}, \qquad (3.3.2)$$

where d_v is the degree of vertex $v \in [n]$. Our main result is as follows:

Theorem 3.4 (Degree sequence of $\mathrm{IRG}_n(\kappa_n)$)   *Let* (κ_n) *be a graphical sequence of kernels, with limit κ as described in Definition 3.3(a). For any fixed $k \geq 0$,*

$$N_k(n)/n \overset{\mathbb{P}}{\longrightarrow} \int_{\mathcal{S}} \frac{\lambda(x)^k}{k!} e^{-\lambda(x)} \mu(\mathrm{d}x), \qquad (3.3.3)$$

where $x \mapsto \lambda(x)$ is defined by

$$\lambda(x) = \int_{\mathcal{S}} \kappa(x, y) \mu(\mathrm{d}y). \qquad (3.3.4)$$

Theorem 3.4 is equivalent to the statement that

$$N_k(n)/n \overset{\mathbb{P}}{\longrightarrow} \mathbb{P}(D = k), \qquad (3.3.5)$$

where D has a mixed-Poisson distribution with mixing distribution W_λ given by

$$\mathbb{P}(W_\lambda \leq x) = \mu(\{y \in \mathcal{S} \colon \lambda(y) \leq x\}). \qquad (3.3.6)$$

See Figure 3.2 for examples of the degree distribution in the generalized random graph: we show the degree distribution itself, the size-biased degree distribution, and the degree distribution of a random neighbor of a uniform vertex.

In the remainder of this section we will prove Theorem 3.4. This proof is a good example of how proofs for inhomogeneous random graphs will be carried out later in this text. Indeed, we start by proving Theorem 3.4 for the finite-types case, which is substantially easier. After this, we give a proof in the general case, for which we need to prove results on approximations of sequences of graphical kernels. These approximations apply to *bounded* kernels, and thus we also need to show that unbounded kernels can be well approximated by bounded kernels. It is here that the assumption (3.2.3) is crucially used.

3.3.1 DEGREE SEQUENCE OF FINITE-TYPE CASE

Now we will prove Theorem 3.4 in the finite-type case, for which $\mathcal{S} = [t]$ for some $t < \infty$. Take a vertex v of type s, let d_v be its degree, and let $d_{v,r}$ be the number of edges from v to vertices of type $r \in [t]$. Then, clearly,

$$d_v = \sum_{r \in [t]} d_{v,r}. \qquad (3.3.7)$$

Recall that, in the finite-type case, the edge probability between vertices of types s and r is denoted by $(\kappa_n(s,r) \wedge n)/n$. Further, (3.2.4) implies that $\kappa_n(s,r) \to \kappa(s,r)$ for every $s, r \in [t]$, while (3.2.1) implies that the number n_s of vertices of type s satisfies $\mu_n(s) = n_s/n \to \mu(s)$ for some probability distribution $(\mu(s))_{s \in [t]}$.

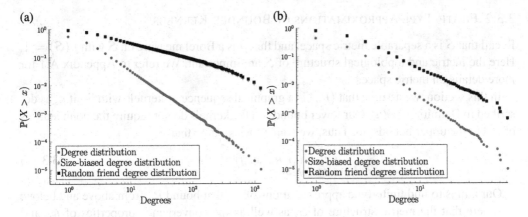

Figure 3.2 Degree distributions in the generalized random graph with $n = 100,000$: (a) $\tau = 2.5$ and (b) $\tau = 3.5$.

Assume that $n \geq \max \kappa$. The random variables $(d_{v,r})_{r \in [t]}$ are independent, and $d_{v,r} \sim$ $\mathrm{Bin}(n_r - \mathbb{1}_{\{s=r\}}, \kappa_n(s,r)/n) \xrightarrow{d} \mathrm{Poi}(\mu(r)\kappa(s,r))$, where n_r is the number of vertices with type r and $\mu(r) = \lim_{n \to \infty} n_r/n$ is the limiting type distribution. Hence

$$d_v \xrightarrow{d} \mathrm{Poi}\Big(\sum_{r \in [t]} \kappa(s,r)\mu(r) \Big) = \mathrm{Poi}(\lambda(s)), \tag{3.3.8}$$

where $\lambda(s) = \int \kappa(s,r)\mu(\mathrm{d}r) = \sum_{r \in [t]} \kappa(s,r)\mu(r)$. Consequently,

$$\mathbb{P}(d_v = k) \to \mathbb{P}(\mathrm{Poi}(\lambda(s)) = k) = \frac{\lambda(s)^k}{k!} e^{-\lambda(s)}. \tag{3.3.9}$$

Let $N_{k,s}(n)$ be the number of vertices in $\mathrm{IRG}_n(\kappa_n)$ of type s with degree k. Then

$$\frac{1}{n}\mathbb{E}[N_{k,s}(n)] = \frac{1}{n}n_s\mathbb{P}(d_v = k) \to \mu(s)\mathbb{P}(\mathrm{Poi}(\lambda(s)) = k). \tag{3.3.10}$$

It is easily checked that $\mathrm{Var}(N_{k,s}(n)) = O(n)$ (see Exercise 3.12). Hence,

$$\frac{1}{n}N_{k,s}(n) \xrightarrow{\mathbb{P}} \mathbb{P}(\mathrm{Poi}(\lambda(s)) = k)\mu(s), \tag{3.3.11}$$

and thus, summing over $s \in [t]$,

$$\frac{1}{n}N_k(n) = \sum_{s \in [t]} \frac{1}{n}N_{k,s}(n) \xrightarrow{\mathbb{P}} \sum_{s \in [t]} \mathbb{P}(\mathrm{Poi}(\lambda(s)) = k)\mu(s) = \mathbb{P}(D = k). \tag{3.3.12}$$

This proves Theorem 3.4 in the finite-type case. $\qquad\square$

In order to prove Theorem 3.4 in the general case, we approximate a sequence of graphical kernels (κ_n) by appropriate regular finite kernels, as we explain in detail in the next subsection.

3.3.2 Finite-Type Approximations of Bounded Kernels

Recall that S is a separable metric space, and that μ is a Borel measure on S with $\mu(S) = 1$. Here the metric and topological structure of S are important. We refer to Appendix A.1 for more details on metric spaces.

In this section, we assume that (κ_n) is a graphical sequence of kernels with limit κ, as described in Definition 3.3(a). Our lower bounds on the kernels do not require the boundedness of κ, but the upper bounds do. Thus, we sometimes assume that

$$\sup_{n \geq 1} \sup_{x,y \in S} \kappa_n(x,y) < \infty, \tag{3.3.13}$$

Our aim is to find finite-type approximations of κ_n that bound κ_n from above and below. It is here that the metric structure of S, as well as the convergence properties of κ_n and a.e.-continuity of $(x,y) \mapsto \kappa(x,y)$ in Definition 3.3, are crucially used:

Proposition 3.5 (Finite-type approximations of general kernels) *If $(\kappa_n)_{n \geq 1}$ is a graphical sequence of kernels on a vertex space $(S, \mu, (x_n)_{n \geq 1})$ with limit κ, then there exist sequences $(\underline{\kappa}_m)_{m \geq 1}$, and $(\overline{\kappa}_m)_{m \geq 1}$ when (3.3.13) holds, of finite-type kernels on the same vertex space $(S, \mu_n, (x_n)_{n \geq 1})$ satisfying the following:*

(a) if κ is irreducible, then so are $\underline{\kappa}_m$ and $\overline{\kappa}_m$ for all large enough m;
(b) $\underline{\kappa}_m(x,y) \nearrow \kappa(x,y)$ for $(\mu \times \mu)$-a.e. $x, y \in S$;
(c) $\overline{\kappa}_m(x,y) \searrow \kappa(x,y)$ for $(\mu \times \mu)$-a.e. $x, y \in S$.

Let us now give some details. We find these finite-type approximations by giving a partition $\mathcal{P}_m$ of S on which $\kappa_n(x,y)$ is almost constant when x and y are inside cells of the partition. Fix $m \geq 1$; this indicates the number of cells in the partition of S. Given a sequence of finite partitions $\mathcal{P}_m = \{A_{m1}, \ldots, A_{mM_m}\}$ of S and an $x \in S$, we define the function $x \mapsto i_m(x)$ by requiring that

$$x \in A_{m,i_m(x)}. \tag{3.3.14}$$

Thus, $i_m(x)$ indicates the cell in $\mathcal{P}_m$ containing x. For $A \subseteq S$, we write $\mathrm{diam}(A) = \sup\{\mathrm{dist}(x,y) \colon x, y \in A\}$, where $\mathrm{dist}(\cdot, \cdot)$ denotes the distance on S. We obtain the following key approximation result:

Lemma 3.6 (Approximating partition) *Fix $m \geq 1$. There exists a sequence of finite partitions $\mathcal{P}_m = \{A_{m1}, \ldots, A_{mM_m}\}$ of S such that:*

(a) each A_{mi} is measurable and $\mu(\partial A_{mi}) = 0$;
(b) for each m, $\mathcal{P}_{m+1}$ refines $\mathcal{P}_m$, i.e., each A_{mi} is a union $\bigcup_{j \in J_{mi}} A_{m+1,j}$ for some set J_{mi};
(c) for almost every $x \in S$, $\mathrm{diam}(A_{m,i_m(x)}) \to 0$ as $m \to \infty$, where $i_m(x)$ is defined by (3.3.14).

Proof This proof is a little technical. When $S = (0,1]$ and μ is continuous, we can take $\mathcal{P}_m$ as the dyadic partition into intervals of length 2^{-m}. If $S = (0,1]$ and μ is arbitrary, then we can do almost the same: only we shift the endpoints of the intervals a little when necessary to avoid point masses of μ.

In general, we can proceed as follows. Let $z_1, z_2, \ldots$ be a dense sequence of points in $\mathcal{S}$. For any z_i, the balls $\mathcal{B}_d(z_i) = \{y \in \mathcal{S} \colon \text{dist}(y, z_i) \leq d\}$, for $d > 0$, have disjoint boundaries, and thus all except at most a countable number of them are μ-continuity sets. Consequently, for every $m \geq 1$, we may choose balls $\mathcal{B}_{mi} = \mathcal{B}_{d_{mi}}(z_i)$ that are μ-continuity sets and have radii satisfying $1/m < d_{mi} < 2/m$. Then, $\bigcup_i \mathcal{B}_{mi} = \mathcal{S}$ and if we define $\mathcal{B}'_{mi} := \mathcal{B}_{mi} \setminus \bigcup_{j<i} \mathcal{B}_{mj}$, we obtain for each m an infinite partition $\{\mathcal{B}'_{mi}\}_{i \geq 1}$ of $\mathcal{S}$ into μ-continuity sets, each with diameter at most $4/m$. To get a finite partition, we choose q_m large enough to ensure that, with $\mathcal{B}'_0 := \bigcup_{i > q_m} \mathcal{B}'_{mi}$, we have $\mu(\mathcal{B}'_{m0}) < 2^{-m}$; then $\{\mathcal{B}'_{mi}\}_{i=0}^{q_m}$ is a partition of $\mathcal{S}$ for each m, with $\text{diam}(\mathcal{B}'_{mi}) \leq 4/m$ for $i \geq 1$.

Finally, we let $\mathcal{P}_m$ consist of all intersections $\bigcap_{l=1}^{m} \mathcal{B}'_{l i_l}$ with $0 \leq i_l \leq q_l$; then conditions (a) and (b) are satisfied. Condition (c) follows from the Borel–Cantelli Lemma: as $\sum_m \mu(\mathcal{B}'_{m0})$ is finite, a.e. x is in finitely many of the sets $\mathcal{B}'_{m0}$. For any such x, if m is large enough then $x \in \mathcal{B}'_{mi}$ for some $i \geq 1$, so the part of $\mathcal{P}_m$ containing x has diameter satisfying $\text{diam}(\mathcal{B}'_{mi}) \leq 4/m$. $\qquad\square$

Now we are ready to complete the proof of Proposition 3.5.

Proof of Proposition 3.5. Recall from Definition 3.3 that a kernel κ is a symmetric measurable function on $\mathcal{S} \times \mathcal{S}$ that is a.e. continuous. Recall also that κ_n is a graphical sequence of kernels, so that it satisfies the convergence properties in (3.2.4). Fixing a sequence of partitions with the properties described in Lemma 3.6, we can define sequences of lower and upper approximations to κ by

$$\underline{\kappa}_m(x, y) = \inf\{\kappa(x', y') \colon x' \in \mathcal{A}_{m, i_m(x)}, y' \in \mathcal{A}_{m, i_m(y)}\}, \tag{3.3.15}$$

$$\overline{\kappa}_m(x, y) = \sup\{\kappa(x', y') \colon x' \in \mathcal{A}_{m, i_m(x)}, y' \in \mathcal{A}_{m, i_m(y)}\}. \tag{3.3.16}$$

We thus replace κ by its infimum or supremum on each $\mathcal{A}_{mi} \times \mathcal{A}_{mj}$. As $\overline{\kappa}_m$ might be $+\infty$, we use it only for bounded κ_n as in (3.3.13). Obviously, $\underline{\kappa}_m$ and $\overline{\kappa}_m$ are constant on $\mathcal{A}_{m,i} \times \mathcal{A}_{m,j}$ for every i, j, so that $\underline{\kappa}_m$ and $\overline{\kappa}_m$ correspond to finite-type kernels (see Exercise 3.14).

By Lemma 3.6(b),

$$\underline{\kappa}_m \leq \underline{\kappa}_{m+1} \quad \text{and} \quad \overline{\kappa}_m \geq \overline{\kappa}_{m+1}. \tag{3.3.17}$$

Furthermore, since κ is almost everywhere continuous then, by Lemma 3.6(c),

$$\underline{\kappa}_m(x, y) \to \kappa(x, y) \quad \text{and} \quad \overline{\kappa}_m(x, y) \to \kappa(x, y) \quad \text{for } (\mu \times \mu)\text{-a.e. } (x, y) \in \mathcal{S}^2. \tag{3.3.18}$$

If (κ_n) is a graphical sequence of kernels with limit κ, then we similarly define

$$\underline{\kappa}_m(x, y) := \inf\{(\kappa \wedge \kappa_n)(x', y') \colon x' \in \mathcal{A}_{m, i_m(x)}, y' \in \mathcal{A}_{m, i_m(y)}, n \geq m\}, \tag{3.3.19}$$

$$\overline{\kappa}_m(x, y) := \sup\{(\kappa \vee \kappa_n)(x', y') \colon x' \in \mathcal{A}_{m, i_m(x)}, y' \in \mathcal{A}_{m, i_m(y)}, n \geq m\}. \tag{3.3.20}$$

By (3.3.17), $\underline{\kappa}_m \leq \underline{\kappa}_{m+1}$, and, by Lemma 3.6(c) and (3.2.4) in Definition 3.3(a),

$$\underline{\kappa}_m(x, y) \nearrow \kappa(x, y) \quad \text{as } m \to \infty, \quad \text{for } (\mu \times \mu)\text{-a.e. } (x, y) \in \mathcal{S}^2. \tag{3.3.21}$$

This proves part (b) of Proposition 3.5. The proof of part (c) is similar. For the irreducibility in part (a), we may assume that μ is irreducible. In fact, $\underline{\kappa}_m$ may be reducible for some

m. We omit the proof that $\underline{\kappa}_m$ can be adapted in such a way that the adapted version is irreducible. $\qquad\qquad\qquad\qquad\qquad\qquad\qquad\qquad\qquad\qquad\qquad\qquad\qquad\quad$ $\square$

Since $\underline{\kappa}_m \leq \kappa$, we can obviously construct our random graph in such a way that all edges in $\mathrm{IRG}_n(\underline{\kappa}_m)$ are also present in $\mathrm{IRG}_n(\kappa_n)$, which we will write as $\mathrm{IRG}_n(\underline{\kappa}_m) \subseteq \mathrm{IRG}_n(\kappa_n)$, and in what follows we will assume this. See also Exercise 3.15. Similarly, we shall assume that $\mathrm{IRG}_n(\overline{\kappa}_m) \supseteq \mathrm{IRG}_n(\kappa_n)$ when κ_n is bounded as in (3.3.13). Moreover, when $n \geq m$,

$$\kappa_n \geq \underline{\kappa}_m, \tag{3.3.22}$$

and we may assume that $\mathrm{IRG}_n(\underline{\kappa}_m) \subseteq \mathrm{IRG}_n(\kappa_n)$. By the convergence of the sequence of kernels (κ_n), we further obtain that the number of edges also converges. Thus, in bounding κ_n, we do not create or destroy too many edges. This provides the starting point of our analysis, which we provide in the following subsection.

3.3.3 Degree Sequences of General Inhomogeneous Random Graphs

Now we are ready to complete the proof of Theorem 3.4 for general sequences of graphical kernels (κ_n). Define $\underline{\kappa}_m$ by (3.3.19). Since we will be only using the lower bounding kernel $\underline{\kappa}_m$ (which always exists), we need not assume that κ_n is bounded.

Let $\varepsilon > 0$ be given. From (3.2.5) and monotone convergence, there is an m such that

$$\iint_{\mathcal{S}^2} \underline{\kappa}_m(x,y)\mu(\mathrm{d}x)\mu(\mathrm{d}y) > \iint_{\mathcal{S}^2} \kappa(x,y)\mu(\mathrm{d}x)\mu(\mathrm{d}y) - \varepsilon. \tag{3.3.23}$$

Recall that $\mathrm{IRG}_n(\underline{\kappa}_m) \subseteq \mathrm{IRG}_n(\kappa_n)$. By (3.2.5) and (3.3.23),

$$\frac{1}{n}|E(\mathrm{IRG}_n(\kappa_n) \setminus \mathrm{IRG}_n(\underline{\kappa}_m))|$$

$$= \frac{1}{n}|E(\mathrm{IRG}_n(\kappa_n))| - \frac{1}{n}|E(\mathrm{IRG}_n(\underline{\kappa}_m))|$$

$$\overset{\mathbb{P}}{\longrightarrow} \frac{1}{2}\iint_{\mathcal{S}^2} \kappa(x,y)\mu(\mathrm{d}x)\mu(\mathrm{d}y) - \frac{1}{2}\iint_{\mathcal{S}^2} \underline{\kappa}_m(x,y)\mu(\mathrm{d}x)\mu(\mathrm{d}y) < \frac{\varepsilon}{2}, \tag{3.3.24}$$

so that, whp, $|E(\mathrm{IRG}_n(\kappa_n) \setminus \mathrm{IRG}_n(\underline{\kappa}_m))| < \varepsilon n$. Let us write $N_k^{(m)}(n)$ for the number of vertices of degree k in $\mathrm{IRG}_n(\underline{\kappa}_m)$ (and $N_k(n)$ for those in $\mathrm{IRG}_n(\kappa_n)$). It follows that, whp,

$$|N_k^{(m)}(n) - N_k(n)| < 2\varepsilon n. \tag{3.3.25}$$

Writing $D^{(m)}$ for the equivalent of D defined using $\underline{\kappa}_m$ in place of κ, we have $N_k^{(m)}(n)/n \overset{\mathbb{P}}{\longrightarrow} \mathbb{P}(D^{(m)} = k)$ by the proof for the finite-type case. Thus, whp,

$$|N_k^{(m)}(n)/n - \mathbb{P}(D^{(m)} = k)| < \varepsilon. \tag{3.3.26}$$

Finally, we have $\mathbb{E}[D] = \int_S \lambda(x)\mu(\mathrm{d}x) = \iint_{S^2} \kappa(x,y)\mu(\mathrm{d}x)\mu(\mathrm{d}y)$. Since $\lambda^{(m)}(x) \leq \lambda(x)$, we can couple the limiting degrees in such a way that $D^{(m)} \leq D$ almost surely, and thus

$$\mathbb{P}(D \neq D^{(m)}) = \mathbb{P}(D - D^{(m)} \geq 1) \leq \mathbb{E}[D - D^{(m)}]$$

$$= \iint_{S^2} \kappa(x,y)\mu(\mathrm{d}x)\mu(\mathrm{d}y) - \iint_{S^2} \underline{\kappa}_m(x,y)\mu(\mathrm{d}x)\mu(\mathrm{d}y) < \varepsilon. \quad (3.3.27)$$

Combining (3.3.25), (3.3.26), and (3.3.27), we see that $|N_k(n)/n - \mathbb{P}(D = k)|) < 4\varepsilon$ whp, as required. $\qquad\square$

3.3.4 Degree Distribution in Inhomogeneous Random Graphs: Discussion

Now that we have identified the limit of the degree distribution, let us discuss its proof as well as some properties of the limiting degree distribution.

Bounded Kernels

First of all, the above proof is exemplary of several proofs that we will use in this chapter as well as in Chapter 6. The current proof is particularly simple, as it makes use only of the *lower bounding* finite-type inhomogeneous random graph, while in many settings we also need the *upper bound*. This upper bound can apply only to *bounded* kernels κ_n as in (3.3.13). As a result, we need to study the effect of bounding κ_n, for example by approximating it by $\kappa_n(x,y) \wedge K$ for large enough K.

Tail Properties of the Degree Distribution

Let $W = W_\lambda$ be the random variable $\lambda(U)$, where U is a random variable on S having distribution μ. Then we can also describe the mixed-Poisson distribution of D as $\mathrm{Poi}(W)$. Under mild conditions, the tail probabilities $\mathbb{P}(D > k)$ and $\mathbb{P}(W > k)$ agree for large k. We state this for the case of power-law tails; many of these results generalize to regularly varying tails. Let $N_{>k}(n)$ be the number of vertices with degree larger than k.

Corollary 3.7 (Power-law tails for the degree sequence)   *Let* (κ_n) *be a graphical sequence of kernels with limit* κ. *Suppose that*

$$\mathbb{P}(W > k) = \mu(\{x \in S : \lambda(x) > k\}) = c_W k^{-(\tau-1)}(1 + o(1)) \quad (3.3.28)$$

as $k \to \infty$, *for some* $c_W > 0$ *and* $\tau > 2$. *Then*

$$N_{>k}(n)/n \xrightarrow{\mathbb{P}} \mathbb{P}(D > k) = c_W k^{-(\tau-1)}(1 + o(1)), \quad (3.3.29)$$

where the first limit is for k *fixed and* $n \to \infty$, *and the second for* $k \to \infty$.

Proof   It suffices to show that $\mathbb{P}(D > k) = c_W k^{-(\tau-1)}(1 + o(1))$; the remaining conclusions then follow from Theorem 3.4. For any $\varepsilon > 0$, as $k \to \infty$,

$$\mathbb{P}(\mathrm{Poi}(W) > k \mid W > (1+\varepsilon)k) \to 1,$$

$$\text{and} \quad \mathbb{P}(\mathrm{Poi}(W) > k \mid W < (1-\varepsilon)k) = o(k^{-(\tau-1)}). \quad (3.3.30)$$

It follows that $\mathbb{P}(D > k) = \mathbb{P}(\mathsf{Poi}(W) > k) = c_w k^{-(\tau-1)}(1 + o(1))$ as $k \to \infty$. Exercise 3.16 asks you to fill in the details of this argument. $\qquad\square$

Corollary 3.7 shows that the general inhomogeneous random graph does include natural cases with power-law degree distributions. Recall that we have already observed in [V1, Theorem 6.7] that this is the case for $\mathrm{GRG}_n(\boldsymbol{w})$ when the weights sequence $\boldsymbol{w}$ is chosen appropriately.

3.4 MULTI-TYPE BRANCHING PROCESSES

In order to study further properties of $\mathrm{IRG}_n(\kappa_n)$, we need to understand the neighborhood structure of vertices. This will be crucially used in the next section, where we study the local convergence properties of $\mathrm{IRG}_n(\kappa_n)$. For simplicity, let us restrict ourselves first to the finite-types case. As we have seen, nice kernels can be arbitrarily well approximated by finite-type kernels, so this should be a good start. Then, for a vertex of type s, the number of neighbors of type r is close to Poisson-distributed with approximate mean $\kappa(s, r)\mu(r)$. Even when we assume independence of the neighborhood structures of different vertices, we still do not arrive at a *classical* branching process as discussed in [V1, Chapter 3]. Instead, we can describe the neighborhood structure with a branching process in which we keep track of the *type* of each vertex. For general κ and μ, we can even have a *continuum* of types. Such branching processes are called *multi-type branching processes*. In this section, we discuss some of the basics of these processes.

3.4.1 MULTI-TYPE BRANCHING PROCESSES WITH FINITELY MANY TYPES

Multi-type branching process can be effectively analyzed using linear algebra in the finite-type case, and functional analysis in the infinite-type case. In order to do so, we first introduce some notation. We will assume that we are in the finite-type case, and denote the number of types by t. We let $\boldsymbol{j} = (j_1, \ldots, j_t) \in \mathbb{N}_0^t$ be a vector of non-negative integers, and denote by $p_{\boldsymbol{j}}^{(s)}$ the probability that an individual of type s gives rise to offspring $\boldsymbol{j}$, i.e., j_s children of type s for all $s \in [t]$. The offsprings of different individuals are all mutually independent. Denote by $Z_{k,r}^{(s)}$ the number of individuals of type r in generation k when starting from a single individual of type s and $\boldsymbol{Z}_k^{(s)} = (Z_{k,1}^{(s)}, \ldots, Z_{k,t}^{(s)})$. We are interested in the survival or extinction of multi-type branching processes, and in the growth of the generation sizes. In the multi-type case, we are naturally led to a matrix set-up.

We now discuss the survival versus extinction of multi-type branching processes. We denote the survival probability of the multi-type branching process when one starts from a single individual of type $s \in [t]$ by

$$\zeta^{(s)} = \mathbb{P}(\boldsymbol{Z}_k^{(s)} \neq \boldsymbol{0} \text{ for all } k), \qquad (3.4.1)$$

and we let $\boldsymbol{\zeta} = (\zeta^{(1)}, \ldots, \zeta^{(t)})$ denote the vector of survival probabilities. Exercise 3.18 identifies the survival probability when starting with a *random* type having distribution $(\mu(s))_{s \in [t]}$. Our first aim is to investigate the condition for $\boldsymbol{\zeta} = \boldsymbol{0}$ to hold.

Multi-Type Branching Processes and Generating Functions

We write $p(j) = (p_j^{(1)}, \ldots, p_j^{(r)})$ and, for $z \in [0,1]^t$, we let

$$G^{(s)}(z) = \sum_j p_j^{(s)} \prod_{r \in [t]} z_r^{j_r} \qquad (3.4.2)$$

be the joint probability generating function of the offspring of an individual of type $s \in [t]$. We write

$$G(z) = (G^{(1)}(z), \ldots, G^{(r)}(z)) \qquad (3.4.3)$$

for the vector of generating functions. We now generalize [V1, Theorem 3.1] to the multi-type case.

Let ζ be the smallest solution in the lexicographic order on $\mathbb{R}^t$ to

$$\zeta = 1 - G(1 - \zeta). \qquad (3.4.4)$$

It turns out that ζ is the vector whose sth component equals the survival probability of $(Z_k^{(s)})_{k \geq 0}$. Define

$$G_k^{(s)}(z) = \mathbb{E}\Big[\prod_{r \in [t]} z_t^{Z_{k,t}^{(s)}} \Big], \qquad (3.4.5)$$

and $G_k(z) = (G_k^{(1)}(z), \ldots, G_k^{(r)}(z))$. Then, $G_{k+1}(z) = G_k(G(z)) = G(G_k(z))$. Since $G_k(0) = (\mathbb{P}(Z_k^{(s)} = 0))_{k \geq 0}$, we also have that $\lim_{k \to \infty} G_k(0)$ is the vector of extinction probabilities, so that $\zeta = 1 - \lim_{k \to \infty} G_k(0)$ is the vector of survival probabilities.

Naturally, the vector of survival probabilities depends sensitively on the type of the ancestor of the branching process. On the other hand, under reasonable assumptions, the *positivity* of the survival probability is independent of the type of the root of the branching process. A necessary and sufficient condition for this property is that, with positive probability, an individual of type s arises as a descendent of an individual of type r for each pair of types s and r. See Exercise 3.19. Exercise 3.20 relates this to the lth power of the mean offspring matrix $(\mathbb{E}[Z_{1,r}^{(s)}])_{s,r \in [t]}$.

We next exclude a case where the branching mechanism is trivial:

Definition 3.8 (Singular multi-type branching processes) We call a multi-type branching process *singular* when G in (3.4.3) equals $G(z) = Mz$ for some matrix M. Otherwise we call the multi-type branching process *non-singular*. ◀

For a singular multi-type branching process, each individual in the branching process has precisely *one* offspring almost surely (see Exercise 3.21). When each individual has precisely one offspring, the multi-type branching process is equivalent to a Markov chain, and the process almost surely survives. Thus, in this case there is no survival versus extinction phase transition. We assume throughout the remainder of this section that the multi-type branching process is non-singular.

3.4.2 SURVIVAL VERSUS EXTINCTION OF MULTI-TYPE BRANCHING PROCESSES

Let $\lambda_{sr} = \mathbb{E}[Z_{1,r}^{(s)}]$ denote the expected number of offspring of type r of a single individual of type s. In analogy to the random graph setting, we write $\lambda_{sr} = \kappa(s,r)\mu(r)$. This can always be done (and in fact in many ways), for example by taking $\kappa(s, r) = t\lambda_{sr}$, $\mu(r) = 1/t$. The current set-up is convenient, however, as it allows the expected number of neighbors in the inhomogeneous random graphs to be matched to the expected number of offspring in the multi-type branching process. For random graphs, $\kappa(s, r) = \kappa(r, s)$ is necessarily symmetric, while for multi-type branching processes this is not necessary. Let $T_\kappa = (\kappa(s, r)\mu(r))_{s,r\in[t]}$ be the matrix of expected number of offspring.

Definition 3.9 (Irreducible and positively regular multi-type branching processes)   We call a multi-type branching process *irreducible* when, for every pair of types $s, r \in [t]$, there exists l such that $(T_\kappa^l)_{s,r} > 0$, where the matrix T_κ^l is the lth power of T_κ. We call a multi-type branching process *positively regular* if there exists l such that $(T_\kappa^l)_{s,r} > 0$ for all $s, r \in [t]$.   ◀

The definition of irreducible multi-type branching processes in Definition 3.9 is closely related to that of irreducible random graph kernels in Definition 3.3. The name irreducibility can be understood since it implies that the Markov chain of the number of individuals of the various types is irreducible.

By the Perron–Frobenius theorem, in the positively regular case, the matrix T_κ has a unique largest eigenvalue equal to $\|T_\kappa\|$ with non-negative left eigenvector x_κ, and the eigenvalue $\|T_\kappa\|$ can be computed as

$$\|T_\kappa\| = \sup_{x:\,\|x\|\leq 1} \|T_\kappa x\|, \quad \text{where} \quad \|x\| = \sqrt{\sum_{s\in[t]} x_s^2}. \tag{3.4.6}$$

Fix $s \in [t]$. We note that, for $k \geq 0$ and all $y \in \mathbb{N}_0^t$,

$$\mathbb{E}[Z_{k+1}^{(s)} \mid Z_k^{(s)} = y] = T_\kappa y, \tag{3.4.7}$$

so that

$$\mathbb{E}[Z_k^{(s)}] = T_\kappa^k e^{(s)}, \tag{3.4.8}$$

where T_κ^k denotes the k-fold application of the matrix T_κ and $e^{(s)}$ is the vector which has 1 at the sth position, and zeros otherwise.

The identities in (3.4.7) and (3.4.8) have several important consequences concerning the phase transitions of multi-type branching processes. First, when $\|T_\kappa\| < 1$,

$$\mathbb{E}[Z_k^{(s)}] \leq \|T_\kappa\|^k \|e^{(s)}\| = \|T_\kappa\|^k, \tag{3.4.9}$$

which vanishes exponentially fast. Therefore, by the Markov inequality ([V1, Theorem 2.17]), a multi-type branching process dies out almost surely. When $\|T_\kappa\| > 1$, on the other hand, the sequence

$$M_k = x_\kappa Z_k^{(s)} \|T_\kappa\|^{-k} \tag{3.4.10}$$

is a non-negative martingale, by (3.4.7) and the fact that x_κ is a left eigenvector with eigenvalue $\|T_\kappa\|$, since $x_\kappa T_\kappa = \|T_\kappa\| x_\kappa$. By the Martingale Convergence Theorem ([V1, Theorem 2.24]), the martingale M_k converges almost surely as $k \to \infty$.

Under some further assumptions on M_k, for example that M_1 has finite second moment, we obtain that $M_k \xrightarrow{a.s.} M_\infty$ and $\mathbb{E}[M_k] \to \mathbb{E}[M_\infty]$. More precisely, there is a multi-type analog of the Kesten–Stigum Theorem ([V1, Theorem 3.10]). Since $\mathbb{E}[M_0] = \boldsymbol{x}_\kappa \mathrm{e}^{(s)} > 0$, we thus have that $Z_k^{(s)}$ grows exponentially with a strictly positive probability, which implies that the survival probability is positive. [V1, Theorem 3.1] can be adapted to show that $Z_k^{(s)} \xrightarrow{\mathbb{P}} 0$ when $\|\boldsymbol{T}_\kappa\| = 1$ in the non-singular case. See, e.g. (Harris, 1963, Sections II.6 and II.7). We conclude that $\zeta > 0$ precisely when $\|\boldsymbol{T}_\kappa\| > 1$ for non-singular and irreducible multi-type branching processes:

Theorem 3.10 (Survival versus extinction of finite-type branching processes) *Let $(\boldsymbol{Z}_k^{(s)})_{k\geq 0}$ be a non-singular positively regular multi-type branching process with offspring matrix $\boldsymbol{T}_\kappa$ on the type space $\mathcal{S} = [t]$. Then the following hold:*

(a) *The survival probability ζ is the largest solution to $\zeta = 1 - G(1 - \zeta)$, and $\zeta > 0$ precisely when $\|\boldsymbol{T}_\kappa\| > 1$.*

(b) *Assume that $\|\boldsymbol{T}_\kappa\| > 1$. Let $\boldsymbol{x}_\kappa$ be the unique positive left eigenvector of $\boldsymbol{T}_\kappa$. Then, as $k \to \infty$, the martingale $M_k = \boldsymbol{x}_\kappa \boldsymbol{Z}_k^{(i)} \|\boldsymbol{T}_\kappa\|^{-k}$ converges almost surely to a non-negative limit on the event of survival precisely when $\mathbb{E}[Z_1^{(s)} \log (Z_1^{(s)})] < \infty$ for all $s \in [t]$, where $Z_1^{(s)} = \|\boldsymbol{Z}_1^{(s)}\|_1 = \sum_{r \in [t]} Z_{1,r}^{(s)}$ is the total number of offspring of a type-s individual.*

3.4.3 Poisson Multi-Type Branching Processes

We now specialize to Poisson multi-type branching processes as these turn out to be the most relevant in the inhomogeneous random graph setting. We call a multi-type branching process *Poisson* when all the numbers of children of each type are independent Poisson random variables. Thus, $\boldsymbol{Z}^{(s)} = (Z_{1,1}^{(s)}, \ldots, Z_{1,t}^{(s)})$ is a vector of independent Poisson random variables with means $(\kappa(s,1)\mu(1), \ldots, \kappa(s,t)\mu(t))$. As we will see later, Poisson multi-type branching processes arise naturally when one is exploring a component of $\mathrm{IRG}_n(\kappa)$ starting at a vertex of type s. This is analogous to the use of the single-type Poisson branching process in the analysis of the Erdős–Rényi random graph $\mathrm{ER}_n(\lambda/n)$, as discussed in detail in [V1, Chapters 4 and 5].

Poisson Multi-Type Branching Processes with Finitely Many Types

For Poisson multi-type branching processes with finitely many types, let $s \in [t]$ be the type of the root, and compute

$$G^{(s)}(\boldsymbol{z}) = \mathbb{E}\Big[\prod_{r\in[t]} z_r^{Z_{1,r}^{(s)}} \Big] = \mathrm{e}^{\sum_{r\in[t]} \kappa(s,r)\mu(r)(z_r-1)} = \mathrm{e}^{(\boldsymbol{T}_\kappa(\boldsymbol{z}-1))_s}. \tag{3.4.11}$$

Thus, the vector of survival probabilities ζ satisfies

$$\zeta = 1 - \mathrm{e}^{-\boldsymbol{T}_\kappa \zeta}, \tag{3.4.12}$$

where, for a vector $\boldsymbol{x}$, $\mathrm{e}^{\boldsymbol{x}}$ denotes the coordinate-wise exponential.

There is a beautiful property of Poisson random variables that allows us to construct a Poisson multi-type branching process in a particularly convenient way. This property follows from the following *Poisson thinning* property:

Lemma 3.11 (Poisson number of multinomial trials)   *Let X have a Poisson distribution with parameter λ. Perform X multinomial trials, where the ith outcome appears with probability p_i for probabilities $(p_i)_{i=1}^{k}$. Consider $(X_i)_{i=1}^{k}$, where X_i denotes the total number of outcomes i. Then $(X_i)_{i=1}^{k}$ is a sequence of* independent *Poisson random variables with parameters $(\lambda p_i)_{i=1}^{k}$.*

Proof   Let $(x_i)_{i=1}^{k}$ denote a sequence of non-negative integers, denote $x = \sum_{i=1}^{k} x_i$ and compute

$$\mathbb{P}((X_i)_{i=1}^{k} = (x_i)_{i=1}^{k}) = \mathbb{P}(X = x)\mathbb{P}((X_i)_{i=1}^{k} = (x_i)_{i=1}^{k} \mid X = x) \qquad (3.4.13)$$

$$= e^{-\lambda}\frac{\lambda^x}{x!}\binom{x}{x_1, x_2, \ldots, x_k} p_1^{x_1} \cdots p_k^{x_k} = \prod_{i=1}^{k} e^{-\lambda p_i}\frac{(\lambda p_i)^{x_i}}{(x_i)!},$$

since $\sum_{i=1}^{k} p_i = 1$. $\square$

By Lemma 3.11, we can alternatively construct a Poisson branching process as follows. For an individual of type s, let its total number of offspring N_s have a Poisson distribution with parameter $\lambda(s) = \sum_{r \in [t]} \kappa(s, r)\mu(r)$. Then give each of the children independently a type r with probability $\kappa(s, r)\mu(r)/\lambda(s)$. Let N_{sr} denote the total number of individuals of type r thus obtained. We conclude that $\boldsymbol{Z}_1^{(s)}$ has the same distribution as $(N_{sr})_{r \in [t]}$.

We now extend the above setting of finite-type Poisson multi-type branching processes to the infinite-type case. Again, we prove results in the infinite-type case by reducing to the finite-type case.

Poisson Multi-Type Branching Processes with Infinitely Many Types

Let κ be a kernel. We define the Poisson multi-type branching processes with kernel κ as follows. Each individual of type $x \in \mathcal{S}$ is replaced in the next generation by a set of individuals distributed as a Poisson process on $\mathcal{S}$ with intensity $\kappa(x, y)\mu(dy)$. Thus, the number of children with types in a subset $\mathcal{A} \subseteq \mathcal{S}$ has a Poisson distribution with mean $\int_{\mathcal{A}} \kappa(x, y)\mu(dy)$, and these numbers are independent for disjoint sets $\mathcal{A}$ and for different individuals; see, e.g., Kallenberg (2002) or Section 1.5.

Let $\zeta_\kappa(x)$ be the survival probability of the Poisson multi-type branching process with kernel κ, starting from a root of type $x \in \mathcal{S}$. Set

$$\zeta_\kappa = \int_{\mathcal{S}} \zeta_\kappa(x)\mu(dx). \qquad (3.4.14)$$

Again, it can be seen, in a way similar to that above, that $\zeta_\kappa > 0$ if and only if $\|T_\kappa\| > 1$, where now the linear operator T_κ is defined, for $f \colon \mathcal{S} \to \mathbb{R}$, by

$$(T_\kappa f)(x) = \int_{\mathcal{S}} \kappa(x, y)f(y)\mu(dy), \qquad (3.4.15)$$

for any (measurable) function f such that this integral is defined (finite or $+\infty$) for a.e. $x \in \mathcal{S}$.

Note that $T_\kappa f$ is defined for every $f \geq 0$, with $0 \leq T_\kappa f \leq \infty$. If $\kappa \in L^1(\mathcal{S} \times \mathcal{S})$, as we assume throughout, then $T_\kappa f$ is also defined for every bounded f. In this case $T_\kappa f \in L^1(\mathcal{S})$ and thus $T_\kappa f$ is finite almost everywhere.

The consideration of multi-type branching processes with a possibly uncountable number of types requires some functional analysis. Similarly to the finite-type case in (3.4.6), we define

$$\|\boldsymbol{T}_\kappa\| = \sup\left\{\|\boldsymbol{T}_\kappa f\| : f \geq 0, \|f\| \leq 1\right\} \leq \infty. \tag{3.4.16}$$

When finite, $\|\boldsymbol{T}_\kappa\|$ is the norm of $\boldsymbol{T}_\kappa$ as an operator on $L^2(\mathcal{S})$; it is infinite if $\boldsymbol{T}_\kappa$ does not define a bounded operator on $L^2(\mathcal{S})$. The norm $\|\boldsymbol{T}_\kappa\|$ is at most the Hilbert–Schmidt norm of $\boldsymbol{T}_\kappa$:

$$\|\boldsymbol{T}_\kappa\| \leq \|\boldsymbol{T}_\kappa\|_{\mathrm{HS}} = \|\kappa\|_{L^2(\mathcal{S}\times\mathcal{S})} = \left(\iint_{\mathcal{S}^2} \kappa(x,y)^2 \mu(\mathrm{d}x)\mu(\mathrm{d}y)\right)^{1/2}. \tag{3.4.17}$$

We also define the non-linear operator Φ_κ by

$$(\Phi_\kappa f)(x) = 1 - \mathrm{e}^{-(\boldsymbol{T}_\kappa f)(x)}, \qquad x \in \mathcal{S}, \tag{3.4.18}$$

for $f \geq 0$. Note that for such an f we have $0 \leq \boldsymbol{T}_\kappa f \leq \infty$, and thus $0 \leq \Phi_\kappa f \leq 1$. We characterize the survival probability $\zeta_\kappa(x)$, and thus ζ_κ, in terms of Φ_κ, showing essentially that the function $x \mapsto \zeta_\kappa(x)$ is the maximal fixed point of the non-linear operator Φ_κ (recall (3.4.12)). Again, the survival probability satisfies that $\zeta_\kappa > 0$ precisely when $\|\boldsymbol{T}_\kappa\| > 1$, recall the finite-types case in Theorem 3.10. This leads us to the following definition:

Definition 3.12 (Super- and subcritical multi-type branching processes)   We call a multi-type branching process *supercritical* when $\|\boldsymbol{T}_\kappa\| > 1$, *critical* when $\|\boldsymbol{T}_\kappa\| = 1$, and *subcritical* when $\|\boldsymbol{T}_\kappa\| < 1$.   ◄

Then, the above discussion can be summarized by saying that a non-singular multi-type branching process survives with positive probability precisely when it is supercritical.

Poisson Branching Processes and Product Kernels: The Rank-1 Case

We continue by studying the rank-1 case, where the kernel is of product structure. Let $\kappa(x,y) = \psi(x)\psi(y)$, so that

$$(\boldsymbol{T}_\kappa f)(x) = \int_\mathcal{S} \psi(x)\psi(y)f(y)\mu(\mathrm{d}y) = \psi(x)\langle\psi, f\rangle_\mu, \tag{3.4.19}$$

where the inner product $\langle f, g\rangle_\mu$ is defined as

$$\langle f, g\rangle_\mu = \int_\mathcal{S} f(x)g(y)\mu(\mathrm{d}y). \tag{3.4.20}$$

In this case, we see that ψ is an eigenvector with eigenvalue

$$\int_\mathcal{S} \psi(y)^2 \mu(\mathrm{d}y) \equiv \|\psi\|^2_{L^2(\mu)}. \tag{3.4.21}$$

Thus, also

$$\|\boldsymbol{T}_\kappa\| = \|\psi\|^2_{L^2(\mu)}, \tag{3.4.22}$$

since ψ is the unique non-negative eigenfunction, and a basis of eigenfunctions can be found by taking a basis in the space orthogonal to ψ (each member of which will have eigenvalue

0). Thus, the rank-1 multi-type branching process is supercritical when $\|\psi\|_{L^2(\mu)}^2 > 1$, critical when $\|\psi\|_{L^2(\mu)}^2 = 1$, and *subcritical* when $\|\psi\|_{L^2(\mu)}^2 < 1$.

The rank-1 case is rather special, and not only since we can explicitly compute the eigenvectors of the operator T_κ. It also turns out that the rank-1 multi-type case reduces to a *single*-type branching process with mixed-Poisson offspring distribution. For this, we recall the construction right below Lemma 3.11. We compute that

$$\lambda(x) = \int_S \psi(x)\psi(y)\mu(dy) = \psi(x)\int_S \psi(y)\mu(dy), \qquad (3.4.23)$$

so that an offspring of an individual of type x receives a mark $y \in \mathcal{A}$ with probability

$$\int_{\mathcal{A}} p(x, dy) = \int_{\mathcal{A}} \frac{\kappa(x,y)\mu(dy)}{\lambda(x)} = \frac{\int_{\mathcal{A}} \psi(x)\psi(y)\mu(dy)}{\psi(x)\int_S \psi(z)\mu(dz)} = \frac{\int_{\mathcal{A}} \psi(y)\mu(dy)}{\int_S \psi(z)\mu(dz)}. \qquad (3.4.24)$$

We conclude that every individual chooses its type *independently* of the type of its parent. This means that this multi-type branching process reduces to a single-type branching process with offspring distribution $\mathsf{Poi}(W_\lambda)$, where

$$\mathbb{P}(W_\lambda \in \mathcal{A}) = \frac{\int_{\mathcal{A}} \psi(y)\mu(dy)}{\int_S \psi(z)\mu(dz)}. \qquad (3.4.25)$$

This makes the rank-1 setting particularly appealing.

Poisson Branching Processes and Sum Kernels

For the sum kernel, the analysis becomes slightly more involved, but can still be solved. Recall that $\kappa(x,y) = \psi(x) + \psi(y)$ for the sum kernel. Anticipating a nice shape of the eigenvalues and eigenvectors, we let $\phi(x) = a\psi(x) + b$, and verify the eigenvalue relation. We compute

$$(T_\kappa\phi)(x) = \int_S \kappa(x,y)\phi(y)\mu(dy) = \int_S [\psi(x) + \psi(y)](a\psi(y) + b))\mu(dy)$$
$$= \psi(x)\big(a\|\psi\|_{L^1(\mu)} + b\big) + \big(a\|\psi\|_{L^2(\mu)}^2 + b\|\psi\|_{L^1(\mu)}\big). \qquad (3.4.26)$$

The eigenvalues and corresponding eigenvectors satisfy

$$(T_\kappa\phi)(x) = \lambda(a\psi(x) + b). \qquad (3.4.27)$$

Solving for a, b, λ leads to $a\|\psi\|_{L^1(\mu)} + b = a\lambda$, $a\|\psi\|_{L^2(\mu)}^2 + b\|\psi\|_{L^1(\mu)} = \lambda b$, so that the vector $(a,b)^T$ is the eigenvector with eigenvalue λ of the matrix

$$\begin{bmatrix} \|\psi\|_{L^1(\mu)} & 1 \\ \|\psi\|_{L^2(\mu)}^2 & \|\psi\|_{L^1(\mu)} \end{bmatrix}. \qquad (3.4.28)$$

Solving this equation leads to eigenvalues

$$\lambda = \|\psi\|_{L^1(\mu)} \pm \|\psi\|_{L^2(\mu)}, \qquad (3.4.29)$$

and the corresponding eigenvectors $\psi(x) \pm \|\psi\|_{L^2(\mu)}$. Clearly, the maximal eigenvalue equals $\lambda = \|\psi\|_{L^2(\mu)} + \|\psi\|_{L^1(\mu)}$, with corresponding $L^2(\mu)$-normalized eigenvector

$$\phi(x) = \frac{\psi(x) + \|\psi\|_{L^2(\mu)}}{2\big(\|\psi\|_{L^2(\mu)}^2 + \|\psi\|_{L^1(\mu)}\|\psi\|_{L^2(\mu)}\big)}. \qquad (3.4.30)$$

All other eigenvectors can be chosen to be orthogonal to ψ and 1, so that this corresponds to a rank-2 setting.

Unimodular Poisson Branching Processes and Notation

In what follows, we will often consider multi-type Poisson branching processes that start from the type distribution μ. Thus, we fix the root $\varnothing$ in the branching-process tree, and give it a random type Q satisfying $\mathbb{P}(Q \in \mathcal{A}) = \mu(\mathcal{A})$, for any measurable $\mathcal{A} \subseteq \mathcal{X}$. This corresponds to the *unimodular* setting, which is important in random graph settings. The idea is that the total number of vertices with types in $\mathcal{A}$ is close to $n\mu(\mathcal{A})$, so that if we pick a vertex uar, it has a type in $\mathcal{A}$ with asymptotic probability equal to $\mu(\mathcal{A})$. Recall Definition 1.26 for unimodular single-type branching processes.

We now introduce some helpful notation along the lines of that in Section 1.5. We let $\mathsf{BP}_{\leq r}$ denote the branching process up to and including generation r, where, for each individual v in the rth generation, we record its type as $Q(v)$. It is convenient to think of the branching-process tree, denoted as BP, as being labeled in the Ulam–Harris way (recall Section 1.5), so that a vertex v in generation r has a label $\varnothing a_1 \cdots a_r$, where $a_i \in \mathbb{N}$. When applied to BP, we denote this process by $(\mathsf{BP}(t))_{t \geq 1}$, where $\mathsf{BP}(t)$ consists of precisely $t+1$ vertices and their types (with $\mathsf{BP}(0)$ equal to the root $\varnothing$ and its type $Q(\varnothing)$). We recall Definitions 1.24 and 1.25 for details.

Monotone Approximations of Kernels

In what follows, we often approximate general kernels by kernels with finitely many types, as described in Proposition 3.5. For monotone sequences, we can prove the following convergence result:

Theorem 3.13 (Monotone approximations of multi-type Poisson branching processes) *Let (κ_n) be a sequence of kernels such that $\kappa_n(x,y) \nearrow \kappa(x,y)$. Let $\mathsf{BP}^{(n)}_{\leq r}$ denote the first r generations of the Poisson multi-type branching process with kernel κ_n and $\mathsf{BP}_{\leq r}$ that of the Poisson multi-type branching process with kernel κ. Then $\mathsf{BP}^{(n)}_{\leq r} \xrightarrow{d} \mathsf{BP}_{\leq r}$. Further, let $\zeta^{(n)}_{\geq k}(x)$ denote the probability that an individual of type x has at least k descendants. Then $\zeta^{(n)}_{\geq k}(x) \nearrow \zeta_{\geq k}(x)$.*

Proof Since $\kappa_n(x,y) \nearrow \kappa(x,y)$, we can write

$$\kappa(x,y) = \sum_{n \geq 1} \Delta \kappa_n(x,y), \quad \text{where} \quad \Delta \kappa_n(x,y) = \kappa_n(x,y) - \kappa_{n-1}(x,y). \quad (3.4.31)$$

We can represent this by a sum of independent Poisson multi-type processes with intensities $\Delta \kappa_n(x,y)$ and can associate a label n with each individual that arises from $\Delta \kappa_n(x,y)$. Then the branching process $\mathsf{BP}^{(n)}_{\leq r}$ is obtained by keeping all vertices with labels at most n, while $\mathsf{BP}_{\leq r}$ is obtained by keeping all vertices. Consequently, $\mathsf{BP}^{(n)}_{\leq r} \xrightarrow{d} \mathsf{BP}_{\leq r}$ follows since $\kappa_n \to \kappa$. Further, $1 - \zeta^{(n)}_{\geq k}(x) = \zeta^{(n)}_{<k}(x) = \mathbb{P}(|\mathsf{BP}^{(n)}_{\leq k}| \geq \bar{k})$, which thus also converges. $\quad\square$

3.5 LOCAL CONVERGENCE FOR INHOMOGENEOUS RANDOM GRAPHS

In this section, we prove the local convergence of $\mathrm{IRG}_n(\kappa_n)$ in general:

Theorem 3.14 (Local convergence of $\mathrm{IRG}_n(\kappa_n)$) *Assume that κ_n is an irreducible graphical kernel converging to some limiting kernel κ. Then $\mathrm{IRG}_n(\kappa_n)$ converges locally in probability to the unimodular multi-type marked branching-process tree $(G, o) \sim \rho$, where*

▷ *the root has type Q with distribution*

$$\rho(Q \in \mathcal{A}) = \mu(\mathcal{A}) \qquad \text{for all} \qquad \mathcal{A} \subseteq \mathcal{S}; \tag{3.5.1}$$

▷ *a vertex of type x independently has offspring distribution $\mathrm{Poi}(\lambda(x))$, with*

$$\lambda(x) = \int_{\mathcal{S}} \kappa(x, y)\mu(\mathrm{d}y), \tag{3.5.2}$$

and each of its offspring receives an independent type with distribution $Q(x)$ given by

$$\rho(Q(x) \in \mathcal{A}) = \frac{\int_{\mathcal{A}} \kappa(x, y)\mu(\mathrm{d}y)}{\int_{\mathcal{S}} \kappa(x, y)\mu(\mathrm{d}y)}. \tag{3.5.3}$$

The proof of Theorem 3.14 follows a familiar pattern: we first prove it for the finite-type case, and then use finite-type approximations to extend the proof to the infinite-type case. We use ρ for the law of the local limit rather than μ, as in Chapter 2, to avoid confusion with the limiting type measure μ appearing in the definition of $\mathrm{IRG}_n(\kappa_n)$.

3.5.1 LOCAL CONVERGENCE: FINITELY MANY TYPES

In order start the proof of (2.4.11) for Theorem 3.14 in the finite-type case, we introduce some notation. Fix a rooted ordered tree $(\mathbf{t}, \boldsymbol{q})$ of r generations, where the vertex $v \in V(\mathbf{t})$ has type $q(v)$ for all $v \in V(\mathbf{t})$. It will be convenient again to think of $\mathbf{t}$ as being labeled in the Ulam–Harris way, so that a vertex v in generation r has label $\varnothing a_1 \cdots a_r$, where $a_i \in \mathbb{N}$ (recall Section 1.5).

We also randomly order the vertices in $[n]$, and let $\bar{B}_r^{(G_n; Q)}(v)$ be the ordered version of $B_r^{(G_n; Q)}(v)$, where the neighbors of each vertex are placed in increasing order of their labels. Let

$$N_{n,r}(\mathbf{t}, \boldsymbol{q}) = \sum_{v \in [n]} \mathbb{1}_{\{\bar{B}_r^{(G_n; Q)}(v) = (\mathbf{t}, \boldsymbol{q})\}} \tag{3.5.4}$$

denote the number of vertices whose ordered local neighborhood up to generation r, including their types, equals $(\mathbf{t}, \boldsymbol{q})$. Here, in $\bar{B}_r^{(G_n; Q)}(v)$, we record the types of the vertices in $\bar{B}_r^{(G_n)}(v)$. Theorem 2.15 implies that in order to prove Theorem 3.14, we need to show that

$$\frac{N_{n,r}(\mathbf{t}, \boldsymbol{q})}{n} \xrightarrow{\mathbb{P}} \rho(\bar{B}_r^{(G, Q)}(o) = (\mathbf{t}, \boldsymbol{q})), \tag{3.5.5}$$

where $(\bar{B}_r^{(G, Q)}(o))_{r \geq 0}$ are the vertex-marked r-neighborhoods of the unimodular branching process $(G, o) \sim \rho$ described in Theorem 3.14, including the types of the tree vertices. Recall that these neighborhoods are ordered trees. This implies the convergence in probability of marked rooted graphs, discussed in Section 2.3.5, and the usual local convergence in probability of the (unmarked) neighborhood follows by summing over the types of the vertices in $\bar{B}_r^{(G_n)}(v)$. Let

$$N_{n,r}(\mathbf{t}) = \sum_{v\in[n]} \mathbb{1}_{\{\bar{B}_r^{(G_n)}(v)=\mathbf{t}\}}. \tag{3.5.6}$$

Then, indeed, since there is only a *finite* number of types, (3.5.5) also implies that $N_{n,r}(\mathbf{t})/n \xrightarrow{\mathbb{P}} \rho(\mathbf{t})$, where, with a slight abuse of notation, we write

$$\rho(\mathbf{t}) = \sum_q \rho(\bar{B}_r^{(G,Q)}(o) = (\mathbf{t},\mathbf{q})) = \rho(\bar{B}_r^{(G)}(o) = \mathbf{t}) \tag{3.5.7}$$

for the probability that the branching process produces a certain marked tree. We can then apply Theorem 2.15.

To prove (3.5.5), we follow the usual pattern of using a *second-moment method*. We first prove that the first moment satisfies $\mathbb{E}[N_{n,r}(\mathbf{t},\mathbf{q})]/n \to \rho(\bar{B}_r^{(G,Q)}(o) = (\mathbf{t},\mathbf{q}))$, after which we prove that $\mathrm{Var}(N_{n,r}(\mathbf{t},\mathbf{q})) = o(n^2)$. Then, (3.5.5) follows by the Chebychev inequality ([V1, Theorem 2.18]).

Local Convergence: First Moment

We start by noting that

$$\frac{1}{n}\mathbb{E}[N_{n,r}(\mathbf{t},\mathbf{q})] = \mathbb{P}(\bar{B}_r^{(G_n;Q)}(o_n) = (\mathbf{t},\mathbf{q})), \tag{3.5.8}$$

where $o_n \in [n]$ is a vertex chosen uar. Our aim is to prove that $\mathbb{P}(\bar{B}_r^{(G_n;Q)}(o_n) = (\mathbf{t},\mathbf{q})) \to \rho(\bar{B}_r^{(G,Q)}(o) = (\mathbf{t},\mathbf{q}))$.

Let us start with the branching process and analyze $\rho(\bar{B}_r^{(G,Q)}(o) = (\mathbf{t},\mathbf{q}))$. Fix a vertex $v \in V(\mathbf{t})$ of type $q(v)$. The probability of obtaining a sequence of d_v children of (ordered) types $(q(v1),\ldots,q(vd_v))$ equals

$$e^{-\lambda(q(v))}\frac{\lambda(q(v))^{d_v}}{d_v!}\prod_{j=1}^{d_v}\frac{\kappa(q(v),q(vj))\mu(q(vj))}{\lambda(q(v))}$$

$$= e^{-\lambda(q(v))}\frac{1}{d_v!}\prod_{j=1}^{d_v}\kappa(q(v),q(vj))\mu(q(vj)), \tag{3.5.9}$$

since we first draw a Poisson $\lambda(q(v))$ number of children, and then assign a type q to each of them with probability $\kappa(q(v),q)\mu(q)/\lambda(q(v))$. This is true independently for all $v \in V(\mathbf{t})$ with $|v| \le r-1$, so that

$$\rho(\bar{B}_r^{(G,Q)}(o) = (\mathbf{t},\mathbf{q})) = \prod_{v\in V(\mathbf{t}):\, |v|\le r-1} e^{-\lambda(q(v))}\frac{1}{d_v!}\prod_{j=1}^{d_v}\kappa(q(v),q(vj))\mu(q(vj)). \tag{3.5.10}$$

For a comparison with the graph exploration, it turns out to be convenient to rewrite this probability slightly. Let $\mathbf{t}_{\le r-1} = \{v\colon |v| \le r-1\}$ denote the vertices in the first $r-1$ generations of $\mathbf{t}$ and let $|\mathbf{t}_{\le r-1}|$ denote its size. We can order the elements of $\mathbf{t}_{\le r-1}$ in their lexicographic or Ulam–Harris ordering as $(v_i)_{i=1}^{|\mathbf{t}_{\leq r-1}|}$ (recall Definition 1.24 in Section 1.5). Then we can write

$$\rho(\bar{B}_r^{(G,Q)}(o) = (\mathbf{t}, \boldsymbol{q})) = \prod_{i=1}^{|\mathbf{t}_{\leq r-1}|} e^{-\lambda(q(v_i))} \frac{1}{d_{v_i}!} \prod_{j=1}^{d_{v_i}} \kappa(q(v_i), q(v_i j))\mu(q(v_i j)). \quad (3.5.11)$$

Let us now turn to $\mathrm{IRG}_n(\kappa_n)$. Fix a vertex $v \in [n]$ of type $q(v)$. Recall that n_q denotes the number of type q vertices. The probability of obtaining a sequence of d_v neighbors of (ordered) types $(q(v1), \ldots, q(vd_v))$ equals

$$\frac{1}{d_v!} \prod_{q \in \mathcal{S}} \left(1 - \frac{\kappa_n(q(v), q)}{n}\right)^{n_q - m_q} \prod_{j=1}^{d_v} \frac{\kappa_n(q(v), q(vj))}{n}[n_{qj} - m_{q(vj)}(j-1)], \quad (3.5.12)$$

where $m_q = \#\{i : q(vi) = q\}$ is the number of type-q vertices in $(q(v1), \ldots, q(vd_v))$ and $m_{q(vj)}(j) = \#\{i \leq j : q(vi) = q\}$ is the number of type-q vertices in $(q(v1), \ldots, q(vj))$. Here, the first factor, $1/d_v!$, arises since we are assigning an ordering on all vertices uar, the second factor, involving the product over $q \in \mathcal{S}$, since all other edges (except for the specified ones) need to be absent, and the third factor, involving the product over $j \in [d_v]$, specifies that the edges to vertices of the (ordered) sequence of types are present. When $n \to \infty$, $\kappa_n(q(v), q) \to \kappa(q(v), q)$ for every $q \in \mathcal{S}$ since $n_q/n \to \mu(q)$, so that

$$\frac{1}{d_v!} \prod_{q \in \mathcal{S}} \left(1 - \frac{\kappa_n(q(v), q)}{n}\right)^{n_q - m_q} \prod_{j=1}^{d_v} \kappa_n(q(v), q(vj))\frac{[n_{qj} - m_{q(vj)}(j-1)]}{n}$$

$$\to e^{-\lambda(q(v))} \frac{1}{d_v!} \prod_{i=1}^{d_v} \kappa(q(v), q(vj))\mu(q(vj)), \quad (3.5.13)$$

as required. The above computation, however, ignores the depletion-of-points effect, that fewer vertices participate in the course of the exploration.

To describe this, recall the lexicographic ordering of the elements in $\mathbf{t}_{\leq r-1}$ as $(v_i)_{i=1}^{|\mathbf{t}_{\leq r-1}|}$, and, for a type q, let $m_q(i) = \#\{j \in [i] : q(v_j) = q\}$ denote the number of type-q individuals in $(\mathbf{t}, \boldsymbol{q})$ encountered up to and including the ith exploration. Then

$$\mathbb{P}(\bar{B}_r^{(G_n;Q)}(o) = (\mathbf{t}, \boldsymbol{q})) = \prod_{i=1}^{|\mathbf{t}_{\leq r-1}|} \frac{1}{d_{v_i}!} \prod_{q \in \mathcal{S}} \left(1 - \frac{\kappa_n(q(v_i), q)}{n}\right)^{n_q - m_q(i-1)} \quad (3.5.14)$$

$$\times \prod_{j=1}^{d_{v_i}} \frac{\kappa_n(q(v_i), q(v_i j))}{n}[n_{qj} - m_{q(vj)}(i+j-1)].$$

As $n \to \infty$, this converges to the rhs of (3.5.11), as required. This completes the proof of (3.5.8), and thus the convergence of the first moment, which, in turn, implies local weak convergence. $\quad \square$

Local Convergence in Probability: Second Moment

Here, we study the second moment of $N_{n,r}(\mathbf{t}, \boldsymbol{q})$, and show that it is close to the first moment squared:

Lemma 3.15 (Concentration of the number of trees) *As* $n \to \infty$,

$$\frac{\text{Var}(N_{n,r}(\mathbf{t}, \mathbf{q})^2)}{n^2} \to 0. \tag{3.5.15}$$

Consequently, $N_{n,r}(\mathbf{t}, \mathbf{q})/n \overset{\mathbb{P}}{\longrightarrow} \rho(\bar{B}_r^{(G,Q)}(o) = (\mathbf{t}, \mathbf{q}))$.

Proof We start by computing

$$\frac{\mathbb{E}[N_{n,r}(\mathbf{t}, \mathbf{q})^2]}{n^2} = \mathbb{P}(\bar{B}_r^{(n,Q)}(o_1) = \bar{B}_r^{(n,Q)}(o_2) = (\mathbf{t}, \mathbf{q})), \tag{3.5.16}$$

where $o_1, o_2 \in [n]$ are two vertices chosen independently and uar from $[n]$.

By Corollary 2.20 and by local weak convergence, as proved above, $\text{dist}_{\text{IRG}_n(\kappa_n)}(o_1, o_2) > 2r$ whp for any r fixed. Thus,

$$\frac{\mathbb{E}[N_{n,r}(\mathbf{t}, \mathbf{q})^2]}{n^2} = \mathbb{P}(\bar{B}_r^{(G_n;Q)}(o_1), \bar{B}_r^{(G_n;Q)}(o_2) = (\mathbf{t}, \mathbf{q}), o_2 \notin B_{2r}^{(G_n)}(o_1)) + o(1). \tag{3.5.17}$$

We now condition on $\bar{B}_r^{(G_n;Q)}(o_1) \simeq (\mathbf{t}, \mathbf{q})$, and write

$$\mathbb{P}(\bar{B}_r^{(G_n;Q)}(o_1), \bar{B}_r^{(G_n;Q)}(o_2) = (\mathbf{t}, \mathbf{q}), o_2 \notin B_{2r}^{(G_n)}(o_1))$$
$$= \mathbb{P}(\bar{B}_r^{(G_n;Q)}(o_2) = (\mathbf{t}, \mathbf{q}) \mid \bar{B}_r^{(G_n;Q)}(o_1) = (\mathbf{t}, \mathbf{q}), o_2 \notin B_{2r}^{(G_n)}(o_1))$$
$$\times \mathbb{P}(\bar{B}_r^{(G_n;Q)}(o_1) = (\mathbf{t}, \mathbf{q}), o_2 \notin B_{2r}^{(G_n)}(o_1)). \tag{3.5.18}$$

We already know that $\mathbb{P}(\bar{B}_r^{(G_n;Q)}(o_1) = (\mathbf{t}, \mathbf{q})) \to \rho(\bar{B}_r^{(G,Q)}(o) = (\mathbf{t}, \mathbf{q}))$, so that also

$$\mathbb{P}(\bar{B}_r^{(G_n;Q)}(o_1) = (\mathbf{t}, \mathbf{q}), o_2 \notin B_{2r}^{(G_n)}(o_1)) \to \rho(\bar{B}_r^{(G,Q)}(o) = (\mathbf{t}, \mathbf{q})). \tag{3.5.19}$$

In Exercise 3.24, you can prove that (3.5.19) does indeed hold.

We next investigate the conditional probability given $\bar{B}_r^{(G_n;Q)}(o_1) = (\mathbf{t}, \mathbf{q})$ and $o_2 \notin B_{2r}^{(G_n)}(o_1)$, by noting that the probability that $\bar{B}_r^{(G_n;Q)}(o_2) = (\mathbf{t}, \mathbf{q})$ is the *same* as the probability that $\bar{B}_r^{(G_n;Q)}(o_2) = (\mathbf{t}, \mathbf{q})$ in $\text{IRG}_{n'}(\kappa_n)$, which is obtained by removing the vertices in $B_r^{(G_n;Q)}(o_1)$, as well as the edges from them, from $\text{IRG}_n(\kappa_n)$. We conclude that the resulting random graph has $n' = n - |V(\mathbf{t})|$ vertices, and $n'_q = n_q - m_q$ vertices of type $q \in [t]$, where m_q is the number of type-q vertices in $(\mathbf{t}, \mathbf{q})$. Further, $\kappa_{n'}(s, r) = \kappa_n(s, r)n'/n$. The whole point is that $\kappa_{n'}(s, r) \to \kappa(s, r)$ and $n'_q/n \to \mu(q)$ still hold. Therefore, we also have

$$\mathbb{P}(\bar{B}_r^{(G_n;Q)}(o_2) = (\mathbf{t}, \mathbf{q}) \mid B_r^{(G_n;Q)}(o_1) = (\mathbf{t}, \mathbf{q}), o_2 \notin B_{2r}^{(G_n)}(o_1))$$
$$\to \rho(\bar{B}_r^{(G,Q)}(o) = (\mathbf{t}, \mathbf{q})). \tag{3.5.20}$$

and we have proved that $\mathbb{E}[N_{n,r}(\mathbf{t}, \mathbf{q})^2]/n^2 \to \rho(\bar{B}_r^{(G,Q)}(o) = (\mathbf{t}, \mathbf{q}))^2$. From this, (3.5.15) follows directly since $\mathbb{E}[N_{n,r}(\mathbf{t}, \mathbf{q})]/n \to \rho(\bar{B}_r^{(G,Q)}(o) = (\mathbf{t}, \mathbf{q}))$. As a result, $N_{n,r}(\mathbf{t}, \mathbf{q})/n \overset{\mathbb{P}}{\longrightarrow} \rho(\bar{B}_r^{(G,Q)}(o) = (\mathbf{t}, \mathbf{q}))$, as required. $\square$

Lemma 3.15 completes the proof of Theorem 3.14 in the finite-type case. $\square$

3.5.2 LOCAL CONVERGENCE: INFINITELY MANY TYPES

We next extend the proof of Theorem 3.14 to the infinite-type case. We follow the strategy in Section 3.3.3.

Fix a general sequence of graphical kernels (κ_n). For $n \geq m$, again define $\underline{\kappa}_m$ by (3.3.19), so that $\kappa_n \geq \underline{\kappa}_m$. Couple the random graphs $\mathrm{IRG}_n(\underline{\kappa}_m)$ and $\mathrm{IRG}_n(\kappa_n)$ in such a way that $E(\mathrm{IRG}_n(\underline{\kappa}_m)) \subseteq E(\mathrm{IRG}_n(\kappa_n))$. Let $\varepsilon' > 0$ be given. Recall (3.3.24), which shows that we can take m so large that the bound

$$|E(\mathrm{IRG}_n(\underline{\kappa}_m))| \leq |E(\mathrm{IRG}_n(\kappa_n))| = \sum_{u \in [n]} (D_u - D_u^{(m)}) \leq \varepsilon' n \qquad (3.5.21)$$

holds whp. We let K denote the maximal degree in $\mathbf{t}$. Let $N_{n,r}^{(m)}(\mathbf{t}, \boldsymbol{q})$ denote $N_{n,r}(\mathbf{t}, \boldsymbol{q})$ for the kernel $\underline{\kappa}_m$ (and keep $N_{n,r}(\mathbf{t}, \boldsymbol{q})$ as in (3.5.4) for the kernel κ_n).

If a vertex v is such that $B_r^{(G_n;Q)}(v) \simeq (\mathbf{t}, \boldsymbol{q})$ in $\mathrm{IRG}_n(\underline{\kappa}_m)$, but not in $\mathrm{IRG}_n(\kappa_n)$, or vice versa, then one vertex in $B_{r-1}^{(G_n;Q)}(v)$ needs to have a different degree in $\mathrm{IRG}_n(\underline{\kappa}_m)$ from that in $\mathrm{IRG}_n(\kappa_n)$. Thus,

$$|N_{n,r}^{(m)}(\mathbf{t}, \boldsymbol{q}) - N_{n,r}(\mathbf{t}, \boldsymbol{q})|$$

$$\leq \sum_{u,v} \mathbb{1}_{\{u \in B_{r-1}^{(G_n;Q)}(v), \bar{B}_r^{(G_n;Q)}(v)=(\mathbf{t},\boldsymbol{q}) \text{ in } \mathrm{IRG}_n(\underline{\kappa}_m)\}} \mathbb{1}_{\{D_u \neq D_u^{(m)}\}}$$

$$+ \sum_{u,v} \mathbb{1}_{\{u \in B_{r-1}^{(G_n;Q)}(v), \bar{B}_r^{(G_n;Q)}(v)=(\mathbf{t},\boldsymbol{q}) \text{ in } \mathrm{IRG}_n(\kappa_n)\}} \mathbb{1}_{\{D_u^{(m)} \neq D_u\}}. \qquad (3.5.22)$$

Recall that the maximal degree of any vertex in $V(\mathbf{t})$ is K. Further, if $\bar{B}_r^{(G_n;Q)}(v) = (\mathbf{t}, \boldsymbol{q})$ and $u \in B_{r-1}^{(G_n;Q)}(v)$, then all the vertices on the path between u and v have degree at most K. Therefore,

$$\sum_v \mathbb{1}_{\{u \in B_{r-1}^{(G_n;Q)}(v), \bar{B}_r^{(G_n;Q)}(v)=(\mathbf{t},\boldsymbol{q}) \text{ in } \mathrm{IRG}_n(\underline{\kappa}_m)\}} \leq \sum_{\ell \leq r-1} K^\ell \leq \frac{K^r - 1}{K - 1}, \qquad (3.5.23)$$

and, in the same way,

$$\sum_v \mathbb{1}_{\{u \in B_{r-1}^{(G_n;Q)}(v), \bar{B}_r^{(G_n;Q)}(v)=(\mathbf{t},\boldsymbol{q}) \text{ in } \mathrm{IRG}_n(\kappa_n)\}} \leq \frac{K^r - 1}{K - 1}. \qquad (3.5.24)$$

We thus conclude that whp

$$|N_{n,r}^{(m)}(\mathbf{t}, \boldsymbol{q}) - N_{n,r}(\mathbf{t}, \boldsymbol{q})| \leq 2\frac{K^r - 1}{K - 1} \sum_{u \in [n]} \mathbb{1}_{\{D_u^{(m)} \neq D_u\}}$$

$$\leq 2\frac{K^r - 1}{K - 1} \sum_{u \in [n]} (D_u^{(m)} - D_u) \leq 2\frac{K^r - 1}{K - 1} \varepsilon' n. \qquad (3.5.25)$$

Taking $\varepsilon' = \varepsilon(K - 1)/[2(K^r - 1)]$, we thus obtain that whp

$$|N_{n,r}^{(m)}(\mathbf{t}, \boldsymbol{q}) - N_{n,r}(\mathbf{t}, \boldsymbol{q})| \leq \varepsilon n, \qquad (3.5.26)$$

as required.

For $N_{n,r}^{(m)}(\mathbf{t}, \mathbf{q})$, we can use Theorem 3.14 in the finite-type case to obtain

$$\frac{1}{n} N_{n,r}^{(m)}(\mathbf{t}, \mathbf{q}) \overset{\mathbb{P}}{\longrightarrow} \rho(\bar{B}_r^{(G,Q,m)}(o) = (\mathbf{t}, \mathbf{q})). \tag{3.5.27}$$

The fact that m can be taken so large that

$$|\rho(\bar{B}_r^{(G,Q,m)}(o) = (\mathbf{t}, \mathbf{q})) - \rho(\bar{B}_r^{(G,Q)}(o) = (\mathbf{t}, \mathbf{q}))| \leq \varepsilon \tag{3.5.28}$$

follows from Theorem 3.13. $\qquad\square$

3.5.3 COMPARISON WITH BRANCHING PROCESSES

In this subsection we describe a beautiful comparison of the neighborhoods of a uniformly chosen vertex in rank-1 inhomogeneous random graphs, such as the generalized random graph, the Chung–Lu model, or the Norros–Reittu model, with a marked branching process. This comparison is particularly pretty when one considers the Norros–Reittu model, in which such neighborhoods are stochastically dominated by a mixed-Poisson branching process. We start by describing the result for the rank-1 setting, after which we extend it to kernels with finitely many types. Recall the rank-1 Poisson branching processes defined in Section 3.4.3, and recall that there such branching processes were shown to be equivalent to one-type mixed-Poisson branching processes.

Stochastic Domination of Connected Components by a Branching Process

We dominate the connected component of a vertex in the Norros–Reittu model by the total progeny of a unimodular branching processes with mixed-Poisson offspring. This domination allows us to control differences, and makes the heuristic argument after Theorem 3.20 below precise.

Define the *mark random variable* or *mark* to be the random variable M with distribution

$$\mathbb{P}(M = m) = w_m/\ell_n, \qquad m \in [n]. \tag{3.5.29}$$

Let $(X_v)_v$ be a collection of independent random variables, where:

(a) the number of children of the root $X_\varnothing$ has a mixed-Poisson distribution with random parameter $w_{M_\varnothing}$, where $M_\varnothing$ is uniformly chosen in $[n]$;

(b) the number of children X_v of a node v in the Ulam–Harris tree has a mixed-Poisson distribution with random parameter w_{M_v}, where $(M_v)_{v \neq \varnothing}$ are iid random marks with distribution (3.5.29) that are independent of $M_\varnothing$.

We call $(X_v, M_v)_v$ a *marked mixed-Poisson branching process* (MMPBP).

Clearly, $w_{M_\varnothing}$ has the same distribution as W_n defined in (1.3.10), while the distribution of w_{M_v} for each v with $|v| \geq 1$ is iid with distribution w_M given by

$$\mathbb{P}(w_M \leq x) = \sum_{m=1}^{n} \mathbb{1}_{\{w_m \leq x\}} \mathbb{P}(M = m) = \frac{1}{\ell_n} \sum_{m=1}^{n} w_m \mathbb{1}_{\{w_m \leq x\}}$$

$$= \mathbb{P}(W_n^\star \leq x) = F_n^\star(x), \tag{3.5.30}$$

where $W_n^\star$ is the *size-biased version* of W_n.

When we are interested only in the *numbers of individuals*, we then obtain a unimodular branching process since the random variables $(X_v)_v$ are independent, and the random variables $(X_v)_{v \neq \varnothing}$ are iid (see Exercise 3.25). However, in what follows, we make explicit use of the marks $(M_v)_{v \neq \varnothing}$, as the complete information $(X_v, M_v)_v$ gives us a way to retrieve the connected component of the vertex $M_\varnothing$, something that would not be possible on the basis of $(X_v)_v$ only.

In order to define the neighborhood exploration in $\mathrm{NR}_n(\boldsymbol{w})$, we introduce a *thinning procedure* that guarantees that we inspect a vertex only once. We think of M_v as being the vertex label in $\mathrm{NR}_n(\boldsymbol{w})$ of the node v in the Ulam–Harris tree, and $X_v = \mathrm{Poi}(w_{M_v})$ as its *potential number of children*. These potential children effectively become children when their marks correspond to vertices in $\mathrm{NR}_n(\boldsymbol{w})$ that have not yet appeared in the exploration. The thinning ensures this. To describe the thinning, we set

(a) $\varnothing$ unthinned; and,

(b) for v with $v \neq \varnothing$, we thin v when either (i) one tree vertex on the (unique) path between the root $\varnothing$ and v has been thinned; or (ii) when $M_v = M_{v'}$ for some unthinned vertex $v' < v$. Here we recall the breadth-first ordering in the Ulam–Harris tree from Definition 1.24 in Section 1.5.

We now make the connection between the thinned marked mixed-Poisson branching process and neighborhood exploration precise:

Proposition 3.16 (Connected components as thinned marked branching processes) *The connected component of a uniformly chosen vertex $\mathscr{C}(o)$ is equal in distribution to the set of vertices $\{M_v \colon v \text{ unthinned}\} \subseteq [n]$ and the edges between them inherited by the marked mixed-Poisson branching process $(X_v, M_v)_v$. Here, $\{M_v \colon v \text{ unthinned}\}$ consists of the marks of unthinned tree nodes encountered in the marked mixed-Poisson branching process up to the end of the exploration. Consequently, the set of vertices at graph distance r from o has the same distribution as*

$$\Big(\{M_v \colon v \text{ unthinned}, |v| = r\}\Big)_{r \geq 0}. \tag{3.5.31}$$

Proof   We prove the two statements simultaneously. By construction, the distribution of o is the same as that of $M_\varnothing$, the mark of the root of the marked mixed-Poisson branching process. We continue by proving that the direct neighbors of the root $\varnothing$ agree in both constructions. In $\mathrm{NR}_n(\boldsymbol{w})$, conditional on $M_\varnothing = l$, the direct neighbors are equal to $\{j \in [n] \setminus \{l\} \colon I_{lj} = 1\}$, where $(I_{lj})_{j \in [n] \setminus \{l\}}$ are independent $\mathrm{Be}(p_{lj})$ random variables with $p_{lj} = 1 - \mathrm{e}^{-w_l w_j / \ell_n}$.

We now prove that the same is true for the marked mixed-Poisson branching process. Conditional on $M_\varnothing = l$, the root has a $\mathrm{Poi}(w_l)$ number of children, where these $\mathrm{Poi}(w_l)$ offspring receive iid marks. We make use of the fundamental "thinning" property of the Poisson distribution in Lemma 3.11. By Lemma 3.11, the random vector $(X_{\varnothing,j})_{j \in [n]}$, where $X_{\varnothing,j}$ is the number of offspring of the root that receive mark j, is a vector of *independent* Poisson random variables with parameters $w_l w_j / \ell_n$. Owing to the thinning, a mark occurs precisely when $X_{\varnothing,j} \geq 1$. Therefore, the mark j occurs, independently for all $j \in [n]$, with probability

$$\mathbb{P}(X_{\varnothing,j} \geq 1) = 1 - \mathbb{P}(X_{\varnothing,j} = 0) = 1 - \mathrm{e}^{-w_l w_j / \ell_n} = p_{jl}^{(\mathrm{NR})}. \tag{3.5.32}$$

This proves that the set of marks of the children of the root in the MMPBD has the same distribution as the set of neighbors of the chosen vertex in $\mathrm{NR}_n(\boldsymbol{w})$.

Next, we look at the number of new elements of $\mathscr{C}(o)$ neighboring a vertex found in the exploration. Fix one such vertex, and let its tree label be $v = \varnothing v_1$. First, condition on $M_v = l$, and assume that v is not thinned. Conditional on $M_v = l$, the number of children of v in the MMPBP has distribution $\mathrm{Poi}(w_l)$. Each of these $\mathrm{Poi}(w_l)$ children receives an iid mark. Let $X_{v,j}$ denote the number of children of v that receive mark j.

By Lemma 3.11, $(X_{v,j})_{j\in[n]}$ is again a vector of independent Poisson random variables with parameters $w_l w_j / \ell_n$. Owing to the thinning, a mark appears within the offspring of individual v precisely when $X_{v,j} \geq 1$, and these events are independent. In particular, for each j that has not appeared as the mark of an unthinned vertex, the probability that it occurs as the child of a vertex having mark l equals $1 - \mathrm{e}^{-w_j w_l / \ell_n} = p_{lj}^{(\mathrm{NR})}$, as required. $\square$

Stochastic Domination by Branching Processes: Finite-Type Case

The rank-1 setting described above is special, since the marks of vertices in the tree are independent random variables in that they do not depend on the type of their parent. However, this is not true in general. We next describe how the result can be generalized. We restrict to the finite-type case for convenience. Further, we let the edge probabilities of our random graph be given by

$$p_{uv} = p_{uv}^{(\mathrm{NR})} = 1 - \mathrm{e}^{-\kappa_n(s_u, s_v)/n}, \tag{3.5.33}$$

where $s_u \in [t]$ is the type of vertex $u \in [n]$ and $\mathcal{S} = [t]$ is the collection of types.

Let us introduce some notation. Recall that n_s denotes the number of vertices of type $s \in [t]$, and write $n_{\leq s} = \sum_{r \leq s} n_r$. Define the intervals $I_s = [n_{\leq s}] \setminus [n_{\leq s-1}]$ (where, by convention, I_0 is the empty set). We note that all vertices in the intervals I_s play the same role, and this is used crucially in the coupling that we present below.

We now describe the cluster exploration of a uniformly chosen vertex $o \in [n]$, which has type s with probability $\mu_n(s) = n_s/n$. To define the cluster of o, as well as the types of the vertices in it, we define the *mark distribution* of a tree vertex of type r to be the random variable $M(r)$ with distribution

$$\mathbb{P}(M(r) = \ell) = \frac{1}{n_r}, \qquad \ell \in I_r. \tag{3.5.34}$$

Let $(X_v, T_v, M_v)_v$ be a collection of random variables, where:

(a) the root $\varnothing$ has type s with probability $\mu_n(s) = n_s/n$, and, given the type s of the root, the number of children $X_\varnothing$ of the root has a mixed-Poisson distribution with random parameter $\lambda_n(s) = \sum_{r\in[t]} \kappa_n(s,r)\mu_n(r)$, where each child v with $|v| = 1$ of $\varnothing$ independently receives a type T_v, where $T_v = r$ with probability $\kappa_n(s,r)\mu_n(r)/\lambda_n(s)$;

(b) given its type s, the number of children X_v of a tree vertex v has a mixed-Poisson distribution with parameter $\lambda_n(s) = \sum_{r\in[t]} \kappa_n(s,r)\mu_n(r)$, and child vj with $j \geq 1$ of v receives a type T_{vj}, where $T_{vj} = r$ with probability $\kappa_n(s,r)\mu_n(r)/\lambda_n(s)$;

(c) given that a tree vertex v has type r, it receives an independent mark $M_v(r)$ with distribution in (3.5.34).

We call $(X_v, T_v, M_v)_v$ a *marked multi-type Poisson branching process*. Then, the following extension of Proposition 3.16 holds:

Proposition 3.17 (Connected components as thinned marked multi-type branching processes) *The connected component of a vertex $\mathscr{C}(v)$ of type s is equal in distribution to the set of vertices $\{M_v : v\ unthinned\} \subseteq [n]$ and the edges between them inherited by $(X_v, T_v, M_v)_v$. Here, $\{M_v : v\ unthinned\}$ are the marks of unthinned tree vertices encountered in the marked multi-type Poisson branching process up to the end of the exploration. Similarly, the set of vertices at graph distance r from o has the same distribution as*

$$\left(\{M_v : v\ unthinned, |v| = r\}\right)_{r \geq 0}. \tag{3.5.35}$$

The reader is asked to prove Proposition 3.17 in Exercise 3.28.

3.5.4 Local Convergence of Generalized Random Graphs

We close this section by investigating the locally tree-like nature of the generalized random graph. Our main result is as follows:

Theorem 3.18 (Locally tree-like nature of $\mathrm{GRG}_n(w)$) *Assume that Conditions 1.1(a),(b) hold. Then $\mathrm{GRG}_n(w)$ converges locally in probability to the unimodular branching-process tree, with offspring distribution $(p_k)_{k \geq 0}$ given by*

$$p_k = \mathbb{P}(D = k) = \mathbb{E}\left[e^{-W} \frac{W^k}{k!}\right]. \tag{3.5.36}$$

This result also applies to $\mathrm{NR}_n(w)$ and $\mathrm{CL}_n(w)$ under the same conditions.

Theorem 3.18 follows directly from Theorem 3.14. However, we give an alternative proof that relies on the locally tree-like nature of $\mathrm{CM}_n(d)$ proved in Theorem 4.1 below and the relation between $\mathrm{GRG}_n(w)$ and $\mathrm{CM}_n(d)$ discussed in Section 1.3 and Theorem 1.9. This approach is interesting in itself, since it allows for general proofs for $\mathrm{GRG}_n(w)$ by proving the result first for $\mathrm{CM}_n(d)$, and then merely extending it to $\mathrm{GRG}_n(w)$. We frequently rely on such a proof strategy for $\mathrm{GRG}_n(w)$.

3.6 Phase Transition for Inhomogeneous Random Graphs

In this section we discuss the phase transition in $\mathrm{IRG}_n(\kappa_n)$. The main result shows that there is a giant component when the associated multi-type branching process is supercritical (recall Definition 3.12), while otherwise there is not:

Theorem 3.19 (Giant component of IRG) *Let (κ_n) be a sequence of irreducible graphical kernels with limit κ, and let $\mathscr{C}_{\max}$ and $\mathscr{C}_{(2)}$ denote the two largest connected components of $\mathrm{IRG}_n(\kappa_n)$ (breaking ties arbitrarily). Then,*

$$|\mathscr{C}_{\max}|/n \overset{\mathbb{P}}{\longrightarrow} \zeta_\kappa, \tag{3.6.1}$$

and $|\mathscr{C}_{(2)}|/n \overset{\mathbb{P}}{\longrightarrow} 0$. In all cases $\zeta_\kappa < 1$, while $\zeta_\kappa > 0$ precisely when $\|T_\kappa\| > 1$.

Theorem 3.19 is a generalization of the law of large numbers for the largest connected component in [V1, Theorem 4.8] for $\mathrm{ER}_n(\lambda/n)$ (see Exercise 3.31); recall also Theorem 2.34.

We do not give a complete proof of Theorem 3.19 in this chapter. The upper bound follows directly from the local convergence in Theorem 3.14, together with Corollary 2.27. For the lower bound, it suffices to prove this for kernels with finitely many types, by Proposition 3.5. This proof is deferred to Section 6.5.3 in Chapter 6. We close this section by discussing a few examples of Theorem 3.19.

Bipartite Random Graph

We let n be even and take $\mathcal{S} = \{1, 2\}$ and

$$\kappa_n(s, r) = \kappa(s, r) = \lambda \mathbb{1}_{\{s \neq r\}}/2. \tag{3.6.2}$$

Thus, for $u < v$, the edge probabilities p_{uv} given by (3.2.6) are equal to $\lambda/(2n)$ (for $2n > \lambda$) when $u \in [n/2]$ and $v \in [n] \setminus [n/2]$, and equal 0 otherwise.

In this case, $\|T_\kappa\| = \lambda$ with corresponding eigenfunction $f(s) = 1$ for all $s \in \mathcal{S}$. Thus, Theorem 3.19 proves that there is a phase transition at $\lambda = 2$. Furthermore, the function $\zeta_\lambda(s)$ reduces to the single value $\zeta_{\lambda/2}$, which is the survival probability of a Poisson branching process with mean offspring $\lambda/2$. This is not surprising, since the degree of each vertex is $\mathrm{Bin}(n/2, \lambda/n)$, so the bipartite random graph of size n is, in terms of its *local structure*, closely related to the Erdős–Rényi random graph of size $n/2$.

Finite-Type Case

The bipartite random graph can also be viewed as a random graph with two types of vertices (i.e., the vertices $[n/2]$ and $[n] \setminus [n/2]$). We now generalize our results to the finite-type case, in which we have seen that κ_n is equivalent to a $t \times t$ matrix $(\kappa_n(s, r))_{s,r \in [t]}$, where t denotes the number of types. In this case, $\mathrm{IRG}_n(\kappa_n)$ has vertices of t different types (or colors), say n_s vertices of type s, where two vertices of type s and r are joined by an edge with probability $n^{-1}\kappa_n(s, r) \wedge 1$. Exercises 3.29 and 3.30 investigate the phase transition in the finite-type case.

Random Graph with Prescribed Expected Degrees

We next consider the Chung–Lu model or expected-degree random graph, where κ_n is given by (3.2.10), i.e., $\kappa_n(u/n, v/n) = w_u w_v / \mathbb{E}[W_n]$ for all $u, v \in [n]$ with $u \neq v$.

We first assume that Conditions 1.1(a)–(c) hold, so that in particular $\mathbb{E}[W^2] < \infty$, where W has distribution function F. A particular instance of this case is the choice $w_v = [1 - F]^{-1}(v/n)$ for $v \in [n]$ in (1.3.15) when $W \sim F$ and $\mathbb{E}[W] < \infty$. In this case, the sequence (κ_n) converges to κ, where the limit κ is given by (recall (3.2.10))

$$\kappa(x, y) = \psi(x)\psi(y)/\mathbb{E}[W], \tag{3.6.3}$$

where $\psi(x) = [1 - F]^{-1}(x)$. We have already obtained $\|T_\kappa\| = \|\psi\|^2 / \int_\mathcal{S} \psi(x)\mu(dx) = \mathbb{E}[W^2]/\mathbb{E}[W]$. Thus,

$$\|T_\kappa\| = \mathbb{E}[W^2]/\mathbb{E}[W]. \tag{3.6.4}$$

In the case where $\mathbb{E}[W^2] = \infty$, on the other hand, we take $f_\varepsilon(x) = c\psi(x)\mathbb{1}_{\{x \in [\varepsilon, 1]\}}$, where $c = c_\varepsilon$ is such that $\|f_\varepsilon\| = 1$. Then, $\|T_\kappa f_\varepsilon\| \to \infty$, so that $\|T_\kappa\| = \infty$, and $\mathrm{CL}_n(\boldsymbol{w})$ is always supercritical in this regime.

General Rank-1 Setting

Theorem 3.19 identifies the phase transition in $\mathrm{IRG}_n(\kappa_n)$. For the rank-1 setting, we prove some quite strong results, as we now discuss. We denote the *complexity* of a connected component $\mathscr{C}$ by $E(\mathscr{C}) - V(\mathscr{C}) + 1$, which equals the number of edges that need to be removed to turn $\mathscr{C}$ into a tree. We further recall that $v_k(\mathscr{C})$ denotes the number of vertices of degree k in the connected component $\mathscr{C}$. Our main result is as follows:

Theorem 3.20 (Phase transition in generalized random graphs) *Suppose that Conditions 1.1(a),(b) hold and consider the random graphs* $\mathrm{GRG}_n(\boldsymbol{w}), \mathrm{CL}_n(\boldsymbol{w})$ *or* $\mathrm{NR}_n(\boldsymbol{w})$, *letting* $n \to \infty$. *Denote* $p_k = \mathbb{P}(\mathrm{Poi}(W) = k)$ *as defined below* (1.3.22). *Let* $\mathscr{C}_{\max}$ *and* $\mathscr{C}_{(2)}$ *be the largest and second largest components of* $\mathrm{GRG}_n(\boldsymbol{w}), \mathrm{CL}_n(\boldsymbol{w})$, *or* $\mathrm{NR}_n(\boldsymbol{w})$.

(a) *If* $\nu = \mathbb{E}[W^2]/\mathbb{E}[W] > 1$, *then there exist* $\xi \in (0,1), \zeta \in (0,1)$ *such that*

$$|\mathscr{C}_{\max}|/n \xrightarrow{\mathbb{P}} \zeta,$$
$$v_k(\mathscr{C}_{\max})/n \xrightarrow{\mathbb{P}} p_k(1 - \xi^k), \text{ for every } k \geq 0,$$
$$|E(\mathscr{C}_{\max})|/n \xrightarrow{\mathbb{P}} \tfrac{1}{2}\mathbb{E}[W](1 - \xi^2).$$

while $|\mathscr{C}_{(2)}|/n \xrightarrow{\mathbb{P}} 0$ *and* $|E(\mathscr{C}_{(2)})|/n \xrightarrow{\mathbb{P}} 0$. *Further,* $\tfrac{1}{2}\mathbb{E}[W](1 - \xi^2) > \zeta$, *so that the complexity of the giant is linear.*

(b) *If* $\nu = \mathbb{E}[W^2]/\mathbb{E}[W] \leq 1$, *then* $|\mathscr{C}_{\max}|/n \xrightarrow{\mathbb{P}} 0$ *and* $|E(\mathscr{C}_{\max})|/n \xrightarrow{\mathbb{P}} 0$.

The proof of Theorem 3.20, except for the proof of the linear complexity, is deferred to Section 4.3.2 in Chapter 4, where a similar result is proved for the configuration model. By the strong relation between the configuration model and the generalized random graph (recall Theorem 1.9), this result can be seen to imply Theorem 3.20.

Let us discuss some implications of Theorem 3.20, focussing on the supercritical case where $\nu = \mathbb{E}[W^2]/\mathbb{E}[W] > 1$. In this case, the parameter ξ is the extinction probability of a branching process with offspring distribution $p_k^\star = \mathbb{P}(\mathrm{Poi}(W^\star) = k)$, where $W^\star$ is the size-biased version of W. Thus,

$$\xi = G_{\mathrm{Poi}(W^\star)}(\xi) = \mathbb{E}[e^{W^\star(\xi-1)}], \tag{3.6.5}$$

where $G_{\mathrm{Poi}(W^\star)}(s) = \mathbb{E}[s^{\mathrm{Poi}(W^\star)}]$ is the probability generating function of a mixed-Poisson random variable with mixing distribution $W^\star$.

Further, since $v_k(\mathscr{C}_{\max})/n \xrightarrow{\mathbb{P}} p_k(1 - \xi^k)$ and $|\mathscr{C}_{\max}|/n \xrightarrow{\mathbb{P}} \zeta$, it must hold that

$$\zeta = \sum_{k \geq 0} p_k(1 - \xi^k) = 1 - G_D(\xi), \tag{3.6.6}$$

where $G_D(s) = \mathbb{E}[s^D]$ is the probability generating function of $D = \mathrm{Poi}(W)$. We also note that $|E(\mathscr{C}_{\max})|/n \xrightarrow{\mathbb{P}} \eta$, and compute

$$\eta = \tfrac{1}{2} \sum_{k \geq 0} k p_k (1 - \xi^k) = \tfrac{1}{2} \mathbb{E}[W] \sum_{k \geq 0} \frac{k p_k}{\mathbb{E}[W]} (1 - \xi^k)$$

$$= \tfrac{1}{2} \mathbb{E}[W] \big(1 - \xi G_{\mathrm{Poi}(W^\star)}(\xi)\big) = \tfrac{1}{2} \mathbb{E}[W] (1 - \xi^2), \tag{3.6.7}$$

as required.

We now compare the limiting total number of edges with $|\mathscr{C}_{\max}|$. Recall the useful correlation inequality in [V1, Lemma 2.14] that states that $\mathbb{E}[f(X) g(X)] \geq \mathbb{E}[f(X)] \mathbb{E}[g(X)]$ for any non-decreasing functions f and g and random variable X. Applying this to $f(k) = k$ and $g(k) = 1 - \xi^k$, which are both increasing, leads to

$$\sum_{k \geq 0} k p_k (1 - \xi^k) > \sum_{k \geq 0} k p_k \sum_{k \geq 0} (1 - \xi^k) p_k = \mathbb{E}[W] \zeta. \tag{3.6.8}$$

As a result, by (3.6.7),

$$\eta = \tfrac{1}{2} \sum_{k \geq 0} k p_k (1 - \xi^k) > \tfrac{1}{2} \mathbb{E}[W] \zeta. \tag{3.6.9}$$

Thus, the average degree η / ζ in the giant component is strictly larger than the average degree in the entire graph $\mathbb{E}[W]/2$.

We finally show that $\eta > \zeta$, so that the giant has linear complexity. By convexity of $x \mapsto x^{k-1}$ and the fact that $\xi < 1$, for $k \geq 1$, we have

$$\sum_{i=0}^{k-1} \xi^i \leq k(1 + \xi^{k-1})/2, \tag{3.6.10}$$

with strict inequality for $k \geq 3$. Multiply by $1 - \xi$ to obtain

$$1 - \xi^k \leq k(1 - \xi)(1 + \xi^{k-1})/2, \tag{3.6.11}$$

again for every $k \geq 1$, again with strict inequality for $k \geq 3$. Now multiply by p_k, and sum to get, since $p_k > 0$ for all $k > 3$,

$$\sum_k p_k (1 - \xi^k) < (1 - \xi) \sum_k k p_k (1 + \xi^{k-1})/2. \tag{3.6.12}$$

The lhs of (3.6.12) equals ζ by (3.6.6). We next investigate the rhs of (3.6.12). Recall that

$$\sum_k k p_k = \mathbb{E}[W], \tag{3.6.13}$$

and, by (3.6.5),

$$\sum_k \frac{k p_k}{\mathbb{E}[W]} \xi^{k-1} = \xi. \tag{3.6.14}$$

Hence, the rhs of (3.6.12) is

$$(1 - \xi)(\mathbb{E}[W] + \mathbb{E}[W]\xi)/2 = \mathbb{E}[W](1 - \xi^2)/2 = \eta, \tag{3.6.15}$$

which is the limit in probability of $|E(\mathscr{C}_{\max})|/n$. Thus, $\zeta < \eta$. $\qquad\square$

Attack Vulnerability of $\mathrm{CL}_n(\boldsymbol{w})$

Suppose an adversary attacks a network by removing some of its vertices. A *clever* adversary would do this in a smart way; this is often referred to as a *deliberate* attack. On the other hand, the vertices might also be exposed to random failures, which is often referred to as a *random* attack. The results as stated above do not specifically apply to these settings, but they do have intuitive consequences.

We model a deliberate attack as the removal of a proportion of the vertices with highest weights, whereas a random attack is modeled by random removal of the vertices independently with a given fixed probability. One of our aims is to quantify the effect of such attacks, and in particular the difference between random and deliberate attacks. We denote the proportion of remaining vertices by p. We always assume that $\nu > 1$, so that a giant component exists before the attack, and we investigate under what conditions on p and the graph $\mathrm{CL}_n(\boldsymbol{w})$, the giant component remains in existence.

We start by addressing the case of random attack for the $\mathrm{CL}_n(\boldsymbol{w})$ model under Conditions 1.1(a)–(c), where $\mathbb{E}[W^2] < \infty$. One of the difficulties of the above set-up is that we remove *vertices* rather than *edges*, so that the resulting graph is no longer an IRG. In percolation jargon, we are dealing with *site* percolation rather than with bond percolation. We will start by relating the obtained graph to an inhomogeneous random graph.

Note that when we *explore* a connected component of a vertex after a random attack, the vertex may not have been affected by the attack, which has probability p. After this, in the exploration, we always inspect an edge between a vertex that is unaffected by the attack and a vertex of which we do not yet know whether it has been attacked or not. As a result, for random attacks, the probability that it is affected equals p independently of the past randomness. Therefore, it is similar to the random graph where p_{uv} is replaced by $p \times p_{uv}$.

For a branching process, this identification is exact, and we have that $\zeta_{\kappa,p} = p\zeta_{p\kappa}$, where $\zeta_{\kappa,p}$ denotes the survival probability of the unimodular multi-type marked branching-process tree in Theorem 3.14, where additionally each individual in the tree is killed with probability $1 - p$ independently of all other randomness. For $\mathrm{CL}_n(\boldsymbol{w})$, this equality is only asymptotic. In the case where $\mathbb{E}[W^2] < \infty$, so that $\nu < \infty$, this means that there exists a critical value $p_c = 1/\nu$, such that, if $p > p_c$, the giant component persists in $\mathrm{CL}_n(\boldsymbol{w})$, where vertices are removed with probability $1 - p$, while the giant component is destroyed for $p \leq p_c$. Thus, when $\mathbb{E}[W^2] < \infty$, $\mathrm{CL}_n(\boldsymbol{w})$ is *sensitive to random attacks*. When $\mathbb{E}[W^2] = \infty$, on the other hand, $\nu = \infty$ also, so that the giant component persists for *every* $p \in [0, 1)$, and the graph is called *robust to random attacks*. Here we must note that the *size* of the giant component does decrease, since $\zeta_{\kappa,p} < p\zeta_\kappa < \zeta_\kappa$!

To mimic a deliberate attack, we remove a proportion p of vertices with highest weight. For convenience, we assume that $\boldsymbol{w} = (w_1, \ldots, w_n)$ is non-increasing. Then, removing a proportion p of the vertices with highest weight means that $\boldsymbol{w}$ is replaced with $\boldsymbol{w}(p)$, which is equal to $w_v(p) = w_v \mathbb{1}_{\{v > np\}}$, and we denote the resulting edge probabilities by

$$p_{uv}(p) = \max\{1, w_u(p)w_v(p)/\ell_n\}. \tag{3.6.16}$$

In this case, the resulting graph on $[n] \setminus [n(1 - p)]$ is again a Chung–Lu model for which ν is replaced with $\nu(p)$, given by

$$\nu(p) = \mathbb{E}[\psi(U)^2 \mathbb{1}_{\{U > 1-p\}}]/\mathbb{E}[W], \tag{3.6.17}$$

where U is uniform on $[0, 1]$ and we recall that we have written $\psi(u) = [1 - F]^{-1}(u)$. Now, for any distribution function F, $\mathbb{E}[[1 - F]^{-1}(U)^2 \mathbf{1}_{\{U > p\}}] < \infty$, so that, for p sufficiently close to 0, $\nu(p) < 1$ (see Exercise 3.39). Thus, the $\mathrm{CL}_n(\boldsymbol{w})$ model is always sensitive to deliberate attacks.

Phase Transitions in Uniformly Grown Random Graphs and for Sum Kernels

Recall the definition of the uniformly grown random graph in (3.2.12). A vertex v is connected independently with all $u \in [v - 1]$ with probability $p_{uv} = \lambda/v$. This leads to an inhomogeneous random graph with type space $\mathcal{S} = [0, 1]$ and limiting kernel $\kappa(x, y) = \lambda/(x \vee y)$. It is non-trivial to compute $\|T_\kappa\|$, but remarkably this can be done, to yield $\|T_\kappa\| = 4\lambda$, so that a giant exists for all $\lambda > \frac{1}{4}$. We do not give the proof of $\|T_\kappa\| = 4\lambda$ here, and refer to Exercise 3.40 for details. Exercise 3.41 investigates when there is a giant for sum kernels, as in (3.2.13).

3.7 RELATED RESULTS FOR INHOMOGENEOUS RANDOM GRAPHS

In this section, we discuss some related results for inhomogeneous random graphs. While we give intuition about their proofs, we do not include them in full detail.

Largest Subcritical Cluster

For the classical $\mathrm{ER}_n(\lambda/n)$, it is well known that the stronger bound $|\mathscr{C}_{\max}| = \Theta_{\mathbb{P}}(\log n)$ holds in the subcritical case for which $\lambda < 1$ (see [V1, Theorems 4.4 and 4.5]), and that $|\mathscr{C}_{(2)}| = \Theta_{\mathbb{P}}(\log n)$ in the supercritical case for which $\lambda > 1$. These bounds do not always hold in the general framework considered here, but if we add some conditions then we can improve the estimates in Theorem 3.19 for the subcritical case to $O_{\mathbb{P}}(\log n)$:

Theorem 3.21 (Largest subcritical and second largest supercritical clusters) *Consider the inhomogeneous random graph* $\mathrm{IRG}_n(\kappa_n)$, *where* (κ_n) *is a graphical sequence of kernels with limit* κ. *It follows that:*

(a) *if* κ *is subcritical and* $\sup_{x,y,n} \kappa_n(x, y) < \infty$, *then* $|\mathscr{C}_{\max}| = O_{\mathbb{P}}(\log n)$;

(b) *if* κ *is supercritical,* κ *is irreducible, and either*

$$\inf_{x,y,n} \kappa_n(x, y) > 0 \qquad or \qquad \sup_{x,y,n} \kappa_n(x, y) < \infty,$$

then $|\mathscr{C}_{(2)}| = O_{\mathbb{P}}(\log n)$.

When $\lim_{n \to \infty} \sup_{x,y} \kappa_n(x, y) = \infty$, the largest subcritical clusters can have rather different behavior, as we now show for the rank-1 case. Note that, by Theorem 3.19 as well as the fact that $\|T_\kappa\| = \nu = \mathbb{E}[W^2]/\mathbb{E}[W]$, a rank-1 model can be subcritical only when $\mathbb{E}[W^2] < \infty$, i.e., in the case of finite-variance degrees. However, when W has a power-law tail, the highest weight can be much larger than $\log n$. Then the largest subcritical cluster is also much larger than $\log n$:

Theorem 3.22 (Subcritical phase for rank-1 inhomogeneous random graphs) *Let w satisfy Conditions 1.1(a)–(c) with $\nu = \mathbb{E}[W^2]/\mathbb{E}[W] < 1$, and assume further that there exist $\tau > 3$ and $c_2 > 0$ such that, for all $x \geq 1$, the empirical weight distribution satisfies*

$$[1 - F_n](x) \leq c_2 x^{-(\tau-1)}. \tag{3.7.1}$$

Then, for $\mathrm{NR}_n(w)$ *with* $w_{\max} = \max_{v \in [n]} w_v$,

$$|\mathscr{C}_{\max}| = \frac{w_{\max}}{1 - \nu} + o_{\mathbb{P}}(n^{1/(\tau-1)}). \tag{3.7.2}$$

Theorem 3.22 is most interesting in the case where the limiting distribution function F in Condition 1.1 has a power-law tail. For example, for w as in (1.3.15), let F satisfy

$$[1 - F](x) = c_W x^{-(\tau-1)}(1 + o(1)). \tag{3.7.3}$$

Then, $w_{\max} = w_1 = [1 - F]^{-1}(1/n) = (c_W n)^{1/(\tau-1)}(1 + o(1))$. Therefore,

$$|\mathscr{C}_{\max}| = (c_W n)^{1/(\tau-1)}/(1 - \nu) + o_{\mathbb{P}}(n^{1/(\tau-1)}). \tag{3.7.4}$$

Thus, the largest connected component is much larger than for $\mathrm{ER}_n(\lambda/n)$ with $\lambda < 1$.

Theorem 3.22 can be intuitively understood as follows. The connected component of a typical vertex is close to a branching process, so that it is whp bounded, and its expected connected component size is close to $1/(1 - \nu)$. Thus, the best way to obtain a large connected component is to start with a vertex with high weight w_i, and let all of its roughly w_i children be independent branching processes. Therefore, in expectation, each child is connected to another $1/(1 - \nu)$ different vertices, leading to a connected component size of roughly $w_i/(1 - \nu)$. This is clearly largest when $w_i = \max_{j \in [n]} w_j = w_{\max}$, leading to an intuitive explanation of Theorem 3.22.

Theorems 3.21 and 3.22 raise the question what the precise conditions are for $|\mathscr{C}_{\max}|$ to be of order $\log n$. Intuitively, if $w_{\max} \gg \log n$ then $|\mathscr{C}_{\max}| = w_{\max}/(1 - \nu)(1 + o_{\mathbb{P}}(1))$, whereas if $w_{\max} = \Theta(\log n)$ then $|\mathscr{C}_{\max}| = \Theta_{\mathbb{P}}(\log n)$ as well. In Turova (2011), it was proved that $|\mathscr{C}_{\max}|/\log n$ converges in probability to a finite constant when $\nu < 1$ and the weights are iid with distribution function F with $\mathbb{E}[e^{\alpha W}] < \infty$, for some $\alpha > 0$, i.e., *exponential tails* are sufficient.

Large Deviations for the Erdős–Rényi Giant

We close this section by discussing a beautiful and surprising large-deviations result for the size of the giant $|\mathscr{C}_{\max}|$ in $\mathrm{ER}_n(\lambda/n)$:

Theorem 3.23 (Large deviations for the Erdős–Rényi giant) *Consider* $\mathrm{ER}_n(\lambda/n)$ *with* $\lambda > 0$. *Then, for all* $u \in [0, 1]$,

$$\lim_{n \to \infty} \frac{1}{n} \log \mathbb{P}(|\mathscr{C}_{\max}| = \lceil nu \rceil) = -J_\lambda(u), \tag{3.7.5}$$

where, for $u \in [u_k, u_{k-1}]$,

$$J_\lambda(u) = -kum(\lambda u) + ku \log u + (1 - ku) \log(1 - ku)$$
$$+ \lambda u - k(k + 1)\lambda u^2/2, \tag{3.7.6}$$

with $m(u) = \log(1 - e^{-u})$ *and where* $u_0 = 1$ *and, for* $k \geq 1$,

$$u_k = \sup \left\{ u \colon \frac{u}{1 - ku} = 1 - e^{-\lambda u} \right\}. \tag{3.7.7}$$

Note that $u_1 = 1$ for $\lambda \leq 1$, so that, for all $u \in [0, 1]$ and $\lambda \leq 1$,

$$J_\lambda(u) = -um(\lambda u) + u \log u + (1 - u) \log(1 - u) + \lambda u(1 - u). \tag{3.7.8}$$

This function is nicely convex with a unique minimum for $u = 0$; the minimum equals 0, as can be expected.

For $\lambda > 1$, however, the situation is more involved. The function $u \mapsto J_\lambda(u)$ still has a unique minimum for $u = \zeta_\lambda$, where $J_\lambda(\zeta_\lambda) = 0$. However, the function $u \mapsto J_\lambda(u)$ is not convex and has infinitely many non-analyticities. This can be understood as follows. When $u \approx \zeta_\lambda$ or when $u > \zeta_\lambda$, the rate function $J_\lambda(u)$ measures the exponential rate of the event that there exists a connected component of size $\lceil nu \rceil$. However, when u becomes quite small compared with ζ_λ, not only should there be a connected component of size $\lceil nu \rceil$, but also *all other connected components should be smaller than* $\lceil nu \rceil$. Since $\mathrm{ER}_n(\lambda/n)$ with $\mathscr{C}_{\max}$ removed is also an Erdős–Rényi random graph with appropriate parameters, when u is quite small it again becomes exponentially rare that this Erdős–Rényi random graph has a giant that has size at most $\lceil nu \rceil$. When u is very small, we need to iterate this several times, and the k parameter in (3.7.6) measures how many such exponential contributions arise before the graph remaining after the removal of all large components becomes such that its giant has whp size at most $\lceil nu \rceil$.

Interestingly, applying Theorem 3.23 to $u = 1$ also provides the relation

$$\lim_{n \to \infty} \frac{1}{n} \log \mathbb{P}(\mathrm{ER}_n(\lambda/n) \text{ connected}) = -\log(1 - e^{-\lambda}); \tag{3.7.9}$$

see also Exercise 3.42.

3.8 Notes and Discussion for Chapter 3

Notes on Section 3.1

Example 3.1 already points at general stochastic block models.

Notes on Section 3.2

The seminal paper by Bollobás et al. (2007) studied inhomogeneous random graphs in an even more general setting, where the number of vertices in the graph need not be equal to n. In this case, the vertex space is called a *generalized vertex space*. However, we simplify the discussion here slightly by assuming that the number of vertices is always equal to n. An example where the extension to a random number of vertices is crucially used was by Turova and Vallier (2010), who studied an interpolation between percolation and $\mathrm{ER}_n(p)$.

The Dubins model was first investigated by Shepp (1989). The uniformly grown random graph was proposed by Callaway et al. (2001). In their model the graph grows dynamically as follows. At each time step, a new vertex is added. Further, with probability δ, two vertices are chosen uar and joined by an undirected edge. This process is repeated for n time steps, where n is the number of vertices in the graph. Callaway et al. predicted, using physical reasoning, that in the limit of large n, the resulting graph has a giant component precisely when $\delta > \frac{1}{8}$, and the proportion of vertices in the giant component is of the order $e^{-\Theta(1/(\sqrt{8\delta-1}))}$ when $\delta > \frac{1}{8}$ is close to $\frac{1}{8}$. Such behavior is sometimes called an *infinite-order phase transition*. Durrett (2003) discusses this model. Here we discussed the variant of this model that was proposed and analyzed by Bollobás et al. (2005).

Notes on Section 3.3

Theorem 3.4 is a special case of (Bollobás et al., 2007, Theorem 3.13). This section is a minor modification of results in Bollobás et al. (2007). Indeed, Proposition 3.5 is an adapted version of (Bollobás et al., 2007, Lemma 7.3); see also the proof of the irreducibility result in Proposition 3.5(a) there. Lemma 3.6 is (Bollobás et al., 2007, Lemma 7.1).

Notes on Section 3.4

See (Athreya and Ney, 1972, Chapter V) or (Harris, 1963, Chapter III) for more background on multi-type branching processes. We note that the convergence in Theorem 3.10(b) may not hold when starting from an individual of a fixed type. This is, for example, the case when there are two types and individuals only have children of the *other* type. Poisson multi-type branching processes are discussed in detail in (Bollobás et al., 2007, Sections 5 and 6). The Perron–Frobenius Theorem can be found in many places; (Noutsos, 2006, Theorem 2.2) is a convenient reference with many related results.

Notes on Section 3.5

Theorem 3.14 is novel, to the best of our knowledge, even though Bollobás et al. (2007) proved various relations between inhomogeneous random graphs and branching processes. The use of ρ for the law of the local limit rather than μ as in Chapter 2 is in honor of Bollobás et al. (2007), who used ρ for the limiting law of the component size distribution. This proof is closely related to that of Theorem 3.14, even though the result is not phrased in terms of local convergence. Proposition 3.16 appeared first as (Norros and Reittu, 2006, Proposition 3.1), where the connections between $\mathrm{NR}_n(\boldsymbol{w})$ and Poisson branching processes were first exploited to prove the versions of Theorem 6.3 in Chapter 6.

Notes on Section 3.6

Theorem 3.19 is a special case of (Bollobás et al., 2007, Theorem 3.1). The finite-type case of Theorem 3.19 was studied by Söderberg (2002, 2003a,c,b). Theorem 3.20 was taken from Janson and Luczak (2009), where the giant component is investigated for the configuration model. We explain the proof in detail in Section 4.3, where we also prove how the result for the configuration model in Theorem 4.9 can be used to prove Theorem 3.20. Earlier versions for random graphs with given expected degrees or the Chung–Lu model appeared in Chung and Lu (2002b, 2004, 2006b) (see also the monograph Chung and Lu (2006a)). Central limit theorems for the giant component for the rank-1 model as studied in Theorem 3.20 are proved in Janson (2020a) (see also Janson (2020b)). See Section 4.6 for more details. I learned the proof of the linear complexity of the giant in Theorem 3.20 from Svante Janson.

Bollobás et al. (2007) proved various other results concerning the giant component of $\mathrm{IRG}_n(\kappa_n)$. For example, (Bollobás et al., 2007, Theorem 3.9) proved that the giant component of $\mathrm{IRG}_n(\kappa_n)$ is stable in the sense that its size does not change much if we add or delete a small linear number of edges. Note that the edges added or deleted do not have to be random or independent of the existing graph; rather, they can be chosen by an adversary after inspecting the whole of $\mathrm{IRG}_n(\kappa_n)$. More precisely, (Bollobás et al., 2007, Theorem 3.9) shows that, for small enough $\delta > 0$, the giant component of $\mathrm{IRG}_n(\kappa_n)$ in the supercritical regime changes by more than εn vertices if we remove any collection of δn edges.

Notes on Section 3.7

Theorem 3.22 is (Janson, 2008, Corollary 4.4). Theorem 3.23 was proved by O'Connell (1998). Andreis et al. (2021) proved related results on large deviations of component sizes in the Erdős–Rényi random graph, while Andreis et al. (2023) extended this to general inhomogeneous random graphs.

3.9 EXERCISES FOR CHAPTER 3

Exercise 3.1 (Erdős–Rényi random graph)   *Show that, for $S = [0, 1]$ and $p_{uv} = \kappa(u/n, v/n)/n$ with $\kappa \colon [0, 1]^2 \to [0, \infty)$ continuous, the model is the Erdős–Rényi random graph with edge probability λ/n precisely when $\kappa(x, y) = \lambda$. Is this also true when $\kappa \colon [0, 1]^2 \to [0, \infty)$ is not continuous?*

Exercise 3.2 (Lower bound on expected number of edges)   *Show that when $\kappa\colon \mathcal{S}\times\mathcal{S}\to[0,\infty)$ is continuous, then*

$$\liminf_{n\to\infty}\frac{1}{n}\mathbb{E}[|E(\mathrm{IRG}_n(\kappa))|]\geq\frac{1}{2}\iint_{\mathcal{S}^2}\kappa(x,y)\mu(\mathrm{d}x)\mu(\mathrm{d}y),\qquad(3.9.1)$$

so that the lower bound in (3.2.3) generally holds.

Exercise 3.3 (Expected number of edges)   *Show that (3.2.3) holds when $\kappa\colon \mathcal{S}\times\mathcal{S}\to[0,\infty)$ is bounded and continuous.*

Exercise 3.4 (Asymptotic equivalence for general IRGs)   *Prove that the random graphs $\mathrm{IRG}_n(\boldsymbol{p})$ with p_{uv} as in (3.2.6) are asymptotically equivalent to $\mathrm{IRG}_n(\boldsymbol{p})$ with $p_{uv}=p_{uv}^{(\mathrm{NR})}(\kappa_n)$ and to $\mathrm{IRG}_n(\boldsymbol{p})$ with $p_{uv}=p_{uv}^{(\mathrm{GRG})}(\kappa_n)$ when (3.2.8) holds.*

Exercise 3.5 (The Chung–Lu model)   *Prove that when $\kappa\colon [0,1]^2\to[0,\infty)$ is given by*

$$\kappa(x,y)=[1-F]^{-1}(x)[1-F]^{-1}(y)/\mathbb{E}[W],\qquad(3.9.2)$$

then κ is graphical precisely when $\mathbb{E}[W]<\infty$, where W has distribution function F. Further, κ is always irreducible.

Exercise 3.6 (The Chung–Lu model repeated)   *Let $\tilde{w}_v=[1-F]^{-1}(v/n)\sqrt{n\mathbb{E}[W]/\ell_n}$ and $w_v=[1-F]^{-1}(v/n)$ as in (1.3.15). Assume that $w_v^2=o(\ell_n)$. Show that the edge probabilities in $\mathrm{CL}_n(\tilde{w})$ are*

$$\tilde{p}_{uv}=[1-F]^{-1}(i/n)[1-F]^{-1}(i/n)/(n\mathbb{E}[W]).$$

Further, show that $\mathrm{CL}_n(\tilde{w})$ and $\mathrm{CL}_n(w)$ are asymptotically equivalent whenever $(\mathbb{E}[W_n]-\mathbb{E}[W])^2=o(1/n^2)$.

Exercise 3.7 (Definitions 3.2 and 3.3 for the homogeneous bipartite graph)   *Prove that Definitions 3.2 and 3.3 hold for the homogeneous bipartite graph.*

Exercise 3.8 (Examples of homogeneous random graphs)   *Show that the Erdős–Rényi random graph, the homogeneous bipartite random graph, and the stochastic block model are all homogeneous random graphs.*

Exercise 3.9 (Homogeneous bipartite graph)   *Prove that the homogeneous bipartite random graph is a special case of the finite-type case.*

Exercise 3.10 (Irreducibility for the finite-type case)   *Prove that, in the finite-type case, irreducibility follows when there exists an m such that the mth power of the matrix $(\kappa(s,r)\mu(r))_{s,r\in[t]}$ contains no zeros.*

Exercise 3.11 (Graphical limit in the finite-type case)   *Prove that, in the finite-type case, the convergence of μ_n in (3.2.1) holds precisely when, for every type $s\in\mathcal{S}$,*

$$\lim_{n\to\infty}n_s/n=\mu(s).\qquad(3.9.3)$$

Exercise 3.12 (Variance of number of vertices of degree k and type s)   *Let $\mathrm{IRG}_n(\kappa_n)$ be a finite-type inhomogeneous random graph with graphical sequence of kernels κ_n. Let $N_{k,s}(n)$ be the number of vertices of degree k and type s. Show that $\mathrm{Var}(N_{k,s}(n))=O(n)$.*

Exercise 3.13 (Proportion of isolated vertices in inhomogeneous random graphs)   *Let $\mathrm{IRG}_n(\kappa_n)$ be an inhomogeneous random graph with a graphical sequence of kernels κ_n that converges to κ. Show that the proportion of isolated vertices converges to*

$$\frac{1}{n}N_0(n)\overset{\mathbb{P}}{\longrightarrow}p_0=\int e^{-\lambda(x)}\mu(\mathrm{d}x).\qquad(3.9.4)$$

Conclude that $p_0>0$ when $\int\lambda(x)\mu(\mathrm{d}x)<\infty$.

Exercise 3.14 (Upper and lower bounding finite-type kernels)   *Prove that the kernels $\underline{\kappa}_m$ and $\overline{\kappa}_m$ in (3.3.15) and (3.3.16) are of finite type.*

Exercise 3.15 (Inclusion of graphs for larger κ)   *Let $\kappa'\leq\kappa$ hold a.e. Show that we can couple $\mathrm{IRG}_n(\kappa')$ and $\mathrm{IRG}_n(\kappa)$ in such a way that $\mathrm{IRG}_n(\kappa')\subseteq\mathrm{IRG}_n(\kappa)$.*

Exercise 3.16 (Tails of Poisson variables)   *Use the stochastic domination of Poisson random variables with different parameters, as well as the concentration properties of Poisson variables, to complete the proof of (3.3.30), showing that the tail asymptotics of the weight distribution and that of the mixed-Poisson random variable with that weight agree.*

Exercise 3.17 (Power laws for sum kernels)   *Let $\kappa(x, y) = \psi(x) + \psi(y)$ for a continuous function $\psi \colon [0, 1] \mapsto [0, \infty)$, and let the reference measure μ be uniform on $[0, 1]$. Use Corollary 3.7 to identify when the degree distribution satisfies a power law. How is the tail behavior of D related to that of ψ?*

Exercise 3.18 (Survival probability of individual with random type)   *Consider a multi-type branching process where the root has type s with probability $\mu(s)$ for all $s \in [t]$. Show that the survival probability ζ equals*

$$\zeta = \sum_{s \in [t]} \zeta^{(s)} \mu(s), \tag{3.9.5}$$

where $\zeta^{(s)}$ was defined in (3.4.1).

Exercise 3.19 (Irreducibility of multi-type branching process)   *Show that the positivity of the survival probability $\zeta^{(s)}$ of an individual of type s is independent of the type s when the probability that an individual of type r has a type-s descendant is strictly positive for every $s, r \in [t]$.*

Exercise 3.20 (Irreducibility of multi-type branching process (cont.))   *Prove that the probability that an individual of type s to have a type-r descendant is strictly positive precisely when there exists an l such that $\boldsymbol{T}_\kappa^l(s, r) > 0$, where $\boldsymbol{T}_\kappa(s, r) = \kappa(s, r) \mu(r)$ is the mean offspring matrix.*

Exercise 3.21 (Singularity of multi-type branching process)   *Prove that $\boldsymbol{G}(\boldsymbol{z}) = \mathbf{M} \boldsymbol{z}$ for some matrix $\mathbf{M}$ precisely when each individual in the multi-type branching process has exactly one offspring almost surely.*

Exercise 3.22 (Erdős–Rényi random graph)   *Prove that $\mathrm{NR}_n(\boldsymbol{w}) = \mathrm{ER}_n(\lambda/n)$ when $\boldsymbol{w}$ is constant with $w_v = -n \log(1 - \lambda/n)$ for all $v \in [n]$.*

Exercise 3.23 (Homogeneous Poisson multi-type branching processes)   *In analogy with the homogeneous random graph as defined in (3.2.11), we call a Poisson multi-type branching process homogeneous when the expected offspring of a tree vertex of type x equals $\lambda(x) = \lambda$ for all $x \in S$. Consider a homogeneous Poisson multi-type branching process with parameter λ. Show that the function $\phi(x) = 1$ is an eigenvector of $\boldsymbol{T}_\kappa$ with eigenvalue λ. Conclude that $(Z_j / \lambda^j)_{j \geq 0}$ is a martingale, where $(Z_j)_{j \geq 0}$ denotes the number of individuals in the jth generation, irrespective of the starting distribution.*

Exercise 3.24 (Proof of no-overlap property in (3.5.19))   *Prove that $\mathbb{P}(\bar{B}_r^{(G_n; Q)}(o_1) = (\mathbf{t}, \boldsymbol{q}), o_2 \in B_{2r}^{(G_n; Q)}(o_1)) \to 0$, and conclude that (3.5.19) holds.*

Exercise 3.25 (Unimodular mixed-Poisson branching process)   *Recall the definition of a unimodular branching process in Definition 1.26. Prove that the mixed-Poisson branching process described in (3.5.29) and (3.5.30) is indeed unimodular.*

Exercise 3.26 (Branching process domination of Erdős–Rényi random graph)   *Show that Exercise 3.22 together with Proposition 3.16 imply that $|\mathscr{C}(o)| \preceq T^\star$, where $T^\star$ is the total progeny of a Poisson branching process with mean $-n \log(1 - \lambda/n)$ offspring.*

Exercise 3.27 (Local convergence of $\mathrm{ER}_n(\lambda/n)$)   *Use Theorem 3.18 to show that $\mathrm{ER}_n(\lambda/n)$ converges locally in probability to the Poisson branching process with parameter λ.*

Exercise 3.28 (Coupling to a multi-type Poisson branching process)   *Prove the stochastic relation between multi-type Poisson branching processes and neighborhoods in Norros–Reittu inhomogeneous random graphs in Proposition 3.17 by adapting the proof of Proposition 3.16.*

Exercise 3.29 (Phase transition for $r = 2$)   *Let $\zeta_\kappa^{(1)}$ and $\zeta_\kappa^{(2)}$ denote the survival probabilities of an irreducible multi-type branching process with two types starting from vertices of types 1 and 2, respectively. Give necessary and sufficient conditions for $\zeta_\kappa^{(i)} > 0$ to hold for $i \in \{1, 2\}$.*

Exercise 3.30 (The size of small components in the finite-type case)   *Prove that, in the finite-types case, when (κ_n) converges to a limiting kernel κ, then $\sup_{x,y,n} \kappa_n(x, y) < \infty$ holds, so that the results of Theorem 3.21 apply in the sub- and supercritical cases.*

Exercise 3.31 (Law of large numbers for $|\mathscr{C}_{\max}|$ for $\mathrm{ER}_n(\lambda/n)$)   *Prove that, for the Erdős–Rényi random graph, Theorem 3.19 implies that $|\mathscr{C}_{\max}|/n \overset{\mathbb{P}}{\longrightarrow} \zeta_\lambda$, where ζ_λ is the survival probability of a Poisson branching process with mean-λ offspring.*

Exercise 3.32 (Connectivity of uniformly chosen vertices)   *Suppose we draw two vertices independently and uar from $[n]$ in $\mathrm{IRG}_n(\kappa_n)$. Prove that Theorem 3.20 implies that the probability that the vertices are connected converges to ζ^2.*

Exercise 3.33 (The size of small components for $\mathrm{CL}_n(\boldsymbol{w})$)   *Use Theorem 3.21 to prove that, for $\mathrm{CL}_n(\boldsymbol{w})$ with weights given by (1.3.15) and $1 < \nu < \infty$, the second largest cluster has size $|\mathscr{C}_{(2)}| = O_{\mathbb{P}}(\log n)$ when W has bounded support or is almost surely bounded below by $\varepsilon > 0$ with $\mathbb{E}[W] < \infty$. Further, $|\mathscr{C}_{\max}| = O_{\mathbb{P}}(\log n)$ when W has bounded support and $\nu < 1$. Here W is a random variable with distribution function F.*

Exercise 3.34 (Average degree in two populations)   *Show that the average degree is close to $pm_1 + (1 - p)m_2$ in the setting of Example 3.1 with n_1 vertices of type 1 satisfying $n_1/n \to p$.*

Exercise 3.35 (Phase transition for two populations)   *Show that $\zeta > 0$ precisely when $[pm_1^2 + (1 - p)m_2^2]/[pm_1 + (1-p)m_2] > 1$ in the setting of Example 3.1 with n_1 vertices of type 1 satisfying $n_1/n \to p$.*

Exercise 3.36 (Phase transition for two populations (cont.))   *In the setting of Exercise 3.35, find an example of p, m_1, m_2 where the average degree is less than 1, yet there exists a giant component.*

Exercise 3.37 (Degree sequence of giant component for rank 1)   *Consider $\mathrm{GRG}_n(\boldsymbol{w})$ as in Theorem 3.20. Show that the proportion of vertices of $\mathscr{C}_{\max}$ having degree ℓ is close to $p_\ell(1 - \xi^\ell)/\zeta$.*

Exercise 3.38 (Degree sequence of complement of giant component)   *Consider $\mathrm{GRG}_n(\boldsymbol{w})$ as in Theorem 3.20. Show that when $\xi < 1$, the proportion of vertices outside the giant component $\mathscr{C}_{\max}$ having degree ℓ is close to $p_\ell \xi^\ell/(1 - \zeta)$. Conclude that the degree sequence of the complement of the giant component never satisfies a power law. Can you give an intuitive explanation for this?*

Exercise 3.39 (Finiteness of $\nu(p)$)   *Prove that $\nu(p)$ in (3.6.17) satisfies $\nu(p) < \infty$ for every $p \in (0, 1]$.*

Exercise 3.40 (Phase transition of uniformly grown random graphs)   *Recall the uniformly grown random graph in (3.2.12). Look up the proof that $\|T_\kappa\| = 4\lambda$ in (Bollobás et al., 2007, Section 16.1).*

Exercise 3.41 (Phase transition of sum kernels)   *Recall the inhomogeneous random graph with sum kernel in (3.2.13). When does it have a giant?*

Exercise 3.42 (Connectivity probability of sparse $\mathrm{ER}_n(\lambda/n)$)   *Use Theorem 3.23 to prove that*

$$\lim_{n \to \infty} \frac{1}{n} \log \mathbb{P}(\mathrm{ER}_n(\lambda/n) \text{ connected}) = \log(1 - e^{-\lambda})$$

as in (3.7.9).

CONNECTED COMPONENTS IN CONFIGURATION MODELS

Abstract

In this chapter we investigate the local limit of the configuration model, identify when it has a giant component, and find its size and degree structure. We give two proofs, one based on a "the giant is almost local" argument, and the other based on a continuous-time exploration of the connected components in the configuration model. Further results include its connectivity transition.

4.1 MOTIVATION: RELATING DEGREES TO LOCAL LIMIT AND GIANT

In this chapter we study the connectivity structure of the configuration model. We focus on the local connectivity, by investigating its local limit, as well as the global connectivity, by identifying its giant component and connectivity transition. In inhomogeneous random graphs there always is a positive proportion of vertices that are isolated (recall Exercise 3.13). In many real-world examples, we observe the presence of a giant component (recall Table 3.1). In many of these examples the giant is almost the whole graph and sometimes, by definition, it *is* the whole graph. For example, the Internet needs to be connected in such a way as to allow e-mail messages to be sent between any pair of vertices. In many other real-world examples, though, it is not at all obvious whether, or why, the network is connected. See Figure 4.1 (which is the same as Figure 3.1), and observe that there are quite a few connected networks in the KONECT data base.

Table 4.1 invites us to think about what makes networks (close to) fully connected. We investigate this question here in the context of the configuration model. The advantage of the configuration model is its high flexibility in degree structure, so that all degrees can have at least a certain minimal value. We will see that this can give rise to *connected* random graphs that at the same time remain sparse, as is the case in many real-world networks.

Organization of this Chapter

This chapter is organized as follows. In Section 4.2 we study the local limit of the configuration model. In Section 4.3 we state and prove the law of large numbers for the giant component in the configuration model, thus establishing the giant's phase transition. In Section 4.4 we study the conditions for the configuration model to be connected. In Section 4.5 we state further results on the configuration model. We close this chapter in Section 4.6 with notes and discussion, and with exercises in Section 4.7.

4.2 LOCAL CONVERGENCE OF THE CONFIGURATION MODEL

We start by investigating the locally tree-like nature of the configuration model. Recall the unimodular branching-process tree from Definition 1.26. Our main result is as follows:

Table 4.1 *The rows in the above table represent the following six real-world networks:*
In the California road network, vertices represent intersections or endpoints of roads.
In the Facebook network, vertices represent the users and the edges Facebook friendships.
Hyves was a Dutch social media platform. Vertices represent users, and edges friendships.
The arXiv astro-physics network represents authors of papers within the astro-physics section of
arXiv, where an edge between authors represents that they have co-authored a paper.
In the high-voltage power network in western USA, the vertices represent transformer substations
and generators, and the edges transmission cables.
In the jazz-musicians data set, vertices represent musicians and connections indicate past
collaborations.

Subject	% in giant	Size	Source	Data
California roads	0.9958	1,965,206	Leskovec et al. (2009)	Leskovec and Krevl (2014)
Facebook	0.9991	721m	Ugander et al. (2011)	Ugander et al. (2011)
Hyves	0.996	8,047,530	Corten (2012)	Corten (2012)
arXiv astro-ph	0.9538	18,771	Leskovec et al. (2007)	Kunegis (2017)
US power grid	1	4,941	Watts and Strogatz (1998)	Kunegis (2017)
Jazz-musicians	1	198	Gleiser and Danon (2003)	Kunegis (2017)

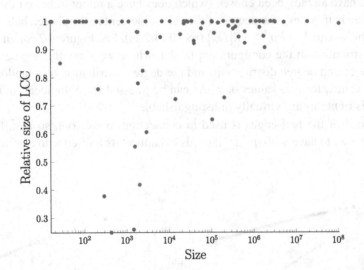

Figure 4.1 Proportion of vertices in the maximal connected component in the 1,203 networks from the KONECT data base. The network is connected when the proportion equals 1.

Theorem 4.1 (Locally tree-like nature of the configuration model) *Assume that Conditions 1.7(a),(b) hold. Then* $\mathrm{CM}_n(\mathbf{d})$ *converges locally in probability to the unimodular branching-process tree* $(G,o) \sim \mu$ *with root offspring distribution* $(p_k)_{k \geq 0}$ *given by* $p_k = \mathbb{P}(D = k)$.

Before starting the proof of Theorem 4.1, let us informally explain the above connection between local neighborhoods and branching processes. We note that the asymptotic off-spring distribution at the root is equal to $(p_k)_{k\geq 0}$, where $p_k = \mathbb{P}(D = k)$ is the asymptotic degree distribution. Indeed, fix $G_n = \mathrm{CM}_n(\boldsymbol{d})$. Then, the probability that a random vertex has degree k is equal to

$$p_k^{(G_n)} = \mathbb{P}(D_n = k) = n_k/n, \tag{4.2.1}$$

where n_k denotes the number of vertices with degree k. By Condition 1.7(a), $p_k^{(G_n)}$ converges to $p_k = \mathbb{P}(D = k)$, for every $k \geq 1$. This explains the offspring of the root of our branching-process approximation.

The offspring distribution of individuals in the first and later generations is given by

$$p_k^\star = \frac{(k+1)p_{k+1}}{\mathbb{E}[D]}. \tag{4.2.2}$$

We now explain this heuristically, by examining the degree of the vertex to which the first half-edge incident to the root is paired. By the uniform matching of half-edges, the probability that a vertex of degree k is chosen is proportional to k. Ignoring the fact that the root and one half-edge have already been chosen (which does have a minor effect on the number of available or free half-edges), the degree of the vertex incident to the chosen half-edge equals k with probability equal to $k p_k^{(G_n)}/\mathbb{E}[D_n]$ (recall (4.2.1)). See Figure 4.2 for an example of the degree distribution in the configuration model, where we show the degree distribution itself, the size-biased degree distribution, and the degree distribution of a random neighbor of a uniform vertex, for two values of τ. As can be guessed from the local limit, the latter two degree distributions are virtually indistinguishable.

However, one of the half-edges is used in connecting to the root, so that, for a vertex incident to the root to have k offspring, it needs to connect its half-edge to a vertex of degree

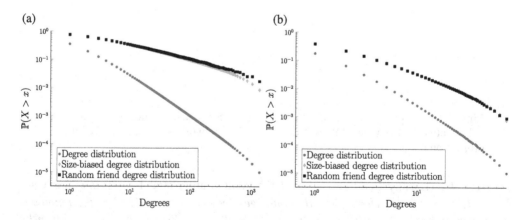

Figure 4.2 Degree distributions in the configuration model with $n = 100,000$ and (a) $\tau = 2.5$; (b) $\tau = 3.5$.

$k + 1$. Therefore, the probability that the offspring, or "forward degree," of any of the direct neighbors of the root is k equals

$$p_k^{\star(G_n)} = \frac{(k+1)p_{k+1}^{(G_n)}}{\mathbb{E}[D_n]}. \tag{4.2.3}$$

Thus, $(p_k^{\star(G_n)})_{k \geq 0}$ can be interpreted as the *forward degree distribution of vertices in the cluster exploration*. When Conditions 1.7(a),(b) hold, we also have $p_k^{\star(G_n)} \to p_k^\star$, where $(p_k^\star)_{k \geq 0}$ is defined in (4.2.2). As a result, we often refer to $(p_k^\star)_{k \geq 0}$ as the *asymptotic forward degree distribution*.

The above heuristic argues that any direct neighbor of the root has a number of forward neighbors with asymptotic law $(p_k^\star)_{k \geq 0}$. However, every time we pair two half-edges, the number of *free* or *available* half-edges decreases by 2. Similarly to the depletion-of-points effect in the exploration of connected components in the Erdős–Rényi random graph $\mathrm{ER}_n(\lambda/n)$, the configuration model $\mathrm{CM}_n(\boldsymbol{d})$ suffers from a *depletion-of-points-and-half-edges* effect. Thus, by iteratively connecting half-edges in a breadth-first way, the offspring distribution changes along the way, which potentially gives trouble.

Luckily, the number of available half-edges is initially $\ell_n - 1$, which is very large when Conditions 1.7(a),(b) hold, since then $\ell_n/n = \mathbb{E}[D_n] \to \mathbb{E}[D] > 0$. Thus, we can pair *many* half-edges before we start noticing that their number decreases. As a result, the degrees of different vertices in the exploration process are *close* to being iid, leading to a branching-process approximation of neighborhoods in the configuration model. In order to prove Theorem 4.1, we need to pair only a *bounded* number of edges, but our approximation extends significantly beyond this.

In order start the proof of Theorem 4.1 based on (2.4.11), we introduce some notation. First, we let $\bar{B}_r^{(G_n)}(v)$ denote the *ordered* version of $B_r^{(G_n)}(v)$, obtained by ordering the half-edges randomly and performing a breadth-first exploration from the smallest to the largest labeled half-edge. We again write $\bar{B}_r^{(G_n)}(v) = \mathbf{t}$ to denote that this ordered neighborhood is equal to the ordered tree $\mathbf{t}$.

Fix a rooted ordered tree $\mathbf{t}$ with r generations, and let

$$N_{n,r}(\mathbf{t}) = \sum_{v \in [n]} \mathbb{1}_{\{\bar{B}_r^{(G_n)}(v)=\mathbf{t}\}} \tag{4.2.4}$$

denote the number of vertices in $G_n = \mathrm{CM}_n(\boldsymbol{d})$ whose ordered local neighborhood up to generation r equals $\mathbf{t}$. By Theorem 2.15, to prove Theorem 4.1, we need to show that

$$\frac{N_{n,r}(\mathbf{t})}{n} \overset{\mathbb{P}}{\longrightarrow} \mu(\bar{B}_r^{(G)}(o) = \mathbf{t}), \tag{4.2.5}$$

where $(G, o) \sim \mu$ denotes the unimodular branching process with root offspring distribution $(p_k)_{k \geq 1}$. Here, we also rely on Theorem 2.8 to see that it suffices to prove (4.2.5) for *trees*, since the unimodular branching-process tree is a tree with probability 1.

To prove (4.2.5), and as we have done before, we use a *second-moment method*. We start by proving that the first moment $\mathbb{E}[N_{n,r}(\mathbf{t})]/n \to \mu(\bar{B}_r^{(G)}(o) \simeq \mathbf{t})$, after which we prove that $\mathrm{Var}(N_{n,r}(\mathbf{t})) = o(n^2)$. Then (4.2.5) follows from the Chebychev inequality [V1, Theorem 2.18].

4.2.1 Proof of Local Convergence of Configuration Model: First Moment

We next relate the neighborhood in a random graph to a branching process where the root has offspring distribution D_n, while all other individuals have offspring distribution $D_n^\star - 1$, where $D_n^\star$ is the size-biased distribution of D_n. Denote this branching process by $(\mathsf{BP}_n(t))_{t \geq 1}$, as described in Section 1.5. In terms of Definition 1.25, we say that a tree vertex is *explored* when we have inspected how many children it has. Here, $\mathsf{BP}_n(t)$ denotes the branching process when we have explored precisely t tree vertices, so that $\mathsf{BP}_n(1)$ denotes the root $\varnothing$ and its neighbors. We explore it in breadth-first order as in Definition 1.25 in Section 1.5.

Clearly, by Conditions 1.7(a),(b), we have $D_n \xrightarrow{d} D$ and $D_n^\star \xrightarrow{d} D^\star$, which implies that $\mathsf{BP}_n(t) \xrightarrow{d} \mathsf{BP}(t)$ for every t finite, where $\mathsf{BP}(t)$ is the restriction of the unimodular branching process (G, o) with root offspring distribution $(p_k)_{k \geq 1}$ to its first t individuals (see Exercise 4.1). Note that, for $\mathbf{t}$ a fixed rooted tree of at most r generations, $B_r^{(G)}(o) \simeq \mathbf{t}$ precisely when $\mathsf{BP}(t_r) \simeq \mathbf{t}$, where t_r denotes the number of vertices in the first $r - 1$ generations in $\mathbf{t}$.

We let $(\mathsf{G}_n(t))_{t \geq 1}$ denote the graph exploration process from a uniformly chosen vertex $o \in [n]$. Here $\mathsf{G}_n(t)$ is the exploration where we have paired precisely $t-1$ half-edges, in the breadth-first manner as described in Definition 1.25, while we also indicate the half-edges incident to the vertices found. Thus, $\mathsf{G}_n(1)$ consists of $o \in [n]$ and its $D_n = d_o$ half-edges, and every further exploration corresponds to the pairing of a half-edge. In particular, from $(\mathsf{G}_n(t))_{t \geq 1}$, we can retrieve $B_r^{(G_n)}(o)$ for every $r \geq 0$, where $G_n = \mathrm{CM}_n(\mathbf{d})$. The following lemma proves that we can couple the graph exploration to the branching process in such a way that $(\mathsf{G}_n(t))_{t \in [m_n]}$ is equal to $(\mathsf{BP}_n(t))_{t \in [m_n]}$ whenever $m_n \to \infty$ sufficiently slowly. In the statement, we write $(\widehat{\mathsf{G}}_n(t), \widehat{\mathsf{BP}}_n(t))_{t \geq 1}$ for the coupling of $(\mathsf{G}_n(t))_{t \in [m_n]}$ and $(\mathsf{BP}_n(t))_{t \in [m_n]}$:

Lemma 4.2 (Coupling graph exploration and branching process) *Subject to Conditions 1.7(a),(b), there exists a coupling $(\widehat{\mathsf{G}}_n(t), \widehat{\mathsf{BP}}_n(t))_{t \geq 1}$ of $(\mathsf{G}_n(t))_{t \geq 1}$ and $(\mathsf{BP}_n(t))_{t \geq 1}$ such that*

$$\mathbb{P}\Big((\widehat{\mathsf{G}}_n(t))_{t \in [m_n]} \neq (\widehat{\mathsf{BP}}_n(t))_{t \in [m_n]} \Big) = o(1), \tag{4.2.6}$$

when $m_n \to \infty$ sufficiently slowly. Consequently, $\mathbb{E}[N_{n,r}(\mathbf{t})]/n \to \mu(\bar{B}_r^{(G)}(o) = \mathbf{t})$.

Remark 4.3 (Extensions) Here we discuss some useful extensions of Lemma 4.2. First, in its proof we will see that any $m_n = o(\sqrt{n/d_{\max}})$ is allowed. Here $d_{\max} = \max_{v \in [n]} d_v$ is the maximal vertex degree in $\mathrm{CM}_n(\mathbf{d})$, which is $o(n)$ when Conditions 1.7(a),(b) hold (compare with Exercise 1.8). Second, Lemma 4.2 can easily be extended to deal with the explorations from *two* sources (o_1, o_2), where we can still take $m_n = o(\sqrt{n/d_{\max}})$ and the two branching processes to which we couple the exploration from two sources, denoted by $(\widehat{\mathsf{BP}}_n^{(1)}(t))_{t \in [m_n]}$ and $(\widehat{\mathsf{BP}}_n^{(2)}(t))_{t \in [m_n]}$, are iid. ◀

Proof of Lemma 4.2. We let the offspring of the root of the branching process $\widehat{D}_n$ be equal to d_o, which is the number of neighbors of the vertex $o \in [n]$ that was chosen uar. By construction $\widehat{D}_n = d_o$, so that also $\widehat{\mathsf{G}}_n(1) = \widehat{\mathsf{BP}}_n(1)$. The proof consists of several parts.

Joint Neighborhood Construction

Fix m. We next explain how to *jointly* construct $(\widehat{\mathsf{G}}_n(t), \widehat{\mathsf{BP}}_n(t))_{t\in[m]}$ *given* that we have already constructed $(\widehat{\mathsf{G}}_n(t), \widehat{\mathsf{BP}}_n(t))_{t\in[m-1]}$. Further, for each half-edge, we record its status as *real* or *ghost*, where the real half-edges correspond to those present in *both* $\widehat{\mathsf{G}}_n(m-1)$ and $\widehat{\mathsf{BP}}_n(m-1)$, while the ghost half-edges are present only in $\widehat{\mathsf{G}}_n(m-1)$ or $\widehat{\mathsf{BP}}_n(m-1)$.

To obtain $\widehat{\mathsf{G}}_n(m)$, we take the first unpaired half-edge x_m in $\widehat{\mathsf{G}}_n(m-1)$. When this half-edge has the ghost status, we draw a uniform *unpaired* half-edge y'_m and then pair x_m to y'_m to obtain $\widehat{\mathsf{G}}_n(m)$, and we give all sibling half-edges of y'_m the ghost status (where we recall that the sibling half-edges of a half-edge y are those half-edges unequal to y that are incident to the same vertex as y).

When the half-edge has the real status, it needs to be paired both in $\widehat{\mathsf{G}}_n(m)$ and $\widehat{\mathsf{BP}}_n(m)$. To obtain $\widehat{\mathsf{G}}_n(m)$, this half-edge needs to be paired with a uniform "free" half-edge, i.e., one that has not been paired so far. For $\widehat{\mathsf{BP}}_n(m)$, this restriction does not hold. We now show how these two choices can be conveniently coupled.

For $\widehat{\mathsf{BP}}_n(m)$, we draw a uniform half-edge y_m from the collection of *all* half-edges, independently of the past randomness. Let U_m denote the vertex to which y_m is incident. We then let the mth individual in $(\widehat{\mathsf{BP}}_n(t))_{t\in[m-1]}$ have precisely $d_{U_m}-1$ children. Note that $d_{U_m}-1$ has the same distribution as $D_n^\star -1$ and, by construction, the collection $\left(d_{U_t}-1\right)_{t\geq 1}$ is iid. This constructs $\widehat{\mathsf{BP}}_n(m)$, except for the statuses of the sibling half-edges incident to U_t, which we describe below.

For $\widehat{\mathsf{G}}_n(m)$, when y_m is still free, i.e., it has not yet been paired in $(\widehat{\mathsf{G}}_n(t))_{t\in[m-1]}$, we let x_m be paired with y_m in $\widehat{\mathsf{G}}_n(m)$; we have thus also constructed $(\widehat{\mathsf{G}}_n(t), \widehat{\mathsf{BP}}_n(t))_{t\in[m]}$. We give all the other half-edges of U_m the status "real" when U_m has not yet appeared in $\widehat{\mathsf{G}}_n(m-1)$, and otherwise we give them the ghost status. The latter case implies that a cycle appears in $(\widehat{\mathsf{G}}_n(t))_{t\in[m]}$. By construction, such a cycle does not occur in $(\widehat{\mathsf{BP}}_n(t))_{t\in[m]}$, where reused vertices are simply repeated several times.

A difference in the coupling arises when y_m has already been paired in $(\widehat{\mathsf{G}}_n(t))_{t\in[m-1]}$, in which case we give all the sibling half-edges of U_t the ghost status. For $\widehat{\mathsf{G}}_n(m)$, we draw a uniform *unpaired* half-edge y'_m and pair x_m with y'_m instead, to obtain $\widehat{\mathsf{G}}_n(m)$, and we give all the sibling half-edges of y'_m the ghost status. Clearly, this might give rise to a difference between $\widehat{\mathsf{G}}_n(m)$ and $\widehat{\mathsf{BP}}_n(m)$.

We continue the above exploration algorithm until it terminates at some time T_n. Since each step pairs exactly one half-edge, we have that $T_n = |E(\mathscr{C}(o))|$, so that $T_n \leq \ell_n/2$ steps. The final result is then $(\widehat{\mathsf{G}}_n(t), \widehat{\mathsf{BP}}_n(t))_{t\in[T_n]}$. At this moment, however, the branching-process tree $(\widehat{\mathsf{BP}}_n(t))_{t\geq 1}$ has *not* been fully explored, since the tree vertices corresponding to ghost half-edges in $(\widehat{\mathsf{BP}}_n(t))_{t\geq 1}$ have not been explored. We complete the tree exploration $(\widehat{\mathsf{BP}}_n(t))_{t\geq 1}$ by iid drawing children of all the ghost tree vertices until the full tree is obtained.

We emphasize that the law of $(\widehat{\mathsf{BP}}_n(t))_{t\geq 1}$ obtained above is not the same as that of $(\mathsf{BP}_n(t))_{t\geq 1}$, since the order in which half-edges are paired is chosen in such a way that $(\widehat{\mathsf{G}}_n(t))_{t\in[T_n]}$ has the same law as the graph exploration process $(\mathsf{G}_n(t))_{t\in[T_n]}$. However, with σ_n the first time that a ghost half-edge is paired, we have that $(\widehat{BP}_n(t))_{t\in[\sigma_n]}$ *does* have the same law as $(BP_n(t))_{t\geq 1}$.

Differences between the Branching Process and Graph Exploration

We now provide bounds on the probability that differences occur between $(\widehat{G}_n(t))_{t\geq 1}$ and $(\widehat{BP}_n(t))_{t\geq 1}$ before time m_n. As discussed above, there are two sources of difference between $(\widehat{G}_n(t))_{t\in[m]}$ and $(\widehat{BP}_n(t))_{t\in[m]}$:

Half-edge reuse. In the above coupling, a half-edge reuse occurs when y_m has already been paired and is being reused in the branching process. As a result, for $(\widehat{G}_n(t))_{t\in[m]}$, we need to redraw y_m to obtain y'_m, which is used instead in $(\widehat{G}_n(t))_{t\in[m]}$;

Vertex reuse. A vertex reuse occurs when $U_m = U_{m'}$ for some $m' < m$. In the above coupling, this means that y_m is a half-edge that has not yet been paired in $(\widehat{G}_n(t))_{t\in[m-1]}$, but it is *incident* to a half-edge that has already been paired in $(\widehat{G}_n(t))_{t\in[m-1]}$. In particular, the vertex U_m to which it is incident has already appeared in $(\widehat{G}_n(t))_{t\in[m-1]}$, and it is being reused in the branching process. In this case, a *copy* of U_m appears in $(\widehat{BP}_n(t))_{t\in[m]}$, while a *cycle* appears in $(\widehat{G}_n(t))_{t\in[m]}$.

We continue by providing a bound on both contributions:

Half-Edge Reuse

At time $m-1$, precisely $2m-1$ half-edges are forbidden for use by $(\widehat{G}_n(t))_{t\in[m]}$. The probability that the half-edge y_m equals one of these half-edges is

$$\frac{2m-1}{\ell_n}. \tag{4.2.7}$$

Hence the expected number of half-edge reuses before time m_n is

$$\sum_{m=1}^{m_n} \frac{2m-1}{\ell_n} = \frac{m_n^2}{\ell_n} = o(1), \tag{4.2.8}$$

when $m_n = o(\sqrt{n})$. The Markov inequality ([V1, Theorem 2.17]) shows that the probability that a half-edge reuse occurs is also $o(1)$ when $m_n = o(\sqrt{n})$.

Vertex Reuse

The probability that vertex v is chosen in the mth draw of $(\widehat{BP}_n(t))_{t\geq 1}$ is equal to d_v/ℓ_n. The probability that vertex v is drawn twice before time m_n is therefore at most

$$\frac{m_n(m_n-1)}{2}\frac{d_v^2}{\ell_n^2}. \tag{4.2.9}$$

The expected number of vertex reuses up to time m_n is thus at most

$$\frac{m_n(m_n-1)}{2\ell_n}\sum_{v\in[n]}\frac{d_v^2}{\ell_n} \leq m_n^2\frac{d_{\max}}{\ell_n} = o(1), \tag{4.2.10}$$

by Condition 1.7(a),(b) when $m_n = o(\sqrt{n/d_{\max}})$. Again the Markov inequality completes the proof.

This completes the coupling part of Lemma 4.2, including the bound on m_n as formulated in Remark 4.3. It is straightforward to check that the exploration can be performed from the two sources (o_1, o_2) independently, thus establishing the required coupling to two independent n-dependent branching processes as claimed in Remark 4.3. □

Completion of the Proof: Convergence of $\mathbb{E}[N_{n,r}(\mathbf{t})]/n$

In order to show that $\mathbb{E}[N_{n,r}(\mathbf{t})]/n \to \mu(\bar{B}_r^{(G)}(o) = \mathbf{t})$, we let t_r denote the number of individuals in the first $r-1$ generations in $\mathbf{t}$ and let $(\mathbf{t}(t))_{t \in [t_r]}$ be its breadth-first exploration as in Definition 1.25. It is here that the fact that both $\bar{B}_r^{(G)}(o)$ and $\mathbf{t}$ are ordered is crucial. Then

$$\mathbb{E}[N_{n,r}(\mathbf{t})]/n = \mathbb{P}((\mathsf{G}_n(t))_{t \in [t_r]} = (\mathbf{t}(t))_{t \in [t_r]}), \qquad (4.2.11)$$

so that

$$
\begin{aligned}
\mathbb{P}(B_r^{(G_n)}(o) \simeq \mathbf{t}) &= \mathbb{P}((\mathsf{G}_n(t))_{t \in [t_r]} = (\mathbf{t}(t))_{t \in [t_r]}) \\
&= \mathbb{P}((\mathsf{BP}_n(t))_{t \in [t_r]} = (\mathbf{t}(t))_{t \in [t_r]}) + o(1) \\
&= \mathbb{P}((\mathsf{BP}(t))_{t \in [t_r]} = (\mathbf{t}(t))_{t \in [t_r]}) + o(1) \\
&= \mu(\bar{B}_r^{(G)}(o) = \mathbf{t}) + o(1), \qquad (4.2.12)
\end{aligned}
$$

where the second equality is (4.2.6) in Lemma 4.2, while the third is the statement that $\mathsf{BP}_n(t) \overset{d}{\longrightarrow} \mathsf{BP}(t)$ for every finite t by Exercise 4.1. This proves the claim. We note that (4.2.12) implies the local weak convergence of $\mathrm{CM}_n(\boldsymbol{d})$. □

4.2.2 Proof of Local Convergence of Configuration Model: Second Moment

We next study the second moment of $N_{n,r}(\mathbf{t})$ and show that it is almost its first moment squared:

Lemma 4.4 (Concentration number of trees) *Subject to Conditions 1.7(a),(b),*

$$\frac{\mathbb{E}[N_{n,r}(\mathbf{t})^2]}{n^2} \to \mu(\bar{B}_r^{(G)}(o) = \mathbf{t})^2. \qquad (4.2.13)$$

Consequently, $N_{n,r}(\mathbf{t})/n \overset{\mathbb{P}}{\longrightarrow} \mu(\bar{B}_r^{(G)}(o) = \mathbf{t})$.

Proof Let $o_1, o_2 \in [n]$ be two vertices chosen uar from $[n]$, independently. We start by computing

$$\frac{\mathbb{E}[N_{n,r}(\mathbf{t})^2]}{n^2} = \mathbb{P}(\bar{B}_r^{(G_n)}(o_1), \bar{B}_r^{(G_n)}(o_2) = \mathbf{t}). \qquad (4.2.14)$$

Recall that $|B_r^{(G_n)}(o_1)| \overset{d}{\longrightarrow} |B_r^{(G)}(o)|$, where $(G, o) \sim \mu$ denotes the local weak limit of $\mathrm{CM}_n(\boldsymbol{d})$ derived above. Since $|B_r^{(G)}(o)|$ is a tight random variable, $o_2 \notin B_{2r}^{(G_n)}(o_1)$ whp (recall Corollary 2.20), so that also

$$\frac{\mathbb{E}[N_{n,r}(\mathbf{t})^2]}{n^2} = \mathbb{P}(\bar{B}_r^{(G_n)}(o_1), \bar{B}_r^{(G_n)}(o_2) = \mathbf{t}, o_2 \notin B_{2r}^{(G_n)}(o_1)) + o(1). \qquad (4.2.15)$$

We now condition on $\bar{B}_r^{(G_n)}(o_1) = \mathbf{t}$, and write

$$\mathbb{P}(\bar{B}_r^{(G_n)}(o_1), \bar{B}_r^{(G_n)}(o_2) = \mathbf{t}, o_2 \notin B_{2r}^{(G_n)}(o_1))$$
$$= \mathbb{P}(\bar{B}_r^{(G_n)}(o_2) = \mathbf{t} \mid \bar{B}_r^{(G_n)}(o_1) = \mathbf{t}, o_2 \notin B_{2r}^{(G_n)}(o_1))$$
$$\times \mathbb{P}(\bar{B}_r^{(G_n)}(o_1) = \mathbf{t}, o_2 \notin B_{2r}^{(G_n)}(o_1)). \tag{4.2.16}$$

We already know that $\mathbb{P}(\bar{B}_r^{(G_n)}(o_1) = \mathbf{t}) \to \mu(\bar{B}_r^{(G)}(o) = \mathbf{t})$, so that also

$$\mathbb{P}(\bar{B}_r^{(G_n)}(o_1) = \mathbf{t}, o_2 \notin B_{2r}^{(G_n)}(o_1)) \to \mu(\bar{B}_r^{(G)}(o) = \mathbf{t}). \tag{4.2.17}$$

In Exercise 4.4, the reader can prove that (4.2.17) does indeed hold.

Conditional on $\bar{B}_r^{(G_n)}(o_1) = \mathbf{t}$ and $o_2 \notin B_r^{(G_n)}(o_1)$, the probability that $\bar{B}_r^{(G_n)}(o_2) = \mathbf{t}$ is the *same* as the probability that $\bar{B}_r^{(G_n)}(o_2) = \mathbf{t}$ in $\mathrm{CM}_{n'}(\boldsymbol{d'})$, which is obtained by removing all vertices in $B_r^{(G_n)}(o_1)$. Thus, since $\bar{B}_r^{(G_n)}(o_1) = \mathbf{t}$, we have that $n' = n - |V(\mathbf{t})|$ and $\boldsymbol{d'}$ is the corresponding degree sequence. The key point is that the degree distribution $\boldsymbol{d'}$ still satisfies Conditions 1.7(a),(b). Therefore, we also have

$$\mathbb{P}(\bar{B}_r^{(G_n)}(o_2) = \mathbf{t} \mid \bar{B}_r^{(G_n)}(o_1) = \mathbf{t}, o_2 \notin B_r^{(G_n)}(o_1)) \to \mu(\bar{B}_r^{(G)}(o) = \mathbf{t}), \tag{4.2.18}$$

and we have proved (4.2.13).

Finally, since $\mathbb{E}[N_{n,r}(\mathbf{t})]/n \to \mu(\bar{B}_r^{(G)}(o) = \mathbf{t})$ and $\mathbb{E}[N_{n,r}(\mathbf{t})^2]/n^2 \to \mu(\bar{B}_r^{(G)}(o) = \mathbf{t})^2$, it follows that $\mathrm{Var}(N_{n,r}(\mathbf{t})/n) \to 0$. Since also $\mathbb{E}[N_{n,r}(\mathbf{t})]/n \to \mu(\bar{B}_r^{(G)}(o) = \mathbf{t})$, we obtain that $N_{n,r}(\mathbf{t})/n \xrightarrow{\mathbb{P}} \mu(\bar{B}_r^{(G)}(o) = \mathbf{t})$, as required. $\qquad\square$

Lemma 4.4 completes the proof of Theorem 4.1. $\qquad\square$

Exercises 4.2 and 4.3 study an adaptation of the above proof for $\mathrm{NR}_n(\boldsymbol{w})$ using Proposition 3.16.

4.2.3 Concentration Inequality Proof using a Revealment Process

In this subsection we provide an alternative proof for the concentration proved in Section 4.2.2 (see in particular Lemma 4.4), based on *concentration techniques*. The proof also improves the local convergence in probability to almost surely, and further will be useful for analyzing uniform random graphs with prescribed degrees, as in the next section.

We label the half-edges arbitrarily as $[\ell_n]$, and pair them one by one, taking the next available half-edge with lowest label. For $t \in [\ell_n/2]$, let $\mathscr{F}_t$ denote the σ-algebra generated by the first t pairings. We define the Doob martingale $(M_t)_{t \geq 1}$ by

$$M_t = \mathbb{E}[N_{n,r}(\mathbf{t}) \mid \mathscr{F}_t], \tag{4.2.19}$$

so that $M_0 = \mathbb{E}[N_{n,r}(\mathbf{t})]$ and $M_{\ell_n/2} = N_{n,r}(\mathbf{t})$. We use the Azuma–Hoeffding inequality [V1, Theorem 2.27] to obtain the concentration of $M_{\ell_n/2} - M_0$. For this, we investigate, for $t \in [\ell_n/2]$,

$$M_t - M_{t-1} = \mathbb{E}[N_{n,r}(\mathbf{t}) \mid \mathscr{F}_t] - \mathbb{E}[N_{n,r}(\mathbf{t}) \mid \mathscr{F}_{t-1}]. \tag{4.2.20}$$

In the first term, we *reveal* one more pairing compared with the second term. We now study the effect of this extra pairing. Let $((x_s, y_s))_{s \in [\ell_n/2]}$ be the pairing conditional on $\mathscr{F}_t$, where we let the pairing $((x_s, y_s))_{s \in [\ell_n/2]}$ be such that $x_s < y_s$. We now construct a pairing $((x'_s, y'_s))_{s \in [\ell_n/2]}$ that has the correct distribution under $\mathscr{F}_{t-1}$, while $((x_s, y_s))_{s \in [\ell_n/2]}$ and $((x'_s, y'_s))_{s \in [\ell_n/2]}$ differ by at most two edges almost surely, i.e., by switching one pairing.

For this, we let $(x_s, y_s) = (x'_s, y'_s)$ for $s \in [t-1]$. Then we let x_t be the half-edge with lowest label that has not been paired yet at time t, and y_t its pair prescribed by $\mathscr{F}_t$. Further, we also let $x'_t = x_t$, and we let y'_t be a pair of x_t chosen *independently* of y_t from the set of available half-edges prescribed by $\mathscr{F}_{t-1}$. Then, clearly, $((x_s, y_s))_{s \in [t]}$ and $((x'_s, y'_s))_{s \in [t]}$ have the correct distributions. We complete the proof by describing how the remaining half-edges can be paired in such a way that at most two edges are different in $((x_s, y_s))_{s \in [\ell_n/2]}$ and $((x'_s, y'_s))_{s \in [\ell_n/2]}$.

Let (x_a, y_b) be the unique pair of $((x_s, y_s))_{s \in [\ell_n/2]}$ such that $y'_t \in \{x_a, y_a\}$. Then, we pair y_t in $((x'_s, y'_s))_{s \in [\ell_n/2]}$ to $\{x_a, y_a\} \setminus \{y'_t\}$. Thus, in $((x'_s, y'_s))_{s \in [\ell_n/2]}$, y_t is paired with y_a when $y'_t = x_a$, and y_t is paired with x_a when $y'_t = y_a$. This means that the pair of edges (x_t, y_t) and (x_a, y_a) in $((x_s, y_s))_{s \in [\ell_n/2]}$ is *switched* to (x_t, y'_t) and either the ordered version of $\{x_a, y_t\}$ or that of $\{y_a, y_t\}$ in $((x'_s, y'_s))_{s \in [\ell_n/2]}$. All other pairs in $((x'_s, y'_s))_{s \in [\ell_n/2]}$ are the same as in $((x_s, y_s))_{s \in [\ell_n/2]}$. Since (x_t, y'_t) is paired independently of (x_t, y_t), the conditional distribution of $((x'_s, y'_s))_{s \in [\ell_n/2]}$ given $\mathscr{F}_{t-1}$ is the same as that of $((x_s, y_s))_{s \in [\ell_n/2]}$ given $\mathscr{F}_{t-1}$, as required.

Let $N'_{n,r}(\mathbf{t})$ be the number of vertices whose r-neighborhood is isomorphic to $\mathbf{t}$ in $((x'_s, y'_s))_{s \in [\ell_n/2]}$. The above coupling gives that

$$M_t - M_{t-1} = \mathbb{E}[N_{n,r}(\mathbf{t}) - N'_{n,r}(\mathbf{t}) \mid \mathscr{F}_t]. \tag{4.2.21}$$

When switching two edges, the number of vertices whose r-neighborhood is isomorphic to $\mathbf{t}$ cannot change by more than $4c$, where $c = \sum_{k=0}^r d^k$ and d is the maximal degree in $\mathbf{t}$. Indeed, the presence of an edge $\{u, v\}$ in the resulting multi-graph G_n affects the event $\{B_r^{(G_n)}(i) \simeq \mathbf{t}\}$ only if there exists a path of length at most r in G_n between i and $\{u, v\}$, the maximal degree along which is at most d. For the given choice of $\{u, v\}$ there are at most $2c$ such values of $i \in [n]$. Since a switch changes two edges, we obtain that $|N_{n,r}(\mathbf{t}) - N'_{n,r}(\mathbf{t})| \leq 4c$. Thus, Azuma–Hoeffding [V1, Theorem 2.27] implies that (with the time variable n in [V1, Theorem 2.27] replaced by $\ell_n/2$)

$$\mathbb{P}(|N_{n,r}(\mathbf{t}) - \mathbb{E}[N_{n,r}(\mathbf{t})]| \geq n\varepsilon) = \mathbb{P}(|M_{\ell_n/2} - M_0| \geq n\varepsilon)$$
$$\leq 2e^{-(n\varepsilon)^2/[16c^2\ell_n]}. \tag{4.2.22}$$

Since this vanishes exponentially, and so is summable, we have proved the following corollary:

Corollary 4.5 (Almost sure local convergence) *The local convergence in Theorem 4.1 in fact occurs almost surely.*

4.2.4 Local Convergence of Uniform Graphs with Prescribed Degrees

We start by investigating the locally tree-like nature of the uniform random graph $\mathrm{UG}_n(\boldsymbol{d})$ with prescribed degrees $\boldsymbol{d}$ (recall Section 1.3.3). Our main result is as follows:

Theorem 4.6 (Locally tree-like nature of uniform graphs with given degrees) *Assume that Conditions 1.7(a),(b) hold. Assume further that the empirical distribution F_n of $\boldsymbol{d}$ satisfies*

$$[1 - F_n](x) \leq c_F x^{-(\tau-1)}, \tag{4.2.23}$$

for some $c_F > 0$, $\tau \in (2, 3)$. Then $\mathrm{UG}_n(\boldsymbol{d})$ converges locally in probability to the unimodular branching-process tree with root offspring distribution $(p_k)_{k \geq 0} = (\mathbb{P}(D = k))_{k \geq 0}$.

We give two proofs of Theorem 4.6. The first proof, below, uses Theorem 1.13 to compare probabilities for $\mathrm{UG}_n(\boldsymbol{d})$ with those in $\mathrm{CM}_n(\boldsymbol{d})$, and then relies on Theorem 4.1. The second proof uses the concentration inequality in (4.2.22). We refer the reader to Section 4.6 for further discussion.

Proof   We rely on Theorem 1.13, for which (4.2.23) provides the assumption. We will compare the neighborhood probabilities in $\mathrm{UG}_n(\boldsymbol{d})$ with those in $\mathrm{CM}_n(\boldsymbol{d})$, and show that these are asymptotically equal. We then use Theorem 4.1 to reach the conclusion.

It is convenient to order all half-edges in $\mathrm{UG}_n(\boldsymbol{d})$ randomly. We then write $\{u \xrightarrow{j} v \text{ in } \mathrm{UG}_n(\boldsymbol{d})\}$ for the event that the jth half-edge incident to u connects to v in $\mathrm{UG}_n(\boldsymbol{d})$. For $\mathrm{CM}_n(\boldsymbol{d})$, we also order the half-edges in $[\ell_n]$ randomly, and we write $\{u \xrightarrow{j} v \text{ in } \mathrm{CM}_n(\boldsymbol{d})\}$ for the event that the jth half-edge incident to u connects to v in $\mathrm{CM}_n(\boldsymbol{d})$.

Fix an ordered tree $\mathbf{t}$ and write $G_n = \mathrm{UG}_n(\boldsymbol{d})$. Let us start by computing $\mathbb{P}(\bar{B}_r^{(G_n)}(o_1) = \mathbf{t})$, where we write $\bar{B}_r^{(G_n)}(o)$ for the *ordered* version of $B_r^{(G_n)}(o)$, in which we order all half-edges incident to o, as well as the *forward* half-edges incident to $v \in B_r^{(G_n)}(o) \setminus \{o\}$, according to their labels. Here we bear in mind that the half-edge at v connecting v to the (unique) vertex closer to the root o is not one of the forward edges.

Recall (1.3.45) in Theorem 1.13. Since the degrees in $\mathbf{t}$ are bounded, we can make the approximation

$$\mathbb{P}(u \sim v \mid \mathcal{E}_U) = (1 + o(1)) \frac{(d_u - |U_u|)(d_v - |U_v|)}{\ell_n}, \qquad (4.2.24)$$

where U_x denotes the set of pairs in U that contain x.

We condition on $u = o$, which occurs with probability $1/n$. We explore $B_r^{(G_n)}(o)$ in a breadth-first manner, starting with the neighbors $(v_i)_{i=1}^{d_u}$ of u. In this setting, $|U_{v_i}| = 0$ for all $i \in [d_u]$, while $d_u - |U_u| = d_u - i + 1$ for the probability of $u \sim v_i$ in (4.2.24). Thus, with $U_{i-1} = \{\{u, v_1\}, \dots, \{u, v_{i-1}\}\}$,

$$\mathbb{P}(u \sim v_i \mid \mathcal{E}_{U_{i-1}}) = (1 + o(1)) \frac{(d_u - i + 1)d_{v_i}}{\ell_n}. \qquad (4.2.25)$$

Taking the product over $i \in [d_u]$, we conclude that

$$\mathbb{P}(\partial B_1^{(G_n)}(u) = \{v_1, \dots, v_{d_u}\}) = (1 + o(1))d_u! \prod_{i=1}^{d_u} \frac{d_{v_i}}{\ell_n}. \qquad (4.2.26)$$

Recalling that $\bar{B}_1^{(G_n)}(u)$ is the *ordered* version of the 1-neighborhood of u and noting that there are $d_u!$ orderings of the edges incident to u, each of them equally likely, we thus obtain that

$$\mathbb{P}(\partial \bar{B}_1^{(G_n)}(u) = (v_1, \ldots, v_{d_u})) = (1 + o(1)) \frac{d_u!}{d_u!} \prod_{j=1}^{d_u} \frac{d_{v_j}}{\ell_n}$$

$$= (1 + o(1)) \prod_{j=1}^{d_u} \frac{d_{v_j}}{\ell_n}. \qquad (4.2.27)$$

Alternatively,

$$\mathbb{P}(u \overset{j}{\rightsquigarrow} v_j \; \forall j \in [d_u] \text{ in } \mathrm{UG}_n(\boldsymbol{d})) = (1 + o(1)) \prod_{j=1}^{d_u} \frac{d_{v_j}}{\ell_n}. \qquad (4.2.28)$$

For $\mathrm{CM}_n(\boldsymbol{d})$, we compute

$$\mathbb{P}(u \overset{j}{\rightsquigarrow} v_j \; \forall j \in [d_u] \text{ in } \mathrm{CM}_n(\boldsymbol{d})) = \prod_{j=1}^{d_u} \frac{d_{v_j}}{\ell_n - 2j + 1}$$

$$= (1 + o(1)) \prod_{j=1}^{d_u} \frac{d_{v_j}}{\ell_n}, \qquad (4.2.29)$$

so that

$$\mathbb{P}(u \overset{j}{\rightsquigarrow} v_j \; \forall j \in [d_u] \text{ in } \mathrm{UG}_n(\boldsymbol{d}))$$
$$= (1 + o(1)) \mathbb{P}(u \overset{j}{\rightsquigarrow} v_j \; \forall j \in [d_u] \text{ in } \mathrm{CM}_n(\boldsymbol{d})). \qquad (4.2.30)$$

This shows that, conditional on $o = u$, the neighborhood sets of u in $\mathrm{UG}_n(\boldsymbol{d})$ can be coupled whp to those in $\mathrm{CM}_n(\boldsymbol{d})$.

We continue by investigating the neighborhood set of another vertex $v \in B_r^{(G_n)}(u)$. For this, we note that one edge has already been used to connect v to $B_r^{(G_n)}(u)$, so there are $d_v - 1$ edges remaining, which we will call *forward edges*. Let v be the sth vertex in $\bar{B}_r^{(G_n)}(u)$, and let $\mathscr{F}_{s-1}$ denote all the information about the edges and vertices that have been explored before vertex v. Then we compute

$$\mathbb{P}(v \overset{j}{\rightsquigarrow} v_j \; \forall j \in [d_v - 1] \text{ in } \mathrm{UG}_n(\boldsymbol{d}) \mid \mathscr{F}_{s-1}) = (1 + o(1)) \prod_{j=1}^{d_v - 1} \frac{d_{v_j}}{\ell_n}, \qquad (4.2.31)$$

where we use that there are $(d_v - 1)!$ orderings of the $d_v - 1$ forward edges. For $\mathrm{CM}_n(\boldsymbol{d})$, we instead compute

$$\mathbb{P}(v \overset{j}{\rightsquigarrow} v_j \; \forall j \in [d_u] \text{ in } \mathrm{CM}_n(\boldsymbol{d}) \mid \mathscr{F}_{s-1}) = \prod_{j=1}^{d_v - 1} \frac{d_{v_j}}{\ell_n - 2j - 2s - 1}$$

$$= (1 + o(1)) \prod_{j=1}^{d_v - 1} \frac{d_{v_j}}{\ell_n}, \qquad (4.2.32)$$

so that, for every $s \geq 1$,

$$\mathbb{P}(v \overset{j}{\leadsto} v_j \; \forall j \in [d_v - 1] \text{ in } \mathrm{UG}_n(\boldsymbol{d}) \mid \mathscr{F}_{s-1})$$
$$= (1 + o(1))\mathbb{P}(v \overset{j}{\leadsto} v_j \; \forall j \in [d_v - 1] \text{ in } \mathrm{CM}_n(\boldsymbol{d}) \mid \mathscr{F}_{s-1}). \qquad (4.2.33)$$

Using (4.2.30) for the neighborhood sets of the root $o = u$, and (4.2.33) repeatedly for all other vertices in $B_r^{(G_n)}(u)$, we conclude that

$$\mathbb{P}(\bar{B}_r^{(G_n)}(o_1) = \mathbf{t} \text{ in } \mathrm{UG}_n(\boldsymbol{d})) = (1 + o(1))\mathbb{P}(\bar{B}_r^{(G_n)}(o_1) = \mathbf{t} \text{ in } \mathrm{CM}_n(\boldsymbol{d})). \qquad (4.2.34)$$

We then use (4.2.12) to conclude that

$$\mathbb{P}(\bar{B}_r^{(G_n)}(o_1) = \mathbf{t} \text{ in } \mathrm{UG}_n(\boldsymbol{d})) = \mu(\bar{B}_r^{(G)}(o) = \mathbf{t}) + o(1), \qquad (4.2.35)$$

as required.

We continue by computing $\mathbb{P}(\bar{B}_r^{(G_n)}(o_1), \bar{B}_r^{(G_n)}(o_2) = \mathbf{t})$, for which we use similar ideas to arrive at

$$\mathbb{P}(\bar{B}_r^{(G_n)}(o_1), \bar{B}_r^{(G_n)}(o_2) = \mathbf{t} \text{ in } \mathrm{UG}_n(\boldsymbol{d}))$$
$$= (1 + o(1))\mathbb{P}(\bar{B}_r^{(G_n)}(o_1), \bar{B}_r^{(G_n)}(o_2) = \mathbf{t} \text{ in } \mathrm{CM}_n(\boldsymbol{d})). \qquad (4.2.36)$$

By (4.2.14), as well as (4.2.13) in Lemma 4.4, we obtain

$$\mathbb{P}(\bar{B}_r^{(G_n)}(o_1), \bar{B}_r^{(G_n)}(o_2) = \mathbf{t} \text{ in } \mathrm{UG}_n(\boldsymbol{d}))$$
$$= \mu(\bar{B}_r^{(G)}(o) = \mathbf{t})^2 + o(1), \qquad (4.2.37)$$

so that

$$\frac{N_{n,r}(\mathbf{t})}{n} = \frac{1}{n}\#\{u \colon \bar{B}_r^{(G_n)}(u) = \mathbf{t} \text{ in } \mathrm{UG}_n(\boldsymbol{d})\} \overset{\mathbb{P}}{\longrightarrow} \mu(\bar{B}_r^{(G)}(o) = \mathbf{t}), \qquad (4.2.38)$$

as required. This completes the proof of Theorem 4.6. $\qquad\qquad\square$

A Concentration Inequality Proof

As for Corollary 4.5, we can use the concentration inequality in (4.2.22) to show that $\mathrm{UG}_n(\boldsymbol{d})$ converges locally almost surely:

Corollary 4.7 (Almost sure local convergence) *The local convergence in Theorem 4.6 in fact occurs almost surely.*

To prove Corollary 4.7, we rely on the following lemma about the probability of the simplicity of $\mathrm{CM}_n(\boldsymbol{d})$, which is interesting in its own right:

Lemma 4.8 (Simplicity probability) *Consider* $\mathrm{CM}_n(\boldsymbol{d})$ *subject to Conditions 1.7(a),(b). Then*

$$\mathbb{P}(\mathrm{CM}_n(\boldsymbol{d}) \text{ simple}) = \mathrm{e}^{-o(n)}. \qquad (4.2.39)$$

Proof We call a vertex *high degree* when its degree is at least K, and *low degree* when its degree is in $[k]$. The quantities K and k will be determined later on. Let

$$d_{\geq K} = \sum_{v \in [n]} d_v \mathbb{1}_{\{d_v \geq K\}}, \qquad d_{[k]} = \sum_{v \in [n]} d_v \mathbb{1}_{\{d_v \in [k]\}} \qquad (4.2.40)$$

be the total numbers of half-edges incident to high- and low-degree vertices, respectively.

We start by pairing the $d_{\geq K}$ half-edges incident to high-degree vertices, and we call their pairing *good* when the half-edges incident to a high-degree vertex are all connected to distinct low-degree vertices. We also call a sub-pairing (i.e., a pairing of a subset of the $d_{\geq K}$ half-edges incident to high-degree vertices) good when the half-edges in it are such that all half-edges incident to the same vertex are paired with distinct vertices.

Let $n_{[k]}$ denote the number of low-degree vertices. Note that, independently of how earlier half-edges have been paired, the probability that the pairing of a half-edge keeps the sub-pairing good is at least $(n_{[k]} - d_{\geq K})/\ell_n \geq \alpha$ for some $\alpha > 0$, when k is such that $n_{[k]} \geq \varepsilon n$ and when K is large enough that $d_{\geq K} \leq \varepsilon n/2$, which we assume from now on.

Let $\mathcal{E}_n$ be the probability that the pairing of half-edges incident to high-degree vertices is good. Then, by the above,

$$\mathbb{P}(\mathcal{E}_n) \geq \alpha^{d_{\geq K}}. \qquad (4.2.41)$$

Now choose $K = K(\varepsilon)$ sufficiently large that $\log(1/\alpha)d_{\geq K} \leq \varepsilon n/2$ for every n. Then we obtain

$$\mathbb{P}(\mathcal{E}_n) \geq e^{-\varepsilon n/2}. \qquad (4.2.42)$$

Having paired the half-edges incident to the high-degree vertices, we pair the remaining half-edges uniformly. Note that $CM_n(d)$ is simple precisely when this pairing produces a simple graph. Since the maximal degree is now bounded by K, the probability of the simplicity of this graph is $\Theta(1) \gg e^{-\varepsilon n/2}$ (recall (1.3.41)). Thus, we arrive at

$$\mathbb{P}(CM_n(d) \text{ simple}) \geq e^{-\varepsilon n}. \qquad (4.2.43)$$

Since $\varepsilon > 0$ is arbitrary, the claim follows, when noting that obviously $\mathbb{P}(CM_n(d) \text{ simple}) \leq 1 = e^{-o(n)}$. $\qquad \square$

Proof of Corollary 4.7. By (4.2.22), and by Lemma 4.8,

$$\mathbb{P}(|N_{n,r}(\mathbf{t}) - \mathbb{E}[N_{n,r}(\mathbf{t})]| \geq n\varepsilon \mid CM_n(d) \text{ simple})$$
$$\leq 2e^{-(n\varepsilon)^2/[16c^2\ell_n]}e^{o(n)} = e^{-\Theta(n)}, \qquad (4.2.44)$$

which completes the proof since $CM_n(d)$, conditioned on simplicity, has the same law as $UG_n(d)$ by (1.3.29) and the discussion below it. $\qquad \square$

4.3 THE GIANT IN THE CONFIGURATION MODEL

In this section we investigate the connected components in the configuration model. Similarly to the Erdős–Rényi random graph, we identify when the configuration model whp has a giant component. Again, this condition has the interpretation that an underlying branching process describing the exploration of a cluster has a strictly positive survival probability.

For a graph G, we recall that $v_k(G)$ denotes the number of vertices of degree k in G and $|E(G)|$ the number of edges. The main result concerning the size and structure of the largest connected components of $CM_n(d)$ is the following:

Theorem 4.9 (Phase transition in $\mathrm{CM}_n(\boldsymbol{d})$)   *Consider* $\mathrm{CM}_n(\boldsymbol{d})$ *subject to Conditions 1.7(a),(b). Assume that* $p_2 = \mathbb{P}(D = 2) < 1$. *Let* $\mathscr{C}_{\max}$ *and* $\mathscr{C}_{(2)}$ *be the largest and second largest connected components of* $\mathrm{CM}_n(\boldsymbol{d})$ *(breaking ties arbitrarily).*

(a) *If* $\nu = \mathbb{E}[D(D-1)]/\mathbb{E}[D] > 1$, *then there exist* $\xi \in [0,1), \zeta \in (0,1]$ *such that*

$$|\mathscr{C}_{\max}|/n \overset{\mathbb{P}}{\longrightarrow} \zeta,$$

$$v_k(\mathscr{C}_{\max})/n \overset{\mathbb{P}}{\longrightarrow} p_k(1 - \xi^k) \quad \text{for every } k \geq 0,$$

$$|E(\mathscr{C}_{\max})|/n \overset{\mathbb{P}}{\longrightarrow} \tfrac{1}{2}\mathbb{E}[D](1 - \xi^2),$$

while $|\mathscr{C}_{(2)}|/n \overset{\mathbb{P}}{\longrightarrow} 0$ *and* $|E(\mathscr{C}_{(2)})|/n \overset{\mathbb{P}}{\longrightarrow} 0$.

(b) *If* $\nu = \mathbb{E}[D(D-1)]/\mathbb{E}[D] \leq 1$, *then* $|\mathscr{C}_{\max}|/n \overset{\mathbb{P}}{\longrightarrow} 0$ *and* $|E(\mathscr{C}_{\max})|/n \overset{\mathbb{P}}{\longrightarrow} 0$.

Consequently, the same result holds for the uniform random graph with degree sequence $\boldsymbol{d}$ *satisfying Conditions 1.7(a),(b), under the extra assumption that* $\sum_{i \in [n]} d_i^2 = O(n)$.

Reformulation in Terms of Branching Processes

We start by interpreting the results in Theorem 4.9 in terms of *branching processes*, as also arising in Section 4.2. As it turns out, we can interpret ζ as the survival probability of the unimodular branching process with root offspring distribution $(p_k)_{k \geq 1}$ that appears in Theorem 4.1 and ξ as the extinction probability of a branching process with offspring distribution $(p_k^\star)_{k \geq 0}$. Thus, ζ satisfies

$$\zeta = \sum_{k \geq 1} p_k(1 - \xi^k), \tag{4.3.1}$$

where ξ satisfies

$$\xi = \sum_{k \geq 0} p_k^\star \xi^k. \tag{4.3.2}$$

Clearly, $\xi = 1$ precisely when

$$\nu = \sum_{k \geq 0} k p_k^\star \leq 1. \tag{4.3.3}$$

By (4.2.2), we can rewrite

$$\nu = \frac{1}{\mathbb{E}[D]} \sum_{k \geq 0} k(k+1)p_{k+1} = \frac{\mathbb{E}[D(D-1)]}{\mathbb{E}[D]}, \tag{4.3.4}$$

which explains the condition on ν in Theorem 4.9(a).

Further, to understand the asymptotics of $v_k(\mathscr{C}_{\max})$, we note that there are $n_k = np_k^{(G_n)} \approx np_k$ vertices with degree k. Each of the k direct neighbors of a vertex of degree k survives with probability close to $1 - \xi$, so that the probability that at least one of them survives is close to $1 - \xi^k$. When one neighbors of the vertex of degree k survives, the vertex itself is part of the giant component, which explains why $v_k(\mathscr{C}_{\max})/n \overset{\mathbb{P}}{\longrightarrow} p_k(1 - \xi^k)$.

Finally, an edge consists of two half-edges, and an edge is part of the giant component precisely when this is true for one vertex incident to it, which occurs with probability

$1 - \xi^2$. There are in total $\ell_n/2 = n\mathbb{E}[D_n]/2 \approx n\mathbb{E}[D]/2$ edges, which explains why $|E(\mathscr{C}_{\max})|/n \xrightarrow{\mathbb{P}} \frac{1}{2}\mathbb{E}[D](1 - \xi^2)$. Therefore, the results in Theorem 4.9 have a simple explanation in terms of the branching-process approximation of the connected component of a uniform vertex in $[n]$ in $\mathrm{CM}_n(\boldsymbol{d})$.

The Condition $\mathbb{P}(D = 2) = p_2 < 1$

Because isolated vertices do not matter, without loss of generality, we may assume that $p_0 = 0$. The case $p_2 = 1$, for which $\nu = 1$, is quite exceptional. We give three examples showing that then quite different behaviors are possible.

Our first example is when $d_v = 2$ for all $v \in [n]$, so we are studying a random 2-regular graph. In this case the components are cycles and the distribution of cycle lengths in $\mathrm{CM}_n(\boldsymbol{d})$ is given by Ewen's sampling formula $\mathrm{ESF}(\frac{1}{2})$; see, e.g., Arratia et al. (2003). This implies that $|\mathscr{C}_{\max}|/n$ converges in distribution to a *non-degenerate* distribution on $[0, 1]$ (Arratia et al., 2003, Lemma 5.7) and not to any constant as in Theorem 4.9. Moreover, the same is true for $|\mathscr{C}_{(2)}|/n$ (and for $|\mathscr{C}_{(3)}|/n, \ldots$), so in this case there are *several* large components.

To see this result intuitively, we note that in the exploration of a cluster we start with one vertex with two half-edges. When pairing a half-edge, it connects to a vertex that again has two half-edges. Therefore, the number of half-edges to be paired is always equal to 2, up to the moment when the cycle is closed, and the cluster is completed. When there are $m = \alpha n$ free half-edges left, the probability of closing up the cycle equals $1/m = 1/(\alpha n)$, and, thus, the time this takes is of order n. A slight extension of this reasoning shows that the time it takes to close a cycle is nT_n, where T_n converges to a limiting non-degenerate random variable (see Exercise 4.5).

Our second example for $p_2 = 1$ is obtained by adding a small number of vertices of degree 1. More precisely, we let $n_1 \to \infty$ be such that $n_1/n \to 0$ and $n_2 = n - n_1$. In this case, components can either be cycles, or strings of vertices with degree 2 terminated with two vertices with degree 1. When $n_1 \to \infty$, it is more likely that a long string of vertices of degree 2 will be terminated by a vertex of degree 1 than by closing the cycle, as for the latter we need to pair to a *unique* half-edge while for the former we have n_1 choices. Therefore, intuitively this implies that $|\mathscr{C}_{\max}| = o_{\mathbb{P}}(n)$ (see Exercise 4.6 for details).

Our third example for $p_2 = 1$ is obtained by instead adding a small number of vertices of degree 4 (i.e., $n_4 \to \infty$ such that $n_4/n \to 0$, and $n_2 = n - n_4$.) We can regard each vertex of degree 4 as *two* vertices of degree 2 that have been identified. Therefore, to obtain $\mathrm{CM}_n(\boldsymbol{d})$ with this degree distribution, we can start from a configuration model having $n' = n + n_4$ vertices, and uniformly identify n_4 pairs of vertices of degree 2. Since the configuration model with $n' = n + n_4$ vertices of degree 2 has many components having size of order n, most of these merge into one giant component upon identification of these pairs. As a result, $|\mathscr{C}_{\max}| = n - o_{\mathbb{P}}(n)$, so there is a giant component containing almost everything (see Exercise 4.7).

We conclude that the case where $p_2 = \mathbb{P}(D = 2) = 1$ is quite sensitive to the precise properties of the degree structure, which are not captured by the limiting distribution $(p_k)_{k \geq 1}$ only. In what follows, we thus ignore the case where $p_2 = 1$.

Organization of the Proof of Theorem 4.9

We give two proofs of the existence and uniqueness of the giant in Theorem 4.9. In the first proof, in Section 4.3.1 we apply the "giant component is almost local" result in Theorems 2.28 and 2.31. In the second proof, in Section 4.3.2, we apply a continuous-time exploration to "uncover" the giant.

4.3.1 "Giant Component is Almost Local" Proof

Setting the Stage for the Proof of Theorem 4.9

Theorem 4.9(b) follows directly from Corollary 2.27 combined with the local convergence in probability in Theorem 4.1 and the fact that, for $\nu \leq 1$ and $p_2 < 1$, the unimodular branching-process tree with root offspring distribution $(p_k)_{k \geq 0}$ given by $p_k = \mathbb{P}(D = k)$ dies out almost surely.

Theorem 4.9(a) follows from Theorems 2.28–2.31, together with the facts that, for the unimodular branching-process tree with root offspring distribution $(p_k)_{k \geq 0}$ given by $p_k = \mathbb{P}(D = k)$, we have

$$\mu(|\mathscr{C}(o)| = \infty, d_o = \ell) = p_\ell(1 - \xi^\ell) \tag{4.3.5}$$

and

$$\mathbb{E}_\mu\left[d_o \mathbb{1}_{\{|\mathscr{C}(o)| = \infty\}}\right] = \mathbb{E}[D](1 - \xi^2). \tag{4.3.6}$$

For the latter, we refer to Exercise 4.8. Thus, it suffices to check the assumptions in Theorems 2.28–2.31. The uniform integrability of $D_n = d_{o_n}^{(G_n)}$ follows from Conditions 1.7(a),(b). For the assumptions in Theorem 2.28, the local convergence in probability follows from Theorem 4.1, so we are left to prove the crucial hypothesis in (2.6.7), which the remainder of the proof does.

We first prove (2.6.7) under the assumption that $d_v \leq b$ for all $v \in [n]$. At the end of the proof, we will lift this assumption. To start with our proof of (2.6.7), applied to $G_n = \mathrm{CM}_n(\boldsymbol{d})$ under the condition that $d_v \leq b$ for all $v \in [n]$, we first use the alternative formulation from Lemma 2.33, and note that (2.6.39) holds for the unimodular branching-process tree with root offspring distribution $(p_k)_{k \geq 0}$ given by $p_k = \mathbb{P}(D = k)$. Thus, instead of proving (2.6.7), it suffices to prove (2.6.38), which we do later.

Recall from (2.6.54) that, with $o_1, o_2 \in [n]$ chosen independently and uar,

$$\frac{1}{n^2}\mathbb{E}\left[\#\{(x, y) \in [n] \times [n] \colon |\partial B_r^{(G_n)}(x)|, |\partial B_r^{(G_n)}(y)| \geq r, x \not\longleftrightarrow y\}\right]$$
$$= \mathbb{P}(|\partial B_r^{(G_n)}(o_1)|, |\partial B_r^{(G_n)}(o_2)| \geq r, o_1 \not\longleftrightarrow o_2). \tag{4.3.7}$$

Thus, (2.6.38) states that

$$\lim_{r \to \infty} \limsup_{n \to \infty} \mathbb{P}(|\partial B_r^{(G_n)}(o_1)|, |\partial B_r^{(G_n)}(o_2)| \geq r, o_1 \not\longleftrightarrow o_2) = 0. \tag{4.3.8}$$

This is what we focus on from now on.

Coupling to Branching Process: Lemma 4.2 and Beyond

We next apply Lemma 4.2, in particular using the two extensions in Remark 4.3. Take an arbitrary $\underline{m}_n = o(\sqrt{n})$; then Remark 4.3 shows that whp we can perfectly couple $(B_k^{(G_n)}(o_1))_{k \leq \underline{k}_n}$, where

$$\underline{k}_n = \inf \{k \colon |B_k^{(G_n)}(o_1)| \geq \underline{m}_n \}, \tag{4.3.9}$$

to a unimodular branching processes $(\mathrm{BP}_k^{(1)})_{k \leq \underline{k}_n}$ with root offspring distribution $(p_k^{(G_n)})_{k \geq 1}$ in (4.2.1). Here we recall the notation below Definition 1.26, where $\mathrm{BP}_k^{(1)}$ was defined to consist of the individuals in the kth generation of this branching process. Since all degrees are bounded, $|B_{\underline{k}_n}^{(G_n)}(o_1)| \leq (1 + b)\underline{m}_n = \Theta(\underline{m}_n)$. Let $\mathcal{C}_n(1)$ denote the event that this perfect coupling happens, so that

$$\mathcal{C}_n(1) = \{(|\partial B_k^{(G_n)}(o_1)|)_{k \leq \underline{k}_n} = (|\mathrm{BP}_k^{(1)}|)_{k \leq \underline{k}_n} \}, \quad \text{and} \quad \mathbb{P}(\mathcal{C}_n(1)) = 1 - o(1). \tag{4.3.10}$$

We can extend the above coupling to deal with vertex o_2, for which we need to explore a little further. For this, we start by defining the necessary notation. Fix $\overline{m}_n \geq \underline{m}_n$, to be determined later. Let

$$\overline{k}_n = \inf \{k \colon |B_k^{(G_n)}(o_2)| \geq \overline{m}_n \}, \tag{4.3.11}$$

and, again since all degree are bounded, $|B_{\overline{k}_n}^{(G_n)}(o_2)| \leq (1 + b)\overline{m}_n = \Theta(\overline{m}_n)$. Further, for $\delta > 0$, we let

$$\begin{aligned} \mathcal{C}_n(2) &= \{(|\partial B_k^{(G_n)}(o_2)|)_{k \leq \underline{k}_n} = (|\mathrm{BP}_k^{(2)}|)_{k \leq \underline{k}_n} \} \\ &\quad \cap \{||\partial B_k^{(G_n)}(o_2)| - |\mathrm{BP}_k^{(2)}|| \leq (\overline{m}_n^2/\ell_n)^{1+\delta} \; \forall k \in [\overline{k}_n] \} \\ &\equiv \mathcal{C}_n(2,1) \cap \mathcal{C}_n(2,2), \end{aligned} \tag{4.3.12}$$

where $\mathcal{C}_n(2, 1)$ and $\mathcal{C}_n(2, 2)$ refer to the events in the first and second line of (4.3.12), respectively. Here $(\mathrm{BP}_k^{(2)})_{k \geq 0}$ is again an n-dependent unimodular branching process independent of $(\mathrm{BP}_k^{(1)})_{k \geq 0}$. With $\underline{m}_n = o(\sqrt{n})$, we will later pick $\overline{m}_n$ such that $\underline{m}_n \overline{m}_n \gg n$, to reach our conclusion. The following lemma shows that also $\mathcal{C}_n(2)$ occurs whp:

Lemma 4.10 (Coupling beyond Lemma 4.2)   *Consider* $\mathrm{CM}_n(\boldsymbol{d})$ *and let* $\overline{m}_n^2/\ell_n \to \infty$. *Then, for every* $\delta > 0$,

$$\mathbb{P}(\mathcal{C}_n(2)) = 1 - o(1). \tag{4.3.13}$$

Proof   The fact that $\mathcal{C}_n(2,1) = \{(|\partial B_k^{(G_n)}(o_2)|)_{k \leq \underline{k}_n} = (|\mathrm{BP}_k^{(2)}|)_{k \leq \underline{k}_n} \}$ occurs whp follows in the same way as in (4.3.10). We thus investigate the bounds in $\mathcal{C}_n(2,1)$ only for $k \in (\underline{k}_n, \overline{k}_n]$.

Define $a_n = (\overline{m}_n^2/\ell_n)^{1+\delta}$ where $\delta > 0$. Let $m_n = (b + 1)\overline{m}_n$ denote the maximal number of vertex explorations needed to explore the k_nth generation. Recalling the notation in Lemma 4.2, $\widehat{\mathsf{G}}_n(m_n)$ and $\widehat{\mathsf{BP}}_n(m_n)$ denote the half-edges and individuals found up to the m_nth step of the exploration starting from o_2; let $\widehat{\mathsf{G}}_n(m_n) \setminus \widehat{\mathsf{BP}}_n(m_n)$ and $\widehat{\mathsf{BP}}_n(m_n) \setminus \widehat{\mathsf{G}}_n(m_n)$ denote the sets of half-edges that are in one, but not the other, exploration. Then

$$\mathcal{C}_n(2,2)^c \subseteq \{|\widehat{\mathsf{G}}_n(m_n) \setminus \widehat{\mathsf{BP}}_n(m_n)| \geq a_n\} \cup \{|\widehat{\mathsf{BP}}_n(m_n) \setminus \widehat{\mathsf{G}}_n(m_n)| \geq a_n\}. \tag{4.3.14}$$

We apply a first-moment method and obtain the bound

$$
\mathbb{P}(\mathcal{C}_n(2,2)^c \cap \mathcal{C}_n(2,1) \cap \mathcal{D}_n)
$$
$$
\leq \frac{1}{a_n} \mathbb{E}\Big[\mathbb{1}_{\mathcal{C}_n(2,1) \cap \mathcal{D}_n} \big| |B_{\bar{k}_n}^{(G_n)}(o_2)| - |BP_{\bar{k}_n}^{(2)}| \big| \Big]
$$
$$
= \frac{1}{a_n} \mathbb{E}\Big[\mathbb{1}_{\mathcal{C}_n(2,1) \cap \mathcal{D}_n} \big(|B_{\bar{k}_n}^{(G_n)}(o_2)| - |BP_{\bar{k}_n}^{(2)}| \big)_+ + \big(|B_{\bar{k}_n}^{(G_n)}(o_2)| - |BP_{\bar{k}_n}^{(2)}| \big)_- \Big]
$$
$$
\leq \frac{1}{a_n} \mathbb{E}\Big[\mathbb{1}_{\mathcal{C}_n(2,1) \cap \mathcal{D}_n} \big(|\widehat{\mathsf{G}}_n(m_n)| - |\widehat{\mathsf{BP}}_n(m_n)| \big)_+
$$
$$
+ \mathbb{1}_{\mathcal{C}_n(2,1) \cap \mathcal{D}_n} \big(|\widehat{\mathsf{G}}_n(m_n)| - |\widehat{\mathsf{BP}}_n(m_n)| \big)_- \Big], \tag{4.3.15}
$$

where the notation is as in Lemma 4.2, $|\widehat{\mathsf{G}}_n(m_n)|$ and $|\widehat{\mathsf{BP}}_n(m_n)|$ denote the number of half-edges and individuals found up to the m_nth step of the exploration starting from o_2, and $m_n = (b+1)\bar{m}_n$, while $x_+ = \max\{0, x\}$ and $x_- = \max\{0, -x\}$.

To bound these expectations, we adapt the proof of Lemma 4.2 to our setting. We start with the first term in (4.3.15), for which we use the exploration up to size $\bar{m}_n$ used in the proof of Lemma 4.2. We note that the only way that $|\widehat{\mathsf{G}}_n(t+1)| - |\widehat{\mathsf{G}}_n(t)|$ can be larger than $|\widehat{\mathsf{BP}}_n(t+1)| - |\widehat{\mathsf{BP}}_n(t)|$ is when a half-edge reuse occurs. This gives rise to a *primary* ghost, which then possibly gives rise to some further *secondary* ghosts, i.e., ghost vertices that are found by pairing a ghost half-edge. Since all degrees are bounded by b, on $\mathcal{C}_n(2,1)$ where the first $\underline{k}_n$ generations are perfectly coupled, the total number of secondary ghosts, together with the single primary ghost, is at most

$$
c_n \equiv \sum_{k=0}^{\bar{k}_n - \underline{k}_n} (b-1)^k = \frac{(b-1)^{\bar{k}_n - \underline{k}_n + 1} - 1}{b - 2}, \tag{4.3.16}
$$

which tends to infinity. On $\mathcal{D}_n$, we can let $\bar{k}_n - \underline{k}_n \to \infty$ arbitrarily slowly. Therefore, on $\mathcal{C}_n(2,1) \cap \mathcal{D}_n$, we have

$$
|\widehat{\mathsf{G}}_n(\bar{m}_n)| - |\widehat{\mathsf{BP}}_n(\bar{m}_n)| \leq c_n \#\{\text{half-edge reuses up to time } \bar{m}_n\}. \tag{4.3.17}
$$

Thus, by (4.2.8),

$$
\mathbb{E}\Big[\mathbb{1}_{\mathcal{C}_n(2,1) \cap \mathcal{D}_n} \big(|\widehat{\mathsf{G}}_n(\bar{m}_n)| - |\widehat{\mathsf{BP}}_n(\bar{m}_n)| \big)_+ \Big] \leq c_n \frac{\bar{m}_n^2}{\ell_n}. \tag{4.3.18}
$$

We continue with the second term in (4.3.15), which is similar. We note that $|\widehat{\mathsf{G}}_n(t+1)| - |\widehat{\mathsf{G}}_n(t)|$ can be smaller than $|\widehat{\mathsf{BP}}_n(t+1)| - |\widehat{\mathsf{BP}}_n(t)|$ when a half-edge reuse occurs, or when a vertex reuse occurs. Thus, again using that the total number of secondary ghosts, together with the single primary ghost, is at most c_n, on $\mathcal{C}_n(2,1) \cap \mathcal{D}_n$,

$$
|\widehat{\mathsf{BP}}_n(\bar{m}_n)| - |\widehat{\mathsf{G}}_n(\bar{m}_n)| \leq c_n \#\{\text{half-edge and vertex reuses up to time } \bar{m}_n\}. \tag{4.3.19}
$$

As a result, by (4.2.8) and (4.2.10),

$$
\mathbb{E}\Big[\mathbb{1}_{\mathcal{C}_n(2,1) \cap \mathcal{D}_n} \big(|\widehat{\mathsf{G}}_n(\bar{m}_n)| - |\widehat{\mathsf{BP}}_n(\bar{m}_n)| \big)_- \Big] \leq c_n \frac{\bar{m}_n^2}{\ell_n}. \tag{4.3.20}
$$

We conclude that

$$\mathbb{P}(\mathcal{C}_n(2)^c) \leq \frac{1}{a_n} 2c_n \frac{\bar{m}_n^2}{\ell_n} = O\left(\left(\frac{\ell_n}{\bar{m}_n^2}\right)^{\delta/2}\right) = o(1), \qquad (4.3.21)$$

when taking $c_n \to \infty$ such that $c_n = o((\bar{m}_n^2/\ell_n)^{\delta/2})$. □

We now define the *successful coupling event* $\mathcal{C}_n$ to be

$$\mathcal{C}_n = \mathcal{C}_n(1) \cap \mathcal{C}_n(2), \qquad \text{so that} \qquad \mathbb{P}(\mathcal{C}_n) = 1 - o(1). \qquad (4.3.22)$$

Neighborhood Growth of the Branching Process

The previous step relates the graph exploration process to the (n-dependent) unimodular pair of branching processes $(\mathsf{BP}_k^{(1)}, \mathsf{BP}_k^{(2)})_{k \leq \underline{k}_n}$. In this step, we investigate the growth of these branching processes in a similar way as for the Erdős–Rényi random graph in Section 2.4.5.

Denote $b_0^{(i)} = |\mathsf{BP}_r^{(i)}|$, which we assume to be at least r (which is true on the event $\mathcal{C}_n \cap \{|\partial B_r^{(G_n)}(o_1)|, |\partial B_r^{(G_n)}(o_2)| \geq r\}$).

Recall that $\nu_n = \mathbb{E}[D_n(D_n - 1)]/\mathbb{E}[D_n]$ denotes the expected forward degree of a uniform half-edge in $G_n = \mathrm{CM}_n(\boldsymbol{d})$, which equals the expected offspring of the branching processes $(\mathsf{BP}_k^{(i)})_{k \geq 0}$. Define

$$\mathcal{D}_n = \{\underline{b}_k^{(i)} \leq |\mathsf{BP}_{r+k}^{(i)}| \leq \bar{b}_k^{(i)} \; \forall i \in [2], \; k \geq 0\}, \qquad (4.3.23)$$

where $(\underline{b}_k^{(i)})_{k \geq 0}$ and $(\bar{b}_k^{(i)})_{k \geq 0}$ satisfy the recursions $\underline{b}_0^{(i)} = \bar{b}_0^{(i)} = b_0^{(i)}$, while, for some $\alpha \in (\frac{1}{2}, 1)$,

$$\underline{b}_{k+1}^{(i)} = \underline{b}_k^{(i)} \nu_n - (\bar{b}_k^{(i)})^\alpha, \qquad \bar{b}_{k+1}^{(i)} = \bar{b}_k^{(i)} \nu_n + (\bar{b}_k^{(i)})^\alpha. \qquad (4.3.24)$$

The following lemma investigates the asymptotics of $(\underline{b}_k^{(i)})_{k \leq k_n - r}$ and $(\bar{b}_k^{(i)})_{k \leq k_n - r}$:

Lemma 4.11 (Asymptotics of $\underline{b}_k^{(i)}$ and $\bar{b}_k^{(i)}$) *Assume that* $\lim_{n \to \infty} \nu_n = \nu > 1$ *and that* $b_0^{(i)} = |\mathsf{BP}_r^{(i)}| \geq r$. *Then there exists an* $A = A_r > 1$ *such that, for all* $k \geq 0$,

$$\bar{b}_k^{(i)} \leq A b_0^{(i)} \nu_n^k, \qquad \underline{b}_k^{(i)} \geq b_0^{(i)} \nu_n^k / A. \qquad (4.3.25)$$

Proof First, obviously, $\bar{b}_k^{(i)} \geq b_0^{(i)} \nu_n^k$. Thus, also using that $\nu_n > 1$,

$$\bar{b}_{k+1}^{(i)} = \bar{b}_k^{(i)} \nu_n + (\bar{b}_k^{(i)})^\alpha \leq \bar{b}_k^{(i)} \nu_n (1 + (\bar{b}_k^{(i)})^{\alpha-1})$$
$$\leq \bar{b}_k^{(i)} \nu_n (1 + r^{-(1-\alpha)} \nu_n^{-(1-\alpha)k}). \qquad (4.3.26)$$

By iteration, this implies the upper bound with A replaced by $\bar{A}_r$, given by

$$\bar{A}_r = \prod_{k \geq 0} (1 + r^{-(1-\alpha)} \nu_n^{-(1-\alpha)k}) < \infty. \qquad (4.3.27)$$

For the lower bound, we use that $\bar{b}_k^{(i)} \leq \bar{A}_r b_0^{(i)} \nu_n^k$ to obtain

$$\underline{b}_{k+1}^{(i)} \geq \underline{b}_k^{(i)} \nu_n - \bar{A}_r^\alpha (b_0^{(i)})^\alpha \nu_n^{\alpha k}. \qquad (4.3.28)$$

We use induction to show that

$$\underline{b}_k^{(i)} \geq a_k b_0^{(i)} \nu_n^k, \qquad (4.3.29)$$

where $a_0 = 1$ and

$$a_{k+1} = a_k - \bar{A}_r^\alpha r^{1-\alpha} \nu_n^{(\alpha-1)k-1}. \qquad (4.3.30)$$

The initialization follows since $\underline{b}_0^{(i)} = b_0^{(i)}$ and $a_0 = 1$. To advance the induction hypothesis, we substitute it to obtain that

$$
\begin{aligned}
\underline{b}_{k+1}^{(i)} &\geq a_k b_0^{(i)} \nu_n^{k+1} - \bar{A}_r^\alpha (b_0^{(i)})^\alpha \nu_n^{\alpha k} \\
&= b_0^{(i)} \nu_n^{k+1} \left(a_k - \bar{A}_r^\alpha (b_0^{(i)})^{\alpha-1} \nu_n^{(\alpha-1)k-1} \right) \\
&\geq b_0^{(i)} \nu_n^{k+1} \left(a_k - \bar{A}_r^\alpha r^{\alpha-1} \nu_n^{(\alpha-1)k-1} \right) = a_{k+1} b_0^{(i)} \nu_n^{k+1},
\end{aligned}
\qquad (4.3.31)
$$

by (4.3.30). Finally, a_k is decreasing, and thus $a_k \searrow a \equiv 1/\underline{A}_r$, where

$$\underline{A}_r = \prod_{k \geq 0} \left(1 - \bar{A}_r^\alpha r^{-(1-\alpha)} \nu_n^{-(1-\alpha)k} \right)^{-1} < \infty$$

for r large enough, so that the claim follows with $A = A_r = \max\{\bar{A}_r, \underline{A}_r\}$. $\qquad\square$

The following lemma shows that $\mathcal{D}_n = \left\{ \underline{b}_k^{(i)} \leq |\mathsf{BP}_{r+k}^{(i)}| \leq \bar{b}_k^{(i)} \; \forall i \in [2], \; k \in [k_n - r] \right\}$ occurs whp when first $n \to \infty$ followed by $r \to \infty$:

Lemma 4.12 ($\mathcal{D}_n$ occurs whp) *Assume that $b_0^{(i)} = |\mathsf{BP}_r^{(i)}| \geq r$. Then*

$$\lim_{r \to \infty} \limsup_{n \to \infty} \mathbb{P}(\mathcal{D}_n) = 1. \qquad (4.3.32)$$

Proof We will show that $\lim_{r \to \infty} \limsup_{n \to \infty} \mathbb{P}(\mathcal{D}_n^c) = 0$. We write

$$\mathbb{P}(\mathcal{D}_n^c) \leq \sum_{k=1}^{k_n - r} \mathbb{P}(\mathcal{D}_{n,k}^c \cap \mathcal{D}_{n,k-1}), \qquad (4.3.33)$$

where

$$\mathcal{D}_{n,k} = \left\{ \underline{b}_k^{(i)} \leq |\mathsf{BP}_{r+k}^{(i)}| \leq \bar{b}_k^{(i)} \; \forall i \in [2] \right\}. \qquad (4.3.34)$$

Note that, when $|\mathsf{BP}_{r+k}^{(i)}| > \bar{b}_k^{(i)}$ and $|\mathsf{BP}_{r+k-1}^{(i)}| \leq \bar{b}_{k-1}^{(i)}$,

$$|\mathsf{BP}_{r+k}^{(i)}| - \nu_n |\mathsf{BP}_{r+k-1}^{(i)}| > \bar{b}_k^{(i)} - \nu_n \bar{b}_{k-1}^{(i)} = (\bar{b}_{k-1}^{(i)})^\alpha \qquad (4.3.35)$$

while when $|\mathsf{BP}_{r+k}^{(i)}| < \underline{b}_k^{(i)}$ and $|\mathsf{BP}_{r+k-1}^{(i)}| \geq \underline{b}_{k-1}^{(i)}$,

$$|\mathsf{BP}_{r+k}^{(i)}| - \nu_n |\mathsf{BP}_{r+k-1}^{(i)}| < \underline{b}_k^{(i)} - \nu_n \underline{b}_{k-1}^{(i)} = -(\bar{b}_{k-1}^{(i)})^\alpha. \qquad (4.3.36)$$

Thus,

$$\mathcal{D}_{n,k}^c \cap \mathcal{D}_{n,k-1} \qquad (4.3.37)$$

$$\subseteq \left\{ \left| |\mathsf{BP}_{r+k}^{(1)}| - \nu_n |\mathsf{BP}_{r+k-1}^{(1)}| \right| \geq (\bar{b}_{k-1}^{(1)})^\alpha \right\} \cup \left\{ \left| |\mathsf{BP}_{r+k}^{(2)}| - \nu_n |\mathsf{BP}_{r+k-1}^{(2)}| \right| \geq (\bar{b}_{k-1}^{(2)})^\alpha \right\}.$$

By the Chebychev inequality, conditional on $\mathcal{D}_{n,k-1}$,

$$\mathbb{P}\left(\left| |\mathsf{BP}_{r+k}^{(i)}| - \nu_n |\mathsf{BP}_{r+k-1}^{(i)}| \right| \geq (\bar{b}_{k-1}^{(i)})^\alpha \mid \mathcal{D}_{n,k-1} \right) \qquad (4.3.38)$$

$$\leq \frac{\mathrm{Var}(|\mathsf{BP}_{r+k}^{(i)}| \mid \mathcal{D}_{n,k-1})}{(\bar{b}_{k-1}^{(i)})^{2\alpha}} \leq \frac{\sigma_n^2 \mathbb{E}[|\mathsf{BP}_{r+k-1}^{(i)}| \mid \mathcal{D}_{n,k-1}]}{(\bar{b}_{k-1}^{(i)})^{2\alpha}} \leq \sigma_n^2 (\bar{b}_{k-1}^{(i)})^{1-2\alpha},$$

where σ_n^2 is the variance of the offspring distribution, given by

$$\sigma_n^2 = \frac{1}{\ell_n} \sum_{v \in [n]} d_v(d_v - 1)^2 - \nu_n^2, \qquad (4.3.39)$$

so that $\sigma_n^2 \leq b(b-1)^2$ is uniformly bounded. Thus, by the union bound for $i \in \{1, 2\}$,

$$\mathbb{P}(\mathcal{D}_{n,k}^c \cap \mathcal{D}_{n,k-1}) \leq 2\sigma_n^2 (\bar{b}_{k-1}^{(i)})^{1-2\alpha}, \qquad (4.3.40)$$

and we conclude that

$$\mathbb{P}(\mathcal{D}_n^c) \leq 2\sigma_n^2 \sum_{k \geq 1} \left((\bar{b}_{k-1}^{(1)})^{1-2\alpha} + (\bar{b}_{k-1}^{(2)})^{1-2\alpha}\right). \qquad (4.3.41)$$

The claim now follows from Lemma 4.11 together with $\alpha \in (\frac{1}{2}, 1)$ and the fact that $\sigma_n^2 \leq b(b-1)^2$ remains uniformly bounded. $\qquad \square$

Completion of the Proof

Recall (4.3.8). Also recall the definition of $\mathcal{C}_n$ in (4.3.22), (4.3.10), and (4.3.12), and that of $\mathcal{D}_n$ in (4.3.23). Let $\mathcal{G}_n = \mathcal{C}_n \cap \mathcal{D}_n$ be the good event. By (4.3.22) and Lemma 4.12,

$$\lim_{r \to \infty} \limsup_{n \to \infty} \mathbb{P}(|\partial B_r^{(G_n)}(o_1)|, |\partial B_r^{(G_n)}(o_2)| \geq r, o_1 \nleftrightarrow o_2; \mathcal{G}_n^c) = 0, \qquad (4.3.42)$$

so that it suffices to investigate $\mathbb{P}(|\partial B_r^{(G_n)}(o_1)|, |\partial B_r^{(G_n)}(o_2)| \geq r, o_1 \nleftrightarrow o_2; \mathcal{G}_n)$.

Our argument below will even show that $\text{dist}_{\text{CM}_n(d)}(o_1, o_2) \leq 2r + \bar{k}_n + \underline{k}_n + 1$ whp on the event that $|\partial B_r^{(G_n)}(o_1)|, |\partial B_r^{(G_n)}(o_2)| \geq r$. On $\mathcal{G}_n$ (recall (4.3.12)),

$$|\partial B_{\bar{k}_n}^{(G_n)}(o_2)| - |\text{BP}_{\bar{k}_n}^{(2)}| \geq -(\overline{m}_n^2/\ell_n)^{1+\delta}. \qquad (4.3.43)$$

Further, on $\mathcal{G}_n$ (recall (4.3.10)),

$$|\partial B_{\underline{k}_n}^{(G_n)}(o_1)| - |\text{BP}_{\underline{k}_n}^{(1)}| = 0. \qquad (4.3.44)$$

When $\partial B_{\underline{k}_n}^{(G_n)}(o_1) \cap \partial B_{\bar{k}_n}^{(G_n)}(o_2) \neq \varnothing$, we have $o_1 \longleftrightarrow o_2$ so this does not contribute to (4.3.42).

On the other hand, when $\partial B_{\underline{k}_n}^{(G_n)}(o_1) \cap \partial B_{\bar{k}_n}^{(G_n)}(o_2) = \varnothing$, by Lemma 4.11 and when $\overline{m}_n^2/\ell_n \to \infty$ sufficiently slowly, $|\partial B_{\underline{k}_n}^{(G_n)}(o_1)| = \Theta_{\mathbb{P}}(\underline{m}_n)$ and $|\partial B_{\bar{k}_n}^{(G_n)}(o_2)| = \Theta_{\mathbb{P}}(\overline{m}_n)$. The same bounds hold for the number of half-edges $Z_{\underline{k}_n}^{(1)}$ and $Z_{\bar{k}_n}^{(2)}$ incident to $\partial B_{\underline{k}_n}^{(G_n)}(o_1)$ and $\partial B_{\bar{k}_n}^{(G_n)}(o_2)$, respectively, since $Z_{\underline{k}_n}^{(1)} \geq |\partial B_{\underline{k}_n+1}^{(G_n)}(o_1)|$ and $Z_{\bar{k}_n}^{(2)} \geq |\partial B_{\bar{k}_n+1}^{(G_n)}(o_2)|$, so that also $Z_{\underline{k}_n}^{(1)} = \Theta_{\mathbb{P}}(\underline{m}_n)$ and $Z_{\bar{k}_n}^{(2)} = \Theta_{\mathbb{P}}(\overline{m}_n)$.

Conditional on having paired some half-edges incident to $\partial B_{\underline{k}_n}^{(G_n)}(o_1)$, each further such half-edge has probability at least $1 - Z_{\bar{k}_n}^{(2)}/\ell_n$ of being paired with a half-edge incident to $\partial B_{\bar{k}_n}^{(G_n)}(o_2)$, thus creating a path between o_1 and o_2. The latter conditional probability is *independent* of the pairing of the earlier half-edges. Thus, the probability that $\partial B_{\underline{k}_n}^{(G_n)}(o_1)$ is not *directly* connected to $\partial B_{\bar{k}_n}^{(G_n)}(o_2)$ is at most

$$\left(1 - \frac{Z_{\bar{k}_n}^{(2)}}{\ell_n}\right)^{Z_{\underline{k}_n}^{(1)}/2}, \qquad (4.3.45)$$

since at least $Z_{\overline{k}_n}^{(1)}/2$ pairings need to be performed. This probability vanishes when $\underline{m}_n \overline{m}_n \gg n$. As a result, as $n \to \infty$,

$$\mathbb{P}(|\partial B_r^{(G_n)}(o_1)|, |\partial B_r^{(G_n)}(o_2)| \geq r, o_1 \not\longleftrightarrow o_2; \mathcal{G}_n) = o(1), \tag{4.3.46}$$

as required. This completes the proof of (2.6.7) for $CM_n(\boldsymbol{d})$ with uniformly bounded degrees, and indeed shows that $\text{dist}_{CM_n(\boldsymbol{d})}(o_1, o_2) \leq 2r + \overline{k}_n + \underline{k}_n + 1$ whp on the event that $|\partial B_r^{(G_n)}(o_1)|, |\partial B_r^{(G_n)}(o_2)| \geq r$.

Degree Truncation to Lift the Uniform Degree Boundedness Condition

We finally lift the assumption that $d_v \leq b$ for all $v \in [n]$ in the proof of (2.6.7), by applying the degree-truncation technique from Theorem 1.11 with b large. This is not necessary when, e.g., $\mathbb{E}[D_n^3] = \Theta(1)$ but it allows us to deal with high variability in the degrees.

By Conditions 1.7(a),(b), we can choose b large enough that

$$\mathbb{P}(D_n^\star > b) = \frac{1}{\ell_n} \sum_{v \in [n]} d_v \mathbb{1}_{\{d_v > b\}} \leq \frac{\varepsilon n}{\ell_n}. \tag{4.3.47}$$

Recall the definition of $CM_{n'}(\boldsymbol{d}')$ in Theorem 1.11 and its proof. This implies that $CM_{n'}(\boldsymbol{d}')$ has at most $(1+\varepsilon)n$ vertices, and that the (at most εn) extra vertices compared with $CM_n(\boldsymbol{d})$ all have degree 1, while the vertices in $[n]$ have degree $d_v' \leq b$. Further, with $\mathscr{C}_{max}'$ the largest connected component in $CM_{n'}(\boldsymbol{d}')$, by (4.3.47), we have

$$|\mathscr{C}_{max}'| \leq |\mathscr{C}_{max}| + \varepsilon n, \tag{4.3.48}$$

so that

$$\mathbb{P}(|\mathscr{C}_{max}| \geq n(\zeta' - 2\varepsilon)) \to 1, \tag{4.3.49}$$

when $|\mathscr{C}_{max}'|/n \xrightarrow{\mathbb{P}} \zeta'$. We denote the limiting parameters of $CM_{n'}(\boldsymbol{d}')$ by $(p_k')_{k \geq 1}, \xi'$ and ζ', and note that, for ε as in (4.3.47), when $\varepsilon \searrow 0$, we have

$$p_k' \to p_k, \qquad \xi' \to \xi, \qquad \zeta' \to \zeta, \tag{4.3.50}$$

so that we can take b sufficiently large that, for all $\varepsilon > 0$,

$$\mathbb{P}(|\mathscr{C}_{max}| \geq n(\zeta - 3\varepsilon)) \to 1. \tag{4.3.51}$$

This proves the required lower bound on $|\mathscr{C}_{max}|$, while the upper bound follows from Corollary 2.27. $\qquad\qquad\qquad\qquad\qquad\qquad\qquad\qquad\qquad\qquad\qquad\qquad\qquad\square$

Remark 4.13 (Small-world properties of $CM_n(\boldsymbol{d})$) We next discuss the consequences of the above proof to the small-world nature of $CM_n(\boldsymbol{d})$, in a similar way to the proof of Theorem 2.36. Here we should consider the use of the degree truncation in Theorem 1.11. Let $o_1, o_2 \in [n]$ be chosen uar. Recall from Theorem 1.11(c) that $\text{dist}_{CM_n(\boldsymbol{d})}(o_1, o_2) \leq$

$\mathrm{dist}_{\mathrm{CM}_{n'}(d')}(o_1, o_2)$. The above "giant is almost local" proof shows that, whp, if $n \to \infty$ followed by $r \to \infty$, then

$$\mathrm{dist}_{\mathrm{CM}_{n'}(d')}(o_1, o_2) \le 2r + \underline{k}_n + \overline{k}_n + 1. \tag{4.3.52}$$

Lemma 4.11 implies the asymptotics $\underline{k}_n = (1 + o_{\mathbb{P}}(1)) \log \underline{m}_n / \log \nu'_n$ and $\overline{k}_n = (1 + o_{\mathbb{P}}(1)) \log \overline{m}_n / \log \nu'_n$, where $\nu'_n = \mathbb{E}[D'_{n'}(D'_{n'} - 1)]/\mathbb{E}[D'_{n'}]$. Thus, on the event $\mathcal{G}_n = \mathcal{C}_n \cap \mathcal{D}_n$,

$$2r + \underline{k}_n + \overline{k}_n + 1 = (1 + o_{\mathbb{P}}(1)) \frac{\log(\underline{m}_n \overline{m}_n)}{\log \nu'_n} = \frac{\log n}{\log \nu'_n}(1 + o_{\mathbb{P}}(1)) \tag{4.3.53}$$

when $\underline{m}_n \overline{m}_n / n \to \infty$ slowly enough. Assume that Conditions 1.7(a)–(c) hold. Then, we can choose $b = b(\varepsilon)$ and n large enough to make $1/\log \nu'_n \le (1 + \varepsilon)/\log \nu$, so that

$$\mathrm{dist}_{\mathrm{CM}_n(d)}(o_1, o_2) \le \frac{\log n}{\log \nu}(1 + \varepsilon). \tag{4.3.54}$$

Such small-world results are explored in more detail in Chapter 7. ◀

4.3.2 CONTINUOUS-TIME EXPLORATION PROOF

In this subsection we give an alternative continuous-time exploration proof of Theorem 4.9, using a clever *randomization scheme* to explore the connected components one by one. This construction is explained in terms of a simple continuous-time algorithm below. The algorithm describes the number of vertices of given degrees that have been found, as well as the total number of unpaired half-edges, at time $t > 0$. It is proved that, when $n \to \infty$, these quantities all converge in probability to *deterministic functions* described in terms of some functions $s \mapsto H(s)$ and $s \mapsto G_D^\star(s)$. In particular, the number of unpaired half-edges in our exploration is given in terms of $s \mapsto H(s)$, so that the first zero of this function gives the size of the giant component. We analyze this algorithm below and show that, when $\zeta > 0$, after a short initial period of exploring small clusters, the giant component is found, and the exploration explores it completely, after which no large component is left. When $\zeta = 0$, however, only small clusters are found. A crucial aspect in the proof resides in how to deal with the depletion-of-points-and-half-edges effect.

Reformulation in Terms of Generating Functions

We start by reformulating the results in Theorem 4.9 in terms of generating functions, which play a crucial role throughout our proof. For $s \in [0, 1]$, let

$$G_D(s) = \sum_{k=1}^{\infty} p_k s^k = \mathbb{E}[s^D] \tag{4.3.55}$$

be the probability generating function for the probability distribution $(p_k)_{k \ge 1}$ given by $p_k = \mathbb{P}(D = k)$. Recall that, for a non-negative random variable D, the random variable $D^\star$ denotes its size-biased distribution. Define further, again for $s \in [0, 1]$,

$$G_D^\star(s) = \mathbb{E}[s^{D^\star}] = \sum_{k \ge 0} p_k^\star s^k = G'_D(s)/\mathbb{E}[D], \tag{4.3.56}$$

$$H(s) = \mathbb{E}[D]s(s - G_D^\star(s)). \tag{4.3.57}$$

Note that $G_D^\star(1) = 1$, and thus $H(0) = H(1) = 0$. Note also that

$$H'(1) = \mathbb{E}[D]\left(1 - \frac{\mathrm{d}}{\mathrm{d}s}G_D^\star(1)\right) = \mathbb{E}[D]\left(1 - \sum_{k \geq 0} kp_k^\star\right)$$

$$= \mathbb{E}[D] - \sum_{k \geq 1} k(k-1)p_k = -\mathbb{E}[D(D-2)]. \tag{4.3.58}$$

For further properties of $s \mapsto H(s)$, see Lemma 4.18 below. We conclude that if $\mathbb{E}[D(D-2)] = \sum_k k(k-2)p_k > 0$ and if $p_0^\star > 0$, then there is a unique $\xi \in (0,1)$ such that $H(\xi) = 0$, or equivalently $G_D^\star(\xi) = \xi$, so that indeed ξ is the extinction probability of the branching process with offspring distribution $(p_k^\star)_{k \geq 0}$ in (4.2.2). Further, $\mathbb{E}[D(D-2)] > 0$ precisely when $\nu = \mathbb{E}[D(D-1)]/\mathbb{E}[D] > 1$. When instead $p_0^\star = 0$, or equivalently $p_1 = 0$, $\xi = 0$ is the unique solution in $[0,1)$ of $H(\xi) = 0$. The functions $s \mapsto H(s)$ and $s \mapsto G_D^\star(s)$ play a central role in our analysis.

Reduction to the Case where $\mathbb{P}(D = 1) = p_1 > 0$

In our proof, it is convenient to assume that $p_1 = \mathbb{P}(D = 1) > 0$. The extinction probability ξ and the survival probability ζ satisfy $\xi = 0$ and $\zeta = 1$ when $p_1 = 0$, which causes technical difficulties in the proof. We now explain how we can reduce the case where $p_1 = 0$ to the case where $p_1 > 0$ in a similar way to that used in Theorem 1.11.

Let $d_{\min} = \min\{k : p_k > 0\}$ be the minimum of the support of the asymptotic degree distribution D. Fix $\varepsilon > 0$, and assume that $\varepsilon < p_{d_{\min}}$. Consider the configuration model with $\tilde{n} = n + (d_{\min} - 1)\varepsilon n$ and degree sequence $\tilde{\boldsymbol{d}} = (\tilde{d}_i)_{i \in [n]}$, with $\tilde{n}_k = n_k$ for all $k > d_{\min}$, $\tilde{n}_{d_{\min}} = n_{d_{\min}} - \varepsilon n$, and $\tilde{n}_1 = d_{\min}\varepsilon n$. This configuration model can be obtained from $\mathrm{CM}_n(\boldsymbol{d})$ by replacing εn vertices of degree $d_{\min}$ by $d_{\min}$ vertices having degree 1, as if we have "forgotten" that these vertices are actually equal.

Clearly, $\mathrm{CM}_n(\boldsymbol{d})$ can be retrieved by identifying εn collections of $d_{\min}$ vertices of degree 1, and collapsing them to a single vertex of degree $d_{\min}$. When $\boldsymbol{d}$ satisfies Conditions 1.7(a),(b), then so does $\tilde{\boldsymbol{d}}$ with limiting degree distribution $\tilde{p}_1 = d_{\min}\varepsilon/(1 + (d_{\min} - 1)\varepsilon)$, $\tilde{p}_{d_{\min}} = (p_{d_{\min}} - \varepsilon)/(1 + (d_{\min} - 1)\varepsilon)$, and $\tilde{p}_k = p_k/(1 + (d_{\min} - 1)\varepsilon)$ for all $k > d_{\min}$. Let $\widetilde{\mathscr{C}}_{\max}$ denote the giant in the configuration model with extra degree-1 vertices. The above procedure satisfies $|\mathscr{C}_{\max}| \geq |\widetilde{\mathscr{C}}_{\max}| - d_{\min}\varepsilon n$. Further, if ζ_ε denotes the limit of $|\widetilde{\mathscr{C}}_{\max}|/\tilde{n}$ for $\tilde{\boldsymbol{d}}$, we have that $\zeta_\varepsilon \to 1$ as $\varepsilon \searrow 0$. As a result, Theorem 4.9 for $\zeta = 1, \xi = 0$ follows from Theorem 4.9 with $p_1 > 0$, for which $\zeta < 1$ and $\xi > 0$. In the remainder of the proof, we therefore without loss of generality assume that $\xi > 0$ and $\zeta < 1$.

Finding the Largest Component

The components of an arbitrary finite graph or multi-graph can be found by a standard breadth-first exploration. Pick an arbitrary vertex v and determine the component of v as follows: include all the neighbors of v in an arbitrary order; then add in the neighbors of the neighbors, and so on, until no more vertices can be added. The vertices included until this moment form the component of v. If there are still vertices left in the graph then pick any such vertex w and repeat the above to determine the second connected component (the component of vertex w). Carry on in this manner until all the components have been found.

The same result can be obtained in the following way, which turns out to be convenient for the exploration of the giant component in the configuration model. Regard each edge as consisting of two half-edges, each half-edge having one endpoint. We label the vertices as sleeping or awake (i.e., used) and the half-edges as sleeping, active, or dead (already paired into edges); the sleeping and active half-edges are also called *living*. We start with all vertices and half-edges sleeping. Pick a vertex and label its half-edges as active. Then take any active half-edge, say x, and find its partner y in the graph; label these two half-edges as dead. Further, if the endpoint of y is sleeping, label it as awake and all other half-edges of the vertex incident to y as active. Repeat as long as there are active half-edges. When there is no active half-edge left, we have obtained the first connected component in the graph. Then start again with another vertex until all components are found.

We apply this algorithm to $\mathrm{CM}_n(\boldsymbol{d})$, revealing its edges during the process. Thus we initially only observe the vertex degrees and the half-edges, not how they are paired with form edges. Hence, each time we need a partner for a half-edge, this partner is uniformly distributed over all other living half-edges. It is here that we are using the specific structure of the configuration model, which simplifies the analysis substantially.

We make the random choices of finding a partner for the half-edges by associating iid random maximal lifetimes E_x to the half-edge x, where E_x has an $\mathsf{Exp}(1)$ distribution. We interpret these lifetimes as *clocks*, and changes in our exploration process occur only when the clock of a half-edge rings. In other words, each half-edge dies spontaneously at rate 1 (unless killed earlier). Each time we need to find a partner for a half-edge x, we then wait until the next living half-edge unequal to x dies and take that one. This process in continuous time can be formulated as an *algorithm*, constructing $\mathrm{CM}_n(\boldsymbol{d})$ and exploring its components simultaneously, as follows. Recall that we start with all vertices and half-edges sleeping. The exploration is then formalized in the following three steps:

Step 1   When there is no active half-edge (as at the beginning), select a sleeping vertex and declare it awake and all its half-edges active. For definiteness, we choose the vertex by choosing a half-edge uar among all sleeping half-edges. When there is no sleeping half-edge left, the process stops; the remaining sleeping vertices are all isolated and we have explored all other components.

Step 2   Pick an active half-edge (which one does not matter) and kill it, i.e., change its status to dead.

Step 3   Wait until the next half-edge dies (spontaneously, as a result of its clock ringing). This half-edge is paired with the one killed in step Step 2 to form an edge of the graph. If the vertex incident to it is sleeping, then we change this vertex to awake and all other half-edges incident to it to active. Repeat from Step 1.

The above randomized algorithm is such that components are created between the successive times at which Step 1 is performed, where we say that Step 1 is performed when there is no active half-edge and, as a result, a new vertex is chosen whose connected component we continue exploring.

The vertices in the component created during one of these intervals between the successive times at which Step 1 is performed are the vertices that are awakened during the interval. Note also that a component is completed and Step 1 is performed exactly when the number of active half-edges is 0 and a half-edge dies at a vertex where all other half-edges (if any) are dead. Below, we investigate the behavior of the key characteristics of the algorithm.

Analysis of the Algorithm for the Configuration Model

We start by introducing the key characteristics of the above exploration algorithm. Let $S(t)$ and $A(t)$ be the numbers of sleeping and active half-edges, respectively, at time t, and let

$$L(t) = S(t) + A(t) \tag{4.3.59}$$

be the number of living half-edges. For definiteness, we define these random functions to be right-continuous.

Let us first look at $L(t)$. We start with $\ell_n = \sum_{i \in [n]} d_i$ half-edges, all sleeping and thus living, but we immediately perform **Step 1** and **Step 2** and kill one of them. Thus, $L(0) = \ell_n - 1$. Next, as soon as a living half-edge dies, we perform **Step 3** and then (instantly) either **Step 2** or both **Step 1** and **Step 2**. Since **Step 1** does not change the number of living half-edges while **Step 2** and **Step 3** each decrease it by 1, the total result is that $L(t)$ is decreased by 2 each time one of the living half-edges dies, except when the last living one dies and the process terminates. Because of this simple dynamics of $t \mapsto L(t)$, we can give sharp asymptotics of $L(t)$ when $n \to \infty$:

Proposition 4.14 (Number of living half-edges) *As $n \to \infty$, for any $t_0 \geq 0$ fixed,*

$$\sup_{0 \leq t \leq t_0} |n^{-1} L(t) - \mathbb{E}[D_n] e^{-2t}| \xrightarrow{\mathbb{P}} 0. \tag{4.3.60}$$

Proof The process $t \mapsto L(t)$ satisfies $L(0) = \ell_n - 1$, and it decreases by 2 at rate $L(t)$. As a result, it is closely related to a *death process*. We study such processes in the following lemma:

Lemma 4.15 (Asymptotics of death processes) *Let $d, \gamma > 0$ be given and let $(N^{(x)}(t))_{t \geq 1}$ be a Markov process such that $N^{(x)}(0) = x$ almost surely, and the dynamics of $t \mapsto (N^{(x)}(t))_{t \geq 1}$ is such that from position y, it jumps down by d at a rate γy. In other words, the waiting time until the next event is $\mathrm{Exp}(1/\gamma y)$ and each jump is of size d downwards. Then, for every $t_0 \geq 0$,*

$$\mathbb{E}\left[\sup_{t \leq t_0} |N^{(x)}(t) - e^{-\gamma dt} x|^2\right] \leq 8d(e^{\gamma dt_0} - 1)x + 8d^2. \tag{4.3.61}$$

Proof The proof follows by distinguishing several cases. First assume that $d = 1$ and that x is an integer. In this case, the process is a standard pure death process taking the values $x, x - 1, x - 2, \ldots, 0$, describing the number of particles alive when the particles die independently at rate $\gamma > 0$. As is well known, and is easily seen by regarding $N^{(x)}(t)$ as the sum of x independent copies of the process $N^{(1)}(t)$, the process $(e^{\gamma t} N^{(x)}(t))_{t \geq 1}$, is a *martingale* starting in x. Furthermore, for every $t \geq 0$, the random variable $N^{(x)}(t)$ has a $\mathrm{Bin}(x, e^{-\gamma t})$ distribution, since each particle (of which there are x) has a probability of dying before time t equal to $e^{-\gamma t}$, and the different particles die independently.

Application of Doob's martingale inequality (recall (1.5.4)), now in continuous time, yields

$$\mathbb{E}\Big[\sup_{t\leq t_0}\big|N^{(x)}(t)-e^{-\gamma t}x\big|^2\Big] \leq \mathbb{E}\Big[\sup_{t\leq t_0}\big|e^{\gamma t}N^{(x)}(t)-x\big|^2\Big] \leq 4\mathbb{E}\Big[\big(e^{\gamma t}N^{(x)}(t_0)-x\big)^2\Big]$$

$$= 4e^{2\gamma t}\mathrm{Var}(N^{(x)}(t_0)) \leq 4(e^{\gamma t_0}-1)x. \tag{4.3.62}$$

This proves the claim when x is integer.

Next, we still assume that $d = 1$, but let $x > 0$ be arbitrary. We can couple the two processes $\big(N^{(x)}(t)\big)_{t\geq 1}$ and $\big(N^{(\lfloor x\rfloor)}(t)\big)_{t\geq 1}$ with different initial values in such a way that whenever the smaller one jumps by 1, so does the other. This coupling keeps

$$\big|N^{(x)}(t)-N^{(\lfloor x\rfloor)}(t)\big| < 1 \tag{4.3.63}$$

for all $t \geq 0$, and thus,

$$\sup_{t\leq t_0}\big|N^{(\lfloor x\rfloor)}(t)-e^{-\gamma t}\lfloor x\rfloor\big| \leq \sup_{t\leq t_0}\big|N^{(x)}(t)-e^{-\gamma t}x\big| + 2, \tag{4.3.64}$$

so that by (4.3.62), in turn,

$$\mathbb{E}\Big[\sup_{t\leq t_0}\big|N^{(x)}(t)-e^{-\gamma t}x\big|^2\Big] \leq 8(e^{\gamma t_0}-1)x + 8. \tag{4.3.65}$$

Finally, for a general $d > 0$, we observe that $N^{(x)}(t)/d$ is a process of the same type with the parameters (γ, d, x) replaced by $(\gamma d, 1, x/d)$, and the general result follows from (4.3.65) and (4.3.62). $\qquad\square$

The proof of Proposition 4.14 follows from Lemma 4.15 with $d = 2, x = \ell_n - 1 = n\mathbb{E}[D_n] - 1$, and $\gamma = 1$. $\qquad\square$

We continue by considering the sleeping half-edges $S(t)$. Let $V_k(t)$ be the number of sleeping vertices of degree k at time t, so that

$$S(t) = \sum_{k=1}^{\infty} kV_k(t). \tag{4.3.66}$$

Note that **Step 2** does not affect sleeping half-edges, and that **Step 3** implies that each sleeping vertex of degree k is eliminated (i.e., awakened) with intensity k, independently of what happens to all other vertices. However, some sleeping vertices eliminated by **Step 1**, which complicates the dynamics of $t \mapsto V_k(t)$.

It is here that the depletion-of-points-and-half-edges effect enters the analysis of the component structure of $\mathrm{CM}_n(\boldsymbol{d})$. This effect is complicated, but we will see that it is quite harmless, as can be understood by noting that we apply **Step 1** only when we have completed the exploration of an entire component. Since we are mainly interested in settings where the giant component is large, we will see that we will not be using **Step 1** very often before having completely explored the giant component. After having completed the exploration of the giant component, we start using **Step 1** again quite frequently, but it will turn out that then it is very unlikely to be exploring any particularly large connected component. Thus, we can have a setting in mind where the number of applications of **Step 1** is quite small. With this intuition in mind, we first ignore the effect of **Step 1** by letting $\tilde{V}_k(t)$ be the number of vertices of degree k such that all its half-edges x have maximal lifetimes $E_x > t$ i.e., none of its k half-edges would have died spontaneously up to time t, assuming they all escaped Step 1. We conclude that, intuitively, the difference between $V_k(t)$ and $\widetilde{V}_k(t)$ can be expected to be insignificant. We thus start by focussing on the dynamics of $(\widetilde{V}_k(t))_{t \geq 1}$, ignoring the effect of Step 1, and later correct for this omission.

For a given half-edge, we call the half-edges incident to the same vertex its *sibling half-edges*. Further, let

$$\widetilde{S}(t) = \sum_{k=1}^{\infty} k \widetilde{V}_k(t) \qquad (4.3.67)$$

denote the number of half-edges whose sibling half-edges have all escaped spontaneous death up to time t. Comparing with (4.3.66), we see that the process $\widetilde{S}(t)$ ignores the effect of Step 1 in an identical way to $\widetilde{V}_k(t)$.

Recall the functions $G_D, G_D^\star$ from (4.3.55) and (4.3.56), and define, for $s \in [0,1]$,

$$h(s) = s\mathbb{E}[D]G_D^\star(s). \qquad (4.3.68)$$

Then, we can identify the asymptotics of $(\widetilde{V}_k(t))_{t \geq 1}$ in a similar way to that in Proposition 4.14:

Lemma 4.16 (Number of living vertices of degree k) *Subject to Conditions 1.7(a),(b), as $n \to \infty$ and for any $t_0 \geq 0$ fixed,*

$$\sup_{t \leq t_0} |n^{-1}\widetilde{V}_k(t) - p_k e^{-kt}| \xrightarrow{\mathbb{P}} 0 \qquad (4.3.69)$$

for every $k \geq 0$, and

$$\sup_{t \leq t_0} \left| n^{-1} \sum_{k=1}^{\infty} \widetilde{V}_k(t) - G_D(e^{-t}) \right| \xrightarrow{\mathbb{P}} 0, \qquad (4.3.70)$$

$$\sup_{t \leq t_0} |n^{-1}\widetilde{S}(t) - h(e^{-t})| \xrightarrow{\mathbb{P}} 0. \qquad (4.3.71)$$

Proof The statement (4.3.69) again follows from Lemma 4.15, now with $\gamma = k$, $x = n_k$ and $d = 1$. We can replace $p_k^{(G_n)} = n_k/n$ by p_k by Condition 1.7(a).

By Condition 1.7(b), the sequence of random variables $(D_n)_{n \geq 1}$ is uniformly integrable, which means that for every $\varepsilon > 0$ there exists $K < \infty$ such that $\sum_{k>K} kn_k/n = \mathbb{E}[D_n \mathbb{1}_{\{D_n > k\}}] < \varepsilon$ for all n. We may further assume (or deduce from Fatou's inequality) that $\sum_{k>K} kp_k < \varepsilon$ and obtain by (4.3.69) that, whp,

$$\sup_{t \leq t_0} |n^{-1}\widetilde{S}(t) - h(e^{-t})| = \sup_{t \leq t_0} \left| \sum_{k=1}^{\infty} k(n^{-1}\widetilde{V}_k(t) - p_k e^{-kt}) \right|$$

$$\leq \sum_{k=1}^{K} k \sup_{t \leq t_0} |n^{-1}\widetilde{V}_k(t) - p_k e^{-kt}| + \sum_{k>K} k \left(\frac{n_k}{n} + p_k \right)$$

$$\leq \varepsilon + \varepsilon + \varepsilon,$$

proving (4.3.71). An almost identical argument yields (4.3.70). $\qquad \square$

Remarkably, the difference between $S(t)$ and $\widetilde{S}(t)$ is easily estimated. The following result can be viewed as the key to why this approach works. Indeed, it gives a *uniform* upper bound on the difference due to the application of **Step 1**:

Lemma 4.17 (Effect of **Step 1**) *Let* $d_{\max} := \max_{v \in [n]} d_v$ *be the maximum degree of* $CM_n(\boldsymbol{d})$. *Then*

$$0 \leq \widetilde{S}(t) - S(t) < \sup_{0 \leq s \leq t} (\widetilde{S}(s) - L(s)) + d_{\max}. \tag{4.3.72}$$

The process $(\widetilde{S}(t))_{t \geq 1}$ runs at scale n (see, e.g., the related statement for $(L(t))_{t \geq 1}$ in Proposition 4.14). Further, $d_{\max} = o(n)$ when Conditions 1.7(a),(b) hold. Finally, one can expect that $\widetilde{S}(s) \leq L(s)$ holds, since the difference is related to the number of active half-edges. Thus, intuitively, $\sup_{0 \leq s \leq t}(\widetilde{S}(s) - L(s)) = \widetilde{S}(0) - L(0) = 0$. We will make that argument precise after we have proved Lemma 4.17. We now conclude that $\widetilde{S}(t) - S(t) = o_{\mathbb{P}}(n)$, and so they have the same limit after rescaling by n. Let us now prove Lemma 4.17:

Proof Clearly, $V_k(t) \leq \widetilde{V}_k(t)$, and thus $S(t) \leq \widetilde{S}(t)$. Furthermore, $\widetilde{S}(t) - S(t)$ increases only as a result of **Step 1**. Indeed, **Step 1** acts to guarantee that $A(t) = L(t) - S(t) \geq 0$, and is only performed when $A(t) = 0$.

If **Step 1** is performed at time t and a vertex of degree $j > 0$ is awakened, then **Step 2** applies instantly and we have $A(t) = j - 1 < d_{\max}$ and consequently

$$\widetilde{S}(t) - S(t) = \widetilde{S}(t) - L(t) + A(t) < \widetilde{S}(t) - L(t) + d_{\max}. \tag{4.3.73}$$

Furthermore, $\widetilde{S}(t) - S(t)$ is never changed by **Step 2** and either unchanged or decreased by **Step 3**. Hence, $\widetilde{S}(t) - S(t)$ does not increase until the next time **Step 1** is performed. Consequently, for any time t, if s was the last time before (or equal to) t that **Step 1** was performed, then $\widetilde{S}(t) - S(t) \leq \widetilde{S}(s) - S(s)$, and the result follows by (4.3.73). $\qquad \square$

Let us now set the stage for taking the limits of $n \to \infty$. Recall that $A(t) = L(t) - S(t)$ denotes the number of awakened vertices, and let

$$\widetilde{A}(t) = L(t) - \widetilde{S}(t) = A(t) - (\widetilde{S}(t) - S(t)) \tag{4.3.74}$$

denote the number of awakened vertices, ignoring the effect of **Step 1**. Thus, $\widetilde{A}(t) \leq A(t)$ since $S(t) \leq \widetilde{S}(t)$. We use $\widetilde{A}(t)$ as a proxy for $A(t)$ similarly to how $\widetilde{S}(t)$ is used as a proxy for $S(t)$.

Recall the definition of $s \mapsto H(s)$ in (4.3.57). By Lemmas 4.14 and 4.16, and the definition that $\widetilde{A}(t) = L(t) - \widetilde{S}(t)$, for any $t_0 \geq 0$, we have

$$\sup_{t \leq t_0} |n^{-1} \widetilde{A}(t) - H(e^{-t})| \overset{\mathbb{P}}{\longrightarrow} 0. \tag{4.3.75}$$

Lemma 4.17 can be rewritten as

$$0 \leq \widetilde{S}(t) - S(t) < -\inf_{s \leq t} \widetilde{A}(s) + d_{\max}. \tag{4.3.76}$$

By (4.3.74) and (4.3.76),

$$\widetilde{A}(t) \leq A(t) < \widetilde{A}(t) - \inf_{s \leq t} \widetilde{A}(s) + d_{\max}, \tag{4.3.77}$$

which, perhaps, illuminates the relation between $A(t)$ and $\widetilde{A}(t)$. Recall that connected components are explored between subsequent zeros of the process $t \mapsto A(t)$. The function $t \mapsto H(e^{-t})$, which acts as the limit of $\widetilde{A}(t)$ (and, as proved below, also of $A(t)$), is strictly positive in $(0, -\log \xi)$ and $H(1) = H(\xi) = 0$. Therefore, we expect $\widetilde{A}(t)$ to be positive for $t \in (0, -\log \xi)$, and, if so, $\inf_{s \le t} \widetilde{A}(s) = 0$. This would prove that indeed $\widetilde{A}(t)$ and $A(t)$ are close on this entire interval, and the exploration on the interval $t \in (0, -\log \xi)$ will turn out to correspond to the exploration of the giant component.

The idea is to continue our algorithm defined by Steps 1–3 until the giant component has been found, which implies that $A(t) > 0$ for the time of exploration of the giant component, and $A(t) = 0$ for the first time we complete the exploration of the giant component, which is $t = -\log \xi$. Thus, the term $\inf_{s \le t} \widetilde{A}(s)$ in (4.3.77) ought to be negligible. When Conditions 1.7(a),(b) hold, we further have that $d_{\max} = o(n)$, so that one can expect $\widetilde{A}(t)$ to be a good approximation to $A(t)$. The remainder of the proof makes this intuition precise. We start by summarizing some useful analytical properties of $s \mapsto H(s)$ on which we will rely later:

Lemma 4.18 (Properties of $s \mapsto H(s)$) *Suppose that Conditions 1.7(a),(b) hold, and let $H(x)$ be given by (4.3.57). Suppose also that $p_2 < 1$.*

(a) *If $\nu = \mathbb{E}[D(D-1)]/\mathbb{E}[D] > 1$ and $p_1 > 0$, then there is a unique $\xi \in (0,1)$ such that $H(\xi) = 0$. Moreover, $H(s) < 0$ for all $s \in (0, \xi)$ and $H(s) > 0$ for all $s \in (\xi, 1)$.*

(b) *If $\nu = \mathbb{E}[D(D-1)]/\mathbb{E}[D] \le 1$, then $H(s) < 0$ for all $s \in (0,1)$.*

Proof As remarked earlier, $H(0) = H(1) = 0$ and $H'(1) = -\mathbb{E}[D(D-2)]$. Furthermore, if we define $\phi(s) := H(s)/s$, then $\phi(s) = \mathbb{E}[D](s - G_D^\star(s))$ is a concave function on $(0, 1]$, and it is strictly concave unless $p_k = 0$ for all $k \ge 3$, in which case $H'(1) = -\mathbb{E}[D(D-2)] = p_1 > 0$. Indeed, $p_1 + p_2 = 1$ when $p_k = 0$ for all $k \ge 3$. Since we assume that $p_2 < 1$, we thus obtain that $p_1 > 0$ in this case.

In case (b), we thus have that ϕ is concave and $\phi'(1) = H'(1) - H(1) \ge 0$, with either the concavity or the inequality strict, and thus $\phi'(s) > 0$ for all $s \in (0, 1)$, whence $\phi(s) < \phi(1) = 0$ for $s \in (0, 1)$.

In case (a), $H'(1) < 0$, and thus $H(s) > 0$ for s close to 1. Further, when $p_1 > 0$, $H'(0) = -h'(0) = -p_1 < 0$, and thus $H(s) \le 0$ for s close to 0. Hence, there is at least one $\xi \in (0, 1)$ with $H(\xi) = 0$ and, since $H(s)/s$ is strictly concave and also $H(1) = 0$, there is at most one such ξ and the result follows. $\qquad\square$

Proof of Theorem 4.9: Preparations

Let $\nu > 1$, let ξ be the zero of H given by Lemma 4.18(a), and let $\theta = -\log \xi$. Then, by Lemma 4.18, $H(e^{-t}) > 0$ for $0 < t < \theta$, and thus $\inf_{t \le \theta} H(e^{-t}) = 0$. Consequently, (4.3.75) implies

$$n^{-1} \inf_{t \le \theta} \widetilde{A}(t) = \inf_{t \le \theta} n^{-1} \widetilde{A}(t) - \inf_{t \le \theta} H(e^{-t}) \overset{\mathbb{P}}{\longrightarrow} 0. \qquad (4.3.78)$$

Further, by Condition 1.7(b), $d_{\max} = o(n)$, and thus $d_{\max}/n \to 0$. Therefore, (4.3.76) and (4.3.78) yield

$$\sup_{t \le \theta} n^{-1} |A(t) - \widetilde{A}(t)| = \sup_{t \le \theta} n^{-1} |\widetilde{S}(t) - S(t)| \overset{\mathbb{P}}{\longrightarrow} 0. \qquad (4.3.79)$$

Thus, by (4.3.75),

$$\sup_{t \leq \theta} |n^{-1}A(t) - H(e^{-t})| \overset{\mathbb{P}}{\longrightarrow} 0. \tag{4.3.80}$$

This is the work horse of our argument. By Lemma 4.18, we know that $t \mapsto H(e^{-t})$ is positive on $(0, -\log \xi)$ when $\nu > 1$. Thus, exploration in the interval $(0, -\log \xi)$ will find the giant component. In particular, we need to show that whp no large connected component is found before or after this interval (showing that the giant is unique), and we need to investigate the properties of the giant, in terms of its number of edges, vertices of degree k, etc. We now provide these details.

Let $0 < \varepsilon < \theta/2$. Since $H(e^{-t}) > 0$ on the compact interval $[\varepsilon, \theta - \varepsilon]$, (4.3.80) implies that $A(t)$ remains whp positive on $[\varepsilon, \theta - \varepsilon]$, and thus we have not started exploring a new component in this interval.

On the other hand, again by Lemma 4.18(a), $H(e^{-(\theta+\varepsilon)}) < 0$ and (4.3.75) implies that $n^{-1}\tilde{A}(\theta + \varepsilon) \overset{\mathbb{P}}{\longrightarrow} H(e^{-(\theta+\varepsilon)})$, while $A(\theta + \varepsilon) \geq 0$. Thus, with $\Delta = |H(e^{-\theta-\varepsilon})|/2 > 0$, whp

$$\tilde{S}(\theta + \varepsilon) - S(\theta + \varepsilon) = A(\theta + \varepsilon) - \tilde{A}(\theta + \varepsilon) \geq -\tilde{A}(\theta + \varepsilon) > n\Delta, \tag{4.3.81}$$

while (4.3.79) yields that $\tilde{S}(\theta) - S(\theta) < n\Delta$ whp. Consequently, whp $\tilde{S}(\theta+\varepsilon) - S(\theta+\varepsilon) > \tilde{S}(\theta) - S(\theta)$, so whp **Step 1** is performed at least once between the times θ and $\theta + \varepsilon$.

Let T_1 be the last time that **Step 1** was performed before time $\theta/2$. Let T_2 be the next time that **Step 1** is performed (by convention, $T_2 = \infty$ if such a time does not exist). We have shown that, for any $\varepsilon > 0$ and whp, $0 \leq T_1 \leq \varepsilon$ and $\theta - \varepsilon \leq T_2 \leq \theta + \varepsilon$. In other words, $T_1 \overset{\mathbb{P}}{\longrightarrow} 0$ and $T_2 \overset{\mathbb{P}}{\longrightarrow} \theta$. We conclude that we have found one component that is explored between time $T_1 \overset{\mathbb{P}}{\longrightarrow} 0$ and time $T_2 \overset{\mathbb{P}}{\longrightarrow} \theta$. This is our candidate for the giant component, and we continue to study its properties, i.e., its size, its number of edges, and its number of vertices of degree k. These properties are stated separately in the next proposition, so that we are able to reuse them later on:

Proposition 4.19 (Connected component properties) *Let $T_1^\star$ and $T_2^\star$ be two random times when **Step 1** is performed, with $T_1^\star \leq T_2^\star$, and assume that $T_1^\star \overset{\mathbb{P}}{\longrightarrow} t_1$ and $T_2^\star \overset{\mathbb{P}}{\longrightarrow} t_2$ where $0 \leq t_1 \leq t_2 \leq \theta < \infty$. If $\mathscr{C}^\star$ is the union of all components explored between $T_1^\star$ and $T_2^\star$ then*

$$v_k(\mathscr{C}^\star)/n \overset{\mathbb{P}}{\longrightarrow} p_k(e^{-kt_1} - e^{-kt_2}), \quad k \geq 0, \tag{4.3.82}$$

$$|\mathscr{C}^\star|/n \overset{\mathbb{P}}{\longrightarrow} G_D(e^{-t_1}) - G_D(e^{-t_2}), \tag{4.3.83}$$

$$|E(\mathscr{C}^\star)|/n \overset{\mathbb{P}}{\longrightarrow} \tfrac{1}{2}h(e^{-t_1}) - \tfrac{1}{2}h(e^{-t_2}). \tag{4.3.84}$$

In particular, if $t_1 = t_2$, then $|\mathscr{C}^\star|/n \overset{\mathbb{P}}{\longrightarrow} 0$ and $|E(\mathscr{C}^\star)| \overset{\mathbb{P}}{\longrightarrow} 0$.

Below, we apply Proposition 4.19 to $T_1 = o_{\mathbb{P}}(1)$ and $T_2 = \theta + o_{\mathbb{P}}(1)$. We can identify the values of the above constants for $t_1 = 0$ and $t_2 = \theta$ as $e^{-t_1} = 1, e^{-t_2} = \xi, G_D(e^{-t_1}) = 1$, $G_D(e^{-t_2}) = 1 - \zeta, h(e^{-t_1}) = 2\mathbb{E}[D], h(e^{-t_2}) = 2\mathbb{E}[D]\xi^2$ (see Exercise 4.9).

By Proposition 4.19 and Exercise 4.9, Theorem 4.9(a) follows if we can prove that the connected component found between times T_1 and T_2 is indeed the giant component. This will be proved after we complete the proof of Proposition 4.19:

Proof The set of vertices $\mathscr{C}^*$ contains all vertices awakened in the interval $[T_1^*, T_2^*)$ and no others, and thus (writing $V_k(t-) = \lim_{s \nearrow t} V_k(s)$)

$$v_k(\mathscr{C}^*) = V_k(T_1^*-) - V_k(T_2^*-), \qquad k \geq 1. \tag{4.3.85}$$

Since $T_2^* \xrightarrow{\mathbb{P}} t_2 \leq \theta$ and H is continuous, we obtain that

$$\inf_{t \leq T_2^*} H(e^{-t}) \xrightarrow{\mathbb{P}} \inf_{t \leq t_2} H(e^{-t}) = 0, \tag{4.3.86}$$

where the latter equality follows since $H(1) = 0$. Now, (4.3.75) and (4.3.76) imply, in analogy with (4.3.78) and (4.3.79), that $n^{-1} \inf_{t \leq T_2^*} \widetilde{A}(t) \xrightarrow{\mathbb{P}} 0$ and thus also

$$\sup_{t \leq T_2^*} n^{-1} |\widetilde{S}(t) - S(t)| \xrightarrow{\mathbb{P}} 0. \tag{4.3.87}$$

Since $\widetilde{V}_k(t) \geq V_k(t)$ for every k and $t \geq 0$,

$$\widetilde{V}_k(t) - V_k(t) \leq k^{-1} \sum_{\ell=1}^{\infty} \ell(\widetilde{V}_\ell(t) - V_\ell(t)) = k^{-1}(\widetilde{S}(t) - S(t)), \quad k \geq 1. \tag{4.3.88}$$

Hence (4.3.87) implies that $\sup_{t \leq T_2^*} |\widetilde{V}_k(t) - V_k(t)| = o_{\mathbb{P}}(n)$ for every $k \geq 1$. Consequently, using Lemma 4.16, for $j = 1, 2$, we have

$$V_k(T_j^*-) = \widetilde{V}_k(T_j^*-) + o_{\mathbb{P}}(n) = np_k e^{-kT_j^*} + o_{\mathbb{P}}(n) = np_k e^{-kt_j} + o_{\mathbb{P}}(n), \tag{4.3.89}$$

and (4.3.82) follows by (4.3.85). Similarly, using $\sum_{k=0}^{\infty} (\widetilde{V}_k(t) - V_k(t)) \leq \widetilde{S}(t) - S(t)$,

$$|\mathscr{C}^*| = \sum_{k=1}^{\infty} (V_k(T_1^*-) - V_k(T_2^*-)) = \sum_{k=1}^{\infty} (\widetilde{V}_k(T_1^*-) - \widetilde{V}_k(T_2^*-)) + o_{\mathbb{P}}(n) \tag{4.3.90}$$

$$= nG_D(e^{-T_1^*}) - nG_D(e^{-T_2^*}) + o_{\mathbb{P}}(n),$$

and

$$2|E(\mathscr{C}^*)| = \sum_{k=1}^{\infty} k(V_k(T_1^*-) - V_k(T_2^*-)) = \sum_{k=1}^{\infty} k(\widetilde{V}_k(T_1^*-) - \widetilde{V}_k(T_2^*-)) + o_{\mathbb{P}}(n)$$

$$= nh(e^{-T_1^*}) - nh(e^{-T_2^*}) + o_{\mathbb{P}}(n). \tag{4.3.91}$$

Thus, (4.3.83) and (4.3.84) follow from the convergence $T_i^* \xrightarrow{\mathbb{P}} t_i$ and the continuity of $t \mapsto G_D(e^{-t})$ and $t \mapsto h(e^{-t})$. $\qquad\square$

We are now ready to complete the Proof of Theorem 4.9.

Completion of the proof of Theorem 4.9

Let $\mathscr{C}'_{\max}$ be the component created at time T_1 and explored until time T_2, where we recall that T_1 is the last time Step 1 was performed before time $\theta/2$ and we let T_2 be the next time it is performed if that did occur and let $T_2 = \infty$ otherwise. Then $T_1 \xrightarrow{\mathbb{P}} 0$ and $T_2 \xrightarrow{\mathbb{P}} \theta$. The cluster $\mathscr{C}'_{\max}$ is our candidate for the giant component $\mathscr{C}_{\max}$, and we next prove that indeed it is, whp, the largest connected component.

By Proposition 4.19, with $t_1 = 0$ and $t_2 = \theta$,

$$|v_k(\mathscr{C}'_{\max})|/n \xrightarrow{\mathbb{P}} p_k(1 - e^{-k\theta}), \tag{4.3.92}$$

$$|\mathscr{C}'_{\max}|/n \xrightarrow{\mathbb{P}} G_D(1) - G_D(e^{-\theta}) = 1 - G_D(\xi), \tag{4.3.93}$$

$$|E(\mathscr{C}'_{\max})|/n \xrightarrow{\mathbb{P}} \frac{1}{2}(h(1) - h(e^{-\theta})) = \frac{1}{2}(h(1) - h(\xi)) = \frac{\mathbb{E}[D]}{2}(1 - \xi^2), \tag{4.3.94}$$

using Exercise 4.9. We have found one large component $\mathscr{C}'_{\max}$ with the claimed numbers of vertices and edges. It remains to show that whp there is no other large component. The basic idea is that *if* there exists another component that has at least $\eta\ell_n$ half-edges in it, then it should have a reasonable chance of actually being found quickly. Since we can show that the probability of finding a large component before T_1 or after T_2 is small, there just cannot be any other large connected component. Let us now make this intuition precise.

No early large component. Here, we first show that it is unlikely that a large component different from $\mathscr{C}'_{\max}$ is found before time T_1. For this, let $\eta > 0$, and apply Proposition 4.19 to $T_0 = 0$ and T_1, where T_1 has been defined to be the last time **Step 1** was performed before time $\theta/2$. Then, since $T_1 \xrightarrow{\mathbb{P}} 0$, the total number of vertices and edges in *all* components found before $\mathscr{C}'_{\max}$, i.e., before time T_1, is $o_{\mathbb{P}}(n)$. Hence, recalling that $\ell_n = \Theta(n)$ by Condition 1.7(b),

$$\mathbb{P}(\text{a component } \mathscr{C} \text{ with } |E(\mathscr{C})| \geq \eta\ell_n \text{ is found before } \mathscr{C}'_{\max}) \to 0. \tag{4.3.95}$$

We conclude that whp no component containing at least $\eta\ell_n$ half-edges is found *before* $\mathscr{C}'_{\max}$ is found.

No late large component. In order to study the probability of finding a component containing at least $\eta\ell_n$ edges *after* $\mathscr{C}'_{\max}$ is found, we start by letting T_3 be the first time after time T_2 that **Step 1** is performed. Since $\widetilde{S}(t) - S(t)$ increases by at most $d_{\max} = o(n)$ each time **Step 1** is performed, we obtain from (4.3.87) that

$$\sup_{t \leq T_3}(\widetilde{S}(t) - S(t)) \leq \sup_{t \leq T_2}(\widetilde{S}(t) - S(t)) + d_{\max} = o_{\mathbb{P}}(n). \tag{4.3.96}$$

Comparing this with (4.3.81), for every $\varepsilon > 0$ and whp we have that $\theta + \varepsilon > T_3$. Since also $T_3 > T_2 \xrightarrow{\mathbb{P}} \theta$, it follows that $T_3 \xrightarrow{\mathbb{P}} \theta$. If $\mathscr{C}'$ is the component created between times T_2 and T_3, then Proposition 4.19 applied to T_2 and T_3 yields $|\mathscr{C}'|/n \xrightarrow{\mathbb{P}} 0$ and $|E(\mathscr{C}')| \xrightarrow{\mathbb{P}} 0$.

On the other hand, if there existed a component $\mathscr{C} \neq \mathscr{C}'_{\max}$ in $\mathrm{CM}_n(\boldsymbol{d})$ with at least $\eta\ell_n$ edges that had not been found before $\mathscr{C}'_{\max}$, then with probability at least η the vertex chosen at random by **Step 1** at time T_2 to start the component $\mathscr{C}'$ would belong to $\mathscr{C}$. If this occurred, we would clearly have that $\mathscr{C} = \mathscr{C}'$. Consequently,

$$\mathbb{P}(\text{a component } \mathscr{C} \text{ with } |E(\mathscr{C})| \geq \eta\ell_n \text{ is found after } \mathscr{C}'_{\max})$$
$$\leq \eta^{-1}\mathbb{P}(|E(\mathscr{C}')| \geq \eta\ell_n) \to 0, \tag{4.3.97}$$

since $|E(\mathscr{C}')| \xrightarrow{\mathbb{P}} 0$.

Completion of the proof of Theorem 4.9(a). Combining (4.3.95) and (4.3.97), we see that whp there is no connected component except $\mathscr{C}'_{\max}$ that has at least $\eta\ell_n$ edges. As a result, we must have that $\mathscr{C}'_{\max} = \mathscr{C}_{\max}$, where $\mathscr{C}_{\max}$ is the largest connected component.

Further, again whp, $|E(\mathscr{C}_{(2)})| < \eta\ell_n$. Consequently, the results for $\mathscr{C}_{\max}$ follow from (4.3.92)–(4.3.94). We have further shown that $|E(\mathscr{C}_{(2)})|/\ell_n \overset{\mathbb{P}}{\longrightarrow} 0$, which implies that $|E(\mathscr{C}_{(2)})|/n \overset{\mathbb{P}}{\longrightarrow} 0$ and $|\mathscr{C}_{(2)}|/n \overset{\mathbb{P}}{\longrightarrow} 0$ because $\ell_n = \Theta(n)$ and $|\mathscr{C}_{(2)}| \le |E(\mathscr{C}_{(2)})| + 1$. This completes the proof of Theorem 4.9(a). $\qquad\square$

Completion of the proof of Theorem 4.9(b). The proof of Theorem 4.9(b) is similar to the last step in the proof for Theorem 4.9(a). Indeed, let $T_1 = 0$ and let T_2 be the next time that Step 1 is performed, and let $T_2 = \infty$ if this does not occur. Then

$$\sup_{t \le T_2} |A(t) - \widetilde{A}(t)| = \sup_{t \le T_2} |\widetilde{S}(t) - S(t)| \le 2d_{\max} = o(n). \tag{4.3.98}$$

For every $\varepsilon > 0$, $n^{-1}\widetilde{A}(\varepsilon) \overset{\mathbb{P}}{\longrightarrow} H(\mathrm{e}^{-\varepsilon}) < 0$ by (4.3.75) and Lemma 4.18(b), while $A(\varepsilon) \ge 0$, and it follows from (4.3.98) that whp $T_2 < \varepsilon$. Hence, $T_2 \overset{\mathbb{P}}{\longrightarrow} 0$. We apply Proposition 4.19 (which holds in this case too, with $\theta = 0$) and find that if $\mathscr{C}$ is the first component found, then $|E(\mathscr{C})|/n \overset{\mathbb{P}}{\longrightarrow} 0$.

Let $\eta > 0$. If $|E(\mathscr{C}_{\max})| \ge \eta\ell_n$, then the probability that the first half-edge chosen by Step 1 belongs to $\mathscr{C}_{\max}$, and thus $\mathscr{C} = \mathscr{C}_{\max}$, is $2|E(\mathscr{C}_{\max})|/(2\ell_n) \ge \eta$, and hence,

$$\mathbb{P}(|E(\mathscr{C}_{\max})| \ge \eta\ell_n) \le \eta^{-1}\mathbb{P}(|E(\mathscr{C})| \ge \eta\ell_n) \to 0. \tag{4.3.99}$$

The results follows since $\ell_n = \Theta(n)$ by Condition 1.7(b) and $|\mathscr{C}_{\max}| \le |E(\mathscr{C}_{\max})| + 1$. This completes the proof of Theorem 4.9(b), and thus that of Theorem 4.9. $\qquad\square$

4.3.3 GIANT COMPONENT OF RELATED RANDOM GRAPHS

In this subsection we extend the results of Theorem 4.9 to related models, such as uniform simple random graphs with a given degree sequence, as well as generalized random graphs. Recall that $\mathrm{UG}_n(\boldsymbol{d})$ denotes a uniform simple random graph with degrees $\boldsymbol{d}$ (see Section 1.3.4 and [V1, Section 7.5]). The results in Theorem 4.9 also hold for $\mathrm{UG}_n(\boldsymbol{d})$ when we assume that Conditions 1.7(a)–(c) hold:

Theorem 4.20 (Phase transition in $\mathrm{UG}_n(\boldsymbol{d})$)   *Let $\boldsymbol{d}$ satisfy Conditions 1.7(a)–(c). Then Theorem 4.9 extends to the* uniform *simple graph with degree sequence $\boldsymbol{d}$.*

Proof   By [V1, Corollary 7.17], and since $\boldsymbol{d} = (d_v)_{v\in[n]}$ satisfies Conditions 1.7(a)–(c), any event $\mathcal{E}_n$ that occurs whp for $\mathrm{CM}_n(\boldsymbol{d})$ also occurs whp for $\mathrm{UG}_n(\boldsymbol{d})$. By Theorem 4.9, the event $\mathcal{E}_n = \{||\mathscr{C}_{\max}|/n - \zeta| \le \varepsilon\}$ occurs whp for $\mathrm{CM}_n(\boldsymbol{d})$, so it also holds whp for $\mathrm{UG}_n(\boldsymbol{d})$. The proof for the other properties is identical. $\qquad\square$

Note that it is not obvious how to extend Theorem 4.20 to the case where $\nu = \infty$, which we discuss now:

Theorem 4.21 (Giant in uniform graph with given degrees for $\nu = \infty$)   *Consider $\mathrm{UG}_n(\boldsymbol{d})$, where the degrees $\boldsymbol{d}$ satisfy Conditions 1.7(a),(b), and assume that there exists $\tau \in (2,3)$ such that, for every $x \ge 1$,*

$$[1 - F_n](x) \le c_F x^{-(\tau-1)}. \tag{4.3.100}$$

Then, Theorem 4.9 extends to the uniform *simple graph with degree sequence $\boldsymbol{d}$.*

Sketch of proof. We do not present the entire proof, but rather sketch it. We will show that, for every $\varepsilon > 0$, there exists $\delta = \delta(\varepsilon) > 0$ such that

$$\mathbb{P}\left(\left|\,|\mathscr{C}_{\max}| - \zeta n\right| > \varepsilon n\right) \leq e^{-\delta n}, \tag{4.3.101}$$

and

$$\mathbb{P}\left(\left|v_k(\mathscr{C}_{\max}) - p_k(1 - \xi^k)n\right| > \varepsilon n\right) \leq e^{-\delta n}. \tag{4.3.102}$$

This *exponential concentration* is quite convenient, as it allows us to extend the result to the setting of uniform random graphs by conditioning $\mathrm{CM}_n(\boldsymbol{d})$ to be simple. Indeed, by Lemma 4.8, it follows that the result also holds for the uniform simple random graph $\mathrm{UG}_n(\boldsymbol{d})$ when Conditions 1.7(a),(b) hold. In Exercise 4.10 below, the reader is invited to fill in the details of the proof of Theorem 4.21. □

We refer to Section 4.5 for a further discussion of (4.3.102). There, we discuss approximations for $\mathbb{P}(\mathrm{CM}_n(\boldsymbol{d})$ simple) under conditions such as (4.3.100).

We next prove Theorem 3.19 for rank-1 inhomogeneous random graphs, as already stated in Theorem 3.20 and as restated here for convenience:

Theorem 4.22 (Phase transition in rank-1 random graphs) *Let $\boldsymbol{w}$ satisfy Condition 1.1(a)–(c). Then the results in Theorem 4.9 also hold for $\mathrm{GRG}_n(\boldsymbol{w})$, $\mathrm{CL}_n(\boldsymbol{w})$, and $\mathrm{NR}_n(\boldsymbol{w})$.*

Proof Let d_v be the degree of vertex $v \in [n]$ in $\mathrm{GRG}_n(\boldsymbol{w})$ defined in [V1, (1.3.18)], where we use a small letter to avoid confusion with D_n, which is the degree of a *uniform* vertex in $[n]$. By [V1, Theorem 7.18], the law of $\mathrm{GRG}_n(\boldsymbol{w})$ conditioned on the degrees $\boldsymbol{d}$ and $\mathrm{CM}_n(\boldsymbol{d})$ conditioned on being simple agree (recall also Theorem 1.4). By Theorem 1.9, $(d_v)_{v \in [n]}$ satisfies Conditions 1.7(a)–(c) in probability. Then, by [V1, Theorem 7.18] and Theorem 4.9, the results in Theorem 4.9 also hold for $\mathrm{GRG}_n(\boldsymbol{w})$. By [V1, Theorem 6.20], the same result applies to $\mathrm{CL}_n(\boldsymbol{w})$, and, by [V1, Exercise 6.39], also to $\mathrm{NR}_n(\boldsymbol{w})$. □

Unfortunately, when $\nu = \infty$ we cannot rely on the fact that, by [V1, Theorem 7.18], the law of $\mathrm{GRG}_n(\boldsymbol{w})$ conditioned on the degrees $\boldsymbol{d}$ and $\mathrm{CM}_n(\boldsymbol{d})$ conditioned on being simple agree. Indeed, when $\nu = \infty$, the probability that $\mathrm{CM}_n(\boldsymbol{d})$ is simple vanishes. Therefore, we instead rely on a *truncation argument* to extend Theorem 4.22 to the case where $\nu = \infty$. It is here that the monotonicity of $\mathrm{GRG}_n(\boldsymbol{w})$ in terms of the edge probabilities can be used rather conveniently:

Theorem 4.23 (Phase transition in $\mathrm{GRG}_n(\boldsymbol{w})$) *Let $\boldsymbol{w}$ satisfy Conditions 1.1(a),(b). Then, the results in Theorem 4.9 also hold for $\mathrm{GRG}_n(\boldsymbol{w})$, $\mathrm{CL}_n(\boldsymbol{w})$, and $\mathrm{NR}_n(\boldsymbol{w})$.*

Proof We prove only that $|\mathscr{C}_{\max}|/n \xrightarrow{\mathbb{P}} \zeta$; the other statements in Theorem 4.9 can be proved in a similar fashion (see Exercises 4.11 and 4.12 below). We prove Theorem 4.23 only for $\mathrm{NR}_n(\boldsymbol{w})$, the proofs for $\mathrm{GRG}_n(\boldsymbol{w})$ and $\mathrm{CL}_n(\boldsymbol{w})$ being similar. The required upper bound $|\mathscr{C}_{\max}|/n \leq \zeta + o_{\mathbb{P}}(1)$ follows by the local convergence in probability in Theorem 3.14 and Corollary 2.27.

For the lower bound, we bound $\mathrm{NR}_n(\boldsymbol{w})$ from below by a random graph with edge probabilities

$$p_{uv}^{(K)} = 1 - e^{-(w_u \wedge K)(w_v \wedge K)/\ell_n}. \tag{4.3.103}$$

Therefore, we also have $|\mathscr{C}_{\max}| \succeq |\mathscr{C}_{\max}^{(K)}|$, where $\mathscr{C}_{\max}^{(K)}$ is the largest connected component in the inhomogeneous random graph with edge probabilities $(p_{uv}^{(K)})_{u,v \in [n]}$. Let

$$w_v^{(K)} = (w_v \wedge K) \frac{1}{\ell_n} \sum_{u \in [n]} (w_u \wedge K), \qquad (4.3.104)$$

so that the edge probabilities in (4.3.103) correspond to the Norros–Reittu model with weights $(w_v^{(K)})_{v \in [n]}$. It is not hard to see that, when Condition 1.1(a) holds for $(w_v)_{v \in [n]}$, Conditions 1.1(a)–(c) hold for $(w_v^{(K)})_{v \in [n]}$, where the limiting random variable equals $W^{(K)} = (W \wedge K) \mathbb{E}[(W \wedge K)] / \mathbb{E}[W]$. Therefore, Theorem 4.22 applies to $(w_v^{(K)})_{v \in [n]}$. We deduce that $|\mathscr{C}_{\max}^{(K)}|/n \xrightarrow{\mathbb{P}} \zeta^{(K)}$, which is the survival probability of the two-stage mixed-Poisson branching process with mixing variable $W^{(K)}$. Since $\zeta^{(K)} \to \zeta$ when $K \to \infty$, we conclude that $|\mathscr{C}_{\max}|/n \xrightarrow{\mathbb{P}} \zeta$. $\qquad \square$

4.4 CONNECTIVITY OF CONFIGURATION MODELS

Assume that $\mathbb{P}(D = 2) < 1$. By Theorem 4.9, we see that $|\mathscr{C}_{\max}|/n \xrightarrow{\mathbb{P}} 1$ when $\mathbb{P}(D \geq 2) = 1$, as in this case the survival probability ζ of the local limit equals 1. In this section, we investigate the conditions under which $\mathrm{CM}_n(\boldsymbol{d})$ is whp *connected*, i.e., $\mathscr{C}_{\max} = [n]$ and $|\mathscr{C}_{\max}| = n$. Our main result shows that this occurs whp when $d_{\min} = \min_{v \in [n]} d_v \geq 3$:

Theorem 4.24 (Connectivity of $\mathrm{CM}_n(\boldsymbol{d})$) *Assume that Conditions 1.7(a),(b) hold. Further, assume that $d_v \geq 3$ for every $v \in [n]$. Then*

$$\mathbb{P}(\mathrm{CM}_n(\boldsymbol{d}) \text{ disconnected}) = o(1). \qquad (4.4.1)$$

If Condition 1.7(a) holds with $p_1 = p_2 = 0$, then $\nu \geq 2 > 1$ is immediate, so we are always in the supercritical regime. Also, $\zeta = 1$ when $p_1 = p_2 = 0$, since survival of the unimodular branching-process tree occurs with probability 1. Therefore, Theorem 4.9 implies that the largest connected component has size $n(1 + o_{\mathbb{P}}(1))$ when Conditions 1.7(a),(b) hold. Theorem 4.24 extends this to the statement that $\mathrm{CM}_n(\boldsymbol{d})$ is whp *connected*.

Theorem 4.24 yields an important difference between the generalized random graph and the configuration model, also from a practical point of view. Indeed, for the generalized random graph to be whp connected, the degrees must tend to infinity. This has already been observed for $\mathrm{ER}_n(p)$ in [V1, Theorem 5.8]. The configuration model can be connected while the average degree is bounded. Many real-world networks are connected, which makes the configuration model often more suitable than inhomogeneous random graphs from this perspective (recall Table 4.1 and Figure 4.1).

Proof The proof is based on a relatively simple counting argument. We recall that a *configuration* denotes a pairing of all the half-edges. We note that the probability of a configuration equals $1/(\ell_n - 1)!!$. On the event that $\mathrm{CM}_n(\boldsymbol{d})$ is disconnected, there exists a set of vertices $\mathcal{I} \subseteq [n]$ with $|\mathcal{I}| \leq \lfloor n/2 \rfloor$ such that all half-edges incident to vertices in $\mathcal{I}$ are paired *only* with half-edges incident to other vertices in $\mathcal{I}$. For $\mathcal{I} \subseteq [n]$, we let $\ell_n(\mathcal{I})$ denote the total degree of $\mathcal{I}$, i.e.,

$$\ell_n(\mathcal{I}) = \sum_{i \in \mathcal{I}} d_i. \qquad (4.4.2)$$

Since $d_{\min} \geq 3$, we can use Theorem 4.9 to conclude that most edges are in $\mathscr{C}_{\max}$, and $\mathcal{I} \neq \mathscr{C}_{\max}$. Therefore, $\ell_n(\mathcal{I}) = o(\ell_n) = o(n)$, and we may, without loss of generality, assume that $\ell_n(\mathcal{I}) \leq \ell_n/2$. We denote by $\mathcal{E}_n$ the event that there exists a collection of connected components $\mathcal{I}$ consisting of $|\mathcal{I}| \leq \lfloor n/2 \rfloor$ vertices for which the sum of degrees is at most $\ell_n(\mathcal{I}) \leq \ell_n/2$, so that $\mathcal{E}_n$ occurs whp, i.e.

$$\mathbb{P}(\mathcal{E}_n^c) = o(1). \tag{4.4.3}$$

Clearly, in order for the half-edges incident to vertices in $\mathcal{I}$ to be paired only to other half-edges incident to vertices in $\mathcal{I}$, $\ell_n(\mathcal{I})$ needs to be even. The number of configurations for which this happens is bounded above by

$$(\ell_n(\mathcal{I}) - 1)!!(\ell_n(\mathcal{I}^c) - 1)!!. \tag{4.4.4}$$

As a result,

$$\mathbb{P}(\mathrm{CM}_n(\boldsymbol{d}) \text{ disconnected}; \mathcal{E}_n) \leq \sum_{\mathcal{I} \subseteq [n]} \frac{(\ell_n(\mathcal{I}) - 1)!!(\ell_n(\mathcal{I}^c) - 1)!!}{(\ell_n - 1)!!}$$

$$= \sum_{\mathcal{I} \subseteq [n]} \prod_{j=1}^{\ell_n(\mathcal{I})/2} \frac{\ell_n(\mathcal{I}) - 2j + 1}{\ell_n - 2j + 1}, \tag{4.4.5}$$

where the sum over $\mathcal{I} \subseteq [n]$ is restricted to sets $\mathcal{I}$ for which $1 \leq |\mathcal{I}| \leq \lfloor n/2 \rfloor$ and $\ell_n(\mathcal{I}) \leq \ell_n/2$ is even. Exercise 4.13 uses (4.4.5) to bound the probability of the existence of an isolated vertex (i.e., a vertex with only self-loops).

Define

$$f(x) = \prod_{j=1}^{x} \frac{2x - 2j + 1}{\ell_n - 2j + 1}, \tag{4.4.6}$$

so that

$$\mathbb{P}(\mathrm{CM}_n(\boldsymbol{d}) \text{ disconnected}; \mathcal{E}_n) \leq \sum_{\mathcal{I} \subseteq [n]} f(\ell_n(\mathcal{I})/2). \tag{4.4.7}$$

We can rewrite

$$f(x) = \frac{\prod_{j=1}^{x}(2x - 2j + 1)}{\prod_{j=1}^{x}(\ell_n - 2j - 1)} = \frac{\prod_{i=0}^{x-1}(2i + 1)}{\prod_{k=0}^{x-1}(\ell_n - 2k + 1)} = \prod_{j=0}^{x-1} \frac{2j + 1}{\ell_n - 2j + 1}, \tag{4.4.8}$$

where we set $j = x - i$ and $j = k + 1$ in the second equality. Thus, for $x \leq \ell_n/4$, $x \mapsto f(x)$ is decreasing because

$$\frac{f(x+1)}{f(x)} = \frac{2x + 1}{\ell_n - 2x + 1} \leq 1. \tag{4.4.9}$$

Since $\ell_n(\mathcal{I}) \leq \ell_n/2$, we also have that $\ell_n(\mathcal{I})/2 \leq \ell_n/4$, so that $f(\ell_n(\mathcal{I})/2) \leq f(a)$ for any $a \leq \ell_n(\mathcal{I})/2$. Now, since $d_i \geq 3$ for every $i \in [n]$ and $\ell_n(\mathcal{I}) \leq \ell_n/2$ is *even*,

$$\ell_n(\mathcal{I}) \geq 2\lceil 3|\mathcal{I}|/2 \rceil, \tag{4.4.10}$$

which depends only on the number of vertices in $\mathcal{I}$. There are precisely $\binom{n}{m}$ ways of choosing m vertices out of $[n]$, so that, with $m = |\mathcal{I}|$,

$$\mathbb{P}(\mathrm{CM}_n(\boldsymbol{d}) \text{ disconnected}; \mathcal{E}_n) \leq \sum_{\mathcal{I} \subseteq [n]} f(\lceil 3|\mathcal{I}|/2\rceil) = \sum_{m=1}^{\lfloor n/2 \rfloor} \binom{n}{m} f(\lceil 3m/2\rceil). \quad (4.4.11)$$

Define

$$h_n(m) = \binom{n}{m} f(\lceil 3m/2\rceil), \quad (4.4.12)$$

so that

$$\mathbb{P}(\mathrm{CM}_n(\boldsymbol{d}) \text{ disconnected}; \mathcal{E}_n) \leq \sum_{m=1}^{\lfloor n/2 \rfloor} h_n(m). \quad (4.4.13)$$

We are left with studying $h_n(m)$. For this, we write

$$\frac{h_n(m+1)}{h_n(m)} = \frac{n-m}{m+1} \frac{f(\lceil 3(m+1)/2\rceil)}{f(\lceil 3m/2\rceil)}. \quad (4.4.14)$$

Note that, for m odd,

$$\frac{f(\lceil 3(m+1)/2\rceil)}{f(\lceil 3m/2\rceil)} = \frac{f((3m+1)/2+1)}{f((3m+1)/2)} = \frac{3m+2}{\ell_n - 3m}. \quad (4.4.15)$$

while, for m even,

$$\frac{f(\lceil 3(m+1)/2\rceil)}{f(\lceil 3m/2\rceil)} = \frac{f(3m/2+2)}{f(3m/2)} = \frac{3m+3}{\ell_n - 3m - 1} \frac{3m+1}{\ell_n - 3m + 1}. \quad (4.4.16)$$

Thus, for m odd and using $\ell_n \geq 3n$,

$$\frac{h_n(m+1)}{h_n(m)} = \frac{n-m}{m+1} \frac{3m+2}{\ell_n - 3m} \leq \frac{3(n-m)}{\ell_n - 3m} \leq 1, \quad (4.4.17)$$

while, for m even and using $\ell_n \geq 3n$,

$$\frac{h_n(m+1)}{h_n(m)} = \frac{n-m}{m+1} \frac{3m+3}{\ell_n - 3m - 1} \frac{3m+1}{\ell_n - 3m + 1} \leq \frac{3m+1}{\ell_n - 3m - 1}. \quad (4.4.18)$$

Thus, we obtain that, for $m \leq n/2$ and since $\ell_n \geq 3n$, there exists a $c > 0$ such that

$$\frac{h_n(m+1)}{h_n(m)} \leq 1 + \frac{c}{n}. \quad (4.4.19)$$

We conclude that, for $m \leq n/2$ such that for $m \geq 3$,

$$h_n(m) = h_n(3) \prod_{j=3}^{m-1} \frac{h_n(j+1)}{h_n(j)} \leq h_n(3) \prod_{j=3}^{\lfloor n/2 \rfloor} (1 + c/n)$$

$$\leq h_n(3)(1 + c/n)^{\lfloor n/2 \rfloor} \leq h_n(3) e^{c/2}, \quad (4.4.20)$$

so that

$$\mathbb{P}(\mathrm{CM}_n(\boldsymbol{d}) \text{ disconnected}; \mathcal{E}_n) \leq \sum_{m=1}^{\varepsilon n} h_n(m) \leq h_n(1) + h_n(2) + \sum_{m=3}^{\lfloor n/2 \rfloor} h_n(m)$$
$$\leq h_n(1) + h_n(2) + n h_n(3) e^{c/2}/2. \qquad (4.4.21)$$

By Exercises 4.13–4.15, $h_n(1), h_n(2) = O(1/n)$, so we are left with computing $h_n(3)$. For this, we note that $\lceil 3m/2 \rceil = 5$ when $m = 3$, so that

$$h_n(3) = \binom{n}{3} f(5) = \frac{9!! n(n-1)(n-2)}{6(\ell_n - 1)(\ell_n - 3)(\ell_n - 5)(\ell_n - 7)(\ell_n - 9)}$$
$$= O(1/n^2). \qquad (4.4.22)$$

As a result, $n h_n(3) = O(1/n)$. We conclude that

$$\mathbb{P}(\mathrm{CM}_n(\boldsymbol{d}) \text{ disconnected}; \mathcal{E}_n) = O(1/n), \qquad (4.4.23)$$

which, together with (4.4.3), completes the proof. □

The bound in (4.4.23) is stronger than required, and even suggests that $\mathbb{P}(\mathrm{CM}_n(\boldsymbol{d})$ disconnected$) = O(1/n)$. However, our proof falls short of this, since we started with the assumption that the non-giant (collection of) component(s) $\mathcal{I}$ satisfies $\ell_n(\mathcal{I}) \leq \ell_n/2$. For this, in turn, we used Theorem 4.9 to conclude that the complementary probability is $o(1)$. In the following theorem we will use the degree-truncation argument in Section 1.3.3, see in particular Remark 1.12, to substantially improve upon Theorem 4.24:

Theorem 4.25 (Connectivity of $\mathrm{CM}_n(\boldsymbol{d})$) *Assume that $d_v \geq 3$ for every $v \in [n]$. Then*

$$\mathbb{P}(\mathrm{CM}_n(\boldsymbol{d}) \text{ disconnected}) = O(1/n). \qquad (4.4.24)$$

Theorem 4.25 improves upon Theorem 4.24 in that there are no restrictions on the degree other than $d_{\min} \geq 3$ (not even the degree regularity in Conditions 1.7(a),(b)) and that the probability of disconnection is $O(1/n)$. Exercise 4.16 uses Theorem 4.25 to prove that

$$\mathbb{P}(\mathrm{CM}_n(\boldsymbol{d}) \text{ disconnected}) = \Theta(1/n) \qquad (4.4.25)$$

when $\mathbb{P}(D = 3) > 0$, and Exercise 4.17 does the same for $\mathbb{P}(D = 4) > 0$.

Proof We start by using Remark 1.12 with $b = 3$, which means that we may assume that all degrees are in $\{3, 4, 5\}$. Indeed, the construction leading to Remark 1.12 splits vertices of high degrees into (possibly several) vertices of degree lying in the interval $[b, 2b)$, which equals $\{3, 4, 5\}$ for $b = 3$. Further, when the graph after splitting is connected, it certainly must have been before splitting. We slightly abuse notation, and keep on writing $\mathrm{CM}_n(\boldsymbol{d})$ for the graph after degree truncation.

We then follow the proof of Theorem 4.24, and now define $\mathcal{I}$ as the collection of components that satisfies $\ell_n(\mathcal{I}) \leq \ell_n/2$. It should be remarked that in this case we cannot rely upon Theorem 4.9, which implies (4.4.3). Theorem 4.9 was used to show that $\ell_n(\mathcal{I}) \leq \ell_n/2$ and $|\mathcal{I}| \leq n/2$ whp. The fact that $\ell_n(\mathcal{I}) \leq \ell_n/2$ was used in (4.4.9) to show that $x \mapsto f(x)$ is decreasing for the appropriate x, and this still holds. The fact that $|\mathcal{I}| \leq n/2$ was used to restrict the sum over m in (4.4.11) and the formulas that followed it, which we can now no longer use, and thus we need an alternative argument.

In the current setting, since the degrees are all in $\{3, 4, 5\}$,

$$3|\mathcal{I}| \leq \ell_n(\mathcal{I}) \leq \ell_n/2, \tag{4.4.26}$$

so that $m \leq \ell_n/6$. Following the proof of Theorem 4.24 up to (4.4.13), we thus arrive at

$$\mathbb{P}(CM_n(\boldsymbol{d}) \text{ disconnected}) \leq \sum_{m=1}^{\lfloor \ell_n/6 \rfloor} h_n(m). \tag{4.4.27}$$

The bound in (4.4.17) remains unchanged since it did not rely on $m \leq n/2$, while, for $m \leq \ell_n/6$, (4.4.18) can be bounded as follows:

$$\frac{3m+1}{\ell_n - 3m - 1} \leq 1 + O(1/n). \tag{4.4.28}$$

As a result, both (4.4.17) and (4.4.18) remain valid, proving that $h_n(m+1)/h_n(m) \leq 1 + c/n$. We conclude that the proof can be completed as for Theorem 4.24. $\qquad\square$

The above proof is remarkably simple, and requires very little of the precise degree distribution to be satisfied except for $d_{\min} \geq 3$. In what follows, we investigate what happens when this fails. We first continue by showing that $CM_n(\boldsymbol{d})$ is with positive probability *disconnected* when n_1, the number of vertices of degree 1, satisfies $n_1 \gg n^{1/2}$:

Proposition 4.26 (Disconnectivity of $CM_n(\boldsymbol{d})$ when $n_1 \gg n^{1/2}$) *Let Conditions 1.7(a),(b) hold, and assume that $n_1 \gg n^{1/2}$. Then*

$$\lim_{n \to \infty} \mathbb{P}(CM_n(\boldsymbol{d}) \text{ connected}) = 0. \tag{4.4.29}$$

Proof We note that $CM_n(\boldsymbol{d})$ is disconnected when there are two vertices of degree 1 whose half-edges are paired with each other. When the half-edges of two vertices of degree 1 are paired with each other, we say that a *2-pair* is created. Then, since after i pairings of degree-1 vertices to higher-degree vertices, there are $\ell_n - n_1 - i + 1$ half-edges incident to higher-degree vertices, out of a total of $\ell_n - 2i + 1$ unpaired half-edges, we have

$$\mathbb{P}(CM_n(\boldsymbol{d}) \text{ contains no 2-pair}) = \prod_{i=1}^{n_1} \frac{\ell_n - n_1 - i + 1}{\ell_n - 2i + 1}$$

$$= \prod_{i=1}^{n_1} \left(1 - \frac{n_1 - i}{\ell_n - 2i + 1}\right). \tag{4.4.30}$$

For each $i \geq 1$,

$$1 - \frac{n_1 - i}{\ell_n - 2i + 1} \leq 1 - \frac{n_1 - i}{\ell_n} \leq e^{-(n_1 - i)/\ell_n}, \tag{4.4.31}$$

so that we arrive at

$$\mathbb{P}(CM_n(\boldsymbol{d}) \text{ contains no 2-pair}) \leq \prod_{i=1}^{n_1} e^{-(n_1 - i)/\ell_n}$$

$$= e^{-n_1(n_1 - 1)/[2\ell_n]} = o(1), \tag{4.4.32}$$

since $\ell_n = \Theta(n)$ and $n_1 \gg n^{1/2}$. $\qquad\square$

We close this section by showing that the probability that $CM_n(\boldsymbol{d})$ is connected is strictly smaller than 1 when $p_2 > 0$:

Proposition 4.27 (Disconnectivity of $CM_n(d)$ when $p_2 > 0$) *Let Conditions 1.7(a),(b) hold, and assume that $p_2 > 0$. Then,*

$$\limsup_{n\to\infty} \mathbb{P}(CM_n(d) \text{ connected}) < 1. \qquad (4.4.33)$$

Proof We perform a second-moment method on the number $P_n(2)$ of connected components consisting of two vertices of degree 2. We compute

$$\mathbb{E}[P_n(2)] = \frac{2n_2(n_2 - 1)}{2(\ell_n - 1)(\ell_n - 3)}, \qquad (4.4.34)$$

since there are $n_2(n_2 - 1)/2$ pairs of vertices of degree 2, and the probability that a fixed pair forms a connected component is equal to $2/(\ell_n - 1)(\ell_n - 3)$. By Conditions 1.7(a),(b), which imply that $n_2/n \to p_2$,

$$\mathbb{E}[P_n(2)] \to p_2^2/\mathbb{E}[D]^2 \equiv \lambda_2. \qquad (4.4.35)$$

By assumption, $p_2 > 0$, so that also $\lambda_2 > 0$. By investigating the higher factorial moments, and using [V1, Theorem 2.6], it follows that $P_n(2) \xrightarrow{d} \text{Poi}(\lambda_2)$, so that

$$\mathbb{P}(CM_n(d) \text{ disconnected}) \geq \mathbb{P}(P_n(2) > 0) \to 1 - e^{-\lambda_2} > 0, \qquad (4.4.36)$$

as required. The proof that [V1, Theorem 2.6] can be applied is Exercise 4.18. $\qquad\square$

Almost-Connectivity for $d_{\min} \geq 2$

We close this section with a detailed result on the size of the giant when $d_{\min} \geq 2$:

Theorem 4.28 (Almost-connectivity of $CM_n(d)$ when $p_1 = 0$) *Consider $CM_n(d)$ where the degrees d satisfy Conditions 1.7(a),(b), and assume that $p_2 \in (0, 1)$. Also assume that $d_v \geq 2$ for every $v \in [n]$. Then*

$$n - |\mathscr{C}_{\max}| \xrightarrow{d} \sum_{k \geq 2} k X_k, \qquad (4.4.37)$$

where $(X_k)_{k \geq 2}$ are independent Poisson random variables with parameters $\lambda_k = \lambda^k 2^{k-1}/k$ with $\lambda = p_2/\mathbb{E}[D]$. Consequently,

$$\mathbb{P}(CM_n(d) \text{ connected}) \to e^{-\sum_{k \geq 2}(2\lambda^2)^k/(2k)} \in (0, 1). \qquad (4.4.38)$$

Rather than giving the complete proof of Theorem 4.28, we give a sketch of it:

Sketch of proof of Theorem 4.28. Let $P_n(k)$ denote the number of k-cycles consisting of degree-2 vertices, for $k \geq 2$. Obviously, every vertex in such a cycle is not part of the giant component, so that

$$n - |\mathscr{C}_{\max}| \geq \sum_{k \geq 2} k P_n(k). \qquad (4.4.39)$$

A multivariate moment method allows one to prove that $(P_n(k))_{k \geq 2} \xrightarrow{d} (X_k)_{k \geq 2}$, where $(X_k)_{k \geq 2}$ are independent Poisson random variables with parameters (see Exercise 4.19)

$$\lambda_2^k/(2k) = \lim_{n\to\infty} \mathbb{E}[P_n(k)].$$

In order to complete the argument, two approaches are possible (and have been used in the literature). First, Federico and van der Hofstad (2017) used counting arguments to show that as soon as a connected component has at least one vertex v of degree $d_v \geq 3$, then it is whp part of the giant component $\mathscr{C}_{\max}$. This then proves that (4.4.39) is whp an equality. See also Exercise 4.20.

Alternatively, and more in the style of Łuczak (1992), one can pair up all the half-edges incident to vertices of degree 2, and then realize that the graph, after pairing of all these degree-2 vertices, is again a configuration model with a changed degree distribution. The cycles consisting of only degree-2 vertices will be removed, so that we need only to consider the contribution of pairing strings of degree-2 vertices to vertices of degrees at least 3. If both ends of the string are *each* connected to two *distinct* vertices of degrees d_s, d_t at least 3, then we can imagine this string to correspond to a single vertex of degree $d_s + d_t - 2 \geq 4$, which is sufficiently large.

Unfortunately, it is also possible that the string of degree-2 vertices is connected to the *same* vertex u of degree $d_u \geq 3$, thus possibly reducing the degree by 2. When $d_u \geq 5$, there are still at least three half-edges remaining at u. Thus, we need only to consider the case where we create a cycle of vertices of degree 2 with one vertex u in it of degree $d_u = 3$ or $d_u = 4$, respectively, which corresponds to vertices of remaining degree 1 or 2, respectively. In Exercise 4.21, the reader is asked to prove that there is a bounded number of such cycles.

We conclude that it suffices to extend the proof of Theorem 4.24 to the setting where there is a *bounded* number of vertices of degrees 1 and 2. The above argument can be repeated for the degree-2 vertices. We can deal with the degree-1 vertices in a similar way. Pairing the degree-1 vertices again leads to vertices of remaining degree at least $3 - 1 = 2$ after the pairing, which is fine when the remaining degree is at least 3; otherwise they can be dealt with in the same way as the other degree-2 vertices. We refrain from giving more details. $\square$

4.5 RELATED RESULTS FOR CONFIGURATION MODELS

In this section we discuss related results on connected components for the configuration model. We start by discussing the subcritical behavior of the configuration model.

Largest Subcritical Cluster

When $\nu < 1$, so that in particular $\mathbb{E}[D^2] < \infty$, the largest connected component for $\mathrm{CM}_n(\boldsymbol{d})$ is closely related to the largest degree:

Theorem 4.29 (Subcritical phase for $\mathrm{CM}_n(\boldsymbol{d})$) *Let $\boldsymbol{d}$ satisfy Conditions 1.7(a)–(c) with $\nu = \mathbb{E}[D(D-1)]/\mathbb{E}[D] < 1$. Suppose further that there exists $\tau > 3$ and $c_2 > 0$ such that, for all $x \geq 1$,*

$$[1 - F_n](x) \leq c_2 x^{-(\tau-1)}. \tag{4.5.1}$$

Then, for $\mathrm{CM}_n(\boldsymbol{d})$ with $d_{\max} = \max_{j \in [n]} d_j$,

$$|\mathscr{C}_{\max}| = \frac{d_{\max}}{1 - \nu} + o_{\mathbb{P}}(n^{1/(\tau-1)}). \tag{4.5.2}$$

Theorem 4.29 is closely related to Theorem 3.22 for $\mathrm{GRG}_n(\boldsymbol{w})$. In fact, we can use Theorem 4.29 to prove Theorem 3.22, see Exercise 4.22. Note that the result in (4.5.2) is most interesting when $d_{\max} = \Theta(n^{1/(\tau-1)})$, as it would be in the case when the degrees obey a power law with exponent τ (for example, in the case where the degrees are iid). This is the only case where Theorem 4.29 is sharp (see Exercise 4.23). When $d_{\max} = o(n^{1/(\tau-1)})$, Theorem 4.29 implies that $|\mathscr{C}_{\max}| = o_{\scriptscriptstyle\mathbb{P}}(n^{1/(\tau-1)})$, a less precise result.

The intuition behind Theorem 4.29 is that from the vertex of maximal degree, there are $d_{\max}$ half-edges that can reach more vertices. Since the random graph is subcritical, one can use Theorem 4.1 to prove that the tree rooted at any half-edge incident to the vertex of maximal degree converges in distribution to a subcritical branching process. Further, the trees rooted at different half-edges are close to being independent. Thus, by the law of large numbers, one can expect that the total number of vertices in these $d_{\max}$ trees is close to $d_{\max}$ times the expected size of a single tree, which is $1/(1 - \nu)$. This explains the result in Theorem 4.29. Part of this intuition is made precise in Exercises 4.25 and 4.26. Exercises 4.27 and 4.28 investigate conditions under which $|\mathscr{C}_{\max}| = d_{\max}/(1-\nu)(1+o_{\scriptscriptstyle\mathbb{P}}(1))$ might, or might not, hold.

Number of Simple Scale-Free Graphs with Given Degrees

We next discuss the number of simple graphs when the degree distribution has infinite variance, and thus no longer satisfies Condition 1.7(c). The main result is as follows:

Theorem 4.30 (Number of simple graphs: infinite-variance degrees) *Let* $\boldsymbol{d} = (d_i)_{i\in[n]}$ *satisfy Conditions 1.7(a),(b), as well as either* $\mathbb{P}(D_n = k) \leq ck^{-\tau}$ *for some* $\tau > \frac{5}{2}$, *or* $\mathbb{P}(D_n > k) \leq ck^{-(\tau-1)}$ *for some* $\tau > 1 + \sqrt{3}$. *Then the number of simple graphs having degree sequence* $\boldsymbol{d}$ *equals*

$$\frac{(\ell_n - 1)!!}{\prod_{v\in[n]} d_v!} \exp\left\{ -\frac{\ell_n}{2} + \frac{\mathbb{E}[D_n^2]}{2\mathbb{E}[D_n]} + \frac{3}{4} + \sum_{1\leq u<v\leq n} \log(1 + d_u d_v/\ell_n) + o(1) \right\}.$$

More general results exist, including cases when $d_{\max} \gg \sqrt{n}$; see the discussion in Section 4.6. Theorem 4.30 also holds when $\mathbb{E}[D_n^2] = o(n^{1/8})$. The difficulty in the proof of Theorem 4.30 is that it allows for degree sequences for which $d_{\max} \gg \sqrt{n}$. In this case, there are *many* multi-edges between vertices of degree of order $d_{\max}$ in $\mathrm{CM}_n(\boldsymbol{d})$, and the conditioning on being simple thus has a dramatic effect.

As a consequence of Theorem 4.30, we obtain that, subject to its assumptions,

$$\mathbb{P}(\mathrm{CM}_n(\boldsymbol{d}) \text{ simple}) = \exp\left\{ -\frac{\ell_n}{2} + \frac{\mathbb{E}[D_n^2]}{2\mathbb{E}[D_n]} + \frac{3}{4} + \sum_{1\leq u<v\leq n} \log(1 + d_u d_v/\ell_n) + o(1) \right\};$$

$$(4.5.3)$$

recall (4.2.39) for a more general, but weaker, estimate. In Exercise 4.29 the reader can show that (4.5.3) is indeed $\mathrm{e}^{-o(n)}$ under the conditions of Theorem 4.30.

Densest Subgraph Problem

We next consider the densest subgraph problem on $\mathrm{CM}_n(\boldsymbol{d})$, as a nice example of how local convergence methods can be used effectively to prove highly non-trivial and non-local results. For a graph $G = (V(G), E(G))$, we let H be a subgraph when $V(H) \subseteq V(G)$, and $E(H) = \{\{u, v\} \in E(G) \colon u, v \in V(H)\}$. We then let

$$\kappa(G) = \max_{\varnothing \neq H \subseteq G} \frac{|E(H)|}{|V(H)|} \tag{4.5.4}$$

be the density of the densest subgraph of G. It is far from obvious that the asymptotics of $\kappa(G_n)$ can be described using local convergence methodologies, but a deep relation exists:

Theorem 4.31 (Densest subgraph of sparse CM)   *Consider* $\mathrm{CM}_n(\boldsymbol{d})$ *subject to Condition 1.7(a), and assume that* $\mathbb{P}(D = 1) < 1$ *as well as that there exists* $\theta > 0$ *such that*

$$\sup_{n \geq 1} \mathbb{E}[e^{\theta D_n}] < \infty. \tag{4.5.5}$$

Then $\kappa(\mathrm{CM}_n(\boldsymbol{d})) \overset{\mathbb{P}}{\longrightarrow} \kappa(\mu)$, *where* μ *is the law of the unimodular branching process with root offspring distribution* $(p_k)_{k \geq 1}$ *with* $p_k = \mathbb{P}(D = k)$, *and* $\kappa(\mu)$ *is defined in* (4.5.9) *below.*

Theorem 4.31 describes the convergence of the edge density of the densest subgraph of $\mathrm{CM}_n(\boldsymbol{d})$ as well as the fact that its limit is a functional of the local limit, as described in Theorem 4.1. Theorem 4.31 holds under a strong degree assumption, in the sense that the degree distribution has *exponentially small* tails. It is unclear whether Theorem 4.31 remains valid when (4.5.5) fails. We refer to the notes and discussion in Section 4.6 for more details. Exercise 4.30 shows that Condition 1.7(a) and (4.5.5) imply that Conditions 1.7(b),(c) hold.

Let us now shed some light on the how the proof of Theorem 4.31 can be related to local convergence. This proof is beautiful, while at the same time also technically demanding. The proof highlights how this link can be used to define $\kappa(\mu)$, as well as to establish the convergence of $\kappa(\mathrm{CM}_n(\boldsymbol{d}))$ to it. This connection is through *load balancing* problems.

Let $G = (V(G), E(G))$ be a finite, simple, undirected graph. As before, we write $\vec{E}(G)$ for the set of directed edges, formed by replacing each edge $\{u, v\} \in E(G)$ by the two directed edges (u, v) and (v, u). An *allocation* on G is a map $\theta \colon \vec{E}(G) \to [0, 1]$ satisfying $\theta(u, v) + \theta(v, u) = 1$ for every $\{u, v\} \in E(G)$. The *load* induced by θ at a vertex $o \in V(G)$ is given by

$$\partial \theta(o) := \sum_{v \colon \{v, o\} \in E(G)} \theta(o, v). \tag{4.5.6}$$

An allocation θ is *balanced* when, for every $(u, v) \in \vec{E}(G)$, $\partial \theta(u) < \partial \theta(v)$ implies that $\theta(u, v) = 0$.

When we are thinking of each edge as carrying a unit amount of load, an allocation needs to be chosen that distributes load over its endpoints in such a way that the total load is as balanced as possible across the graph. Thus, a balanced allocation optimizes this allocation problem, in that a balanced θ minimizes $\sum_{v \in V(G)} f(\partial \theta(v))$ either over *some* strictly convex $f \colon [0, 1] \to [0, \infty)$, or over *all* convex $f \colon [0, 1] \to [0, \infty)$.

From now on, we let θ denote a balanced allocation. Remarkably, it can be seen that $\partial\theta(v)$ measures the *local density* of G at $v \in V(G)$. In particular, in terms of this load balancing problem,

$$\kappa(G) = \max_{v \in V(G)} \partial\theta(v). \tag{4.5.7}$$

We are left with studying the vector $(\partial\theta(v))_{v \in [n]}$. Denote the *empirical load distribution* by

$$\mathcal{L}_G(\mathcal{A}) = \frac{1}{|V(G)|} \sum_{v \in V(G)} \mathbb{1}_{\{\partial\theta(v) \in \mathcal{A}\}}, \tag{4.5.8}$$

for every Borel set $\mathcal{A} \subseteq [0, \infty)$. When G_n converges locally, one would also expect that $\mathcal{L}_{G_n}(\mathcal{A}) \to \mathcal{L}(\mathcal{A})$ for some limiting measure $\mathcal{L}$. This indeed turns out to be true (but is technically quite challenging). In fact, it turns out that if G_n converges locally to $(G, o) \sim \mu$ then $\mathcal{L} = \mathcal{L}_\mu$. In terms of $\mathcal{L}$, we have the characterization that

$$\kappa(\mu) = \sup\{t \in \mathbb{R}: \mathcal{L}[t, \infty) > 0\}. \tag{4.5.9}$$

Unfortunately, this is not the end of the story. Indeed, by the above, one would expect that $\kappa(G_n) \xrightarrow{\mathbb{P}} \kappa(\mu)$ if G_n converges locally in probability to $(G, o) \sim \mu$. This, however, is far from obvious as the graph parameter $\kappa(G)$ is too sensitive to be controlled only by local convergence. Indeed, let G_n converge locally, and add a disjoint clique K_{m_n} to G_n of size $m_n = o(n)$ to obtain G_n^+. Then, obviously, $\kappa(G_n^+) = \max\{\kappa(G_n), (m_n - 1)/2\}$. Thus, the precise structure of the graph G_n is highly relevant, and $\mathrm{CM}_n(\boldsymbol{d})$ under the condition (4.5.5) turns out to be "nice enough."

We do not prove Theorem 4.31 but instead indicate how (4.5.5) can be used to show that $\kappa(\mathrm{CM}_n(\boldsymbol{d}))$ is bounded. This proceeds in four key steps.

In the *first step*, we investigate the number of edges N_S between vertices in a set $S \subseteq [n]$, and show that N_S is stochastically bounded by a binomial random variable with mean d_S^2/m, where $d_S = \sum_{v \in S} d_v$ is the total degree of the set S. This can be seen by pairing the half-edges one by one, giving priority to the half-edges incident to vertices in S. Let $(X_t)_{t \geq 1}$ denote the Markov chain that describes the number of edges with both endpoints in S after t pairings. Then, conditioning on $(X_s)_{s=1}^t$, the probability that $X_{t+1} = X_t + 1$ is

$$\frac{(d_S - X_t - t - 1)_+}{\ell_n - 2t - 1} \leq \frac{d_S - t - 1}{\ell_n - 2t - 1} \mathbb{1}_{\{t \leq d_S\}} \leq \frac{d_S}{\ell_n} \mathbb{1}_{\{t \leq d_S\}}. \tag{4.5.10}$$

This in fact shows that N_S is stochastically dominated by a $\mathrm{Bin}(d_S, d_S/\ell_n)$ variable.

In the *second step*, the tail N_S is bounded, using that

$$\mathbb{P}(\mathrm{Bin}(n, p) \geq r) \leq \binom{n}{r} p^r \leq \frac{(np)^r}{r!} = \mathbb{E}[\mathrm{Bin}(n, p)]^r/r!. \tag{4.5.11}$$

Thus,

$$\mathbb{P}(N_S \geq r) \leq \frac{1}{r! \ell_n^r} d_S^{2r} \leq \left(\frac{2r}{\theta^2 \ell_n}\right)^r \prod_{s \in S} e^{\theta d_s}, \tag{4.5.12}$$

using the crude bounds $x^{2r} \leq (2r)!e^x$ for $x = d_{\mathcal{S}}\theta$, and $(2r)!/r! \leq (2r)^r$. As a result, with $X_{k,r}$ denoting the number of subgraphs in $\mathrm{CM}_n(\boldsymbol{d})$ with k vertices and at least r edges,

$$\mathbb{E}[X_{k,r}] \leq \sum_{|\mathcal{S}|=k} \mathbb{P}(N_{\mathcal{S}} \geq r) \leq \Big(\frac{2r}{\theta^2 \ell_n}\Big)^r \sum_{|\mathcal{S}|=k} \prod_{s \in \mathcal{S}} e^{\theta d_s}$$

$$\leq \Big(\frac{2r}{\theta^2 \ell_n}\Big)^r \frac{1}{k!} \Big(\sum_{v \in [n]} e^{\theta d_v}\Big)^k \leq \Big(\frac{2r}{\theta^2 \ell_n}\Big)^r \Big(\frac{\mathrm{e}}{k} \sum_{v \in [n]} e^{\theta d_v}\Big)^k, \qquad (4.5.13)$$

since $k! \geq (k/\mathrm{e})^k$. We can rewrite the resulting bound slightly more conveniently. Denote

$$\alpha = \sup_{n \geq 1} \mathbb{E}[D_n], \qquad \lambda = \sup_{n \geq 1} \mathbb{E}[e^{\theta D_n}], \qquad (4.5.14)$$

and pick $\theta > 0$ small enough that $\lambda < \infty$. It is here that (4.5.5) is crucially used. Then,

$$\mathbb{E}[X_{k,r}] \leq \Big(\frac{2r}{\theta^2 \alpha n}\Big)^r \Big(\frac{\mathrm{e}\lambda n}{k}\Big)^k. \qquad (4.5.15)$$

In the *third step*, we first note that, for any set $\mathcal{S} \subseteq [n]$ with $|\mathcal{S}| \geq n\delta$, the edge density of $\mathcal{S}$ is at most

$$\frac{d_{\mathcal{S}}}{2|\mathcal{S}|} \leq \frac{\ell_n}{2\delta n} = \frac{\mathbb{E}[D_n]}{2\delta}, \qquad (4.5.16)$$

which remains uniformly bounded. Thus, to show that $\kappa(\mathrm{CM}_n(\boldsymbol{d}))$ is uniformly bounded, it suffices to analyze sets of size at most δn. For $\delta \in (0,1)$ and $t > 1$, we then let $Z_{\delta,t}$ denote the number of subsets $\mathcal{S}$ with $|\mathcal{S}| \leq \delta n$ and $|E(\mathcal{S})| \geq t|\mathcal{S}|$ in $\mathrm{CM}_n(\boldsymbol{d})$. We next show that there exists a $\delta > 0$ such that, for every $t > 1$, there exists a $\chi < \infty$ such that

$$\mathbb{E}[Z_{\delta,t}] \leq \chi\Big(\frac{\log n}{n}\Big)^{t-1}. \qquad (4.5.17)$$

In particular, $Z_{\delta,t} = 0$ whp, so that the density of the densest subgraph is bounded by $(1 + \varepsilon)(1 \wedge \mathbb{E}[D]/(2\delta))$. In order to see (4.5.17), we note that

$$\mathbb{E}[Z_{\delta,t}] = \sum_{k=1}^{\delta n} \mathbb{E}[X_{k,\lceil kt \rceil}]. \qquad (4.5.18)$$

By (4.5.15),

$$\mathbb{E}[X_{k,\lceil kt \rceil}] \leq \Big(\frac{2\lceil kt \rceil}{\theta^2 \alpha k}\Big)^{\lceil kt \rceil} (\mathrm{e}\lambda)^k \Big(\frac{k}{n}\Big)^{\lceil kt \rceil - k} \leq f(k/n)^k, \qquad (4.5.19)$$

where we define

$$f(\delta) = \Big(1 \vee \frac{2(t+1)}{\theta^2 \alpha}\Big)^{t+1} (\mathrm{e}\lambda)\delta^{t-1}. \qquad (4.5.20)$$

We choose $\delta \in (0,1)$ small enough that $f(\delta) < 1$. Note that $\delta \mapsto f(\delta)$ is increasing, so that, for every $1 \leq m \leq \delta n$,

$$\mathbb{E}[Z_{\delta,t}] = \sum_{k=1}^{m} f(m/n)^k + \sum_{k=m+1}^{\delta n} f(\delta)^k$$

$$\leq \frac{f(m/n)}{1-f(m/n)} + \frac{f(\delta)^m}{1-f(\delta)}. \tag{4.5.21}$$

Finally, choose $m = c \log n$ with c fixed. Then $f(m/n)$ is of order $(\log n/n)^{t-1}$, while $f(\delta)^m \ll (\log n/n)^{t-1}$ when c is large enough. This proves (4.5.17).

The *fourth step* concludes the proof. The bound in (4.5.17) shows that $\kappa(\mathrm{CM}_n(\boldsymbol{d}))$ remains uniformly bounded. Further, it also shows that *either* there is a set of size at least δn whose density is at least t, *or* $Z_{\delta,t} > 0$. The latter occurs with vanishing probability for the appropriate $\delta > 0$, so that whp there is a set of size at least δn whose density is at least t. The fact that such high-density sets must be large is crucial to go from the convergence of $\mathcal{L}_{G_n}$ to $\mathcal{L}_\mu$ (which follows from local convergence) to that of $\kappa(\mathrm{CM}_n(\boldsymbol{d}))$ to $\kappa(\mu)$ (which, as we have seen, generally does not follow from local convergence). Indeed, local convergence has direct implications on the local properties of only a *positive proportion* of vertices, so problems might arise in this convergence should the maximum in (4.5.7) be carried by a vanishing proportion of vertices.

4.6 NOTES AND DISCUSSION FOR CHAPTER 4

Notes on Section 4.2

Theorems 4.1 and 4.6 are classical results, and have appeared in various guises throughout the literature. For example, Dembo and Montanari (2010b) crucially relied on it to identify the limiting pressure for the Ising model on the configuration model; see also Dembo and Montanari (2010a) and Bordenave (2016) for more detailed discussions. Bordenave and Caputo (2015) proved that the neighborhoods in the configuration model satisfy a large-deviation principle at speed n. This in particular implies that the probability that $\mathrm{CM}_n(\boldsymbol{d})$ contains the wrong number of r-neighborhoods of a specific type decays exponentially, as in (4.2.22) but with the right constant in the exponent. The conditions posed by Bordenave and Caputo (2015) are substantially stronger than those in Theorem 4.6 in that they assume that $d_{\max}$ is uniformly bounded. The results in Bordenave and Caputo (2015) also apply to the Erdős–Rényi random graph. The technique for obtaining the concentration inequality in (4.2.22) was pioneered by Wormald (1999). Lemma 4.8 is (Bollobás and Riordan, 2015, Lemma 21).

Notes on Section 4.3

The "giant is almost local" proof of Theorem 4.9 in Section 4.3.1 was adapted from van der Hofstad (2021). The continuous-time exploration proof in Section 4.3.2 was adapted from Janson and Luczak (2009), who, in turn, generalize the results by Molloy and Reed (1995, 1998). The results by Molloy and Reed (1995, 1998) are not phrased in terms of branching processes, which makes them a bit more difficult to grasp. We also refer to Bollobás and Riordan (2015), who gave an alternative proof using branching-process approximations on the exploration of the giant component. They also provided the extension Theorem 4.21 of Theorem 4.20, by showing that the probability of a deviation of order εn of $v_k(\mathscr{C}_{\max})$ is exponentially small for the configuration model. Since the probability of simplicity in $\mathrm{CM}_n(\boldsymbol{d})$ for power-law degrees with infinite variance degree is not exponentially small (recall Lemma 4.8), this implies Theorem 4.21. In their statement of the main result implying Theorem 4.21, Bollobás and Riordan (2015) used a condition slightly different from (4.3.100), namely, that there exists a $p > 1$ such that

$$\mathbb{E}[D_n^p] \to \mathbb{E}[D^p] < \infty. \tag{4.6.1}$$

It is straightforward to show that (4.6.1) for some $p > 1$ holds precisely when (4.3.100) holds for some $\tau > 2$. See Exercise 4.31.

The sharpest results for $n_2 = n(1 - o(1))$ are in Federico (2023), to which we refer for details. There, Federico proved the results in Exercises 4.6 and 4.7 and derived the *exact* asymptotics of $n - |\mathscr{C}_{\max}|$.

Barbour and Röllin (2019) proved a central limit theorem for the giant in Theorem 4.9, where the asymptotics of the variance already had already been identified by Ball and Neal (2017). Janson (2020a) (see also Janson (2020b)) lifted the simplicity condition under Condition 1.7(a)–(c) using switchings, so that the results extend to uniform random graphs with prescribed degrees, as conjectured in Barbour and Röllin (2019). Janson and Luczak (2008) proved related central limit theorems for the k-core in the configuration model.

Notes on Section 4.4

The results concerning the connectivity of $\mathrm{CM}_n(\boldsymbol{d})$ are folklore. A version of Theorem 4.24 can be found in (Chatterjee and Durrett, 2009, Lemma 1.2). We could not find the precise version stated in Theorems 4.24 and 4.25. Theorem 4.28 was proved by Federico and van der Hofstad (2017). This paper also allowed for a number of vertices n_1 of degree 1 satisfying $n_1 = \rho\sqrt{n}$. Earlier versions include the results by Łuczak (1992) for $d_{\min} \geq 2$ and by Wormald (1981), who proved r-connectivity when $d_{\min} = r$ (meaning that the removal of any set of $r - 1$ vertices keeps the graph connected). A result related to Theorem 4.25 can be found in Ruciński and Wormald (2002) for a different class of almost d-regular random graphs, where edges are added one by one until the addition of any further edge creates a vertex of degree $d + 1$. Such random graphs can, however, be quite far from uniform (Molloy et al. (2022)).

Notes on Section 4.5

Theorem 4.29 was proved by Janson (2008). Theorem 1.1 in that paper shows that $|\mathscr{C}_{\max}| \leq An^{1/(\tau-1)}$ when the power-law upper bound in (4.5.1) holds, while Theorem 1.3 in the same paper gives the asymptotic statement. Further, Remark 1.4 states that the jth largest cluster has size $d_{(j)}/(1 - \nu) + o_{\mathbb{P}}(n^{1/(\tau-1)})$.

Theorem 4.30 was proved by Gao and Wormald (2016), to which we refer for further discussion. The most general result available is their Theorem 6.

Theorem 4.31 was proved by Anantharam and Salez (2016); see in particular Theorem 3 in that paper. In the case of the Erdős–Rényi random graph, the results in Anantharam and Salez (2016) proved the conjectures by Hajek (1990), who established the link of load balancing problems to the densest subgraph problem. In particular, (4.5.7) is (Hajek, 1990, Corollary 7). See also Hajek (1996) for a discussion of load balancing problems on infinite graphs. The proof of the boundedness of $\kappa(\mathrm{CM}_n(\boldsymbol{d}))$ under the exponential moment condition in (4.5.5) was taken from (Anantharam and Salez, 2016, Section 11). It is unclear what the minimal condition is that guarantees that $\kappa(\mathrm{CM}_n(\boldsymbol{d})) \xrightarrow{\mathbb{P}} \kappa(\mu)$. For $\mathrm{CM}_n(\boldsymbol{d})$ with infinite-variance degrees, it is not hard to see that $\kappa(\mathrm{CM}_n(\boldsymbol{d})) \xrightarrow{\mathbb{P}} \infty$. Indeed, subject to Conditions 1.7(a),(b), whp there exists a clique of increasing size when there is a growing number of vertices of degree much larger than $\sqrt{n}$, as we will discuss in more detail in Chapter 7.

4.7 EXERCISES FOR CHAPTER 4

Exercise 4.1 (Convergence of n-dependent branching process) *Assume that Conditions 1.7(a),(b) hold. Prove that $D_n^\star \xrightarrow{d} D^\star$, and conclude that $\mathrm{BP}_n(t) \xrightarrow{d} \mathrm{BP}(t)$ for every t finite, where the branching processes $(\mathrm{BP}_n(t))_{t \geq 1}$ and $(\mathrm{BP}(t))_{t \geq 1}$ were defined in Section 4.2.1.*

Exercise 4.2 (Coupling to n-dependent branching process for $\mathrm{NR}_n(\boldsymbol{w})$) *Use Proposition 3.16 to adapt the coupling to an n-dependent branching process in Lemma 4.2, as well as Remark 4.3, to $\mathrm{NR}_n(\boldsymbol{w})$.*

Exercise 4.3 (Local convergence of $\mathrm{NR}_n(\boldsymbol{w})$) *Use the solution of Exercise 4.2 to show that $\mathrm{NR}_n(\boldsymbol{w})$ converges locally in probability to a mixed-Poisson unimodular branching process with mixing distribution W when Conditions 1.1(a),(b) hold.*

Exercise 4.4 (Proof of no-overlap property in (4.2.17)) *Subject to the conditions in Theorem 4.1, prove that $\mathbb{P}(B_r^{(G_n)}(o_1) \simeq \mathbf{t}, o_2 \in B_{2r}^{(G_n)}(o_1)) \to 0$, and conclude that the no-overlap property in (4.2.17) holds.*

Exercise 4.5 (Component size of vertex 1 in a 2-regular graph) *Consider $\mathrm{CM}_n(\boldsymbol{d})$ where all degrees are equal to 2, i.e., $n_2 = n$. Let $\mathscr{C}(1)$ denote the size of the connected component of vertex 1. Show that*

$$|\mathscr{C}(1)|/n \xrightarrow{d} T, \tag{4.7.1}$$

where $\mathbb{P}(T \leq x) = 1 - \sqrt{1-x}$.

Exercise 4.6 (Component size in a 2-regular graph with some degree-1 vertices) *Consider $\mathrm{CM}_n(\boldsymbol{d})$ with $n_1 \to \infty$ with $n_1/n \to 0$, and $n_2 = n - n_1$. Let $\mathscr{C}(1)$ denote the size of the connected component of vertex 1. Show that*

$$|\mathscr{C}(1)|/n \xrightarrow{\mathbb{P}} 0. \tag{4.7.2}$$

Exercise 4.7 (Component size in a 2-regular graph with some degree-4 vertices) *Consider $\mathrm{CM}_n(\boldsymbol{d})$ with $n_4 \to \infty$ with $n_4/n \to 0$, and $n_2 = n - n_4$. Let $\mathscr{C}(1)$ denote the the size of the connected component of vertex 1. Show that*

$$|\mathscr{C}(1)|/n \xrightarrow{\mathbb{P}} 1. \tag{4.7.3}$$

Exercise 4.8 (Expected degree giant in $\mathrm{CM}_n(\boldsymbol{d})$) *Prove that $\mathbb{E}_\mu\left[d_o \mathbb{1}_{\{|\mathscr{C}(o)|=\infty\}}\right] = \mathbb{E}[D](1 - \xi^2)$ as claimed in (4.3.6), where μ is the law of the unimodular branching-process tree with root offspring distribution $(p_k)_{k\geq 0}$ given by $p_k = \mathbb{P}(D = k)$.*

Exercise 4.9 (Limiting constants in Theorem 4.9) *Recall the constants $t_1 = 0$ and $t_2 = \theta = -\log \xi$, where ξ is the zero of H given by Lemma 4.18(a). Prove that for $t_1 = 0$ and $t_2 = \theta$, $\mathrm{e}^{-t_1} = 1, \mathrm{e}^{-t_2} = \xi$, $G_D(\mathrm{e}^{-t_1}) = 1$, $G_D(\mathrm{e}^{-t_2}) = 1 - \zeta$, $h(\mathrm{e}^{-t_1}) = 2\mathbb{E}[D]$, and $h(\mathrm{e}^{-t_2}) = 2\mathbb{E}[D]\xi^2$, where, for $\theta = \infty$, e^{-t_2} should be interpreted as 0.*

Exercise 4.10 (Giant in $\mathrm{UG}_n(\boldsymbol{d})$ for $\nu = \infty$) *Combine (4.2.39) and (4.3.101)–(4.3.102) to complete the proof of the identification of the giant in $\mathrm{UG}_n(\boldsymbol{d})$ for $\nu = \infty$ in Theorem 4.21.*

Exercise 4.11 (Number of degree-k vertices in giant $\mathrm{NR}_n(\boldsymbol{w})$) *Let $\boldsymbol{w}$ satisfy Conditions 1.1(a),(b). Adapt the proof of $|\mathscr{C}_{\max}|/n \xrightarrow{\mathbb{P}} \zeta$ in Theorem 4.23 to show that $v_k(\mathscr{C}_{\max})/n \xrightarrow{\mathbb{P}} p_k(1 - \xi^k)$ for $\mathrm{NR}_n(\boldsymbol{w})$.*

Exercise 4.12 (Number of edges in giant $\mathrm{NR}_n(\boldsymbol{w})$) *Let $\boldsymbol{w}$ satisfy Conditions 1.1(a),(b). Use Exercise 4.11 to show that $|E(\mathscr{C}_{\max})|/n \xrightarrow{\mathbb{P}} \frac{1}{2}\mathbb{E}[W](1 - \xi^2)$.*

Exercise 4.13 (Isolated vertex in $\mathrm{CM}_n(\boldsymbol{d})$) *Use (4.4.5) to show that, when $d_v \geq 3$ for all $v \in [n]$,*

$$\mathbb{P}(\exists \text{ isolated vertex in } \mathrm{CM}_n(\boldsymbol{d})) \leq \frac{3n}{(2\ell_n - 1)(2\ell_n - 3)}. \tag{4.7.4}$$

Exercise 4.14 (Isolated vertex (Cont.)) *Use (4.4.11) to reprove Exercise 4.13. Hence, the bound in (4.4.11) is quite sharp.*

Exercise 4.15 (Connected component of size 2) *Use (4.4.11) to prove that, when $d_v \geq 3$ for all $v \in [n]$,*

$$\mathbb{P}(\exists \text{ component of size 2 in } \mathrm{CM}_n(\boldsymbol{d})) \leq \frac{15n(n-1)}{(2\ell_n - 1)(2\ell_n - 3)(2\ell_n - 5)}. \tag{4.7.5}$$

Exercise 4.16 (Lower bound on probability $\mathrm{CM}_n(\boldsymbol{d})$ disconnected) *Show that*

$$\mathbb{P}(\mathrm{CM}_n(\boldsymbol{d}) \text{ disconnected}) \geq c/n$$

for some $c > 0$ when $\mathbb{P}(D = 3) > 0$ and $\mathbb{E}[D] < \infty$.

Exercise 4.17 (Lower bound on probability $\mathrm{CM}_n(\boldsymbol{d})$ disconnected) *Show that*

$$\mathbb{P}(\mathrm{CM}_n(\boldsymbol{d}) \text{ disconnected}) \geq c/n$$

for some $c > 0$ when $\mathbb{P}(D = 4) > 0$ and $\mathbb{E}[D] < \infty$.

Exercise 4.18 (Factorial moments of $P_n(2)$) *Consider* $\mathrm{CM}_n(\boldsymbol{d})$ *subject to Conditions 1.7(a),(b), and assume that $p_2 > 0$. Let $P_n(2)$ denote the number of 2-cycles consisting of two vertices of degree 2. Prove that, for every $k \geq 1$ and with $\lambda_2 = p_2^2/\mathbb{E}[D]^2$,*

$$\mathbb{E}[(P_n(2))_k] \to \lambda_2^k, \tag{4.7.6}$$

where we recall that $x_k = x(x-1)\cdots(x-k+1)$. Conclude that $P_n(2) \xrightarrow{d} \mathrm{Poi}(\lambda_2)$.

Exercise 4.19 (Cycles in $\mathrm{CM}_n(\boldsymbol{d})$) *Let $P_n(k)$ denote the number of k-cycles consisting of degree-2 vertices, for $k \geq 2$. Let $\lambda = p_2/\mathbb{E}[D]$. Use the multivariate moment method to prove that $(P_n(k))_{k\geq 2} \xrightarrow{d} (X_k)_{k\geq 2}$, where $(X_k)_{k\geq 2}$ are independent Poisson random variables with parameters*

$$\lambda^k 2^{k-1}/k = \lim_{n\to\infty} \mathbb{E}[P_n(k)].$$

Exercise 4.20 ($\mathscr{C}_{\max}$ when $d_{\min} = 2$) *Consider* $\mathrm{CM}_n(\boldsymbol{d})$ *with $d_{\min} = 2$ and assume that $\mathbb{P}(D \geq 3) > 0$. Show that Theorem 4.28 holds if $\mathbb{P}(\exists v \colon d_v \geq 3$ and $v \notin \mathscr{C}_{\max}) = o(1)$.*

Exercise 4.21 (Cycles of degree 2 vertices with one other vertex) *Subject to Conditions 1.7(a),(b) and $d_{\min} \geq 2$, show that the expected number of cycles consisting of vertices of degree 2 with a starting and ending vertex of degree k converges to*

$$\frac{k(k-1)p_k}{2\mathbb{E}[D]^2} \sum_{\ell \geq 1} (2p_2/\mathbb{E}[D])^\ell.$$

Exercise 4.22 (Subcritical power-law $\mathrm{GRG}_n(\boldsymbol{w})$ in Theorem 3.22) *Use the size of the largest connected component in the subcritical power-law $\mathrm{CM}_n(\boldsymbol{d})$ in Theorem 4.29, combined with Theorem 1.9, to identify the largest connected component in the subcritical power-law $\mathrm{GRG}_n(\boldsymbol{w})$ in Theorem 3.22.*

Exercise 4.23 (Sharp asymptotics in Theorem 4.29) *Recall the setting of the largest connected subcritical component in $\mathrm{CM}_n(\boldsymbol{d})$ in Theorem 4.29. Prove that $|\mathscr{C}_{\max}| = d_{\max}/(1-\nu)(1+o_{\mathbb{P}}(1))$ precisely when $d_{\max} = \Theta(n^{1/(\tau-1)})$. Prove that $|\mathscr{C}_{\max}| = d_{\max}/(1-\nu)(1+o_{\mathbb{P}}(1))$ precisely when $d_{\max} = \Theta(n^{1/(\tau-1)})$.*

Exercise 4.24 (Sub-polynomial subcritical clusters) *Use Theorem 4.29 to prove that $|\mathscr{C}_{\max}| = o_{\mathbb{P}}(n^\varepsilon)$ for every $\varepsilon > 0$ when (4.5.1) holds for every $\tau > 1$. Thus, when the maximal degree is sub-polynomial in n, also the size of the largest connected component is.*

Exercise 4.25 (Single tree asymptotics in Theorem 4.29) *Assume that the conditions in Theorem 4.29 hold. Use Theorem 4.1 to prove that the tree rooted at any half-edge incident to the vertex of maximal degree converges in distribution to a subcritical branching process with expected total progeny $1/(1-\nu)$.*

Exercise 4.26 (Two-tree asymptotics in Theorem 4.29) *Assume that the conditions in Theorem 4.29 hold. Use the local convergence in Theorem 4.1 to prove that the two trees rooted at any pair of half-edges incident to the vertex of maximal degree jointly converge in distribution to two* independent *subcritical branching processes with expected total progeny $1/(1-\nu)$.*

Exercise 4.27 (Theorem 4.29 when $d_{\max} = o(\log n)$) *Assume that the subcritical conditions in Theorem 4.29 hold, so that $\nu < 1$. Suppose that $d_{\max} = o(\log n)$. Do you expect $|\mathscr{C}_{\max}| = d_{\max}/(1-\nu)(1+o_{\mathbb{P}}(1))$ to hold? Note: No proof is expected; a reasonable argument will suffice.*

Exercise 4.28 (Theorem 4.29 when $d_{\max} \gg \log n$) *Assume that the subcritical conditions in Theorem 4.29 hold, so that $\nu < 1$. Suppose that $d_{\max} \gg \log n$. Do you expect $|\mathscr{C}_{\max}| = d_{\max}/(1-\nu)(1+o_{\mathbb{P}}(1))$ to hold? Note: No proof is expected; a reasonable argument will suffice.*

Exercise 4.29 (Probability of simplicity in Theorem 4.30) *Subject to the conditions in Theorem 4.30, show that (4.5.3) implies that $\mathbb{P}(\mathrm{CM}_n(\boldsymbol{d})$ simple$) = \mathrm{e}^{-o(n)}$, as proved in Lemma 4.8.*

Exercise 4.30 (Exponential moments) *Show that Condition 1.7(a) and $\sup_{n\geq 1} \mathbb{E}[\mathrm{e}^{\theta D_n}] < \infty$ as in (4.5.5) imply that $\mathbb{E}[D_n^p] \to \mathbb{E}[D^p]$ for every $p > 0$. Conclude that then also Conditions 1.7(b)-(c) hold.*

Exercise 4.31 (Moment versus tails) *Show that $\mathbb{E}[D_n^p] \to \mathbb{E}[D^p] < \infty$ for some $p > 1$ precisely when $[1 - F_n](x) \leq c_F x^{-(\tau-1)}$ for all $x \geq 1$ and some $\tau > 2$.*

CONNECTED COMPONENTS IN PREFERENTIAL ATTACHMENT MODELS

Abstract

In this chapter we investigate the connectivity structure of preferential attachment models. We start by discussing an important tool: exchangeable random variables and their distribution as described in de Finetti's Theorem. We apply these results to Pólya urn schemes, which, in turn, we use to describe the distribution of the degrees in preferential attachment models.

It turns out that Pólya urn schemes can also be used to describe the local limit of preferential attachment models. A crucial ingredient is the fact that the edges in the Pólya urn representation are *conditionally* independent, given the appropriate randomness. The resulting local limit is the *Pólya point tree*, a specific multi-type branching process with continuous types.

5.1 MOTIVATION: CONNECTIONS AND LOCAL DEGREE STRUCTURE

The models discussed so far share the property that they are *static* and their edge-connection probabilities are close to being *independent*. As discussed at great length in [V1, Chapter 8], see also Section 1.3.5, preferential attachment models were invented for their *dynamic* structure: since edges incident to younger vertices connect to older vertices in a way that favours *high-degree vertices*, preferential attachment models develop power-law degree distributions. This intuitive dynamics comes at the expense of creating *dynamic* models in which edge-connection probabilities are hard to compute. As a result, we see that proofs for preferential attachment models are generally substantially harder than those for inhomogeneous random graphs and configuration models.

In this chapter, we explain how this difference can be overcome, to some extent, by realizing that the degree evolution in preferential attachment models can be described in terms of *exchangeable* random variables. Because of this, we can describe these models in terms of independent edges, given some appropriate extra randomness.

Organization of this Chapter

We start in Section 5.2 by discussing *exchangeable random variables*, and their fascinating properties. Indeed, de Finetti's Theorem implies that infinite sequences of exchangeable random variables are, conditional on the appropriate randomness, independent and identically distributed. We continue in Section 5.3 by stating local convergence for preferential attachment models and setting the stage for its proof. A major result here is the finite-graph Pólya urn description of the preferential attachment model, which states that its edges are *conditionally independent* given the appropriate randomness. The proof of local convergence is completed in Section 5.4 using various martingale, coupling, and Poisson-process techniques. In Section 5.5 we investigate the connectivity of preferential attachment models. Section 5.6 highlights some further results for preferential attachment models. We close this chapter in Section 5.7 with notes and discussion and in Section 5.8 with exercises.

5.2 EXCHANGEABLE RANDOM VARIABLES AND PÓLYA URN SCHEMES

In this section we discuss the distribution of infinite sequences of exchangeable random variables and their applications to Pólya urn schemes.

De Finetti's Theorem for Exchangeable Random Variables

We start by defining the conditions for sequences of random variables to be exchangeable:

Definition 5.1 (Exchangeable random variables) A finite sequence of random variables $(X_i)_{i=1}^n$ is called *exchangeable* when the distribution of $(X_i)_{i=1}^n$ is the same as that of $(X_{\sigma(i)})_{i=1}^n$ for any permutation $\sigma \colon [n] \to [n]$. An infinite sequence $(X_i)_{i\geq 1}$ is called *exchangeable* when $(X_i)_{i=1}^n$ is exchangeable for every $n \geq 1$. ◀

The notion of exchangeability is rather strong and implies for example that the distribution of X_i is the same for every i (see Exercise 5.1) as well as that (X_i, X_j) have the same distribution for every $i \neq j$.

Clearly, when a sequence of random variables is iid then it is also exchangeable (see Exercise 5.2). A second example arises when we take a sequence of random variables that are iid *conditionally* on some random variables. An example could be a sequence of Bernoulli random variables that are iid conditional on their success probability U but U itself is random. This is called a *mixture of iid random variables*. Remarkably, the distribution of an infinite sequence of exchangeable random variables is *always* such a mixture. This is the content of de Finetti's Theorem, which we state and prove here in the case where $(X_i)_{i\geq 1}$ are indicator variables:

Theorem 5.2 (De Finetti's Theorem) *Let $(X_i)_{i\geq 1}$ be an infinite sequence of exchangeable random variables, and assume that $X_i \in \{0, 1\}$. Then there exists a random variable U with $\mathbb{P}(U \in [0, 1]) = 1$ such that, for all $n \geq 1$ and $k \in [n]$,*

$$\mathbb{P}(X_1 = \cdots = X_k = 1, X_{k+1} = \cdots = X_n = 0) = \mathbb{E}[U^k(1 - U)^{n-k}]. \qquad (5.2.1)$$

The theorem of de Finetti (Theorem 5.2) states that an infinite exchangeable sequence of indicators has the same distribution as an independent Bernoulli sequence with a *random* success probability U. Thus, the different elements of the sequence are *not* independent, but their dependence enters only through the success probability U.

The proof of Theorem 5.2 can be relatively easily extended to more general settings, for example, when X_i takes on a *finite* number of values. Since we are relying on Theorem 5.2 only for indicator variables, we refrain from stating this more general version.

Define S_n to be the number of ones in $(X_i)_{i=1}^n$, i.e.,

$$S_n = \sum_{k=1}^n X_k. \qquad (5.2.2)$$

Then Theorem 5.2 is equivalent to the statement that

$$\mathbb{P}(S_n = k) = \mathbb{E}\Big[\mathbb{P}\big(\mathsf{Bin}(n, U) = k\big)\Big]. \qquad (5.2.3)$$

The reader is asked to prove (5.2.3) in Exercise 5.4. Equation (5.2.3) also allows us to compute the distribution of U. Indeed, when we suppose that

$$\lim_{n \to \infty} \mathbb{P}(S_n \in (an, bn)) = \int_a^b f(u)du, \tag{5.2.4}$$

where f is a density, then (5.2.3) implies that f is in fact the density of the random variable U. This is useful in applications of de Finetti's Theorem (Theorem 5.2). Equation (5.2.4) follows by noting that $S_n/n \xrightarrow{a.s.} U$ by the strong law of large numbers applied to the conditional law given U. In Exercise 5.3, you are asked to fill in the details.

Proof of Theorem 5.2. The proof makes use of Helly's Theorem, which states that any sequence of *bounded* random variables has a weakly converging subsequence. We fix $m \geq n$ and condition on S_m to write

$$\mathbb{P}(X_1 = \cdots = X_k = 1, X_{k+1} = \cdots = X_n = 0) \tag{5.2.5}$$
$$= \sum_{j=k}^m \mathbb{P}(X_1 = \cdots = X_k = 1, X_{k+1} = \cdots = X_n = 0 \mid S_m = j)\mathbb{P}(S_m = j).$$

By exchangeability, and conditional on $S_m = j$, each sequence $(X_i)_{i=1}^m$ containing precisely j ones is equally likely. There are precisely $\binom{m}{j}$ such sequences, and precisely $\binom{m-n}{j-k}$ of them start with k ones followed by $n - k$ zeros. Thus,

$$\mathbb{P}(X_1 = \cdots = X_k = 1, X_{k+1} = \cdots = X_n = 0 \mid S_m = j) = \frac{\binom{m-n}{j-k}}{\binom{m}{j}}. \tag{5.2.6}$$

Writing $(m)_k = m(m-1) \cdots (m-k+1)$ for the kth falling factorial of m, we therefore arrive at

$$\mathbb{P}(X_1 = \cdots = X_k = 1, X_{k+1} = \cdots = X_n = 0)$$
$$= \sum_{j=k}^m \frac{(j)_k (m-j)_{n-k}}{(m)_n} \mathbb{P}(S_m = j). \tag{5.2.7}$$

When $m \to \infty$ and for k and n with $k \leq n$ fixed,

$$\frac{(j)_k (m-j)_{n-k}}{(m)_n} = \left(\frac{j}{m}\right)^k \left(1 - \frac{j}{m}\right)^{n-k} + o(1). \tag{5.2.8}$$

Equation (5.2.8) can be seen by splitting between $j > \varepsilon m$ and $j \leq \varepsilon m$ for $\varepsilon > 0$ arbitrarily small. For the former $(j)_k = j^k(1 + o(1))$, while for the latter $(j)_k \leq (\varepsilon m)^k$ and $(m-j)_{n-k}/(m)_n \leq m^{-k}$.

Recall that $S_m = j$, so that

$$\mathbb{P}(X_1 = \cdots = X_k = 1, X_{k+1} = \cdots = X_n = 0) = \lim_{m \to \infty} \mathbb{E}[Y_m^k(1 - Y_m)^{n-k}], \tag{5.2.9}$$

where $Y_m = S_m/m$. Note that it is here that we make use of the fact that $(X_i)_{i \geq 1}$ is an *infinite* exchangeable sequence of random variables. Equation (5.2.9) is the point of departure for the completion of the proof.

We have that $Y_m \in [0,1]$ since $S_m \in [0,m]$, so that the sequence of random variables $(Y_m)_{m \geq 1}$ is bounded. By Helly's Theorem, it thus contains a weakly converging subsequence, i.e., there exists a subsequence $(Y_{m_l})_{l \geq 1}$ with $\lim_{l \to \infty} m_l = \infty$ and a random variable U such that $Y_{m_l} \xrightarrow{d} U$. Since the random variable $Y_m^k(1 - Y_m)^{n-k}$ is uniformly bounded for each k, n, Lebesgue's Dominated Convergence Theorem ([V1, Theorem A.1]) gives that

$$
\begin{aligned}
\lim_{m \to \infty} \mathbb{E}\left[Y_m^k(1 - Y_m)^{n-k}\right] &= \lim_{l \to \infty} \mathbb{E}\left[Y_{m_l}^k(1 - Y_{m_l})^{n-k}\right] \\
&= \mathbb{E}\left[U^k(1 - U)^{n-k}\right].
\end{aligned}
\tag{5.2.10}
$$

This completes the proof. Yet a careful reader may wonder whether the above proof on the basis of subsequences is enough. Indeed, it is possible that *another* subsequence $(Y_{m_l'})_{l \geq 1}$ with $\lim_{l \to \infty} m_l' = \infty$ has a different limiting random variable V such that $Y_{m_l'} \xrightarrow{d} V$. However, from (5.2.9) we then conclude that $\mathbb{E}\left[V^k(1 - V)^{n-k}\right] = \mathbb{E}\left[U^k(1 - U)^{n-k}\right]$ for every k, n. In particular, $\mathbb{E}[V^k] = \mathbb{E}[U^k]$ for every $k \geq 0$. Since the random variables U, V are almost surely bounded by 1, and have the same moments, they also have the same distribution. We conclude that $Y_{m_l} \xrightarrow{d} U$ for *every* subsequence $(m_l)_{l \geq 1}$ along which $(Y_{m_l})_{l \geq 1}$ converges, and this is equivalent to $Y_m \xrightarrow{d} U$.   □

The theorem of de Finetti implies that if X_k and X_n are coordinates of an infinite exchangeable sequence of indicators then they are *positively correlated*; see Exercise 5.5. Thus, it is impossible for infinite exchangeable sequences of indicator variables to be negatively correlated, which is somewhat surprising.

In the proof of de Finetti's Theorem, it is imperative that the sequence $(X_i)_{i \geq 1}$ is *infinite*. This is not merely a technicality of the proof. Rather, there are finite exchangeable sequences of random variables for which the equality (5.2.1) does *not* hold. Indeed, take an urn filled with b blue and r red balls, and draw balls successively without replacement. Thus, the urn is sequentially being depleted, and it will be empty after the $(b + r)$th ball is drawn. Let X_i denote the indicator that the ith ball drawn is blue. Then, clearly, the sequence $(X_i)_{i=1}^{r+b}$ is exchangeable. However,

$$
\begin{aligned}
\mathbb{P}(X_1 = X_2 = 1) &= \frac{b(b-1)}{(b+r)(b+r-1)} \\
&< \left(\frac{b}{b+r}\right)^2 = \mathbb{P}(X_1 = 1)\mathbb{P}(X_2 = 1),
\end{aligned}
\tag{5.2.11}
$$

so that X_1 and X_2 are *negatively* correlated.

Pólya Urn Schemes

An important application of de Finetti's Theorem (Theorem 5.2) arises in so-called *Pólya urn schemes*. An urn contains a number of balls. We start with $B_0 = b_0$ blue balls and $R_0 = r_0$ red balls at time $n = 0$. Let $W_b, W_r : \mathbb{N} \to (0, \infty)$ be two weight functions. Then, at time $n + 1$, the probability of drawing a blue ball, conditional on the number B_n of blue balls at time n, is proportional to the weight of the blue balls at time n, i.e., it is

$$\frac{W_b(B_n)}{W_b(B_n) + W_r(R_n)}. \tag{5.2.12}$$

After drawing a ball, it is replaced together with a *second* ball of the same color; we denote this Pólya urn scheme by $((B_n, R_n))_{n \geq 0}$. Naturally, since we always replace one ball by two balls, the total number of balls $B_n + R_n = b_0 + r_0 + n$ is deterministic.

In this section, we restrict to the case where there exist $a_r, a_b > 0$ such that

$$W_b(k) = a_b + k, \qquad W_r(k) = a_r + k, \tag{5.2.13}$$

i.e., both weight functions are *linear* with the same slope, but possibly a different intercept. Our main result concerning Pólya urn schemes is the following theorem:

Theorem 5.3 (Limit theorem for linear Pólya urn schemes) *Let $((B_n, R_n))_{n \geq 0}$ be a Pólya urn scheme starting with $(B_0, R_0) = (b_0, r_0)$ balls of each color, and with linear weight functions W_b and W_r as in (5.2.13) for some $a_r, a_b > 0$. Then, as $n \to \infty$,*

$$\frac{B_n}{B_n + R_n} \xrightarrow{a.s.} U, \tag{5.2.14}$$

where U has a Beta distribution with parameters $a = b_0 + a_b$ and $b = r_0 + a_r$, and, for all $k \leq n$,

$$\mathbb{P}(B_n = b_0 + k) = \mathbb{E}\Big[\mathbb{P}\big(\mathrm{Bin}(n, U) = k\big)\Big]. \tag{5.2.15}$$

Before proving Theorem 5.3, let us comment on its remarkable content. Clearly, the number of blue balls B_n is *not* a binomial random variable, as early draws of blue balls reinforce the proportion of blue balls in the end. However, (5.2.15) states that we can *first* draw a random variable U and then, *conditionally* on that random variable, the number of blue balls *is* binomial. This is an extremely useful perspective, as we will see later on. The urn conditioned on the limiting variable U is sometimes called a Pólya urn with *strength U*, and Theorem 5.3 implies that this is a mere binomial experiment given the strength. The variables $a = b_0 + a_b$ and $b = r_0 + a_r$ of the Beta distributions indicate the *initial weights* of each of the two colors.

Proof of Theorem 5.3. We start with the almost sure convergence in (5.2.14). Let $M_n = (B_n + a_b)/(B_n + R_n + a_b + a_r)$. Note that

$$\begin{aligned}
\mathbb{E}[M_{n+1} \mid (B_l)_{l=1}^n] &= \frac{1}{B_{n+1} + R_{n+1} + a_b + a_r} \mathbb{E}[(B_{n+1} + a_b) \mid B_n] \\
&= \frac{1}{B_{n+1} + R_{n+1} + a_b + a_r} \Big[B_n + a_b + \frac{B_n + a_b}{B_n + R_n + a_b + a_r}\Big] \\
&= \frac{B_n + a_b}{B_{n+1} + R_{n+1} + a_b + a_r} \Big[\frac{B_n + R_n + a_b + a_r + 1}{B_n + R_n + a_b + a_r}\Big] \\
&= \frac{B_n + a_b}{B_n + R_n + a_b + a_r} = M_n, \tag{5.2.16}
\end{aligned}$$

since $B_{n+1} + R_{n+1} + a_b + a_r = B_n + R_n + a_b + a_r + 1$. As a result, $(M_n)_{n \geq 0}$ is a non-negative martingale, and thus converges almost surely to some random variable U by the Martingale Convergence Theorem ([V1, Theorem 2.24]).

We continue by identifying the limiting random variable in (5.2.14), which will follow from (5.2.15). Let X_n denote the indicator that the nth ball drawn is blue. We first show that $(X_n)_{n \geq 1}$ is an infinite exchangeable sequence. Note that

$$B_n = b_0 + \sum_{j=1}^{n} X_j, \qquad R_n = r_0 + \sum_{j=1}^{n} (1 - X_j) = r_0 + b_0 + n - B_n. \qquad (5.2.17)$$

Now, for any sequence $(x_t)_{t=1}^{n}$,

$$\mathbb{P}\left((X_t)_{t=1}^{n} = (x_t)_{t=1}^{n}\right) = \prod_{t=1}^{n} \frac{W_b(b_{t-1})^{x_t} W_r(r_{t-1})^{1-x_t}}{W_b(b_{t-1}) + W_r(r_{t-1})}, \qquad (5.2.18)$$

where $b_t = b_0 + \sum_{j=1}^{t} x_j$ and $r_t = r_0 + b_0 + t - b_t$. Denote $k = \sum_{t=1}^{n} x_t$. Then, by (5.2.13) and (5.2.17),

$$\prod_{t=1}^{n} (W_b(b_{t-1}) + W_r(r_{t-1})) = \prod_{t=1}^{n} (b_0 + r_0 + a_b + a_r + t - 1), \qquad (5.2.19)$$

while

$$\prod_{t=1}^{n} W_b(b_{t-1})^{x_t} = \prod_{m=0}^{k-1} (b_0 + a_b + m), \qquad \prod_{t=1}^{n} W_r(r_{t-1})^{1-x_t} = \prod_{j=0}^{n-k-1} (r_0 + a_r + j). \qquad (5.2.20)$$

Thus, we arrive at

$$\mathbb{P}\left((X_t)_{t=1}^{n} = (x_t)_{t=1}^{n}\right) = \frac{\prod_{m=0}^{k-1}(b+m) \prod_{j=0}^{n-k-1}(r+j)}{\prod_{t=0}^{n-1}(b+r+t)}, \qquad (5.2.21)$$

where $b = b_0 + a_b$ and $r = r_0 + a_r$. In particular, (5.2.21) does not depend on the *order* in which the elements of $(x_t)_{t=1}^{n}$ appear, so that the sequence $(X_n)_{n \geq 1}$ is an infinite exchangeable sequence. de Finetti's Theorem (Theorem 5.2) implies that $(X_n)_{n \geq 1}$ is a mixture of Bernoulli random variables with a random success probability U, and we are left to compute the distribution of U. We also observe that the distribution of U depends only on b_0, r_0, a_b, a_r through $b = b_0 + a_b$ and $r = r_0 + a_r$.

To identify the law of U, we verify (5.2.4). For fixed $0 \leq k \leq n$, there are $\binom{n}{k}$ sequences of k ones and $n - k$ zeros. Each sequence has the same probability, given by (5.2.21). Thus,

$$\begin{aligned}
\mathbb{P}(S_n = k) &= \binom{n}{k} \frac{\prod_{m=0}^{k-1}(b+m) \prod_{j=0}^{n-k-1}(r+j)}{\prod_{t=0}^{n-1}(b+r+t)} \\
&= \frac{\Gamma(n+1)}{\Gamma(k+1)\Gamma(n-k+1)} \times \frac{\Gamma(k+b)}{\Gamma(b)} \times \frac{\Gamma(n-k+r)}{\Gamma(r)} \times \frac{\Gamma(b+r)}{\Gamma(n+b+r)} \\
&= \frac{\Gamma(b+r)}{\Gamma(r)\Gamma(b)} \times \frac{\Gamma(k+b)}{\Gamma(k+1)} \times \frac{\Gamma(n-k+r)}{\Gamma(n-k+1)} \times \frac{\Gamma(n+1)}{\Gamma(n+b+r)}. \qquad (5.2.22)
\end{aligned}$$

For k and $n - k$ large, by [V1, (8.3.9)],

$$\mathbb{P}(S_n = k) = \frac{\Gamma(b+r)}{\Gamma(r)\Gamma(b)} \frac{k^{b-1}(n-k)^{r-1}}{n^{b+r-1}}(1 + o(1)). \tag{5.2.23}$$

Taking $k = \lceil un \rceil$ (recall (5.2.4)), this leads to

$$\lim_{n\to\infty} n\mathbb{P}(S_n = \lceil un \rceil) = \frac{\Gamma(b+r)}{\Gamma(r)\Gamma(b)} u^{b-1}(1 - u)^{r-1}, \tag{5.2.24}$$

which is the density of a Beta distribution with parameters b and r; the convergence is uniform for $u \in (a, b)$ for all $a, b \in (0, 1)$. It is not hard to show from (5.2.24) that (5.2.4) holds with $f(u)$ the right-hand side of (5.2.24) (see Exercise 5.6). $\qquad\square$

Multiple Urn Extensions

We next explain how Theorem 5.3 can be inductively extended to urns with *several* colors of balls. This is essential in the analysis of preferential attachment models, where we need a large number of urns.

Assume that we have an urn with ℓ colors $(C_i(n))_{i\in[\ell],n\geq0}$, where $C_i(n)$ denotes the number of balls of color i at time n. Again, we restrict to the setting where the weight functions in the urn are affine, i.e., there exist $(a_i)_{i\in[\ell]}$ such that

$$W_i(k) = a_i + k. \tag{5.2.25}$$

We assume that the Pólya urn starts with k_i balls of color i, and that a ball is drawn according to the weights $W_i(C_i(n))$ for $i \in [\ell]$, after which it is replaced by two balls of the same color. For $j \in [\ell]$, we let $a_{[j,\ell]} = \sum_{i=j}^{\ell} a_i$ and $C_{[j,\ell]}(n) = \sum_{i=j}^{\ell} C_i(n)$. We first view the balls of color 1 and the other colors as a two-type urn. Thus,

$$\frac{C_1(n)}{n} \xrightarrow{a.s.} U_1, \tag{5.2.26}$$

where U_1 has a Beta distribution with parameters $a = k_1 + a_1$ and $b = k_{[2,\ell]} + a_{[2,\ell]}$. This highlights the proportion of balls of color 1, but groups all other balls together as one "combined" color. This combined color takes a proportion $1 - U_1$ of the balls. Now, the times that a "combined" color ball is drawn again forms a (multi-type) Pólya urn scheme, now with the colors $2, \ldots, \ell$. This implies that

$$\frac{C_2(n)}{n} \xrightarrow{a.s.} U_2(1 - U_1), \tag{5.2.27}$$

where U_2 is independent of U_1 and has a Beta distribution with parameters $a = k_2 + a_2$ and $b = k_{[3,\ell]} + a_{[3,\ell]}$. Repeating the procedure gives that

$$\frac{C_i(n)}{n} \xrightarrow{a.s.} U_i \prod_{j=1}^{i-1}(1 - U_j), \tag{5.2.28}$$

where U_i is independent of $(U_1, \ldots, U_{i-1})$ and has a Beta distribution with parameters $a = k_i + a_i$ and $b = k_{[i,\ell]} + a_{[i,\ell]}$. This gives not only an extension of Theorem 5.3 to urns with multiple colors but also an appealing independence structure of the limits.

Applications to Relative Sizes in Scale-Free Trees

We close this section by discussing applications of Pólya urn schemes to scale-free trees. We start at time $n = 2$ with an initial graph consisting of two vertices of which vertex 1 has degree d_1 and vertex 2 has degree d_2. Needless to say, in order for the initial graph to be possible, we need $d_1 + d_2$ to be even, and the graph may contain self-loops and multiple edges. After this, we successively attach vertices to older vertices with probability proportional to the degree plus $\delta > -1$. We do not allow for self-loops in the growth of the trees, so that the structures connected to vertices 1 and 2 are trees (but the entire structure is not when $d_1 + d_2 > 2$). This is a generalization of $(\mathrm{PA}_n^{(1,\delta)}(b))_{n \geq 2}$, in which we are more flexible in choosing the initial graph. The model for $(\mathrm{PA}_n^{(1,\delta)}(b))_{n \geq 1}$ arises when $d_1 = d_2 = 2$ (see Exercise 5.8). For $(\mathrm{PA}_n^{(1,\delta)}(d))_{n \geq 1}$, $d_1 = d_2 = 1$ is the most relevant (recall from Section 1.3.5 that $(\mathrm{PA}_n^{(1,\delta)}(d))_{n \geq 1}$ starts at time 1 with two vertices and one edge between them).

We decompose the growing tree into two trees. For $i = 1, 2$, we let $T_i(n)$ be the tree of vertices that are closer to vertex i than to vertex $3 - i$. Thus, the tree $T_2(n)$ consists of those vertices for which the path in the tree from the vertex to vertex 1 passes through vertex 2, and $T_1(n)$ consists of the remainder of the scale-free tree. Let $S_i(n) = |T_i(n)|$ denote the number of vertices in $T_i(n)$. Clearly, $S_1(n) + S_2(n) = n$, which is the total number of vertices in the tree at time n. We can apply Theorem 5.3 to describe the relative sizes of $T_1(n)$ and $T_2(n)$:

Theorem 5.4 (Tree decomposition for scale-free trees)   *For scale-free trees with initial degrees $d_1, d_2 \geq 1$, as $n \to \infty$,*

$$\frac{S_1(n)}{n} \xrightarrow{a.s.} U, \tag{5.2.29}$$

where U has a Beta distribution with parameters $a = (d_1 + \delta)/(2 + \delta)$ and $b = (d_2 + \delta)/(2 + \delta)$, and

$$\mathbb{P}(S_1(n) = k) = \mathbb{E}\Big[\mathbb{P}\big(\mathrm{Bin}(n - 1, U) = k - 1\big)\Big]. \tag{5.2.30}$$

By Theorem 5.4, we can decompose a scale-free tree into two disjoint scale-free trees each of which contains an almost surely positive proportion of the vertices.

Proof   The evolution of $(S_1(n))_{n \geq 2}$ can be viewed as a Pólya urn scheme. Indeed, when $S_1(n) = s_1(n)$, the probability of attaching the $(n + 1)$th vertex to $T_1(n)$ is equal to

$$\frac{(2s_1(n) + d_1 - 2) + \delta s_1(n)}{(2s_1(n) + d_1 - 2) + \delta s_1(n) + 2(s_2(n) + d_2) + \delta s_2(n)}, \tag{5.2.31}$$

since the number of vertices in $T_i(n)$ equals $S_i(n)$, while the total degree of $T_i(n)$ equals $(2S_i(n) + d_i - 2)$. We can rewrite this as

$$\frac{s_1(n) + (d_1 - 2)/(2 + \delta)}{s_1(n) + s_2(n) + (d_1 + d_2 - 4)/(2 + \delta)}, \tag{5.2.32}$$

which is equal to (5.2.12) in the case (5.2.13) when $r_0 = b_0 = 1$ and $a_b = (d_1 - 2)/(2 + \delta)$, $a_r = (d_2 - 2)/(2 + \delta)$. Therefore, Theorem 5.4 follows directly from Theorem 5.3. $\qquad \square$

We continue by adapting the above argument to the size of the connected component of, or subtree containing, vertex 1 in $\mathrm{PA}_n^{(1,\delta)}(a)$ (recall Section 1.3.5), which we denote by $S_1'(n)$:

Theorem 5.5 (Tree decomposition for preferential attachment trees)   *For* $\mathrm{PA}_n^{(1,\delta)}(a)$, *as* $n \to \infty$,

$$\frac{S_1'(n)}{n} \xrightarrow{a.s.} U', \tag{5.2.33}$$

where U' *has a mixed Beta distribution with random parameters* $a = I + 1$ *and* $b = 1 + (1 + \delta)/(2 + \delta)$, *and where, for* $\ell \geq 2$,

$$\mathbb{P}(I = \ell) = \mathbb{P}(\text{first vertex that is not connected to vertex 1 is vertex } \ell). \tag{5.2.34}$$

Consequently,

$$\mathbb{P}(S_1(n) = k) = \mathbb{E}\Big[\mathbb{P}(\mathrm{Bin}(n - 1, U') = k - 1)\Big]. \tag{5.2.35}$$

Proof   By construction, all vertices in $[I - 1]$ are in the subtree containing vertex 1. We note that $S_1'(n) = n$ for all $n < I$, and $S_1'(I) = I - 1$, while the subtree containing vertex I consists only of the vertex I, i.e., $S_2'(I) = 1$. For $n \geq I + 1$, the evolution of $(S_1'(n))_{n\geq2}$ can be viewed as a Pólya urn scheme. Indeed, when $S_1'(n) = s_1'(n)$, the probability of attaching the $(n + 1)$th vertex to the tree rooted at vertex 1 is equal to

$$\frac{(2 + \delta)s_1'(n)}{(2 + \delta)n + 1 + \delta}. \tag{5.2.36}$$

We can rewrite this as

$$\frac{s_1'(n)}{n + (1 + \delta)/(2 + \delta)} = \frac{(s_1'(n) - I) + I}{(n - I) + I + (1 + \delta)/(2 + \delta)}, \tag{5.2.37}$$

which is equal to (5.2.12) in the case (5.2.13) when $b_0 = r_0 = 1$ and $a_b = I, a_r = (1 + \delta)/(2 + \delta)$. Therefore, Theorem 5.5 follows directly from Theorem 5.3.   □

Applications to Relative Degrees in Scale-Free Trees

We continue by discussing an application of Pólya urn schemes to relative initial degrees. For this, we fix an integer $k \geq 2$ and consider only times $n \geq k$ at which an edge is attached to one of the k initial vertices. We work with $(\mathrm{PA}_n^{(1,\delta)}(a))_{n\geq1}$, so that we start at time $n = 1$ with one vertex having one self-loop, after which we successively attach vertices to older vertices with probabilities proportional to the degree plus $\delta > -1$, allowing for self-loops. The main result is as follows:

Theorem 5.6 (Relative degrees in scale-free trees)   *For* $(\mathrm{PA}_n^{(1,\delta)}(a))_{n\geq1}$, *as* $n \to \infty$,

$$\frac{D_k(n)}{D_{[k]}(n)} \xrightarrow{a.s.} \psi_k, \tag{5.2.38}$$

where $D_{[k]}(n) = D_1(n) + \cdots + D_k(n)$ *and* $\psi_k \sim \mathrm{Beta}(1 + \delta, (k - 1)(2 + \delta))$.

By Theorem 1.17, $D_k(n)n^{-1/(2+\delta)} \xrightarrow{a.s.} \xi_k$, where ξ_k is positive almost surely by the argument in the proof of [V1, Theorem 8.14]. It thus follows from Theorem 5.6 that $\psi_k = \xi_k/(\xi_1 + \cdots + \xi_k)$. We conclude that Theorem 5.6 allows us to identify properties of the law of the limiting degrees.

Proof of Theorem 5.6. Denote the sequence of stopping times $(\tau_k(n))_{n\geq 2k-1}$, by $\tau_k(2k-1) = k-1$, where

$$\tau_k(n) = \inf\{t \colon D_{[k]}(t) = n\}, \tag{5.2.39}$$

i.e., $\tau_k(n)$ is the time where the total degree of vertices $[k]$ equals n. The initial condition $\tau_k(2k-1) = k-1$ is chosen such that the half-edge incident to vertex k is already considered to be present at time $k-1$, but the receiving end of that edge is not. This guarantees that also the attachment of the edge of vertex k is properly taken into account.

Note that $\tau_k(n) < \infty$ for every n, since $D_j(n) \xrightarrow{a.s.} \infty$ as $n \to \infty$ for every j. Moreover, since $\tau_k(n) \xrightarrow{a.s.} \infty$ as $n \to \infty$,

$$\lim_{n\to\infty} \frac{D_k(n)}{D_{[k]}(n)} = \lim_{n\to\infty} \frac{D_k(\tau_k(n))}{D_{[k]}(\tau_k(n))} = \lim_{n\to\infty} \frac{D_k(\tau_k(n))}{n}. \tag{5.2.40}$$

Now, the random variables $((D_k(\tau_k(n)), D_{[k-1]}(\tau_k(n))))_{n\geq 2k-1}$ form a Pólya urn scheme, with $D_k(\tau_k(2k-1)) = 1$, and $D_{[k-1]}(\tau_k(2k-1)) = 2k-2$. The edge at time $\tau_k(n)$ is attached to vertex k with probability

$$\frac{D_k(\tau_k(n)) + \delta}{n + k\delta}, \tag{5.2.41}$$

which is the probability of a Pólya urn scheme having linear weights as in (5.2.13) with $a_b = \delta, a_r = (k-1)\delta, b_0 = 1$, and $r_0 = 2(k-1)$. Thus, the statement follows from Theorem 5.3. □

Theorem 5.6 is easily extended to $(\mathrm{PA}_n^{(1,\delta)}(b))_{n\geq 1}$:

Theorem 5.7 (Relative degrees in scale-free trees)   *For* $(\mathrm{PA}_n^{(1,\delta)}(b))_{n\geq 1}$, *as* $n \to \infty$,

$$\frac{D_k(n)}{D_{[k]}(n)} \xrightarrow{a.s.} \psi_k', \tag{5.2.42}$$

where $\psi_k' \sim \mathrm{Beta}(1+\delta, (2k-1)+(k-1)\delta)$ *for* $k \geq 3$, *and* $\psi_2' \sim \mathrm{Beta}(2+\delta, 2+\delta)$.

The dynamics for $(\mathrm{PA}_n^{(1,\delta)}(b))_{n\geq 1}$ are slightly different than those of $(\mathrm{PA}_n^{(1,\delta)}(a))_{n\geq 1}$, since $\mathrm{PA}_n^{(1,\delta)}(b)$ does not allow for self-loops in the growth of the tree. Indeed, now the random variables $((D_k(\tau_k(n)), D_{[k-1]}(\tau_k(n))))_{n\geq 2k}$ form a Pólya urn scheme, starting with $D_k(\tau_k(2k)) = 1$, and $D_{[k-1]}(\tau_k(2k)) = 2k-1$. The edge at time $\tau_k(n)$ is attached to vertex k with probability

$$\frac{D_k(\tau_k(n)) + \delta}{n + k\delta}, \tag{5.2.43}$$

which are the probabilities of a Pólya urn scheme in (5.2.12) in the linear weight case in (5.2.13) when $a_b = \delta, a_r = (k-1)\delta, b_0 = 1$, and $r_0 = 2k-1$. The setting is a little different for $k = 2$, since vertex 3 attaches to vertices 1 and 2 with equal probability, so that $\psi_2' \sim \text{Beta}(2 + \delta, 2 + \delta)$. Thus, again the statement follows from Theorem 5.3. See Exercise 5.10 for the complete proof. We conclude that, even though $(\text{PA}_n^{(1,\delta)}(a))_{n \geq 1}$ and $(\text{PA}_n^{(1,\delta)}(b))_{n \geq 1}$ have the *same* asymptotic degree distribution, the limiting degree ratios in Theorems 5.6 and 5.7 *are* different.

5.3 LOCAL CONVERGENCE OF PREFERENTIAL ATTACHMENT MODELS

In this section we study the local limit of preferential attachment models, which is a more difficult subject that of inhomogeneous random graphs or configuration models. Indeed, it turns out that the local limit is *not* described by a homogeneous unimodular branching process but rather by an inhomogeneous multi-type branching process.

5.3.1 LOCAL CONVERGENCE OF PAMS WITH FIXED NUMBER OF EDGES

In this subsection we study the local limit of the preferential attachment model, using Pólya urn schemes. We start with $(\text{PA}_n^{(m,\delta)}(d))_{n \geq 1}$, as this model turns out to be the simplest for local limits due to its close relation to Pólya urn schemes. In Section 5.4.4, we discuss the related models $(\text{PA}_n^{(m,\delta)}(a))_{n \geq 1}$ and $(\text{PA}_n^{(m,\delta)}(b))_{n \geq 1}$.

Recall from Section 1.3.5 that the graph starts at time 1 with two vertices having m edges between them. Let τ_k be the kth time that an edge is added to either vertex 1 or 2. The relation to Pólya urn schemes can be informally explained by noting that the random variable $k \mapsto D_1(\tau_k)/(D_1(\tau_k) + D_2(\tau_k))$ can be viewed as the proportion of type-1 vertices in a Pólya urn starting with m balls of types 1 and 2, respectively. Application of de Finetti's Theorem shows that $D_1(\tau_k)/(D_1(\tau_k) + D_2(\tau_k))$ converges almost surely to a certain Beta distribution, which we denote by ψ. What is particularly nice about this description is that the random variable $D_1(\tau_k)$ has *exactly* the same distribution as m plus a $\text{Bin}(k, \psi)$ distribution, i.e., *conditional on* ψ, $D_1(\tau_k)$ is a sum of iid random variables. In the graph context, the Pólya urn description becomes more daunting. However, a description in terms of Beta random variables can again be given.

Definition of the Pólya Point Tree

We will show that, asymptotically, the neighborhood of a random vertex o_n in $\text{PA}_n^{(m,\delta)}(d)$ is a multi-type branching process, in which every vertex has a type that is closely related to the *age* of the vertex. Thus, as for inhomogeneous random graphs and the configuration model, the local limit of $\text{PA}_n^{(m,\delta)}(d)$ is again a tree, but it is not *homogeneous*. This tree has degrees with a Poissonian component to them, as for inhomogeneous random graphs, owing to the younger vertices that connect to the vertex of interest, but also a deterministic component in that every vertex is connected to m older vertices.

To keep track of these two aspects, we give the nodes in the tree a type that consists of its *age* as well as a label indicating the *relative age wrt its parent*. Indeed, each node has two different classes of children, labeled O and Y. The children labeled with O are the *older* neighbors to which one of the initial m edges of the parent is connected, while children labeled Y are *younger* nodes, which use one of their m edges to connect to their parent. Since we are interested in the asymptotic neighborhood of a uniform vertex, the age of the root, which corresponds to the limit of o_n/n, is a uniform random variable on $[0, 1]$. In order to describe its immediate neighbors, we have to consider to which older nodes of label o the root is connected, as well as the number and ages of the younger nodes, of label Y. After this, we again have to consider the number of Y- and o-labeled children that its children have, etc.

Let us now describe these constructs in detail. We define a multi-type branching process, called the *Pólya point tree*. Its nodes are labeled by finite words, using the Ulam–Harris labeling of trees in Section 1.5, as $w = w_1 w_2 \cdots w_l$, each carrying an *age* as well as a *label* Y or o denoting whether the child is *younger* or *older* than its parent in the tree.

The root $\varnothing$ has *age* $U_\varnothing$, where $U_\varnothing$ is chosen uar in $[0, 1]$. The root is special, and has no label in $\{Y, o\}$, since it has no parent. Having discussed the root of the tree, we now construct the remainder of the tree by recursion.

In the recursion step, we assume that the Ulam–Harris word w (recall Section 1.5) and the corresponding age variable $A_w \in [0, 1]$ have been chosen in a previous step. For $j \geq 1$, let wj be the jth child of w, and set

$$
m_-(w) = \begin{cases} m & \text{if } w \text{ is the root or of label o,} \\ m - 1 & \text{if } w \text{ is of label Y.} \end{cases} \tag{5.3.1}
$$

The intuition behind (5.3.1) is that $m_-(w)$ equals the number of older children of w, which equals m when w is older than its parent, and $m - 1$ when w is younger than its parent.

Recall that a Gamma distribution with parameters r and λ has the density given in (1.5.1). Let Γ have a Gamma distribution with parameters $r = m + \delta$ and $\lambda = 1$, and let $\Gamma^\star$ be the size-biased version of Y, which has a Gamma distribution with parameters $r = m + \delta + 1$ and $\lambda = 1$ (see Exercise 5.11). We then take

$$
\Gamma_w \sim \begin{cases} \Gamma & \text{if } w \text{ is the root or of label Y,} \\ \Gamma^\star & \text{if } w \text{ is of label o,} \end{cases} \tag{5.3.2}
$$

independently of everything else.

Let $w1, \ldots, wm_-(w)$ be the children of w having label o, and let their ages $A_{w1}, \ldots, A_{wm_-(w)}$ be given by

$$
A_{wj} = U_{wj}^{1/\chi} A_w, \tag{5.3.3}
$$

where $(U_{wj})_{j=1}^{m_-(w)}$ are iid uniform random variables on $[0, 1]$ that are independent of everything else, and let

$$
\chi = \frac{m + \delta}{2m + \delta}. \tag{5.3.4}
$$

Further, let $(A_{w(m_-(w))+j)})_{j \geq 1}$ be the (ordered) points of a Poisson point process on $[A_w, 1]$ with intensity

$$
\rho_w(x) = \frac{\Gamma_w}{\tau - 1} \frac{x^{1/(\tau-1)-1}}{A_w^{1/(\tau-1)}}, \tag{5.3.5}
$$

where we recall that $\tau = 3 + \delta/m$ by (1.3.63), and the nodes $(w(m_-(w) + j))_{j \geq 1}$ have label Y. The children of w are the nodes wj with labels o and Y.

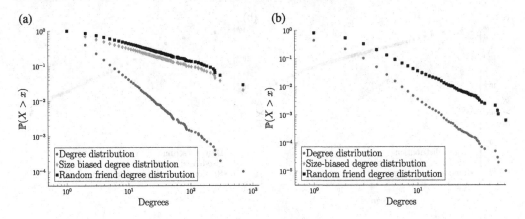

Figure 5.1 Degree distributions in the preferential attachment model with $n = 10,000$, $m = 2$, and (a) $\delta = -1, \tau = 2.5$; (b) $\delta = 1, \tau = 3.5$.

The above random tree is known as the *Pólya point tree*. The Pólya point tree is a multi-type discrete-time branching process, where the type of a node w is equal to the pair (a_w, t_w), with $a_w \in [0, 1]$ corresponding to the age of the vertex, and $t_w \in \{Y, O\}$ to its label. Thus, the type space $\mathcal{S} = [0, 1] \times \{Y, O\}$ of the multi-type branching process is *continuous*.

Let us discuss the offspring structure of the above process. Obviously, there are finitely many children of label O. Further, note that $1/(\tau - 1) = m/(2m + \delta) > 0$, so the intensity ρ_w in (5.3.5) of the Poisson process is integrable. Thus every vertex in the random tree has almost surely *finitely* many children.

With the above description in hand, we are ready to state our main result concerning local convergence of $\mathrm{PA}_n^{(m,\delta)}(d)$:

Theorem 5.8 (Local convergence of preferential attachment models) *Fix $m \geq 1$ and $\delta > -m$. The preferential attachment model $\mathrm{PA}_n^{(m,\delta)}(d)$ converges locally in probability to the Pólya point tree.*

See Figures 5.1 and 5.2 for examples of the various degree distributions in the preferential attachment model, where we plot the degree distribution itself, the size-biased degree distribution, and the degree distribution of a random neighbor of a uniform vertex. Contrary to the generalized random graphs and the configuration model (recall Figures 3.2 and 4.2), the latter two degree distributions are *different*, particularly for small values of n, as in Figure 5.1, even though their power-law exponents *do* agree.

Extensions of Theorem 5.8 to other models, including those where self-loops are allowed, are given in Section 5.4.4 below. These results show that Theorem 5.8 is quite robust to minor changes in the model definition, and that it also applies to $\mathrm{PA}_n^{(m,\delta)}(a)$ and $\mathrm{PA}_n^{(m,\delta)}(b)$ with the *same* limit. We refer to the discussion in Section 5.7 for more details and also the history of Theorem 5.8.

The proof of Theorem 5.8 is organized as follows. We start in Section 5.3.2 by investigating consequences of Theorem 5.8 for the degree structure of $\mathrm{PA}_n^{(m,\delta)}(d)$ (or any other graph having the same local limit). In Section 5.3.3 we prove that $\mathrm{PA}_n^{(m,\delta)}(d)$ can be

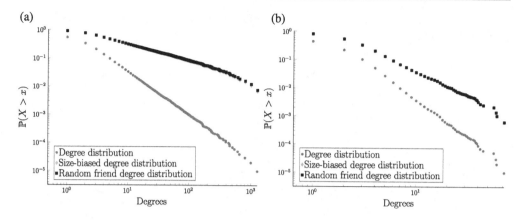

Figure 5.2 Degree distributions in the preferential attachment model with $n = 100,000$, $m = 2$, and (a) $\delta = -1, \tau = 2.5$; (b) $\delta = 1, \tau = 3.5$.

represented in terms of *conditionally independent edges* by relying on a Pólya urn description. The remainder of the proof of local convergence is deferred to Section 5.4.

5.3.2 CONSEQUENCES OF LOCAL CONVERGENCE: DEGREE STRUCTURE

Before turning to the proof of Theorem 5.8, we use it to describe some properties of the degrees of vertices in $\mathrm{PA}_n^{(m,\delta)}(d)$. This is done in the following lemma:

Lemma 5.9 (Degree sequence of $\mathrm{PA}_n^{(m,\delta)}(d)$) *Let $D_o(n)$ be the degree at time n of a vertex chosen uar from $[n]$ in $\mathrm{PA}_n^{(m,\delta)}(d)$. Then,*

$$\mathbb{P}(D_o(n) = k) \to p_k = \frac{2m+\delta}{m} \frac{\Gamma(m+2+\delta+\delta/m)}{\Gamma(m+\delta)} \frac{\Gamma(k+\delta)}{\Gamma(k+3+\delta+\delta/m)}, \quad (5.3.6)$$

and, with $D'_o(n)$ the degree at time n of one of the m older neighbors of o,

$$\mathbb{P}(D'_o(n) = k) \qquad\qquad\qquad (5.3.7)$$
$$\to p'_k = \frac{2m+\delta}{m^2} \frac{\Gamma(m+2+\delta+\delta/m)}{\Gamma(m+\delta)} \frac{(k-m+1)\Gamma(k+1+\delta)}{\Gamma(k+4+\delta+\delta/m)}.$$

Further, let $P_k(n)$ denote the proportion of vertices of degree k in $\mathrm{PA}_n^{(m,\delta)}(d)$, and $P'_k(n)$ the proportion of degree-k older neighbors. Then $P_k(n) \overset{\mathbb{P}}{\longrightarrow} p_k$ and $P'_k(n) \overset{\mathbb{P}}{\longrightarrow} p'_k$.

Note that the limiting degree distribution in (5.3.6) is equal to that for $\mathrm{PA}_n^{(m,\delta)}(a)$ in (1.3.60), again exemplifying that the details of the model have little influence on the limiting degree sequence. It is not hard to see from Lemma 5.9 that

$$p_k = c_{m,\delta} k^{-\tau}(1 + O(1/k)), \qquad p'_k = c'_{m,\delta} k^{-(\tau-1)}(1 + O(1/k)), \qquad (5.3.8)$$

for some constants $c_{m,\delta}$ and $c'_{m,\delta}$ and with $\tau = 3 + \delta/m$ (see Exercise 5.12). We conclude that there is a form of *size biasing* in that older neighbors of a uniform vertex have a limiting degree distribution that satisfies a power law (like the degree of the random vertex itself), but with an exponent that is one lower than that of the vertex itself (recall Figures 5.1 and

5.2). Exercises 5.13–5.15 study the joint distribution (D, D') and various conditional power laws.

Proof of Lemma 5.9 subject to Theorem 5.8. We note that local convergence in probability implies the convergence of the degree distribution. It thus suffices to study the distribution of the degree of the root in the Pólya point tree. We first condition on the age $A_\varnothing = U_\varnothing$ of the root of the Pólya point tree, where $U_\varnothing$ is standard uniform. Let D be the degree of the root. Conditioning on $A_\varnothing = a$, the degree D is m plus a Poisson variable with parameter

$$\frac{\Gamma_\varnothing}{a^{1/(\tau-1)}(\tau-1)} \int_a^1 x^{1/(\tau-1)-1} \mathrm{d}x = \Gamma_\varnothing \frac{1 - a^{1/(\tau-1)}}{a^{1/(\tau-1)}} \equiv \Gamma_\varnothing \kappa(a), \tag{5.3.9}$$

where $\Gamma_\varnothing$ is a Gamma variable with parameters $r = m + \delta$ and $\lambda = 1$. Thus, taking the expectation wrt $\Gamma_\varnothing$, we obtain

$$\begin{aligned}
\mathbb{P}(D = k \mid A_\varnothing = a) &= \int_0^\infty \mathbb{P}(D = k \mid A_\varnothing = a, \Gamma_\varnothing = y) \frac{y^{m+\delta-1}}{\Gamma(m+\delta)} e^{-y} \mathrm{d}y \\
&= \int_0^\infty e^{-y\kappa(a)} \frac{(y\kappa(a))^{k-m}}{(k-m)!} \frac{y^{m+\delta-1}}{\Gamma(m+\delta)} e^{-y} \mathrm{d}y \\
&= \frac{\kappa(a)^{k-m}}{(1+\kappa(a))^{k-m+m+\delta}} \frac{\Gamma(k+\delta)}{(k-m)!\Gamma(m+\delta)} \\
&= (1 - a^{1/(\tau-1)})^{k-m} a^{(m+\delta)/(\tau-1)} \frac{\Gamma(k+\delta)}{(k-m)!\Gamma(m+\delta)}, \tag{5.3.10}
\end{aligned}$$

where we use $\kappa(a)/(1 + \kappa(a)) = 1 - a^{1/(\tau-1)}$ and $1/(1 + \kappa(a)) = a^{1/(\tau-1)}$. We thus conclude that

$$\begin{aligned}
\mathbb{P}(D = k) &= \int_0^1 \mathbb{P}(D = k \mid A_\varnothing = a) \mathrm{d}a \\
&= \int_0^1 (1 - a^{1/(\tau-1)})^{k-m} a^{(m+\delta)/(\tau-1)} \frac{\Gamma(k+\delta)}{(k-m)!\Gamma(m+\delta)} \mathrm{d}a. \tag{5.3.11}
\end{aligned}$$

Recall that

$$\int_0^1 u^{p-1}(1-u)^{q-1} \mathrm{d}u = \frac{\Gamma(p)\Gamma(q)}{\Gamma(p+q)}. \tag{5.3.12}$$

Using the integral transform $u = a^{1/(\tau-1)}$, for which $\mathrm{d}a = (\tau-1)u^{2-\tau}\mathrm{d}u$, we arrive at

$$\begin{aligned}
\mathbb{P}(D = k) &= (\tau-1)\frac{\Gamma(k+\delta)}{(k-m)!\Gamma(m+\delta)} \int_0^1 (1-u)^{k-m} u^{m+\delta+1+\delta/m} \mathrm{d}u \\
&= (\tau-1)\frac{\Gamma(k+\delta)}{(k-m)!\Gamma(m+\delta)} \frac{\Gamma(k-m+1)\Gamma(m+2+\delta+\delta/m)}{\Gamma(k+3+\delta+\delta/m)} \\
&= (\tau-1)\frac{\Gamma(k+\delta)\Gamma(m+2+\delta+\delta/m)}{\Gamma(m+\delta)\Gamma(k+3+\delta+\delta/m)}. \tag{5.3.13}
\end{aligned}$$

Since $\tau - 1 = (2m + \delta)/m$ by (1.3.63), this proves (5.3.6).

We next extend this to the convergence in distribution of $D'_o(n)$, for which we again note that local convergence implies the convergence of the degree distribution of neighbors of the root, and so in particular of $D'_o(n)$. It thus suffices to study the distribution of the degree of a uniform neighbor of the root in the Pólya point tree. We first condition on the age $A_\varnothing = U_\varnothing$ of the root of the Pólya point tree, where $U_\varnothing$ is standard uniform, and recall that the age $A_{\varnothing 1}$ of one of the m older vertices to which $\varnothing$ is connected has distribution $A_{\varnothing 1} = U_{\varnothing 1}^{1/\chi} A_\varnothing$, where $U_{\varnothing 1}$ is uniform on $[0, 1]$ and $1/\chi = (\tau - 1)/(\tau - 2)$ by (5.3.4). Let D' be the degree of vertex $\varnothing 1$. By (5.3.5), conditioning on $A_{\varnothing 1} = b$, the degree D' is m plus a Poisson variable with parameter

$$\frac{\Gamma_{\varnothing 1}}{b^{1/(\tau-1)}(\tau-1)} \int_b^1 x^{1/(\tau-1)-1} \mathrm{d}x = \Gamma_{\varnothing 1} \frac{1 - b^{1/(\tau-1)}}{b^{1/(\tau-1)}} \equiv \Gamma_{\varnothing 1} \kappa(b), \qquad (5.3.14)$$

where $\Gamma_{\varnothing 1}$ is a Gamma variable with parameters $r = m + 1 + \delta$ and $\lambda = 1$.

Thus, taking expectations with respect to $\Gamma_{\varnothing 1}$, we obtain as before

$$\mathbb{P}(D' = k \mid A_{\varnothing 1} = b) = \int_0^\infty \mathbb{P}(D' = k \mid A_{\varnothing 1} = b, \Gamma_{\varnothing 1} = y) \frac{y^{m+1+\delta}}{\Gamma(m+2+\delta)} e^{-y} \mathrm{d}y$$

$$= \int_0^\infty e^{-y\kappa(b)} \frac{(y\kappa(b))^{k-m}}{(k-m)!} \frac{y^{m+\delta}}{\Gamma(m+1+\delta)} e^{-y} \mathrm{d}y$$

$$= \frac{\kappa(b)^{k-m}}{(1+\kappa(b))^{k-m+m+1+\delta}} \frac{\Gamma(k+1+\delta)}{(k-m)!\Gamma(m+1+\delta)} \qquad (5.3.15)$$

$$= \frac{\Gamma(k+1+\delta)}{(k-m)!\Gamma(m+1+\delta)} (1 - b^{1/(\tau-1)})^{k-m} b^{(m+1+\delta)/(\tau-1)},$$

where we again use that $\kappa(b)/(1+\kappa(b)) = 1 - b^{1/(\tau-1)}$ and $1/(1+\kappa(b)) = b^{1/(\tau-1)}$.

We next use that $A_{\varnothing 1} = U_{\varnothing 1}^{(\tau-2)/(\tau-1)} A_\varnothing$, where $A_\varnothing$ is uniform on $[0, 1]$. Recall that the vector $(A_\varnothing, U_{\varnothing 1})$ has density 1 on $[0, 1]^2$. Define the random vector $(A_\varnothing, A_{\varnothing 1}) = (A_\varnothing, U_{\varnothing 1}^{(\tau-2)/(\tau-1)} A_\varnothing)$, so that $(A_\varnothing, A_{\varnothing 1})$ has joint density on $\{(a, b): b \leq a\}$ given by

$$f_{(A_\varnothing, A_{\varnothing 1})}(a, b) = \frac{\tau - 2}{\tau - 1} a^{-(\tau-2)/(\tau-1)} b^{-1/(\tau-1)}. \qquad (5.3.16)$$

We thus conclude that $\mathbb{P}(D' = k)$ equals

$$\frac{\tau-2}{\tau-1} \int_0^1 a^{-(\tau-1)/(\tau-2)} \int_0^a b^{-1/(\tau-1)} \mathbb{P}(D' = k \mid A_{\varnothing 1} = b) \mathrm{d}b \, \mathrm{d}a$$

$$= \frac{\tau-2}{\tau-1} \frac{\Gamma(k+1+\delta)}{(k-m)!\Gamma(m+1+\delta)} \qquad (5.3.17)$$

$$\times \int_0^1 a^{-(\tau-2)/(\tau-1)} \int_0^a (1 - b^{1/(\tau-1)})^{k-m} b^{(m+1+\delta)/(\tau-1)-1/(\tau-1)} \mathrm{d}b \, \mathrm{d}a$$

$$= (\tau-2)(\tau-1) \frac{\Gamma(k+1+\delta)}{(k-m)!\Gamma(m+1+\delta)} \int_0^1 \int_0^u (1-v)^{k-m} v^{m+1+\delta+\delta/m} \mathrm{d}v \, \mathrm{d}u,$$

where we have now used the integral transform $u = a^{\tau-1}$ and $v = b^{\tau-1}$. Recall (5.3.12). Interchanging the integrals over u and v thus leads to the conclusion that $\mathbb{P}(D' = k)$ equals

$$(\tau - 2)(\tau - 1)\frac{\Gamma(k+1+\delta)}{(k-m)!\Gamma(m+1+\delta)} \int_0^1 (1-v)^{k-m+1}v^{m+1+\delta+\delta/m}dv$$

$$= (\tau - 2)(\tau - 1)\frac{\Gamma(k+1+\delta)}{(k-m)!\Gamma(m+1+\delta)}\frac{\Gamma(m+2+\delta+\delta/m)\Gamma(k-m+2)}{\Gamma(k+4+\delta+\delta/m)}$$

$$= \frac{(2m+\delta)}{m^2}\frac{\Gamma(m+2+\delta+\delta/m)}{\Gamma(m+\delta)}\frac{(k-m+1)\Gamma(k+1+\delta)}{\Gamma(k+4+\delta+\delta/m)}, \tag{5.3.18}$$

as required. □

5.3.3 FINITE-GRAPH PÓLYA VERSION OF PREFERENTIAL ATTACHMENT MODELS

The proof of Theorem 5.8 relies crucially on the exchangeability and applications of de Finetti's Theorem (Theorem 5.2). The crucial observation is that de Finetti's Theorem can be used to give an *equivalent* formulation of $\mathrm{PA}_n^{(m,\delta)}(d)$ that relies on independent random variables. We now explain this *finite-graph Pólya version* of $\mathrm{PA}_n^{(m,\delta)}(d)$.

We start by introducing the necessary notation. Let $(\psi_j)_{j\geq 1}$ be independent Beta random variables with parameters $\alpha = m + \delta, \beta_j = (2j-3)m + \delta(j-1)$, i.e.,

$$\psi_j \sim \mathsf{Beta}\big(m+\delta, (2j-3)m + \delta(j-1)\big). \tag{5.3.19}$$

Define

$$\varphi_j^{(n)} = \psi_j \prod_{i=j+1}^n (1-\psi_i), \qquad S_k^{(n)} = \sum_{j=1}^k \varphi_j^{(n)} = \prod_{i=k+1}^n (1-\psi_i). \tag{5.3.20}$$

Here the latter equality follows simply by induction on $k \geq 1$ (see Exercise 5.16). Finally, let $I_k^{(n)} = [S_{k-1}^{(n)}, S_k^{(n)})$. We now construct a graph as follows:

▷ conditional on $\psi_1, \ldots, \psi_n$, choose $(U_{k,i})_{k\in[n],i\in[m]}$ as a sequence of independent random variables, with $U_{k,i}$ chosen uar from the (random) interval $[0, S_{k-1}^{(n)}]$;
▷ for $k \in [n]$ and $j < n$, join two vertices j and k if $j < k$ and $U_{k,i} \in I_j^{(n)}$ for some $i \in [m]$ (with multiple edges between j and k if there are several such i).

Call the resulting random multi-graph on $[n+1]$ the *finite-size Pólya graph of size n*. The main result for $\mathrm{PA}_n^{(m,\delta)}(d)$ is as follows:

Theorem 5.10 (Finite-graph Pólya version of $\mathrm{PA}_n^{(m,\delta)}(d)$) *Fix $m \geq 1$ and $\delta > -m$. Then, the distribution of $\mathrm{PA}_n^{(m,\delta)}(d)$ is the same as that of the finite-size Pólya graph of size n.*

The importance of Theorem 5.10 is that the edges in the finite-size Pólya graph are *independent* conditional on the Beta variables $(\psi_k)_{k\geq 1}$, in a similar way as for (5.2.15) in Theorem 5.3. This independence makes explicit computations possible. Exercises 5.18–5.17, for example, use Theorem 5.10 to derive properties of the number of multiple edges in $\mathrm{PA}_n^{(m,\delta)}(d)$ for $m = 2$.

In terms of the above Pólya point tree, the proof shows that the Gamma variables that define the "strengths" Γ_w are inherited from the Beta random variables $(\psi_k)_{k\in[n]}$, while the age variables A_w are inherited from the random variables $(S_k^{(n)})_{k \in [n]}$ (see Lemmas 5.17 and 5.18 below).

Remark 5.11 (Different starting graph)   For $\mathrm{PA}_n^{(m,\delta)}(d)$, we note that $\sum_{v \in [n]} D_v(n) = 2m(n-1)$ and that it has n vertices, since the graph at time 1 consists of two vertices with m edges between them. However, it is sometimes more convenient to deal with a model where the graph at time 2 consists of two vertices with $2m$ edges between them, so that the sum of the degrees at time n equals $2mn$ and there are n vertices. In particular, for $m = 1$, this gives rise to $(\mathrm{PA}_n^{(1,\delta)}(b))_{n \geq 1}$. The proof shows that Theorem 5.10 remains true almost verbatim, but now with $(\psi_j)_{j \in [n]}$ replaced by $(\psi_j')_{j \in [n]}$ given by $\psi_1' = 1, \psi_2' \sim \mathsf{Beta}(2m+\delta, 2m+\delta)$, while, for $j \geq 3$,

$$\psi_j' \sim \mathsf{Beta}\big(m + \delta, (2j - 1)m + \delta(j - 1)\big). \tag{5.3.21}$$

Recall also Theorem 5.7. The above changes affect the finite-size Pólya graph in only a minor way. ◄

Let us give some insight into the proof of Theorem 5.10, after which we give two full proofs. The first proof relies on Pólya urn methods and the second on a direct computation.

For the Pólya urn proof, we rely on the fact that there is a close connection between the preferential attachment model and the Pólya urn model, in the following sense. Every new connection that a vertex gains can be represented by a new ball added to the urn corresponding to that vertex, as in Theorems 5.6 and 5.7. As time progresses, the number of urns corresponding to the vertices changes, which is a major complication. As it turns out, however, the attachment probabilities are *consistent*, which allows the Pólya urn description to be extended to this setting of increasing numbers of urns. Let us now make this intuition precise.

Pólya urn proof of Theorem 5.10. Let us consider first a two-urn model, where the number of balls in one urn represents the degree of a particular vertex k, and the number of balls in the other represents the sum of the degrees of the vertices $[k-1]$ as in Theorems 5.6 and 5.7. We start this process at the point when $n = k$, and k has connected to precisely m vertices in $[k-1]$. Note that at this point, by the structure of $\mathrm{PA}_n^{(m,\delta)}(d)$, the urn representing the degree of vertex k has m balls while the other urn, corresponding to the vertices in $[k-1]$, has $(2k-3)m$ balls.

Consider a time in the evolution of $\mathrm{PA}_n^{(m,\delta)}(d)$ when we have $n - 1 \geq k$ old vertices, and $i - 1$ edges between the new vertex n and $[k-1]$. Assume that at this point the degree of k is d_k and the sum of the degrees of the vertices in $[k-1]$ is $d_{[k-1]}$. The probability that the ith edge from n to $[n-1]$ is attached to k is then

$$\frac{d_k + \delta}{2m(n-1) + (1+\delta)(i-1)}, \tag{5.3.22}$$

while the probability that it is connected to a vertex in $[k-1]$ is equal to

$$\frac{d_{[k-1]} + \delta(k-1)}{2m(n-1) + (1+\delta)(i-1)}. \tag{5.3.23}$$

Thus, conditioned on connecting to $[k]$, the probability that the ith edge from n to $[n-1]$ is attached to k is $(d_k + \delta)/(k\delta + d_{[k]})$, while the probability that the ith edge from n to $[n-1]$ is attached to $[k-1]$ is $(d_{[k-1]} + \delta(k-1))/(k\delta + d_{[k]})$.

Taking into account that the two urns start with m and $(2k-3)m$ balls, respectively, we see that the evolution of the two bins is a Pólya urn with strengths ψ_k and $1 - \psi_k$, where $\psi_k \sim \text{Beta}(m + \delta, (2k-3)m + \delta(k-1))$ (recall Theorem 5.3). We next use this to complete the proof of Theorem 5.10, where we use induction. Indeed, using the two-urn process as an inductive input, we construct the finite-size Pólya graph defined in Theorem 5.10 in a similar way as for the Pólya urns with multiple colors in (5.2.28).

Let $X_t \in [0, \lceil t/m \rceil]$ be the vertex receiving the tth edge (the other endpoint of this edge being the vertex $\lceil t/m \rceil + 1$). For $t \in [m]$, X_t is deterministic (and equal to 1 since we start at time 1 with two vertices and m edges between them); however, beginning at $t = m + 1$, we have a two-urn model, starting with m balls in each urn. As shown above, the two urns can be described as Pólya urns with strengths $1 - \psi_2$ and ψ_2. Once $t > 2m$, X_t can take three values but, conditioned on $X_t \leq 2$, the process continues to be a two-urn model with strengths $1 - \psi_2$ and ψ_2.

To determine the probability of the event that $X_t \leq 2$, we now use the above two-urn model with $k = 3$, which gives that the probability of the event $X_t \leq 2$ is $1 - \psi_3$, at least as long as $t \leq 3m$. Combining these two-urn models, we get a three-urn model with strengths $(1 - \psi_2)(1 - \psi_3)$, $\psi_2(1 - \psi_3)$, and ψ_3. Again, this model remains valid for $t > 3m$, as long as we condition on $X_t \leq 3$. Continuing inductively, we see that the sequence X_t evolves in stages:

▷ For $t \in [m]$, the variable X_t is deterministic: $X_t = 1$.
▷ For $t = m + 1, \ldots, 2m$, the distribution of $X_t \in \{1, 2\}$ is described by a two-urn model with strengths $1 - \psi_2$ and ψ_2, where $\psi_2 \sim \text{Beta}(m + \delta, m + \delta)$.
▷ In general, for $t = m(k-1) + 1, \ldots, km$, the distribution of $X_t \in [k]$ is described by a k-urn model with strengths

$$\varphi_j^{(k)} = \psi_j \prod_{i=j+1}^{k} (1 - \psi_i), \qquad j \in [k]. \tag{5.3.24}$$

Here the Beta-variable ψ_k in (5.3.19) is chosen at the beginning of the kth stage, independently of the previously chosen strengths $\psi_1, \ldots, \psi_{k-1}$ (for convenience, we set $\psi_1 = 1$).

Note that the random variables $\varphi_j^{(k)}$ can be expressed in terms of the random variables introduced in Theorem 5.10 as follows. By (5.3.20), $S_k^{(n)} = \prod_{j=k+1}^{n} (1 - \psi_j)$. This implies that $\varphi_j^{(n)} = \psi_j S_j^{(n)}$, which relates the strengths $\varphi_j^{(n)}$ to the random variables defined right before Theorem 5.10, and shows that the process derived above is indeed the process given in the theorem. □

We next give a direct proof of Theorem 5.10, which is of independent interest as it also indicates how the conditional independence of edges can be used effectively:

Direct proof of Theorem 5.10. In what follows, we let PA'_n denote the finite-size Pólya graph of size n. Our aim is to show that $\mathbb{P}(\mathrm{PA}'_n = G) = \mathbb{P}(\mathrm{PA}_n^{(m,\delta)}(d) = G)$ for any graph G. Here, we think of G as being a directed and edge-labeled graph, where every vertex has out-degree m and the out-edges are labeled as $[m]$. Thus, the out-edges are the edges from young to old. Indeed, recall from Section 1.3.5 that the graph starts at time 1 with two vertices and m edges between them. The vertex set of $\mathrm{PA}_n^{(m,\delta)}(d)$ is $[n]$. In the proof, it is convenient to denote the labeled edge set of G as $\vec{E}(G) = \{(u, v_j(u), j) \colon u \in [n], j \in [m]\}$, where $v_j(u) < u$ is the vertex to which the jth edge of u is attached in G. We can assume that $v_j(2) = 1$ for all $j \in [m]$, since $\mathrm{PA}_n^{(m,\delta)}(d)$ starts at time 1 with two vertices having m edges between them.

Fix an edge-labeled graph G for which $\mathbb{P}(\mathrm{PA}_n^{(m,\delta)}(d) = G) > 0$. On the one hand, we can compute directly that

$$\mathbb{P}(\mathrm{PA}_n^{(m,\delta)}(d) = G) = \prod_{u \in [3,n], j \in [m]} \frac{d_{v_j(u)}^{(G)}(u) + \delta}{2m(u-2) + j - 1 + \delta(u-1)}, \tag{5.3.25}$$

where $[3, n] = \{3, \ldots, n\}$. Note that, for every $s \in [n]$,

$$\prod_{u \in [3,n], j \in [m] \colon v_j(u) = s} (d_{v_j(u)}^{(G)}(u) + \delta) = \prod_{i=0}^{d_s^{(G)} - m - 1} (i + m + \delta), \tag{5.3.26}$$

with the empty product defined as 1. Therefore,

$$\mathbb{P}(\mathrm{PA}_n^{(m,\delta)}(d) = G) \tag{5.3.27}$$

$$= \prod_{s \in [n]} \prod_{i=0}^{d_s^{(G)} - m - 1} (i + m + \delta) \prod_{u \in [3,n], j \in [m]} \frac{1}{2m(u-2) + j - 1 + \delta(u-1)}.$$

Thus, we are left with showing that $\mathbb{P}(\mathrm{PA}'_n = G)$ is equal to the rhs of (5.3.27).

To identify $\mathbb{P}(\mathrm{PA}'_n = G)$, it is convenient to condition on the Beta variables $(\psi_j)_{j \in [n]}$. We denote the conditional measure by $\mathbb{P}_n$; i.e., for every event $\mathcal{E}$,

$$\mathbb{P}_n(\mathcal{E}) = \mathbb{P}(\mathcal{E} \mid (\psi_j)_{j \in [n]}). \tag{5.3.28}$$

The advantage of this measure is that now edges are *conditionally independent*, which allows us to give an exact formula for the probability of a certain graph occurring. We start by computing the edge probabilities under $\mathbb{P}_n$, where we recall that $\{u \overset{j}{\rightsquigarrow} v\}$ is the event that the jth edge of u connects to v:

Lemma 5.12 (Edge probabilities in PA'_n conditioned on Beta variables) *Fix $m \geq 1$ and $\delta > -m$, and consider PA'_n. For any $u > v$ and $j \in [m]$,*

$$\mathbb{P}_n(u \overset{j}{\rightsquigarrow} v) = \psi_v (1 - \psi)_{(v,u)}, \tag{5.3.29}$$

where, for $A \subseteq [n]$,

$$(1 - \psi)_A = \prod_{a \in A} (1 - \psi_a). \tag{5.3.30}$$

Proof Recall the construction between (5.3.19) and Theorem 5.10. When we condition on $(\psi_j)_{j\in[n]}$, the only randomness left is that in the uniform random variables $(U_{k,i})_{k\in[n],i\in[m]}$, where $U_{k,i}$ is uniform on $[0, S_{k-1}^{(n)}]$. Then, $u \overset{j}{\rightsquigarrow} v$ occurs precisely when $U_{u,j} \in I_v^{(n)} = [S_{v-1}^{(n)}, S_v^{(n)})$, which occurs with $\mathbb{P}_n$-probability equal to $|I_v^{(n)}|/S_{u-1}^{(n)}$. Note that

$$|I_v^{(n)}| = S_v^{(n)} - S_{v-1}^{(n)} = (1-\psi)_{[v+1,n]} - (1-\psi)_{[v,n]} = \psi_v(1-\psi)_{(v,n]}, \qquad (5.3.31)$$

while, from (5.3.20),

$$S_{u-1}^{(n)} = \prod_{i=u}^{n}(1-\psi_i) = (1-\psi)_{[u,n]}. \qquad (5.3.32)$$

Taking the ratio yields, with $u > v$,

$$\mathbb{P}_n(u \overset{j}{\rightsquigarrow} v) = \psi_v \frac{(1-\psi)_{(v,n]}}{(1-\psi)_{[u,n]}} = \psi_v(1-\psi)_{(v,u)}, \qquad (5.3.33)$$

as required. $\qquad\qquad\square$

In the finite-size Pólya graph PA_n', different edges are *conditionally independent*, so that we obtain the following corollary:

Corollary 5.13 (Graph probabilities in PA_n' conditioned on Beta variables) *Fix $m \geq 1$ and $\delta > -m$, and consider PA_n'. For any edge-labeled multi-graph G,*

$$\mathbb{P}_n(\mathrm{PA}_n' = G) = \prod_{s=2}^{n} \psi_s^{p_s}(1-\psi_s)^{q_s}, \qquad (5.3.34)$$

where $p_s = p_s^{(G)}$ and $q_s = q_s^{(G)}$ are given by

$$p_s = d_s^{(G)} - m, \qquad q_s = \sum_{u\in[3,n]}\sum_{j\in[m]} \mathbb{1}_{\{s\in(v_j(u),u)\}}. \qquad (5.3.35)$$

Proof Multiply the factors $\mathbb{P}_n(u \overset{j}{\rightsquigarrow} v_j(u))$ for every labeled edge $(u, v_j(u), j) \in \vec{E}(G)$, and collect the powers of ψ_s and $1 - \psi_s$. $\qquad\square$

We note that p_s equals the number of edges in the graph G that point towards s. This is relevant, since in (5.3.29) in Lemma 5.12 every older vertex v in an edge receives a factor ψ_v. Further, again by (5.3.29), there are factors $1 - \psi_s$ for every $s \in (v, u)$ and all edges (v, u), so q_s counts how many factors $1 - \psi_s$ occur.

When taking expectations wrt $(\psi_v)_{v\in[n]}$ into account, by Corollary 5.13, we obtain expectations of the form $\mathbb{E}[\psi^p(1-\psi)^q]$, where $\psi \sim \mathrm{Beta}(\alpha,\beta)$ and $p, q \geq 0$. These are computed in the following lemma:

Lemma 5.14 (Expectations of powers of Beta variables) *For all $p, q \in \mathbb{N}$ and $\psi \sim \mathrm{Beta}(\alpha,\beta)$,*

$$\mathbb{E}[\psi^p(1-\psi)^q] = \frac{(\alpha+p-1)_p(\beta+q-1)_q}{(\alpha+\beta+p+q-1)_{p+q}}, \qquad (5.3.36)$$

where, as before, $(x)_m = x(x-1)\cdots(x-m+1)$ denotes the mth falling factorial of x.

Proof A direct computation based on the density of a Beta random variable in (1.5.2) yields

$$\mathbb{E}[\psi^p(1-\psi)^q] = \frac{B(\alpha+p,\beta+q)}{B(\alpha,\beta)} = \frac{\Gamma(\alpha+\beta)}{\Gamma(\alpha)\Gamma(\beta)}\frac{\Gamma(\alpha+p)\Gamma(\beta+q)}{\Gamma(\alpha+\beta+p+q)}$$
$$= \frac{(\alpha+p-1)_p(\beta+q-1)_q}{(\alpha+\beta+p+q-1)_{p+q}}. \tag{5.3.37}$$

$\square$

The above computation, when applied to Corollary 5.13, leads to the following expression for the probability of observing a particular edge-labeled multi-graph G:

Corollary 5.15 (Graph probabilities in PA'_n) *Fix $m \geq 1$ and $\delta > -m$, and consider PA'_n. For any edge-labeled multi-graph G,*

$$\mathbb{P}(\mathrm{PA}'_n = G) = \prod_{s=2}^{n-1} \frac{(\alpha+p_s-1)_{p_s}(\beta_s+q_s-1)_{q_s}}{(\alpha+\beta_s+p_s+q_s-1)_{p_s+q_s}}, \tag{5.3.38}$$

where $\alpha = m + \delta, \beta_s = (2s-3)m + \delta(s-1)$, and $p_s = p_s^{(G)}$ and $q_s = q_s^{(G)}$ are defined in (5.3.35).

Note that the contribution for $s = n$ equals 1, since $p_n^{(G)} = d_n^{(G)} - m = 0$ almost surely in PA'_n. Corollary 5.15 allows us to complete the direct proof of Theorem 5.10:

Corollary 5.16 (Graph probabilities in PA'_n and $\mathrm{PA}_n^{(m,\delta)}(d)$) *Fix $m \geq 1$ and $\delta > -m$, and consider PA'_n and $\mathrm{PA}_n^{(m,\delta)}(d)$. For any edge-labeled multi-graph G,*

$$\mathbb{P}(\mathrm{PA}'_n = G) = \mathbb{P}(\mathrm{PA}_n^{(m,\delta)}(d) = G), \tag{5.3.39}$$

where $\alpha = m + \delta, \beta_s = (2s-3)m + \delta(s-1)$, and $p_s = p_s^{(G)}$ and $q_s = q_s^{(G)}$ are defined in (5.3.35). Consequently, Corollaries 5.13 and 5.15 also hold for $\mathrm{PA}_n^{(m,\delta)}(d)$.

Proof We evaluate (5.3.38) in Corollary 5.15 explicitly. Since $\alpha = m + \delta$ and $p_s = d_s^{(G)} - m$,

$$(\alpha+p_s-1)_{p_s} = \prod_{i=0}^{d_s^{(G)}-m-1}(i+m+\delta), \tag{5.3.40}$$

so that

$$\prod_{s=2}^{n-1}(\alpha+p_s-1)_{p_s} = \prod_{s=2}^{n-1}\prod_{i=0}^{d_s^{(G)}-m-1}(i+m+\delta), \tag{5.3.41}$$

which produces the first product in (5.3.27) except for the $s = 1$ factor.

We next identify the other factors, for which we start by analyzing $q_s^{(G)}$ as follows:

$$q_s^{(G)} = \sum_{u\in[3,n]}\sum_{j\in[m]}\mathbb{1}_{\{s\in(v_j(u),u)\}} = \sum_{u\in[3,n]}\sum_{j\in[m]}\left(\mathbb{1}_{\{s\in(v_j(u),n]\}} - \mathbb{1}_{\{s\in[u,n]\}}\right). \tag{5.3.42}$$

We can use

$$\sum_{u\in[3,n]}\sum_{j\in[m]}\mathbb{1}_{\{v_j(u)\in[s-1]\}} = d_{[s-1]}^{(G)} - m(s-1), \tag{5.3.43}$$

since the lhs counts the in-edges in $[s-1]$ except for those from vertex 2 to vertex 1, while $d_{[s-1]}^{(G)}$ counts all in- and out-edges, and there are exactly $m(s-2)$ out-edges in $[s-1]$ and m edges from vertex 2 to vertex 1. Further, note that, for $s \in [n]$,

$$\sum_{u \in [3,n]} \sum_{j \in [m]} \mathbb{1}_{\{s \in [u,n]\}} = m \sum_{u \in [3,s]} \mathbb{1}_{\{s \in [u,n]\}} = m(s-2). \tag{5.3.44}$$

Thus,

$$q_s^{(G)} = d_{[s-1]}^{(G)} - m(2s-3). \tag{5.3.45}$$

As a result, by (5.3.35) and the recursions

$$p_s^{(G)} + q_s^{(G)} = q_{s+1}^{(G)} + m, \qquad \alpha + \beta_s = \beta_{s+1} - m, \tag{5.3.46}$$

we obtain

$$(\alpha + \beta_s + p_s + q_s - 1)_{p_s+q_s} = (\beta_{s+1} + q_{s+1} - 1)_{q_{s+1}+m}$$
$$= (\beta_{s+1} + q_{s+1} - 1)_{q_{s+1}} (\beta_{s+1} - 1)_m. \tag{5.3.47}$$

Therefore, by (5.3.45) and since $\beta_s = (2s-3)m + \delta(s-1)$,

$$\prod_{s=2}^{n-1} \frac{(\beta_s + q_s - 1)_{q_s}}{(\alpha + \beta_s + p_s + q_s - 1)_{p_s+q_s}}$$

$$= \prod_{s=2}^{n-1} \frac{1}{(\beta_{s+1} - 1)_m} \prod_{s=2}^{n-1} \frac{(\beta_s + q_s - 1)_{q_s}}{(\beta_{s+1} + q_{s+1} - 1)_{q_{s+1}}}$$

$$= (\delta + d_1^{(G)} - 1)_{d_1^{(G)} - m} \prod_{s=2}^{n-1} \frac{1}{(m(2s-1) + \delta s - 1)_m}. \tag{5.3.48}$$

Indeed, the starting value in the telescoping product equals, again by (5.3.45), with $\beta_2 = m + \delta$ and $q_2 = d_1^{(G)} - m$,

$$(\beta_2 + q_2 - 1)_{q_2} = (\delta + d_1^{(G)} - 1)_{d_1^{(G)} - m}, \tag{5.3.49}$$

and, since $q_n = d_{[n-1]}^{(G)} - m(2n+1) = 0$, the ending value equals

$$(\beta_n + q_n - 1)_{q_n} = 1. \tag{5.3.50}$$

The first factor in (5.3.48) produces the missing $s = 1$ factor in (5.3.27), as compared with (5.3.41). For the second factor in (5.3.48), we compute that, with $u = s + 1 \in [3, n]$,

$$\prod_{s=2}^{n-1} (m(2s-1) + \delta s - 1)_m = \prod_{u \in [3,n]} \prod_{i \in [m]} (\delta(u-1) + m(2u-3) - i)$$

$$= \prod_{u \in [3,n], j \in [m]} (2m(u-2) + j - 1 + \delta(u-1)), \tag{5.3.51}$$

where $j = m - i + 1$, as required. $\qquad \square$

5.4 PROOF OF LOCAL CONVERGENCE FOR PREFERENTIAL ATTACHMENT MODELS

In this section we complete the proof of the local convergence of preferential attachment models to the Pólya point tree in Theorem 5.8. This section is organized as follows. In Section 5.4.1 we discuss some necessary preliminaries, such as the convergence of the rescaled Beta variables to Gamma variables and the regularity properties of the Pólya point tree. Our local convergence proof again relies on a second-moment method for the number of vertices whose ordered r-neighborhood agrees with a specific ordered tree $\mathbf{t}$. We investigate the first moment of the subgraph counts in Section 5.4.2 and handle the second moment in Section 5.4.3. We close in Section 5.4.4 by discussing the local convergence of related preferential attachment models.

5.4.1 LOCAL CONVERGENCE: PRELIMINARIES

In this subsection we set the stage for the local convergence proof. We start by analyzing the asymptotics of $(\psi_k)_{k \in [n]}$ and $(S_k^{(n)})_{k \in [n]}$.

Asymptotics of $(\psi_k)_{k \in [n]}$ and $(S_k^{(n)})_{k \in [n]}$

We start by analyzing the random variables in Theorem 5.10, to prepare us for proving local convergence in Theorem 5.8. The next lemma describes the asymptotics of ψ_k for k large:

Lemma 5.17 (Gamma asymptotics of Beta variables) *As $k \to \infty$, $k\psi_k \xrightarrow{d} \Gamma$, where Γ has a Gamma distribution with parameters $r = m + \delta$ and $\lambda = 2m + \delta$. More precisely, take $f_k(x)$ such that $\mathbb{P}(\psi_k \leq f_k(x)) = \mathbb{P}(\chi_k \leq x)$, where χ_k has a Gamma distribution with $r = m + \delta$ and $\lambda = 1$ (so that $\Gamma \overset{d}{=} \chi_k/(2m + \delta)$). Then, for every $\varepsilon > 0$, there exists $K = K_\varepsilon \geq 1$ sufficiently large such that, for all $k \geq K$ and $x \leq (\log k)^2$,*

$$\frac{1-\varepsilon}{k(2m+\delta)}x \leq f_k(x) \leq \frac{1+\varepsilon}{k(2m+\delta)}x. \tag{5.4.1}$$

Further, $\chi_k \leq (\log k)^2$ for all $k \geq K$, with probability at least $1 - \varepsilon$.

The construction in Lemma 5.17 implies that $\psi_k \overset{d}{=} f_k(\chi_k)$, which allows for a convenient coupling between the Beta and Gamma random variables.

Proof Fix $x \geq 0$. We use Stirling's formula as in [V1, (8.3.9)] to compute that

$$\mathbb{P}(k\psi_k \leq x) = \frac{\Gamma\big(m + \delta + (2k-3)m + \delta(k-1)\big)}{\Gamma\big(m+\delta\big)\Gamma\big((2k-3)m + \delta(k-1)\big)}$$

$$\times \int_0^{x/k} u^{m+\delta-1}(1-u)^{(2k-3)m+\delta(k-1)-1}\mathrm{d}u \tag{5.4.2}$$

$$= (1 + o(1))\frac{[(2m+\delta)k]^{m+\delta}}{\Gamma(m+\delta)k^{m+\delta}} \int_0^x v^{m+\delta-1}(1-v/k)^{(2k-3)m+\delta(k-1)-1}\mathrm{d}v.$$

For every $u > 0$, $(1 - u/k)^{(2k-3)m+\delta(k-1)-1} \to e^{-u(2m+\delta)}$, so that dominated convergence implies that

$$\mathbb{P}(k\psi_k \leq x) \to \int_0^x \frac{(2m+\delta)^{m+\delta} u^{m+\delta-1}}{\Gamma(m+\delta)} e^{-u(2m+\delta)} \mathrm{d}u, \tag{5.4.3}$$

as required.

We continue by proving the upper bound in (5.4.1). Let $\chi_k' = \chi_k/[(2k-3)m + \delta(k-1) - 1]$. We prove below that $\psi_k \preceq \chi_k'$, or, for every $x \geq 0$,

$$\mathbb{P}(\psi_k \leq x) \geq \mathbb{P}(\chi_k' \leq x). \tag{5.4.4}$$

This proves the upper bound in (5.4.1) for $k \geq K$ and K large enough. The inequality in (5.4.4) is obviously true for $x \geq 1$, so we will assume that $x \in [0, 1)$ from now on. Then we can write, with $b = (2k-3)m + \delta(k-1) - 1$,

$$\mathbb{P}(\chi_k' \leq x) = \frac{\mathbb{E}[\mathbb{1}_{\{\psi_k \leq x\}} e^{-b\psi_k}(1 - \psi_k)^{-b}]}{\mathbb{E}[e^{-b\psi_k}(1 - \psi_k)^{-b}]}. \tag{5.4.5}$$

Note that $y \mapsto e^{-by}(1-y)^{-b}$ is increasing, while $y \mapsto \mathbb{1}_{\{y \leq x\}}$ is decreasing, so that, by the correlation inequality in [V1, Lemma 2.14],

$$\frac{\mathbb{E}[\mathbb{1}_{\{\psi_k \leq x\}} e^{-b\psi_k}(1-\psi_k)^{-b}]}{\mathbb{E}[e^{-b\psi_k}(1-\psi_k)^{-b}]} \leq \frac{\mathbb{E}[\mathbb{1}_{\{\psi_k \leq x\}}]\mathbb{E}[e^{-b\psi_k}(1-\psi_k)^{-b}]}{\mathbb{E}[e^{-b\psi_k}(1-\psi_k)^{-b}]} = \mathbb{P}(\psi_k \leq x), \tag{5.4.6}$$

as required.

We continue with the lower bound in (5.4.1), and now instead aim to prove that, for $x \leq (\log k)^2/b$ and again with $b = (2k-3)m + \delta(k-1) - 1$,

$$\mathbb{P}(\psi_k \leq (1-\varepsilon)x) \leq \mathbb{P}(\chi_k' \leq xb_k). \tag{5.4.7}$$

We now write

$$\begin{aligned}
\mathbb{P}(\psi_k \leq (1-\varepsilon)x) &= \frac{\int_0^{(1-\varepsilon)xb} y^{\alpha_k - 1}(1 - y/b)^b \mathrm{d}y}{\int_0^b y^{\alpha_k - 1}(1 - y/b)^b \mathrm{d}y} \\
&= \frac{\mathbb{E}[\mathbb{1}_{\{\chi_k' \leq (1-\varepsilon)xb\}} e^{\chi_k'}(1 - \chi_k'/b)^{-b}]}{\mathbb{E}[\mathbb{1}_{\{\chi_k' \leq b\}} e^{\chi_k'}(1 - \chi_k'/b)^{-b}]} \\
&= \frac{\mathbb{E}[\mathbb{1}_{\{\chi_k' \leq (1-\varepsilon)xb\}} e^{\chi_k'}(1 - \chi_k'/b)^{-b} \mid \chi_k' \leq b]}{\mathbb{E}[e^{\chi_k'}(1 - \chi_k'/b)^{-b} \mid \chi_k' \leq b]}.
\end{aligned} \tag{5.4.8}$$

Note that $y \mapsto \mathbb{1}_{\{y \leq (1-\varepsilon)xb\}}$ is non-increasing, and $y \mapsto e^y(1-y/b)^{-b}$ is non-decreasing. Therefore, again by [V1, Lemma 2.14],

$$\mathbb{P}(\psi_k \leq (1-\varepsilon)x) \leq \mathbb{P}(\chi_k' \leq (1-\varepsilon)xb_k \mid \chi_k' \leq b). \tag{5.4.9}$$

Thus, for the lower bound in (5.4.1), it suffices to show that, for all k large enough, and for $x \leq (\log k)^2/b$,

$$\mathbb{P}(\chi_k' \leq (1-\varepsilon)xb \mid \chi_k' \leq b) \leq \mathbb{P}(\chi_k' \leq xb). \tag{5.4.10}$$

In turn, this follows from the statement that, for $x \leq (\log k)^2/b$,

$$e(x) \equiv \mathbb{P}(\chi_k' \leq xb)\mathbb{P}(\chi_k' \leq b) - \mathbb{P}(\chi_k' \leq (1-\varepsilon)xb) \geq 0. \tag{5.4.11}$$

Note that $e(x)$ can be simplified to

$$e(x) = \mathbb{P}(xb(1-\varepsilon) < \chi'_k \le bx) - \mathbb{P}(\chi'_k > b)\mathbb{P}(\chi'_k \le xb). \tag{5.4.12}$$

We bound the first term on the rhs of (5.4.12) from below as follows:

$$\mathbb{P}(xb(1-\varepsilon) < \chi'_k \le bx)$$
$$= \int_{xb(1-\varepsilon)}^{bx} \frac{y^{m+\delta-1}}{\Gamma(m+\delta)} e^{-y} dy \ge \frac{[xb(1-\varepsilon)]^{m+\delta-1}}{\Gamma(m+\delta)} \int_{xb(1-\varepsilon)}^{bx} e^{-y} dy$$
$$= \frac{[xb(1-\varepsilon)]^{m+\delta-1}}{\Gamma(m+\delta)} [e^{-xb(1-\varepsilon)} - e^{-bx}], \tag{5.4.13}$$

while the second term on the rhs of (5.4.12) is bounded from above by

$$\mathbb{P}(\chi'_k > b) \le e^{-b/2} \, \mathbb{E}[e^{\chi'_k/2}] \le 2^{m+\delta} e^{-b/2}, \tag{5.4.14}$$

and

$$\mathbb{P}(\chi'_k \le xb) = \int_0^{xb} \frac{y^{m+\delta-1}}{\Gamma(m+\delta)} e^{-y} dy \le \frac{[xb]^{m+\delta-1}}{\Gamma(m+\delta)} \int_0^{xb} e^{-y} dy$$
$$\le \frac{[xb]^{m+\delta-1}}{\Gamma(m+\delta)}. \tag{5.4.15}$$

Substitution yields that

$$e(x) \ge \frac{[xb]^{m+\delta-1}}{\Gamma(m+\delta)} [(1-\varepsilon)^{m+\delta-1}(e^{-xb(1-\varepsilon)} - e^{-bx}) - 2^{m+\delta} e^{-b/2}], \tag{5.4.16}$$

which, for any $\varepsilon \in (0,1)$, is non-negative for all $x < \frac{1}{3}$, say, and $b = (2k-3)m + \delta(k-1)$ sufficiently large. This is much more than is needed.

We complete the proof by showing that $\chi_k \le (\log k)^2$ for all $k \ge K$ with probability at least $1 - \varepsilon$, for which we note that

$$\mathbb{P}(\chi_k \ge (\log k)^2) \le \mathbb{E}[e^{\chi_k/2}] \, e^{-(\log k)^2/2} = 2^{m+\delta} e^{-(\log k)^2/2}, \tag{5.4.17}$$

which is summable in k, so that the union bound yields the claim. $\qquad\square$

By Lemma 5.17, we see that indeed the Beta random variables $(\psi_k)_{k \in [n]}$ give rise to the Gamma random variables in (5.3.2), which explains their appearance in the Pólya point tree.

We continue by analyzing the asymptotics for the random variables $(S_k^{(n)})_{k \in [n]}$:

Proposition 5.18 (Asymptotics of $S_k^{(n)}$)   *Recall that* $\chi = (m+\delta)/(2m+\delta)$. *For every* $\varepsilon > 0$, *there exist* $\eta > 0$ *and* $K < \infty$ *such that, for all* $n \ge K$ *and with probability at least* $1 - \varepsilon$,

$$\max_{k \in [n]} \left| S_k^{(n)} - \left(\frac{k}{n}\right)^\chi \right| \le \eta, \tag{5.4.18}$$

and

$$\max_{k \in [n] \setminus [K]} \left| S_k^{(n)} - \left(\frac{k}{n}\right)^\chi \right| \le \varepsilon \left(\frac{k}{n}\right)^\chi. \tag{5.4.19}$$

Proof We will give the intuition behind Proposition 5.18. We recall from (5.3.20) that $S_k^{(n)} = \prod_{i=k+1}^{n}(1 - \psi_i)$, where $(\psi_k)_{k \in [n]}$ are independent random variables. We write

$$\log S_k^{(n)} = \sum_{i=k}^{n} \log(1 - \psi_i) = \sum_{i=k}^{n} \mathbb{E}[\log(1 - \psi_i)] + \sum_{i=k}^{n}(\log(1 - \psi_i) - \mathbb{E}[\log(1 - \psi_i)]).$$
$$(5.4.20)$$

Note that $(M_n)_{n \geq k}$, with $M_n = \sum_{i=k}^{n}(\log(1 - \psi_i) - \mathbb{E}[\log(1 - \psi_i)])$, is a martingale. Thus, by Kolmogorov's inequality in (1.5.5), and all $t \geq k$,

$$\mathbb{P}\Big(\sup_{n=k}^{t} \Big| \sum_{i=k}^{n}(\log(1-\psi_i) - \mathbb{E}[\log(1-\psi_i)]) \Big| \geq \varepsilon \Big) \leq \varepsilon^{-2} \sum_{i=k}^{t} \mathrm{Var}(\log(1-\psi_i)). \quad (5.4.21)$$

Using that $\log(1 - x) \leq x/(1 - x)$ for all $x \in [0, 1]$ and Lemma 5.14, we obtain the bound

$$\mathrm{Var}(\log(1 - \psi_i)) \leq \mathbb{E}[(\log(1 - \psi_i))^2] \leq \mathbb{E}\Big[\frac{\psi_i^2}{(1 - \psi_i)^2}\Big] = O(i^{-2}), \quad (5.4.22)$$

so that, for all $t \geq n$,

$$\mathbb{P}\Big(\sup_{n=k}^{t} \Big| \sum_{i=k}^{n}(\log S_k^{(n)} - \mathbb{E}[\log(1 - \psi_i)]) \Big| \geq \varepsilon \Big) \leq \frac{C}{\varepsilon^2} \sum_{i \geq k} \frac{1}{i^2}, \quad (5.4.23)$$

which can be made small by letting $k \geq K$ and K be large. This shows that the random part in (5.4.20) is whp small for $k \geq K$.

To compute the asymptotics of the deterministic first part in (5.4.20), now using that $x \leq \log(1 - x) \leq x + x^2/(1 - x)$ for all $x \in [0, 1]$,

$$0 \leq \sum_{i=k}^{n}(\mathbb{E}[\log(1 - \psi_i)] - \mathbb{E}[\psi_i]) \leq \sum_{i=k}^{n} \mathbb{E}\Big[\frac{\psi_i^2}{1 - \psi_i}\Big] \leq C \sum_{i \geq k} \frac{1}{i^2}, \quad (5.4.24)$$

which can again be made small when $k \geq K$ with K large. Further, by Lemma 5.14, with $\alpha = m + \delta$ and $\beta_i = (2i - 3)m + \delta(i - 1)$, we have

$$\sum_{i=k}^{n} \mathbb{E}[\psi_i] = \sum_{i=k}^{n} \frac{m + \delta}{(2i - 3)m + \delta(i - 1)} = \frac{m + \delta}{2m + \delta} \log(n/k) + O(1/k) \quad (5.4.25)$$
$$= \chi \log(n/k) + O(1/k),$$

since $\sum_{i=k}^{n} 1/i = \log(n/k) + O(1/k)$. We conclude that

$$\log S_k^{(n)} = \chi \log(n/k) + O(1/k), \quad (5.4.26)$$

which completes the proof of (5.4.19). The proof of (5.4.18) follows easily, since (5.4.19) is stronger for $k \geq K$, while $\mathbb{E}[S_k^{(n)}] = o(1)$ for $k \in [K]$. We omit further details. $\qquad\square$

Density of the Pólya Point Tree

Our proof of Theorem 5.8 proves a stronger result. Indeed, first, it proves *marked* local convergence in probability, where the marks are the vertex types in $[0, 1] \times \{\mathrm{Y}, \mathrm{O}\}$, and we recall Definition 2.10 in Section 2.3.5 for the definition of marked local convergence. Second, we will prove that the probability that the vertex labels in $B_r^{(G_n)}(o)$, rescaled by appropriate powers of $1/n$, are equal to some fixed positive values converges to the *joint density* of the ages in the Pólya point tree. For this, it is useful to have a description of this joint density.

Before we can formulate our main result concerning this joint density, we will introduce some further notation. Recall the definition of the Pólya point tree in Section 5.3.1 and also the Poisson intensities in (5.3.5) and the corresponding Gamma variables in (5.3.2). Below, we write $x \mapsto \rho_w(x; \Gamma_w)$ for the Poisson intensity in (5.3.5) conditioned on the Gamma variable Γ_w.

Fix an ordered tree $\mathbf{t}$, and let (G, o) be the Pólya point tree. In what follows, it is useful to regard $B_r^{(G)}(v)$ as a rooted *edge-marked* graph, where an edge receives a label in $[m]$ corresponding to the label of the directed edge (directed from young to old) that gives rise to that edge (in either possible direction). Thus, in the pre-limiting preferential attachment model, the edge $\{u, v\}$ receives label j when $u \overset{j}{\rightsquigarrow} v$ or when $v \overset{j}{\rightsquigarrow} u$.

We denote this marked ordered neighborhood as $\bar{B}_r^{(G)}(o)$. The edge labels are *almost* contained in the ordered tree $\mathbf{t}$ but not quite, since when a vertex has label $\textsc{y}$ it is unclear which edge of its parent gave rise to this connection, and, together with the $m-1$ edge labels of its older children, these edge labels should equal $[m]$. We will slightly abuse notation, and also write $\mathbf{t}$ for this edge-labeled tree, and we will write $\bar{B}_r^{(G)}(o) = \mathbf{t}$ when the two graphs are the same as edge-labeled trees.

For $(a_w)_{w \in V(\mathbf{t})} \in [0, 1]^{|V(\mathbf{t})|}$, we define $f_{\mathbf{t}}\big((a_w)_{w \in V(\mathbf{t})}\big)$ to be the density of the ages in the Pólya point tree when the ordered r-neighborhood $\bar{B}_r^{(G)}(o)$ in the Pólya point tree equals $\mathbf{t}$. Thus,

$$\mu(\bar{B}_r^{(G)}(o) = \mathbf{t}, A_w \in a_w \mathrm{d}a_w \ \forall w \in V(\mathbf{t})) = f_{\mathbf{t}}\big((a_w)_{w \in V(\mathbf{t})}\big) \prod_{w \in V(\mathbf{t})} \mathrm{d}a_w. \quad (5.4.27)$$

Note that $(a_w)_{w \in V(\mathbf{t})} \mapsto f_{\mathbf{t}}\big((a_w)_{w \in V(\mathbf{t})}\big)$ is a sub-probability measure, as it need not integrate to 1. We let $\bar{\mathbf{t}}$ denote a rooted vertex- and edge-marked tree, where the vertex labels corresponding to the ages of the nodes are in $[0, 1]$ and the edge labels are in $[m]$. Thus,

$$\bar{\mathbf{t}} = \big(\mathbf{t}, (a_w)_{w \in V(\mathbf{t})}\big), \quad (5.4.28)$$

where $a_w \in [0, 1]$ is the age of $w \in V(\mathbf{t})$. The following proposition identifies the density $f_{\mathbf{t}}\big((a_w)_{w \in V(\mathbf{t})}\big)$ in (5.4.27), which corresponds to the density of the ages in the Pólya point tree when the edge-marked neighborhood equals $\mathbf{t}$:

Proposition 5.19 (Joint density of the Pólya point tree) *The density* $f_{\mathbf{t}}\big((a_w)_{w \in V(\mathbf{t})}\big)$ *in* (5.4.27) *satisfies*

$$f_{\mathbf{t}}\big((a_w)_{w \in V(\mathbf{t})}\big) = \mathbb{E}[g_{\mathbf{t}}\big((a_w)_{w \in V(\mathbf{t})}; (\chi_w)_{w \in V(\mathbf{t})}\big)], \quad (5.4.29)$$

where $(\chi_w)_{w \in V(\mathbf{t})}$ *are iid Gamma variables with parameters* $r = m + \delta$ *and* $\lambda = 1$, *and, with* $d_w^{(\mathrm{in})}(\bar{\mathbf{t}}) = \#\{\{v, w\} \in E(\mathbf{t}) \colon a_v > a_w\}$ *the in-degree of* w *in* $\bar{\mathbf{t}}$,

$$g_{\mathbf{t}}\big((a_w)_{w \in V(\mathbf{t})}; (\chi_w)_{w \in V(\mathbf{t})}\big) \quad (5.4.30)$$

$$= \prod_{w \in V(\mathbf{t})} \left(\frac{\chi_w}{2m + \delta}\right)^{d_w^{(\mathrm{in})}(\bar{\mathbf{t}})} \prod_{w \in V^\circ(\mathbf{t})} e^{-\int_{a_w}^1 \rho_w(\mathrm{d}t; \chi_w)} \prod_{(w, w\ell) \in E(\mathbf{t})} \frac{1}{(a_w \wedge a_{w\ell})^{1-\chi}(a_w \vee a_{w\ell})^\chi},$$

where $V^\circ(\mathbf{t})$ denotes the set of vertices in the tree $\mathbf{t}$ that are at a distance strictly smaller than r from the root.

Proof The proof is split into several steps. We start by removing the size-biasing of the Gamma variables in (5.3.2).

Un-Size-Biasing the Gamma Variables

Recall the Gamma variables present in (5.3.2). Denote the conditional density in (5.4.27) given the random variables $(\Gamma_w)_{w\in V(\mathbf{t})}$ by $f_{\mathbf{t}}((a_w)_{w\in V(\mathbf{t})}; (\Gamma_w)_{w\in V(\mathbf{t})})$, so that

$$f_{\mathbf{t}}((a_w)_{w\in V(\mathbf{t})}) = \mathbb{E}[f_{\mathbf{t}}((a_w)_{w\in V(\mathbf{t})}; (\Gamma_w)_{w\in V(\mathbf{t})})]. \tag{5.4.31}$$

Now recall the size-biasing in (5.3.2), present for all individuals of label o. In terms of these random variables, note that, for each function $h\colon \mathbb{R} \mapsto \mathbb{R}$, and using that $\mathbb{E}[Y] = m + \delta$,

$$\mathbb{E}[h(Y^\star)] = \mathbb{E}\Big[\frac{Y}{\mathbb{E}[Y]}h(Y)\Big] = \mathbb{E}\Big[\frac{Y}{m+\delta}h(Y)\Big]. \tag{5.4.32}$$

Thus, with $(\chi_w)_{w\in V(\mathbf{t})}$ a collection of iid Gamma random variables with parameters $m + \delta$ and 1,

$$f_{\mathbf{t}}((a_w)_{w\in V(\mathbf{t})}) = \mathbb{E}\Big[\prod_{w\in V(\mathbf{t})} \Big(\frac{\chi_w}{m+\delta}\Big)^{\mathbb{1}_{\{\text{label}(w)=\text{O}\}}} f_{\mathbf{t}}((a_w)_{w\in V(\mathbf{t})}; (\chi_w)_{w\in V(\mathbf{t})})\Big], \tag{5.4.33}$$

where we write label(w) for the label of w. We claim that $g_{\mathbf{t}}((a_w)_{w\in V(\mathbf{t})}; (\chi_w)_{w\in V(\mathbf{t})})$ in (5.4.30) is given by

$$g_{\mathbf{t}}((a_w)_{w\in V(\mathbf{t})}; (\chi_w)_{w\in V(\mathbf{t})})$$
$$= \prod_{w\in V(\mathbf{t})} \Big(\frac{\chi_w}{m+\delta}\Big)^{\mathbb{1}_{\{\text{label}(w)=\text{O}\}}} f_{\mathbf{t}}((a_w)_{w\in V(\mathbf{t})}; (\chi_w)_{w\in V(\mathbf{t})}), \tag{5.4.34}$$

which then completes the proof of (5.4.29).

To prove (5.4.34), we recall again that the vertices have labels in $\{\text{Y}, \text{O}\}$ indicating their relative age compared with their parent. Let $d_w(\bar{\mathbf{t}})$ be the number of younger children of w in $V(\bar{\mathbf{t}})$, i.e.,

$$d_w(\bar{\mathbf{t}}) = \#\{\ell\colon w\ell \in V(\mathbf{t}), a_{w\ell} > a_w\}. \tag{5.4.35}$$

Then $d_w(\bar{\mathbf{t}}) = d_w^{(\text{in})}(\bar{\mathbf{t}})$ when w has label Y, while $d_w(\bar{\mathbf{t}}) = d_w^{(\text{in})}(\bar{\mathbf{t}}) - 1$ when w has label O. We can then rewrite the first factor on the rhs of (5.4.30) as follows:

$$\prod_{w\in V(\mathbf{t})} \Big(\frac{\chi_w}{2m+\delta}\Big)^{d_w^{(\text{in})}(\bar{\mathbf{t}})} = \prod_{w\in V(\mathbf{t})} \Big(\frac{\chi_w}{2m+\delta}\Big)^{\mathbb{1}_{\{\text{label}(w)=\text{O}\}}} \prod_{w\in V(\mathbf{t})} \Big(\frac{\chi_w}{2m+\delta}\Big)^{d_w(\bar{\mathbf{t}})}. \tag{5.4.36}$$

Thus, to prove (5.4.34), we need to show that

$$
\begin{aligned}
f_{\mathbf{t}}\big((a_w)_{w\in V(\mathbf{t})};(\chi_w)_{w\in V(\mathbf{t})}\big) \\
= \prod_{w\in V(\mathbf{t})}\left(\frac{\chi_w}{2m+\delta}\right)^{d_w(\bar{\mathbf{t}})}\left(\frac{m+\delta}{2m+\delta}\right)^{\mathbb{1}_{\{\text{label}(w)=\mathrm{o}\}}} \\
\times \prod_{w\in V^{\circ}(\mathbf{t})}\mathrm{e}^{-\int_{a_w}^{1}\rho_w(\mathrm{d}t;\chi_w)}\prod_{(w,w\ell)\in E(\mathbf{t})}\frac{1}{(a_w\wedge a_{w\ell})^{1-\chi}(a_w\vee a_{w\ell})^{\chi}}.
\end{aligned}
\tag{5.4.37}
$$

Bringing in the Densities: Older Children

Recall from (5.3.1) that each vertex $w\in V(\mathbf{t})$ has $m_-(w)$ older children of label o. By (5.3.3), the density of the age $A_{w\ell}$ of the child $w\ell$ of w with age a_w is given by

$$
\begin{aligned}
f(a_{w\ell};a_w) &= \chi a_{w\ell}^{\chi-1}a_w^{-\chi} = \frac{m+\delta}{2m+\delta}a_{w\ell}^{-(1-\chi)}a_w^{-\chi} \\
&= \frac{m+\delta}{2m+\delta}\frac{1}{(a_w\wedge a_{w\ell})^{1-\chi}(a_w\vee a_{w\ell})^{\chi}},
\end{aligned}
\tag{5.4.38}
$$

for $a_{w\ell}\in[0,a_w]$. These ages are iid, so that the joint density of the o children of w is

$$
\prod_{\ell=1}^{m_-(w)}\frac{m+\delta}{2m+\delta}\frac{1}{(a_w\wedge a_{w\ell})^{1-\chi}(a_w\vee a_{w\ell})^{\chi}}.
\tag{5.4.39}
$$

The $m+\delta$ factors cancel the inverse power of $m+\delta$ on the rhs of (5.4.34).

Bringing in the Densities: Younger Children and Poisson Processes

Recall the density of the points of an inhomogeneous Poisson processes with intensity Λ on $[0,\infty)$ from (1.5.6) in Section 1.5. This observation implies that, for each $w\in V^{\circ}(\mathbf{t})$, the density of the ages of its younger children having label y equals

$$
\mathrm{e}^{-\int_{a_w}^{1}\rho_w(\mathrm{d}t;\chi_w)}\prod_{\ell:\,a_{w\ell}>a_w}\rho_w(a_{w\ell};\chi_{w\ell}),
\tag{5.4.40}
$$

and $a_{w\ell}>a_w$ for all $\ell>m_-(w)$. Here, we also note that $a_{w\ell}<a_{w(\ell+1)}$ for all w and $\ell>m_-(w)$ such that $w\ell, w(\ell+1)\in V(\mathbf{t})$). By (5.3.5),

$$
\begin{aligned}
\rho_w(x;\chi_w) &= \frac{\chi_w}{\tau-1}\frac{x^{1/(\tau-1)-1}}{a_w^{1/(\tau-1)}} = \frac{\chi_w}{\tau-1}a_{w\ell}^{1/(\tau-1)-1}a_w^{-1/(\tau-1)} \\
&= \chi_w\frac{m}{2m+\delta}a_{w\ell}^{-\chi}a_w^{-(1-\chi)} = \chi_w\frac{m}{2m+\delta}\frac{1}{(a_w\wedge a_{w\ell})^{1-\chi}(a_w\vee a_{w\ell})^{\chi}}.
\end{aligned}
\tag{5.4.41}
$$

Since the number of ℓ-values with $a_{w\ell}>a_w$ in (5.4.40) equals $d_w(\bar{\mathbf{t}})$, this leads to

$$
\mathrm{e}^{-\int_{a_w}^{1}\rho_w(\mathrm{d}t;\chi_w)}\chi_w^{d_w(\bar{\mathbf{t}})}\left(\frac{m}{2m+\delta}\right)^{d_w(\bar{\mathbf{t}})}\prod_{\ell:\,a_{w\ell}>a_w}\frac{1}{(a_w\wedge a_{w\ell})^{1-\chi}(a_w\vee a_{w\ell})^{\chi}}.
\tag{5.4.42}
$$

When we recall that each ℓ-value with $a_{w\ell} > a_w$ is assigned an edge label in $[m]$, which occurs independently with probability $1/m$, the density of the edge-labeled younger children of w is given by

$$
e^{-\int_{a_w}^1 \rho_w(dt;\chi_w)} \chi_w^{d_w(\bar{t})} \left(\frac{1}{2m+\delta}\right)^{d_w(\bar{t})} \prod_{\ell:\, a_{w\ell} > a_w} \frac{1}{(a_w \wedge a_{w\ell})^{1-\chi}(a_w \vee a_{w\ell})^{\chi}}. \quad (5.4.43)
$$

Multiplying Out

We multiply (5.4.39) and (5.4.43) to obtain that the density of the ages of the children of w for each $w \in V(\mathbf{t})$ is given by

$$
\chi_w^{d_w(\bar{t})} e^{-\int_{a_w}^1 \rho_w(dt;\chi_w)} \left(\frac{1}{2m+\delta}\right)^{d_w(\bar{t})} \left(\frac{m+\delta}{2m+\delta}\right)^{m_-(w)} \prod_{\ell:\, w\ell \in V(\mathbf{t})} \frac{1}{(a_w \wedge a_{w\ell})^{1-\chi}(a_w \vee a_{w\ell})^{\chi}}.
$$

The above holds for all $w \in V^\circ(\mathbf{t})$, i.e., all w that are at a distance strictly smaller than r from the root $\varnothing$. We next multiply over all such w, to obtain

$$
f_{\mathbf{t}}\big((a_w)_{w\in V(\mathbf{t})}; (\chi_w)_{w\in V(\mathbf{t})}\big) = \prod_{w\in V^\circ(\mathbf{t})} e^{-\int_{a_w}^1 \rho_w(dt;\chi_w)} \left(\frac{\chi_w}{2m+\delta}\right)^{d_w(\bar{t})} \left(\frac{m+\delta}{2m+\delta}\right)^{m_-(w)}
$$

$$
\times \prod_{(w,w\ell)\in E(\mathbf{t})} \frac{1}{(a_w \wedge a_{w\ell})^{1-\chi}(a_w \vee a_{w\ell})^{\chi}}. \quad (5.4.44)
$$

Since

$$
\sum_{w\in V^\circ(\mathbf{t})} m_-(w) = \sum_{w\in V(\mathbf{t})} \mathbb{1}_{\{\text{label}(w)=o\}}, \quad (5.4.45)
$$

this is indeed the same as the rhs of (5.4.37). This proves (5.4.34), and thus (5.4.29). $\qquad \square$

Regularity Properties of the Pólya Point Tree

We next discuss some properties of the Pólya point tree, showing that all random variables and intensities used in within the r-neighborhood of the root satisfy uniform bounds whp (recall (5.3.5)):

Lemma 5.20 (Regularity of Pólya point tree) *Consider the Pólya point tree $(G, \varnothing)$. Fix $r \geq 1$ and $\varepsilon > 0$. Then there exist constants $\eta > 0$ and $K < \infty$ such that, with probability at least $1 - \varepsilon$,*

(a) $B_r^{(G)}(\varnothing) \leq K$;
(b) $A_w \geq \eta$ for all $w \in B_r^{(G)}(\varnothing)$;
(c) $\Gamma_w \leq K$ and $\rho_w(\cdot) \leq K$ for all $w \in B_r^{(G)}(\varnothing)$;
(d) $\min_{w,w'\in B_r^{(G)}(\varnothing)} |A_w - A_{w'}| \geq \eta$.

Proof The proof of this lemma is standard, and can be obtained, for example, by induction on r or by using Proposition 5.19. The last bound follows from the continuous nature of the random variables A_w, which implies that $A_w \neq A_{w'}$ for all distinct pairs w, w', so that any finite number are pairwise separated by at least η for an appropriate $\eta = \eta(\varepsilon)$ with probability at least $1 - \varepsilon$. $\qquad \square$

5.4.2 LOCAL CONVERGENCE: FIRST MOMENT

As in the proof of Proposition 5.19, it will be useful to regard $B_r^{(G_n)}(v)$ as a rooted *edge-marked* graph, where an edge receives a label in $[m]$ corresponding to the label of the directed edge that gives rise to it (in either possible direction). Thus, the edge $\{u, v\}$ receives label j when $u \xrightarrow{j} v$ or when $v \xrightarrow{j} u$. We denote this marked ordered neighborhood as $\bar{B}_r^{(G_n)}(v)$. Let

$$N_{n,r}(\mathbf{t}) = \sum_{v \in [n]} \mathbb{1}_{\{\bar{B}_r^{(G_n)}(v) = \mathbf{t}\}}. \tag{5.4.46}$$

With $B_r^{(G)}(\varnothing)$ the r-neighborhood of $\varnothing$ in the Pólya point tree (which in itself is also ordered), we aim to show that

$$\frac{N_{n,r}(\mathbf{t})}{n} \xrightarrow{\mathbb{P}} \mu(B_r^{(G)}(\varnothing) = \mathbf{t}), \tag{5.4.47}$$

where, again, $B_r^{(G)}(\varnothing) = \mathbf{t}$ denotes that the ordered trees $B_r^{(G)}(\varnothing)$ and $\mathbf{t}$ agree, and μ denotes the law of the Pólya point tree.

Proving the convergence of $N_{n,r}(\mathbf{t})$ is much harder than for inhomogeneous random graphs and configuration models, considered in Theorems 3.14 and 4.1, respectively, as the type of a vertex is crucial in determining the number and types of its children, and the type space is *continuous*.

We start with the first moment, for which we note that

$$\mathbb{E}[N_{n,r}(\mathbf{t})/n] = \mathbb{P}(\bar{B}_r^{(G_n)}(o_n) = \mathbf{t}), \tag{5.4.48}$$

where $o_n \in [n]$ is a uniform vertex. We do this using an *explicit* computation, alike the one used in the direct proof of Theorem 5.10. In fact, we will prove a stronger statement, in which we also study the vertex labels in $\bar{B}_r^{(G_n)}(o_n)$ and compare it with the density in Proposition 5.19. Let us introduce some necessary notation.

Recall the definition of the rooted vertex- and edge-marked tree $\bar{\mathbf{t}}$ in (5.4.28), where the vertex labels were in $[0, 1]$ and the edge labels in $[m]$. We fix a tree $\mathbf{t}$ of height exactly r. We let the vertex $v_w = \lceil na_w \rceil \in [n]$ correspond to the node in the tree having age a_w. With a slight abuse of notation, we also write $\bar{B}_r^{(G_n)}(o_n) = \bar{\mathbf{t}}$ to denote that the vertices, edges, and edge labels in $\bar{B}_r^{(G_n)}(o_n)$ are given by those in $\bar{\mathbf{t}}$. Note that this is rather different from $B_r^{(G_n)}(o_n) \simeq \mathbf{t}$ as defined in Definition 2.3, where $\mathbf{t}$ was unlabeled and we were investigating whether $B_r^{(G_n)}(o_n)$ and $\mathbf{t}$ are isomorphic, and even different from $\bar{B}_r^{(G_n)}(o_n) = \mathbf{t}$ as in (5.4.46), where only the *edges* receive marks, and not the vertices. The definition of $\bar{B}_r^{(G_n)}(o) = \bar{\mathbf{t}}$ used here is tailor-made to study the local convergence of $\mathrm{PA}_n^{(m,\delta)}(d)$ as a *marked* graph, where the vertex marks denote the vertex labels or ages of the vertices and also the edges receive marks in $[m]$.

Let $\bar{\mathbf{t}} = (\mathbf{t}, (v_w)_{w \in V(\mathbf{t})})$ be the vertex-marked version of $\mathbf{t}$, now with $v_w = \lceil na_w \rceil \in [n]$ denoting the vertex label of the tree node w in $\mathbf{t}$ (instead of the age as in (5.4.28)). Below, we write $v \in V(\bar{\mathbf{t}})$ to indicate that there exists a $w \in V(\mathbf{t})$ with $v_w = v$. Also, let $\partial V(\bar{\mathbf{t}})$ denote the vertices at distance exactly r from the root of $\bar{\mathbf{t}}$, and let $V^\circ(\bar{\mathbf{t}}) = V(\bar{\mathbf{t}}) \setminus \partial V(\bar{\mathbf{t}})$ denote the restriction of $\bar{\mathbf{t}}$ to all vertices at distance at most $r - 1$ from its root. The main result of this section is the following theorem:

Theorem 5.21 (Marked local weak convergence) *Fix $m \geq 1$ and $\delta > -m$, and consider $G_n = \mathrm{PA}_n^{(m,\delta)}(d)$. Uniformly for $v_w \geq \varepsilon n$ and $\hat{\chi}_{v_w} \leq K$ for all $w \in V(t)$, where $\hat{\chi}_v = f_v(\psi_v)$ and when $(v_w)_{w \in V(t)}$ are all distinct,*

$$\mathbb{P}\big(\bar{B}_r^{(G_n)}(o_n) = \bar{t} \mid (\psi_{v_w})_{w \in V(t)}\big)$$
$$= (1 + o_{\mathbb{P}}(1))\frac{1}{n^{|V(t)|}}g_t\big((v_w/n)_{w \in V(t)}; (\hat{\chi}_{v_w})_{w \in V(t)}\big). \tag{5.4.49}$$

Consequently, if $(\chi_w)_{w \in V(t)}$ is an iid sequence of Gamma variables with parameters $2m + \delta$ and 1,

$$\mathbb{E}[g_t((a_w)_{w \in V(t)}; (\chi_w)_{w \in V(t)})] = f_t((a_w)_{w \in V(t)}), \tag{5.4.50}$$

and thus $\mathrm{PA}_n^{(m,\delta)}(d)$ converges to the Pólya point tree in the marked local weak sense.

By Lemma 5.17, $(\hat{\chi}_v)_{v \in [n]}$ are iid Gamma variables with parameters $m + \delta$ and 1. The coordinates of the sequence $(\chi_w)_{w \in V(t)}$ defined by

$$\chi_w = \hat{\chi}_{v_w} \tag{5.4.51}$$

are indeed iid when the $(v_w)_{w \in V(t)}$ are distinct. By (5.4.50), the relation (5.4.49) can be seen as a *density theorem* for the densities of the ages of vertices in r-neighborhoods.

Note that the type space of the Pólya point tree equals $\mathcal{S} = \{Y, O\} \times [0, \infty)$ (except for the root, which has a type only in $[0, 1]$). However, the $\{Y, O\}$-components of the types are *deterministic* when one knows the ages in $\bar{B}_r^{(G_n)}(o_n)$, so these do not need to receive much attention in what follows.

We prove Theorem 5.21 below. The main ingredient to the proof is Proposition 5.22, which gives an explicit description for the lhs of (5.4.49):

Proposition 5.22 (Law of vertex- and edge-marked neighborhoods in $\mathrm{PA}_n^{(m,\delta)}(d)$) *Fix $m \geq 2$ and $\delta > -m$, and consider $G_n = \mathrm{PA}_n^{(m,\delta)}(d)$. Let $\bar{t} = (t, (v_w)_{w \in V(t)})$ be a rooted vertex- and edge-marked tree with root o_n. Fix $\bar{t}$ such that $(v_w)_{w \in V(t)}$ are all distinct, with the oldest vertex having age at least εn. Then, for all $(\psi_v)_{v \in V(\bar{t})}$ such that $\psi_v \leq K/v$ for all $v \in V(\bar{t})$, as $n \to \infty$,*

$$\mathbb{P}\Big(\bar{B}_r^{(G_n)}(o_n) = \bar{t} \mid (\psi_v)_{v \in V(\bar{t})}\Big)$$
$$= \frac{1 + o_{\mathbb{P}}(1)}{n} \prod_{v \in V(\bar{t})} \psi_v^{p'_v} \prod_{v \in V^\circ(\bar{t})} \exp\Big(-(2m+\delta)n\psi_v(v/n)^\chi(1 - (v/n)^{1-\chi})\Big)$$
$$\times \prod_{s \in [n] \setminus V(\bar{t})} \frac{(\beta_s + q'_s - 1)_{q'_s}}{(\alpha + \beta_s + q'_s - 1)_{q'_s}}, \tag{5.4.52}$$

where, if $u \sim s$ denotes that u and s are neighbors in $\bar{t}$,

$$p'_s = \mathbb{1}_{\{s \in V(\bar{t})\}} \sum_{u \in V(\bar{t})} \mathbb{1}_{\{u \sim s, u > s\}}, \tag{5.4.53}$$

$$q'_s = \sum_{u \in V(\bar{t})} \sum_{j \in [m]} \mathbb{1}_{\{s \in (v_j(u), u)\}}. \tag{5.4.54}$$

Proof We start by analyzing the conditional law of $\bar{B}_r^{(G_n)}(o_n)$ given all $(\psi_v)_{v\in[n]}$. After this, we take the expectation wrt ψ_v for $v \notin \bar{B}_r^{(G_n)}(o_n)$ to get the claim.

Computing the Conditional Law of $\bar{B}_r^{(G_n)}(o_n)$ Given $(\psi_v)_{v\in[n]}$

We start by introducing some useful notation. Recall (5.3.29). Define, for $u > v$, the edge probability

$$P_{u,v} = \psi_v \prod_{s\in(v,u)} (1 - \psi_s). \tag{5.4.55}$$

We first condition on *all* $(\psi_v)_{v\in[n]}$ and use Lemma 5.12 to obtain, for a vertex-marked edge-labeled tree $\bar{t}$,

$$\mathbb{P}_n(\bar{B}_r^{(G_n)}(o_n) = \bar{t}) = \frac{1}{n} \prod_{v\in V(\bar{t})} \psi_v^{p_v'} \prod_{s=2}^{n} (1 - \psi_s)^{q_s'}$$

$$\times \prod_{v\in V^\circ(\bar{t})} \prod_{u,j:\, u\xrightarrow{j}v} [1 - P_{u,v}], \tag{5.4.56}$$

where the $1/n$ is due to the uniform choice of the root, the first double product is due to all the required edges to ensure that $\bar{B}_r^{(G_n)}(o_n) \subseteq \bar{t}$, while the second double product is due to all the other edges, which must be absent, so that $\bar{B}_r^{(G_n)}(o_n)$ really equals $\bar{t}$.

No-Further-Edge Probability

We continue by analyzing the second line in (5.4.56), which, for clarity, we call the *no-further-edge probability*. First of all, since we are exploring the r-neighborhood of o, the only edges that are not allowed are of the form $u \xrightarrow{j} v$, where $v \in V^\circ(\bar{t})$ and $u > v$, i.e., they are younger vertices than those in $V^\circ(\bar{t})$ that do not form edges in $\bar{t}$.

Recall that the minimal age of a vertex in $\bar{t}$ is εn. Further, by Lemma 5.17, with overwhelming probability, $\psi_v \leq (\log n)^2/n$ for all $v \geq \varepsilon n$. In particular, $P_{u,v}$ is small uniformly in $v \in V(\bar{t})$ and $u > v$. Since there are only finitely many elements in $V^\circ(\bar{t})$, we can thus approximate as follows:

$$\prod_{v\in V^\circ(\bar{t})} \prod_{u,j:\, u\xrightarrow{j}v} [1 - P_{u,v}] = (1 + o_{\mathbb{P}}(1)) \prod_{v\in V^\circ(\bar{t})} \prod_{u,j:\, u>v} [1 - P_{u,v}]. \tag{5.4.57}$$

We can make the further approximation

$$\prod_{u,j:\, u>v} [1 - P_{u,v}] = e^{\Theta(1)\sum_{u,j:\, u\in(v,n]} P_{u,v}^2} \exp\left(-\sum_{u,j:\, u>v} P_{u,v}\right). \tag{5.4.58}$$

Then we can compute, using (5.3.29) and (5.3.20),

$$\sum_{u,j:\, u\in(v,n]} P_{u,v} = m\psi_v \sum_{u\in(v,n]} \prod_{s\in(v,u)} (1 - \psi_s)$$

$$= m\psi_v \sum_{u\in(v,n]} \frac{S_v^{(n)}}{S_u^{(n)}}. \tag{5.4.59}$$

We will take $v = \lceil sn \rceil$ for some $s \in [\varepsilon, 1]$. By Lemma 5.18, $\sup_{s \in [\varepsilon, 1]} |S_{ns}^{(n)} - s^\chi| \xrightarrow{\mathbb{P}} 0$. Thus, we also have

$$\sup_{s \in [\varepsilon, 1]} \left| \sum_{u \in (sn, n]} 1/S_u^{(n)} - \int_s^1 t^{-\chi} dt \right| \xrightarrow{\mathbb{P}} 0. \tag{5.4.60}$$

We conclude that

$$\frac{m}{n \psi_{sn} S_{sn}^{(n)}} \sum_{u \in (sn, n]} P_{u,sn} \xrightarrow{\mathbb{P}} m \int_s^1 t^\chi dt = \frac{m}{1 - \chi} [1 - s^{1-\chi}]$$

$$= (2m + \delta)[1 - s^{1-\chi}]. \tag{5.4.61}$$

As a result,

$$m \sum_{u \in (sn, n]} P_{u,v} = (1 + o_{\mathbb{P}}(1)) n \psi_{sn} S_{sn}^{(n)} (2m + \delta)[1 - s^{1-\chi}]$$

$$= (1 + o_{\mathbb{P}}(1))(sn \psi_{sn}) s^{\chi-1} (2m + \delta)[1 - s^{1-\chi}]$$

$$= (1 + o_{\mathbb{P}}(1))(2m + \delta)(sn \psi_{sn}) \kappa(v/n), \tag{5.4.62}$$

where we recall that $\kappa(u) = [1 - u^{1/(\tau-1)}]/u^{1/(\tau-1)}$ from (5.3.9).

Further, by Lemma 5.17, $(v \psi_v)_{v \in V^\circ(\mathbf{t})}$ converges in distribution to a sequence of independent Gamma random variables, and thus, for all $v \in V^\circ(\mathbf{t})$,

$$\sum_{u,j:\, u \in (v, n]} P_{u,v}^2 \xrightarrow{\mathbb{P}} 0. \tag{5.4.63}$$

Therefore,

$$\prod_{v \in V(\mathbf{t})} \prod_{u,j:\, u \not\to v}^j [1 - P_{u,v}] = (1 + o_{\mathbb{P}}(1)) \prod_{v \in V^\circ(\mathbf{t})} e^{-(2m+\delta)(v\psi_v)\kappa(v/n)}. \tag{5.4.64}$$

Conclusion of the Proof

We next use Lemma 5.14 to take the expectation wrt ψ_s for all $s \notin V(\mathbf{t})$, to obtain

$$\mathbb{P}\Big(\bar{B}_r(o_n) = \bar{\mathbf{t}} \,\Big|\, (\psi_j)_{j \in V(\mathbf{t})} \Big)$$

$$= \frac{1 + o_{\mathbb{P}}(1)}{n} \prod_{v \in V(\mathbf{t})} \psi_v^{p_v'} \prod_{v \in V^\circ(\mathbf{t})} e^{-(2m+\delta)(v\psi_v)\kappa(v/n)} (1 - \psi_v)^{q_v'}$$

$$\times \prod_{s \in [n] \setminus V(\mathbf{t})} \frac{(\alpha + p_s' - 1)_{p_s'} (\beta_s + q_s' - 1)_{q_s'}}{(\alpha + \beta_s + p_s' + q_s' - 1)_{p_s' + q_s'}}. \tag{5.4.65}$$

Since q_v' is uniformly bounded, we also have $(1 - \psi_v)^{q_v'} \xrightarrow{\mathbb{P}} 1$ for all $v \in V(\mathbf{t})$. Finally, note that $p_s' = 0$ for all $s \in [n] \setminus V(\mathbf{t})$. $\qquad\square$

We are now ready to complete the proof of Theorem 5.21:

Proof of Theorem 5.21. We collect the different factors on the rhs of (5.4.52) in Proposition 5.22, and compare the result with that in Proposition 5.19.

Coupling of Beta and Gamma Variables

First, we use the construction in Lemma 5.17 and take $(\hat{\chi}_k)_{k \geq 1}$ as the sequence of iid Gamma $r = m + \delta$ and $\lambda = 1$ variables. Let $\psi_k = f_k(\hat{\chi}_k)$, so that $(\psi_k)_{k \geq 2}$ are independent $\text{Beta}(m + \delta, (2k - 3)m + \delta(k - 1))$ variables, as required (recall (5.3.19)). Together with (5.4.51), this provides the coupling between the Gamma and Beta variables.

In order to relate this to the Gamma variables in the description of the Pólya point tree (recall (5.3.2)–(5.3.5)), we need to look into the precise structure of the rhs of (5.4.52), as some of its ingredients give rise to the size-biasing in (5.3.2). In the proof below, we restrict to the $(\hat{\chi}_k)_{k \geq 1}$ for which $\hat{\chi}_v \leq K$ for all $v \in V(\bar{t})$, which occurs whp for K large by Lemma 5.20 and since $v > \varepsilon n$.

Product of Beta Variables

We next analyze the first term on the rhs of (5.4.52), under the above coupling. Note that $p_v' = d_{w_v}^{(\text{in})}(t)$ for $v \in V(\bar{t})$, where $d_w^{(\text{in})}(t)$ is the in-degree of w in $\bar{t}$, i.e., the number of younger neighbors of w in $V(\bar{t})$, and w_v is the vertex in $V(t)$ corresponding to $v \in V(\bar{t})$. Under the above coupling, and the fact that $v \geq \varepsilon n$ for all $v \in V(\bar{t})$, we first observe that

$$\psi_v = (1 + o_{\mathbb{P}}(1)) \frac{\hat{\chi}_v}{v(2m + \delta)}. \tag{5.4.66}$$

Thus,

$$\prod_{v \in V(\bar{t})} \psi_v^{p_v'} = (1 + o_{\mathbb{P}}(1)) \prod_{v \in V(\bar{t})} \left(\frac{\hat{\chi}_v}{v(2m + \delta)} \right)^{d_{w_v}^{(\text{in})}(t)}, \tag{5.4.67}$$

where the error term is *uniform* on the event that $\hat{\chi}_v \leq K$ for all $v \in V(\bar{t})$.

Limit of the No-Further-Edge Probability

We continue by analyzing the second term on the rhs of (5.4.52), and note that, under the coupling of $(\psi_k)_{k \geq 2}$ and $(\hat{\chi}_k)_{k \geq 2}$, for $v = an$, we have

$$(2m + \delta)(an\psi_{\lceil an \rceil})\kappa(a) = (1 + o_{\mathbb{P}}(1)) \int_a^1 \rho_a(dt; \hat{\chi}_{an}), \tag{5.4.68}$$

where we recall ρ_w from (5.3.5), and its conditional form given Γ_w denoted by $\rho_w(x; \Gamma_w)$. For $w \in V(t)$, let v_w be such that $a_w n = v_w$ and write $\chi_w = \hat{\chi}_{v_w}$. Using (5.4.66), this leads to

$$\prod_{v \in V^{\circ}(\bar{t})} e^{-(2m+\delta)(v\psi_v)\kappa(v/n)} = (1 + o_{\mathbb{P}}(1)) \prod_{v \in V^{\circ}(\bar{t})} e^{-\int_{v/n}^1 \rho_{w_v}(dt; \hat{\chi}_{v_w})}$$

$$= (1 + o_{\mathbb{P}}(1)) \prod_{w \in V^{\circ}(t)} e^{-\int_{a_w}^1 \rho_w(dt; \chi_w)}. \tag{5.4.69}$$

Product of Falling Factorials

We now analyze the product of falling factorials on the second line of the rhs of (5.4.52). Note that $q_s' = 0$ for all $s \in [\varepsilon n]$ by (5.4.54), while $q_s' \leq |V(\bar{t})|$ is uniformly bounded.

Therefore,

$$\frac{(\beta_s + q'_s - 1)_{q'_s}}{(\alpha + \beta_s + q'_s - 1)_{q'_s}} = (1 + \Theta(\beta_s^{-2}))(1 - \frac{\alpha}{\beta_s})^{q'_s}$$

$$= (1 + \Theta(\beta_s^{-2}))e^{-\alpha q'_s/\beta_s}. \tag{5.4.70}$$

Thus, since $q'_s = 0$ for all $s \in [\varepsilon n]$, we also have

$$\prod_{s \in [n] \backslash V(\bar{t})} \frac{(\beta_s + q'_s - 1)_{q'_s}}{(\alpha + \beta_s + q'_s - 1)_{q'_s}} = (1 + o(1))e^{-\alpha \sum_{s \in [n] \backslash V(\bar{t})} q'_s/\beta_s},$$

where the error term is uniformly bounded. We recall that the edges in our edge-marked tree $\bar{t}$ are given by $\vec{E}(\bar{t}) = \{(u, v_j(u), j) \colon u \in [n], j \in [m]\}$. We use (5.4.54) and $\beta_s = (2s - 3)m + \delta(s - 1)$ to write

$$\sum_{s \in [n] \backslash V(\bar{t})} q'_s/\beta_s = \sum_{(u, v_j(u), j) \in \vec{E}(\bar{t})} \sum_{s \in [n] \backslash V(\bar{t})} \frac{\mathbb{1}_{\{s \in (v, u)\}}}{\beta_s}$$

$$= \sum_{(u, v_j(u), j) \in \vec{E}(\bar{t})} \sum_{s \in [n] \backslash V(\bar{t})} \frac{\mathbb{1}_{\{s \in (v, u)\}}}{(2s - 3)m + \delta(s - 1)}. \tag{5.4.71}$$

Since $v \geq \varepsilon n$ for all $v \in V(\bar{t})$,

$$\sum_{s \in [n] \backslash V(\bar{t})} \frac{\mathbb{1}_{\{s \in (v, u)\}}}{(2s - 3)m + \delta(s - 1)} = O(1/v) + \sum_{s \in [n]} \frac{\mathbb{1}_{\{s \in (v, u)\}}}{(2m + \delta)s}$$

$$= O(1/v) + \frac{\log(u/v)}{2m + \delta}. \tag{5.4.72}$$

As a result, since $\alpha = m + \delta$ and $\chi = (m + \delta)/(2m + \delta)$,

$$\prod_{s \in [n] \backslash V(\bar{t})} \frac{(\beta_s + q'_s - 1)_{q'_s}}{(\alpha + \beta_s + q'_s - 1)_{q'_s}} = (1 + o(1)) \prod_{(u, v_j(u), j) \in \vec{E}(\bar{t})} e^{-\chi \log(u/v)}$$

$$= (1 + o(1)) \prod_{(u, v_j(u), j) \in \vec{E}(\bar{t})} \left(\frac{v}{u}\right)^{\chi}. \tag{5.4.73}$$

Note that $(u, v_j(u), j) \in \vec{E}(\bar{t})$ when there exists $w \in V(t)$ and ℓ such that $(w, w\ell) \in E(t)$, so that

$$\prod_{(u, v_j(u), j) \in \vec{E}(\bar{t})} \left(\frac{v}{u}\right)^{\chi} = \prod_{(w, w\ell) \in E(t)} \left(\frac{v_w \wedge v_{w\ell}}{v_w \vee v_{w\ell}}\right)^{\chi}. \tag{5.4.74}$$

Collecting Terms and Proof of (5.4.50)

We combine (5.4.67), and (5.4.73) and (5.4.74), to arrive at

$$\prod_{v \in V(\bar{t})} \psi_v^{p_v'} \prod_{s \in [n] \setminus V(\bar{t})} \frac{(\beta_s + q_s' - 1)_{q_s'}}{(\alpha + \beta_s + q_s' - 1)_{q_s'}} \tag{5.4.75}$$

$$= (1 + o_{\mathbb{P}}(1)) \prod_{v \in V(\bar{t})} \left(\frac{\hat{\chi}_v}{2m + \delta} \right)^{d_{w_v}^{(in)}(\bar{t})} \prod_{(w, w\ell) \in E(t)} \frac{1}{(v_w \wedge v_{w\ell})^{1-\chi} (v_w \vee v_{w\ell})^{\chi}}.$$

Combining this further with (5.4.69) and using (5.4.30) in Proposition 5.19, we obtain (5.4.49) with $(\chi_w)_{w \in V(t)} = (\hat{\chi}_{\lceil na_w \rceil})_{w \in V(t)}$, which is indeed an iid sequence of Gamma$(m + \delta, 1)$ random variables when the $(\lceil na_w \rceil)_{w \in V(t)}$ are distinct, and we recall (5.4.30).

Conclusion of the Proof of Theorem 5.21

Finally, we prove that $PA_n^{(m,\delta)}(d)$ converges to the Pólya point tree in the marked local weak convergence sense. For this, we note that the rhs of (5.4.49) has full measure when integrating over $(a_w)_{w \in V(t)}$ followed by summing over the ordered Ulam–Harris trees t. Further, the Pólya point tree is regular, as proved in Lemma 5.20. Thus, whp $\bar{B}_r^{(G_n)}(o_n)$ also has the same regularity properties, so that the assumptions that $\min V(\bar{t}) > \varepsilon n$ and $\chi_v \leq K$ hold whp for $\varepsilon > 0$ small and K large.

Recall the definition of marked local weak convergence from Definition 2.10. Recall that the type space is $[0, \infty) \times \{Y, O\}$. We put the Euclidean distance on $[0, 1]$ for the ages of the vertices, and the discrete distance on $\{Y, O\}$, meaning that $d(s, t) = \mathbb{1}_{\{s \neq t\}}$ for $s, t \in \{Y, O\}$. Then, for fixed t, (5.4.49) can be summed over all v_w such that $|v_w/n - a_w| \leq 1/r$ with $v_w \in [n]$ equal to the vertex that corresponds to $w \in V(t)$, to obtain that

$$\mathbb{P}(\bar{B}_r^{(G_n)}(o_n) = t, |v_w/n - a_w| \leq 1/r \; \forall w \in V(t))$$
$$\to \mu(\bar{B}_r^{(G)}(o) = t, |A_w - a_w| \leq 1/r \; \forall w \in V(t)). \tag{5.4.76}$$

Denote

$$N_{n,r}(t, (a_w)_{w \in V(t)}) = \#\{v : \bar{B}_r^{(G_n)}(v) = t, |v_w - a_w| \leq 1/r \; \forall w \in V(t)\}. \tag{5.4.77}$$

Then, (5.4.76) shows that

$$\frac{1}{n} \mathbb{E}\left[N_{n,r}(t, (a_w)_{w \in V(t)}) \right] \to \mu(B_r^{(G)}(o) = t, |A_w - a_w| \leq 1/r \; \forall w \in V(t)). \tag{5.4.78}$$

In turn, this shows that the claimed marked local weak convergence holds. $\quad\square$

5.4.3 LOCAL CONVERGENCE: SECOND MOMENT

In this subsection we complete the proof of the marked local convergence in probability. Below, we say that the two rooted vertex- and edge-marked trees $\bar{t}_1$ and $\bar{t}_2$ have distinct and disjoint vertex sets when (a) the vertices in $(v_w)_{w \in V(\bar{t}_1)}$, as well as those in $(v_w)_{w \in V(\bar{t}_2)}$, are distinct; and (b) there is no vertex appearing in $(v_w)_{w \in V(\bar{t}_1)}$ as well as in $(v_w)_{w \in V(\bar{t}_2)}$. The main result is the following theorem:

Theorem 5.23 (Marked local convergence) *Fix $m \geq 1$ and $\delta > -m$, and consider* $G_n = \mathrm{PA}_n^{(m,\delta)}(d)$. *Let $\bar{t}_1$ and $\bar{t}_2$ be two rooted vertex- and edge-marked trees with distinct and disjoint vertex sets and with root vertices o_1 and o_2, respectively. Uniformly for $v_w > \varepsilon n$ and $\chi_{v_w} \leq K$ for all $w \in V(t_1) \cup V(t_2)$, with $\hat{\chi}_v = f_v(\psi_v)$,*

$$\mathbb{P}\big(\bar{B}_r^{(G_n)}(o_1) = \bar{t}_1, \bar{B}_r^{(G_n)}(o_2) = \bar{t}_2 \mid (\psi_{v_w})_{w \in V(t_1) \cup V(t_2)}\big) \tag{5.4.79}$$

$$= \frac{1 + o_{\mathbb{P}}(1)}{n^{|V(t)|}} g_{t_1}\big((v_w/n)_{w \in V(t_1)}; (\hat{\chi}_{v_w})_{w \in V(t_1)}\big) g_{t_2}\big((v_w/n)_{w \in V(t_2)}; (\hat{\chi}_{v_w})_{w \in V(t_2)}\big).$$

Consequently, $\mathrm{PA}_n^{(m,\delta)}(d)$ converges locally in probability in the marked sense to the Pólya point tree.

Theorem 5.23 proves a result that is stronger than the local convergence in probability claimed in Theorem 5.8, in several ways. Foremost, Theorem 5.23 proves *marked* local convergence, so that also the *ages* of the vertices in the r-neighborhood converge to those in the Pólya point tree. Further, Theorem 5.23 establishes a *local density* limit theorem for the vertex marks.

The proof of Theorem 5.23 follows that of Theorem 5.21, so we can be more succinct. We have the following characterization of the conditional law of the vertex- and edge-marked versions of $\bar{B}_r^{(G_n)}(o_1)$ and $\bar{B}_r^{(G_n)}(o_2)$, where $G_n = \mathrm{PA}_n^{(m,\delta)}(d)$, which is a generalization of Proposition 5.22 to two neighborhoods:

Proposition 5.24 (Law of neighborhoods in $\mathrm{PA}_n^{(m,\delta)}(d)$) *Fix $m \geq 1$ and $\delta > -m$, and consider $G_n = \mathrm{PA}_n^{(m,\delta)}(d)$. Let $\bar{t}_1$ and $\bar{t}_2$ be two rooted vertex- and edge-marked trees with distinct and disjoint vertex sets and root vertices o_1 and o_2, respectively. Uniformly for $v_w > \varepsilon n$ and $\chi_{v_w} \leq K$ for all $w \in V(t_1) \cup V(t_2)$, where $\chi_v = f_v(\psi_v)$, as $n \to \infty$,*

$$\mathbb{P}\Big(\bar{B}_r^{(G_n)}(o_1) = \bar{t}_1, \bar{B}_r^{(G_n)}(o_2) = \bar{t}_2 \mid (\psi_v)_{v \in V(\bar{t}_1) \cup V(\bar{t}_2)}\Big)$$

$$= (1 + o_{\mathbb{P}}(1)) \prod_{v \in V(\bar{t}_1) \cup V(\bar{t}_2)} \psi_v^{p'_v} \prod_{v \in V^\circ(\bar{t}_1) \cup V^\circ(\bar{t}_2)} e^{-(2m+\delta)(v\psi_v)\kappa(v/n)}$$

$$\times \prod_{s \in [n] \setminus (V(\bar{t}_1) \cup V(\bar{t}_2))} \frac{(\beta_s + q'_s - 1)_{q'_s}}{(\alpha + \beta_s + q'_s - 1)_{q'_s}}, \tag{5.4.80}$$

where now

$$p'_s = \mathbb{1}_{\{s \in V(\bar{t}_1) \cup V(\bar{t}_2)\}} \sum_{u \in V(\bar{t}_1) \cup V(\bar{t}_2)} \mathbb{1}_{\{u \sim s, u > s\}}, \tag{5.4.81}$$

$$q'_s = \sum_{u,v \in V(\bar{t}_1) \cup V(\bar{t}_2)} \mathbb{1}_{\{u \xrightarrow{j_u} v\}} \mathbb{1}_{\{s \in (v,u)\}}. \tag{5.4.82}$$

Proof We first condition on *all* $(\psi_v)_{v \in [n]}$ and use Lemma 5.12 to obtain, for two trees $\bar{t}_1$ and $\bar{t}_2$, as in (5.4.56),

$$\mathbb{P}_n(\bar{B}_r(o_1) = \bar{t}_1, \bar{B}_r(o_2) = \bar{t}_2) = \frac{1}{n^2} \prod_{v \in V(\bar{t}_1) \cup V(\bar{t}_2)} \psi_v^{p'_v} \prod_{s=2}^{n} (1 - \psi_s)^{q'_s}$$

$$\times \prod_{v \in V^\circ(\bar{t}_1) \cup V^\circ(\bar{t}_2)} \prod_{\substack{u,j:\ u \overset{j}{\not\sim} v}} [1 - P_{u,v}], \qquad (5.4.83)$$

where the factor $1/n^2$ is due to the uniform choices of the vertices o_1, o_2, the first double product is due to all the edges required to ensure that $\bar{B}_r^{(G_n)}(o_i) \subseteq \bar{t}_i$ for $i \in \{1,2\}$, while the second double product is due to the edges that are not allowed to be there, so that $\bar{B}_r^{(G_n)}(o_i)$ really equals $\bar{t}_i$ for $i \in \{1,2\}$. The remainder of the proof follows the steps in the proof of Proposition 5.22, and is omitted. $\qquad \square$

We continue by proving Theorem 5.23. This proof follows that of Theorem 5.21, using Proposition 5.24 instead of Proposition 5.22. The major difference is that in Proposition 5.24, we assume that $V(\bar{t}_1)$ and $V(\bar{t}_2)$ are *disjoint*, which we prove to be the case whp next.

Disjoint Neighborhoods

We note that, as in Corollary 2.20,

$$\mathbb{P}(\bar{B}_r^{(G_n)}(o_1) \cap \bar{B}_r^{(G_n)}(o_2) \neq \varnothing) = 1 - \mathbb{E}\left[|B_{2r}^{(G_n)}(o_1)|/n\right] = 1 - o(1),$$

by dominated convergence and since $|B_{2r}^{(G_n)}(o_1)|$ is a tight sequence of random variables by Theorem 5.21. Thus, we may restrict our analysis to $\bar{t}_1$ and $\bar{t}_2$ for which $V(\bar{t}_1)$ and $V(\bar{t}_2)$ are disjoint, which we assume from this point onwards.

Completion of the Proof of Theorem 5.23

Recall $N_{n,r}(t, (a_w)_{w \in V(t)})$ from (5.4.77). Note that, for disjoint $V(\bar{t}_1)$ and $V(\bar{t}_2)$, the two factors on the rhs of (5.4.79) are independent. By taking the expectation in (5.4.79) and using (5.4.50), we obtain

$$\mathbb{P}(\bar{B}_r^{(G_n)}(o_1) = \bar{t}_1, \bar{B}_r^{(G_n)}(o_2) = \bar{t}_2)$$

$$= (1 + o(1)) \frac{1}{n^{|V(t_1)| + |V(t_2)|}} g_{t_1}((v_w/n)_{w \in V(t_1)}) g_{t_2}((v_w/n)_{w \in V(t_2)}). \qquad (5.4.84)$$

As in the proof of Theorem 5.21, (5.4.84) implies that

$$\frac{1}{n^2} \mathbb{E}\left[N_{n,r}(t, (a_w)_{w \in V(t)})^2\right] \to \mu(B_r^{(G)}(o) = t, |A_w - a_w| \leq 1/r \ \forall w \in V(t))^2. \quad (5.4.85)$$

In turn, this implies that

$$\frac{1}{n} N_{n,r}(t, (a_w)_{w \in V(t)}) \xrightarrow{\mathbb{P}} \mu(B_r^{(G)}(o) = t, |A_w - a_w| \leq 1/r \ \forall w \in V(t)), \quad (5.4.86)$$

which implies the claimed marked local convergence in probability. $\qquad \square$

Offspring Operator of the Pólya Point Tree

We close this subsection by discussing the offspring operator of the local limit. The Pólya point tree, arising as the local limit in Theorems 5.8 and 5.23, is a multi-type branching process (recall Section 3.4). There, we saw that the offspring operator is an important functional for the behavior of such multi-type branching processes. This operator is defined as follows.

Let $s, t \in \{\text{Y}, \text{O}\}$ and $x, y \in [0, 1]$. Let $\kappa\big((x, s), ([0, y], t)\big)$ denote the expected number of children of type in $([0, y], t)$ of an individual of type (x, s), and let

$$\kappa\big((x, s), (y, t)\big) = \frac{\mathrm{d}}{\mathrm{d}y}\kappa\big((x, s), ([0, y], t)\big) \tag{5.4.87}$$

denote the integral kernel of the offspring operator of the Pólya point tree multi-type branching process. Then $\kappa\big((x, s), (y, t)\big)$ is computed in the following lemma:

Lemma 5.25 (Offspring operator of Pólya point tree) *For all $s, t \in \{\text{Y}, \text{O}\}$ and $x, y \in [0, 1]$,*

$$\kappa\big((x, s), (y, t)\big) = \frac{c_{st}\big(\mathbb{1}_{\{x > y, t = \text{O}\}} + \mathbb{1}_{\{x < y, t = \text{Y}\}}\big)}{(x \vee y)^{\chi}(x \wedge y)^{1-\chi}}, \tag{5.4.88}$$

with

$$c_{st} = \begin{cases} \frac{m(m+\delta)}{2m+\delta} & \text{for } st = \text{OO}, \\ \frac{m(m+1+\delta)}{2m+\delta} & \text{for } st = \text{OY}, \\ \frac{(m-1)(m+\delta)}{2m+\delta} & \text{for } st = \text{YO}, \\ \frac{m(m+\delta)}{2m+\delta} & \text{for } st = \text{YY}. \end{cases} \tag{5.4.89}$$

Proof Recall (5.3.3). Note that $U^{1/\chi}$ has density $\chi y^{\chi-1}$. Thus, for $[a, b] \subseteq [0, x]$, and with $\chi = (m + \delta)/(2m + \delta)$,

$$\kappa\big((x, \text{O}), ([a, b], \text{O})\big) = m\mathbb{E}\left[\int_{a/x}^{b/x} \chi y^{\chi-1}\mathrm{d}y\right] = m\frac{b^{\chi} - a^{\chi}}{x^{\chi}}. \tag{5.4.90}$$

Further, by (5.3.5), for $[a, b] \subseteq [x, 1]$ and noting that $1/(\tau - 1) = m/(2m + \delta) = 1 - \chi$, we have

$$\begin{aligned} \kappa\big((x, \text{O}), ([a, b], \text{Y})\big) &= (\tau - 1)\mathbb{E}\left[\text{Poi}\Big(\Gamma^{\star}\int_{a}^{b} \frac{y^{1/(\tau-1)-1}}{x^{1/(\tau-1)}}\mathrm{d}y\Big)\right] \\ &= \mathbb{E}[\Gamma^{\star}]\frac{b^{1/(\tau-1)} - a^{1/(\tau-1)}}{x^{1/(\tau-1)}} \\ &= (m + 1 + \delta)\frac{b^{1-\chi} - a^{1-\chi}}{x^{1-\chi}}. \end{aligned} \tag{5.4.91}$$

Similarly, for $[a, b] \subseteq [0, x]$,

$$\kappa\big((x, \text{Y}), ([a, b], \text{O})\big) = (m - 1)\mathbb{E}\left[\int_{a/x}^{b/x} \chi y^{\chi-1}\mathrm{d}y\right] = (m - 1)\frac{b^{\chi} - a^{\chi}}{x^{\chi}}, \tag{5.4.92}$$

while, for $[a, b] \subseteq [x, 1]$,

$$\begin{aligned} \kappa\big((x, \text{Y}), ([a, b], \text{Y})\big) &= (\tau - 1)\mathbb{E}\left[\text{Poi}\Big(\Gamma\int_{a}^{b} \frac{y^{1/(\tau-1)-1}}{x^{1/(\tau-1)}}\mathrm{d}y\Big)\right] \\ &= \mathbb{E}[\Gamma]\frac{b^{1/(\tau-1)} - a^{1/(\tau-1)}}{x^{1/(\tau-1)}} \\ &= (m + \delta)\frac{b^{1-\chi} - a^{1-\chi}}{x^{1-\chi}}. \end{aligned} \tag{5.4.93}$$

We now use (5.4.87), and see that all terms in $\kappa\big((x,s),(y,t)\big)$ are of the form of the rhs of (5.4.88), with

$$
c_{st} = \begin{cases}
m\chi & \text{for } st = \text{OO}, \\
(m+1+\delta)(1-\chi) & \text{for } st = \text{OY}, \\
(m-1)\chi & \text{for } st = \text{YO}, \\
(m+\delta)(1-\chi) & \text{for } st = \text{YY}.
\end{cases}
\tag{5.4.94}
$$

Substituting $\chi = (m+\delta)/(2m+\delta)$ yields (5.4.89). $\square$

5.4.4 LOCAL CONVERGENCE OF RELATED MODELS

In this subsection we discuss the local convergence of two related models. The main result is the following theorem:

Theorem 5.26 (Local convergence of related preferential attachment models)   *Fix $m \geq 1$ and $\delta > -m$. The preferential attachment models $\mathrm{PA}_n^{(m,\delta)}(a)$ and $\mathrm{PA}_n^{(m,\delta)}(b)$ converge locally in probability to the Pólya point tree.*

Proof   We do not present the entire proof but, rather, explain how the proof of Theorem 5.8 can be adapted. References can be found in the notes in Section 5.7. The proof of Theorem 5.8 has two main steps, the first being the fixed-graph Pólya urn representation in Theorem 5.10, which is the crucial starting point of the analysis. In the second step, this representation is used to perform the second-moment method for the marked neighborhood counts. In the present proof, we focus on the first part, as this is the most sensitive to minor changes in the model.

Finite-Graph Pólya Version of $\mathrm{PA}_{mn}^{(1,\delta/m)}(a)$ and $\mathrm{PA}_{mn}^{(1,\delta/m)}(b)$

We start with $\mathrm{PA}_n^{(m,\delta)}(b)$. We recall that $\mathrm{PA}_n^{(m,\delta)}(b)$ can be obtained from $\mathrm{PA}_{mn}^{(1,\delta/m)}(b)$ by collapsing the vertices $[mv] \setminus [m(v-1)]$ in $\mathrm{PA}_{mn}^{(1,\delta/m)}(b)$ into vertex v in $\mathrm{PA}_n^{(m,\delta)}(b)$. When we do this collapsing, all edges are also collapsed. Thus, the event $u \overset{j}{\rightsquigarrow} v$ in $\mathrm{PA}_n^{(m,\delta)}(b)$ is equivalent to $m(u-1)+j \overset{1}{\rightsquigarrow} [mv] \setminus [m(v-1)]$ in $\mathrm{PA}_{mn}^{(1,\delta/m)}(b)$. Further, for $m=1$, $\mathrm{PA}_{mn}^{(1,\delta/m)}(b)$ and $\mathrm{PA}_{mn}^{(1,\delta/m)}(d)$ are the same, except that $\mathrm{PA}_{mn}^{(1,\delta/m)}(b)$ starts with two vertices with two edges between them, while $\mathrm{PA}_{mn}^{(1,\delta/m)}(d)$ starts with two vertices with one edge between them. This different starting graph was addressed in Remark 5.11, where it was explained that the finite-graph Pólya version is changed only in a minor way. We can thus use the results obtained thus far, together with a collapsing procedure, to obtain the Pólya urn description of $\mathrm{PA}_{mn}^{(1,\delta/m)}(b)$.

For $\mathrm{PA}_n^{(m,\delta)}(a)$, we also use that it can be obtained from $\mathrm{PA}_{mn}^{(1,\delta/m)}(a)$ by collapsing the vertices $[mv] \setminus [m(v-1)]$ in $\mathrm{PA}_{mn}^{(1,\delta/m)}(a)$ into vertex v in $\mathrm{PA}_n^{(m,\delta)}(a)$. However, $\mathrm{PA}_{mn}^{(1,\delta/m)}(a)$ is not quite the same as $\mathrm{PA}_{mn}^{(1,\delta/m)}(d)$. Instead, we use the description in Theorem 5.6 and compare it with that in Theorem 5.7 to see that now the Beta variables are given by $(\psi_j')_{j\in[n]}$ with $\psi_1' = 1$, and, for $j \geq 2$,

$$
\psi_j' \sim \mathrm{Beta}\big(1 + \delta/m, 2(j-1) + (j-1)\delta/m\big).
\tag{5.4.95}
$$

Also, the definitions in (5.3.20) need to be updated. This affects the finite-graph Pólya for $m = 1$ only in a minor way.

Effect of Collapsing Vertices

The above gives us finite-graph Pólya versions of $\mathrm{PA}_{mn}^{(1,\delta/m)}(a)$ and $\mathrm{PA}_{mn}^{(1,\delta/m)}(b)$. These allow us to describe the local limit for $m = 1$ as before. However, in order to go to $\mathrm{PA}_n^{(m,\delta)}(a)$ and $\mathrm{PA}_n^{(m,\delta)}(b)$, we need to investigate the effect of the collapsing procedure on the local limit. Now each vertex has m older neighbors rather than 1 as for $m = 1$, and this effect is easily understood. Note in particular that $\tau = 3 + \delta/m$ is the same in $\mathrm{PA}_{mn}^{(1,\delta/m)}(a)$ and $\mathrm{PA}_{mn}^{(1,\delta/m)}(b)$, respectively, compared with $\mathrm{PA}_n^{(m,\delta)}(a)$ and $\mathrm{PA}_n^{(m,\delta)}(b)$, respectively, so the distribution of the ages of the older vertices is also the same in both.

A more challenging difference arises in the description of the Poisson processes of younger neighbors in (5.3.5). We use the fact that the sum of independent Poisson processes is again a Poisson process, where the intensity of the sum is the sum of the individual intensities. Thus, the intensity after collapsing becomes

$$\rho_w(x) = \frac{(\tau - 1)x^{1/(\tau-1)-1}}{A_w^{1/(\tau-1)}} \sum_{i=1}^{m} \Gamma_{w,i}, \tag{5.4.96}$$

where $(\Gamma_{w,i})_{i\in[m]}$ are iid Gamma variables with parameters given in (5.3.2) for $m = 1$ and with δ replaced by δ/m. Recall that the sum of iid Gamma parameters with parameters $(r_i)_{i\in[m]}$ and scale parameter $\lambda = 1$ is again Gamma distributed, now with parameter $r = \sum_{i=1}^{m} r_i$ and scale parameter $\lambda = 1$. Recall that either $r_i = 1 + \delta/m$ or $r_i = 1 + \delta/m + 1$ and that there is no $r_i = 1 + \delta/m + 1$ when w has label Y, while there is exactly one i with $r_i = 1 + \delta/m + 1$ when i has label O. Thus, we see that $\sum_{i=1}^{m} \Gamma_{w,i}$ has a Gamma distribution with parameters $r = m + \delta$ and $\lambda = 1$ when the label of w is Y, while it has parameters $r = m + \delta + 1$ and $\lambda = 1$ when the label of w is O, as in (5.3.2) for $m \geq 2$. We refrain from giving more details. $\qquad\square$

5.5 CONNECTIVITY OF PREFERENTIAL ATTACHMENT MODELS

In this section we investigate the connectivity of $\mathrm{PA}_n^{(m,\delta)}(a)$. Recall that $\mathrm{PA}_n^{(m,\delta)}(b)$ and $\mathrm{PA}_n^{(m,\delta)}(d)$ are by construction connected, so $\mathrm{PA}_n^{(m,\delta)}(a)$ is special in this respect. We start by describing the connectivity when $m = 1$, which is special. For $m = 1$, the number of connected components N_n of $\mathrm{PA}_n^{(1,\delta)}$ has distribution given by

$$N_n = I_1 + I_2 + \cdots + I_n, \tag{5.5.1}$$

where I_i is the indicator that the ith edge forms a self-loop, so that $(I_i)_{i\geq 1}$ are independent indicator variables with

$$\mathbb{P}(I_i = 1) = \frac{1+\delta}{(2+\delta)(i-1)+1+\delta}. \tag{5.5.2}$$

It is not hard to see that this implies that $N_n/\log n$ converges in probability to $(1+\delta)/(2+\delta) < 1$, so that whp there exists a largest connected component of size at least $n/\log n$. As a result, whp $\mathrm{PA}_n^{(1,\delta)}$ is not connected, but has a few connected components which are almost all quite large. We do not elaborate more on the connectivity properties for $m = 1$ and instead leave the asymptotics of the number of connected components as Exercises 5.26 and 5.27.

For $m \geq 2$, the situation is entirely different since then $\mathrm{PA}_n^{(m,\delta)}(a)$ *is* connected whp at all sufficiently large times:

Theorem 5.27 (Connectivity of $\mathrm{PA}_n^{(m,\delta)}(a)$ for $m \geq 2$) *Fix $m \geq 2$. Then, with high probability for T large, $\mathrm{PA}_n^{(m,\delta)}(a)$ is connected for all $n \geq T$.*

Proof Again, we let N_n denote the number of connected components of $\mathrm{PA}_n^{(m,\delta)}(a)$. We note that $I_n = N_n - N_{n-1} = 1$ precisely when all m edges of vertex n are attached to vertex n. Thus,

$$\mathbb{P}(I_n = 1) = \prod_{j \in [m]} \frac{2e - 1 + \delta}{(2m + \delta)n + (2j - 1 + \delta)}. \tag{5.5.3}$$

For $m \geq 2$,

$$\sum_{n=2}^{\infty} \mathbb{P}(I_n = 1) < \infty, \tag{5.5.4}$$

so that, almost surely, $I_n = 1$ occurs only *finitely* often. As a result, $\lim_{n \to \infty} N_n < \infty$ almost surely since

$$N_n \leq 1 + \sum_{n=2}^{\infty} I_n. \tag{5.5.5}$$

This implies that, for $m \geq 2$, $\mathrm{PA}_n^{(m,\delta)}(a)$ almost surely contains only finitely many connected components. Further, $\mathrm{PA}_n^{(m,\delta)}(a)$ has a *positive* probability of being disconnected at a certain time $n \geq 2$ (see Exercise 5.30 below). However, for $m \geq 2$, $I_n = N_n - N_{n-1}$ can be negative, since the edges of the nth vertex can be attached to two *distinct* connected components. We will see that this happens whp, which explains why $N_n = 1$ whp for n large, as we show next.

We first fix $K \geq 1$ large. Then, with probability converging to 1 as $K \to \infty$,

$$\sum_{n=K}^{\infty} \mathbb{1}_{\{N_n > N_{n-1}\}} = 0. \tag{5.5.6}$$

We condition on $\sum_{n=K}^{\infty} \mathbb{1}_{\{N_n > N_{n-1}\}} = 0$, so that no new connected components are formed after time K, and the number of connected components can only decrease in time. Let $\mathscr{F}_s$ denote the σ-algebra generated by $(\mathrm{PA}_n^{(m,\delta)}(a))_{n=1}^{s}$. We are left with proving that, for n sufficiently large, the vertices in $[K]$ are whp all connected in $\mathrm{PA}_n^{(m,\delta)}(a)$.

The proof proceeds in two steps. We show that, if $N_n \geq 2$ and n is large, $\mathbb{P}(N_{2n} - N_n \leq -1 \mid \mathscr{F}_n)$ is uniformly bounded from below. Indeed, we condition on $\mathscr{F}_n$ for which $N_n \geq 2$ and $N_n \leq K$. Then, using $N_n \leq K$, $\mathrm{PA}_n^{(m,\delta)}(a)$ must have one connected component of size at least n/K, while every other component has at least one vertex in it, and its degree is at least m. Fix $s \in [2n] \setminus [n]$. Then, the probability that the first edge of $v_s^{(m)}$ connects to the connected component of size at least n/K, while the second connects to the connected component of size at least 1, is, conditional on $\mathscr{F}_{s-1}$, at least

$$\frac{m+\delta}{2(2m+\delta)n} \frac{(m+\delta)n/K}{2(2m+\delta)n} \geq \frac{\varepsilon}{n}, \tag{5.5.7}$$

for some $\varepsilon > 0$ and uniformly in s. Thus, conditional on $\mathscr{F}_n$, the probability that this happens for at least one $s \in [2n] \setminus [n]$ is at least

$$1 - \left(1 - \frac{\varepsilon}{n}\right)^n \geq \eta > 0, \tag{5.5.8}$$

uniformly for every n. Thus, when $N_n \geq 2$, $\mathbb{P}(N_{2n} - N_n \leq -1 \mid \mathscr{F}_n) \geq \eta$. As a result, $N_n \xrightarrow{a.s.} 1$, so that $N_T = 1$ for some $T < \infty$ almost surely. Without loss of generality, we can take $T \geq K$. When $\sum_{n=K}^{\infty} \mathbb{1}_{\{N_n > N_{n-1}\}} = 0$, if $N_T = 1$ for some T then $N_n = 1$ for *all* $n \geq T$. This proves that $\mathrm{PA}_n^{(m,\delta)}(a)$ is whp connected for all $n \geq T$, where T is large, which implies Theorem 5.27. $\qquad\square$

Exercise 5.31 investigates the all-time connectivity of $(\mathrm{PA}_n^{(m,\delta)}(a))_{n\geq 1}$.

5.6 FURTHER RESULTS FOR PREFERENTIAL ATTACHMENT MODELS

Local Limit of Preferential Attachment Models with Random Out-Degrees

We start by discussing a result concerning a preferential attachment model where vertex i has m_i edges to connect to the graph at time $i - 1$. Each of these edges is connected to a vertex in $[i]$ with a probability proportional to the degree plus δ, as usual. There are various possible choices for the model (corresponding to $\mathrm{PA}_n^{(m,\delta)}(a)$, $\mathrm{PA}_n^{(m,\delta)}(b)$, and $\mathrm{PA}_n^{(m,\delta)}(d)$), but these all behave similarly, so we will be less precise.

We assume that $(m_i)_{i\in[n]}$ is an iid sequence and that $\delta > -\inf\{d : \mathbb{P}(m_i = d) > 0\}$ is such that $m_i + \delta > 0$ almost surely. We call this the *preferential attachment model with random out-degrees*. Here, we describe the local limit of this model:

Theorem 5.28 (Local convergence of preferential attachment models: random out-degrees) *Let $(m_i)_{i\geq 1}$ be iid out-degrees, and fix $\delta > -\inf\{d : \mathbb{P}(m_i = d) > 0\}$. The preferential attachment model with random out-degrees converges locally in probability to the Pólya point tree with random degrees.*

The Pólya point tree with random degrees arises from the Pólya point tree by letting the initial degrees be random. We use the model related to $\mathrm{PA}_n^{(1,\delta)}(a)$, so that now blocks of vertices of *random length* are being collapsed into single vertices in this model, and the δ corresponding to the ith block is equal to δ/m_i. Interestingly, and similarly to many related random graph models, there is an explicit size-biasing present in this model. Indeed, the out-degrees of nodes of label o in the tree have distribution

$$\frac{m+\delta}{\mathbb{E}[M]+\delta}\mathbb{P}(M = m), \tag{5.6.1}$$

where M is the random variable of which $(m_i)_{i\geq 1}$ are iid copies, while the out-degrees of nodes of label y in the tree have the size-biased distribution

$$\frac{m}{\mathbb{E}[M]}\mathbb{P}(M = m). \tag{5.6.2}$$

For $\delta = 0$ only, the two are the same.

Local Structure of Bernoulli Preferential Attachment Model

Recall the Bernoulli preferential attachment model $(\mathrm{BPA}_n^{(f)})_{n\geq 1}$ defined in Section 1.3.5. Here we investigate the local structure of this model. A special case is $(\mathrm{BPA}_n^{(f)})_{n\geq 1}$ with an affine attachment function f, i.e., the setting where there exist $\gamma, \beta > 0$ such that

$$f(k) = \gamma k + \beta. \qquad (5.6.3)$$

Owing to the attachment rules in (1.3.66), the model does not satisfy the rescaling property that the model with cf has the same law as the model with f for any $c > 0$. In fact, it turns out that the parameter $\gamma > 0$ (which, by convention, is always taken to be 1 for $(\mathrm{PA}_n^{(m,\delta)})_{n\geq 1}$) is now the parameter that determines the tail behavior of the degree distribution (recall Exercise 1.23). In Exercises 5.23 and 5.24, the reader is invited to compute the average degree of this affine model, as well as the number of edges added at time n for large n.

The case of general attachment functions $k \mapsto f(k)$ is more delicate to describe. We start by introducing some notation. We call a preferential attachment function $f \colon \mathbb{N}_0 \mapsto (0, \infty)$ *concave* when

$$f(0) \leq 1 \qquad \text{and} \qquad \Delta f(k) := f(k+1) - f(k) < 1 \quad \text{for all } k \geq 0. \qquad (5.6.4)$$

Concavity implies the existence of the limit

$$\gamma := \lim_{k\to\infty} \frac{f(k)}{k} = \min_{k\geq 0} \Delta f(k). \qquad (5.6.5)$$

The following result investigates the proportion of vertices $v \in [n]$ whose connected component $\mathscr{C}(v)$ at time n has size k:

Theorem 5.29 (Component sizes of Bernoulli preferential attachment models) *Let f be a concave attachment function. The Bernoulli preferential attachment model with conditionally independent edges $\mathrm{BPA}_n^{(f)}$ satisfies that, for every $k \geq 1$,*

$$\frac{1}{n}\#\{v \in [n] \colon |\mathscr{C}(v)| = k\} \overset{\mathbb{P}}{\longrightarrow} \mu(|\mathcal{T}| = k), \qquad (5.6.6)$$

where $|\mathcal{T}|$ is the total progeny of an appropriate multi-type branching process.

The limiting multi-type branching process in Theorem 5.29 is such that $\mu(|\mathcal{T}| = k) > 0$ for all $k \geq 1$, so that $\mathrm{BPA}_n^{(f)}$ is disconnected whp. While Theorem 5.29 does not *quite* prove the local convergence of $\mathrm{BPA}_n^{(f)}$, it is strongly related. See Section 5.7 for a more detailed discussion. Thus, in what follows, we discuss the proof as if the theorem does yield local convergence.

We next describe the local limit and degree evolution in more detail. We start with two main building blocks. Let $(Z_t)_{t\geq 0}$ be a pure-birth Markov process with birth rate $f(k)$ when it is in state k, starting from $Z_0 = 0$ (i.e., it jumps from k to $k+1$ at rate $f(k)$). Further, for $\sigma \geq 0$, let $(Z_t^{[\sigma]} - \mathbb{1}_{[\sigma,\infty)}(t))_{t\geq 0}$ be the process $(Z_t)_{t\geq 0}$ *conditioned on having a jump at time σ.*

Let $\mathcal{S} := \{\mathrm{Y}\} \times \mathbb{R} \cup (\{\mathrm{o}\} \times [0, \infty)) \times \mathbb{R}$ be the type space, considering the *label* as an element in $\{\mathrm{Y}\} \cup (\{\mathrm{o}\} \times [0, \infty))$ and the *location* as being in $\mathbb{R}$. It turns out that the location of a vertex $t \in \mathbb{N}$ in $\mathrm{BPA}_n^{(f)}$ corresponds to $\log(t/n)$, and we allow for $t > n$ in our description. Individuals of label Y correspond to individuals that are *younger* than their parent in the tree. Individuals of label (o, σ) correspond to individuals that are *older* than their parent, and for them we need to record the relative location of an individual compared with its parent (roughly corresponding to the log of the *ratio* of their ages).

The local limit is a multi-type branching process with the following properties and offspring distributions. The root has label Y and location $-E$, where E is a standard exponential random variable with parameter 1; this variable corresponds to $\log(U)$ with U uniform in $[0, 1]$. A particle of label Y at location x has younger children of label Y with relative locations at the jumps of the process $(Z_t)_{t \geq 0}$, so that their locations are equal to $x + \pi_i$, where π_i is the ith jump of $(Z_t)_{t \geq 0}$. A particle of label Y at location x has older children with labels $(o, -\pi_i)$, where $(\pi_i)_{i \geq 0}$ are the points in a Poisson point process on $(-\infty, 0]$ with intensity measure given by

$$e^t \mathbb{E}[f(Z_{-t})] \, dt, \tag{5.6.7}$$

their locations being $x + \pi_i$. A particle of label (o, σ) generates offspring

▷ of labels in $o \times [0, \infty)$ in the same manner as for a parent of label Y;

▷ of label Y with locations at the jumps of $(Z_t^{[\sigma]} - \mathbb{1}_{[\sigma, \infty)}(t))_{t \geq 0}$ plus x.

The above describes the evolution of the branching process limit for *all* times. This can be interesting when investigating, e.g., the degree evolutions and graph structures of the vertices in $[n]$ at all times $t \geq n$. However, for local convergence, we are interested only in the subgraph of vertices in $[n]$, so that only vertices with a *negative* location matter. Thus, finally, we *kill* all particles with location $x > 0$ together with their entire tree of descendants. This describes the local limit of $\mathrm{BPA}_n^{(f)}$, and $|\mathcal{T}|$ is the total progeny of this multi-type branching process.

We next explain in more detail how the above multi-type branching process arises. An important ingredient is the birth process $(Z_t)_{t \geq 0}$. This process can be related to the in-degree evolution in $(\mathrm{BPA}_n^{(f)})_{n \geq 1}$ as follows. Recall from (1.3.66) that $D_i^{(\mathrm{in})}(n)$ denotes the in-degree of vertex i at time n. Note that the $(D_i^{(\mathrm{in})}(n))_{n \geq 0}$ are independent growth processes, and

$$\mathbb{P}(D_i^{(\mathrm{in})}(n+1) - D_i^{(\mathrm{in})}(n) = 1 \mid \mathrm{BPA}_n^{(f)}) = \frac{f(D_i^{(\mathrm{in})}(n))}{n}, \tag{5.6.8}$$

where $D_i^{(\mathrm{in})}(i) = 0$. Now consider $(Z_t)_{t \geq 0}$, and note that, for $\varepsilon > 0$ small,

$$\mathbb{P}(Z_{t+\varepsilon} - Z_t \mid Z_t) = \varepsilon f(Z_t) + o(\varepsilon). \tag{5.6.9}$$

Fix

$$t_n = \sum_{k=1}^n \frac{1}{k}. \tag{5.6.10}$$

Then,

$$\mathbb{P}(Z_{t_n} - Z_{t_{n-1}} \mid Z_{t_{n-1}}) = (t_n - t_{n-1}) f(Z_{t_{n-1}}) + o(t_n - t_{n-1})$$
$$= \frac{f(Z_{t_{n-1}})}{n} + o(1/n), \tag{5.6.11}$$

so that the evolution of $(D_i^{(\mathrm{in})}(n))_{n>i}$ is *almost* the same as that of $(Z_{t_n})_{n>i}$ when also $Z_{t_i} = 0$. Further, at the times k where a jump occurs in $(Z_{t_k})_{k>i}$, there is an edge between vertex i and vertex k. The latter has location $\log(k/n) \approx t_k - t_n$, and the vertex to which vertex i connects is younger and thus has label Y. The above gives a nice description of the children of the root that have label Y.

The situation changes a little when we consider an individual of type (o, σ). Indeed, when considering its younger children, *we know already* that one younger child is present with a location $x+\sigma$. This means that the birth process of younger children is $(Z_t^{[\sigma]} - \mathbb{1}_{[\sigma,\infty)}(t))_{t\geq 0}$ instead of $(Z_t)_{t\geq 0}$, where we see that the offspring distribution depends on the jump σ of the individual of label (o, σ).

We close this discussion by explaining how the children having labels in $\{\mathrm{o}\} \times [0, \infty)$ arise. Note that these children are the same for individuals having label Y as well as for those having label (o, σ) with $\sigma \geq 0$. Since the connection decisions are *independent* and edge probabilities are small for n large, the number of connections to vertices in the range $[a, b]n$ are close to a Poisson random variable with an appropriate parameter, thus leading to an appropriate Poisson process. The expected number of neighbors of vertex qn with ages in $[a, b]n$ is roughly

$$\sum_{i=an}^{bn} \mathbb{E}\left[\frac{f(D_i^{(\mathrm{in})}(qn))}{qn}\right] \approx \frac{1}{q}\int_a^b \mathbb{E}[f(D_{un}^{(\mathrm{in})}(qn))]\mathrm{d}u \approx \frac{1}{q}\int_a^b \mathbb{E}[f(Z_{\log(q/u)})]\mathrm{d}u$$

$$= \int_{\log(q/a)}^{\log(q/b)} \mathrm{e}^{-t}\mathbb{E}[f(Z_t)]\mathrm{d}t = \int_{\log(a/q)}^{\log(b/q)} \mathrm{e}^t\mathbb{E}[f(Z_{-t})]\mathrm{d}t, \tag{5.6.12}$$

as in (5.6.7). When the age is in $[a, b]n$, the location is in $[\log(a), \log(b)]$, so the change in location compared with q is in $[\log(a/q), \log(b/q)]$. This explains how the children with labels in $\{\mathrm{o}\} \times [0, \infty)$ arise, and completes our discussion of the local structure of $\mathrm{BPA}_n^{(f)}$.

Giant Component for Bernoulli Preferential Attachment

The preferential attachment model $\mathrm{PA}_n^{(m,\delta)}(a)$ turns out to be *connected* whp, as we have discussed in Section 5.5. This, however, is not true for $\mathrm{BPA}_n^{(f)}$. Here, we describe the existence of the giant component in this model:

Theorem 5.30 (Existence of a giant component: linear case)   *If $f(k) = \gamma k + \beta$ for some $0 \leq \gamma < 1$ and $0 < \beta \leq 1$, then $\mathrm{BPA}_n^{(f)}$ has a unique giant component if and only if*

$$\gamma \geq \tfrac{1}{2} \qquad or \qquad \beta > \frac{(\tfrac{1}{2} - \gamma)^2}{1 - \gamma}. \tag{5.6.13}$$

Consequently, if $\mathscr{C}_{\max}$ and $\mathscr{C}_{(2)}$ denote the largest and second largest connected components in $\mathrm{BPA}_n^{(f)}$,

$$|\mathscr{C}_{\max}|/n \xrightarrow{\mathbb{P}} \zeta, \qquad |\mathscr{C}_{(2)}|/n \xrightarrow{\mathbb{P}} 0, \tag{5.6.14}$$

where $\zeta > 0$ precisely when (5.6.13) holds.

We next state a more general theorem about concave attachment functions. For this, we introduce some notation. Define the increasing functions M, respectively, M^Y and $M^{(O,\sigma)}$, by

$$M(t) = \int_0^t e^{-s}\mathbb{E}[f(Z_s)]ds, \qquad M^Y(t) = \mathbb{E}[Z_t], \tag{5.6.15}$$

$$M^{(O,\sigma)}(t) = \mathbb{E}[Z_t \mid \Delta Z_\sigma = 1] - \mathbb{1}_{[\sigma,\infty)}(t) \qquad \text{for } \sigma \in [0,\infty). \tag{5.6.16}$$

Next, define a linear operator A_α on the Banach space $C(\mathcal{S})$ of continuous, bounded functions on the type space $\mathcal{S} := \{Y\} \cup (\{O\} \times [0,\infty])$ by

$$(A_\alpha g)(\sigma) = \int_0^\infty g(O,t)e^{\alpha t}dM(t) + \int_0^\infty g(Y)e^{-\alpha t}dM^\sigma(t), \quad \text{for } \sigma \in \mathcal{S}. \tag{5.6.17}$$

The operator A_α should be thought of as describing the expected offspring of vertices of different types, as explained in more detail below. The main result on the existence of a giant component in the preferential attachment model with conditionally independent edges is the following theorem:

Theorem 5.31 (Existence of a giant component) *No giant component exists in* $\mathrm{BPA}_n^{(f)}$ *if and only if there exists* $0 < \alpha < 1$ *such that* A_α *is a compact operator with spectral radius* $\rho(A_\alpha) \leq 1$. *Equivalently,* $|\mathscr{C}_{\max}|/n \overset{\mathbb{P}}{\longrightarrow} \zeta$ *and* $|\mathscr{C}_{(2)}|/n \overset{\mathbb{P}}{\longrightarrow} 0$, *where* $\zeta = 0$ *precisely when there exists* $0 < \alpha < 1$ *such that* A_α *is a compact operator with spectral radius* $\rho(A_\alpha) \leq 1$.

In general, $\zeta = 1 - \sum_{k \geq 1} \mu(|\mathcal{T}| = k)$, as described in Theorem 5.29. However, this description does not immediately allow for an explicit computation of ζ, so that it is hard to decide when $\zeta > 0$. It turns out that A_α is a well-defined compact operator if and only if $(A_\alpha 1)(0) < \infty$. When thinking of A_α as the reproduction operator, the spectral radius $\rho(A_\alpha)$ indicates whether the multi-type branching process has a positive survival probability. Thus, the extinction condition $\rho(A_\alpha) \leq 1$ should be thought of as the equivalent of the usual condition $\mathbb{E}[X] \leq 1$ for the extinction of a discrete single-type branching process.

5.7 Notes and Discussion for Chapter 5

Notes on Section 5.2

The proof of Theorem 5.2 has been adapted from Ross (1996). More extensive discussions on exchangeable random variables and their properties can be found in Aldous (1985) and Pemantle (2007), the latter focussing on random walks with self-interaction, where exchangeability is a crucial tool. There has been a lot of work on urn schemes, also in cases where the weight functions are not linear with equal slope, when the limits can be seen to obey rather different characteristics. See, e.g., (Athreya and Ney, 1972, Chapter 9). For results on the relation between degrees in various random trees, including plane trees, which are closely related to preferential attachment trees, we refer to Janson (2004). Collevecchio et al. (2013) discussed the relation between the degrees of vertices and preferential attachment models.

Notes on Section 5.3

The multi-type branching process local limit in Theorem 5.8 was established by Berger et al. (2014) for preferential attachment models with a fixed number of outgoing edges per vertex. Berger et al. (2014) only treated the case where $\delta \geq 0$; the more recent extension to $\delta > -m$ is due to Garavaglia et al. (2022) and was first developed in the context of this book. This is due to the fact that Berger et al. (2014) viewed the attachment probabilities as a *mixture* between attaching uniformly and according to the degree. We, instead, rely on the Pólya urn description, which works for all $\delta > -m$. (Berger et al., 2014, Theorem 2.2) proved local weak convergence for $\delta \geq 0$. Theorem 5.8 stated local convergence in probability to the Pólya point tree for all $\delta > -m$. Local convergence in probability for $\delta \geq 0$ follows from the convergence in probability of appropriate subgraph counts in (Berger et al., 2014, Lemma 2.4). We refrain from discussing this issue further.

The Pólya-urn proof of Theorem 5.10 follows (Berger et al., 2014, Section 3.1) closely, apart from the fact that we do not rely on the mixture of choosing a vertex uniformly and according to degree. The direct proof of Theorem 5.10 is novel and also appeared in a more general setting in Garavaglia et al. (2022).

Berger et al. (2014) also studied two related settings, one where the edges are attached independently (i.e., without an intermediate update of the degrees while attaching the m edges incident to the newest vertex), and a conditional model in which the edges are attached to *distinct* vertices. This shows that the result is quite robust, as Theorem 5.26 also indicates.

A related version of Theorem 5.10 for $\delta = 0$ was proved by Bollobás and Riordan (2004a) in terms of a pairing representation. This applies to $\mathrm{PA}_n^{(m,\delta)}(a)$ with $\delta = 0$. Another related version of Theorem 5.10 is proved in Rudas et al. (2007) and applies to general preferential attachment trees with $m = 1$. Its proof relies on a continuous-time embedding in terms of continuous-time branching processes. We further refer to Lo (2021) for results on the local convergence of preferential attachment trees with additive fitness.

Notes on Section 5.4

The preliminaries in Lemma 5.17, Proposition 5.18 and Lemma 5.20 follow the proofs of the respective results by Berger et al. (2014) closely, even though we make some simplifications. The remainder of the proof of Theorem 5.8 has been developed for this book, and is close in spirit to that in Garavaglia et al. (2022), where the setting with iid out-degrees is also studied.

Notes on Section 5.5

These results are novel.

Notes on Section 5.6

The preferential attachment model with random out-degrees was introduced by Deijfen et al. (2009), who studied its degree structure. Its local limit was derived in Garavaglia et al. (2022), where Theorem 5.28 is proved.

Theorem 5.29 is (Dereich and Mörters, 2013, Theorem 1.9). The proof of (Dereich and Mörters, 2013, Theorem 1.9) relied on a clever coupling argument; see, e.g., Proposition 6.1 in that paper for a statement that $|\mathscr{C}(o_n)| \wedge c_n$ can be, whp, successfully coupled to $|\mathcal{T}| \wedge c_n$ when $c_n \to \infty$ sufficiently slowly. Here $o_n \in [n]$ is chosen uar. It seems to us that this statement can straightforwardly be adapted to the statement that $B_r^{(G_n)}(o_n)$ can, whp, be successfully coupled to $B_r^{(G)}(\varnothing)$ for all r such that $|B_r^{(G)}(\varnothing)| \leq c_n$. Here $(G, \varnothing)$ denotes the local limit of the Bernoulli preferential attachment model. See also (Dereich and Mörters, 2013, Lemma 6.6) for a related coupling result. This would provide local weak convergence.

For local convergence in probability, a variance computation is needed to perform the second-moment method. In (Dereich and Mörters, 2013, Proposition 7.1), such a variance computation was performed for the number of vertices whose cluster size is at least c_n, and its proof relied on an exploration process argument. Again, it seems to us that this statement can straightforwardly be adapted to a second-moment method for $N_n(\mathbf{t})$, counting the number of $v \in [n]$ for which $B_r^{(G_n)}(v)$ is isomorphic to an ordered tree $\mathbf{t}$.

The results on the giant component for preferential attachment models with conditionally independent edges in Theorems 5.30 amd 5.31 were proved in Dereich and Mörters (2013). Theorem 5.30 is (Dereich and Mörters, 2013, Proposition 1.3). Theorem 5.31 is (Dereich and Mörters, 2013, Theorem 1.1). The proof uses a nice sprinkling argument, where first f is replaced with $(1 - \varepsilon)f$ and concentration is proved for the number of vertices having connected components of size at least c_n, and then the remaining edges are "sprinkled" to obtain the full law of large numbers. See (Dereich and Mörters, 2013, Section 8) where this argument is worked out nicely.

The notation used by Dereich and Mörters (2013) for Theorem 5.30 is slightly different from ours. Dereich and Mörters (2009) proved that their model obeys an asymptotic power law with exponent $\tau = 1 + 1/\gamma = 3 + \delta/m$, so that γ intuitively corresponds to $\gamma = m/(2m + \delta)$. As a result, $\gamma \geq \frac{1}{2}$ corresponds to $\delta \leq 0$ and $\tau \in (2, 3]$, which is also precisely the setting where the configuration model always has a giant component (recall Theorem 4.9). The parameter γ plays a crucial role in the analysis, as can already be observed in Theorem 5.30. The fact that A_α is a well-defined compact operator if and only if $(A_\alpha 1)(0) < \infty$ is (Dereich and Mörters, 2013, Lemma 3.1). We refer to Dereich and Mörters (2009, 2011) for related results on Bernoulli preferential attachment models.

5.8 EXERCISES FOR CHAPTER 5

Exercise 5.1 (Stationarity of exchangeable sequences) *Show that the marginal distribution of X_i is the same as that of X_1 if $(X_i)_{i=1}^n$ are exchangeable. Show also that the distribution of (X_i, X_j), for $j \neq i$, is the same as that of (X_1, X_2).*

Exercise 5.2 (Iid sequences are exchangeable) *Show that $(X_i)_{i \geq 1}$ forms an infinite sequence of exchangeable random variables if $(X_i)_{i \geq 1}$ are iid.*

Exercise 5.3 (Limiting density in de Finetti's Theorem) *Use de Finetti's Theorem (Theorem 5.2) to prove that $S_n/n \xrightarrow{a.s.} U$, where U appears in (5.2.1). Use this to prove (5.2.4).*

Exercise 5.4 (Number of ones in $(X_i)_{i=1}^n$) *Prove that $\mathbb{P}(S_n = k) = \mathbb{E}\big[\mathbb{P}\big(\mathsf{Bin}(n, U) = k\big)\big]$ in (5.2.3) follows from de Finetti's Theorem (Theorem 5.2).*

Exercise 5.5 (Positive correlation of exchangeable random variables) *Let $(X_i)_{i \geq 1}$ be an infinite sequence of exchangeable random variables. Prove that*

$$\mathbb{P}(X_k = X_n = 1) \geq \mathbb{P}(X_k = 1)\mathbb{P}(X_n = 1). \tag{5.8.1}$$

Prove that equality holds if and only if $(X_k)_{k \geq 1}$ are iid.

Exercise 5.6 (Limiting density of mixing distribution for Pólya urn schemes) *Show that (5.2.24) identifies the limiting density in (5.2.4).*

Exercise 5.7 (Uniform recursive trees) *A uniform recursive tree is obtained by starting with a single vertex, and successively attaching the $(n + 1)$th vertex to a uniformly chosen vertex in $[n]$. Prove that, for uniform recursive trees, the tree decomposition in Theorem 5.4 is such that*

$$\frac{S_1(n)}{S_1(n) + S_2(n)} \xrightarrow{a.s.} U, \tag{5.8.2}$$

where U is uniform on $[0, 1]$. Use this to prove that $\mathbb{P}(S_1(n) = k) = 1/n$ for each $k \in [n]$.

Exercise 5.8 (Scale-free trees) *Recall the model studied in Theorem 5.4, where at time $n = 2$, we start with two vertices of which vertex 1 has degree d_1 and vertex 2 has degree d_2. After this, we successively attach vertices to older vertices with probabilities proportional to the degree plus $\delta > -1$ as in (1.3.64). Show that the model for $(\mathrm{PA}_n^{(1,\delta)}(b))_{n \geq 1}$, for which the graph at time $n = 2$ consists of two vertices joined by two edges, arises when $d_1 = d_2 = 2$. What does Theorem 5.4 imply for $(\mathrm{PA}_n^{(1,\delta)}(b))_{n \geq 1}$?*

Exercise 5.9 (Relative degrees of vertices 1 and 2) *Use Theorem 5.6 to compute $\lim_{n \to \infty} \mathbb{P}(D_2(n) \geq x D_1(n))$ for $(\mathrm{PA}_n^{(1,\delta)}(a))_{n \geq 1}$.*

Exercise 5.10 (Proof of Theorem 5.7) *Complete the proof of Theorem 5.7 on the relative degrees in scale-free trees for $(\mathrm{PA}_n^{(1,\delta)}(b))_{n \geq 1}$ by adapting the proof of Theorem 5.6.*

Exercise 5.11 (Size-biased Gamma is again Gamma) *Let X have a Gamma distribution with shape parameter r and scale parameter λ. Show that its size-biased version X^* has a Gamma distribution with shape parameter $r + 1$ and scale parameter λ.*

Exercise 5.12 (Power-law exponents in $\mathrm{PA}_n^{(m,\delta)}(d)$) *Use Lemma 5.9 to prove the power-law relations in (5.3.8) of the asymptotic degree and neighbor degree distributions in $\mathrm{PA}_n^{(m,\delta)}(d)$, and identify the constants $c_{m,\delta}$ and $c'_{m,\delta}$ appearing in them.*

Exercise 5.13 (Joint law (D, D') for $\mathrm{PA}_n^{(m,\delta)}(d)$ (Berger et al., 2014, Lemma 5.3))   *Adapt the proof of Lemma 5.9 to show that, for $j \geq m$ and $k \geq m + 1$,*

$$\mathbb{P}(D = j, D' = k) = \frac{2m + \delta}{m^2} \frac{\Gamma(k + 1 + \delta)}{(k - m)!\Gamma(m + 1 + \delta)} \frac{\Gamma(j + \delta)}{(j - m)!\Gamma(m + \delta)}$$

$$\times \int_0^1 \int_v^1 (1 - v)^{k-m} v^{m+1+\delta+\delta/m}(1 - u)^{j-m} u^{m+\delta} du dv. \tag{5.8.3}$$

Exercise 5.14 (Conditional law D' given D for $\mathrm{PA}_n^{(m,\delta)}(d)$ (Berger et al., 2014, Lemma 5.3))   *Use Lemma 5.9 and Exercise 5.13 to show that, for fixed $j \geq m$ and as $k \to \infty$, there exists $C_j > 0$ such that*

$$\mathbb{P}(D' = k \mid D = j) = C_j k^{-(2+\delta/m)}(1 + o(1)). \tag{5.8.4}$$

Exercise 5.15 (Joint law (D, D') for $\mathrm{PA}_n^{(m,\delta)}(d)$ (Berger et al., 2014, Lemma 5.3))   *Use Lemma 5.9 and Exercise 5.13 to show that, for fixed $k \geq m + 1$ and as $j \to \infty$, there exists $\tilde{C}_j > 0$ such that*

$$\mathbb{P}(D = j \mid D' = k) = \tilde{C}_j j^{-(4+\delta+\delta/m)}(1 + o(1)). \tag{5.8.5}$$

Exercise 5.16 (Simple form of $S_k^{(n)}$ in (5.3.20))   *Prove, using induction on k, that the equality $S_k^{(n)} = \prod_{i=k+1}^n (1 - \psi_i)$ in (5.3.20) holds.*

Exercise 5.17 (Multiple edges and Theorem 5.10)   *Fix $m = 2$. Let M_n denote the number of edges in $\mathrm{PA}_n^{(m,\delta)}(d)$ that need to be removed so that no multiple edges remain. Use Theorem 5.10 to show that, conditional on $(\psi_k)_{k \geq 1}$, the sequence $(M_{n+1} - M_n)_{n \geq 2}$ is an independent sequence with*

$$\mathbb{P}\big(M_{n+1} - M_n = 1 \mid (\psi_k)_{k \geq 1}\big) = \sum_{k=1}^n \left(\frac{\varphi_k}{S_n^{(n)}}\right)^2. \tag{5.8.6}$$

Exercise 5.18 (Multiple edges and Theorem 5.10 (cont.))   *Fix $m = 2$. Let M_n denote the number of edges in $\mathrm{PA}_n^{(m,\delta)}(d)$ that need to be removed so that no multiple edges remain, as in Exercise 5.17. Use Exercise 5.17 to show that*

$$\mathbb{E}[M_{n+1} - M_n] = \sum_{k=1}^n \mathbb{E}\left[\left(\frac{\varphi_k}{S_n^{(n)}}\right)^2\right]. \tag{5.8.7}$$

Exercise 5.19 (Multiple edges and Theorem 5.10 (cont.))   *Fix $m = 2$. Let M_n denote the number of edges in $\mathrm{PA}_n^{(m,\delta)}(d)$ that need to be removed so that no multiple edges remain, as in Exercise 5.17. Compute*

$$\mathbb{E}\left[\left(\frac{\varphi_k}{S_n^{(n)}}\right)^2\right].$$

Exercise 5.20 (Multiple edges and Theorem 5.10 (cont.))   *Fix $m = 2$. Let M_n denote the number of multiple edges in $\mathrm{PA}_n^{(m,\delta)}(d)$, as in Exercise 5.18, and fix $\delta > -1$. Use Exercise 5.19 to show that $\mathbb{E}[M_n]/\log n \to c$, and identify $c > 0$. What happens when $\delta \in (-2, -1)$?*

Exercise 5.21 (Almost sure limit of normalized product of ψ's)   *Let $(\psi_j)_{j \geq 1}$ be independent Beta random variables with parameters $\alpha = m + \delta, \beta = (2j - 3)m + \delta(j - 1)$ as in (5.3.19). Fix $k \geq 1$. Prove that $(M_n(k))_{n \geq k+1}$, where*

$$M_n(k) = \prod_{j=k+1}^n \frac{1 - \psi_j}{\mathbb{E}[1 - \psi_j]}, \tag{5.8.8}$$

is a multiplicative positive martingale. Thus, $M_n(k)$ converges almost surely by the Martingale Convergence Theorem ([V1, Theorem 2.24]).

Exercise 5.22 (Almost sure limit of normalized product of $S_k^{(n)}$)   *Use Exercise 5.21, combined with the fact that $\prod_{j=k+1}^n \mathbb{E}[1 - \psi_j] = c_k (k/n)^\chi (1 + o(1))$ for some $c_k > 0$, to conclude that $(k/n)^\chi \prod_{j=k+1}^n (1 - \psi_j) = (k/n)^\chi S_k^{(n)}$ converges almost surely for fixed k.*

Exercise 5.23 (Recursion formula for total edges in affine $\mathrm{BPA}_t^{(f)}$)   *Consider the affine $\mathrm{BPA}_n^{(f)}$ with $f(k) = \gamma k + \beta$. Derive a recursion formula for $\mathbb{E}[|E(\mathrm{BPA}_n^{(f)})|]$, where we recall that $|E(\mathrm{BPA}_n^{(f)})|$ is the total number of edges in $\mathrm{BPA}_n^{(f)}$. Identify μ such that $\mathbb{E}[|E(\mathrm{BPA}_n^{(f)})|]/n \to \mu$.*

Exercise 5.24 (Number of edges per vertex in affine $\mathrm{BPA}_t^{(f)}$)   *Consider the affine $\mathrm{BPA}_n^{(f)}$ with $f(k) = \gamma k + \beta$, as in Exercise 5.23. Argue that $|E(\mathrm{BPA}_n^{(f)})|/n \xrightarrow{\mathbb{P}} \mu$.*

Exercise 5.25 (Degree of last vertex in affine $\mathrm{BPA}_t^{(f)}$)   *Use the conclusion of Exercise 5.24 to show that $D_n(n) \xrightarrow{d} \mathrm{Poi}(\mu)$.*

Exercise 5.26 (CLT for number of connected components for $m = 1$)   *Show that the number of connected components N_n in $\mathrm{PA}_n^{(1,\delta)}$ satisfies a central limit theorem when $n \to \infty$, with equal asymptotic mean and variance given by*

$$\mathbb{E}[N_n] = \frac{1+\delta}{2+\delta} \log n (1 + o(1)), \qquad \mathrm{Var}(N_n) = \frac{1+\delta}{2+\delta} \log n (1 + o(1)). \qquad (5.8.9)$$

Exercise 5.27 (Number of connected components for $m = 1$)   *Use Exercise 5.26 to show that the number of connected components N_n in $\mathrm{PA}_n^{(1,\delta)}$ satisfies $N_n / \log n \xrightarrow{\mathbb{P}} (1+\delta)/(2+\delta)$.*

Exercise 5.28 (Number of self-loops in $\mathrm{PA}_n^{(m,\delta)}$)   *Fix $m \geq 1$ and $\delta > -m$. Use a similar analysis to that in Exercise 5.26 to show that the number of self-loops S_n in $\mathrm{PA}_n^{(m,\delta)}(a)$ satisfies $S_n / \log n \xrightarrow{\mathbb{P}} (m+1)(m+\delta)/[2(2m+\delta)]$.*

Exercise 5.29 (Number of self-loops in $\mathrm{PA}_n^{(m,\delta)}(b)$)   *Fix $m \geq 1$ and $\delta > -m$. Use a similar analysis to that in Exercise 5.28 to show that the number of self-loops S_n in $\mathrm{PA}_n^{(m,\delta)}(b)$ satisfies $S_n / \log n \xrightarrow{\mathbb{P}} (m-1)(m+\delta)/[2(2m+\delta)]$.*

Exercise 5.30 (All-time connectivity for $(\mathrm{PA}_n^{(m,\delta)}(a))_{n \geq 1}$)   *Fix $m \geq 2$. Show that the probability that $(\mathrm{PA}_n^{(m,\delta)}(a))_{n \geq 1}$ is connected for all times $n \geq 1$ equals $\mathbb{P}(I_n = 0 \,\forall n \geq 2)$, where we recall that I_n is the indicator that all the m edges of vertex n create self-loops.*

Exercise 5.31 (All-time connectivity for $(\mathrm{PA}_n^{(m,\delta)}(a))_{n \geq 1}$ (cont.))   *Fix $m \geq 2$. Show that the probability that $(\mathrm{PA}_n^{(m,\delta)}(a))_{n \geq 1}$ is connected for all times $n \geq 1$ is in $(0,1)$.*

Part III

Small-World Properties of Random Graphs

Summary of Part II

So far, we have considered the simplest connectivity properties possible. We focused on vertex degrees in Volume 1, and in Part II of this book we extended this to the local structure of the random graphs involved as well as the existence and uniqueness of a macroscopic connected, or *giant*, component. We can summarize the results obtained in the following *meta theorem*:

Meta Theorem A. (Existence and uniqueness of giant component) *In a random graph model with power-law degrees having power-law exponent* τ, *there is a unique giant component when* $\tau \in (2, 3)$, *while, when* $\tau > 3$, *there is a unique giant component only when the degree structure in the graph exceeds a certain precise threshold.*

The above means, informally, that the existence of the giant component is quite robust when $\tau \in (2, 3)$, while it is not when $\tau > 3$. This informally extends even to the random removal of edges, exemplifying the robust nature of the giant when $\tau \in (2, 3)$. These results make the general philosophy that "random graphs with similar degree characteristics behave alike" precise, at least at the level of the existence and robustness of a giant component. The precise condition guaranteeing the existence of the giant varies, but generally amounts to the survival of the local limit.

Overview of Part III

In Part III we aim to extend the discussion of the similarity of random graphs to their *small-world characteristics*, by investigating distances within the giant component. We focus on two settings:

▷ the *typical distance* in a random graph, which means the graph distance between most pairs of vertices, as characterized by the graph distance between two uniformly chosen vertices, conditioned on their being connected; and

▷ the *maximal distance* in a random graph, as characterized by its diameter.

In more detail, Part III is organized as follows. We study distances in general inhomogeneous random graphs in Chapter 6 and those in the configuration model, as well the closely related uniform random graph with prescribed degrees, in Chapter 7. In the last chapter of this part, Chapter 8, we study distances in the preferential attachment model.

SMALL-WORLD PHENOMENA IN INHOMOGENEOUS RANDOM GRAPHS

Abstract

In this chapter we investigate the small-world structure in rank-1 and general inhomogeneous random graphs. For this, we develop *path-counting techniques* that are interesting in their own right.

6.1 MOTIVATION: DISTANCES AND BRANCHING PROCESSES

In this chapter we investigate the small-world properties of inhomogeneous random graphs. It turns out that, when such inhomogeneous random graphs contain a giant, they are also *small worlds*, in the sense that typical distances in them grow at most as the logarithm of the network size. In some cases, for example when the variance of the weight distribution is infinite in rank-1 models, their typical distances are even much smaller than the logarithm of the network size, turning these random graphs into *ultra-small* worlds.

These results closely resemble typical distances in real-world networks. Indeed, see Figure 6.1 for the median (or 50%percentile) of typical distances in the KONECT data base, as compared with the logarithm of the network size. We see that graph distances are, in general, quite small. However, it is unclear whether these real-world examples correspond to small-world or ultra-small-world behavior.

Organization of this Chapter

We start in Section 6.2 by discussing results on the small-world phenomenon in inhomogeneous random graphs. We state results for general inhomogeneous random graphs and then specialize to rank-1 inhomogeneous random graphs, for which we can characterize their ultra-small-world structure in more detail.

The proofs for the main results are in Sections 6.3–6.5. In Section 6.3 we prove lower bounds on typical distances. In Section 6.4 we prove the corresponding upper bounds in the doubly logarithmic regime, and in Section 6.5 we discuss path-counting techniques to obtain the logarithmic upper bound for $\tau > 3$. In Section 6.6 we discuss related results for distances in inhomogeneous random graphs, including their diameter. We close the chapter with notes and discussion in Section 6.7 and exercises in Section 6.8.

6.2 SMALL-WORLD PHENOMENA IN INHOMOGENEOUS RANDOM GRAPHS

In this section we consider the distances between vertices of $\mathrm{IRG}_n(\kappa_n)$, where, as usual, (κ_n) is a graphical sequence of kernels with limit κ.

Recall that we write $\mathrm{dist}_G(u, v)$ for the graph distance between the vertices $u, v \in [n]$ in a graph G having vertex set $V(G) = [n]$. Here the graph distance between u and v is the minimum number of edges in the graph G in all paths from u to v. Further, by convention,

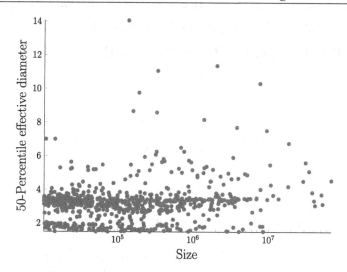

Figure 6.1 Median of typical distances in the 727 networks of size larger than 10,000 from the KONECT data base.

we let $\text{dist}_G(u, v) = \infty$ when u, v are in different connected components. We define the *typical distance* to be $\text{dist}_G(o_1, o_2)$, where o_1, o_2 are two vertices that are chosen uar from the vertex set $[n]$.

It is possible that no path connecting o_1 and o_2 exists; then $\text{dist}_{\text{IRG}_n(\kappa_n)}(o_1, o_2) = \infty$. By Theorem 3.19, $\mathbb{P}(\text{dist}_{\text{IRG}_n(\kappa_n)}(o_1, o_2) = \infty) \to 1 - \zeta^2 > 0$, since $\zeta < 1$ (see Exercise 3.32). In particular, when $\zeta = 0$, which is equivalent to $\nu = \|T_\kappa\| \leq 1$, $\mathbb{P}(\text{dist}_{\text{IRG}_n(\kappa_n)}(o_1, o_2) = \infty) \to 1$. Therefore, in our main results, we condition on o_1 and o_2 being connected, and consider only cases where $\zeta > 0$.

Logarithmic Typical Distances in Inhomogeneous Random Graphs

We start by discussing the logarithmic asymptotics of the typical distance in the case where $\nu = \|T_\kappa\| \in (1, \infty)$. The main result on typical distances in $\text{IRG}_n(\kappa_n)$ is as follows:

Theorem 6.1 (Typical distances in inhomogeneous random graphs)   *Consider* $\text{IRG}_n(\kappa_n)$, *where* (κ_n) *is a graphical sequence of kernels with limit* κ *for which* $\nu = \|T_\kappa\| \in (1, \infty)$. *Let* $\varepsilon > 0$ *be fixed. Then, with* o_1, o_2 *chosen independently and uar from* $[n]$,

(a) if $\sup_{x,y,n} \kappa_n(x, y) < \infty$ *then*

$$\mathbb{P}(\text{dist}_{\text{IRG}_n(\kappa_n)}(o_1, o_2) \leq (1 - \varepsilon) \log_\nu n) = o(1); \tag{6.2.1}$$

(b) if κ *is irreducible then*

$$\mathbb{P}(\text{dist}_{\text{IRG}_n(\kappa_n)}(o_1, o_2) \leq (1 + \varepsilon) \log_\nu n) = \zeta_\kappa^2 + o(1). \tag{6.2.2}$$

Consequently, when $\sup_{x,y,n} \kappa_n(x, y) < \infty$ *and* κ *is irreducible, conditional on* $o_1 \longleftrightarrow o_2$,

$$\frac{\text{dist}_{\text{IRG}_n(\kappa_n)}(o_1, o_2)}{\log_\nu n} \xrightarrow{\mathbb{P}} 1. \tag{6.2.3}$$

In the terminology of [V1, Section 1.4], Theorem 6.1(b) implies that $\mathrm{IRG}_n(\kappa_n)$ is a *small world* when κ is irreducible and $\nu = \|T_\kappa\| \in (1, \infty)$. Theorem 6.1(a) shows that the graph distances are of order $\Theta_\mathbb{P}(\log n)$ when $\sup_{x,y,n} \kappa_n(x, y) < \infty$, so that $\mathrm{IRG}_n(\kappa_n)$ is not an ultra-small world. When $\|T_\kappa\| = \infty$, a truncation argument can be used to prove that $\mathrm{dist}_{\mathrm{IRG}_n(\kappa_n)}(o_1, o_2) = o_\mathbb{P}(\log n)$, but its exact asymptotics is unclear. See Exercise 6.2.

The intuition behind Theorem 6.1 is that, by (3.4.7) and (3.4.8), a Poisson multi-type branching process with kernel κ has neighborhoods that grow exponentially, i.e., the number of vertices at distance k grows like $\|T_\kappa\|^k$. Thus, if we are to examine the distance between two vertices o_1 and o_2 chosen uar from $[n]$ then we need to explore the neighborhood of vertex o_1 up to the moment that it "catches" vertex o_2. For this to happen, the neighborhood must have size of order n, so that we need $\|T_\kappa\|^k = \nu^k \sim n$, i.e., $k = k_n \sim \log_\nu n$. However, proving such a fact is quite tricky, since there are far fewer possible further vertices to explore when the neighborhood has size proportional to n. The proof overcomes this fact by exploring from the *two* vertices o_1 and o_2 *simultaneously* up to the first moment that their neighborhoods share a common vertex, since then the shortest path is obtained. It turns out that shared vertices start appearing when the neighborhoods have size roughly $\sqrt{n}$. At this moment, the neighborhood exploration is still quite close to that in the local branching-process limit. Since $\|T_\kappa\|^r = \nu^r \sim \sqrt{n}$ when $r = r_n \sim \frac{1}{2} \log_\nu n$, this still predicts that distances are close to $2r_n \sim \log_\nu n$.

We next specialize to rank-1 inhomogeneous random graphs, where we also investigate in more detail what happens when $\nu = \infty$ in the case where the degree power-law exponent τ satisfies $\tau \in (2, 3)$.

Distances in Rank-1 IRGs with Finite-Variance Weights

We continue by investigating the behavior of $\mathrm{dist}_{\mathrm{NR}_n(w)}(o_1, o_2)$ for $\mathrm{NR}_n(w)$ in the case where the weights have finite variance:

Theorem 6.2 (Typical distances in rank-1 random graphs with finite-variance weights) *Consider* $\mathrm{NR}_n(w)$*, where the weights* $w = (w_v)_{v \in [n]}$ *satisfy Conditions 1.1(a)–(c) with* $\nu > 1$*. Then, with* o_1, o_2 *chosen independently and uar from* $[n]$ *and conditional on* $o_1 \longleftrightarrow o_2$,

$$\frac{\mathrm{dist}_{\mathrm{NR}_n(w)}(o_1, o_2)}{\log n} \xrightarrow{\mathbb{P}} \frac{1}{\log \nu}. \tag{6.2.4}$$

The same result applies, under identical conditions, to $\mathrm{GRG}_n(w)$ *and* $\mathrm{CL}_n(w)$.

Theorem 6.2 can be seen as a special case of Theorem 6.1. However, in Theorem 6.1(a) we require that κ_n is *bounded*. In the setting of Theorem 6.2 this would imply that the maximal weight $w_{\max} = \max_{v \in [n]} w_v$ is uniformly bounded in n. Theorem 6.2 does not require this strong assumption.

We give a complete proof of Theorem 6.2 in Sections 6.3 and 6.5 below. There, we also use the ideas in the proof of Theorem 6.2 to prove Theorem 6.1. The intuition behind Theorem 6.2 is closely related to that behind Theorem 6.1. For rank-1 models, in particular, $\|T_\kappa\| = \nu = \mathbb{E}[W^2]/\mathbb{E}[W]$, which explains the relation between Theorems 6.1 and 6.2. Exercise 6.1 investigates typical distances for the Erdős–Rényi random graph.

Theorem 6.2 leaves open what happens when $\nu = \infty$. We can use Theorem 6.2 to show that $\mathrm{dist}_{\mathrm{NR}_n(w)}(o_1, o_2) = o_{\mathbb{P}}(\log n)$ (see Exercise 6.2). We next study the case where the weights have an asymptotic power-law distribution with $\tau \in (2, 3)$.

Distances in Rank-1 IRGs with Infinite-Variance Weights

We continue to study typical distances in rank-1 random graphs, in the case where the degrees obey a power law with degree exponent τ satisfying $\tau \in (2, 3)$. It turns out that the scaling of $\mathrm{dist}_{\mathrm{NR}_n(w)}(o_1, o_2)$ depends sensitively on the precise way in which $\nu_n = \mathbb{E}[W_n^2]/\mathbb{E}[W_n] \to \infty$, where W_n is the weight of a uniform vertex. Below, we assume that the vertex weights obey an asymptotic power law with exponent τ satisfying $\tau \in (2, 3)$. We later discuss what happens in related settings, for example when $\tau = 3$.

Recall from (1.3.10) that $F_n(x)$ denotes the proportion of vertices having weight at most x. Infinite-variance weights correspond to settings where $F_n(x) \approx F(x)$ and F has power-law tails having tail exponent $\tau - 1$ with $\tau \in (2, 3)$. In our setting, however, we need to know that $x \mapsto F_n(x)$ is already close to such a power law. For this, we assume that there exists a $\tau \in (2, 3)$ such that, for all $\delta > 0$, there exists $c_1 = c_1(\delta)$ and $c_2 = c_2(\delta)$ such that, uniformly in n,

$$c_1 x^{-(\tau-1+\delta)} \leq [1 - F_n](x) \leq c_2 x^{-(\tau-1-\delta)}, \tag{6.2.5}$$

where the upper bound holds for *all* $x \geq 1$, while the lower bound is required to hold only for $1 \leq x \leq n^{\beta}$ for some $\beta > \frac{1}{2}$.

The assumption in (6.2.5) precisely is what we need, and it states that $[1 - F_n](x)$ obeys power-law type bounds for appropriate values of x. Note that the lower bound in (6.2.5) cannot be valid for *all* x, since $F_n(x) > 0$ implies that $F_n(x) \geq 1/n$ so that the lower and upper bounds in (6.2.5) are contradictory when $x \gg n^{1/(\tau-1)}$. Thus, the lower bound can hold only for $x = O(n^{1/(\tau-1)})$. When $\tau \in (2, 3)$, we have that $1/(\tau-1) \in (\frac{1}{2}, 1)$, and we need the lower bound to hold only for $x \leq n^{\beta}$ for *some* $\beta \in (\frac{1}{2}, 1)$. Exercises 6.3 and 6.4 give simpler conditions for (6.2.5) in special cases, such as iid weights.

The main result on graph distances in the case of infinite-variance weights is as follows:

Theorem 6.3 (Typical distances in rank-1 random graphs with infinite-variance weights) *Consider* $\mathrm{NR}_n(w)$, *where the weights* $w = (w_v)_{v \in [n]}$ *satisfy Conditions 1.1(a),(b) and* (6.2.5). *Then, with* o_1, o_2 *chosen independently and uar from* $[n]$ *and conditional on* $o_1 \longleftrightarrow o_2$,

$$\frac{\mathrm{dist}_{\mathrm{NR}_n(w)}(o_1, o_2)}{\log \log n} \xrightarrow{\mathbb{P}} \frac{2}{|\log(\tau-2)|}. \tag{6.2.6}$$

The same result applies, under identical conditions, to $\mathrm{GRG}_n(w)$ *and* $\mathrm{CL}_n(w)$.

Theorem 6.3 implies that $\mathrm{NR}_n(w)$, with w satisfying (6.3.21), is an ultra-small world.

See Figure 6.2 for a simulation of the typical distances in $\mathrm{GRG}_n(w)$ with $\tau = 2.5$ and $\tau = 3.5$, respectively, where the distances are noticeably smaller in the ultra-small setting with $\tau = 2.5$ compared with the small-world case with $\tau = 3.5$.

In the next two sections we prove Theorems 6.2 and 6.3. The main tool to study typical distances in $\mathrm{NR}_n(w)$ is by comparison with branching processes. For $\tau > 3$, the branching-process approximation has *finite mean*, and we can make use of the martingale limit results

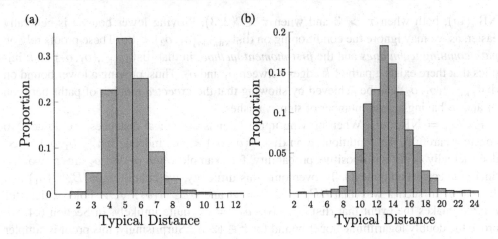

Figure 6.2 Typical distances between 2,000 pairs of vertices in the generalized random graph with $n = 100,000$, for (a) $\tau = 2.5$ and (b) $\tau = 3.5$.

for the number of individuals in generation k as $k \to \infty$, so that this number grows exponentially as ν^k. This explains the logarithmic growth of the typical distances. When $\tau \in (2,3)$, on the other hand, the branching process has infinite mean. In this case, the number of individuals in generation k, conditional on survival of the branching process, grows *super-exponentially* in k, which explains why the typical distances grow doubly logarithmically. See Section 7.4, where this is explained in more detail in the context of the configuration model.

The super-exponential growth implies that a path between two vertices typically passes through vertices with larger and larger weights as we move away from the starting and ending vertices. Thus, starting from the first vertex $o_1 \in [n]$, the path connecting o_1 to o_2 uses vertices whose weights first grow until the midpoint of the path is reached, and then decrease again to reach o_2. This can be understood by noting that the probability that a vertex with weight w is not connected to any vertex with weight larger than $y > w$ in $\mathrm{NR}_n(\boldsymbol{w})$ is

$$\exp\left\{ -\sum_{v \in [n]} w w_v \mathbb{1}_{\{w_v > y\}} / \ell_n \right\} = \exp\left\{ - w[1 - F_n^\star](y) \right\}, \qquad (6.2.7)$$

where $F_n^\star(y) = \sum_{i \in [n]} w_i \mathbb{1}_{\{w_i \leq y\}} / \ell_n$ is the distribution function of $W_n^\star$, to be introduced in (6.3.24) below. When (6.2.5) holds, it follows that $[1 - F_n^\star](y)$ is close to $y^{-(\tau-2)}$; the size-biasing increases the power by 1 (recall Lemma 1.23). Therefore, the probability that a vertex with weight w is not connected to any vertex with weight larger than $y > w$ in $\mathrm{NR}_n(\boldsymbol{w})$ is approximately $e^{-cwy^{-(\tau-2)}}$ for some $c > 0$. For w large, this probability is small when $y \ll w^{1/(\tau-2)}$. Thus, a vertex of weight w is whp connected to a vertex of weight approximately $w^{1/(\tau-2)}$, where $1/(\tau-2) > 1$ for $\tau \in (2,3)$.

Organization of the Proofs of Small-World Results

The proofs of Theorems 6.1, 6.2, and 6.3 are organized as follows. Below, in the rank-1 setting, we focus on $\mathrm{NR}_n(\boldsymbol{w})$ but all proofs will be performed simultaneously for $\mathrm{GRG}_n(\boldsymbol{w})$ and $\mathrm{CL}_n(\boldsymbol{w})$ as well. In Section 6.3 we prove lower bounds on the typical distance in

$\mathrm{NR}_n(w)$, both when $\tau > 3$ and when $\tau \in (2,3)$. Proving lower bounds is generally easier, as we may ignore the conditioning on $\mathrm{dist}_{\mathrm{NR}_n(w)}(o_1, o_2) < \infty$. These proofs rely on *path-counting techniques* and the *first-moment method*, in that $\mathrm{dist}_{\mathrm{NR}_n(w)}(o_1, o_2) = k$ implies that there exists a path of k edges between o_1 and o_2. Thus, proving a lower bound on $\mathrm{dist}_{\mathrm{NR}_n(w)}(o_1, o_2)$ can be achieved by showing that the *expected* number of paths between o_1 and o_2 having a given number of steps vanishes.

Fix $G_n = \mathrm{NR}_n(w)$. When proving upper bounds on typical distances, we do need to consider carefully the conditioning on $\mathrm{dist}_{G_n}(o_1, o_2) < \infty$. Indeed, $\mathrm{dist}_{G_n}(o_1, o_2) = \infty$ does actually occur with positive probability, for example when o_1 and o_2 are in two distinct connected components. To overcome this difficulty, we condition on $B_r^{(G_n)}(o_1)$ and $B_r^{(G_n)}(o_2)$ in such a way that $\partial B_r^{(G_n)}(o_1) \neq \varnothing$ and $\partial B_r^{(G_n)}(o_2) \neq \varnothing$ hold, which, for r large, makes the event that $\mathrm{dist}_{\mathrm{NR}_n(w)}(o_1, o_2) = \infty$ quite unlikely. In Section 6.4, we prove the doubly logarithmic upper bound for $\tau \in (2,3)$. Surprisingly, this proof is simpler than that for logarithmic distances, primarily because we know that the shortest paths for $\tau \in (2,3)$ generally go from lower-weight vertices to higher-weight ones, until the hubs are reached, and then they go back.

In Section 6.5 we investigate the *variance* of the number of paths between sets of vertices in $\mathrm{NR}_n(w)$, using an intricate path-counting method that estimates the sum, over pairs of paths, of the probability that they are *both* occupied. For this, the precise *joint topology* of these pairs of paths is crucial. We use a second-moment method to show that, under the conditional laws, given $B_r^{(G_n)}(o_1)$ and $B_r^{(G_n)}(o_2)$ such that $\partial B_r^{(G_n)}(o_1) \neq \varnothing$ and $\partial B_r^{(G_n)}(o_2) \neq \varnothing$, whp there is a path of appropriate length linking $\partial B_r^{(G_n)}(o_1)$ and $\partial B_r^{(G_n)}(o_2)$. This proves the logarithmic upper bound when $\tau > 3$. In each of our proofs, we formulate the precise results as separate theorems and prove them under conditions that are slightly weaker than those in Theorems 6.1, 6.2, and 6.3.

6.3 TYPICAL-DISTANCE LOWER BOUNDS IN INHOMOGENEOUS RANDOM GRAPHS

In this section we prove lower bounds on typical distances. In Section 6.3.1 we prove the lower bound in Theorem 6.1(a), first in the setting of Theorem 6.2; this is followed by the proof of Theorem 6.1(a). In Section 6.3.2 we prove the doubly logarithmic lower bound on distances for infinite-variance degrees for $\mathrm{NR}_n(w)$ in Theorem 6.3.

6.3.1 LOGARITHMIC LOWER BOUND DISTANCES FOR FINITE-VARIANCE DEGREES

In this subsection we prove a logarithmic lower bound on the typical distance in $\mathrm{NR}_n(w)$:

Theorem 6.4 (Logarithmic lower bound on typical distances $\mathrm{NR}_n(w)$) *Assume that*

$$\limsup_{n \to \infty} \nu_n = \nu \in (1, \infty), \qquad where \qquad \nu_n = \frac{\mathbb{E}[W_n^2]}{\mathbb{E}[W_n]} = \frac{\sum_{v \in [n]} w_v^2}{\sum_{v \in [n]} w_v}. \qquad (6.3.1)$$

Then, for any $\varepsilon > 0$, with o_1, o_2 chosen independently and uar from $[n]$,

$$\mathbb{P}(\mathrm{dist}_{\mathrm{NR}_n(w)}(o_1, o_2) \leq (1 - \varepsilon)\log_\nu n) = o(1). \qquad (6.3.2)$$

The same result applies, under identical conditions, to $\mathrm{GRG}_n(w)$ and $\mathrm{CL}_n(w)$.

Proof The idea behind the proof of Theorem 6.4 is that it is quite unlikely for a path containing far fewer than $\log_\nu n$ edges to exist. In order to show this, we use a first-moment bound and show that the *expected* number of occupied paths connecting the two vertices chosen uar from $[n]$ having length at most $(1 - \varepsilon) \log_\nu n$ is $o(1)$. We will now fill in the details.

We set $k_n = \lceil (1 - \varepsilon) \log_\nu n \rceil$. Then, conditioning on o_1, o_2 gives

$$\mathbb{P}(\text{dist}_{\text{NR}_n(w)}(o_1, o_2) \leq k_n) = \frac{1}{n^2} \sum_{u,v \in [n]} \sum_{k=0}^{k_n} \mathbb{P}(\text{dist}_{\text{NR}_n(w)}(u, v) = k)$$

$$= \frac{1}{n} + \frac{1}{n^2} \sum_{u,v \in [n]: \, u \neq v} \sum_{k=1}^{k_n} \mathbb{P}(\text{dist}_{\text{NR}_n(w)}(u, v) = k). \quad (6.3.3)$$

In this section and in Section 6.5, we make use of *path-counting techniques* (see in particular Section 6.5.1). Here, we show that short paths are unlikely to exist by giving upper bounds on the expected number of paths of various types. In Section 6.5.1 we give bounds on the *variance* of the number of paths of various types, so as to show that long paths are quite likely to exist. Such variance bounds are quite challenging, and here we give some basics to highlight the main ideas in a much simpler setting.

Definition 6.5 (Paths in inhomogeneous random graphs) Let us fix $k \geq 1$. A *path* $\vec{\pi} = (\pi_0, \ldots, \pi_k)$ of length k between vertices u and v is a sequence of vertices connecting $\pi_0 = u$ to $\pi_k = v$. We call a path $\vec{\pi}$ *self-avoiding* when it visits every vertex at most once, i.e., $\pi_i \neq \pi_j$ for every $i \neq j$. Let $\mathcal{P}_k(u, v)$ denote the set of k-step self-avoiding paths between vertices u and v, and let $\mathcal{P}_k(u)$ denote the set of k-step self-avoiding paths starting from u. We say that $\vec{\pi}$ is *occupied* when all edges in $\vec{\pi}$ are occupied in $\text{NR}_n(w)$. ◀

See Figure 6.3 for an example of a 12-step self-avoiding path between u and v.

When $\text{dist}_{\text{NR}_n(w)}(u, v) = k$, there must be a path of length k such that all edges $\{\pi_l, \pi_{l+1}\}$ are occupied in $\text{NR}_n(w)$, for $l = 0, \ldots, k - 1$. The probability that the edge $\{\pi_l, \pi_{l+1}\}$ is occupied in $\text{NR}_n(w)$ is equal to

$$1 - e^{-w_{\pi_l} w_{\pi_{l+1}}/\ell_n} \leq w_{\pi_l} w_{\pi_{l+1}}/\ell_n. \quad (6.3.4)$$

For $\text{CL}_n(w)$ and $\text{GRG}_n(w)$, an identical upper bound holds, which explains why the proof of Theorem 6.4 for $\text{NR}_n(w)$ applies verbatim to those models. By the union bound or Boole's inequality,

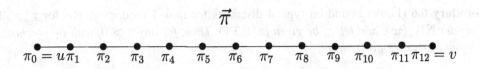

$$\vec{\pi}$$

$$\pi_0 = u \quad \pi_1 \quad \pi_2 \quad \pi_3 \quad \pi_4 \quad \pi_5 \quad \pi_6 \quad \pi_7 \quad \pi_8 \quad \pi_9 \quad \pi_{10} \quad \pi_{11} \quad \pi_{12} = v$$

Figure 6.3 A 12-step self-avoiding path $\vec{\pi}$ connecting vertices u and v.

$$\mathbb{P}(\mathrm{dist}_{\mathrm{NR}_n(w)}(u,v)=k) \leq \mathbb{P}(\exists \vec{\pi} \in \mathcal{P}_k(u,v) \colon \vec{\pi} \text{ occupied in } \mathrm{NR}_n(w))$$

$$\leq \sum_{\vec{\pi} \in \mathcal{P}_k(u,v)} \mathbb{P}(\vec{\pi} \text{ occupied in } \mathrm{NR}_n(w)). \qquad (6.3.5)$$

For any path $\vec{\pi} \in \mathcal{P}_k(u,v)$,

$$\mathbb{P}(\vec{\pi} \text{ occupied}) = \prod_{s=0}^{k-1} \mathbb{P}(\{\pi_l,\pi_{l+1}\} \text{ occupied in } \mathrm{NR}_n(w)) \leq \prod_{l=0}^{k-1} \frac{w_{\pi_l} w_{\pi_{l+1}}}{\ell_n}$$

$$= \frac{w_{\pi_0} w_{\pi_k}}{\ell_n} \prod_{l=1}^{k-1} \frac{w_{\pi_l}^2}{\ell_n} = \frac{w_u w_v}{\ell_n} \prod_{l=1}^{k-1} \frac{w_{\pi_l}^2}{\ell_n}. \qquad (6.3.6)$$

Therefore

$$\mathbb{P}(\mathrm{dist}_{\mathrm{NR}_n(w)}(u,v)=k) \leq \frac{w_u w_v}{\ell_n} \sum_{\vec{\pi} \in \mathcal{P}_k(u,v)} \prod_{l=1}^{k-1} \frac{w_{\pi_l}^2}{\ell_n}$$

$$\leq \frac{w_u w_v}{\ell_n} \prod_{l=1}^{k-1} \left(\sum_{\pi_l \in [n]} \frac{w_{\pi_l}^2}{\ell_n} \right) = \frac{w_u w_v}{\ell_n} \nu_n^{k-1}, \qquad (6.3.7)$$

where ν_n is defined in (6.3.1). We conclude that

$$\mathbb{P}(\mathrm{dist}_{\mathrm{NR}_n(w)}(o_1,o_2) \leq k_n) \leq \frac{1}{n} + \frac{1}{n^2} \sum_{u,v \in [n]} \sum_{k=1}^{k_n} \frac{w_u w_v}{\ell_n} \nu_n^{k-1} = \frac{1}{n} + \frac{\ell_n}{n^2} \sum_{k=1}^{k_n} \nu_n^{k-1}$$

$$= \frac{1}{n} + \frac{\ell_n}{n^2} \frac{\nu_n^{k_n}-1}{\nu_n-1}. \qquad (6.3.8)$$

By (6.3.1), $\limsup_{n\to\infty} \nu_n = \nu \in (1,\infty)$, so that, for n large enough, $\nu_n \geq \nu - \delta > 1$ while $\ell_n/n = \mathbb{E}[W_n] \to \mathbb{E}[W] < \infty$. Thus, since $\nu \mapsto (\nu^k - 1)/(\nu - 1)$ is increasing for every integer $k \geq 0$,

$$\mathbb{P}(\mathrm{dist}_{\mathrm{NR}_n(w)}(o_1,o_2) \leq k_n) \leq O((\nu+\delta)^{k_n}/n) = o(1), \qquad (6.3.9)$$

when $\delta = \delta(\varepsilon) > 0$ is chosen such that $(1-\varepsilon)/\log(\nu+\delta) < 1$, since $k_n = \lceil (1-\varepsilon)\log_\nu n \rceil$. This completes the proof of Theorem 6.4. $\qquad \square$

The condition (6.3.1) is slightly weaker than Condition 1.1(c), which is assumed in Theorem 6.2, as shown in Exercises 6.5 and 6.6. Exercise 6.7 extends the proof of Theorem 6.4 to show that $(\mathrm{dist}_{\mathrm{NR}_n(w)}(o_1,o_2) - \log n/\log\nu_n)_-$ is tight, where we write $(x)_- = \max\{-x, 0\}$.

We close this section by extending the above result to settings where ν_n is not necessarily bounded, the most interesting case being $\tau = 3$:

Corollary 6.6 (Lower bound on typical distances for rank-1 random graphs for $\tau = 3$) *Consider* $\mathrm{NR}_n(w)$, *and let* ν_n *be given in* (6.3.1). *Then, for any* $\varepsilon > 0$, *with* o_1, o_2 *chosen independently and uar from* $[n]$,

$$\mathbb{P}(\mathrm{dist}_{\mathrm{NR}_n(w)}(o_1,o_2) \leq (1-\varepsilon)\log_{\nu_n} n) = o(1). \qquad (6.3.10)$$

The same result applies, under identical conditions, to $\mathrm{GRG}_n(w)$ *and* $\mathrm{CL}_n(w)$.

The proof of Corollary 6.6 is left as Exercise 6.8. In the case where $\tau = 3$ and $[1 - F_n](x)$ is, for a large range of x values, of order x^{-2} (which is stronger than $\tau = 3$), it can be expected that $\nu_n = \Theta(\log n)$, so that, in that case,

$$\mathbb{P}\Big(\text{dist}_{\text{NR}_n(w)}(o_1, o_2) \leq (1 - \varepsilon)\frac{\log n}{\log \log n}\Big) = o(1). \tag{6.3.11}$$

Exercise 6.9 investigates the situation where $\tau = 3$. Exercise 6.10 considers the case $\tau \in (2, 3)$, where Corollary 6.6 unfortunately does not give particularly interesting results.

Lower Bound on Typical Distances for General IRGs: Proof of Theorem 6.1(a)

The proof of the upper bound in Theorem 6.1(a) is closely related to that in Theorem 6.4. Note that

$$\mathbb{P}(\text{dist}_{\text{IRG}_n(\kappa_n)}(u, v) = k) \leq \sum_{\pi_1,\ldots,\pi_{k-1} \in [n]} \prod_{l=0}^{k-1} \frac{\kappa_n(x_{\pi_l}, x_{\pi_{l+1}})}{n}, \tag{6.3.12}$$

where, by convention, $\pi_0 = u, \pi_k = v$ and we can restrict the vertices to be *distinct*. Therefore,

$$\mathbb{P}(\text{dist}_{\text{IRG}_n(\kappa_n)}(o_1, o_2) = k) \leq \frac{1}{n^{k+2}} \sum_{\pi_0, \pi_1, \ldots, \pi_k \in [n]} \prod_{l=0}^{k-1} \kappa_n(x_{\pi_l}, x_{\pi_{l+1}}). \tag{6.3.13}$$

If the above $(k + 1)$-dimensional discrete integral were to be replaced by a continuous integral and κ_n by its limit κ then we would arrive at

$$\frac{1}{n} \int_S \cdots \int_S \prod_{l=0}^{k-1} \kappa(x_l, x_{l+1}) \prod_{i=0}^{k} \mu(\mathrm{d}x_i) = \frac{1}{n}\|T_\kappa^k \mathbb{1}\|_1, \tag{6.3.14}$$

which is bounded from above by $\frac{1}{n}\|T_\kappa\|^k$. Repeating the bound in (6.3.9) would then prove that, when $\nu = \|T_\kappa\| > 1$,

$$\mathbb{P}(\text{dist}_{\text{IRG}_n(\kappa_n)}(o_1, o_2) \leq (1 - \varepsilon)\log_\nu n) = o(1). \tag{6.3.15}$$

However, in the general case, it is not so easy to replace the $(k + 1)$-fold discrete sum in (6.3.13) by a $(k + 1)$-fold integral. We next explain how nevertheless this can be done, starting with the finite-type case.

In the finite-type case, where the types are given by $[t]$, (6.3.12) turns into

$$\mathbb{P}(\text{dist}_{\text{IRG}_n(\kappa)}(o_1, o_2) = k) \leq \frac{1}{n} \sum_{i_0,\ldots,i_k \in [t]} \frac{n_{i_0}}{n} \prod_{l=0}^{k-1} \kappa^{(n)}(i_l, i_{l+1}) \frac{n_{i_{l+1}}}{n}, \tag{6.3.16}$$

where n_i denotes the number of vertices of type $i \in [t]$ and where the probability that there exists an edge between vertices of types i and j is equal to $\kappa^{(n)}(i, j)/n$.

Under the conditions in Theorem 6.1(a), we have $\mu_i^{(n)} = n_i/n \to \mu(i)$ and $\kappa^{(n)}(i, j) \to \kappa(i, j)$ as $n \to \infty$. This also implies that $\|T_{\kappa_n}\| \to \nu$, where ν is largest eigenvalue of the matrix $\mathbf{M} = (\mathbf{M}_{ij})_{i,j \in [t]}$ with $\mathbf{M}_{ij} = \kappa(i,j)\mu(j)$. Denoting $\mathbf{M}_{ij}^{(n)} = \kappa^{(n)}(i,j)n_j/n \to \mathbf{M}_{ij}$, we obtain

$$\mathbb{P}(\text{dist}_{\text{IRG}_n(\kappa)}(o_1,o_2) = k) \leq \frac{1}{n}\langle(\boldsymbol{\mu}^{(n)})^T, [\mathbf{M}^{(n)}]^k \mathbf{1}\rangle, \tag{6.3.17}$$

where $\mathbf{1}$ is the all-1s vector. Obviously, since there are $t < \infty$ types,

$$\langle(\boldsymbol{\mu}^{(n)})^T, [\mathbf{M}^{(n)}]^k \mathbf{1}\rangle \leq \|\mathbf{M}^{(n)}\|^k \|\boldsymbol{\mu}^{(n)}\| \|\mathbf{1}\| \leq \|\mathbf{M}^{(n)}\|^k \sqrt{t}. \tag{6.3.18}$$

Thus,

$$\mathbb{P}(\text{dist}_{\text{IRG}_n(\kappa)}(o_1,o_2) = k) \leq \frac{\sqrt{t}}{n}\|\mathbf{M}^{(n)}\|^k. \tag{6.3.19}$$

We conclude that

$$\mathbb{P}(\text{dist}_{\text{IRG}_n(\kappa_n)}(o_1,o_2) \leq (1-\varepsilon)\log_{\nu_n} n) = o(1), \tag{6.3.20}$$

where $\nu_n = \|\mathbf{M}^{(n)}\| \to \nu$. This proves Theorem 6.1(a) in the finite-type setting.

We next extend the proof of Theorem 6.1(a) to the infinite-type setting. Assume that the conditions in Theorem 6.1(a) hold. Recall the bound in (3.3.20), which bounds κ_n from above by $\bar{\kappa}_m$, which is of finite type. Then, use the fact that $\|T_{\bar{\kappa}_m}\| \searrow \|T_\kappa\| = \nu > 1$ to conclude that $\mathbb{P}(\text{dist}_{\text{IRG}_n(\kappa_n)}(o_1,o_2) \leq (1-\varepsilon)\log_\nu n) = o(1)$ holds under the conditions of Theorem 6.1(a). This completes the proof of Theorem 6.1(a). $\qquad\square$

6.3.2 Doubly Logarithmic Lower Bound for Infinite-Variance Weights

In this subsection we prove a doubly logarithmic lower bound on the typical distances of rank-1 random graphs for $\tau \in (2,3)$. The main result we prove is the following theorem:

Theorem 6.7 (Log log lower bound on typical distances in rank-1 random graphs) *Consider* $\text{NR}_n(\boldsymbol{w})$, *where the weights* $\boldsymbol{w} = (w_v)_{v \in [n]}$ *satisfy Condition 1.1(a), and suppose that there exist* $\tau \in (2,3)$ *and* $c_2 < \infty$ *such that, for all* $x \geq 1$,

$$[1 - F_n](x) \leq c_2 x^{-(\tau-1)}. \tag{6.3.21}$$

Then, for every $\varepsilon > 0$, *with* o_1, o_2 *chosen independently and uar from* $[n]$,

$$\mathbb{P}\left(\text{dist}_{\text{NR}_n(\boldsymbol{w})}(o_1,o_2) \leq (1-\varepsilon)\frac{2\log\log n}{|\log(\tau-2)|}\right) = o(1). \tag{6.3.22}$$

The same result applies, under identical conditions, to $\text{GRG}_n(\boldsymbol{w})$ *and* $\text{CL}_n(\boldsymbol{w})$.

Below, we rely on the fact that (6.3.21) implies that the degree $D_n = d_o$ of a uniform vertex is such that $(D_n)_{n \geq 1}$ is uniformly integrable, so that Condition 1.1(b) follows from Condition 1.1(a) and (6.3.21).

We follow the proof of Theorem 6.4 as closely as possible. The problem with that proof is that, under the condition in (6.3.21), ν_n is too large. Indeed, Exercise 6.10 shows that the lower bound obtained in Corollary 6.6 is a constant, which is not very useful. What goes wrong in that argument is that vertices with extremely high weights provide the main contribution to ν_n, and hence to the upper bound as in (6.3.8). However, this argument completely ignores the fact that it is quite *unlikely* that such a high-weight vertex appears in a path. Indeed, as argued in (6.2.7), the probability that a vertex with weight w is directly connected to a vertex having weight at least y is at most

$$\sum_{v \in [n]} \frac{ww_v \mathbb{1}_{\{w_v > y\}}}{\ell_n} = w[1 - F_n^{\star}](y), \qquad (6.3.23)$$

where

$$F_n^{\star}(x) = \frac{1}{\ell_n} \sum_{i \in [n]} w_i \mathbb{1}_{\{w_i \leq x\}}, \qquad (6.3.24)$$

so that (6.3.23) is small when y is too large. The main contribution to ν_n, on the other hand, comes from vertices having maximal weight of the order $n^{1/(\tau-1)}$.

This problem is resolved by a suitable truncation argument on the weights of vertices in occupied paths, which effectively removes these high-weight vertices. Therefore, instead of obtaining $\nu_n = \sum_{v \in [n]} w_v^2/\ell_n$, we obtain a version of this sum *restricted to vertices having a relatively small weight*. Effectively, this means that we split the space of all paths into *good paths*, i.e., paths that avoid high-weight vertices, and *bad paths*, which are paths that use high-weight vertices.

We now present the details of this argument. We again start from

$$\mathbb{P}(\text{dist}_{\text{NR}_n(w)}(o_1, o_2) \leq k_n) = \frac{1}{n} + \frac{1}{n^2} \sum_{u,v \in [n]: \, u \neq v} \mathbb{P}(\text{dist}_{\text{NR}_n(w)}(u, v) \leq k_n). \quad (6.3.25)$$

When $\text{dist}_{\text{NR}_n(w)}(u, v) \leq k_n$, there exists an occupied path $\vec{\pi} \in \mathcal{P}_k(u, v)$ for some $k \leq k_n$.

We fix an *increasing* sequence of numbers $(b_l)_{l \geq 0}$ that serve as truncation values for the weights of vertices along our occupied path. We determine the precise values of $(b_l)_{l \geq 0}$, which is a quite delicate procedure, below.

Definition 6.8 (Good and bad paths)   Fix $k \geq 1$. Recall the definitions of k-step self-avoiding paths $\mathcal{P}_k(u, v)$ and $\mathcal{P}_k(u)$ from Definition 6.5. We say that a path $\vec{\pi} \in \mathcal{P}_k(u, v)$ is *good* when $w_{\pi_l} \leq b_l \wedge b_{k-l}$ for every $l \in [k]$, and *bad* otherwise. Let $\mathcal{GP}_k(u, v)$ be the set of good paths in $\mathcal{P}_k(u, v)$, and let

$$\mathcal{BP}_k(u) = \{\vec{\pi} \in \mathcal{P}_k(u): w_{\pi_k} > b_k, w_{\pi_l} \leq b_l \, \forall l < k\} \qquad (6.3.26)$$

denote the set of *bad paths* of length k starting in u. ◀

The condition $w_{\pi_l} \leq b_l \wedge b_{k-l}$ for every $l = 0, \ldots, k$ is equivalent to the statement that $w_{\pi_l} \leq b_l$ for $l \leq k/2$, while $w_{\pi_l} \leq b_{k-l}$ for $k/2 \leq l \leq k$. Thus, b_l provides an upper bound on the weight of the lth and $(k-l)$th vertices of the occupied path, ensuring that the weights in it cannot be too large. See Figure 6.4 for a visualization of a good path and the bounds on the weight of its vertices.

Let

$$\mathcal{E}_k(u, v) = \{\exists \vec{\pi} \in \mathcal{GP}_k(u, v): \vec{\pi} \text{ occupied}\} \qquad (6.3.27)$$

denote the event that there exists a good path of length k between u and v.

Let $\mathcal{F}_k(u)$ be the event that there exists a bad path of length k starting from u, i.e.,

$$\mathcal{F}_k(u) = \{\exists \vec{\pi} \in \mathcal{BP}_k(u): \vec{\pi} \text{ occupied}\}. \qquad (6.3.28)$$

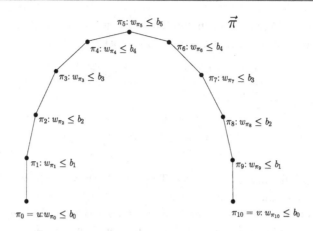

Figure 6.4 A 10-step good path connecting $\pi_0 = u$ and $\pi_{10} = v$ and the upper bounds on the weight of its vertices. Vertices with large weights are higher in the figure.

Then, since $\mathrm{dist}_{\mathrm{NR}_n(w)}(u, v) \leq k_n$ implies that there either is a good path between vertices u and v, or a bad path starting in u or in v, for $u \neq v$,

$$\{\mathrm{dist}_{\mathrm{NR}_n(w)}(u, v) \leq k_n\} \subseteq \bigcup_{k=1}^{k_n} \Big(\mathcal{F}_k(u) \cup \mathcal{F}_k(v) \cup \mathcal{E}_k(u, v)\Big), \qquad (6.3.29)$$

so that, by Boole's inequality and for $u \neq v$,

$$\mathbb{P}(\mathrm{dist}_{\mathrm{NR}_n(w)}(u, v) \leq k_n) \leq \sum_{k=1}^{k_n} \big[\mathbb{P}(\mathcal{F}_k(u)) + \mathbb{P}(\mathcal{F}_k(v)) + \mathbb{P}(\mathcal{E}_k(u, v))\big]. \qquad (6.3.30)$$

In order to estimate the probabilities $\mathbb{P}(\mathcal{F}_k(u))$ and $\mathbb{P}(\mathcal{E}_k(u, v))$, we introduce some notation. For $b \geq 0$, define the *truncated second moment*

$$\nu_n(b) = \frac{1}{\ell_n} \sum_{i \in [n]} w_i^2 \mathbb{1}_{\{w_i \leq b\}} \qquad (6.3.31)$$

to be the restriction of ν_n to vertices of weight at most b, and recall that $F_n^\star(x)$ from (6.3.24) denotes the distribution function of $W_n^\star$, the size-biased version of W_n. The following lemma gives bounds on $\mathbb{P}(\mathcal{F}_k(u))$ and $\mathbb{P}(\mathcal{E}_k(u, v))$ in terms of the tail distribution function $1 - F_n^\star$ and $\nu_n(b)$, which, in turn, we bound using Lemmas 1.23 and 1.22, respectively:

Lemma 6.9 (Truncated path probabilities) *For every $k \geq 1$, $(b_l)_{l \geq 0}$ with $b_l \geq 0$ and $l \mapsto b_l$ non-decreasing, in $\mathrm{NR}_n(\boldsymbol{w})$, $\mathrm{CL}_n(\boldsymbol{w})$, and $\mathrm{GRG}_n(\boldsymbol{w})$,*

$$\mathbb{P}(\mathcal{F}_k(u)) \leq w_u[1 - F_n^\star](b_k) \prod_{l=1}^{k-1} \nu_n(b_l) \qquad (6.3.32)$$

and

$$\mathbb{P}(\mathcal{E}_k(u,v)) \leq \frac{w_u w_v}{\ell_n} \prod_{l=1}^{k-1} \nu_n(b_l \wedge b_{k-l}). \tag{6.3.33}$$

When $b_l = \infty$ for each l, the bound in (6.3.33) equals that obtained in (6.3.7).

Proof   We start by proving (6.3.32). By Boole's inequality,

$$\mathbb{P}(\mathcal{F}_k(u)) = \mathbb{P}(\exists \vec{\pi} \in \mathcal{BP}_k(u) : \vec{\pi} \text{ occupied in } \mathrm{NR}_n(\boldsymbol{w}))$$

$$\leq \sum_{\vec{\pi} \in \mathcal{BP}_k(u)} \mathbb{P}(\vec{\pi} \text{ occupied in } \mathrm{NR}_n(\boldsymbol{w})). \tag{6.3.34}$$

By (6.3.6), (6.3.31), and (6.3.24),

$$\mathbb{P}(\mathcal{F}_k(u)) \leq \sum_{\vec{\pi} \in \mathcal{BP}_k(u)} \frac{w_u w_{\pi_k}}{\ell_n} \prod_{l=1}^{k-1} \frac{w_{\pi_l}^2}{\ell_n}$$

$$\leq w_u \sum_{\pi_k : \, w_{\pi_k} > b_k} \frac{w_{\pi_k}}{\ell_n} \times \prod_{l=1}^{k-1} \sum_{\pi_l : \, w_{\pi_l} \leq b_l} \frac{w_{\pi_l}^2}{\ell_n}$$

$$= w_u [1 - F_n^\star](b_k) \prod_{l=1}^{k-1} \nu_n(b_l). \tag{6.3.35}$$

The same bound applies to $\mathrm{CL}_n(\boldsymbol{w})$ and $\mathrm{GRG}_n(\boldsymbol{w})$.

The proof of (6.3.33) is similar. Indeed, by (6.3.6),

$$\mathbb{P}(\mathcal{E}_k(u,v)) \leq \sum_{\vec{\pi} \in \mathcal{GP}_k(u,v)} \frac{w_u w_v}{\ell_n} \prod_{l=1}^{k-1} \frac{w_{\pi_l}^2}{\ell_n} \leq \frac{w_u w_v}{\ell_n} \prod_{l=1}^{k-1} \nu_n(b_l \wedge b_{k-l}), \tag{6.3.36}$$

since $w_{\pi_l} \leq b_l \wedge b_{k-l}$. Now follow the steps in the proof of (6.3.32). Again the same bound applies to $\mathrm{CL}_n(\boldsymbol{w})$ and $\mathrm{GRG}_n(\boldsymbol{w})$. $\square$

In order to apply Lemma 6.9 effectively, we use Lemmas 1.22 and 1.23 to derive bounds on $[1 - F_n^\star](x)$ and $\nu_n(b)$:

Lemma 6.10 (Bounds on sums)   *Suppose that the weights $\boldsymbol{w} = (w_v)_{v \in [n]}$ satisfy Condition 1.1(a) and that there exist $\tau \in (2,3)$ and c_2 such that, for all $x \geq 1$, (6.3.21) holds. Then, there exists a constant $c_2^\star > 0$ such that, for all $x \geq 1$,*

$$[1 - F_n^\star](x) \leq c_2^\star x^{-(\tau-2)}, \tag{6.3.37}$$

and there exists a $c_\nu > 0$ such that, for all $b \geq 1$,

$$\nu_n(b) \leq c_\nu b^{3-\tau}. \tag{6.3.38}$$

Proof   The bound in (6.3.37) follows from Lemma 1.23, and the bound in (6.3.38) from (1.4.13) in Lemma 1.22 with $a = 2 > \tau - 1$ when $\tau \in (2,3)$. For both lemmas, the assumptions follow from (6.3.21). See Exercise 6.13 below for the bound on $\nu_n(b)$ in (6.3.38). $\square$

With Lemmas 6.9 and 6.10 in hand, we are ready to choose $(b_l)_{l \geq 0}$ and to complete the proof of Theorem 6.7:

Proof of Theorem 6.7. Take $k_n = \lceil 2(1 - \varepsilon) \log \log n / |\log(\tau - 2)| \rceil$. By (6.3.25) and (6.3.29),

$$\mathbb{P}(\mathrm{dist}_{\mathrm{NR}_n(w)}(o_1, o_2) \leq k_n) \leq \frac{1}{n} + \sum_{k=1}^{k_n} \left[\frac{2}{n} \sum_{u \in [n]} \mathbb{P}(\mathcal{F}_k(u)) + \frac{1}{n^2} \sum_{u,v \in [n]:\, u \neq v} \mathbb{P}(\mathcal{E}_k(u, v)) \right],$$

$$(6.3.39)$$

where the term $1/n$ is due to $o_1 = o_2$ for which $\mathrm{dist}_{\mathrm{NR}_n(w)}(o_1, o_2) = 0$. We use Lemmas 6.9 and 6.10 to provide bounds on $\mathbb{P}(\mathcal{F}_k(u))$ and $\mathbb{P}(\mathcal{E}_k(u, v))$. These bounds are quite similar.

We first describe how we choose the truncation values $(b_l)_{l \geq 0}$ in such a way that $[1 - F_n^\star](b_k)$ is small enough to make $\mathbb{P}(\mathcal{F}_k(u))$ small, and, for this choice of $(b_l)_{l \geq 0}$, we show that the contribution due to $\mathbb{P}(\mathcal{E}_k(u, v))$ is small. This means that it is quite unlikely that u or v is connected to a vertex at distance k with too high a weight, i.e., with a weight at least b_k. At the same time, it is also unlikely that there is a good path $\vec{\pi} \in \mathcal{P}_k(u, v)$ whose weights are all small, i.e., for which $w_{\pi_k} \leq b_k$ for every $k \leq k_n$, because k is too small to achieve this.

By Lemma 6.9, we wish to choose b_k in such a way that

$$\frac{1}{n} \sum_{u \in [n]} \mathbb{P}(\mathcal{F}_k(u)) \leq \frac{\ell_n}{n} [1 - F_n^\star](b_k) \prod_{l=0}^{k-1} \nu_n(b_l) \qquad (6.3.40)$$

is small. Below (6.2.7), it was argued that we should choose b_k such that $b_k \approx b_{k-1}^{1/(\tau-2)}$. In order to make the contribution due to $\mathbb{P}(\mathcal{F}_k(u))$ small, however, we will take b_k somewhat larger. We now make this argument precise.

We take $\delta \in (0, \tau - 2)$ sufficiently small and let

$$a = 1/(\tau - 2 - \delta) > 1. \qquad (6.3.41)$$

Take $b_0 = e^A$ for some constant $A \geq 0$ sufficiently large, and define $(b_l)_{l \geq 0}$ recursively by

$$b_l = b_{l-1}^a, \qquad \text{which implies that} \qquad b_l = b_0^{a^l} = e^{A(\tau-2-\delta)^{-l}}. \qquad (6.3.42)$$

We will start from (6.3.30). By Lemma 6.9, we obtain an upper bound on $\mathbb{P}(\mathcal{F}_k(u))$ in terms of factors $\nu_n(b_l)$ and $[1 - F_n^\star](b_k)$, which are bounded in Lemma 6.10. We start by applying the bound on $\nu_n(b_l)$ to obtain

$$\prod_{l=1}^{k-1} \nu_n(b_l) \leq \prod_{l=1}^{k-1} c_\nu b_l^{3-\tau} = c_\nu^{k-1} e^{A(3-\tau) \sum_{l=1}^{k-1} a^l}$$

$$\leq c_\nu^{k-1} e^{A(3-\tau)a^k/(a-1)} = c_\nu^{k-1} b_k^{(3-\tau)/(a-1)}. \qquad (6.3.43)$$

Combining (6.3.43) with the bound on $[1 - F_n^\star](b_k)$ in Lemma 6.10 yields, for $k \geq 1$,

$$\mathbb{P}(\mathcal{F}_k(u)) \leq c_2^\star w_u c_\nu^{k-1} b_k^{-(\tau-2)+(3-\tau)/(a-1)}. \qquad (6.3.44)$$

Since $3 - \tau + \delta < 1$ when $\tau \in (2, 3)$ and $\delta \in (0, \tau - 2)$, we have

$$(\tau - 2) - (3 - \tau)/(a - 1) = (\tau - 2) - (3 - \tau)(\tau - 2 - \delta)/(3 - \tau + \delta)$$

$$= \delta/(3 - \tau + \delta) > \delta, \qquad (6.3.45)$$

so that, for $k \geq 1$,

$$\mathbb{P}(\mathcal{F}_k(u)) \leq c_2^\star w_u c_\nu^{k-1} b_k^{-\delta}. \tag{6.3.46}$$

As a result, for each $\delta > 0$,

$$\frac{1}{n} \sum_{u \in [n]} \sum_{k=1}^{k_n} \mathbb{P}(\mathcal{F}_k(u)) \leq \frac{1}{n} \sum_{u \in [n]} c_2^\star w_u \sum_{k \geq 1} c_\nu^{k-1} b_k^{-\delta} = O(1) \sum_{k \geq 1} c_\nu^{k-1} b_k^{-\delta} \leq \varepsilon, \tag{6.3.47}$$

by (6.3.42), taking $A = A(\delta, \varepsilon)$ sufficiently large.

Similarly, since $b_l \geq 1$, by (6.3.43) and the fact that $l \mapsto b_l$ is non-decreasing,

$$\mathbb{P}(\mathcal{E}_k(u, v)) \leq \frac{w_u w_v}{\ell_n} \prod_{l=1}^{k-1} \nu_n(b_l \wedge b_{k-l}) \leq \frac{w_u w_v}{\ell_n} c_\nu^{k-1} b_{\lceil k/2 \rceil}^{2(3-\tau)/(a-1)}, \tag{6.3.48}$$

so that, again using that $l \mapsto b_l$ is non-decreasing,

$$\sum_{k=1}^{k_n} \frac{1}{n^2} \sum_{u,v \in [n]} \mathbb{P}(\mathcal{E}_k(u, v)) \leq \frac{1}{n^2} \sum_{k=1}^{k_n} \sum_{u,v \in [n]} \frac{w_u w_v}{\ell_n} c_\nu^{k-1} b_{\lceil k/2 \rceil}^{2(3-\tau)/(a-1)}$$

$$\leq \frac{\ell_n}{n^2} k_n c_\nu^{k_n-1} b_{\lceil k_n/2 \rceil}^{2(3-\tau)/(a-1)}, \tag{6.3.49}$$

by (6.3.42). We complete the proof by analyzing this bound.

Recall that $k \leq k_n = \lceil 2(1 - \varepsilon) \log\log n / |\log(\tau - 2)| \rceil$. Take $\delta = \delta(\varepsilon) > 0$ small enough that $(\tau - 2 - \delta)^{-(k_n+1)/2} \leq (\log n)^{1-\varepsilon/4}$. Then, by (6.3.42),

$$b_{\lceil k_n/2 \rceil} \leq e^{A(\tau-2-\delta)^{-(k_n+1)/2}} \leq e^{A(\log n)^{1-\varepsilon/4}}, \tag{6.3.50}$$

and we conclude that

$$\sum_{k=1}^{k_n} \frac{1}{n^2} \sum_{u,v \in [n]} \mathbb{P}(\mathcal{E}_k(u, v)) \leq \frac{\ell_n}{n^2} k_n c_\nu^{k_n} \exp\left(2A(3 - \tau)(\log n)^{1-\varepsilon/4}\right) = o(1), \tag{6.3.51}$$

since $k_n = O(\log\log n)$ and $\ell_n/n^2 = \Theta(1/n)$. This completes the proof of Theorem 6.7. $\qquad\square$

6.4 DOUBLY LOGARITHMIC UPPER BOUND FOR INFINITE-VARIANCE WEIGHTS

In this section we prove the doubly logarithmic upper bound on typical distances in the case where the asymptotic weight distribution has infinite variance. Throughout this section, we assume that there exist $\tau \in (2, 3)$, $\beta > \frac{1}{2}$ and $c_1 > 0$ such that, uniformly in n and $x \leq n^\beta$,

$$[1 - F_n](x) \geq c_1 x^{-(\tau-1)}. \tag{6.4.1}$$

The bound in (6.4.1) corresponds to the lower bound in (6.2.5). The main result in this section is the following theorem:

Theorem 6.11 (Doubly logarithmic upper bound on typical distance for $\tau \in (2,3)$) *Consider* $\mathrm{NR}_n(\boldsymbol{w})$, *where the weights* $\boldsymbol{w} = (w_v)_{v\in[n]}$ *satisfy Conditions 1.1(a),(b) and (6.4.1). Then, for every $\varepsilon > 0$, with o_1, o_2 chosen independently and uar from $[n]$,*

$$\mathbb{P}\left(\mathrm{dist}_{\mathrm{NR}_n(\boldsymbol{w})}(o_1, o_2) \leq \frac{2(1+\varepsilon)\log\log n}{|\log(\tau-2)|} \,\Big|\, \mathrm{dist}_{\mathrm{NR}_n(\boldsymbol{w})}(o_1, o_2) < \infty\right) \to 1. \quad (6.4.2)$$

The same result applies, under identical conditions, to $\mathrm{GRG}_n(\boldsymbol{w})$ *and* $\mathrm{CL}_n(\boldsymbol{w})$.

The proof of Theorem 6.11 is organized as follows. We start by showing that the vertices of weight larger than n^β are all connected to one another. Thus, these giant-weight vertices whp form a complete graph or *clique*. In the second step, we prove a doubly logarithmic upper bound on the distance between a vertex and the set of giant-weight vertices. The latter bound holds only when the vertex is in the giant component, a fact that we need to take into account carefully. In the final step, we complete the proof of Theorem 6.11.

Giant-Weight Vertices Form a (Near-)Clique

Recall the definition of $\beta > \frac{1}{2}$ used in (6.4.1). Let

$$\mathrm{Giant}_n = \{a \in [n]: w_a \geq n^\beta\} \quad (6.4.3)$$

denote the set of giant-weight vertices. Further, for $A \subseteq [n]$ and a graph G_n, we say that A forms a *clique* in G_n when the edges $a_1 a_2$ are occupied for all $a_1, a_2 \in A$. The next lemma shows that, whp, Giant_n forms a (near-)clique, in that it forms a clique whp for $\mathrm{NR}_n(\boldsymbol{w})$ and $\mathrm{CL}_n(\boldsymbol{w})$, while the diameter is at most 2 for $\mathrm{GRG}_n(\boldsymbol{w})$:

Lemma 6.12 (Giant-weight vertices form (near-)clique) *Consider* $\mathrm{NR}_n(\boldsymbol{w})$ *under the conditions of Theorem 6.11,*

$$\mathbb{P}(\mathrm{Giant}_n \text{ does not form a clique in } \mathrm{NR}_n(\boldsymbol{w})) \leq n^2 e^{-n^{2\beta}/\ell_n}. \quad (6.4.4)$$

The same result applies, under identical conditions, to $\mathrm{CL}_n(\boldsymbol{w})$, *while for* $\mathrm{GRG}_n(\boldsymbol{w})$ *the diameter of* Giant_n *is at most 2.*

Proof Let $a_1, a_2 \in \mathrm{Giant}_n$, so that $w_{a_1}, w_{a_2} \geq n^\beta$. There are at most $|\mathrm{Giant}_n|^2 \leq n^2$ pairs of vertices in Giant_n, so that

$$\mathbb{P}(\mathrm{Giant}_n \text{ does not form clique in } \mathrm{NR}_n(\boldsymbol{w}))$$
$$\leq n^2 \max_{a_1, a_2 \in \mathrm{Giant}_n} \mathbb{P}(a_1 a_2 \text{ vacant in } \mathrm{NR}_n(\boldsymbol{w})). \quad (6.4.5)$$

The edge $a_1 a_2$ is vacant with probability

$$\mathbb{P}(a_1 a_2 \text{ vacant in } \mathrm{NR}_n(\boldsymbol{w})) = e^{-w_{a_1} w_{a_2}/\ell_n} \leq e^{-n^{2\beta}/\ell_n}, \quad (6.4.6)$$

since $w_a \geq n^\beta$ for every $a \in \mathrm{Giant}_n$. Multiplying out gives the result. For $\mathrm{CL}_n(\boldsymbol{w})$, $\mathbb{P}(a_1 a_2 \text{ vacant}) = 0$ since $p_{a_1 a_2}^{(\mathrm{CL})} = w_{a_1} w_{a_2}/\ell_n \wedge 1 = 1$ when $w_{a_1}, w_{a_2} \geq n^\beta$ with $\beta > \frac{1}{2}$, so the same proof applies. For $\mathrm{GRG}_n(\boldsymbol{w})$, we need to strengthen this analysis slightly. Indeed, for $\mathrm{GRG}_n(\boldsymbol{w})$, for all $a_1, a_2 \in \mathrm{Giant}_n$,

$$\mathbb{P}(a_1 a_2 \text{ occupied in } \mathrm{GRG}_n(\boldsymbol{w})) \geq \frac{n^{2\beta}}{\ell_n + n^{2\beta}} = 1 - \Theta(n^{1-2\beta}) \geq \frac{1}{2}, \quad (6.4.7)$$

and the edge statuses are independent random variables. As a result, the diameter of Giant_n is bounded by the diameter of $\text{ER}_n(p)$ with $p = \frac{1}{2}$. Thus, it suffices to prove that the diameter of $\text{ER}_n(\frac{1}{2})$ is whp bounded by 2. For this, we note that, with $p = \frac{1}{2}$,

$$\mathbb{P}(\text{diam}(\text{ER}_n(p)) > 2) \leq n^2 \mathbb{P}(\text{dist}_{\text{ER}_n(p)}(1, 2) > 2). \tag{6.4.8}$$

The event $\{\text{dist}_{\text{ER}_n(p)}(1, 2) > 2\}$ implies that not all 2-step paths between 1 and 2 are occupied, so that, by the independence of these 2-step paths,

$$\mathbb{P}(\text{dist}_{\text{ER}_n(p)}(1, 2) > 2) = (1 - \tfrac{1}{4})^{n-2}, \tag{6.4.9}$$

which, combined with (6.4.8), completes the proof. $\square$

Connections to Giant_n Occur at Doubly Logarithmic Distances

Fix $r \geq 1$ large. We next show that vertices that survive up to distance r have a high probability of connecting to Giant_n using a path of at most $(1 + \varepsilon)\frac{\log\log n}{|\log(\tau-2)|}$ edges:

Proposition 6.13 (Connecting to Giant_n) *Consider* $\text{NR}_n(\boldsymbol{w})$ *under the conditions of Theorem 6.11. Let* $u \in [n]$ *be such that* $w_u > 1$. *Then, there exist* $c, c_1^* > 0$ *and* $\eta > 0$ *such that*

$$\mathbb{P}\Big(\text{dist}_{\text{NR}_n(\boldsymbol{w})}(u, \text{Giant}_n) \geq (1+\varepsilon)\frac{\log\log n}{|\log(\tau-2)|}\Big) \leq ce^{-c_1^* w_u^\eta}. \tag{6.4.10}$$

Consequently, if $\mathcal{W}_r(u) = \sum_{k \in \partial B_r^{(G_n)}(u)} w_k$ *denotes the weight of vertices at graph distance r from u and* $G_n = \text{NR}_n(\boldsymbol{w})$,

$$\mathbb{P}\Big(\text{dist}_{\text{NR}_n(\boldsymbol{w})}(\partial B_r^{(G_n)}(u), \text{Giant}_n) \geq (1+\varepsilon)\frac{\log\log n}{|\log(\tau-2)|} \,\Big|\, B_r^{(G_n)}(u)\Big) \leq ce^{-c_1^* \mathcal{W}_r(u)^\eta}. \tag{6.4.11}$$

The same results apply, under identical conditions, to $\text{GRG}_n(\boldsymbol{w})$ *and* $\text{CL}_n(\boldsymbol{w})$.

Proof We start by proving (6.4.10). The bound in (6.4.10) is trivial unless w_u is large, which we assume from now on. Without loss of generality, we may assume that the weights $(w_1, \ldots, w_n)$ are non-increasing.

Let $x_0 = u$, and define, recursively,

$$x_\ell = \min\{v \in [n]: x_{\ell-1}v \text{ occupied in } \text{NR}_n(\boldsymbol{w})\}. \tag{6.4.12}$$

Thus, x_ℓ is the maximal-weight neighbor of $x_{\ell-1}$ in $\text{NR}_n(\boldsymbol{w})$. We stop the above recursion when $w_{x_\ell} \geq n^\beta$, since then $x_\ell \in \text{Giant}_n$. Recall the heuristic approach below (6.2.7), which shows that a vertex with weight w is whp connected to a vertex with weight $w^{1/(\tau-2)}$. We now make this precise.

We take $a = 1/(\tau - 2 + \delta)$, where we choose $\delta > 0$ small enough that $a > 1$. By (6.3.24),

$$\mathbb{P}(w_{x_{\ell+1}} < w_{x_\ell}^a \mid (x_s)_{s \leq \ell}) = \exp\Big\{ -w_{x_\ell} \sum_{v \in [n]} \mathbb{1}_{\{w_v \geq w_{x_\ell}^a\}} w_v/\ell_n \Big\}$$

$$= \exp\Big\{ -w_{x_\ell}[1 - F_n^\star](w_{x_\ell}^a) \Big\}. \tag{6.4.13}$$

We split the argument depending on whether $w_{x_\ell}^a \leq n^\beta$. First, when $w_{x_\ell}^a \leq n^\beta$, by (6.4.1) and uniformly for $x \leq n^\beta$,

$$[1 - F_n^\star](x) \geq \frac{xn}{\ell_n}[1 - F_n](x) \geq c_1^\star x^{-(\tau-2)}, \qquad (6.4.14)$$

where, for n large enough, we can take $c_1^* = c_1/(2\mathbb{E}[W])$. Therefore

$$\mathbb{P}(w_{x_{\ell+1}} < w_{x_\ell}^a \mid (x_s)_{s \leq \ell}) \leq \exp\{-c_1^\star w_{x_\ell}^{1-(\tau-2)a}\} \leq \exp\{-c_1^\star w_{x_\ell}^\delta\}, \qquad (6.4.15)$$

since $a = 1/(\tau - 2 + \delta) > 1$ so that $1 - (\tau - 2)a = a\delta > \delta$.

Second, when $w_{x_\ell}^a > n^\beta$ but $w_{x_\ell} < n^\beta$, we can use (6.4.14) for $x = n^\beta$ to obtain

$$\mathbb{P}(w_{x_{\ell+1}} < n^\beta \mid (x_s)_{s \leq \ell}) \leq \exp\{-c_1^\star w_{x_\ell} n^{-\beta(\tau-2)}\} \leq \exp\{-c_1^\star n^{\beta[1-(\tau-2)]/a}\}$$
$$\leq \exp\{-c_1^\star n^{\beta\delta/a}\}. \qquad (6.4.16)$$

Therefore, in both cases, and with $\eta = \delta(\beta/a \wedge 1)$,

$$\mathbb{P}(w_{x_{\ell+1}} < n^\beta \wedge w_{x_\ell}^a \mid (x_s)_{s \leq \ell}) \leq e^{-c_1^\star w_{x_\ell}^\eta}. \qquad (6.4.17)$$

As a result, when x_ℓ is such that w_{x_ℓ} is quite large, whp, we have $w_{x_{\ell+1}} \geq w_{x_\ell}$. This produces, whp, a short path to Giant_n. We now investigate the properties of this path.

Let the recursion stop at some integer time k. The key observation is that, when this occurs, we must have that $w_{x_{\ell+1}} > w_{x_\ell}^a$ for each $\ell \leq k-1$ where k is such that $w_{x_{k-1}} \in [n^{\beta/a}, n^\beta]$, and at the same time $w_{x_k} \geq n^\beta$. Then, we conclude that the following facts are true:

(1) $w_{x_\ell} \geq w_{x_0}^{a^\ell} = w_u^{a^\ell}$ for every $\ell \leq k-1$;
(2) $\text{dist}_{\text{NR}_n(w)}(u, \text{Giant}_n) \leq k$.

By (1), $w_{x_{k-1}} \geq w_u^{a^{k-1}}$, and $w_{x_{k-1}} \in [n^{\beta/a}, n^\beta]$. Therefore, $w_u^{a^{k-1}} \leq n^\beta$, which, in turn, implies that

$$a^{k-1} \leq \beta \log n, \qquad \text{or} \qquad k - 1 \leq (\log\log n + \log\beta)(\log a). \qquad (6.4.18)$$

Let $k_n = \lceil (1+\varepsilon) \log\log n / |\log(\tau-2)| \rceil$. By (1) and (2), when $\text{dist}_{\text{NR}_n(w)}(u, \text{Giant}_n) > k_n$ occurs, then there must exist an $\ell \leq k_n$ such that $w_{x_{\ell+1}} \leq n^\beta \wedge w_{x_\ell}^a$. We conclude that

$$\mathbb{P}\left(\text{dist}_{\text{NR}_n(w)}(u, \text{Giant}_n) \geq k_n\right)$$

$$\leq \sum_{\ell=0}^{k_n} \mathbb{P}(w_{x_{\ell+1}} \leq n^\beta \wedge w_{x_\ell}^a, w_{x_s} > n^\beta \wedge w_{x_{s-1}}^a \, \forall s \leq \ell)$$

$$\leq \sum_{\ell=0}^{k_n} \mathbb{E}[\mathbb{1}_{\{w_{x_s} > n^\beta \wedge w_{x_s}^a \, \forall s \leq \ell\}} \mathbb{P}(w_{x_{\ell+1}} \leq w_{x_\ell}^a \mid (x_s)_{s \leq \ell})]$$

$$\leq \sum_{\ell=0}^{k_n} \mathbb{E}[e^{-c_1^\star w_{x_\ell}^\eta}] \leq \sum_{\ell=0}^{k_n} e^{-c_1^\star w_u^{\delta a^\ell}} \leq c e^{-c_1^\star w_u^\eta}. \qquad (6.4.19)$$

This completes the proof of (6.4.10).

The proof of (6.4.11) is similar, and is achieved by conditioning on $B_r^{(G_n)}(u)$ and by noting that we can interpret $\partial B_r^{(G_n)}(u)$ as a single vertex having weight given by

$\mathcal{W}_r(u) = \sum_{v \in \partial B_r^{(G_n)}(u)} w_v$. Indeed, for $\mathrm{NR}_n(\boldsymbol{w})$, there is an edge between v and z with probability $1 - e^{-w_v w_z / \ell_n}$, which is the same as the probability that a Poisson random variable with parameter $w_v w_z / \ell_n$ is at least 1. Thus, there is an edge between z and $\partial B_r^{(G_n)}(u)$ when at least one of these edges is present, the probability of which equals the probability that a Poisson random variable with parameter $w_z \sum_{v \in \partial B_r^{(G_n)}(u)} w_v / \ell_n = w_z \mathcal{W}_r(u) / \ell_n$ is at least 1. It is here that we use the relation of the edge probabilities in $\mathrm{NR}_n(\boldsymbol{w})$ and Poisson random variables.

By [V1, (6.8.12) and (6.8.13)], the edge probabilities $p_{uv}^{(\mathrm{CL})}$ satisfy $p_{uv}^{(\mathrm{CL})} \geq p_{uv}^{(\mathrm{NR})}$ for all $u, v \in [n]$, so the results immediately carry over to $\mathrm{CL}_n(\boldsymbol{w})$. For $G_n = \mathrm{GRG}_n(\boldsymbol{w})$, however, for all $z, v \in [n] \setminus V(B_r^{(G_n)}(u))$, we have

$$\mathbb{P}(zv \in E(G_n) \mid B_r^{(G_n)}(u)) = 1 - (1 - p_{zv}^{(\mathrm{GRG})}) \geq 1 - e^{-p_{zv}^{(\mathrm{GRG})}}$$
$$= 1 - e^{-w_z w_v / (\ell_n + w_v w_z)}. \tag{6.4.20}$$

This probability obeys bounds similar to those for $\mathrm{NR}_n(\boldsymbol{w})$ as long as $w_v w_z = o(n)$. Since we are applying this to $v = x_{\ell+1}$ and $z = x_\ell$, we have that $w_{x_{\ell+1}} \leq n^\beta$ and $w_{x_\ell} \leq n^{\beta/a}$, so that $w_{x_{\ell+1}} w_{x_\ell} \leq n^{\beta(1+1/a)}$. Then, we can choose (β, a) such that $\beta > \frac{1}{2}$ while at the same time $\beta(1 + 1/a) = \beta(\tau - 1 + \delta) < 1$, which is possible since $\tau - 1 \in (1, 2)$. Now the proof can be followed as for $\mathrm{NR}_n(\boldsymbol{w})$. $\qquad\square$

Completion of the Proof of Theorem 6.11

To prove the upper bound in Theorem 6.11, for $\varepsilon \in (0, 1)$ we take

$$k_n = \left\lceil \frac{(1 + \varepsilon) \log \log n}{|\log (\tau - 2)|} \right\rceil, \tag{6.4.21}$$

so that it suffices to show that, for every $\varepsilon > 0$,

$$\lim_{n \to \infty} \mathbb{P}(\mathrm{dist}_{\mathrm{NR}_n(w)}(o_1, o_2) \leq 2k_n \mid \mathrm{dist}_{\mathrm{NR}_n(w)}(o_1, o_2) < \infty) = 1. \tag{6.4.22}$$

Since

$$\mathbb{P}(\mathrm{dist}_{\mathrm{NR}_n(w)}(o_1, o_2) \leq 2k_n \mid \mathrm{dist}_{\mathrm{NR}_n(w)}(o_1, o_2) < \infty)$$
$$= \frac{\mathbb{P}(\mathrm{dist}_{\mathrm{NR}_n(w)}(o_1, o_2) \leq 2k_n)}{\mathbb{P}(\mathrm{dist}_{\mathrm{NR}_n(w)}(o_1, o_2) < \infty)}, \tag{6.4.23}$$

this follows from the two bounds

$$\limsup_{n \to \infty} \mathbb{P}(\mathrm{dist}_{\mathrm{NR}_n(w)}(o_1, o_2) < \infty) \leq \zeta^2, \tag{6.4.24}$$

$$\liminf_{n \to \infty} \mathbb{P}(\mathrm{dist}_{\mathrm{NR}_n(w)}(o_1, o_2) \leq 2k_n) \geq \zeta^2, \tag{6.4.25}$$

where $\zeta = \mu(|\mathscr{C}(o)| = \infty) > 0$ is the survival probability of the branching-process approximation to the neighborhoods of $\mathrm{NR}_n(\boldsymbol{w})$, as identified in Theorem 3.18. For (6.4.24), we make the following split, for some $r \geq 1$:

$$\mathbb{P}(\mathrm{dist}_{\mathrm{NR}_n(w)}(o_1, o_2) < \infty)$$
$$\leq \mathbb{P}(|\partial B_r^{(G_n)}(o_1)| > 0, |\partial B_r^{(G_n)}(o_2)| > 0, \mathrm{dist}_{\mathrm{NR}_n(w)}(o_1, o_2) > 2r)$$
$$+ \mathbb{P}(\mathrm{dist}_{\mathrm{NR}_n(w)}(o_1, o_2) \leq 2r). \tag{6.4.26}$$

To prove (6.4.25), we fix $r \geq 1$ and write

$$
\begin{aligned}
\mathbb{P}(\mathrm{dist}_{\mathrm{NR}_n(w)}(o_1, o_2) &\leq 2k_n) \\
&\geq \mathbb{P}(2r < \mathrm{dist}_{\mathrm{NR}_n(w)}(o_1, o_2) \leq 2k_n) \\
&\geq \mathbb{P}\big(\mathrm{dist}_{\mathrm{NR}_n(w)}(o_i, \mathrm{Giant}_n) \leq k_n, i = 1, 2, \mathrm{dist}_{\mathrm{NR}_n(w)}(o_1, o_2) > 2r\big) \\
&\geq \mathbb{P}(|\partial B_r^{(G_n)}(o_1)| > 0, |\partial B_r^{(G_n)}(o_1)| > 0, \mathrm{dist}_{\mathrm{NR}_n(w)}(o_1, o_2) > 2r) \\
&\quad - 2\mathbb{P}(\mathrm{dist}_{\mathrm{NR}_n(w)}(o_1, \mathrm{Giant}_n) > k_n, |\partial B_r^{(G_n)}(o_1)| > 0). \quad (6.4.27)
\end{aligned}
$$

The first terms in (6.4.26) and (6.4.27) are the same. By Corollary 2.19(b), this term satisfies

$$
\begin{aligned}
\mathbb{P}(|\partial B_r^{(G_n)}(o_1)| > 0, |\partial B_r^{(G_n)}(o_1)| &> 0, \mathrm{dist}_{\mathrm{NR}_n(w)}(o_1, o_2) > 2r) \\
&= \mathbb{P}(|\partial B_r^{(G_n)}(o)| > 0)^2 + o(1), \quad (6.4.28)
\end{aligned}
$$

which converges to ζ^2 when $r \to \infty$.

We are left with showing that the second terms in (6.4.26) and (6.4.27) vanish when first $n \to \infty$ followed by $r \to \infty$. By Corollary 2.20, $\mathbb{P}(\mathrm{dist}_{\mathrm{NR}_n(w)}(o_1, o_2) \leq 2r) = o(1)$, which completes the proof of (6.4.24).

For the second term in (6.4.27), we condition on $B_r^{(G_n)}(o_1)$, and use that $\partial B_r^{(G_n)}(o_1)$ is measurable wrt $B_r^{(G_n)}(o_1)$, to obtain

$$
\begin{aligned}
\mathbb{P}\big(\mathrm{dist}_{\mathrm{NR}_n(w)}(o_1, \mathrm{Giant}_n) &> k_n, |\partial B_r^{(G_n)}(o_1)| > 0\big) \\
&= \mathbb{E}\Big[\mathbb{1}_{\{|\partial B_r^{(G_n)}(o_1)| > 0\}} \mathbb{P}\big(\mathrm{dist}_{\mathrm{NR}_n(w)}(o_1, \mathrm{Giant}_n) > k_n \mid B_r^{(G_n)}(o_1)\big)\Big]. \quad (6.4.29)
\end{aligned}
$$

By Proposition 6.13, with $\mathcal{W}_r(o_1) = \sum_{k \in \partial B_r^{(G_n)}(o_1)} w_k$,

$$
\mathbb{P}\big(\mathrm{dist}_{\mathrm{NR}_n(w)}(o_1, \mathrm{Giant}_n) > k_n \mid B_r^{(G_n)}(o_1)\big) \leq c e^{-c_1^\star \mathcal{W}_r(o_1)^\eta}. \quad (6.4.30)
$$

By the local convergence in probability in Theorem 3.18 and Conditions 1.1(a),(b), we obtain that, with $(G, o) \sim \mu$ the local limit of $\mathrm{NR}_n(w)$,

$$
\mathcal{W}_r(o_1) \overset{d}{\longrightarrow} \sum_{i=1}^{|\partial B_r^{(G)}(o_1)|} W_i^\star, \quad (6.4.31)
$$

where $(W_i^\star)_{i \geq 1}$ are iid random variables with distribution function $F^\star$. Therefore,

$$
\mathcal{W}_r(o_1) \overset{\mathbb{P}}{\longrightarrow} \infty \quad (6.4.32)
$$

when first $n \to \infty$ followed by $r \to \infty$, and we use that $|\partial B_r^{(G)}(o_1)| \overset{\mathbb{P}}{\longrightarrow} \infty$ as $r \to \infty$, since $|\partial B_r^{(G)}(o_1)| > 0$. As a result, when first $n \to \infty$ followed by $r \to \infty$,

$$
\mathbb{1}_{\{|\partial B_r^{(G_n)}(o_1)| > 0\}} \mathbb{P}\big(\mathrm{dist}_{\mathrm{NR}_n(w)}(o_1, \mathrm{Giant}_n) > k_n \mid B_r^{(G_n)}(o_1)\big) \overset{\mathbb{P}}{\longrightarrow} 0. \quad (6.4.33)
$$

By Lebesgue's Dominated Convergence Theorem [V1, Theorem A.1] this implies

$$
\mathbb{E}\Big[e^{-c_1^\star \mathcal{W}_r(o_1)^\eta} \mathbb{1}_{\{|\partial B_r^{(G_n)}(o_1)| > 0\}}\Big] \to 0, \quad (6.4.34)
$$

when first $n \to \infty$ followed by $r \to \infty$. This proves (6.4.25), and thus completes the proof of the upper bound in Theorem 6.3 for $\mathrm{NR}_n(w)$. The proofs for $\mathrm{GRG}_n(w)$ and $\mathrm{CL}_n(w)$ are similar, and are left as Exercise 6.14. $\qquad\square$

6.5 LOGARITHMIC UPPER BOUND FOR FINITE-VARIANCE WEIGHTS

In this section we give the proof of the logarithmic upper bound typical distances in rank-1 random graphs with finite-variance weights stated in Theorem 6.2. For this, we use the second-moment method to show that whp there exists a path of at most $(1 + \varepsilon) \log_\nu n$ edges between o_1 and o_2, when o_1 and o_2 are such that $\partial B_r^{(G_n)}(o_1), \partial B_r^{(G_n)}(o_2) \neq \varnothing$ for $G_n = \mathrm{CL}_n(\boldsymbol{w})$. This proves Theorem 6.2 for $\mathrm{CL}_n(\boldsymbol{w})$.

The extensions to $\mathrm{NR}_n(\boldsymbol{w})$ and $\mathrm{GRG}_n(\boldsymbol{w})$ follow by asymptotic equivalence of these graphs, as discussed in [V1, Section 6.7]. Even though this shows that $\mathrm{NR}_n(\boldsymbol{w}), \mathrm{CL}_n(\boldsymbol{w})$, and $\mathrm{GRG}_n(\boldsymbol{w})$ all behave similarly, for our second-moment methods, we will need to be especially careful about the model with which we are working.

To apply the second-moment method, we give a bound on the *variance* of the number of paths of given lengths using path-counting techniques. This section is organized as follows. In Section 6.5.1 we highlight our path-counting techniques. In Section 6.5.2 we apply these methods to give upper bounds on typical distances for finite-variance weights. We also investigate the case where $\tau = 3$, for which we prove that typical distances are bounded by $\log n / \log \log n$ under appropriate conditions.

The above path-counting methods can also be used to study general inhomogeneous random graphs, as discussed in Section 6.5.3, where we prove Theorem 6.1(b) and use its proof ideas to complete the proof of the law of large numbers for the giant in Theorem 3.19.

6.5.1 PATH-COUNTING TECHNIQUES

In this subsection we study path-counting techniques in the context of inhomogeneous random graphs. We consider an inhomogeneous random graph on the vertices $\mathcal{I}$ with edge probabilities $p_{ij} = u_i u_j$, for some weights $(u_i)_{i \in \mathcal{I}}$ satisfying $u_i \leq 1$ for all $i \in \mathcal{I}$. Throughout this section, G_n will denote this random graph.

We obtain $\mathrm{CL}_n(\boldsymbol{w})$ by taking $u_i = w_i / \sqrt{\ell_n}$ and $\mathcal{I} = [n]$, and we assume that $w_u w_v / \ell_n \leq 1$ for all $u, v \in [n]$. The latter turns out to be a consequence of Conditions 1.1(a)–(c). Since $\mathrm{NR}_n(\boldsymbol{w})$ and $\mathrm{GRG}_n(\boldsymbol{w})$ are closely related to $\mathrm{CL}_n(\boldsymbol{w})$ when Conditions 1.1(a)–(c) hold, this suffices for our purposes. We add the flexibility of choosing $\mathcal{I} \subseteq [n]$, since sometimes it is convenient to exclude a subset of vertices, such as the high-weight vertices, from our second-moment computations.

For $a, b \in \mathcal{I}$ and $k \geq 1$, let

$$N_k(a, b) = \#\{\vec{\pi} \in \mathcal{P}_k(a, b): \vec{\pi} \text{ occupied in } G_n\} \qquad (6.5.1)$$

denote the number of self-avoiding paths of length k between the vertices a and b, where we recall that a path $\vec{\pi}$ is self-avoiding when it visits every vertex at most once (see Definition 6.5). Let

$$n_k(a, b) = \mathbb{E}[N_k(a, b)] \qquad (6.5.2)$$

denote the expected number of occupied paths of length k connecting a and b. Define

$$\bar{n}_k(a, b) = u_a u_b \left(\sum_{i \in \mathcal{I} \setminus \{a, b\}} u_i^2 \right)^{k-1}, \qquad \underline{n}_k(a, b) = u_a u_b \left(\sum_{i \in \mathcal{I}_{a,b,k}} u_i^2 \right)^{k-1}, \qquad (6.5.3)$$

where $\mathcal{I}_{a,b,k}$ is the subset of $\mathcal{I}$ in which a and b, as well as the $k-1$ vertices with highest weights, have been removed. In Section 6.3 we proved implicitly an upper bound on $\mathbb{E}[N_k(a,b)]$ of the form (see also Exercise 6.15)

$$n_k(a,b) = \mathbb{E}[N_k(a,b)] \leq \bar{n}_k(a,b). \tag{6.5.4}$$

In this section we prove that $\underline{n}_k(a,b)$ is a lower bound on $n_k(a,b)$ and use related bounds to prove a variance bound on $N_k(a,b)$.

Before stating our main result, we introduce some further notation. Let

$$\nu_{\mathcal{I}} = \sum_{i \in \mathcal{I}} u_i^2, \qquad \gamma_{\mathcal{I}} = \sum_{i \in \mathcal{I}} u_i^3 \tag{6.5.5}$$

denote the sums of squares and third powers of $(u_i)_{i \in \mathcal{I}}$, respectively. Our aim is to show that whp paths of length k exist between the vertices a and b for an appropriate choice of k. We do this by applying a second-moment method to $N_k(a,b)$, for which we need a lower bound on $\mathbb{E}[N_k(a,b)]$ and an upper bound on $\mathrm{Var}(N_k(a,b))$ such that $\mathrm{Var}(N_k(a,b)) = o(\mathbb{E}[N_k(a,b)]^2)$ (recall [V1, Theorem 2.18]), as in the next proposition, which is interesting in its own right:

Proposition 6.14 (Variance of numbers of paths) *For any $k \geq 1$, $a, b \in \mathcal{I}$ and $(u_i)_{i \in \mathcal{I}}$,*

$$\mathbb{E}[N_k(a,b)] \geq \underline{n}_k(a,b), \tag{6.5.6}$$

while, assuming that $\nu_{\mathcal{I}} > 1$,

$$\mathrm{Var}(N_k(a,b))$$
$$\leq n_k(a,b) + \bar{n}_k(a,b)^2 \left(\frac{\gamma_{\mathcal{I}} \nu_{\mathcal{I}}^2}{\nu_{\mathcal{I}} - 1} \left(\frac{1}{u_a} + \frac{1}{u_b} \right) + \frac{\gamma_{\mathcal{I}}^2 \nu_{\mathcal{I}}}{u_a u_b (\nu_{\mathcal{I}} - 1)^2} + e_k \right), \tag{6.5.7}$$

where

$$e_k = \left(1 + \frac{\gamma_{\mathcal{I}}}{u_a \nu_{\mathcal{I}}} \right) \left(1 + \frac{\gamma_{\mathcal{I}}}{u_b \nu_{\mathcal{I}}} \right) \frac{\nu_{\mathcal{I}}}{\nu_{\mathcal{I}} - 1} \left(e^{2k^3 \gamma_{\mathcal{I}}^2 / \nu_{\mathcal{I}}^3} - 1 \right). \tag{6.5.8}$$

Remark 6.15 (Path-counting and existence of k-step paths) Path-counting methods are highly versatile. While in Proposition 6.14 we focus on Chung–Lu-type inhomogeneous random graphs, we will apply them to general inhomogeneous random graphs with finitely many types in Section 6.5.3 and to the configuration model in Section 7.3.3. For such applications, we need to slightly modify our bounds, particularly those in Lemma 6.18, owing to a slightly altered dependence structure between the occupation statuses of distinct paths. ◀

We apply Proposition 6.14 in cases where $\mathbb{E}[N_k(a,b)] = n_k(a,b) \to \infty$, by taking $\mathcal{I}$ to be a large subset of $[n]$ and u_i to equal $w_i / \sqrt{\ell_n}$ for $\mathrm{CL}_n(\boldsymbol{w})$. In this case, $\nu_{\mathcal{I}} \approx \nu_n \approx \nu > 1$. In our applications of Proposition 6.14, the ratio $\bar{n}_k(a,b) / \underline{n}_k(a,b)$ will be bounded, and $k^3 \gamma_{\mathcal{I}}^2 / \nu_{\mathcal{I}}^3 = o(1)$, so that the term involving e_k is an error term. The starting and ending vertices $a, b \in \mathcal{I}$ will correspond to a *union of vertices* in $[n]$ of quite large size; this relies on the local limit stated in Theorem 3.18. As a result, $\gamma_{\mathcal{I}} / u_a$ and $\gamma_{\mathcal{I}} / u_b$ are typically small, so that also

$$\frac{\mathrm{Var}(N_k(a,b))}{\mathbb{E}[N_k(a,b)]^2} \approx \frac{\gamma_{\mathcal{I}} \nu_{\mathcal{I}}^2}{\nu_{\mathcal{I}} - 1} \left(\frac{1}{u_a} + \frac{1}{u_b} \right) + \frac{\gamma_{\mathcal{I}}^2 \nu_{\mathcal{I}}}{u_a u_b (\nu_{\mathcal{I}} - 1)^2} \tag{6.5.9}$$

is small. As a result, whp there exists a path of k steps, as required. The choice of a, b, and $\mathcal{I}$ is quite delicate, which explains why we formulated Proposition 6.14 in such generality.

We next prove Proposition 6.14, which, in particular for (6.5.7), requires some serious combinatorial arguments.

Proof of Proposition 6.14. Recall Definition 6.5 and that $N_k(a, b)$ is a sum of indicators:

$$N_k(a, b) = \sum_{\vec{\pi} \in \mathcal{P}_k(a,b)} \mathbb{1}_{\{\vec{\pi} \text{ occupied in } G_n\}}, \tag{6.5.10}$$

where G_n is our Chung–Lu-type inhomogeneous random graph with edge probabilities $p_{ij} = u_i u_j$. If all the indicators in (6.5.10) were *independent*, the upper bound $n_k(a, b)$ on $\mathrm{Var}(N_k(a, b))$ in (6.5.7) would hold. The second term on the rhs of (6.5.7) thus accounts for the positive dependence between the indicators of two paths being occupied.

We start by proving (6.5.6), which is relatively straightforward. By (6.5.10),

$$\mathbb{E}[N_k(a, b)] = \sum_{\vec{\pi} \in \mathcal{P}_k(a,b)} \mathbb{P}(\vec{\pi} \text{ occupied in } G_n) = \sum_{\vec{\pi} \in \mathcal{P}_k(a,b)} \prod_{l=0}^{k} u_{\pi_l} u_{\pi_{l+1}}$$

$$= u_{\pi_0} u_{\pi_k} \sum_{\vec{\pi} \in \mathcal{P}_k(a,b)} \prod_{l=1}^{k-1} u_{\pi_l}^2. \tag{6.5.11}$$

Every $\vec{\pi} \in \mathcal{P}_k(a, b)$ starts at $\pi_0 = a$ and ends at $\pi_k = b$. Further, since $\vec{\pi}$ is assumed to be self-avoiding,

$$\sum_{\vec{\pi} \in \mathcal{P}_k(a,b)} \prod_{l=1}^{k-1} u_{\pi_l}^2 = \sum_{\pi_1,\ldots,\pi_{k-1} \in \mathcal{I} \setminus \{a,b\}}^{*} \prod_{l=1}^{k-1} u_{\pi_l}^2, \tag{6.5.12}$$

where we recall that $\sum_{\pi_1,\ldots,\pi_r \in \mathcal{I}}^{*}$ denotes a sum over *distinct* indices. Each sum over π_j yields a factor that is at least $\sum_{i \in \mathcal{I}_{a,b,k}} u_i^2$, which proves (6.5.6).

To compute $\mathrm{Var}(N_k(a, b))$, we again start from (6.5.10), which yields

$$\mathrm{Var}(N_k(a, b)) = \sum_{\vec{\pi}, \vec{\rho} \in \mathcal{P}_k(a,b)} \left[\mathbb{P}(\vec{\pi}, \vec{\rho} \text{ occupied}) - \mathbb{P}(\vec{\pi} \text{ occ.}) \mathbb{P}(\vec{\rho} \text{ occ.}) \right], \tag{6.5.13}$$

where we abbreviate $\{\vec{\pi} \text{ occupied in } G_n\}$ to $\{\vec{\pi} \text{ occupied}\}$ or $\{\vec{\pi} \text{ occ.}\}$ when no confusion can arise.

For $\vec{\pi}, \vec{\rho}$, we denote the edges that the paths $\vec{\pi}$ and $\vec{\rho}$ have in common by $\vec{\pi} \cap \vec{\rho}$. The occupation statuses of $\vec{\pi}$ and $\vec{\rho}$ are independent precisely when $\vec{\pi} \cap \vec{\rho} = \varnothing$, so that

$$\mathrm{Var}(N_k(a, b)) \leq \sum_{\substack{\vec{\pi}, \vec{\rho} \in \mathcal{P}_k(a,b) \\ \vec{\pi} \cap \vec{\rho} \neq \varnothing}} \mathbb{P}(\vec{\pi}, \vec{\rho} \text{ occupied}). \tag{6.5.14}$$

Define $\vec{\rho} \setminus \vec{\pi}$ to be the edges in $\vec{\rho}$ that are *not* part of $\vec{\pi}$, so that

$$\mathbb{P}(\vec{\pi}, \vec{\rho} \text{ occupied}) = \mathbb{P}(\vec{\pi} \text{ occupied}) \mathbb{P}(\vec{\rho} \text{ occupied} \mid \vec{\pi} \text{ occupied})$$

$$= \prod_{l=0}^{k} u_{\pi_l} u_{\pi_{l+1}} \prod_{e \in \vec{\rho} \setminus \vec{\pi}} u_{\underline{e}} u_{\overline{e}}, \tag{6.5.15}$$

where, for an edge $e = \{x, y\}$, we write $\bar{e} = x, \underline{e} = y$. For $\vec{\pi} = \vec{\rho}$,

$$\mathbb{P}(\vec{\pi}, \vec{\rho} \text{ occupied}) = \mathbb{P}(\vec{\pi} \text{ occupied}), \tag{6.5.16}$$

and this contributes $n_k(a, b)$ to $\text{Var}(N_k(a, b))$. Thus, from now on, we consider $(\vec{\pi}, \vec{\rho})$ such that $\vec{\pi} \neq \vec{\rho}$ and $\vec{\pi} \cap \vec{\rho} \neq \varnothing$.

The probability in (6.5.15) needs to be summed over all possible pairs of paths $(\vec{\pi}, \vec{\rho})$ with $\vec{\pi} \neq \vec{\rho}$ that share at least one edge. In order to do this effectively, we introduce some notation.

Let $l = |\vec{\pi} \cap \vec{\rho}|$ denote the number of edges in $\vec{\pi} \cap \vec{\rho}$, so that $l \geq 1$ precisely when $\vec{\pi} \cap \vec{\rho} \neq \varnothing$. Note that $l \in [k - 2]$, since $\vec{\pi}$ and $\vec{\rho}$ are distinct self-avoiding paths of length k between the same vertices a and b. Let $k - l = |\vec{\rho} \setminus \vec{\pi}| \geq 2$ be the number of edges in $\vec{\rho}$ that are not part of $\vec{\pi}$.

Let m denote the number of connected subpaths in $\vec{\rho} \setminus \vec{\pi}$, so that $m \geq 1$ whenever $\vec{\pi} \neq \vec{\rho}$. Since $\pi_0 = \rho_0 = a$ and $\pi_k = \rho_k = b$, these subpaths start and end in vertices along the path $\vec{\pi}$. We can thus view the subpaths in $\vec{\rho} \setminus \vec{\pi}$ as *excursions* of the path $\vec{\rho}$ from the walk $\vec{\pi}$. By construction, between two excursions there is at least one edge that $\vec{\pi}$ and $\vec{\rho}$ have in common. We next characterize this excursion structure:

Definition 6.16 ((Edge-)shapes of pairs of paths)   Let m be the number of connected subpaths in $\vec{\rho} \setminus \vec{\pi}$. We define the *shape* of the pair $(\vec{\pi}, \vec{\rho})$ by

$$\text{Shape}(\vec{\pi}, \vec{\rho}) = (\vec{x}_{m+1}, \vec{s}_m, \vec{t}_m, \vec{o}_{m+1}, \vec{r}_{m+1}), \tag{6.5.17}$$

where

(1) $\vec{x}_{m+1} \in \mathbb{N}_0^{m+1}$, where $x_j \geq 0$ is the length of the subpath in $\vec{\rho} \cap \vec{\pi}$ in between the $(j - 1)$th and the jth subpath of $\vec{\pi} \setminus \vec{\rho}$. Here $x_1 \geq 0$ is the number of common edges in the subpath of $\vec{\rho} \cap \vec{\pi}$ that contains a, while $x_{m+1} \geq 0$ is the number of common edges in the subpath of $\vec{\rho} \cap \vec{\pi}$ that contains b. For $j \in \{2, \dots, m\}$, $x_j \geq 1$;

(2) $\vec{s}_m \in \mathbb{N}^m$, where $s_j \geq 1$ is the number of edges in the jth subpath of $\vec{\pi} \setminus \vec{\rho}$;

(3) $\vec{t}_m \in \mathbb{N}^m$, where $t_j \geq 1$ is the number of edges in the jth subpath of $\vec{\rho} \setminus \vec{\pi}$;

(4) $\vec{o}_{m+1} \in [m + 1]^{m+1}$, where o_j is the order of the jth common subpath in $\vec{\rho} \cap \vec{\pi}$ of the path $\vec{\pi}$ in $\vec{\rho}$, e.g., $o_2 = 5$ means that the second subpath that $\vec{\pi}$ has in common with $\vec{\rho}$ is the fifth subpath that $\vec{\rho}$ has in common with $\vec{\pi}$. Note that $o_1 = 1$ and $o_{m+1} = m + 1$, since $\vec{\pi}$ and $\vec{\rho}$ start and end in a and b, respectively;

(5) $\vec{r}_{m+1} \in \{0, 1\}^{m+1}$, where r_j describes the direction in which the jth common subpath in $\vec{\rho} \cap \vec{\pi}$ of the path $\vec{\pi}$ is traversed by $\vec{\rho}$, with $r_j = 1$ when the direction is the same for $\vec{\pi}$ and $\vec{\rho}$ and $r_j = 0$ otherwise. Thus, $r_1 = r_{m+1} = 1$.   ◀

The information in $\text{Shape}(\vec{\pi}, \vec{\rho})$ in Definition 6.16 is precisely what is needed to piece together the topology of the two paths, except for information about the *vertices* involved in $\vec{\pi}$ and $\vec{\rho}$. The subpaths in Definition 6.16 of $\vec{\rho} \setminus \vec{\pi}$ avoid the *edges* in $\vec{\pi}$ but may contain vertices that appear in $\vec{\pi}$. This explains why we call the shapes *edge*-shapes. See Figure 6.5 for an example of a pair of paths $(\vec{\pi}, \vec{\rho})$ and its corresponding shape.

We next discuss properties of shapes and use shapes to analyze $\text{Var}(N_k(a, b))$ further. Recall that $l = |\vec{\pi} \cap \vec{\rho}|$ denotes the number of common edges in $\vec{\pi}$ and $\vec{\rho}$, and m the number of connected subpaths in $\vec{\rho} \setminus \vec{\pi}$. Then

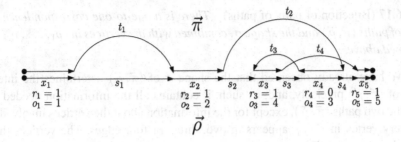

Figure 6.5 An example of a pair of paths $(\vec{\pi}, \vec{\rho})$ and its corresponding shape.

$$\sum_{j=1}^{m+1} x_j = l, \qquad \sum_{j=1}^{m} s_j = \sum_{j=1}^{m} t_j = k - l. \tag{6.5.18}$$

Let $\mathrm{Shape}_{m,l}$ denote the set of shapes corresponding to pairs of paths $(\vec{\pi}, \vec{\rho})$ with m excursions and l common edges so that (6.5.18) holds. Then,

$$\mathrm{Var}(N_k(a,b)) \leq n_k(a,b) + \sum_{l=1}^{k-2} \sum_{m=1}^{k-l} \sum_{\sigma \in \mathrm{Shape}_{m,l}} \sum_{\substack{\vec{\pi}, \vec{\rho} \,\in\, \mathcal{P}_k(a,b) \\ \mathrm{Shape}(\vec{\pi}, \vec{\rho})\,=\,\sigma}} \mathbb{P}(\vec{\pi}, \vec{\rho} \text{ occupied}). \tag{6.5.19}$$

We continue by investigating the structure of the *vertices* in $(\vec{\pi}, \vec{\rho})$. Fix a pair of paths $(\vec{\pi}, \vec{\rho})$ such that $\mathrm{Shape}(\vec{\pi}, \vec{\rho}) = \sigma$ for some $\sigma \in \mathrm{Shape}_{m,l}$. There are $k + 1$ vertices in $\vec{\pi}$. Every subpath of $\vec{\rho} \setminus \vec{\pi}$ starts and ends in a vertex that is also in $\vec{\pi}$. There are m connected subpaths in $\vec{\rho} \setminus \vec{\pi}$ and $l = |\vec{\pi} \cap \vec{\rho}|$ common edges, so that there are at most $k - l - m$ extra vertices in $\vec{\rho} \setminus \vec{\pi}$. We conclude that the union of paths $\vec{\pi} \cup \vec{\rho}$ visits at most $2k + 1 - l - m$ distinct vertices and thus at most $2k - 1 - l - m$ vertices unequal to a or b.

Vertex a is in $1 + \delta_{x_1,0}$ edges and vertex b is in $1 + \delta_{x_{m+1},0}$ edges. Of the other $k - 1$ vertices in $\vec{\pi}$, precisely $2m - \delta_{x_1,0} - \delta_{x_{m+1},0}$ are in three edges, the remaining $k - 1 - 2m + \delta_{x_1,0} + \delta_{x_{m+1},0}$ vertices are in two or four edges. The remaining $k - l - m$ vertices in $\vec{\rho} \setminus \vec{\pi}$ that are not in $\vec{\pi}$ are in two edges. By construction $\vec{\pi}$ and $\vec{\rho}$ are self-avoiding, so the $k + 1$ vertices in $\vec{\pi}$, and those in $\vec{\rho}$, are distinct. In contrast, the $k - l - m$ vertices in $\vec{\rho} \setminus \vec{\pi}$ may intersect those of $\vec{\pi}$.

We summarize the vertex information of $\vec{\pi}$ and $\vec{\rho}$ in the vector $(v_1, \ldots, v_{2k-1-l-m}) \in \mathcal{I}^{2k-1-l-m}$ denoting the vertices in the union of $\vec{\pi}$ and $\vec{\rho}$ that are unequal to a or b. We order these vertices as follows:

▷ the vertices $(v_1, \ldots, v_{2m-a_1-a_{m+1}})$ are in three edges, in the same order as their appearance in $\vec{\pi}$, where we denote $a_1 = \delta_{x_1,0}, a_{m+1} = \delta_{x_{m+1},0}$;

▷ the vertices $(v_{2m-a_1-a_{m+1}+1}, \ldots, v_{k-1})$ are the ordered vertices in $\vec{\pi}$ that are not in three edges and are unequal to a or b, listed in the same order as in $\vec{\pi}$;

▷ the vertices $(v_k, \ldots, v_{2k-1-l-m})$ are the ordered vertices in $\vec{\rho}$ that are not in three edges and are unequal to a or b, listed in the same order as in $\vec{\rho}$.

Thus, vertices that are in four edges in $\vec{\pi} \cup \vec{\rho}$ occur twice in $(v_1, \ldots, v_{2k-1-l-m})$. The vector $(v_1, \ldots, v_{2k-1-l-m})$ is precisely the missing information to reconstruct $(\vec{\pi}, \vec{\rho})$ from σ :

Lemma 6.17 (Bijection of pairs of paths)   *There is a one-to-one correspondence between the pairs of paths $(\vec{\pi}, \vec{\rho})$ and the shape σ, combined with the vertices in $(v_1, \ldots, v_{2k-1-l-m})$ as described above.*

Proof   We have already observed that the shape σ of $(\vec{\pi}, \vec{\rho})$ determines the intersection structure of $(\vec{\pi}, \vec{\rho})$ precisely, and, as such, it contains all the information needed to piece together the two paths $(\vec{\pi}, \vec{\rho})$, except for the information about the vertices involved in these paths. Every vertex in $\vec{\pi} \cup \vec{\rho}$ appears in two, three, or four edges. The vertices that occur in three edges occur at the start of $(v_1, \ldots, v_{2k-1-l-m})$ and the other vertices are those in $\vec{\pi} \setminus \vec{\rho}$ and $\vec{\rho} \setminus \vec{\pi}$, respectively. The above ordering ensures that we can uniquely determine where these vertices are located along the paths $\vec{\pi}$ and $\vec{\rho}$. $\square$

Fix the pair of paths $(\vec{\pi}, \vec{\rho})$ for which $\mathrm{Shape}(\vec{\pi}, \vec{\rho}) = \sigma$ for some $\sigma \in \mathrm{Shape}_{m,l}$, and recall that $a_1 = \delta_{x_1,0}$, $a_{m+1} = \delta_{x_{m+1},0}$. Then, by (6.5.15) and Lemma 6.17,

$$\mathbb{P}(\vec{\pi}, \vec{\rho} \text{ occupied}) = u_a^{1+a_1} u_b^{1+a_{m+1}} \overbrace{\prod_{s=1}^{2m-a_1-a_{m+1}} u_{v_s}^3}^{2m-a_1-a_{m+1}} \overbrace{\prod_{t=2m-a_1-a_{m+1}+1}^{2k-1-l-m} u_{v_t}^2}^{2k-1-l-m}. \qquad (6.5.20)$$

Fix $\sigma \in \mathrm{Shape}_{m,l}$. We bound from above the sum over $\vec{\pi}, \vec{\rho} \in \mathcal{P}_k(a,b)$ such that $\mathrm{Shape}(\vec{\pi}, \vec{\rho}) = \sigma$ by summing (6.5.20) over *all* $(v_1, \ldots, v_{2k-1-l-m}) \in \mathcal{I}^{2k-1-l-m}$, to obtain

$$\sum_{\substack{\vec{\pi}, \vec{\rho} \in \mathcal{P}_k(a,b) \\ \mathrm{Shape}(\vec{\pi}, \vec{\rho}) = \sigma}} \mathbb{P}(\vec{\pi}, \vec{\rho} \text{ occupied})$$

$$\leq u_a u_b \gamma_{\mathcal{I}}^{2m} \nu_{\mathcal{I}}^{2k-1-3m-l} \left(\frac{u_a \nu_{\mathcal{I}}}{\gamma_{\mathcal{I}}} \right)^{\delta_{x_1,0}} \left(\frac{u_b \nu_{\mathcal{I}}}{\gamma_{\mathcal{I}}} \right)^{\delta_{x_{m+1},0}}$$

$$= \bar{n}_k(a,b)^2 \gamma_{\mathcal{I}}^{2(m-1)} \nu_{\mathcal{I}}^{-3(m-1)-l} \left(\frac{\gamma_{\mathcal{I}}}{u_a \nu_{\mathcal{I}}} \right)^{1-\delta_{x_1,0}} \left(\frac{\gamma_{\mathcal{I}}}{u_b \nu_{\mathcal{I}}} \right)^{1-\delta_{x_{m+1},0}}. \qquad (6.5.21)$$

Therefore, we arrive at

$$\mathrm{Var}(N_k(a,b)) \leq n_k(a,b) + \bar{n}_k(a,b)^2 \sum_{l=1}^{k-2} \sum_{m=1}^{k-l} \gamma_{\mathcal{I}}^{2(m-1)} \nu_{\mathcal{I}}^{-3(m-1)-l}$$

$$\times \sum_{\sigma \in \mathrm{Shape}_{m,l}} \left(\frac{\gamma_{\mathcal{I}}}{u_a \nu_{\mathcal{I}}} \right)^{1-\delta_{x_1,0}} \left(\frac{\gamma_{\mathcal{I}}}{u_b \nu_{\mathcal{I}}} \right)^{1-\delta_{x_{m+1},0}}. \qquad (6.5.22)$$

Equation (6.5.22) is our first main result on $\mathrm{Var}(N_k(a,b))$, and we are left with investigating the combinatorial nature of the sums over the shapes. We continue to bound the number of shapes in the following lemma:

Lemma 6.18 (Number of shapes)   *Fix $m \geq 1$ and $l \leq k - 2$.*

(a) *For $m = 1$, the number of shapes in $\mathrm{Shape}_{m,l}$ with fixed $a_1 = \delta_{x_1,0}$, $a_{m+1} = \delta_{x_{m+1},0}$ equals l when $a_1 = a_{m+1} = 0$, 1 when $a_1 + a_{m+1} = 1$, and 0 when $a_1 = a_{m+1} = 1$.*

(b) *For $m \geq 2$, the number of shapes in $\mathrm{Shape}_{m,l}$ with fixed $a_1 = \delta_{x_1,0}$, $a_{m+1} = \delta_{x_{m+1},0}$ is bounded by*

$$2^{m-1}(m-1)! \binom{k-l-1}{m-1}^2 \binom{l}{m-a_1-a_{m+1}}. \qquad (6.5.23)$$

Consequently, for all $m \geq 2$,

$$|\text{Shape}_{m,l}| \leq k \frac{(2k^3)^{m-1}}{(m-1)!}. \qquad (6.5.24)$$

Proof Since $r_1 = r_{m+1} = 1$, there are 2^{m-1} directions in which the common parts of the pairs of paths can be traversed. Since there are m distinct parts, there are $m + 1$ common parts (where the first and last common parts might contain no edges). The first part contains vertex a and the last part contains vertex b. Thus, there are $(m - 1)!$ orders $\vec{o}_{m+1}$ of the common parts when we have fixed the directions in which the paths can be traversed.

Recall that $\mathbb{N}$ denotes the positive integers and $\mathbb{N}_0 = \mathbb{N} \cup \{0\}$ the non-negative integers. In counting the number of $\vec{x}_{m+1}, \vec{s}_m, \vec{t}_m$, we will repeatedly use the fact that there are $\binom{a+b-1}{b-1}$ possible sequences $(y_1, \ldots, y_b) \in \mathbb{N}_0^b$ such that $\sum_{j=1}^b y_j = a$. This can be seen by representing a as a sequence of a ones and $b - 1$ zeros. There are $\binom{a+b-1}{b-1}$ such sequences. Then, there is a bijection to sequences $(y_1, \ldots, y_b) \in \mathbb{N}_0^b$ with $\sum_{j=1}^b y_j = a$ found by letting y_i be the number of ones in between the $(i - 1)$th and ith chosen zero. Similarly, there are $\binom{a-1}{b-1}$ possible sequences $(y_1, \ldots, y_b) \in \mathbb{N}^b$ such that $\sum_{j=1}^b y_j = a$, since we can apply the previous equality to $(y_1 - 1, \ldots, y_b - 1) \in \mathbb{N}_0^b$.

Using the above, we continue to count the number of shapes. The number of vectors $(s_1, \ldots, s_m) \in \mathbb{N}^m$ such that $s_j \geq 1$ and $\sum_{j=1}^m s_j = k - l$ equals $\binom{k-l-1}{m-1}$. The same applies to $(t_1, \ldots, t_m) \in \mathbb{N}^m$ such that $t_j \geq 1$ and $\sum_{j=1}^m t_j = k - l$.

In counting the number of possible $\vec{x}_{m+1}$ such that $\sum_{j=1}^{m+1} x_j = l$, we need to count their numbers separately for $x_1 = 0$ and $x_1 \geq 1$, and for $x_{m+1} = 0$ and $x_{m+1} \geq 1$. When $m = 1$, the number is zero when $x_1 = x_2 = 0$, since $x_1 = x_2 = 0$ implies that the paths share no edges. Recall a_1, a_{m+1}, and suppose that $m - a_1 - a_{m+1} \geq 0$. Then, there are

$$\binom{l}{m-a_1-a_{m+1}}$$

possible choices of $\vec{x}_{m+1}$ with fixed $a_1 = \delta_{x_1,0}, a_{m+1} = \delta_{x_{m+1},0}$. The claims in part (a), as well as that in (6.5.23) in part (b), follow by multiplying these bounds on the number of choices for $\vec{r}_{m+1}, \vec{o}_{m+1}, \vec{s}_m, \vec{t}_m$ and $\vec{x}_{m+1}$.

To prove (6.5.24) in part (b), we continue by obtaining the bound

$$\binom{k-l-1}{m-1}^2 (m-1)! = \frac{1}{(m-1)!} \left(\frac{(k-l-1)!}{(k-l-m)!} \right)^2 \leq \frac{k^{2(m-1)}}{(m-1)!}, \qquad (6.5.25)$$

and, using that $\binom{a}{b} \leq a^b/b!$ and $l \leq k$, we then have

$$\binom{l}{m-a_1-a_{m+1}} \leq \frac{l^{m-a_1-a_{m+1}}}{(m-a_1-a_{m+1})!} \leq k^m. \qquad (6.5.26)$$

Therefore, the number of shapes in $\mathrm{Shape}_{m,l}$ is bounded, for each $l \geq 1$ and $m \geq 2$, by

$$2^{m-1} \frac{k^{2(m-1)}}{(m-1)!} k^m = k \frac{(2k^3)^{m-1}}{(m-1)!}, \tag{6.5.27}$$

as required.　　□

We are now ready to complete the proof of Proposition 6.14:

Proof of Proposition 6.14. By (6.5.22) and applying Lemma 6.18, it suffices to show that the sum of

$$|\mathrm{Shape}_{m,l}| \times \left(\frac{2\gamma_{\mathcal{I}}^2}{\nu_{\mathcal{I}}^3}\right)^{m-1} \nu_{\mathcal{I}}^{-l} \left(\frac{\gamma_{\mathcal{I}}}{u_a \nu_{\mathcal{I}}}\right)^{1-a_1} \left(\frac{\gamma_{\mathcal{I}}}{u_b \nu_{\mathcal{I}}}\right)^{1-a_{m+1}} \tag{6.5.28}$$

over $l \in [k-2]$, $m \in [k-l]$, and $a_1, a_{m+1} \in \{0,1\}$ (where, by convention, $\binom{l}{-1} = 0$), is bounded by the contribution in parentheses in the second term in (6.5.7).

We start with $m = 1$, for which we obtain that the sum of (6.5.28) over the other variables $l \in [k-2]$ and $a_1, a_{m+1} \in \{0,1\}$ equals

$$\gamma_{\mathcal{I}}\left(\frac{1}{u_a} + \frac{1}{u_b}\right) \sum_{l=1}^{\infty} \nu_{\mathcal{I}}^{-(l-1)} + \frac{\gamma_{\mathcal{I}}^2}{u_a u_b \nu_{\mathcal{I}}} \sum_{l=1}^{\infty} l \nu_{\mathcal{I}}^{-(l-1)}$$

$$= \frac{\gamma_{\mathcal{I}} \nu_{\mathcal{I}}^2}{\nu_{\mathcal{I}} - 1}\left(\frac{1}{u_a} + \frac{1}{u_b}\right) + \frac{\gamma_{\mathcal{I}}^2 \nu_{\mathcal{I}}}{u_a u_b (\nu_{\mathcal{I}} - 1)^2}, \tag{6.5.29}$$

where we use that, for $a \in [0,1)$,

$$\sum_{l=0}^{\infty} a^{-l} = a/(1-a), \qquad \sum_{l=0}^{\infty} l a^{-(l-1)} = a^2/(1-a)^2. \tag{6.5.30}$$

The terms in (6.5.29) are the first two terms that are multiplied by $\bar{n}_k(a,b)^2$ on the rhs of (6.5.7).

This leaves us to bound the contribution when $m \geq 2$. Since (6.5.24) is independent of l, we can start by summing (6.5.28) over $l \in [k]$ and over $a_1, a_{m+1} \in \{0,1\}$, to obtain a bound of the form (recall (6.5.8))

$$k\left(1 + \frac{\gamma_{\mathcal{I}}}{u_a \nu_{\mathcal{I}}}\right)\left(1 + \frac{\gamma_{\mathcal{I}}}{u_b \nu_{\mathcal{I}}}\right) \frac{\nu_{\mathcal{I}}}{\nu_{\mathcal{I}} - 1} \sum_{m \geq 2} \frac{(2k^3)^{m-1}}{(m-1)!}\left(\frac{\gamma_{\mathcal{I}}^2}{\nu_{\mathcal{I}}^3}\right)^{m-1}$$

$$= k\left(1 + \frac{\gamma_{\mathcal{I}}}{u_a \nu_{\mathcal{I}}}\right)\left(1 + \frac{\gamma_{\mathcal{I}}}{u_b \nu_{\mathcal{I}}}\right) \frac{\nu_{\mathcal{I}}}{\nu_{\mathcal{I}} - 1}\left(e^{2k^3 \gamma_{\mathcal{I}}^2 / \nu_{\mathcal{I}}^3} - 1\right) = e_k. \tag{6.5.31}$$

After multiplication with $\bar{n}_k(a,b)^2$, the term in (6.5.31) is the same as the last term appearing on the rhs of (6.5.7). Summing the bounds in (6.5.29) and (6.5.31) proves (6.5.7).　□

Exercises 6.16–6.20 study various consequences of our path-counting techniques. In the next subsection, we use Proposition 6.14 to prove lower bounds on graph distances.

6.5.2 LOGARITHMIC DISTANCE BOUNDS FOR FINITE-VARIANCE WEIGHTS

In this subsection we prove that, when Conditions 1.1(a)–(c) hold, two uniform vertices that are conditioned to be connected are whp within a distance $(1 + \varepsilon) \log_{\nu} n$:

Theorem 6.19 (Logarithmic upper bound on typical distances for finite-variance weights) *Consider* $\mathrm{NR}_n(\boldsymbol{w})$*, where the weights* $\boldsymbol{w} = (w_v)_{v \in [n]}$ *satisfy Conditions 1.1(a)–(c) with* $\nu = \mathbb{E}[W^2]/\mathbb{E}[W] \in (1, \infty)$*. Then, for any* $\varepsilon > 0$*, with* o_1, o_2 *chosen independently and uar from* $[n]$,

$$\mathbb{P}(\mathrm{dist}_{\mathrm{NR}_n(\boldsymbol{w})}(o_1, o_2) \leq (1 + \varepsilon)\log_\nu n \mid \mathrm{dist}_{\mathrm{NR}_n(\boldsymbol{w})}(o_1, o_2) < \infty) = 1 + o(1). \quad (6.5.32)$$

The same result applies, under identical conditions, to $\mathrm{GRG}_n(\boldsymbol{w})$ *and* $\mathrm{CL}_n(\boldsymbol{w})$.

Theorem 6.19 provides the upper bound on the typical distances that matches Theorem 6.4, and together these two theorems prove Theorem 6.2. The remainder of this subsection is devoted to the proof of Theorem 6.19.

Organization of the Proof of Theorem 6.19

In the proof of Theorem 6.4 it is convenient to work with $\mathrm{CL}_n(\boldsymbol{w})$, since Proposition 6.14 is designed for that setting. As mentioned before, for $\mathrm{GRG}_n(\boldsymbol{w})$ and $\mathrm{NR}_n(\boldsymbol{w})$ the results will follow from the asymptotic equivalence arguments in [V1, Section 6.7]. Indeed, [V1, Corollary 6.20 and (6.8.13)] imply that $\mathrm{GRG}_n(\boldsymbol{w})$ and $\mathrm{NR}_n(\boldsymbol{w})$ are asymptotically equivalent to $\mathrm{CL}_n(\boldsymbol{w})$ when Conditions 1.1(a)–(c) hold. Thus, we fix $G_n = \mathrm{CL}_n(\boldsymbol{w})$ from now on.

We prove Theorem 6.19 by combining a *branching-process comparison* of local neighborhoods, as given by Theorem 3.18, to a *second-moment method* as in Proposition 6.14 regarding the number of paths of a given length. More precisely, we fix $r \geq 1$ large, recall that $B_r^{(G_n)}(o_1)$ and $B_r^{(G_n)}(o_2)$ denote the rooted graphs of vertices at distance at most r from o_1 and o_2 respectively, and let $\partial B_r^{(G_n)}(o_1)$ and $\partial B_r^{(G_n)}(o_2)$ denote the sets of vertices at distance precisely equal to r.

We condition on $B_r^{(G_n)}(o_1)$ and $B_r^{(G_n)}(o_2)$ such that $\partial B_r^{(G_n)}(o_1) \neq \emptyset$ and $\partial B_r^{(G_n)}(o_2) \neq \emptyset$. By the local convergence in Theorem 3.18, the probability of the latter event is close to ζ_r^2, where $\zeta_r = \mu(\partial B_r^{(G)}(o) \neq \emptyset)$ is the probability that the local limit $(G, o) \sim \mu$ of $\mathrm{CL}_n(\boldsymbol{w})$ survives to generation r. Corollary 2.19 proves the asymptotic independence of the neighborhoods of o_1 and o_2, respectively. Then $\zeta_r \searrow \zeta$ when $r \to \infty$, and, since $\nu > 1$, conditional on $|\partial B_r^{(G)}(o)| > 0$, we have $|\partial B_r^{(G)}(o)| \geq M$ whp, for any M and as $r \to \infty$. This explains the branching-process approximation.

We now state the precise branching-process approximation result that we rely upon, and link the second-moment method for the number of paths, as proved in Proposition 6.14, to our setting. We take $u_i = w_i/\sqrt{\ell_n}$,

$$a = \partial B_r^{(G_n)}(o_1), \qquad b = \partial B_r^{(G_n)}(o_2), \qquad (6.5.33)$$

and, for some $\varepsilon > 0$,

$$u_a = \frac{1 - \varepsilon}{\sqrt{\ell_n}} \sum_{v \in \partial B_r^{(G_n)}(o_1)} w_v = (1 - \varepsilon)\mathcal{W}_r(o_1)/\sqrt{\ell_n},$$

$$(6.5.34)$$

$$u_b = \frac{1 - \varepsilon}{\sqrt{\ell_n}} \sum_{v \in \partial B_r^{(G_n)}(o_2)} w_v = (1 - \varepsilon)\mathcal{W}_r(o_2)/\sqrt{\ell_n}.$$

The $1 - \varepsilon$ factors in (6.5.34) are due to the fact that the edge probabilities in the graph on $\{a, b\} \cup [n] \setminus (B_r^{(G_n)}(o_1) \cup B_r^{(G_n)}(o_2))$ are not *exactly* of the form $p_{ij} = u_i u_j$. Indeed, for $i, j \in \{a, b\}$, the edge probabilities are slightly different. When Conditions 1.1(a)–(c) hold, however, the bound *almost* holds for a and b, which explains the factors $1 - \varepsilon$.

We formalize the above ideas in the following lemma:

Lemma 6.20 (Branching-process approximation) *Under the conditions of Theorem 6.19, as $n \to \infty$,*

$$
(\mathcal{W}_r(o_1), \mathcal{W}_r(o_2)) \xrightarrow{d} \left(\sum_{j=1}^{Z_r^{(1)}} W^{\star(1)}(j), \sum_{j=1}^{Z_r^{(2)}} W^{\star(2)}(j) \right), \qquad (6.5.35)
$$

where $(Z_m^{(1)}, Z_m^{(2)})_{m \geq 0}$ are the generation sizes of two independent unimodular branching processes as in Theorem 3.18, and $(W^{\star(1)}(j))_{j \geq 1}$ and $(W^{\star(2)}(j))_{j \geq 1}$ are two independent sequences of iid random variables with distribution $F^\star$.

Proof It is now convenient to start with $G_n = \mathrm{NR}_n(\boldsymbol{w})$. By Corollary 2.19, $|\partial B_r^{(G_n)}(o_1)|$ and $|\partial B_r^{(G_n)}(o_2)|$ jointly converge in distribution to $(Z_r^{(1)}, Z_r^{(2)})$, which are independent generation sizes of the local limit of $\mathrm{NR}_n(\boldsymbol{w})$ as in Theorem 3.18. Each of the individuals in $\partial B_r^{(G_n)}(o_1)$ and $\partial B_r^{(G_n)}(o_2)$ receives a mark M_i with weight w_{M_i}. By Proposition 3.16, these marks are iid random variables conditioned to be unthinned, where whp no vertex in $B_r^{(G_n)}(o_1) \cup B_r^{(G_n)}(o_2)$ is thinned. Then, $\mathcal{W}_r(o_i) = \sum_{j=1}^{|\partial B_r^{(G_n)}(o_i)|} W_n^{\star(i)}(j)$, where $(W_n^{\star(i)}(j))_{j \geq 1}$ are iid copies of $W_n^\star$. By Conditions 1.1(a),(b), $W_n^\star \xrightarrow{d} W^\star$, so that also $\mathcal{W}_r(o_i) \xrightarrow{d} \sum_{j=1}^{|\partial B_r^{(G)}(o_i)|} W^{\star(i)}(j)$.

The joint convergence follows in a similar fashion, now using local convergence in probability. As discussed before, the above results extend trivially to $\mathrm{GRG}_n(\boldsymbol{w})$ and $\mathrm{CL}_n(\boldsymbol{w})$ by asymptotic equivalence. $\qquad \square$

Second-Moment Method and Path Counting

We again focus on $G_n = \mathrm{CL}_n(\boldsymbol{w})$. Fix $k = k_n = \lceil (1 + \varepsilon) \log_\nu n \rceil - 2r$. We next present the details of the second-moment method, which shows that whp, on the event that $\partial B_r^{(G_n)}(o_1) \neq \varnothing$ and $\partial B_r^{(G_n)}(o_2) \neq \varnothing$, there exists a path of length k_n connecting $\partial B_r^{(G_n)}(o_1)$ and $\partial B_r^{(G_n)}(o_2)$. This ensures that, on the event that $\partial B_r^{(G_n)}(o_1) \neq \varnothing$ and $\partial B_r^{(G_n)}(o_2) \neq \varnothing$, the event $\mathrm{dist}_{\mathrm{NR}_n(w)}(o_1, o_2) \leq k_n$ occurs whp.

To show that $\mathrm{dist}_{\mathrm{NR}_n(w)}(o_1, o_2) \leq k_n - 2r$ occurs whp, we take $u_i = w_i / \sqrt{\ell_n}$. We aim to apply Proposition 6.14, for which we fix $K \geq 1$ sufficiently large and take $a = \partial B_r^{(G_n)}(o_1)$, $b = \partial B_r^{(G_n)}(o_2)$, and

$$
\mathcal{I}_{a,b} = \{i \in [n] : w_i \leq K\} \setminus (B_r^{(G_n)}(o_1) \cup B_r^{(G_n)}(o_2)). \qquad (6.5.36)
$$

In order to apply Proposition 6.14, we start by relating the random graph obtained by restricting $\mathrm{CL}_n(\boldsymbol{w})$ to the vertex set $\mathcal{I}_{a,b}$ to the model on the vertex set $\mathcal{I}_{a,b} \cup \{a, b\}$ with edge probabilities $p_{ij} = u_i u_j$ with $u_i = w_i / \sqrt{\ell_n}$ for $i \in \mathcal{I}_{a,b}$ and u_a, u_b given by (6.5.34). For this, we note that for $i, j \in \mathcal{I}_{a,b}$, this equality holds by definition of $\mathrm{CL}_n(\boldsymbol{w})$. We next take $i = a$ and $j \in \mathcal{I}_{a,b}$; the argument for $i = b$ and $j \in \mathcal{I}_{a,b}$ is identical.

The conditional probability that $j \in \mathcal{I}_{a,b}$ is connected to at least one vertex in $\partial B_r^{(G_n)}(o_1)$, given $B_r^{(G_n)}(o_1)$, equals

$$1 - \prod_{v \in \partial B_r^{(G_n)}(o_1)} \left(1 - \frac{w_v w_j}{\ell_n}\right). \tag{6.5.37}$$

Since, for all $x_i \in [0, 1]$,

$$\prod_i (1 - x_i) \leq 1 - \sum_i x_i + \frac{1}{2} \sum_{i \neq j} x_i x_j,$$

we obtain that

$$1 - \prod_{v \in \partial B_r^{(G_n)}(o_1)} \left(1 - \frac{w_v w_j}{\ell_n}\right) \geq \sum_{v \in \partial B_r^{(G_n)}(o_1)} \frac{w_v w_j}{\ell_n} - \sum_{v_1, v_2 \in \partial B_r^{(G_n)}(o_1)} \frac{w_{v_1} w_{v_2} w_j^2}{2\ell_n^2}$$

$$\geq \frac{\mathcal{W}_r(o_1) w_j}{\ell_n} - \frac{\mathcal{W}_r(o_1)^2 w_j^2}{2\ell_n^2}. \tag{6.5.38}$$

By Conditions 1.1(a)–(c), $w_j = o(\sqrt{n})$ (recall Exercise 1.8), and $\mathcal{W}_r(o_1)$ is a tight sequence of random variables (see also Lemma 6.21 below), so that, whp for any $\varepsilon > 0$,

$$1 - \prod_{v \in \partial B_r^{(G_n)}(o_1)} \left(1 - \frac{w_v w_j}{\ell_n}\right) \geq (1 - \varepsilon) \frac{\mathcal{W}_r(o_1) w_j}{\ell_n}. \tag{6.5.39}$$

With the choices in (6.5.34), we see that our graph is bounded below by that studied in Proposition 6.14. By the above description, it is clear that all our arguments will be *conditional*, given $B_r^{(G_n)}(o_1)$ and $B_r^{(G_n)}(o_2)$. For this, we define $\mathbb{P}_r$ to be the conditional distribution given $B_r^{(G_n)}(o_1)$ and $B_r^{(G_n)}(o_2)$, and we let $\mathbb{E}_r$ and Var_r be the corresponding conditional expectation and variance.

In order to apply Proposition 6.14, we investigate the quantities appearing in it:

Lemma 6.21 (Parameters in path counting) *Under the conditions of Theorem 6.19, and conditioning on $B_r^{(G_n)}(o_1)$ and $B_r^{(G_n)}(o_2)$ with $\partial B_r^{(G_n)}(o_1) \neq \varnothing, \partial B_r^{(G_n)}(o_2) \neq \varnothing$, and with $a = \partial B_r^{(G_n)}(o_1), b = \partial B_r^{(G_n)}(o_2)$, for $k = k_n = \lceil (1 + \varepsilon) \log_\nu n \rceil - 2r$,*

$$n_k(a, b) \xrightarrow{\mathbb{P}} \infty, \qquad \bar{n}_k(a, b) = (1 + o_{\mathbb{P}}(1)) \underline{n}_k(a, b), \tag{6.5.40}$$

and, as $n \to \infty$,

$$\frac{\mathrm{Var}_r(N_k(a, b))}{\mathbb{E}_r[N_k(a, b)]^2} \leq \frac{K\nu^2}{\nu - 1}\left(\frac{1}{\sqrt{\ell_n} u_a} + \frac{1}{\sqrt{\ell_n} u_b}\right) + \frac{K^2 \nu^2}{(\nu - 1)\ell_n u_a u_b} + o_{\mathbb{P}}(1). \tag{6.5.41}$$

Proof By (6.5.3),

$$\underline{n}_k(a, b) = u_a u_b \nu_{\mathcal{I}_{a,b,k}}^{k-1}, \qquad \text{and} \qquad \frac{\bar{n}_k(a, b)}{\underline{n}_k(a, b)} = \left(\frac{\nu_{\mathcal{I}_{a,b}}}{\nu_{\mathcal{I}_{a,b,k}}}\right)^{k-1}. \tag{6.5.42}$$

We start by investigating $\nu_{\mathcal{I}}$. Denote

$$\nu(K) = \frac{\mathbb{E}[W^2 \mathbf{1}_{\{W \leq K\}}]}{\mathbb{E}[W]}. \tag{6.5.43}$$

Then, by (6.5.36) and since $B_r^{(G_n)}(o_1)$ and $B_r^{(G_n)}(o_2)$ contain a finite number of vertices,

$$\nu_{\mathcal{I}_{a,b}} \xrightarrow{\mathbb{P}} \nu(K). \tag{6.5.44}$$

The same applies to $\nu_{\mathcal{I}_{a,b,k}}$. Then, with $K > 0$ chosen sufficently large that $\nu(K) \geq \nu - \varepsilon/2$ and with $k = k_n = \lceil (1 + \varepsilon) \log_\nu n \rceil - 2r$,

$$\underline{n}_k(a, b) = u_a u_b \nu_{\mathcal{I}_{a,b,k}}^{k-1} = \frac{\mathcal{W}_r(o_1) \mathcal{W}_r(o_2)}{\ell_n} n^{(1+\varepsilon) \log \nu_{\mathcal{I}_{a,b,k}} / \log \nu - 1} \xrightarrow{\mathbb{P}} \infty, \tag{6.5.45}$$

when K and n are large that $(1 + \varepsilon)\nu(K)/\nu > 1$. This proves the first property in (6.5.40).

To prove the second property in (6.5.40), we note that the set $\mathcal{I}_{a,b,k}$ is obtained from $\mathcal{I}_{a,b}$ by removing the k vertices with highest weight. Since $w_i \leq K$ for all $i \in \mathcal{I}$ (recall (6.5.36)), $\nu_{\mathcal{I}_{a,b}} \leq \nu_{\mathcal{I}_{a,b,k}} + kK/\ell_n$. Since $k \leq A \log n$, we therefore arrive at

$$\frac{\bar{n}_k(a, b)}{\underline{n}_k(a, b)} \leq \left(1 + kK/(\ell_n \nu_{\mathcal{I}_{a,b,k}})\right)^{k-1} \leq e^{k^2 K/(\ell_n \nu_{\mathcal{I}_{a,b,k}})} \xrightarrow{\mathbb{P}} 1, \tag{6.5.46}$$

as required.

To prove (6.5.41), we rely on Proposition 6.14. We have already shown that $n_k(a, b) = \mathbb{E}_r[N_k(a, b)] \xrightarrow{\mathbb{P}} \infty$, so that the first term on the rhs of (6.5.7) is $o_{\mathbb{P}}(\mathbb{E}_r[N_k(a, b)]^2)$. Further, by (6.5.36),

$$\gamma_{\mathcal{I}} \leq \nu_{\mathcal{I}}(\max_{i \in \mathcal{I}} u_i) \leq \frac{\nu_{\mathcal{I}} K}{\sqrt{\ell_n}}, \tag{6.5.47}$$

so that, for $k \leq A \log n$ with $A > 1$ fixed,

$$\left(1 + \frac{\gamma_{\mathcal{I}}}{u_a \nu_{\mathcal{I}}}\right)\left(1 + \frac{\gamma_{\mathcal{I}}}{u_b \nu_{\mathcal{I}}}\right) k(e^{2k^3 \gamma_{\mathcal{I}}^2 / \nu_{\mathcal{I}}^3} - 1) = o_{\mathbb{P}}(1). \tag{6.5.48}$$

Substituting these bounds into (6.5.41) and using (6.5.40) yields the claim. $\square$

Completion of the Proof of Theorem 6.19

Now we are are ready to complete the proof of Theorem 6.19. Recall that $k_n = k_n(\varepsilon) = \lceil (1 + \varepsilon) \log_\nu n \rceil$, and fix $G_n = \mathrm{CL}_n(\boldsymbol{w})$. We need to show that

$$\mathbb{P}(k_n < \mathrm{dist}_{\mathrm{CL}_n(\boldsymbol{w})}(o_1, o_2) < \infty) = o(1). \tag{6.5.49}$$

Indeed, (6.5.49) implies that $\mathbb{P}(\mathrm{dist}_{\mathrm{CL}_n(\boldsymbol{w})}(o_1, o_2) > k_n \mid \mathrm{dist}_{\mathrm{CL}_n(\boldsymbol{w})}(o_1, o_2) < \infty) = o(1)$, since $\mathbb{P}(\mathrm{dist}_{\mathrm{CL}_n(\boldsymbol{w})}(o_1, o_2) < \infty) \to \zeta^2 > 0$ by Theorem 3.20.

We rewrite

$$\mathbb{P}(k_n < \mathrm{dist}_{\mathrm{CL}_n(\boldsymbol{w})}(o_1, o_2) < \infty) \tag{6.5.50}$$
$$= \mathbb{P}(k_n < \mathrm{dist}_{\mathrm{CL}_n(\boldsymbol{w})}(o_1, o_2) < \infty, \partial B_r^{(G_n)}(o_1) \neq \varnothing, \partial B_r^{(G_n)}(o_2) \neq \varnothing)$$
$$\leq \mathbb{P}(N_{k_n-2r}(\partial B_r^{(G_n)}(o_1), \partial B_r^{(G_n)}(o_2)) = 0, \partial B_r^{(G_n)}(o_1) \neq \varnothing, \partial B_r^{(G_n)}(o_2) \neq \varnothing)$$
$$\leq \mathbb{E}\left[\mathbb{P}_r(N_{k_n-2r}(\partial B_r^{(G_n)}(o_1), \partial B_r^{(G_n)}(o_2)) = 0)\mathbb{1}_{\{\partial B_r^{(G_n)}(o_1) \neq \varnothing, \partial B_r^{(G_n)}(o_2) \neq \varnothing\}}\right],$$

where we recall that $\mathbb{P}_r$ is the conditional distribution given $B_r^{(G_n)}(o_1)$ and $B_r^{(G_n)}(o_2)$.

By Lemma 6.21 and the Chebychev inequality [V1, Theorem 2.18], the conditional probability of $\{\mathrm{dist}_{\mathrm{CL}_n(w)}(o_1, o_2) > k_n\}$, given $B_r^{(G_n)}(o_1), B_r^{(G_n)}(o_2)$, is at most

$$\frac{\mathrm{Var}_r(N_{k_n - 2r}(a, b))}{\mathbb{E}_r[N_{k_n - 2r}(a, b)]^2} \leq \frac{K\nu^2}{\nu - 1}\left(\frac{1}{\sqrt{\ell_n}u_a} + \frac{1}{\sqrt{\ell_n}u_b}\right) + \frac{K^2\nu^2}{(\nu - 1)\ell_n u_a u_b} + o_{\mathbb{P}}(1). \quad (6.5.51)$$

When $\partial B_r^{(G_n)}(o_1) \neq \varnothing$ and $\partial B_r^{(G_n)}(o_2) \neq \varnothing$, by (6.5.34) and as $n \to \infty$,

$$\frac{1}{\sqrt{\ell_n}u_a} + \frac{1}{\sqrt{\ell_n}u_b} \xrightarrow{\mathbb{P}} \left((1 - \varepsilon)\sum_{j=1}^{Z_r^{(1)}} W^{\star(1)}(j)\right)^{-1} + \left((1 - \varepsilon)\sum_{j=1}^{Z_r^{(2)}} W^{\star(2)}(j)\right)^{-1} \xrightarrow{\mathbb{P}} 0,$$
$$(6.5.52)$$

when $r \to \infty$. Therefore, with first $n \to \infty$ followed by $r \to \infty$,

$$\mathbb{P}_r\left(N_{k - 2r}(a, b) = 0 \mid \partial B_r^{(G_n)}(o_1) \neq \varnothing, \partial B_r^{(G_n)}(o_2) \neq \varnothing\right) \xrightarrow{\mathbb{P}} 0, \quad (6.5.53)$$

and, by Lebesgue's Dominated Convergence Theorem [V1, Theorem A.1],

$$\mathbb{P}(\mathrm{dist}_{\mathrm{CL}_n(w)}(o_1, o_2) > k_n, \partial B_r^{(G_n)}(o_1) \neq \varnothing, \partial B_r^{(G_n)}(o_2) \neq \varnothing) \to 0, \quad (6.5.54)$$

when first $n \to \infty$ followed by $r \to \infty$, which completes the proof. $\qquad\square$

Distances for the Critical Case $\tau = 3$

When $\tau = 3$, w_i is approximately $c(n/i)^{1/2}$. It turns out that this changes the typical distances only by a doubly logarithmic factor:

Theorem 6.22 (Typical distances in critical $\tau = 3$ case) *Consider* $\mathrm{NR}_n(w)$, *where the weights* $w = (w_v)_{v \in [n]}$ *satisfy Condition 1.1(a),(b), and consider that there exist constants* $c_2 > c_1 > 0$ *and* $\beta > 0$ *such that, for all* $x \leq n^\beta$,

$$[1 - F_n](x) \geq c_1/x^2, \quad (6.5.55)$$

and, for all $x \geq 0$,

$$[1 - F_n](x) \leq c_2/x^2. \quad (6.5.56)$$

Then, with o_1, o_2 *chosen independently and uar from* $[n]$, *and conditional on* $o_1 \longleftrightarrow o_2$,

$$\frac{\mathrm{dist}_{\mathrm{NR}_n(w)}(o_1, o_2) \log \log n}{\log n} \xrightarrow{\mathbb{P}} 1. \quad (6.5.57)$$

The same result applies, under identical conditions, to $\mathrm{GRG}_n(w)$ *and* $\mathrm{CL}_n(w)$.

The lower bound in Theorem 6.22 has been stated already in Corollary 6.6. The upper bound can be proved using the path-counting techniques in Proposition 6.14 and adaptations of it. We now sketch this proof.

Again fix $G_n = \mathrm{CL}_n(w)$. Let $\eta \in (0, 1)$ and let

$$\beta_n = e^{\nu_n^{1-\eta}}. \quad (6.5.58)$$

Define the *core* of $\mathrm{CL}_n(w)$ as follows:

$$\mathrm{Core}_n = \{i \in [n]: w_i \geq \beta_n\}. \quad (6.5.59)$$

The proof of Theorem 6.22 follows from the following two lemmas:

Lemma 6.23 (Typical distances in core)   *Under the conditions of Theorem 6.22, let $o'_1, o'_2 \in$ Core$_n$ be chosen with probabilities proportional to their weights, i.e.,*

$$\mathbb{P}(o'_i = j) = \frac{w_j}{\sum_{v \in \text{Core}_n} w_v}, \tag{6.5.60}$$

and let H'_n be the graph distance between o'_1, o'_2 in Core$_n$. Then, for any $\varepsilon > 0$, there exists an $\eta \in (0, 1)$ in (6.5.58) such that

$$\mathbb{P}\left(H'_n \leq \frac{(1 + \varepsilon) \log n}{\log \log n}\right) \to 1. \tag{6.5.61}$$

Lemma 6.24 (From periphery to core)   *Under the conditions of Theorem 6.22, let o_1, o_2 be two vertices chosen uar from $[n]$. Then, for any $\eta > 0$ in (6.5.58),*

$$\mathbb{P}(\text{dist}_{\text{CL}_n(w)}(o_1, \text{Core}_n) \leq \nu_n^{1-\eta}, \text{dist}_{\text{CL}_n(w)}(o_2, \text{Core}_n) \leq \nu_n^{1-\eta}) \to \zeta^2. \tag{6.5.62}$$

Further, $\text{CL}_n(w), \text{GRG}_n(w),$ and $\text{NR}_n(w)$ are asymptotically equivalent when restricted to the edges in $[n] \times \{v : w_v \leq \beta_n\}$ for any $\beta_n = o(\sqrt{n})$.

Proof of Theorem 6.22 subject to Lemmas 6.23 and 6.24. To see that Lemmas 6.23 and 6.24 imply Theorem 6.22, we note that

$$\text{dist}_{\text{CL}_n(w)}(o_1, o_2) \leq \text{dist}_{\text{CL}_n(w)}(o_1, \text{Core}_n) + \text{dist}_{\text{CL}_n(w)}(o_2, \text{Core}_n)$$
$$+ \text{dist}_{\text{CL}_n(w)}(o'_1, o'_2), \tag{6.5.63}$$

where $o'_1, o'_2 \in$ Core$_n$ are the vertices in Core$_n$ found first in the breadth-first search from o_1 and o_2, respectively. By the asymptotic equivalence of $\text{CL}_n(w), \text{GRG}_n(w),$ and $\text{NR}_n(w)$ on $[n] \times \{v : w_v \leq \beta_n\}$, stated in Lemma 6.24, whp $\text{dist}_{\text{CL}_n(w)}(o_1, \text{Core}_n) = \text{dist}_{\text{NR}_n(w)}(o_1, \text{Core}_n)$ and $\text{dist}_{\text{CL}_n(w)}(o_2, \text{Core}_n) = \text{dist}_{\text{NR}_n(w)}(o_2, \text{Core}_n)$, so we can work with $\text{NR}_n(w)$ outside Core$_n$. Then, by Proposition 3.16, $o'_1, o'_2 \in$ Core$_n$ are chosen with probabilities proportional to their weights, as assumed in Lemma 6.23.

Fix $k_n = \lceil (1 + \varepsilon) \log n / \log \log n \rceil$. We conclude that, when n is sufficiently large that $\nu_n^{1-\eta} \leq \varepsilon k_n / 4$,

$$\mathbb{P}(\text{dist}_{\text{CL}_n(w)}(o_1, o_2) \leq k_n)$$
$$\geq \mathbb{P}(\text{dist}_{\text{CL}_n(w)}(o_i, \text{Core}_n) \leq \nu_n^{1-\eta}, i = 1, 2) \tag{6.5.64}$$
$$\times \mathbb{P}(\text{dist}_{\text{CL}_n(w)}(o'_1, o'_2) \leq (1 - \varepsilon/2)k_n \mid \text{dist}_{\text{CL}_n(w)}(o_i, \text{Core}_n) \leq \nu_n^{1-\eta}, i = 1, 2).$$

By Lemma 6.24, the first probability converges to ζ^2 and by Lemma 6.23 the second probability converges to 1. We conclude that

$$\mathbb{P}\left(\text{dist}_{\text{CL}_n(w)}(o_1, o_2) \leq (1 + \varepsilon) \frac{\log n}{\log \log n}\right) \to \zeta^2. \tag{6.5.65}$$

Since also $\mathbb{P}(\text{dist}_{\text{CL}_n(w)}(o_1, o_2) < \infty) \to \zeta^2$, this completes the proof of Theorem 6.22. $\square$

The proofs of Lemmas 6.23 and 6.24 follow from path-counting techniques similar to those carried out earlier. Exercises 6.21–6.24 complete the proof of Lemma 6.23. Exercise 6.25 asks you to verify the asymptotic equivalence stated in Lemma 6.24, while Exercise 6.26 asks you to give the proof of (6.5.62) in Lemma 6.24.

6.5.3 DISTANCES AND GIANTS FOR INHOMOGENEOUS RANDOM GRAPHS

In this subsection we use the path-counting techniques in Section 6.5.1 to give some missing proofs for general inhomogeneous random graphs. We assume that (κ_n) is a graphical sequence of kernels with limit κ that is irreducible, and with $\nu = \|T_\kappa\| \in (1, \infty)$. We start by proving Theorem 6.1(b), and then we use it to prove Theorem 3.19.

Logarithmic Upper Bound on Distances in IRGs in Theorem 6.1(b)

Without loss of generality, we may assume that κ_n is bounded, i.e., $\sup_n \sup_{x,y} \kappa_n(x, y) < \infty$. Indeed, we can always stochastically dominate an $\mathrm{IRG}_n(\kappa_n)$ with an unbounded κ_n from below by an $\mathrm{IRG}_n(\kappa_n)$ with a bounded kernel that approximates it arbitrarily well. Since graph distances increase by decreasing κ_n, if we prove Theorem 6.1(b) in the bounded case, then the unbounded case will follow immediately. Further, we can approximate a bounded κ_n from above by a finite-type kernel. Therefore, it now suffices to prove Theorem 6.1(b) for a kernel of finite type.

Our arguments make use of the first extension of the path-counting arguments in Proposition 6.14 that is beyond the rank-1 setting. This shows again the versatility of the path-counting methods. The key is that the expected number of *shortcuts* from a path $\vec{\pi}$ of length ℓ is approximately bounded by $\|T_{\kappa_n}\|^\ell / n$, which is small when ℓ is not too large. This allows us to handle the complicated sums over shapes that occur beyond the rank-1 setting.

Let us set up the finite-types case. For $u, v \in [n]$, we let $\kappa_n(u, v) = \kappa^{(n)}(i_u, i_v)$, where $i_u \in [t]$ denotes the *type* of vertex $u \in [n]$. We assume that, for all $i, j \in [t]$,

$$\lim_{n \to \infty} \kappa^{(n)}(i, j) = \kappa(i, j), \tag{6.5.66}$$

and, for all $i \in [t]$,

$$\lim_{n \to \infty} \mu_n(i) = \lim_{n \to \infty} \frac{1}{n} \#\{v \in [n]: i_v = i\} = \mu(i). \tag{6.5.67}$$

In this case, $\|T_{\kappa_n}\|$ is the largest eigenvalue of the matrix $\mathbf{M}_{i,j}^{(n)} = \kappa^{(n)}(i, j)\mu_n(j)$, which converges to the largest eigenvalue of the matrix $\mathbf{M}_{i,j} = \kappa(i, j)\mu(j)$ and equals $\nu = \|T_\kappa\| \in (1, \infty)$, by assumption. Without loss of generality, we may assume that $\mu(i) > 0$ for all $i \in [t]$. This sets the stage for our analysis.

Fix $G_n = \mathrm{IRG}_n(\kappa_n)$; fix $r \geq 1$ and assume that $\partial B_r^{(G_n)}(o_1), \partial B_r^{(G_n)}(o_2) \neq \varnothing$. We will prove that

$$\mathbb{P}\big(\mathrm{dist}_{\mathrm{IRG}_n(\kappa_n)}(o_1, o_2) \leq (1 + \varepsilon) \log_\nu n \mid B_r^{(G_n)}(o_1), B_r^{(G_n)}(o_2)\big) = 1 + o_\mathbb{P}(1). \tag{6.5.68}$$

We follow the proof of Theorem 6.19 and rely on path-counting techniques. We again take

$$a = \partial B_r^{(G_n)}(o_1), \qquad b = \partial B_r^{(G_n)}(o_2), \tag{6.5.69}$$

and

$$\mathcal{I}_{a,b} = [n] \setminus (B_r^{(G_n)}(o_1) \cup B_r^{(G_n)}(o_2)). \tag{6.5.70}$$

Recall from (6.5.1) that $N_k(a, b)$ denotes the number of k-step occupied self-avoiding paths connecting a and b.

We aim to use the second-moment method for $N_k(a,b)$, for which we need to investigate the mean and variance of $N_k(a,b)$. Let $\mathbb{P}_r$ denote the conditional probability given $B_r^{(G_n)}(o_1)$ and $B_r^{(G_n)}(o_2)$, and let $\mathbb{E}_r$ and Var_r denote the corresponding conditional expectation and variance. We compute

$$\mathbb{E}_r[N_k(a,b)] = \sum_{\vec{\pi} \in \mathcal{P}_k(a,b)} \mathbb{P}(\vec{\pi} \text{ occupied in } G_n) = \sum_{\pi \in \mathcal{P}_k(a,b)} \prod_{l=0}^{k} \frac{\kappa_n(\pi_l, \pi_{l+1})}{n}$$

$$\leq \frac{1}{n} \langle \boldsymbol{x}, \boldsymbol{T}_{\kappa_n}^k \boldsymbol{y} \rangle, \tag{6.5.71}$$

where $\boldsymbol{x} = (x_i)_{i=1}^t$, and $\boldsymbol{y} = (y_i)_{i=1}^t$, with x_i the number of type-i vertices in $\partial B_r^{(G_n)}(o_1)$ and y_i the number of type-i vertices in $\partial B_r^{(G_n)}(o_2)$, respectively. An identical lower bound holds with an extra factor $(\underline{\mu}_n - k/n)/\underline{\mu}_n$, where $\underline{\mu}_n = \min_{j \in [t]} \mu_n(j) \to \min_{j \in [t]} \mu(j) > 0$, by assumption.

Recall the notation and results in Section 3.4, and in particular Theorem 3.10(b). The types of o_1 and o_2 are asymptotically independent, and the probability that o_1 has type j is equal to $\mu_n(j)$, which converges to $\mu(j)$. On the event that the type of o_1 equals j_1, the vector of the numbers of individuals in $\partial B_r^{(G_n)}(o_1)$ converges in distribution to $(Z_r^{(1,j_1)}(i))_{i \in [t]}$, which, by Theorem 3.10(b), is close to $M_\infty \boldsymbol{x}_\kappa(i)$ for some strictly positive random variable M_∞. We conclude that

$$\boldsymbol{x} \xrightarrow{d} \boldsymbol{Z}_r^{(1,j_1)}, \qquad \boldsymbol{y} \xrightarrow{d} \boldsymbol{Z}_r^{(2,j_2)}, \tag{6.5.72}$$

where the limiting branching processes are independent. Equation (6.5.72) replaces the convergence in Lemma 6.20 for $\mathrm{GRG}_n(\boldsymbol{w})$.

We conclude that, for $k = k_n = \lceil (1+\varepsilon) \log_\nu n \rceil$, conditioning on $B_r^{(G_n)}(o_1)$ and $B_r^{(G_n)}(o_2)$ such that $a = \partial B_r^{(G_n)}(o_1)$, and $b = \partial B_r^{(G_n)}(o_2)$, with $\partial B_r^{(G_n)}(o_1), \partial B_r^{(G_n)}(o_2) \neq \varnothing$, we have

$$\mathbb{E}_r[N_k(a,b)] \xrightarrow{\mathbb{P}} \infty. \tag{6.5.73}$$

This completes the analysis of the first moment.

We next analyze $\mathrm{Var}_r(N_k(a,b))$ and prove that $\mathrm{Var}_r(N_k(a,b))/\mathbb{E}_r[N_k(a,b)]^2 \xrightarrow{\mathbb{P}} \infty$. Here our proof will be slightly more sketchy.

Recall (6.5.19). In this case we compute, as in (6.5.15) (see also (6.5.20)),

$$\mathbb{P}(\vec{\pi}, \vec{\rho} \text{ occupied}) = \mathbb{P}(\vec{\pi} \text{ occupied})\mathbb{P}(\vec{\rho} \text{ occupied} \mid \vec{\pi} \text{ occupied}). \tag{6.5.74}$$

Now recall the definition of a *shape* in (6.5.17), in Definition 6.16. Fix $\sigma \in \mathrm{Shape}_{m,l}$ and $\vec{\rho} \in \mathcal{P}_k(a,b)$ with $\mathrm{Shape}(\vec{\pi}, \vec{\rho}) = \sigma$. The factor $\mathbb{P}(\vec{\rho} \text{ occupied} \mid \vec{\pi} \text{ occupied})$, summed out over the free vertices of $\vec{\rho}$ (i.e., those that are not also vertices in $\vec{\pi}$) gives rise to m factors of the form $\boldsymbol{T}_{\kappa_n}^{t_i}(i_{\pi_{u_i}}, i_{\pi_{v_i}})/n$, for $i \in [m]$ and some vertices π_{u_i} and π_{v_i} in the path $(\pi_i)_{i=0}^k$. We use that, uniformly in $q \geq 1$,

$$\frac{1}{n} \max_{i,j \in [t]} \boldsymbol{T}_{\kappa_n}^q(i,j) \leq \frac{C}{n} \|\boldsymbol{T}_{\kappa_n}\|^q. \tag{6.5.75}$$

Thus, for every of the m subpaths of length t_i we obtain a factor $\frac{C}{n}\|T_{\kappa_n}\|^{t_i}$. Using that $\sum_{i=1}^{m} t_i = k - l$, by (6.5.18), we arrive at

$$\sum_{\substack{\vec{\pi},\vec{\rho} \in \mathcal{P}_k(a,b) \\ \text{Shape}(\vec{\pi},\vec{\rho}) = \sigma}} \mathbb{P}(\vec{\pi}, \vec{\rho} \text{ occupied})$$

$$\leq \mathbb{E}_r[N_k(a,b)] \prod_{i=1}^{m} \frac{C}{n}\|T_{\kappa_n}\|^{t_i} \leq \mathbb{E}_r[N_k(a,b)]\|T_{\kappa_n}\|^{k-l}\left(\frac{C}{n}\right)^m. \quad (6.5.76)$$

This replaces (6.5.21). The proof can now be completed in an identical way to that of (6.5.7) combined with that of (6.5.49) in the proof of Theorem 6.19. We omit further details. □

Concentration of the Giant in Theorem 3.19

By Theorem 3.14, we know that $\text{IRG}_n(\kappa_n)$ converges locally in probability. This immediately implies the upper bound on the giant component in Theorem 3.19, as was explained below Theorem 3.19. We now use the previous proof on typical distances to prove the concentration of the giant. We do this by applying Theorem 2.28, for which we need to verify that (2.6.7) holds. It turns out to be more convenient to verify condition (2.6.38).

Again, it suffices to consider the finite-type setting. In our argument we will combine the path-counting methods from Proposition 6.14 with the "giant is almost local" method. Exercise 6.27 investigates a direct proof of (2.6.38) as in Section 2.6.4.

We rewrite the expectation appearing in (2.6.38) as

$$\frac{1}{n^2}\mathbb{E}\left[\#\{x,y \in [n]: |\partial B_r^{(G_n)}(x)|, |\partial B_r^{(G_n)}(y)| \geq r, x \not\leftrightarrow y\}\right]$$

$$= \mathbb{P}(|\partial B_r^{(G_n)}(o_1)|, |\partial B_r^{(G_n)}(o_2)| \geq r, o_1 \not\leftrightarrow o_2). \quad (6.5.77)$$

We condition on $B_r^{(G_n)}(o_1)$ and $B_r^{(G_n)}(o_2)$, and note that the events $\{|\partial B_r^{(G_n)}(o_1)| \geq r\}$ and $\{|\partial B_r^{(G_n)}(o_2)| \geq r\}$ are measurable with respect to $B_r^{(G_n)}(o_1)$ and $B_r^{(G_n)}(o_2)$, to obtain

$$\mathbb{P}(|\partial B_r^{(G_n)}(o_1)|, |\partial B_r^{(G_n)}(o_2)| \geq r, o_1 \not\leftrightarrow o_2)$$

$$= \mathbb{E}\left[\mathbb{1}_{\{|\partial B_r^{(G_n)}(o_1)|,|\partial B_r^{(G_n)}(o_2)|\geq r\}}\mathbb{P}(o_1 \not\leftrightarrow o_2 \mid B_r^{(G_n)}(o_1), B_r^{(G_n)}(o_2))\right]. \quad (6.5.78)$$

In the proof of Theorem 6.1(b) we showed that, on $\{\partial B_r^{(G_n)}(o_1), \partial B_r^{(G_n)}(o_2) \neq \varnothing\}$,

$$\mathbb{P}(\text{dist}_{\text{IRG}_n(\kappa_n)}(o_1,o_2) \leq (1+\varepsilon)\log_\nu n \mid B_r^{(G_n)}(o_1), B_r^{(G_n)}(o_2)) = 1 - o_\mathbb{P}(1), \quad (6.5.79)$$

when first $n \to \infty$ followed by $r \to \infty$. In particular, on $\{\partial B_r^{(G_n)}(o_1), \partial B_r^{(G_n)}(o_2) \neq \varnothing\}$,

$$\mathbb{P}(o_1 \not\leftrightarrow o_2 \mid B_r^{(G_n)}(o_1), B_r^{(G_n)}(o_2)) = o_\mathbb{P}(1), \quad (6.5.80)$$

again when first $n \to \infty$ followed by $r \to \infty$. Since also

$$\mathbb{P}(o_1 \not\leftrightarrow o_2 \mid B_r^{(G_n)}(o_1), B_r^{(G_n)}(o_2)) \leq 1,$$

the Dominated Convergence Theorem [V1, Theorem A.1] completes the proof of (2.6.38) for $\text{IRG}_n(\kappa_n)$, as required. □

6.6 RELATED RESULTS ON DISTANCES IN INHOMOGENEOUS RANDOM GRAPHS

In this section we discuss some related results for inhomogeneous random graphs. While we give some intuition about their proofs, we do not include them in full detail.

Diameter in Inhomogeneous Random Graphs

Recall that the *diameter* $\mathrm{diam}(G)$ of the graph G equals the maximal finite graph distance between any pair of vertices, i.e.,

$$\mathrm{diam}(G) = \max_{u,v:\ \mathrm{dist}_G(u,v)<\infty} \mathrm{dist}_G(u,v). \tag{6.6.1}$$

See Figure 6.6 for the diameters of networks in the KONECT data base. While there are some networks with quite large diameters (often corresponding to road or other spatial networks), the diameters in the majority of the networks are quite small.

We next investigate the diameter of an $\mathrm{IRG}_n(\kappa_n)$, which tends to be much larger than the typical distances owing to the long thin lines that are distributed as a $\mathrm{IRG}_n(\kappa_n)$ with a subcritical κ_n by a duality principle for $\mathrm{IRG}_n(\kappa_n)$. Before we state the results, we introduce the notion of the *dual kernel*:

Definition 6.25 (Dual kernel for $\mathrm{IRG}_n(\kappa_n)$)   Let (κ_n) be a sequence of supercritical kernels with limit κ. The limiting *dual kernel* is the kernel $\widehat{\kappa}$ defined by $\widehat{\kappa}(x,y) = \kappa(x,y)$ with *reference measure* $\mathrm{d}\widehat{\mu}(x) = (1 - \zeta_\kappa(x))\mu(\mathrm{d}x)$. Note that this reference measure integrates to $1 - \zeta_\kappa > 0$, not to 1.   ◀

The dual kernel describes the graph that remains after the *removal of the giant component*. Here, the reference measure $\widehat{\mu}$ measures the structure of the types of vertices in the graph. Indeed, a vertex x is in the giant component with probability $\zeta_\kappa(x)$; if in fact it is in the giant then it must be removed. Thus, $\widehat{\mu}$ describes the proportion of vertices, of various types, that are outside the giant component. As before, we define the operator $T_{\widehat{\kappa}}$ by

$$(T_{\widehat{\kappa}}f)(x) = \int_{\mathcal{S}} \widehat{\kappa}(x,y)f(y)\mathrm{d}\widehat{\mu}(y) = \int_{\mathcal{S}} \kappa(x,y)f(y)[1 - \zeta_\kappa(x)]\mu(\mathrm{d}y), \tag{6.6.2}$$

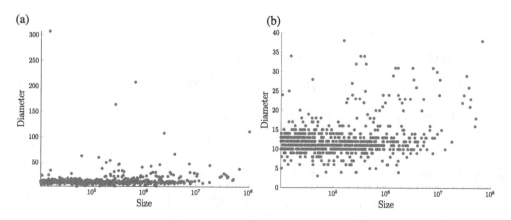

Figure 6.6 (a) Diameters of the 727 networks of size larger than 10,000, from the KONECT data base, and (b) the 721 diameters that are at most 40.

and we write $\|T_{\widehat{\kappa}}\|$ for

$$\|T_{\widehat{\kappa}}\| = \sup\left\{\|T_{\widehat{\kappa}}f\|_{\widehat{\mu}} : f \geq 0, \|f\|_{\widehat{\mu}} = 1\right\}, \tag{6.6.3}$$

where

$$\|f\|_{\widehat{\mu}}^2 = \int_S f^2(x)\widehat{\mu}(\mathrm{d}x). \tag{6.6.4}$$

The following theorem describes the diameter in terms of the above notation:

Theorem 6.26 (Diameter of $\mathrm{IRG}_n(\kappa_n)$ in the finite-types case) *Let (κ_n) be a sequence of kernels with limit κ, which has finitely many types. If $0 < \|T_\kappa\| < 1$ then*

$$\frac{\mathrm{diam}(\mathrm{IRG}_n(\kappa_n))}{\log n} \xrightarrow{\mathbb{P}} \frac{1}{\log(1/\|T_\kappa\|)} \tag{6.6.5}$$

as $n \to \infty$. If $\|T_\kappa\| > 1$ and κ is irreducible then

$$\frac{\mathrm{diam}(\mathrm{IRG}_n(\kappa_n))}{\log n} \xrightarrow{\mathbb{P}} \frac{2}{\log(1/\|T_{\widehat{\kappa}}\|)} + \frac{1}{\log\|T_\kappa\|}, \tag{6.6.6}$$

where $\widehat{\kappa}$ is the dual kernel to κ.

If we compare Theorem 6.26 with Theorem 6.2 then we see that the diameter has the same scaling as the typical distance when $\|T_\kappa\| < \infty$, but that $\mathrm{diam}(\mathrm{IRG}_n(\kappa_n))/\log n$ converges in probability to a strictly larger limit than the one when $\mathrm{dist}_{\mathrm{IRG}_n(\kappa_n)}(o_1, o_2)/\log n$ is conditioned on being finite. This effect is particularly noticeable in the case of rank-1 models with $\tau \in (2,3)$, where, conditional on its being finite, $\mathrm{dist}_{\mathrm{IRG}_n(\kappa_n)}(o_1, o_2)/\log\log n$ converges in probability to a finite limit, while $\mathrm{diam}(\mathrm{IRG}_n(\kappa_n))/\log n$ converges to a non-zero limit. This can be explained by noticing that the diameter in $\mathrm{IRG}_n(\kappa_n)$ is due to very thin lines of length of order $\log n$. Since these lines involve only very few vertices, they do not contribute to $\mathrm{dist}_{\mathrm{IRG}_n(\kappa_n)}(o_1, o_2)$ but they do contribute to $\mathrm{diam}(\mathrm{IRG}_n(\kappa_n))$. This is another argument for why we prefer to work with typical distances rather than the diameter. Exercise 6.28 investigates the consequences for $\mathrm{ER}_n(\lambda/n)$.

We do not prove Theorem 6.26 here. For $\mathrm{GRG}_n(w)$, it also follows from Theorem 7.19 below, which states a related result for the configuration model.

Distance Fluctuations for Finite-Variance Weights

We continue by studying the *fluctuations* of the typical distance when $\mathbb{E}[W^2] < \infty$. We impose a slightly stronger condition on the distribution function F of W, namely, that there exists a $\tau > 3$ and $c > 0$ such that

$$1 - F(x) \leq cx^{-(\tau-1)}. \tag{6.6.7}$$

Equation (6.6.7) implies that the degrees have finite variance; see Exercise 6.29.

Theorem 6.27 (Limit law for the typical distance in $\mathrm{NR}_n(w)$) *Consider $\mathrm{NR}_n(w)$, where the weights $w = (w_v)_{v \in [n]}$ are given by $w_v = [1 - F]^{-1}(v/n)$ as in (1.3.15), with F*

satisfying (6.6.7), and let $\nu = \mathbb{E}[W^2]/\mathbb{E}[W] > 1$. *For* $k \geq 1$, *define* $a_k = \lfloor \log_\nu k \rfloor - \log_\nu k \in (-1, 0]$. *Then, there exist random variables* $(R_a)_{a \in (-1,0]}$ *with*

$$\limsup_{K \to \infty} \sup_{a \in (-1,0]} \mathbb{P}(|R_a| < K) = 1 \tag{6.6.8}$$

such that, as $n \to \infty$ *and for all* $k \in \mathbb{Z}$, *with* o_1, o_2 *chosen independently and uar from* $[n]$,

$$\mathbb{P}\big(\mathrm{dist}_{\mathrm{NR}_n(w)}(o_1, o_2) - \lfloor \log_\nu n \rfloor = k \mid \mathrm{dist}_{\mathrm{NR}_n(w)}(o_1, o_2) < \infty\big)$$
$$= \mathbb{P}(R_{a_n} = k) + o(1). \tag{6.6.9}$$

The same result applies, under identical conditions, to $\mathrm{GRG}_n(w)$ *and* $\mathrm{CL}_n(w)$.

While Theorem 6.1 implies that $\mathrm{dist}_{\mathrm{NR}_n(w)}(o_1, o_2)/\log n \xrightarrow{\mathbb{P}} 1/\log \nu$ conditional on o_1 and o_2 being connected, Theorem 6.27 implies that the fluctuations of $\mathrm{dist}_{\mathrm{NR}_n(w)}(o_1, o_2)$ around $\log_\nu n$ remain uniformly bounded in probability.

The random variables $(R_a)_{a \in (-1,0]}$ can be determined in terms of the limit law in the branching-process approximation of the neighborhoods of $\mathrm{NR}_n(w)$, as in Theorem 3.18. These random variables depend sensitively on a, and this has the implication that although $\big(\mathrm{dist}_{\mathrm{NR}_n(w)}(o_1, o_2) - \lfloor \log_\nu n \rfloor\big)_{n \geq 2}$ is a *tight* sequence of random variables, it does *not* weakly converge. See Exercises 6.30–6.32 for further properties of this sequence of random variables.

Distances in Generalized Random Graphs for $\tau = 3$

We next investigate what happens at the critical value of τ, which is $\tau = 3$, when we allow for an additional logarithmic correction to the weight distribution:

Theorem 6.28 (Critical $\tau = 3$ case: interpolation)   *Consider* $\mathrm{GRG}_n(w)$, *where the iid weights* $(w_v)_{v \in [n]}$ *satisfy that, as* $w \to \infty$,

$$\mathbb{P}(w_1 > w) = w^{-2}(\log w)^{2\alpha + o(1)}, \tag{6.6.10}$$

for some $\alpha > -\frac{1}{2}$. *Consider two vertices* o_1, o_2 *chosen independently and uar from* $[n]$. *Then, conditional on* $o_1 \longleftrightarrow o_2$,

$$\mathrm{dist}_{\mathrm{GRG}_n(w)}(o_1, o_2) = (1 + o_{\mathbb{P}}(1)) \frac{1}{1 + 2\alpha} \frac{\log n}{\log \log n}. \tag{6.6.11}$$

Theorem 6.28 shows how the addition of powers of the logarithm changes typical distances; note that the degree distribution has tails identical to the weight distribution in (6.6.10). It can be compared with Theorem 6.22, which is more general in terms of the weight sequence (they do not need to be iid), yet less general in terms of the logarithmic correction in (6.6.10). Exercise 6.33 investigates the lower bound in Theorem 6.28 for all $\alpha > -\frac{1}{2}$, while Exercise 6.34 shows a logarithmic lower bound for $\alpha < -\frac{1}{2}$.

6.7 Notes and Discussion for Chapter 6

Notes on Section 6.2

Theorem 6.1 is a simplified version of (Bollobás et al., 2007, Theorem 3.14). A first version of Theorem 6.1 was proved by Chung and Lu (2002a, 2003) for the random graph with prescribed expected degrees

$CL_n(w)$, in the case of *admissible* deterministic weights. We refer to (Chung and Lu, 2003, p. 94) for the definition of admissible weight sequences.

Theorem 6.2 has a long history, and many versions of it have been proved in the literature. We refer the reader to Chung and Lu (2002a, 2003) for the Chung–Lu model, and van den Esker et al. (2008) for its extensions to the Norros–Reittu model and the generalized random graph.

Theorem 6.3 for the random graph with prescribed expected degrees, or Chung–Lu model, was first proved by Chung and Lu (2002a, 2003), in the case of deterministic weights $w_v = c(n/v)^{1/(\tau-1)}$ having average degree strictly greater than 1 and maximum weight m satisfying $\log m \gg \log n/\log\log n$. These restrictions were lifted in (Durrett, 2007, Theorem 4.5.2). Indeed, the bound on the average degree is not necessary, since, for $\tau \in (2,3)$, $\nu = \infty$ and therefore the IRG is *always* supercritical. An upper bound as in Theorem 6.3 for the Norros–Reittu model with iid weights was proved by Norros and Reittu (2006). Theorem 6.3 has been proved in many versions, both fully as well as in partial forms; see, e.g., Norros and Reittu (2006); Chung and Lu (2002a, 2003); Dereich et al. (2012).

Notes on Section 6.3

As far as we are aware, the proof of Theorem 6.4 is new in the present context. Similar arguments have often been used, though, to prove lower bounds on distances in various situations.

The truncated first-moment method in the proof of Theorem 6.7 was inspired by Dereich et al. (2012).

Notes on Section 6.4

Theorem 6.11 is novel in its precise form, and also its proof is different from those in the literature. See the notes of Section 6.2 for the relevant references.

Notes on Section 6.5

The path-counting techniques in Proposition 6.14 are novel. They were inspired by the path-counting techniques used by Eckhoff et al. (2013) for smallest-weight problems on the complete graph where many of the counting arguments appeared. Related proofs for the upper bound on typical distances in rank-1 random graphs with $\nu < \infty$ as in Theorem 6.19 often rely on branching-process approximations up to a generation $r = r_n \to \infty$.

Notes on Section 6.6

Theorem 6.26 is a special case of (Bollobás et al., 2007, Theorem 3.16). In the special case of $ER_n(\lambda/n)$, it extends a previous result of Chung and Lu (2001) that proves logarithmic asymptotics of $\mathrm{diam}(ER_n(\lambda/n))$, and it negatively answers a question of Chung and Lu (2001). Related results for the configuration model, which also imply results for the generalized random graph, can be found in Fernholz and Ramachandran (2007). See also Theorem 7.19 below. See Riordan and Wormald (2010) for additional results and branching-process proofs. There, the case of $ER_n(\lambda/n)$ with $\lambda > 1$ such that $\lambda - 1 \gg n^{-1/3}$ was also discussed. Similar results were derived by Ding et al. (2011, 2010).

Theorem 6.27 is proved more generally in van den Esker et al. (2008), both in the case of iid weights as well as for deterministic weights under a mild further condition on the distribution function.

Theorem 6.28 was proved by Dereich et al. (2017).

6.8 EXERCISES FOR CHAPTER 6

Exercise 6.1 (Typical distances in $ER_n(\lambda/n)$ revisited) *Fix $\lambda > 1$. Use either Theorem 6.1 or Theorem 6.2 to prove that, with o_1, o_2 chosen independently and uar from $[n]$, and conditional on $o_1 \longleftrightarrow o_2$,*

$$\mathrm{dist}_{ER_n(\lambda/n)}(o_1, o_2)/\log n \overset{\mathbb{P}}{\longrightarrow} 1/\log \lambda.$$

Exercise 6.2 (Convergence in probability of typical distance in $IRG_n(\kappa_n)$) *Suppose that the graphical sequence of kernels $(\kappa_n)_{n \geq 1}$ converges to κ, where κ is irreducible and $\|T_\kappa\| = \infty$. Prove that Theorem 3.19 together with Theorem 6.1(b) imply that, with o_1, o_2 chosen independently and uar from $[n]$, and conditional on $o_1 \longleftrightarrow o_2$,*

$$\mathrm{dist}_{IRG_n(\kappa_n)}(o_1, o_2)/\log n \overset{\mathbb{P}}{\longrightarrow} 0. \tag{6.8.1}$$

Exercise 6.3 (Power-law tails in key example of deterministic weights)   *Let w be defined as $w_v = [1 - F]^{-1}(v/n)$ as in (1.3.15), and assume that F satisfies*

$$1 - F(x) = x^{-(\tau-1)} L(x), \tag{6.8.2}$$

where $\tau \in (2, 3)$ and $x \mapsto L(x)$ is slowly varying. Prove that, for all $\delta > 0$, there exists $c_1 = c_1(\delta)$ and $c_2 = c_2(\delta)$ such that, uniformly in n,

$$c_1 x^{-(\tau-1+\delta)} \leq [1 - F_n](x) \leq c_2 x^{-(\tau-1-\delta)}, \tag{6.8.3}$$

where the lower bound holds uniformly in $x \leq n^\beta$ for some $\beta > \frac{1}{2}$, as stated in (6.2.5).

Exercise 6.4 (Power-law tails for iid weights)   *Fix iid weights $w = (w_v)_{v \in [n]}$ with distribution F satisfying (6.8.2) with $\tau \in (2, 3)$, and where $x \mapsto L(x)$ is slowly varying. Prove that (6.2.5) holds with probability converging to 1.*

Exercise 6.5 (Conditions (6.3.1) and Condition 1.1)   *Show that (6.3.1) holds when there is precisely one vertex with weight $w_1 = \sqrt{n}$, whereas $w_v = \lambda > 1$ for all $v \in [n] \setminus \{1\}$, but Condition 1.1(c) does not.*

Exercise 6.6 (Conditions (6.3.1) and Condition 1.1)   *In the setting of Exercise 6.5, argue that the upper bound derived in Theorem 6.4 is not sharp, since the vertex with weight $w_1 = \sqrt{n}$ can occur at most once in a self-avoiding path.*

Exercise 6.7 (Lower bound on fluctuations)   *Adapt the proof of the lower bound on typical distances in the finite-variance weight setting in Theorem 6.4 to show that, for every ε, we can find a constant $K = K(\varepsilon) > 0$ such that*

$$\mathbb{P}\big(\text{dist}_{\text{NR}_n(w)}(o_1, o_2) \leq \log_{\nu_n} n - K\big) \leq \varepsilon. \tag{6.8.4}$$

Conclude that if $\log \nu_n = \log \nu + O(1/\log n)$, then the same statement holds with $\log_\nu n$ replacing $\log_{\nu_n} n$.

Exercise 6.8 (Proof of Corollary 6.6)   *Adapt the proof of the lower bound on typical distances in the finite-variance weight setting in Theorem 6.4 to prove Corollary 6.6.*

Exercise 6.9 (Lower bound on typical distances for $\tau = 3$)   *Let $w_v = c\sqrt{n/v}$, so that $\tau = 3$. Prove that $\nu_n / \log n$ converges as $n \to \infty$. Use Corollary 6.6 to obtain that, for any $\varepsilon > 0$,*

$$\mathbb{P}\Big(\text{dist}_{\text{NR}_n(w)}(o_1, o_2) \leq (1 - \varepsilon) \frac{\log n}{\log \log n}\Big) = o(1). \tag{6.8.5}$$

Exercise 6.10 (Lower bound on typical distances for $\tau \in (2, 3)$)   *Let $w_v = c(n/v)^{1/(\tau-1)}$ with $\tau \in (2, 3)$. Prove that there exists a constant $c' > 0$ such that $\nu_n \geq c' n^{(3-\tau)/(\tau-1)}$. Show that Corollary 6.6 implies that $\text{dist}_{\text{NR}_n(w)}(o_1, o_2) \geq (\tau - 1)/(\tau - 3)$ whp in this case. How useful is this bound?*

Exercise 6.11 (Convergence in probability of typical distance in $\text{IRG}_n(\kappa_n)$)   *Suppose that the graphical sequence of kernels (κ_n) satisfies $\sup_{x,y,n} \kappa_n(x, y) < \infty$, where the limit κ is irreducible and $\nu = \|T_\kappa\| > 1$. Prove that Theorems 3.19 and 6.1(a),(b) imply that, conditional on $\text{dist}_{\text{IRG}_n(\kappa_n)}(o_1, o_2) < \infty$,*

$$\text{dist}_{\text{IRG}_n(\kappa_n)}(o_1, o_2)/\log n \xrightarrow{\mathbb{P}} 1/\log \nu. \tag{6.8.6}$$

Exercise 6.12 (Distance between fixed vertices)   *Show that (6.3.30) and Lemma 6.9 imply that, for all $a, b \in [n]$ with $a \neq b$,*

$$\mathbb{P}(\text{dist}_{\text{NR}_n(w)}(a, b) \leq k_n) \leq \frac{1}{n} + \frac{w_a w_b}{\ell_n} \sum_{k=1}^{k_n} \prod_{l=1}^{k-1} \nu_n(b_l \wedge b_{k-l})$$

$$+ (w_a + w_b) \sum_{k=1}^{k_n} [1 - F_n^\star](b_k) \prod_{l=1}^{k} \nu_n(b_l). \tag{6.8.7}$$

Exercise 6.13 (Bound on truncated forward degree $\nu_n(b)$)   *Assume that (6.3.21) holds. Prove the bound on $\nu_n(b)$ in (6.3.38) by combining (1.4.12) in Lemma 1.22 with $\ell_n = \Theta(n)$ by Conditions 1.1(a),(b).*

Exercise 6.14 (Ultra-small distances for $\mathrm{CL}_n(\boldsymbol{w})$ and $\mathrm{GRG}_n(\boldsymbol{w})$) *Complete the proof of the doubly logarithmic upper bound on typical distances in Theorem 6.11 for $\mathrm{CL}_n(\boldsymbol{w})$ and $\mathrm{GRG}_n(\boldsymbol{w})$.*

Exercise 6.15 (Upper bound on the expected number of paths) *Consider an inhomogeneous random graph with edge probabilities $p_{ij} = u_i u_j$ for $(u_i)_{i \in [n]} \in [0, 1]^n$. Prove (6.5.4), which states that*

$$\mathbb{E}[N_k(a, b)] \leq u_a u_b \Big(\sum_{i \in \mathcal{I} \setminus \{a, b\}} u_i^2 \Big)^{k-1}.$$

Exercise 6.16 (Variance of two-paths) *Consider an inhomogeneous random graph with edge probabilities $p_{ij} = u_i u_j$ for $(u_i)_{i \in [n]} \in [0, 1]^n$. Prove that $\mathrm{Var}(N_k(a, b)) \leq \mathbb{E}[N_k(a, b)]$ for $k = 2$.*

Exercise 6.17 (Variance of three-paths) *Consider an inhomogeneous random graph with edge probabilities $p_{ij} = u_i u_j$ for $(u_i)_{i \in [n]} \in [0, 1]^n$. Compute $\mathrm{Var}(N_3(a, b))$ explicitly, and compare it with the bound in (6.5.7).*

Exercise 6.18 (Connections between sets in $\mathrm{NR}_n(\boldsymbol{w})$) *Let $A, B \subseteq [n]$ be two disjoint sets of vertices. Prove that*

$$\mathbb{P}(A \text{ directly connected to } B \text{ in } \mathrm{NR}_n(\boldsymbol{w})) = 1 - e^{-w_A w_B / \ell_n}, \tag{6.8.8}$$

where $w_A = \sum_{a \in A} w_a$ is the weight of A.

Exercise 6.19 (Expectation of paths between sets in $\mathrm{ER}_n(\lambda/n)$) *Consider $\mathrm{ER}_n(\lambda/n)$. Fix $A, B \subseteq [n]$ with $A \cap B = \varnothing$, and let $N_k(A, B)$ denote the number of self-avoiding paths of length k connecting A to B (where a path connecting A and B avoids A and B except for the starting point and endpoint). Show that, for $k(|A| + |B|)/n = o(1)$,*

$$\mathbb{E}[N_k(A, B)] = \lambda^k |A||B| \left(1 - \frac{|A| + |B|}{n} \right)^k (1 + o(1)). \tag{6.8.9}$$

Exercise 6.20 (Variance on path counts for $\mathrm{ER}_n(\lambda/n)$ (cont.)) *In the setting of Exercise 6.19, use Proposition 6.14 to bound the variance of $N_k(A, B)$, and prove that*

$$N_k(A, B)/\mathbb{E}[N_k(A, B)] \xrightarrow{\mathbb{P}} 1 \tag{6.8.10}$$

when $|A|, |B| \to \infty$ with $|A| + |B| = o(n/k)$ and $k = \lceil \log_\lambda n \rceil$.

Exercise 6.21 (Logarithmic bound for ν_n when $\tau = 3$) *Define*

$$a = o_1', \qquad b = o_2', \qquad \mathcal{I} = \{i \in [n] \colon w_i \in [K, \sqrt{\beta_n}]\}, \tag{6.8.11}$$

where o_1', o_2' are independent copies from the sized-biased distribution in (6.5.60). Prove that $\tau = 3$ in the form of (6.5.55) and (6.5.56) implies that $\nu_{\mathcal{I}} \geq c \log \beta_n$ for some $c > 0$ and all n sufficiently large. It may be helpful to use

$$\frac{1}{n} \sum_{i \in \mathcal{I}} w_i^2 = \mathbb{E}[W_n^2 \mathbb{1}_{\{W_n \in [K, \sqrt{\beta_n}]\}}] = 2 \int_0^{\sqrt{\beta_n}} x [F_n(\sqrt{\beta_n}) - F_n(x \vee K)] \mathrm{d}x. \tag{6.8.12}$$

Exercise 6.22 (Expected number of paths within Core_n diverges) *Recall the setting in (6.8.11) in Exercise 6.21. Fix $\eta > 0$. Prove that*

$$\mathbb{E}[N_k(a, b)] \to \infty$$

for $a = o_1', b = o_2'$, and $k = \lceil (1 + \eta) \log n / \log \nu_n \rceil$.

Exercise 6.23 (Concentration of number of paths within Core_n) *Recall the setting in (6.8.11) in Exercise 6.21. Prove that*

$$\mathrm{Var}(N_k(a, b))/\mathbb{E}[N_k(a, b)]^2 \to 0$$

for $a = o_1', b = o_2'$, and $k = \lceil (1 + \eta) \log n / \log \nu_n \rceil$.

Exercise 6.24 (Concentration of number of paths within Core_n)   *Complete the proof of Lemma 6.23 on the basis of Exercises 6.21–6.23.*

Exercise 6.25 (Asymptotic equivalence in Lemma 6.24)   *Recall the conditions of Theorem 6.22. Prove that $\mathrm{CL}_n(\boldsymbol{w}), \mathrm{GRG}_n(\boldsymbol{w})$, and $\mathrm{NR}_n(\boldsymbol{w})$ are asymptotically equivalent when restricted to the edges in $[n] \times \{v \colon w_v \leq \beta_n\}$ for any $\beta_n = o(\sqrt{n})$. Hint: Use the asymptotic equivalence in [V1, Theorem 6.18] for general inhomogeneous random graphs.*

Exercise 6.26 (Completion of the proof of Lemma 6.24)   *Complete the proof of (6.5.62) in Lemma 6.24 by adapting the arguments in (6.5.51)–(6.5.54).*

Exercise 6.27 (Concentration of the giant in IRGs)   *In Section 6.5.3, Theorem 3.19 for finite-type inhomogeneous random graphs was proved using a path-counting method based on Theorem 6.1(b). Give a direct proof of the "giant is almost local" condition in (2.6.38) by adapting the argument in Section 2.6.4 for the Erdős–Rényi random graph. You may assume that $\mu(s) > 0$ for every $s \in [t]$ for which $\mu_n(s) > 0$.*

Exercise 6.28 (Diameter of $\mathrm{ER}_n(\lambda/n)$)   *Recall the asymptotics of the diameter in $\mathrm{IRG}_n(\kappa_n)$ in Theorem 6.26. For $\mathrm{ER}_n(\lambda/n)$, show that $\|\boldsymbol{T}_\kappa\| = \lambda$ and $\|\boldsymbol{T}_{\widehat{\kappa}}\| = \mu_\lambda$, where μ_λ is the dual parameter in [V1, (3.6.6)], so that Theorem 6.26 becomes*

$$\frac{\mathrm{diam}(\mathrm{ER}_n(\lambda/n))}{\log n} \overset{\mathbb{P}}{\longrightarrow} \frac{2}{\log(1/\mu_\lambda)} + \frac{1}{\log \lambda}. \tag{6.8.13}$$

Exercise 6.29 (Finite variance of degrees when (6.6.7) holds)   *Prove that (6.6.7) implies that $\mathbb{E}[W^2] < \infty$. Use this to prove that the degrees have uniformly bounded variance when (6.6.7) holds.*

Exercise 6.30 (Tightness of centered typical distances in $\mathrm{NR}_n(\boldsymbol{w})$)   *Prove that, under the conditions of Theorem 6.27, and conditional on $\mathrm{dist}_{\mathrm{NR}_n(\boldsymbol{w})}(o_1, o_2) < \infty$, the sequence $\big(\mathrm{dist}_{\mathrm{NR}_n(\boldsymbol{w})}(o_1, o_2) - \lfloor \log_\nu n \rfloor\big)_{n \geq 2}$ is tight.*

Exercise 6.31 (Non-convergence of centered typical distances in $\mathrm{NR}_n(\boldsymbol{w})$)   *Prove that, under the conditions of Theorem 6.27, and conditional on $\mathrm{dist}_{\mathrm{NR}_n(\boldsymbol{w})}(o_1, o_2) < \infty$, the sequence $\mathrm{dist}_{\mathrm{NR}_n(\boldsymbol{w})}(o_1, o_2) - \lfloor \log_\nu n \rfloor$ does not weakly converge when the distribution of R_a depends continuously on a and when there are $a, b \in (-1, 0]$ such that the distribution of R_a is not equal to that of R_b.*

Exercise 6.32 (Extension of Theorem 6.27 to $\mathrm{GRG}_n(\boldsymbol{w})$ and $\mathrm{CL}_n(\boldsymbol{w})$)   *Use [V1, Theorem 6.18] to prove that Theorem 6.27 holds verbatim for $\mathrm{GRG}_n(\boldsymbol{w})$ and $\mathrm{CL}_n(\boldsymbol{w})$ when (6.6.7) holds. Hint: Use asymptotic equivalence.*

Exercise 6.33 (Extension lower bound Theorem 6.28 to all $\alpha > -\frac{1}{2}$)   *Consider $\mathrm{NR}_n(\boldsymbol{w})$ with weights $\boldsymbol{w}$ satisfying*

$$\mathbb{P}(W_n > x) = x^{-2}(\log x)^{2\alpha + o(1)}, \tag{6.8.14}$$

for all $x \leq n^\varepsilon$ for some $\varepsilon > 0$, where $W_n = w_o$ is the weight of a uniform vertex in $[n]$. Prove the lower bound in Theorem 6.28 for all $\alpha > -\frac{1}{2}$.

Exercise 6.34 (Extension of the lower bound in Theorem 6.28 to $\alpha < -\frac{1}{2}$)   *Consider $\mathrm{NR}_n(\boldsymbol{w})$ as in Exercise 6.33, but now with $\alpha < -\frac{1}{2}$. Let $\nu = \mathbb{E}[W^2]/\mathbb{E}[W] < \infty$. Prove that the lower bound in Theorem 6.28 is replaced by*

$$\mathbb{P}(\mathrm{dist}_{\mathrm{GRG}_n(\boldsymbol{w})}(o_1, o_2) \leq (1 - \varepsilon) \log_\nu n) = o(1). \tag{6.8.15}$$

SMALL-WORLD PHENOMENA IN CONFIGURATION MODELS

Abstract

In this chapter we investigate the distance structure of the configuration model by investigating its typical distances and its diameter. We adapt the path-counting techniques in Section 6.5 to the configuration model, and obtain typical distances from the "giant is almost local" proof. To understand the ultra-small distances for infinite-variance degree configuration models, we investigate the generation growth of infinite-mean branching processes. The relation to branching processes informally leads to the *power-iteration technique*, which allows one to deduce typical distance results in a relatively straightforward way.

7.1 MOTIVATION: DISTANCES AND DEGREE STRUCTURE

In this chapter we investigate graph distances in the configuration model. We start with a motivating example.

Motivating Example

Recall Figure 1.5(a), in which graph distances in the Autonomous Systems (AS) graph in the Internet, also called *AS counts*, are shown. A relevant question is whether such a histogram can be *predicted* by the graph distances in a random graph model a having similar degree structure and size to the AS graph. Figure 7.1 compares simulations of the typical distances for $\tau \in (2,3)$ with the distances in the AS graph, with $n = 10,940$ equal to the number of autonomous systems, and $\tau = 2.25$ the best approximation to the degree power-law exponent of the AS graph. We see that the typical distances in the configuration model $\mathrm{CM}_n(\boldsymbol{d})$ and the AS counts are quite close. Further, Figure 7.2 shows the 90% percentile of typical distances in the KONECT data base (recall that Figure 6.1 indicates the median value). We see that this 90th percentile mostly remains relatively small, even for large networks.

Figures 7.1 and 7.2 again raise the question how graph distances depend on the structure of the random graphs and real-world networks in question, such as their size and degree structure. The configuration model is highly flexible, in the sense that it offers complete freedom in the choice of the degree distribution. Thus, we can use the configuration model (CM) to single out the *relation* between graph distances and degree structure, in a similar way to that in which we investigate the giant component size and connectivity as a function of the degree distribution, as discussed in detail in Chapter 4. Finally, we can verify whether graph distances in CM are closely related to those in inhomogeneous random graphs, as discussed in Chapter 6, so as to detect another sign of the wished-for *universality* of structural properties of random graphs with similar degree distributions.

Organization of this Chapter

This chapter is organized as follows. In Section 7.2 we summarize the main results on typical distances in the configuration model. In Section 7.3 we prove these distance results, using path-counting techniques and comparisons with branching processes. We do this by

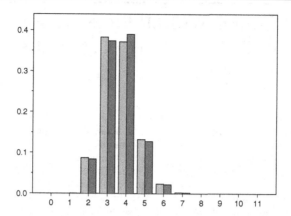

Figure 7.1 Number of AS traversed in hopcount data (lighter gray), and, for comparison, the model (darker gray) with $\tau = 2.25, n = 10,940$.

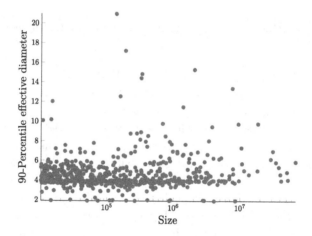

Figure 7.2 90th percentile of typical distances in the 727 networks of size larger than 10,000 from the KONECT data base.

formulating specific theorems, for upper and lower bounds on graph distances, that often hold under slightly weaker conditions. In Section 7.4 we study the generation sizes of infinite-mean branching processes, as these arise in the CM with infinite-variance degrees. These generation sizes heuristically explain the ultra-small nature of CMs in the infinite-variance degree regime. In Section 7.5 we study the diameter of the CM. In Section 7.6, we state further results in the CM. We close the chapter with notes and discussion in Section 7.7 and with exercises in Section 7.8.

7.2 Small-World Phenomenon in Configuration Models

In this section we describe the main results on typical distances in the CM, both in the case of finite-variance degrees and in the case of infinite-variance degrees. These results are proved in the following section.

Distances in Configuration Models with Finite-Variance Degrees

We start by analyzing the typical distances for the configuration model $\mathrm{CM}_n(\boldsymbol{d})$ when Conditions 1.7(a)–(c) hold:

Theorem 7.1 (Typical distances in $\mathrm{CM}_n(\boldsymbol{d})$ for finite-variance degrees) *Consider* $\mathrm{CM}_n(\boldsymbol{d})$ *where the degrees* $\boldsymbol{d} = (d_v)_{v \in [n]}$ *satisfy Conditions 1.7(a)–(c), with* $\nu > 1$. *Then, for* o_1, o_2 *chosen independently and uar from* $[n]$, *and conditioning on* $o_1 \longleftrightarrow o_2$,

$$\mathrm{dist}_{\mathrm{CM}_n(\boldsymbol{d})}(o_1, o_2)/\log n \xrightarrow{\ \mathbb{P}\ } 1/\log \nu. \tag{7.2.1}$$

Theorem 7.1 shows that the typical distances in $\mathrm{CM}_n(\boldsymbol{d})$ are of order $\log_\nu n$, and is thus similar in spirit to Theorem 6.2. We shall see that its proof is also quite similar.

Finite-Mean and Infinite-Variance Degrees

We next study typical distances in configuration models with degrees having finite mean and infinite variance. We start by formulating the precise condition on the degrees with which we will work. This condition is identical to the condition on F_n for $\mathrm{NR}_n(\boldsymbol{w})$ formulated in (6.2.5).

Recall from (1.3.31) that $F_n(x)$ denotes the proportion of vertices having degree at most x. Then, we assume that there exists a $\tau \in (2, 3)$ and that, for all $\delta > 0$, there exist $c_1 = c_1(\delta)$ and $c_2 = c_2(\delta)$ such that, uniformly in n,

$$c_1 x^{-(\tau-1+\delta)} \leq [1 - F_n](x) \leq c_2 x^{-(\tau-1-\delta)}, \tag{7.2.2}$$

where the upper bound holds for every $x \geq 1$, while the lower bound is required to hold only for $1 \leq x \leq n^\beta$, for some $\beta > \frac{1}{2}$. The typical distances of $\mathrm{CM}_n(\boldsymbol{d})$ under the infinite-variance condition in (7.2.2) are identified in the following theorem:

Theorem 7.2 (Typical distances in $\mathrm{CM}_n(\boldsymbol{d})$ for $\tau \in (2, 3)$) *Consider* $\mathrm{CM}_n(\boldsymbol{d})$ *where the degrees* $\boldsymbol{d} = (d_v)_{v \in [n]}$ *satisfy Conditions 1.7(a),(b) and (7.2.2). Then, with* o_1, o_2 *chosen independently and uar from* $[n]$, *and conditioning on* $o_1 \longleftrightarrow o_2$,

$$\frac{\mathrm{dist}_{\mathrm{CM}_n(\boldsymbol{d})}(o_1, o_2)}{\log \log n} \xrightarrow{\ \mathbb{P}\ } \frac{2}{|\log(\tau - 2)|}. \tag{7.2.3}$$

Theorem 7.2 is similar in spirit to Theorem 6.3 for $\mathrm{NR}_n(\boldsymbol{w})$. We will again see that the high-degree vertices play a crucial role in creating short connections and that shortest paths generally pass through the core of high-degree vertices, unlike in the finite-variance case. See Figure 7.3 for a simulation of the typical distances in $\mathrm{CM}_n(\boldsymbol{d})$ with $\tau = 2.5$ and $\tau = 3.5$, respectively, where the distances are again noticeably smaller for the ultra-small setting, with $\tau = 2.5$, as compared with the small-world case, with $\tau = 3.5$. Exercises 7.1–7.3 investigate examples of Theorems 7.1 and 7.2.

Theorems 7.1 and 7.2 leave open many other possible settings, particularly in the critical cases $\tau = 2$ and $\tau = 3$. In these cases it can be expected that the results depend on finer properties of the degrees. We will present some results along the way, when such results follow relatively straightforwardly from our proofs.

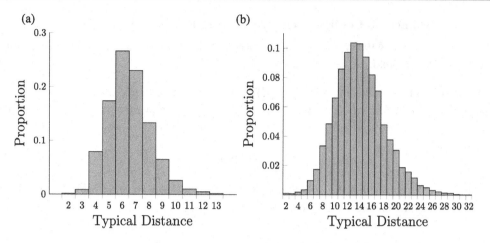

Figure 7.3 Typical distances between 2,000 pairs of vertices in the configuration model with $n = 100,000$ and (a) $\tau = 2.5$ and (b) $\tau = 3.5$.

7.3 Proofs of Small-World Results for the Configuration Model

In this section we give the proofs of Theorems 7.1 and 7.2 describing the small-world properties in $\mathrm{CM}_n(\boldsymbol{d})$. These proofs are adaptations of the proofs of Theorems 6.2 and 6.3, and we focus on the differences in the proofs.

The section is organized as follows. In Section 7.3.1 we give a branching-process approximation for the neighborhoods of a pair of uniform vertices in $\mathrm{CM}_n(\boldsymbol{d})$, using the local convergence in Theorem 4.1 and its proof. In Section 7.3.2 we use path-counting upper bounds to prove lower bounds on typical distances by the first-moment method. In Section 7.3.3 we employ path-counting techniques similar to those in in Section 6.5.1 using second-moment methods, adapted to the CM, where edges are formed by pairing half-edges. We use these to prove logarithmic upper bounds on graph distances. We close in Section 7.3.4 by proving doubly logarithmic upper bounds on typical distances of CMs with infinite-variance degrees. We also discuss the diameter of the *core* of high-degree vertices.

7.3.1 Branching-Process Approximation

In this subsection we summarize some links between the breadth-first exploration in $\mathrm{CM}_n(\boldsymbol{d})$ and branching processes. Recall the coupling of such breadth-first exploration processes in Lemma 4.2, and their extensions to two starting points in Remark 4.3. Corollary 2.19 describes related results assuming local convergence, but the exploration to which we couple in this section involves *unpaired half-edges* incident to vertices at a given distance, rather than the vertices at a given distance themselves as in Corollary 2.19.

Fix $G_n = \mathrm{CM}_n(\boldsymbol{d})$. Let o_1, o_2 be two independent and uniformly chosen vertices in $[n]$. By convention, let $Z_0^{(n;1)} = Z_0^{(n;2)} = 1$. For $r \geq 1$ and $i \in \{1, 2\}$, let $Z_r^{(n;i)}$ denote the number of unpaired half-edges incident to vertices at graph distance $r - 1$ from vertex o_i after exploration of $B_{r-1}^{(G_n)}(o_i)$, so that $Z_1^{(n;i)} = d_{o_i}$. Thus, $Z_r^{(n;i)}$ is obtained after pairing all the $Z_{r-1}^{(n;i)}$ half-edges at distance $r-1$ from the root o_i and counting the number of unpaired sibling half-edges incident to the half-edges chosen.

It would be tempting to think of $Z_r^{(n;i)}$ as $|\partial B_r^{(G_n)}(o_i)|$, and mostly it actually is. The difference is that $|\partial B_r^{(G_n)}(o_i)|$ counts the number of *vertices* at distance r of o_i while $Z_r^{(n;i)}$ counts the number of unpaired half-edges incident to vertices at distance $r-1$. When $B_r^{(G_n)}(o_i)$ is a tree, which happens whp for fixed r, then indeed $Z_r^{(n;i)} = |\partial B_r^{(G_n)}(o_i)|$. However, we do not impose this. It is more convenient to work with $Z_r^{(n;i)}$ than with $|\partial B_r^{(G_n)}(o_i)|$, since we can think of the half-edges counted in $Z_r^{(n;i)}$ as corresponding to a "super-vertex" that encodes much of the information in $\partial B_r^{(G_n)}(o_i)$. In particular, conditional on $B_r^{(G_n)}(o_i)$, half-edges are still paired uar, so the super-vertices can be combined with the vertices in $[n] \setminus (B_r^{(G_n)}(o_1) \cup B_r^{(G_n)}(o_2))$ to form a new configuration model. Such graph surgery procedures are very convenient for configuration models.

The following corollary shows that, for all $r \geq 1$, the processes $(Z_l^{(n;1)}, Z_l^{(n;2)})_{l=0}^r$ are close to two independent unimodular branching processes with root-offspring distribution $(p_k)_{k\geq 1}$ (recall Definition 1.26):

Corollary 7.3 (Coupling of neighborhoods of two vertices)   *Consider* $\mathrm{CM}_n(\boldsymbol{d})$ *where the degrees* $\boldsymbol{d} = (d_v)_{v\in[n]}$ *satisfy Conditions 1.7(a),(b). Let* o_1, o_2 *be chosen independently and uar from* $[n]$. *Then, for every fixed* $r \geq 1$,

$$\mathbb{P}(\mathrm{dist}_{\mathrm{CM}_n(\boldsymbol{d})}(o_1, o_2) \leq 2r) = o(1), \tag{7.3.1}$$

and

$$(Z_l^{(n;1)}, Z_l^{(n;2)})_{l=0}^r \xrightarrow{d} (Z_l^{(1)}, Z_l^{(2)})_{l=0}^r, \tag{7.3.2}$$

where $(Z_l^{(1)}, Z_l^{(2)})_{l\geq 0}$ *are two iid unimodular branching processes with root-offspring distribution* $(p_k)_{k\geq 1}$ *as in the local limit in Theorem 4.1.*

Corollary 7.3 follows from the local convergence in probability in Theorem 4.2, combined with Corollaries 2.19 and 2.20. However, since $Z_r^{(n;i)}$ is not quite $|\partial B_r^{(G_n)}(o_i)|$, a little more work is needed to prove (7.3.2). This is left as Exercise 7.4.

7.3.2 PATH-COUNTING UPPER BOUNDS AND DISTANCE LOWER BOUNDS

In this subsection we present path-counting techniques similar to those in Section 6.3.1, and use them to prove lower bounds on graph distances. Since $\mathrm{CM}_n(\boldsymbol{d})$ is a multi-graph, and not a simple graph like $\mathrm{NR}_n(\boldsymbol{w})$, we adapt the definition of a path in Definition 6.5 to make precise the meaning of a path in $\mathrm{CM}_n(\boldsymbol{d})$. We start by introducing some notation:

Definition 7.4 (Paths in configuration models)   A *path* $\vec{\pi}$ of length k in $\mathrm{CM}_n(\boldsymbol{d})$ is a sequence

$$\vec{\pi} = \{(\pi_0, s_0), (\pi_1, s_1, t_1), \ldots, (\pi_{k-1}, s_{k-1}, t_{k-1}), (\pi_k, t_k)\}, \tag{7.3.3}$$

where $\pi_i \in [n]$ denotes the ith vertex along the path, and $s_i \in [d_{\pi_i}]$ and $t_{i+1} \in [d_{\pi_{i+1}}]$ denote the labels of the half-edges incident to π_i and π_{i+1}, respectively, which are such that s_i and t_{i+1} are paired to create an edge between π_i and π_{i+1}. In particular, multiple edges between π_i and π_{i+1} give rise to *distinct* paths through the same vertices.   ◀

For a path $\vec{\pi}$ as in (7.3.3), we write $\vec{\pi} \subseteq \mathrm{CM}_n(\boldsymbol{d})$ when the path $\vec{\pi}$ in (7.3.3) is present in $\mathrm{CM}_n(\boldsymbol{d})$, so that the half-edge corresponding to s_i is paired with the half-edge corresponding to t_{i+1} for $i = 0, \dots, k-1$. Without loss of generality, we assume throughout that the path $\vec{\pi}$ is *self-avoiding*, i.e., that $\pi_0, \dots, \pi_k$ are distinct vertices.

In this section, we perform first-moment computations on the number of paths present in $\mathrm{CM}_n(\boldsymbol{d})$. In the next section, we perform second-moment methods.

Upper Bounds on the Expected Number of Paths in $\mathrm{CM}_n(\boldsymbol{d})$

For $a, b \in [n], \mathcal{I} \subseteq [n]$, and $k \geq 1$, we let $\mathcal{P}_k(a, b) = \mathcal{P}_k(a, b; \mathcal{I})$ denote the set of k-paths that use only vertices in $\mathcal{I}$, and we let $N_k(a, b) = N_k(a, b; \mathcal{I})$, given by

$$N_k(a, b) = \#\{\vec{\pi} \in \mathcal{P}_k(a, b) \colon \vec{\pi} \subseteq \mathrm{CM}_n(\boldsymbol{d}), \pi_i \in \mathcal{I} \, \forall i \in [k-1]\}, \qquad (7.3.4)$$

denote the number of paths of length k between the vertices a and b. Then, we prove the following upper bound on the expected number of paths connecting a and b:

Proposition 7.5 (Expected numbers of paths) *For any $k \geq 1$, $a, b \in [n]$, and $\boldsymbol{d} = (d_v)_{v \in [n]}$,*

$$\mathbb{E}[N_k(a, b)] \leq \frac{d_a d_b \ell_n}{(\ell_n - 2k + 1)^2} \nu_{\mathcal{I}}^{k-1}, \qquad \text{where} \quad \nu_{\mathcal{I}} = \sum_{i \in \mathcal{I} \setminus \{a, b\}} \frac{d_i(d_i - 1)}{\ell_n}. \qquad (7.3.5)$$

Proof The probability that the path $\vec{\pi}$ in (7.3.3) is present in $\mathrm{CM}_n(\boldsymbol{d})$ is equal to

$$\mathbb{P}(\vec{\pi} \subseteq \mathrm{CM}_n(\boldsymbol{d})) = \prod_{i=1}^{k} \frac{1}{\ell_n - 2i + 1}, \qquad (7.3.6)$$

and the number of paths with fixed vertices $\pi_0, \dots, \pi_k$ is equal to

$$d_{\pi_0} \left(\prod_{i=1}^{k-1} d_{\pi_i}(d_{\pi_i} - 1) \right) d_{\pi_k}. \qquad (7.3.7)$$

Substituting $\pi_0 = a, \pi_k = b$, we arrive at

$$\mathbb{E}[N_k(a, b)] = \frac{d_a d_b}{\ell_n - 2k + 1} \sideset{}{^*}\sum_{\pi_1, \dots, \pi_{k-1}} \prod_{i=1}^{k-1} \frac{d_{\pi_i}(d_{\pi_i} - 1)}{\ell_n - 2i + 1}, \qquad (7.3.8)$$

where the sum is over *distinct* elements of $\mathcal{I} \setminus \{a, b\}$ (as indicated by the asterisk). Let R denote the subset of vertices of $\mathcal{I} \setminus \{a, b\}$ for which $d_i \geq 2$. Then

$$\mathbb{E}[N_k(a, b)] = \frac{d_a d_b}{\ell_n - 2k + 1} \sideset{}{^*}\sum_{\pi_1, \dots, \pi_{k-1} \in R} \prod_{i=1}^{k-1} \frac{d_{\pi_i}(d_{\pi_i} - 1)}{\ell_n - 2i + 1}. \qquad (7.3.9)$$

By an inequality of Maclaurin (Hardy et al., 1988, Theorem 52), for $r = |R|, 2 \leq k \leq r+1$, and any $(a_i)_{i \in R}$ with $a_i \geq 0$, we have

$$\frac{(r - k + 1)!}{r!} \sideset{}{^*}\sum_{\pi_1, \dots, \pi_{k-1} \in R} \prod_{i=1}^{k-1} a_{\pi_i} \leq \left(\frac{1}{r} \sum_{i \in R} a_i \right)^{k-1}. \qquad (7.3.10)$$

Let $a_i = d_i(d_i - 1)$, so that

$$\sum_{i \in R} a_i = \ell_n \nu_{\mathcal{I}}. \tag{7.3.11}$$

Finally, we arrive at

$$\mathbb{E}[N_k(a,b)] \le \frac{d_a d_b}{\ell_n - 2k + 1} (\ell_n \nu_{\mathcal{I}}/r)^{k-1} \prod_{i=1}^{k-1} \frac{(r - i + 1)}{(\ell_n - 2i + 1)}$$

$$\le \frac{d_a d_b}{\ell_n - 2k + 1} \frac{\ell_n}{\ell_n - 2k + 3} \nu_{\mathcal{I}}^{k-1} \prod_{i=0}^{k-2} \frac{(1 - i/r)}{(1 - 2i/\ell_n)}. \tag{7.3.12}$$

Further, $\ell_n = \sum_{v \in [n]} d_v \ge 2r$, so that $1 - i/r \le 1 - 2i/\ell_n$. Substitution yields the required bound. $\qquad\square$

Logarithmic Lower Bound on Typical Distances in the Configuration Model

With Proposition 7.5 in hand, we can immediately prove the lower bound on the typical distance when the degrees have finite second moment (as in Theorem 6.4):

Theorem 7.6 (Logarithmic lower bound on typical distances) *Consider* $\mathrm{CM}_n(d)$. *Assume that*

$$\limsup_{n \to \infty} \nu_n > 1, \quad where \quad \nu_n = \mathbb{E}[D_n(D_n - 1)]/\mathbb{E}[D_n]. \tag{7.3.13}$$

Then, for any $\varepsilon > 0$, with o_1, o_2 chosen independently and uar from $[n]$,

$$\mathbb{P}(\mathrm{dist}_{\mathrm{CM}_n(d)}(o_1, o_2) \le (1 - \varepsilon) \log_{\nu_n} n) = o(1). \tag{7.3.14}$$

We leave the proof of Theorem 7.6, which is almost identical to that of Theorem 6.4, with (7.3.5) in Proposition 7.5 to hand, as Exercise 7.5.

We next investigate the $\tau = 3$ case, where the degree distribution has logarithmic corrections to the power law, as also investigated in Theorem 6.28 for $\mathrm{NR}_n(w)$:

Corollary 7.7 (Critical $\tau = 3$ case: interpolation) *Consider* $\mathrm{CM}_n(d)$ *where the degrees* $d = (d_v)_{v \in [n]}$ *satisfy Conditions 1.7(a),(b), and there exists an α such that, for all $x \ge 1$,*

$$[1 - F_n](x) \le c_2 x^{-2} (\log x)^{2\alpha}. \tag{7.3.15}$$

Let o_1, o_2 be chosen independently and uar from $[n]$. Then, for any $\varepsilon > 0$ and $\alpha > -\frac{1}{2}$,

$$\mathbb{P}\left(\mathrm{dist}_{\mathrm{CM}_n(d)}(o_1, o_2) \le (1 - \varepsilon) \frac{\log n}{(2\alpha + 1) \log \log n} \right) = o(1), \tag{7.3.16}$$

while, for $\alpha < -\frac{1}{2}$ and with $\nu = \lim_{n \to \infty} \nu_n < \infty$,

$$\mathbb{P}\left(\mathrm{dist}_{\mathrm{CM}_n(d)}(o_1, o_2) \le (1 - \varepsilon) \log_\nu n \right) = o(1). \tag{7.3.17}$$

Proof The proof follows from Theorem 7.6 by realizing that (7.3.15) implies that $\nu_n \le C(\log n)^{1+2\alpha}$ for $\alpha > -\frac{1}{2}$, while $(D_n^2)_{n \ge 1}$ is uniformly integrable for $\alpha < -\frac{1}{2}$, so that $\nu_n \to \nu = \mathbb{E}[D(D-1)]/\mathbb{E}[D] < \infty$. The details are left as Exercise 7.6. $\qquad\square$

Doubly Logarithmic Lower Bound for Infinite-Variance Degrees

We next extend the above upper bounds on the expected number of paths in order to deal with the case $\tau \in (2, 3)$, where, similarly to the analysis in Section 6.3.2 for $\mathrm{NR}_n(w)$, we need to *truncate* the degrees occurring in the paths that arise. Our main result is as follows:

Theorem 7.8 (Doubly logarithmic lower bound on typical distances in $\mathrm{CM}_n(d)$) *Consider* $\mathrm{CM}_n(d)$ *where the degrees* $d = (d_v)_{v \in [n]}$ *satisfy Condition 1.7(a) and there exist* $\tau \in (2, 3)$ *and* c_2 *such that, for all* $x \geq 1$,

$$[1 - F_n](x) \leq c_2 x^{-(\tau - 1)}. \tag{7.3.18}$$

Then, for any $\varepsilon > 0$, *with* o_1, o_2 *chosen independently and uar from* $[n]$,

$$\mathbb{P}\left(\mathrm{dist}_{\mathrm{CM}_n(d)}(o_1, o_2) \leq (1 - \varepsilon)\frac{2 \log \log n}{|\log(\tau - 2)|}\right) = o(1). \tag{7.3.19}$$

Proof The proof of Theorem 7.8 is similar to that of Theorem 6.7, and we discuss the changes only. For a fixed set of distinct vertices $(\pi_0, \ldots, \pi_k)$, (7.3.6) and (7.3.7) yield that the probability that there exist edges between π_{i-1} and π_i for all $i \in [k]$ is bounded above by

$$\frac{d_{\pi_0} d_{\pi_k}}{\ell_n - 2k + 1} \prod_{i=1}^{k-1} \frac{d_{\pi_i}(d_{\pi_i} - 1)}{\ell_n - 2i + 1}. \tag{7.3.20}$$

Equation (7.3.20) replaces the similar identity (6.3.6) for $\mathrm{CL}_n(w)$. We see that w_{π_0} and w_{π_k} in (6.3.6) are replaced by d_{π_0} and d_{π_k} in (7.3.20), and, for $i \in [k-1]$, the factors $w_{\pi_i}^2$ in (6.3.6) are replaced by $d_{\pi_i}(d_{\pi_i} - 1)$ in (7.3.20) while the factors ℓ_n in (6.3.6) are replaced by $\ell_n - 2i + 1$ for $i \in [k]$ in (7.3.20).

Define, as in (6.3.31),

$$\nu_n(b) = \frac{1}{\ell_n} \sum_{v \in [n]} d_v(d_i - 1) \mathbb{1}_{\{d_i \leq b\}}. \tag{7.3.21}$$

Then, the arguments in Section 6.3.2 imply that (see in particular Exercise 6.12),

$$\mathbb{P}(\mathrm{dist}_{\mathrm{CM}_n(d)}(u, v) \leq k_n) \leq \frac{d_u d_v}{\ell_n} \sum_{k=1}^{k_n} \frac{\ell_n^k(\ell_n - 2k - 1)!!}{(\ell_n - 1)!!} \prod_{l=1}^{k-1} \nu_n(b_l \wedge b_{k-l}) \tag{7.3.22}$$

$$+ (d_u + d_v) \sum_{k=1}^{k_n} \frac{\ell_n^k(\ell_n - 2k - 1)!!}{(\ell_n - 1)!!} [1 - F_n^\star](b_k) \prod_{l=1}^{k} \nu_n(b_l),$$

i.e., the bound in (6.8.7) is changed by factors $\ell_n^k(\ell_n - 2k - 1)!!/(\ell_n - 1)!!$ in the sum. For $k = O(\log \log n)$ and when Conditions 1.7(a),(b) hold,

$$\frac{\ell_n^k(\ell_n - 2k - 1)!!}{(\ell_n - 1)!!} = \prod_{i=1}^{k} \frac{\ell_n}{\ell_n - 2i + 1} = 1 + O(k^2/\ell_n) = 1 + o(1), \tag{7.3.23}$$

so this change has a negligible effect. Since (6.3.38) in Lemma 6.10 applies under the conditions of Theorem 7.8, we can follow the proof of Theorem 6.7 verbatim. $\square$

7.3.3 PATH-COUNTING LOWER BOUNDS AND RESULTING DISTANCE UPPER BOUNDS

In this subsection we provide upper bounds on typical distances in $\mathrm{CM}_n(\boldsymbol{d})$. We start by using the "giant is almost local" results proved in Section 4.3.1; see in particular Remark 4.13. After this, we continue with path-counting techniques similar to those in Section 6.5.1, focussing on the *variance* of the number of paths in $\mathrm{CM}_n(\boldsymbol{d})$. Such estimates turn out to be extremely versatile and can be used extensively to prove various upper bounds on distances, as we will show in the remainder of the section.

Consequences of the "Giant is Almost Local" Proof

We start by proving the following result using extensions of the "giant is almost local" proof:

Theorem 7.9 (Logarithmic upper bound on graph distances in $\mathrm{CM}_n(\boldsymbol{d})$) *Consider* $\mathrm{CM}_n(\boldsymbol{d})$ *where the degrees* $\boldsymbol{d} = (d_v)_{v \in [n]}$ *satisfy Conditions 1.7(a)–(c) with* $\nu = \mathbb{E}[D(D-1)]/\mathbb{E}[D]$ $\in (1, \infty)$. *Then, for any* $\varepsilon > 0$, *with* o_1, o_2 *chosen independently and uar from* $[n]$,

$$\mathbb{P}(\mathrm{dist}_{\mathrm{CM}_n(\boldsymbol{d})}(o_1, o_2) \le (1+\varepsilon) \log_\nu n \mid \mathrm{dist}_{\mathrm{CM}_n(\boldsymbol{d})}(o_1, o_2) < \infty) = 1 + o(1). \quad (7.3.24)$$

Proof Recall Section 4.3.1, where the *degree-truncation technique* from Theorem 1.11 was used with b sufficiently large. Recall that $\mathrm{CM}_{n'}(\boldsymbol{d}')$ denotes the CM after the degree-truncation method has been applied. Then, for $G_n = \mathrm{CM}_{n'}(\boldsymbol{d}')$ with $d'_{\max} \le b$, the proof shows that, when $|\partial B_r^{(G_n)}(o_1)|, |\partial B_r^{(G_n)}(o_2)| \ge r$, whp also $\mathrm{dist}_{\mathrm{CM}_{n'}(\boldsymbol{d}')}(o_1, o_2) \le \log_{\nu'_n} n(1 + o_\mathbb{P}(1))$ (recall Remark 4.13). Here $\nu'_n = \mathbb{E}[D'_n(D'_n - 1)]/\mathbb{E}[D'_n]$ denotes the expected forward degree of a uniform half-edge in $\mathrm{CM}_{n'}(\boldsymbol{d}')$.

We next relate this to the bound in (7.3.24). We note that $\nu'_n \to \nu'$ when Conditions 1.7(a),(b) hold, where, by construction,

$$\nu' = \frac{\mathbb{E}[D'(D'-1)]}{\mathbb{E}[D']} = \frac{\mathbb{E}[(D \wedge b)((D \wedge b) - 1)]}{\mathbb{E}[D]}. \quad (7.3.25)$$

The latter equality holds since $\sum_{v \in [n']} d'_v = \sum_{v \in [n]} d_v$, and $d'_v = d_v \wedge b$ for $v \in [n]$, while $d'_v = 1$ for $v \in [n'] \setminus [n]$. Thus, we also have

$$\sum_{v \in [n']} d'_v(d'_v - 1) = \sum_{v \in [n]} d'_v(d'_v - 1) = \sum_{v \in [n]} (d_v \wedge b)((d_v \wedge b) - 1). \quad (7.3.26)$$

Taking the limit as $n \to \infty$, and using Condition 1.7(a), proves (7.3.25).

By (7.3.25), $\nu' = \nu'(b) \nearrow \nu$ for $b \to \infty$. Thus, whp and for any $\varepsilon > 0$,

$$\mathrm{dist}_{\mathrm{CM}_{n'}(\boldsymbol{d}')}(o_1, o_2) \le (1+\varepsilon) \log_\nu n. \quad (7.3.27)$$

Since also $\mathbb{P}(\mathrm{dist}_{\mathrm{CM}_n(\boldsymbol{d})}(o_1, o_2) < \infty) \to \zeta^2$ and $\mathbb{P}(\mathrm{dist}_{\mathrm{CM}_{n'}(\boldsymbol{d}')}(o_1, o_2) < \infty) \to (\zeta')^2$, where $\zeta' = \zeta'(b) \to \zeta$ when $b \to \infty$, this gives the first proof of the upper bound in Theorem 7.9. $\square$

Exercise 7.7 shows that the analysis in Section 4.3.1 can be performed without the degree-truncation argument of Theorem 1.11 when $\limsup_{n \to \infty} \mathbb{E}[D_n^3] < \infty$. Exercise 7.8 extends the condition to $\limsup_{n \to \infty} \mathbb{E}[D_n^p] < \infty$ for some $p > 2$.

Second-Moment Method for the Number of Paths in the Configuration Model

We next extend the first-moment bounds on the number of paths in $CM_n(d)$ in Section 7.3.2 to second-moment bounds. This allows us to give a second proof of Theorem 7.9, as well as proofs of several other useful and interesting results.

We start by setting up the notation. Let $\mathcal{I} \subseteq [n]$ be a subset of the vertices. Fix $k \geq 1$, and define

$$\bar{n}_k(a,b) = \frac{\ell_n^k(\ell_n - 2k - 1)!!}{(\ell_n - 1)!!} \frac{d_a d_b}{\ell_n} \left(\sum_{i \in \mathcal{I} \setminus \{a,b\}} \frac{d_i(d_i - 1)}{\ell_n} \right)^{k-1}, \tag{7.3.28}$$

$$\underline{n}_k(a,b) = \frac{d_a d_b}{\ell_n} \left(\sum_{i \in \mathcal{I}_{a,b,k}} \frac{d_i(d_i - 1)}{\ell_n} \right)^{k-1}, \tag{7.3.29}$$

where $\mathcal{I}_{a,b,k}$ is the subset of $\mathcal{I}$ in which a and b, as well as the $k - 1$ indices with highest degrees, have been removed. Let

$$\nu_{\mathcal{I}} = \frac{1}{\ell_n} \sum_{i \in \mathcal{I}} d_i(d_i - 1), \qquad \gamma_{\mathcal{I}} = \frac{1}{\ell_n^{3/2}} \sum_{i \in \mathcal{I}} d_i(d_i - 1)(d_i - 2). \tag{7.3.30}$$

The following proposition replaces the similar Proposition 6.14 for $CL_n(w)$, which was crucial in deriving lower bounds on typical distances:

Proposition 7.10 (Variance of number of paths) *For any $k \geq 1$, $a, b \in \mathcal{I}$,*

$$\mathbb{E}[N_k(a,b)] \geq \underline{n}_k(a,b). \tag{7.3.31}$$

Further, assuming that $\nu_{\mathcal{I}} > 1$,

$$\mathrm{Var}(N_k(a,b)) \leq n_k(a,b)$$
$$+ \bar{n}_k(a,b)^2 \left(\frac{\gamma_{\mathcal{I}} \nu_{\mathcal{I}}^2}{\nu_{\mathcal{I}} - 1} \left(\frac{1}{d_a} + \frac{1}{d_b} \right) + \frac{\gamma_{\mathcal{I}}^2 \nu_{\mathcal{I}}}{d_a d_b(\nu_{\mathcal{I}} - 1)^2} + e_k' \right), \tag{7.3.32}$$

where

$$e_k' = \left(\prod_{i=1}^{k} \frac{\ell_n - 2i + 1}{\ell_n - 2i - 2k + 1} - 1 \right)$$
$$+ k \frac{\ell_n^{2k}(\ell_n - 4k - 1)!!}{(\ell_n - 1)!!} \left(1 + \frac{\gamma_{\mathcal{I}}}{d_a \nu_{\mathcal{I}}} \right) \left(1 + \frac{\gamma_{\mathcal{I}}}{d_b \nu_{\mathcal{I}}} \right) \frac{\nu_{\mathcal{I}}}{\nu_{\mathcal{I}} - 1} \left(e^{2k^3 \gamma_{\mathcal{I}}^2 / \nu_{\mathcal{I}}^3} - 1 \right). \tag{7.3.33}$$

Proof The proof of (7.3.31) follows immediately from (7.3.8), together with the fact that $1/(\ell_n - 2i + 1) \geq 1/\ell_n$.

For the proof of (7.3.32), we follow the proof of (6.5.7), and discuss the differences only. We recall that

$$N_k(a,b) = \sum_{\vec{\pi} \in \mathcal{P}_k(a,b)} \mathbb{1}_{\{\vec{\pi} \subseteq CM_n(d)\}} \tag{7.3.34}$$

is the number of paths $\vec{\pi}$ of length k between the vertices a and b, where a path was defined in (7.3.3). Since $N_k(a,b)$ is a sum of indicators, its variance can be written as follows:

$$\text{Var}(N_k(a,b))$$
$$= \sum_{\vec{\pi},\vec{\rho} \in \mathcal{P}_k(a,b)} \left[\mathbb{P}(\vec{\pi},\vec{\rho} \subseteq \text{CM}_n(\boldsymbol{d})) - \mathbb{P}(\vec{\pi} \subseteq \text{CM}_n(\boldsymbol{d}))\mathbb{P}(\vec{\rho} \subseteq \text{CM}_n(\boldsymbol{d})) \right]. \quad (7.3.35)$$

Equation (7.3.35) replaces (6.5.13) for $\text{NR}_n(\boldsymbol{w})$.

We next proceed to prove (7.3.32), with e_k' defined in (7.3.33). We say that two paths $\vec{\pi}$ and $\vec{\rho}$ are *disjoint* when they use distinct sets of half-edges. Thus, it is possible that the vertex sets $\{\pi_1,\ldots,\pi_{k-1}\}$ and $\{\rho_1,\ldots,\rho_{k-1}\}$ have a non-empty intersection, but then the half-edges leading in and out of the joint vertices for $\vec{\pi}$ and $\vec{\rho}$ must be distinct. For $\text{NR}_n(\boldsymbol{w})$, pairs of paths using different edges are *independent*, so that these pairs do not contribute to $\text{Var}(N_k(a,b))$. For $\text{CM}_n(\boldsymbol{d})$, however, for disjoint pairs of paths $\vec{\pi}$ and $\vec{\rho}$,

$$\mathbb{P}(\vec{\pi},\vec{\rho} \subseteq \text{CM}_n(\boldsymbol{d})) = \prod_{i=1}^{k} \frac{\ell_n - 2i + 1}{\ell_n - 2i - 2k + 1} \mathbb{P}(\vec{\pi} \subseteq \text{CM}_n(\boldsymbol{d}))\mathbb{P}(\vec{\rho} \subseteq \text{CM}_n(\boldsymbol{d})).$$
$$(7.3.36)$$

Summing the quantity

$$\left(\prod_{i=1}^{k} \frac{\ell_n - 2i + 1}{\ell_n - 2i - 2k + 1} - 1 \right) \mathbb{P}(\vec{\pi} \subseteq \text{CM}_n(\boldsymbol{d}))\mathbb{P}(\vec{\rho} \subseteq \text{CM}_n(\boldsymbol{d})) \quad (7.3.37)$$

over all $\vec{\pi}$ and $\vec{\rho}$ gives rise to the first contribution to e_k'. For the other contributions, we follow the proof of (6.5.13) for $\text{NR}_n(\boldsymbol{w})$, and omit further details. $\qquad\square$

With Proposition 7.10 in hand, we can straightforwardly adapt the proof of Theorem 6.19 to $\text{CM}_n(\boldsymbol{d})$ to prove Theorem 7.9. We leave this proof of Theorem 7.9 as Exercise 7.9.

Critical Case $\tau = 3$ with Logarithmic Corrections

Recall the lower bounds on distances from Corollary 7.7. Upper bounds can be derived in a similar way as for $\text{CL}_n(\boldsymbol{w})$ in the proof of Theorem 6.22 in Section 6.5.2. We refrain from giving more details.

7.3.4 DOUBLY-LOGARITHMIC UPPER BOUNDS FOR INFINITE-VARIANCE DEGREES

In order to prove the doubly logarithmic upper bound on the typical distance for $\text{CM}_n(\boldsymbol{d})$ in Theorem 7.2, we will use a different approach compared to that in the related proof of Theorem 6.11 for $\text{NR}_n(\boldsymbol{w})$. We start by describing the setting. We assume that there exist $\tau \in (2,3)$, $\beta > \frac{1}{2}$, and c_1 such that, uniformly in n and $x \leq n^\beta$,

$$[1 - F_n](x) \geq c_1 x^{-(\tau-1)}. \quad (7.3.38)$$

Our main result is as follows:

Theorem 7.11 (Doubly logarithmic upper bound on typical distance for $\tau \in (2,3)$) *Consider* $\text{CM}_n(\boldsymbol{d})$ *where the degrees* $\boldsymbol{d} = (d_v)_{v \in [n]}$ *satisfy Conditions 1.7(a),(b) and (7.3.38). Then, for any* $\varepsilon > 0$, *with* o_1, o_2 *chosen independently and uar from* $[n]$,

$$\lim_{n\to\infty} \mathbb{P}\left(\mathrm{dist}_{\mathrm{CM}_n(d)}(o_1, o_2) \leq \frac{2(1+\varepsilon)\log\log n}{|\log(\tau-2)|} \,\Big|\, \mathrm{dist}_{\mathrm{CM}_n(d)}(o_1, o_2) < \infty\right) = 1.$$

(7.3.39)

Exercise 7.10 explores a proof of Theorem 7.11, based on the proof for $\mathrm{NR}_n(w)$ in Theorem 6.11, that is an alternative to the proof given below.

Proof This proof of Theorem 7.11 makes precise the statement that vertices of large degree d are directly connected vertices of degree approximately $d^{1/(\tau-2)}$.

Connectivity of Sets

We start by studying the connectivity of sets in $\mathrm{CM}_n(d)$, for which we rely on the following connectivity lemma, which is of independent interest:

Lemma 7.12 (Connectivity sets in $\mathrm{CM}_n(d)$)   *For any two sets of vertices $A, B \subseteq [n]$,*

$$\mathbb{P}(A \text{ not directly connected to } B \text{ in } \mathrm{CM}_n(d)) \leq e^{-d_A d_B/(2\ell_n)},$$

(7.3.40)

where, for any $A \subseteq [n]$,

$$d_A = \sum_{a \in A} d_a$$

(7.3.41)

denotes the total degree of vertices in A.

Lemma 7.12 extends the result in Exercise 6.18 for $\mathrm{NR}_n(w)$ to $\mathrm{CM}_n(d)$.

Proof There are d_A half-edges incident to the set A, which we pair one by one. After having paired k half-edges, all to half-edges that are not incident to B, the probability of pairing the next half-edge to a half-edge that is not incident to B equals

$$1 - \frac{d_B}{\ell_n - 2k + 1} \leq 1 - \frac{d_B}{\ell_n}.$$

(7.3.42)

Some half-edges incident to A may attach to other half-edges incident to A, so that possibly *fewer* than d_A half-edges need to be paired in order to to pair all of them. However, since each pairing uses up at most two half-edges incident to A, we need to pair at least $d_A/2$ half-edges, so that

$$\mathbb{P}(A \text{ not directly connected to } B) \leq \left(1 - \frac{d_B}{\ell_n}\right)^{d_A/2} \leq e^{-d_A d_B/(2\ell_n)},$$

(7.3.43)

where we use $1 - x \leq e^{-x}$. $\square$

Power-Iteration Proof for Ultra-Small Distances

We prove Theorem 7.11 using a technique that we call *power iteration*. Power iteration shows that vertices with large degrees are directly connected to vertices with *even larger* degrees and thus can be used to find a short path to the vertices with the highest degrees in the configuration model.

Fix $G_n = \mathrm{CM}_n(d)$. Assume that the degrees $d = (d_v)_{v\in[n]}$ satisfy Conditions 1.7(a), (b) and (7.3.38). Fix $r \geq 1$, and condition on $B_r^{(G_n)}(o_1)$ and $B_r^{(G_n)}(o_2)$ such that $\partial B_r^{(G_n)}$

$(o_1) \neq \varnothing$ and $\partial B_r^{(G_n)}(o_2) \neq \varnothing$. By Corollary 7.3 $(Z_r^{(n;1)}, Z_r^{(n;2)}) \xrightarrow{d} (Z_r^{(1)}, Z_r^{(2)})$, and $Z_r^{(n;1)}$ and $Z_r^{(n;2)}$ are whp quite large since we are conditioning on $Z_r^{(n;1)} \geq 1$ and $Z_r^{(n;2)} \geq 1$. Fix $C > 1$ large, and note that, by Lemma 7.12, the conditional probability that none of the $Z_r^{(n;i)}$ half-edges is paired to a vertex of degree at least d is at most

$$\exp\left\{ -Z_r^{(n;i)} \sum_{v \in [n]} d_v \mathbb{1}_{\{d_v \geq d\}}/(2\ell_n) \right\}. \tag{7.3.44}$$

We use (7.3.38) to obtain the bound

$$\sum_{v \in [n]} d_v \mathbb{1}_{\{d_v \geq d\}}/(2\ell_n) \geq d[1 - F_n](d)(n/2\ell_n) \geq cd^{2-\tau}, \tag{7.3.45}$$

where $c = c_1/(2 \sup_n \mathbb{E}[D_n])$. With $d = (Z_r^{(n;i)})^{1/(\tau-2+\varepsilon)}$, the probability (7.3.44) is at most $\exp\{-c(Z_r^{(n;i)})^{\varepsilon/(2-\tau+\varepsilon)}\}$. Call the maximal-degree vertex to which one of the $Z_r^{(n;i)}$ half-edges is paired the *first power-iteration vertex*.

We now iterate these ideas. Denote $\underline{u}_k = \underline{u}_k^{(i)} = (Z_r^{(n;i)})^{1/(\tau-2+\varepsilon)^k}$. Then, the probability that the $(k-1)$th power-iteration vertex is not paired to a vertex of degree at least $\underline{u}_k$ is at most $\exp\{-c\underline{u}_{k-1}^{\varepsilon/(2-\tau+\varepsilon)}\}$, and we call the maximum-degree vertex to which the $(k-1)$th power-iteration vertex is paired the kth *power-iteration vertex*.

We iterate this until we reach one of the hubs in $\{v: d_v > n^\beta\}$, where $\beta > \frac{1}{2}$, for which we need at most $k_n^\star$ iterations, with $k_n^\star$ satisfying

$$\underline{u}_{k_n^\star} = (Z_r^{(n;i)})^{1/(\tau-2+\varepsilon)^{k_n^\star}} \geq n^\beta. \tag{7.3.46}$$

The smallest $k_n^\star$ for which this occurs is

$$k_n^\star = \left\lceil \frac{\log\log(n^\beta) - \log\log(Z_r^{(n;i)})}{|\log(\tau - 2 + \varepsilon)|} \right\rceil. \tag{7.3.47}$$

Finally, the probability that power iteration fails from vertex o_i is at most

$$\sum_{k=1}^{\infty} \exp\{-c\underline{u}_{k-1}^{\varepsilon/(2-\tau+\varepsilon)}\} \xrightarrow{\mathbb{P}} 0, \tag{7.3.48}$$

when first $n \to \infty$ followed by $r \to \infty$.

On the event that power iteration succeeds from both vertices o_i, the graph distance of o_i to $\{v: d_v > n^\beta\}$ is at most $k_n^\star + r$. Since $\{v: d_v > n^\beta\}$ is whp a clique in $\mathrm{CM}_n(\boldsymbol{d})$ (recall Exercise 7.11), we conclude that, conditional on $o_1 \longleftrightarrow o_2$, $\mathrm{dist}_{\mathrm{CM}_n(\boldsymbol{d})}(o_1, o_2) \leq 2k_n^\star + 1 + 2r$ whp. $\qquad\square$

7.3.5 The Core and its Diameter

We next discuss the *core* of the configuration model, consisting of those vertices that have degree at least some power of $\log n$, and we will bound the diameter of this core. Fix $\tau \in (2, 3)$. We take $\sigma > 1/(3 - \tau)$ and define the core Core_n of the configuration model to be

$$\mathrm{Core}_n = \{v \in [n]: d_v \geq (\log n)^\sigma\}, \tag{7.3.49}$$

i.e., the set of vertices of degree at least $(\log n)^\sigma$. Then, the diameter of the core is bounded in the following theorem, which is interesting in its own right:

Theorem 7.13 (Diameter of the core) *Consider* $\mathrm{CM}_n(\boldsymbol{d})$ *where the degrees* $\boldsymbol{d} = (d_v)_{v\in[n]}$ *satisfy Conditions 1.7(a),(b) and (7.3.38). For any* $\sigma > 1/(3 - \tau)$*, the diameter of* Core_n *is whp bounded above by*

$$\frac{2\log\log n}{|\log(\tau - 2)|} + 1. \tag{7.3.50}$$

We prove Theorem 7.13 below and start by setting up the notation for it. We note that (7.2.2) implies that, for some $\beta \in (\frac{1}{2}, 1/(\tau - 1))$, we have

$$d_{\max} = \max_{v\in[n]} d_v \geq u_1, \qquad \text{where} \qquad u_1 = n^\beta. \tag{7.3.51}$$

Define

$$\Gamma_1 = \{v \in [n]: d_v \geq u_1\}, \tag{7.3.52}$$

so that $\Gamma_1 \neq \varnothing$. For some constant $C > 0$ to be determined later on, and for $k \geq 2$, we recursively define

$$u_k = C\log n \left(u_{k-1}\right)^{\tau - 2}. \tag{7.3.53}$$

We identify u_k in the following lemma:

Lemma 7.14 (Identification $(u_k)_{k\geq 1}$) *For every* $k \geq 1$,

$$u_k = (C\log n)^{a_k} n^{b_k}, \tag{7.3.54}$$

where

$$b_k = \beta(\tau - 2)^{k-1}, \qquad a_k = [1 - (\tau - 2)^{k-1}]/(3 - \tau). \tag{7.3.55}$$

Proof We note that a_k, b_k satisfy the following recursions for $k \geq 2$,

$$b_k = (\tau - 2)b_{k-1}, \qquad a_k = 1 + (\tau - 2)a_{k-1}, \tag{7.3.56}$$

with initial conditions $b_1 = \beta, a_1 = 0$. Solving the recursions yields our claim. $\qquad\square$

As a consequence of Lemma 7.12, Γ_1 is whp a clique (see Exercise 7.11). Define

$$\Gamma_k = \{v \in [n]: d_v \geq u_k\}. \tag{7.3.57}$$

The key step in the proof of Theorem 7.13 is the following proposition, showing that whp every vertex in Γ_k is connected to a vertex in Γ_{k-1}:

Proposition 7.15 (Connectivity between Γ_{k-1} and Γ_k) *Consider* $\mathrm{CM}_n(\boldsymbol{d})$ *where the degrees* $\boldsymbol{d} = (d_v)_{v\in[n]}$ *satisfy Conditions 1.7(a),(b) and (7.3.38). Fix* $k \geq 2$*, and take* $C > 2\mathbb{E}[D]/c_1$*, with* c_1 *as in (7.3.38). Then the probability that there exists an* $i \in \Gamma_k$ *that is not directly connected to* Γ_{k-1} *in* $\mathrm{CM}_n(\boldsymbol{d})$ *is at most* $n^{-\delta}$*, for some* $\delta > 0$ *that is independent of* k.

Proposition 7.15 implies that a vertex u of degree $d_u \gg 1$ is whp directly connected to a vertex of degree approximately $d_u^{1/(\tau-2)}$. Thus, vertices of high degree are directly connected to vertices of *even higher* degree, where the higher degree is a power larger than 1 of the original degree. This is the phenomenon that we have called "power iteration." Theorem 7.13 can be understood by noting that from a vertex of degree $(\log n)^\sigma$ for some $\sigma > 0$, we need roughly $\log\log n / |\log(\tau-2)|$ power iterations to go to a vertex of degree n^β with $\beta > \frac{1}{2}$.

Proof We note that, by definition,

$$\sum_{v \in \Gamma_{k-1}} d_v \geq u_{k-1}|\Gamma_{k-1}| = u_{k-1}n[1 - F_n](u_{k-1}). \tag{7.3.58}$$

By (7.2.2), and since $k \mapsto u_k$ is decreasing with $u_1 = n^\beta$,

$$[1 - F_n](u_{k-1}) \geq c(u_{k-1})^{1-\tau}. \tag{7.3.59}$$

As a result, we obtain that for every $k \geq 2$,

$$\sum_{v \in \Gamma_{k-1}} d_v \geq cn(u_{k-1})^{2-\tau}. \tag{7.3.60}$$

By (7.3.60) and Lemma 7.12, using Boole's inequality, the probability that there exists a $v \in \Gamma_k$ that is not directly connected to Γ_{k-1} is bounded by

$$n \exp\left\{ - u_k n u_{k-1}[1 - F(u_{k-1})]/(2\ell_n)\right\} \leq n \exp\left\{ - c u_k (u_{k-1})^{2-\tau}/(2\mathbb{E}[D_n])\right\}$$
$$= n \exp\left\{ cC \log n/(2\mathbb{E}[D_n])\right\}$$
$$= n^{1-cC/(2\mathbb{E}[D_n])}, \tag{7.3.61}$$

where we use (7.3.53). By Conditions 1.7(a),(b), $\mathbb{E}[D_n] \to \mathbb{E}[D]$ so that, as $n \to \infty$ and taking $C > 2\mathbb{E}[D]/c$, we obtain the claim for any $\delta < cC/[2\mathbb{E}[D]] - 1$. □

We now complete the proof of Theorem 7.13:

Proof of Theorem 7.13. Fix

$$k_n^\star = \left\lfloor \frac{\log\log n}{|\log(\tau-2)|} \right\rfloor. \tag{7.3.62}$$

As a result of Proposition 7.15 and the fact that $k_n^\star n^{-\delta} = o(1)$, whp every $i \in \Gamma_k$ is directly connected to Γ_{k-1} for *all* $k \leq k_n^\star$. By Exercise 7.11, Γ_1 forms whp a complete graph. As a result, the diameter of $\Gamma_{k_n^\star}$ is at most $2k_n^\star + 1$. Therefore, it suffices to prove that

$$\text{Core}_n \subseteq \Gamma_{k_n^\star}. \tag{7.3.63}$$

By (7.3.53), in turn, this is equivalent to $u_{k_n^\star} \leq (\log n)^\sigma$, for any $\sigma > 1/(3-\tau)$. According to Lemma 7.14,

$$u_{k_n^\star} = (C \log n)^{a_{k_n^\star}} n^{b_{k_n^\star}}. \tag{7.3.64}$$

We note that $n^{b_{k_n^\star}} = \exp\left\{(\tau-2)^{k_n^\star} \log n\right\}$. Since, for $\tau \in (2,3)$,

$$x(\tau-2)^{\log x/|\log(\tau-2)|} = x \times x^{-1} = 1, \tag{7.3.65}$$

we find with $x = \log n$ that $n^{b_{k_n^\star}} \leq e^{1/(\tau-2)}$. Further, $a_k \to 1/(\tau-3)$ as $k \to \infty$, so that $(C \log n)^{a_{k_n^\star}} = (C \log n)^{1/(3-\tau)+o(1)}$. We conclude that

$$u_{k_n^\star} = (\log n)^{1/(3-\tau)+o(1)}, \tag{7.3.66}$$

so that, by choosing n sufficiently large, we can make $1/(3-\tau)+o(1) \leq \sigma$. This completes the proof of Theorem 7.13. $\qquad\qquad\qquad\qquad\qquad\qquad\qquad\qquad\qquad\qquad\qquad\qquad\square$

Exercise 7.12 studies an alternative proof of Theorem 7.11 that proceeds by showing that whp a short path exists between $\partial B_r^{(G_n)}(o_1)$ and Core_n when $\partial B_r^{(G_n)}(o_1)$ is non-empty.

Critical Case $\tau = 2$

We next discuss the critical case where $\tau = 2$ and the degree distribution has logarithmic corrections. We use adaptations of the power-iteration technique.

Let us focus on one specific example, where

$$[1 - F_n](x) \geq \frac{c_1}{x}(\log x)^{-\alpha}, \tag{7.3.67}$$

for all $x \leq n^\beta$ and some $\beta > \frac{1}{2}$. We take $\alpha > 1$, since otherwise $(D_n)_{n \geq 1}$ might not be uniformly integrable. Our main result is as follows:

Theorem 7.16 (Example of ultra-ultra-small distances for $\tau = 2$) *Consider* $\mathrm{CM}_n(\boldsymbol{d})$ *where the degrees* $\boldsymbol{d} = (d_v)_{v \in [n]}$ *satisfy Conditions 1.7(a),(b) and (7.3.67). Fix $r \geq 1$ and let $\underline{u}_0 = r$. Define, for $k \geq 1$, recursively $\underline{u}_k = \exp\{\underline{u}_{k-1}^{(1-\varepsilon)/(\alpha-1)}\}$. Let $k_n^\star = \inf\{k \colon \underline{u}_k \geq n^\beta\}$. Then, with o_1, o_2 chosen independently and uar from $[n]$,*

$$\mathbb{P}(\mathrm{dist}_{\mathrm{CM}_n(\boldsymbol{d})}(o_1, o_2) \leq 2(k_n^\star + r) + 1 \mid \mathrm{dist}_{\mathrm{CM}_n(\boldsymbol{d})}(o_1, o_2) < \infty) \to 1, \tag{7.3.68}$$

when first $n \to \infty$ followed by $r \to \infty$.

Proof We start from the setting discussed just above (7.3.44) and see how power iteration applies in this case. We compute

$$\frac{1}{n}\sum_{v \in [n]} d_v \mathbb{1}_{\{d_v \geq d\}} = \mathbb{E}[D_n \mathbb{1}_{\{D_n \geq d\}}] = \sum_{k \geq 0} \mathbb{P}(D_n \mathbb{1}_{\{D_n \geq d\}} > k)$$

$$= \sum_{k \geq d} \mathbb{P}(D_n > k) = \sum_{k \geq d}[1 - F_n](k). \tag{7.3.69}$$

Using the lower bound in (7.3.67), which is valid until $d = n^\beta$, we obtain

$$\frac{1}{n}\sum_{v \in [n]} d_v \mathbb{1}_{\{d_v \geq d\}} \geq \sum_{k=d}^{n^\beta} \frac{c_1}{k}(\log k)^{-\alpha} \geq c[(\log d)^{1-\alpha} - (\log n^\beta)^{1-\alpha}]$$

$$\geq c(\log d)^{1-\alpha}, \tag{7.3.70}$$

provided that $d \leq n^\varepsilon$ for some $\varepsilon < \beta$.

Therefore, the probability that a vertex of degree d_u is not directly connected to a vertex of degree $d = \exp\{d_u^{(1-\varepsilon)/(\alpha-1)}\}$ is bounded by

$$\exp\{-cd_u(\log d)^{1-\alpha}\} \leq \exp\{-cd_u^\varepsilon\}. \tag{7.3.71}$$

Denote $\underline{u}_0 = r$, and recursively define $\underline{u}_k = \exp\{\underline{u}_{k-1}^{(1-\varepsilon)/(\alpha-1)}\}$ for $k \geq 1$. Again the kth power-iteration vertex is the maximal-degree vertex to which the $(k-1)$th power-iteration vertex is connected.

The probability that the $(k-1)$th power-iteration vertex is not paired to a vertex of degree at least $\underline{u}_k$ is at most $\exp\{-c\underline{u}_{k-1}^{\varepsilon}\}$. Recall that $k_n^{\star} = \inf\{k : \underline{u}_k \geq n^{\beta}\}$ denotes the number of steps needed to reach a hub. Since the set of hubs $\{v : d_v \geq n^{\beta}\}$ is whp a clique, by Exercise 7.11, the probability that $\mathrm{dist}_{\mathrm{CM}_n(d)}(o_1, o_2) > 2(r + k_n^{\star}) + 1$ is at most

$$o_{\mathbb{P}}(1) + 2\sum_{k \geq 1} e^{-c\underline{u}_{k-1}^{\varepsilon}} \to 0, \qquad (7.3.72)$$

when first $n \to \infty$ followed by $r \to \infty$. This completes the proof. $\qquad \square$

Exercises 7.13 and 7.14 investigate the scaling of $k_n^{\star}$, which is *very* small. Exercise 7.15 investigates another example for which $\tau = 2$, in which (7.3.67) is replaced by $[1 - F_n](x) \geq c_1 e^{-c(\log x)^{\gamma}}/x$ for some $c_1, c, \gamma \in (0, 1)$, and all $x \leq n^{\beta}$.

7.4 BRANCHING PROCESSES WITH INFINITE MEAN

Recall the branching-process limit for neighborhoods in $\mathrm{CM}_n(d)$ in Corollary 7.3. When $\tau \in (2, 3)$, the branching processes $(Z_j^{(1)})_{j \geq 0}$ and $(Z_j^{(2)})_{j \geq 0}$ are well defined but have *infinite mean* in generations 2, 3, etc. In this section we give a scaling result for the generation sizes for branching processes with infinite mean. This result is crucial to describe the *fluctuations* of the typical distances in $\mathrm{CM}_n(d)$, and it also allows us to understand how ultra-small distances of order $\log \log n$ arise. The main result in this section is as follows:

Theorem 7.17 (Branching processes with infinite mean) *Let $(Z_k)_{k \geq 0}$ be a branching process with offspring distribution $Z_1 = X$ having distribution function F_X. Assume that there exist $\alpha \in (0, 1)$ and a non-negative non-increasing function $x \mapsto \gamma(x)$ such that*

$$x^{-\alpha - \gamma(x)} \leq 1 - F_X(x) \leq x^{-\alpha + \gamma(x)}, \qquad \textit{for large } x, \qquad (7.4.1)$$

where $x \mapsto \gamma(x)$ satisfies

(a) $x \mapsto x^{\gamma(x)}$ is non-decreasing,
(b) $\int_0^{\infty} \gamma(e^{e^x})\, dx < \infty$, or, equivalently, $\int_e^{\infty} \frac{\gamma(y)}{y \log y}\, dy < \infty$.

Then $\alpha^k \log(Z_k \vee 1) \xrightarrow{a.s.} Y$ as $k \to \infty$, with $\mathbb{P}(Y = 0)$ equal to the extinction probability of $(Z_k)_{k \geq 0}$.

In the analysis for the configuration model, we take $\alpha = \tau - 2$, as α corresponds to the tail exponent of the *size-biased* random variable $D^{\star}$ (recall Lemma 1.23). Theorem 7.17 covers the case where the branching process has a heavy-tailed offspring distribution. Indeed, it is not hard to show that Theorem 7.17 implies that $\mathbb{E}[X^s] = \infty$ for every $s > \alpha \in (0, 1)$ (see Exercise 7.16 below).

We will not prove Theorem 7.17 in full generality. Rather, we prove it in the simpler, yet still quite general, case in which $\gamma(x) = c(\log x)^{\gamma - 1}$ for some $\gamma \in [0, 1)$ and $c > 0$. See Exercise 7.17 to see that this case indeed satisfies the assumptions in Theorem 7.17.

In the proof of Theorem 7.17 for the special case $\gamma(x) = c(\log x)^{\gamma-1}$, we rely on regularly varying functions. Note, however, that (7.4.1) does not assume that $x \mapsto 1 - F_x(x)$ is regularly varying (meaning that $x^\alpha[1 - F_x(x)]$ is slowly varying; recall Definition 1.19). Thus, instead, we work with the special slowly varying functions

$$1 - F_x^\pm(x) = x^{-\alpha \pm \gamma(x)}. \tag{7.4.2}$$

Proof of Theorem 7.17 for $\gamma(x) = c(\log x)^{\gamma-1}$. The proof is divided into five main steps.

The Split

We first assume that $\mathbb{P}(Z_1 \geq 1) = 1$, so that the survival probability equals 1. Define

$$M_k = \alpha^k \log(Z_k \vee 1). \tag{7.4.3}$$

For $i \geq 1$, we define

$$Y_i = \alpha^i \log\left(\frac{(Z_i \vee 1)}{(Z_{i-1} \vee 1)^{1/\alpha}}\right). \tag{7.4.4}$$

We make the split

$$M_k = Y_1 + Y_2 + \cdots + Y_k. \tag{7.4.5}$$

From this split, it is clear that the almost sure convergence of M_k follows when the sum $\sum_{i=0}^\infty Y_i$ converges, which is the case when, in turn,

$$\sum_{i=1}^\infty \mathbb{E}[|Y_i|] < \infty. \tag{7.4.6}$$

Inserting Normalization Sequences

We next investigate $\mathbb{E}[|Y_i|]$. We prove by induction on i that there exist constants $\kappa < 1$ and $K > 0$ such that

$$\mathbb{E}[|Y_i|] \leq K\kappa^i. \tag{7.4.7}$$

For $i = 0$, this follows from the fact that, when (7.4.1) holds, the random variable $Y_1 = \alpha \log(Z_1 \vee 1)$ has bounded expectation. This initializes the induction hypothesis. We next turn to the advancement of the induction hypothesis. For this, we recall the definition of u_n in [V1, (2.6.7)], as

$$u_n = \inf\{x: 1 - F_x(x) \leq 1/n\}, \tag{7.4.8}$$

which indicates the order of magnitude of $\max_{i=1}^n X_i$. Here $(X_i)_{i=1}^n$ are iid random variables with distribution function F_x. We also rely on $u_n^\pm$, which are the u_ns corresponding to $F_x^\pm$ in (7.4.2). Obviously,

$$u_n^- \leq u_n \leq u_n^+. \tag{7.4.9}$$

Then we define

$$U_i = \alpha^i \log\left(\frac{u_{Z_{i-1}\vee 1}}{(Z_{i-1} \vee 1)^{1/\alpha}}\right), \qquad V_i = \alpha^i \log\left(\frac{Z_i \vee 1}{u_{Z_{i-1}\vee 1}}\right), \tag{7.4.10}$$

so that $Y_i = U_i + V_i$ and

$$\mathbb{E}[|Y_i|] \leq \mathbb{E}[|U_i|] + \mathbb{E}[|V_i|]. \tag{7.4.11}$$

We bound each of these terms separately.

Bounding the Normalizing Constants

In this step we analyze the normalizing constants $n \mapsto u_n$, assuming (7.4.1), and use this analysis, as well as the induction hypothesis, to bound $\mathbb{E}[|U_i|]$.

When (7.4.1) holds, and since $\lim_{x \to \infty} \gamma(x) = 0$, there exists a constant $C_\varepsilon \geq 1$ such that, for all $n \geq 1$,

$$u_n \leq C_\varepsilon n^{1/\alpha + \varepsilon}. \tag{7.4.12}$$

This gives a first bound on $n \mapsto u_n$. We next substitute this bound into (7.4.1) and use that $x \mapsto x^{\gamma(x)}$ is non-decreasing together with $\gamma(x) = (\log x)^{\gamma - 1}$, to obtain

$$1 + o(1) = n[1 - F_X(u_n)] \geq n\left[u_n^{-\alpha - \gamma(u_n)}\right]$$
$$\geq n\left[u_n^{-\alpha} \exp\left\{\log\left(C_\varepsilon n^{1/\alpha + \varepsilon}\right)^\gamma\right\}\right]. \tag{7.4.13}$$

In turn, this implies that there exists a constant $c > 0$ such that

$$u_n \leq n^{1/\alpha} e^{c(\log n)^\gamma}. \tag{7.4.14}$$

In a similar way, we can show the matching lower bound $u_n \geq n^{1/\alpha} e^{-c(\log n)^\gamma}$. As a result,

$$\mathbb{E}[|U_i|] \leq c\alpha^i \mathbb{E}[(\log(Z_{i-1} \vee 1))^\gamma]. \tag{7.4.15}$$

Using the concavity of $x \mapsto x^\gamma$ for $\gamma \in [0, 1)$, as well as Jensen's inequality, we arrive at

$$\mathbb{E}[|U_i|] \leq c\alpha^i \left(\mathbb{E}[(\log(Z_{i-1} \vee 1))]\right)^\gamma = \alpha^{i(1-\gamma)} \mathbb{E}[M_{i-1}]^\gamma. \tag{7.4.16}$$

By (7.4.5) and (7.4.7), which implies that $\mathbb{E}[M_{i-1}] \leq K\kappa/(1 - \kappa)$, we arrive at

$$\mathbb{E}[|U_i|] \leq \alpha^{i(1-\gamma)} c\left(\frac{K\kappa}{1 - \kappa}\right)^\gamma, \tag{7.4.17}$$

so that (7.4.7) follows for U_i, with $\kappa = \alpha^{1-\gamma} < 1$ and K replaced by $c\left(\frac{K\kappa}{1-\kappa}\right)^\gamma$. An identical argument implies that

$$\mathbb{E}[\log(u_{Z_{i-1} \vee 1}^+ / u_{Z_{i-1} \vee 1}^-)] \leq \alpha^{i(1-\gamma)} c\left(\frac{K\kappa}{1 - \kappa}\right)^\gamma. \tag{7.4.18}$$

Logarithmic Moment of an Asymptotically Stable Random Variable

In this step, which is the most technical, we bound $\mathbb{E}[|V_i|]$. We note that by [V1, Theorem 2.33] and for Z_i quite large, the random variable $(Z_i \vee 1)/(u_{Z_{i-1} \vee 1})$ should be close to being a stable random variable. We first add and subtract a convenient additional term, and write

$$\mathbb{E}[|V_i|] = \left(\mathbb{E}[|V_i|] - 2\mathbb{E}[\log(u_{Z_{i-1} \vee 1}^+ / u_{Z_{i-1} \vee 1}^-)]\right)$$
$$+ 2\mathbb{E}[\log(u_{Z_{i-1} \vee 1}^+ / u_{Z_{i-1} \vee 1}^-)]. \tag{7.4.19}$$

The latter term is bounded in (7.4.18). For the first term, we will rely on stochastic domination results in terms of $1 - F_X^{\pm}$ in (7.4.2).

We make use of the relation to stable distributions by obtaining the bound

$$\mathbb{E}\big[|V_i|\big] - 2\mathbb{E}\big[\log(u_{Z_{i-1}\vee 1}^+/u_{Z_{i-1}\vee 1}^-)\big]\big)$$

$$\leq \alpha^i \sup_{m \geq 1} \Big\{ \mathbb{E}\big[|\log(S_m/u_m)|\big] - 2\log(u_m^+/u_m^-) \Big\}, \tag{7.4.20}$$

where $S_m = X_1 + \cdots + X_m$, and $(X_i)_{i=1}^m$ are iid copies of the offspring distribution X. Our aim is to prove that there exists a constant $C > 0$ such that, for all $m \geq 1$,

$$\mathbb{E}\big[|\log(S_m/u_m)|\big] - 2\log(u_m^+/u_m^-) \leq C. \tag{7.4.21}$$

In order to prove (7.4.21), we note that it suffices to obtain the bounds

$$\mathbb{E}\big[(\log(S_m/u_m))_+\big] - \log(u_m^+/u_m^-) \leq C_+, \tag{7.4.22}$$

$$\mathbb{E}\big[(\log(S_m/u_m))_-\big] - \log(u_m^+/u_m^-) \leq C_-, \tag{7.4.23}$$

where, for $x \in \mathbb{R}$, $x_+ = \max\{x, 0\}$ and $x_- = \max\{-x, 0\}$. Since $|x| = x_+ + x_-$, we then obtain (7.4.21) with $C = C_+ + C_-$.

We start by proving (7.4.23). Let $S_m^- = \sum_{i=1}^m X_i^-$, where X_i^- has distribution function F_X^-. We note that $(\log x)_- = \log(x^{-1} \vee 1)$, so that

$$\mathbb{E}\big[(\log(S_m/u_m))_-\big] - \log(u_m^+/u_m^-)$$

$$= \mathbb{E}\big[\log(u_m/(S_m \wedge u_m))\big] - \log(u_m^+/u_m^-)$$

$$\leq \mathbb{E}\big[\log(u_m/(S_m^- \wedge u_m^-))\big] - \log(u_m^+/u_m^-)$$

$$\leq \mathbb{E}\big[\log(u_m^-/(S_m^- \wedge u_m^-))\big], \tag{7.4.24}$$

where $x \wedge y = \min\{x, y\}$ and we have used (7.4.9). The random variables X_i^- have a regularly varying tail, so that we can use extreme-value theory in order to bound the above quantity.

The function $x \mapsto \log((u_m/(x \wedge u_m))$ is non-increasing, and, since $S_m^- \geq X_{(m)}^-$ where $X_{(m)}^- = \max_{1 \leq i \leq m} X_i^-$, we arrive at

$$\mathbb{E}\big[\log(u_m^-/(S_m^- \wedge u_m^-))\big] \leq \mathbb{E}\big[\log(u_m^-/(X_{(m)}^- \wedge u_m^-))\big]. \tag{7.4.25}$$

We next use that, for $x \geq 1$, $x \mapsto \log(x)$ is concave, so that, for every $s \geq 0$,

$$\mathbb{E}\big[\log(u_m^-/(X_{(m)}^- \wedge u_m^-))\big] = \frac{1}{s}\mathbb{E}\big[\log((u_m^-/(X_{(m)}^- \wedge u_m^-))^s)\big]$$

$$\leq \frac{1}{s}\log\Big(\mathbb{E}\big[(u_m^-/(X_{(m)}^- \wedge u_m^-))^s\big]\Big)$$

$$\leq \frac{1}{s} + \frac{1}{s}\log\Big((u_m^-)^s\mathbb{E}\big[(X_{(m)}^-)^{-s}\big]\Big), \tag{7.4.26}$$

where, in the last step, we have made use of the fact that $u_m^-/(x \wedge u_m^-) \leq 1 + u_m^-/x$.

Now rewrite $X_{(m)}^{-s}$ as $(-Y_{(m)})^s$, where $Y_j = -1/X_j^-$ and $Y_{(m)} = \max_{1 \leq j \leq m} Y_j$. Clearly, $Y_j \in [-1, 0]$ since $X_i^- \geq 1$, so that $\mathbb{E}[(-Y_1)^s] < \infty$. Also, $u_m^- Y_{(m)} = -u_m/X_{(m)}^-$

converges in distribution to $-E^{-1/\alpha}$, where E is exponential with mean 1, so it follows from (Pickands III, 1968, Theorem 2.1) that, as $m \to \infty$,

$$\mathbb{E}\left[(u_m^- Y_{(m)})^s\right] \to \mathbb{E}[E^{-s/\alpha}] < \infty, \tag{7.4.27}$$

when $s < \alpha$, as required.

We proceed by proving (7.4.22), which is a slight adaptation of the above argument. Let $S_m^+ = \sum_{i=1}^m X_i^+$, where X_i^+ has distribution function F_x^+. Now we make use of the fact that $(\log x)_+ = \log(x \vee 1) \le 1 + x$ for $x \ge 0$, so that we need to obtain the bound, using (7.4.9),

$$\mathbb{E}\left[\log\left(S_m \vee u_m/u_m\right)\right] - \log(u_m^+/u_m^-)$$
$$\le \mathbb{E}\left[\log\left(S_m \vee u_m/u_m^+\right)\right] \le \mathbb{E}\left[\log\left(S_m^+ \vee u_m^+/u_m^+\right)\right]$$
$$\le \frac{1}{s}\mathbb{E}\left[\log\left((S_m^+ \vee u_m^+/u_m^+))^s\right)\right] \le \frac{1}{s} + \frac{1}{s}\log\mathbb{E}\left[(S_m^+/u_m^+)^s\right]. \tag{7.4.28}$$

We again rely on the theory of random variables with a regularly varying tail, of which X_i^+ is an example. The discussion in (Hall, 1981, page 565 and Corollary 1) yields, for $s < \alpha$, $\mathbb{E}[S_m^s] = \mathbb{E}[|S_m|^s] \le 2^{s/2}\lambda_s(m)$, for some function $\lambda_s(m)$ depending on s, m, and F_X. Using the discussion in (Hall, 1981, p. 564), we have that $\lambda_s(m) \le C_s m^{s/\alpha} l(m^{1/\alpha})^s$, where $l(\cdot)$ is a slowly varying function. With some more effort, it can be shown that we can replace $m^{s/\alpha} l(m^{1/\alpha})^s$ by $(u_m^+)^s$, which gives

$$\log\mathbb{E}\left[\left(\frac{S_m^+}{u_m^+}\right)^s\right] \le C_s, \tag{7.4.29}$$

and which together with (7.4.28) proves (7.4.22) with $C_+ = 1/s + 2^{s/2}C_s/s$.

Completion of the Proof of Theorem 7.17 when $X \ge 1$

Combining (7.4.11) with (7.4.17) and (7.4.20), (7.4.21), we arrive at

$$\mathbb{E}[|Y_i|] \le 3c\alpha^{i(1-\gamma)}\left(\frac{K\kappa}{1-\kappa}\right)^\gamma + C\alpha^i \le K\kappa^i, \tag{7.4.30}$$

when we take $\kappa = \alpha^{1-\gamma}$ and take K to be sufficiently large, for example $K \ge 2C$ and $K \ge 6c\left(\frac{K\kappa}{1-\kappa}\right)^\gamma$. This completes the proof when the offspring distribution X satisfies $X \ge 1$. The proof for $X \ge 0$ uses [V1, Theorem 3.12]; see Exercise 7.21. $\square$

We finally state some properties of the almost sure limit Y of $(\alpha^k \log(Z_k \vee 1))_{k \ge 0}$, of which we omit a proof:

Theorem 7.18 (Limiting variable for infinite-mean branching processes) *Under the conditions of Theorem 7.17, with Y the almost sure limit of $\alpha^k \log(Z_k \wedge 1)$,*

$$\lim_{x \to \infty} \frac{\log \mathbb{P}(Y > x)}{x} = -1. \tag{7.4.31}$$

Theorem 7.18 can be understood from the fact that, by (7.4.3) and (7.4.4),

$$Y = \sum_{i=1}^\infty Y_i, \quad \text{where} \quad Y_1 = \alpha \log(Z_1 \vee 1). \tag{7.4.32}$$

By (7.4.1),

$$\mathbb{P}(Y_1 > x) = \mathbb{P}(Z_1 > e^{x^{1/\alpha}}) = e^{-x(1+o(1))}, \tag{7.4.33}$$

which shows that Y_1 satisfies (7.4.31). The equality in (7.4.32) together with (7.4.4) suggests that the tails of Y_1 are equal to those of Y, which heuristically explains (7.4.31). Exercise 7.23 gives an example where the limit Y is *exactly* exponential, so that the asymptotics in Theorem 7.18 is exact. The key behind this argument is the fact that Y in Theorem 7.17 satisfies the distributional equation

$$Y \stackrel{d}{=} \alpha \max_{i=1}^{X} Y_i, \tag{7.4.34}$$

where $(Y_i)_{i \geq 1}$ are iid copies of Y that are independent of X (see Exercise 7.22).

Intuition Behind Ultra-Small Distances using Theorem 7.17

We now use the results in Theorem 7.17 to explain how we can understand the ultra-small distances in the configuration model for $\tau \in (2, 3)$ in Theorem 7.11. Fix $G_n = \mathrm{CM}_n(\boldsymbol{d})$, and suppose that the degrees satisfy the assumptions in Theorem 7.11. First, note that

$$\mathbb{P}(\mathrm{dist}_{\mathrm{CM}_n(\boldsymbol{d})}(o_1, o_2) = k) = \frac{1}{n}\mathbb{E}[|\partial B_k^{(G_n)}(o_1)|]. \tag{7.4.35}$$

This suggests that $\mathrm{dist}_{\mathrm{CM}_n(\boldsymbol{d})}(o_1, o_2)$ should be closely related to the value of k (if any exists) such that $\mathbb{E}[|\partial B_k^{(G_n)}(o_1)|] = \Theta(n)$. By the branching-process approximation (as discussed precisely in Corollary 7.3),

$$|\partial B_k^{(G_n)}(o_1)| = Z_k^{(n;1)} \approx Z_k^{(1)} \approx \exp\left\{(\tau - 2)^{-k}Y^{(1)}(1 + o_{\mathbb{P}}(1))\right\} \tag{7.4.36}$$

(the first approximation being true for fairly small k, but not necessarily for large k), this suggests that $\mathrm{dist}_{\mathrm{CM}_n(\boldsymbol{d})}(o_1, o_2) \approx k_n$, where, by Theorem 7.17,

$$\Theta(n) = Z_{k_n}^{(1)} = \exp\left\{(\tau - 2)^{-k_n}Y^{(1)}(1 + o_{\mathbb{P}}(1))\right\}, \tag{7.4.37}$$

which in turn suggests that $\mathrm{dist}_{\mathrm{CM}_n(\boldsymbol{d})}(o_1, o_2) \approx \log\log n / |\log(\tau-2)|$. Of course, for such values the branching-process approximation may fail miserably, and in fact it does. This is exemplified by the fact that the rhs of (7.4.37) can become much larger than n, which is clearly impossible for $|\partial B_{k_n}^{(G_n)}(o_1)|$.

More intriguingly, we see that the proposed typical distances are a factor 2 too small compared with Theorem 7.2. The reason is that the double-exponential growth can clearly no longer be valid when $Z_k^{(n;1)}$ becomes too large, and thus, $Z_k^{(n;1)}$ must be far away from $Z_k^{(1)}$ in this regime. The whole problem is that we are using the branching-process approximation well beyond its "expiry date."

Hence, let us try this again but, rather than using it for *one* neighborhood, let us use the branching-process approximation from two sides. Now we rely on the statement that

$$\mathbb{P}(\mathrm{dist}_{\mathrm{CM}_n(\boldsymbol{d})}(o_1, o_2) \leq 2k) = \mathbb{P}(\partial B_k^{(G_n)}(o_1) \cap \partial B_k^{(G_n)}(o_2) \neq \varnothing). \tag{7.4.38}$$

Again using (7.4.36), we see that

$$\log|\partial B_k^{(G_n)}(o_i)| \approx (\tau - 2)^{-k}Y^{(i)}(1 + o_{\mathbb{P}}(1)), \qquad i \in \{1, 2\},$$

where $Y^{(1)}$ and $Y^{(2)}$ are independent copies of the random variable Y in Theorem 7.17. We see that $|\partial B_k^{(G_n)}(o_1)|$ and $|\partial B_k^{(G_n)}(o_2)|$ grow roughly at the same pace, and, in particular, we have $|\partial B_k^{(G_n)}(o_i)| = n^{\Theta(1)}$ roughly at the same time, namely, when $k \approx \log\log n / |\log(\tau - 2)|$. Thus, we conclude that

$$\mathrm{dist}_{\mathrm{CM}_n(d)}(o_1, o_2) \approx 2\log\log n / |\log(\tau - 2)|,$$

as rigorously proved in Theorem 7.2. We will see in more detail in Theorem 7.25 that the above growth from two sides does allow for better branching-process approximations.

An Insightful Lower Bound

Suppose, for simplicity, that the branching offspring random variable X satisfies $\mathbb{P}(X \geq 1) = 1$, so that the infinite-mean branching process does not die out. Then, we can construct a lower bound on the size of the kth generation by recursively taking the child having the largest offspring. More precisely, we let $Q_0 = 1$, and, given $Q_{k-1} = q_{k-1}$, we let Q_k denote the maximal offspring of the q_{k-1} individuals in the $(k-1)$th generation, so that Q_k counts individuals in the kth generation. We call $(Q_k)_{k \geq 0}$ the *maximum process*, while we could aptly call $(Z_k)_{k \geq 0}$ the *sum process*. We note that $Q_1 = Z_1$, while Q_2 is the maximum number of children of individuals in the first generation.

The process $(Q_k)_{k \geq 0}$ is a Markov chain (see Exercise 7.24), and

$$\mathbb{P}(Q_k > q \mid Q_{k-1} = q_{k-1}) = 1 - \mathbb{P}(Q_k \leq q \mid Q_{k-1} = q_{k-1}) = 1 - F_X(q)^{q_{k-1}}. \quad (7.4.39)$$

We can write, similarly to (7.4.5) and by Exercise 7.25,

$$\alpha^k \log Q_k = \sum_{i=1}^{k} \alpha^i \log(Q_i / Q_{i-1}^{1/\alpha}), \quad (7.4.40)$$

which suggests that

$$\alpha^k \log Q_k \xrightarrow{a.s.} Q_\infty. \quad (7.4.41)$$

Exercise 7.26 investigates the convergence in (7.4.41) in more detail.

The double-exponential growth of the maximum process for infinite-mean branching processes under conditions as in Theorem 7.17, together with the fact that the conditional dependence in such processes for configuration models is quite small, give the key intuition behind the upper bound in Theorem 7.11.

7.5 DIAMETER OF THE CONFIGURATION MODEL

We continue the discussion of distances in the configuration model by investigating the *diameter* in the model.

7.5.1 DIAMETER OF THE CONFIGURATION MODEL: LOGARITHMIC CASE

Before stating the main result, we introduce some notation. Recall that $G_D^\star(x)$ is the probability generating function of $p^\star = (p_k^\star)_{k \geq 0}$ defined in (4.2.2) (note also (4.3.56)). We recall that ξ is the extinction probability of the branching process with offspring distribution $p^\star$, and further define

$$\mu = \frac{d}{dz} G^\star_D(z)\Big|_{z=\xi} = \sum_{k \geq 0} k \xi^{k-1} p^\star_k = \sum_{k \geq 1} k(k+1)\xi^{k-1} p_{k+1}/\mathbb{E}[D]. \qquad (7.5.1)$$

When $\xi < 1$, we also have that $\mu < 1$. Then, the main result is as follows:

Theorem 7.19 (Diameter of the configuration model) *Consider* $\mathrm{CM}_n(\boldsymbol{d})$ *where the degrees* $\boldsymbol{d} = (d_v)_{v \in [n]}$ *satisfy Conditions 1.7(a),(b). Assume that* $\mathbb{E}[D_n^2] \to \mathbb{E}[D^2] \in (0, \infty) \cup \{\infty\}$, *where* $\nu = \mathbb{E}[D(D-1)]/\mathbb{E}[D] > 1$. *Assume further that* $n_1 = 0$ *when* $p_1 = 0$, *and that* $n_2 = 0$ *when* $p_2 = 0$. *Then,*

$$\frac{\mathrm{diam}(\mathrm{CM}_n(\boldsymbol{d}))}{\log n} \xrightarrow{\;\mathbb{P}\;} \frac{1}{\log \nu} + 2\frac{\mathbb{1}_{\{p_1 > 0\}}}{|\log \mu|} + \frac{\mathbb{1}_{\{p_1 = 0, p_2 > 0\}}}{|\log p^\star_1|}. \qquad (7.5.2)$$

For finite-variance degrees, we note that, by Theorems 7.1 and 7.19, the diameter of the configuration model is *strictly* larger than the typical distance, except when $p_1 = p_2 = 0$. In the latter case, the degrees are at least 3, so that thin lines, consisting of degree-2 vertices connected to each other, are not possible and the configuration model is whp *connected* (recall Theorem 4.24). By [V1, Corollary 7.17] (recall also the discussion around (1.3.41)), Theorem 7.19 also applies to uniform random graphs with a given degree sequence, when the degrees have finite second moment, as in the examples below.

We also remark that Theorem 7.19 applies not only to the finite-variance case but also to the finite-mean and infinite-variance case. In the latter case, the diameter is of order $\log n$ unless $p_1 = p_2 = 0$, in which case Theorem 7.19 implies that the diameter is $o_{\mathbb{P}}(\log n)$. We will discuss the latter case in more detail in Theorem 7.20 below.

Random Regular Graphs

Let r be the degree of the random regular graph, where $r \geq 3$. By [V1, Corollary 7.17] (recall also (1.3.41)), the diameter of a random regular r-graph has whp the same asymptotics as the diameter of $\mathrm{CM}_n(\boldsymbol{d})$, where $d_v = r$ for all $v \in [n]$. Thus, $p_r = 1$ and $p_k = 0$ for any $k \neq r$. We assume that nr is even, so that the degree sequence is graphical. It is not hard to see that all the assumptions of Theorem 7.19 are satisfied. Moreover, $\nu = r - 1$. When $r \geq 3$, we thus obtain that

$$\frac{\mathrm{diam}(\mathrm{CM}_n(\boldsymbol{d}))}{\log n} \xrightarrow{\;\mathbb{P}\;} \frac{1}{\log(r-1)}. \qquad (7.5.3)$$

When $r = 2$, on the other hand, the graph is *critical*, so that there is no unique giant component. Since $\nu = 1$, we have that $\mathrm{diam}(\mathrm{CM}_n(\boldsymbol{d})) \gg \log n$. This is quite reasonable, since the graph consists of a collection of *cycles*. The diameter of such a graph is equal to half the longest cycle. Exercises 7.27 and 7.28 explore the length of the longest cycle in a random 2-regular graph, and the consequence of having a long cycle on the diameter.

Erdős–Rényi Random Graph

We next study the diameter of $\mathrm{ER}_n(\lambda/n)$. We let $\lambda > 1$. By [V1, Theorem 5.12], Conditions 1.7(a),(b) hold with $p_k = e^{-\lambda}\lambda^k/k!$. Also, μ in (7.5.1) equals $\mu = \mu_\lambda$, the dual parameter in [V1, (3.6.6)] (see Exercise 7.29).

Again, we make essential use of [V1, Theorem 7.18] (recall also Theorem 1.4 and the discussion below (1.3.29)), which relates the configuration model and the generalized random graph. We note that $\mathrm{ER}_n(\lambda/n)$ is the same as $\mathrm{GRG}_n(\boldsymbol{w})$, where $w_v = n\lambda/(n - \lambda)$ for all $v \in [n]$ (recall [V1, Exercise 6.1]).

Clearly, $\boldsymbol{w} = (n\lambda/(n - \lambda))_{v \in [n]}$ satisfies Conditions 1.1(a)–(c), so that the degree sequence of $\mathrm{ER}_n(\lambda/n)$ also satisfies Conditions 1.7(a)–(c), where the convergence holds in probability (recall [V1, Theorem 5.12]). From the above identifications and using [V1, Theorem 7.18], we find that

$$\frac{\mathrm{diam}(\mathrm{ER}_n(\lambda/n))}{\log n} \xrightarrow{\mathbb{P}} \frac{1}{\log \lambda} + \frac{2}{|\log \mu_\lambda|}. \tag{7.5.4}$$

This identifies the diameter of the Erdős–Rényi random graph, for which Theorem 7.19 agrees with Theorem 6.26. Exercise 7.30 investigates the diameter of $\mathrm{GRG}_n(\boldsymbol{w})$.

Sketch of the Proof of Theorem 7.19

We recognize the term $\log_\nu n$ as corresponding to the typical distances in the graph (recall Theorem 7.1). Except when $p_1 = p_2 = 0$, there is a correction to this term, which is due to deep, yet thin, neighborhoods. These thin neighborhoods act as "traps," from which it takes much longer than usual to escape. The diameter is typically realized as the distance between two traps. The deepest traps turn out to be $\log n/|\log \mu|$ deep when $p_1 > 0$ and $\log n/[2|\log p_1^\star|]$ deep when $p_1 = 0$ and $p_2 > 0$. Thus, let us look at traps in the configuration model in more detail.

When $p_1 > 0$, yet $\nu > 1$, the outside of the giant component is a subcritical graph, which locally looks like a subcritical branching process, owing to the duality principle. Thus, in the complement of the giant, there exist trees of size up to $\Theta(\log n)$, and the maximal diameter of such trees is close to $\log n/|\log \mu|$, where now μ is the dual parameter. Of course, these are not the components of maximal diameter, but they turn out to be closely related, as we discuss next.

In the supercritical case, we can view the giant as consisting of the 2-core and all the trees that hang off it. The 2-*core* is the maximal subgraph of the giant for which every vertex has degree at least 2. It turns out that the trees that hang off the 2-core have a very similar law as the subcritical trees outside of the giant. Therefore, the maximal height of such trees is again of the order $\log n/|\log \mu|$. Asymptotically, the diameter in $\mathrm{CM}_n(\boldsymbol{d})$ is formed between pairs of vertices for which the diameter of the tree that they are in when the 2-core is cut away is largest, thus giving rise to the $\log n/|\log \mu|$ contribution, whereas the distances between the two vertices closest to them in the 2-core is close to $\log_\nu n$. This can be understood by realizing that for *most* vertices these trees have a bounded height, so that the typical distances in the 2-core and in the giant are close for most vertices.

The above intuition does not give the right answer when $p_1 = 0$ yet $p_2 > 0$. Assume that $n_1 = 0$. Then, the giant and the 2-core are close to each other, so that the above argument does not apply. Instead, it turns out that the diameter is realized by the diameter of the 2-core, which is close to $\log n[1/\log \nu + 1/|\log p_1^\star|]$. Indeed, the 2-core contains long paths of degree-2 vertices. The longest such paths has length that is close to $\log n/|\log p_1^\star|$. Therefore, the largest distance between a vertex inside this long path and the ends of the path is close to $\log n/[2|\log p_1^\star|]$. Now, it turns out that pairs of such vertices realize the asymptotic diameter, which explains why the diameter is close to $\log n[1/\log \nu + 1/|\log p_1^\star|]$.

Finally, we discuss what happens when $p_1 = p_2 = 0$. In this case, the assumption in Theorem 7.19 implies that $n_1 = n_2 = 0$, so that $d_{\min} \geq 3$. Then, $\mathrm{CM}_n(\boldsymbol{d})$ is whp connected (recall Theorem 4.24), and the 2-core is the graph itself. Also, there cannot be any long thin parts of the giant, since every vertex has degree at least 3 so that local neighborhoods grow exponentially with overwhelming probability. Therefore, the graph distances and the diameter have the same asymptotics, as proved by Theorem 7.19 when $d_{\min} \geq 3$.

The above case-distinctions explain the intuition behind Theorem 7.19. This intuition is far from a proof; see Section 7.7 for references.

7.5.2 Diameter of Infinite-Variance Configuration Models: Log Log Case

We next use the diameter of the core in Theorem 7.13 to study the diameter of $\mathrm{CM}_n(\boldsymbol{d})$ when $\tau \in (2,3)$. Note that the diameter is equal to a positive constant times $\log n$, by Theorem 7.19 when $p_1 + p_2 > 0$. Therefore, we turn to the case where $p_1 = p_2 = 0$ and $\tau \in (2,3)$. When $d_{\min} \geq 3$, we know by Theorem 4.24 that $\mathrm{CM}_n(\boldsymbol{d})$ is whp connected. The main result about the diameter in this case is as follows:

Theorem 7.20 (Diameter of $\mathrm{CM}_n(\boldsymbol{d})$ for $\tau \in (2,3)$) *Consider* $\mathrm{CM}_n(\boldsymbol{d})$ *where the degrees* $\boldsymbol{d} = (d_v)_{v \in [n]}$ *satisfy Conditions 1.7(a),(b) and (7.2.2). Assume further that* $d_{\min} = \min_{v \in [n]} d_v \geq 3$ *and* $p_{d_{\min}} = \mathbb{P}(D = d_{\min}) > 0$. *Then,*

$$\frac{\mathrm{diam}(\mathrm{CM}_n(\boldsymbol{d}))}{\log \log n} \overset{\mathbb{P}}{\longrightarrow} \frac{2}{|\log(\tau - 2)|} + \frac{2}{\log(d_{\min} - 1)}. \qquad (7.5.5)$$

When comparing Theorem 7.20 with Theorem 7.2, we see that for $d_{\min} \geq 3$ the diameter is of the same doubly logarithmic order as the typical distance, but the constant differs.

As we have already seen in the sketch of the proof of Theorem 7.19, the value of the diameter is due to pairs of vertices that have small or thin local neighborhoods. Such neighborhoods act as local "traps." When $p_1 + p_2 > 0$, these thin neighborhoods are close to lines, and they can be of logarithmic length in n. When $d_{\min} \geq 3$, however, we see that even the thinnest possible neighborhoods are bounded below by a binary tree. Indeed, by assumption, there is a positive proportion of vertices of degree $d_{\min}$. As a result, below we will see that the expected number of vertices whose $(1 - \varepsilon) \log \log n / \log(d_{\min} - 1)$ neighborhood contains only minimal-degree vertices tends to infinity (see Lemma 7.22). The minimal path between two such vertices then consists of three parts: the two paths from the two vertices to leave their minimally connected neighborhood, and the path between the boundaries of these minimally connected neighborhoods. The latter path has length $2 \log \log n / |\log(\tau - 2)|$, as in Theorem 7.2. This explains why there is an extra term $2 \log \log n / \log(d_{\min} - 1)$ in Theorem 7.20 as compared with Theorem 7.2.

We next sketch the proof of Theorem 7.20. We start with the lower bound, which is the easier part.

Proof Sketch for the Lower Bound on the Diameter

Fix $G_n = \mathrm{CM}_n(\boldsymbol{d})$ under the conditions in Theorem 7.20, for which we introduce the notion of minimally k-connected vertices:

Definition 7.21 (Minimally k-connected vertices)   We call a vertex v *minimally k-connected* when all $i \in B_k^{(G_n)}(v)$ satisfy $d_i = d_{\min}$. Let M_k denote the number of minimally k-connected vertices.   ◄

To prove the lower bound on the diameter, we show that $M_k \overset{\mathbb{P}}{\longrightarrow} \infty$ for $k = \lceil (1 - \varepsilon) \log \log n / \log (d_{\min} - 1) \rceil$. Following this, we show that two minimally k-connected vertices o_1, o_2 are such that whp the distance between $\partial B_k^{(G_n)}(o_1)$ and $\partial B_k^{(G_n)}(o_2)$ is at least $2 \log \log n / |\log (\tau - 2)|$:

Lemma 7.22 (Moments of number of minimally k-connected vertices)   *Let* $\mathrm{CM}_n(\mathbf{d})$ *satisfy* $d_{\min} \geq 3$, $n_{d_{\min}} > d_{\min}(d_{\min} - 1)^{k-1}$. *For* $k_n \leq (1 - \varepsilon) \log \log n / \log (d_{\min} - 1)$,

$$\mathbb{E}[M_{k_n}] \to \infty, \quad \text{and} \quad \frac{M_{k_n}}{\mathbb{E}[M_{k_n}]} \overset{\mathbb{P}}{\longrightarrow} 1. \tag{7.5.6}$$

We leave the proof of Lemma 7.22 to Exercises 7.31–7.33. To complete the proof of the lower bound on the diameter, we fix $\varepsilon > 0$ sufficiently small and take

$$k_n^\star = \left\lceil (1 - \varepsilon) \frac{\log \log n}{\log (d_{\min} - 1)} \right\rceil.$$

Clearly,

$$d_{\min}(d_{\min} - 1)^{k_n^\star - 1} \leq (\log n)^{1 - \varepsilon} \leq \ell_n / 8, \tag{7.5.7}$$

so that, in particular, we may use Lemma 7.22.

We conclude that, whp, $M_{k_n^\star} \geq n^{1 - o(1)}$. Since each minimally $k_n^\star$-connected vertex uses up at most

$$1 + \sum_{l=1}^{k_n^\star} d_{\min}(d_{\min} - 1)^{l-1} = n^{o(1)} \tag{7.5.8}$$

vertices of degree $d_{\min}$, whp there must be *at least two* minimally $k_n^\star$-connected vertices whose $k_n^\star$-neighborhoods are disjoint. We fix two such vertices and denote them by v_1 and v_2. We note that v_1 and v_2 have precisely $d_{\min}(d_{\min} - 1)^{k_n^\star - 1}$ unpaired half-edges in $\partial B_{k_n^\star}^{(G_n)}(v_1)$ and $\partial B_{k_n^\star}^{(G_n)}(v_2)$. Let $\mathcal{A}_{12}$ denote the event that v_1, v_2 are minimally $k_n^\star$-connected and their $k_n^\star$-neighborhoods are disjoint.

Conditional on $\mathcal{A}_{12}$, the random graph found by collapsing the half-edges in $\partial B_{k_n^\star}^{(G_n)}(v_1)$ to a single vertex a and the half-edges in $\partial B_{k_n^\star}^{(G_n)}(v_1)$ to a single vertex b is a configuration model on the vertex set $\{a, b\} \cup [n] \setminus (B_{k_n^\star}^{(G_n)}(v_1) \cup B_{k_n^\star}^{(G_n)}(v_2))$, having degrees $\mathbf{d}'$ given by $d_a' = d_b' = d_{\min}(d_{\min} - 1)^{k_n^\star - 1}$ and $d_i' = d_i$ for every $i \in [n] \setminus (B_{k_n^\star}^{(G_n)}(v_1) \cup B_{k_n^\star}^{(G_n)}(v_2))$.

By the truncated first-moment method on paths, performed in the proof of Theorem 7.8 (recall (7.3.22)), it follows that, for any $\varepsilon > 0$,

$$\mathbb{P}\left(\mathrm{dist}_{\mathrm{CM}_n(\mathbf{d})}(\partial B_{k_n^\star}^{(G_n)}(v_1), \partial B_{k_n^\star}^{(G_n)}(v_2)) \leq (1 - \varepsilon) \frac{2 \log \log n}{|\log (\tau - 2)|} \,\Big|\, \mathcal{A}_{12} \right) = o(1). \tag{7.5.9}$$

Therefore, whp,

$$
\mathrm{diam}(\mathrm{CM}_n(\boldsymbol{d})) \geq (1 - \varepsilon)\frac{2\log\log n}{|\log(\tau - 2)|} + 2k_n^\star
$$

$$
= (1 - \varepsilon)\log\log n\Big[\frac{2}{|\log(\tau - 2)|} + \frac{2}{\log(d_{\min} - 1)}\Big]. \qquad (7.5.10)
$$

Since $\varepsilon > 0$ is arbitrary, this suggests the lower bound in Theorem 7.20.

Proof Sketch for the Upper Bound on the Diameter

We finally sketch the proof of the upper bound on the diameter. We aim to prove that, with $k_n^\star = \lceil(1 + \varepsilon)\log\log n[2/|\log(\tau - 2)| + 2/\log(d_{\min} - 1)]\rceil$,

$$
\mathbb{P}(\exists v_1, v_2 \in [n]\colon \mathrm{dist}_{\mathrm{CM}_n(\boldsymbol{d})}(v_1, v_2) \geq k_n^\star) = o(1). \qquad (7.5.11)
$$

We already know that whp $\mathrm{diam}(\mathrm{Core}_n) \leq 2\log\log n/|\log(\tau - 2)| + 1$, by Theorem 7.13, and we assume this from now on. Fix $v_1, v_2 \in [n]$. Then (7.5.11) follows if we can show that

$$
\mathbb{P}(\exists v\colon \mathrm{dist}_{\mathrm{CM}_n(\boldsymbol{d})}(v, \mathrm{Core}_n) \geq (1 + \varepsilon)\log\log n/\log(d_{\min} - 1)) = o(1). \qquad (7.5.12)
$$

For this, we argue that, uniformly in $v \in [n]$,

$$
\mathbb{P}(\mathrm{dist}_{\mathrm{CM}_n(\boldsymbol{d})}(v, \mathrm{Core}_n) \geq (1 + \varepsilon)\log\log n/\log(d_{\min} - 1)) = o(1/n), \qquad (7.5.13)
$$

which would prove (7.5.11).

To prove (7.5.13), it is convenient to explore the neighborhood of the vertex v by pairing up only the first $d_{\min}$ half-edges incident to v and the $d_{\min} - 1$ half-edges incident to any other vertex appearing in the neighborhood. We call this exploration graph the *minimal k-exploration graph*.

One can show that it is quite unlikely that there are many cycles within this exploration graph, so that it is actually close to a tree. Therefore, the number of half-edges on the boundary of this k-exploration graph with $k = k_n^\star = \lceil(1 + \varepsilon/2)\log\log n/\log(d_{\min} - 1)\rceil$ is close to

$$
(d_{\min} - 1)^{k_n^\star} \approx (\log n)^{1+\varepsilon/2}. \qquad (7.5.14)
$$

This is large, but not extremely large. However, one of these vertices is bound to have a quite large degree and thus, from this vertex, it is quite likely that we can connect to Core_n quickly, meaning in $o(\log\log n)$ steps. The main ingredient in the upper bound is the statement that, whp, the minimal $k_n^\star$-exploration tree connects to Core_n whp. We omit further details.

7.6 RELATED RESULTS ON DISTANCES IN CONFIGURATION MODELS

Distances for Infinite-Mean Degrees

We assume that there exist $\tau \in (1, 2)$ and $c > 0$ such that

$$
\lim_{x \to \infty} x^{\tau - 1}[1 - F](x) = c. \qquad (7.6.1)
$$

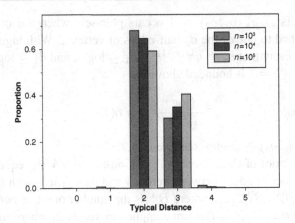

Figure 7.4 Typical distances for $\tau = 1.8$ and $n = 10^3, 10^4, 10^5$.

We will now study the configuration model $\mathrm{CM}_n(\boldsymbol{d})$ where the degrees $\boldsymbol{d} = (d_v)_{v \in [n]}$ are an iid sequence of random variables with distribution F satisfying (7.6.1).

We will make heavy use of the results and notation used in [V1, Theorem 7.23], which we first recall: the random probability distribution $P = (P_i)_{i \geq 1}$ is given by

$$P_i = Z_i/Z, \tag{7.6.2}$$

where $Z_i = \Gamma_i^{-1/(\tau-1)}$ and $\Gamma_i = \sum_{j=1}^i E_i$ with $(E_i)_{i \geq 1}$ an iid sequence of exponential random variables with parameter 1 and where $Z = \sum_{i \geq 1} \Gamma_i^{-1/(\tau-1)}$. The latter is finite almost surely, since $1/(\tau-1) > 1$ for $\tau \in (1,2)$ (see Exercise 7.34).

Recall further that $M_{P,k}$ is a multinomial distribution with parameters k and (random) probabilities $P = (P_i)_{i \geq 1}$. Thus, $M_{P,k} = (B_1, B_2, \ldots)$, where, conditional on $P = (P_i)_{i \geq 1}$, B_i is the number of outcomes i in k independent trials such that each outcome is equal to i with probability P_i.

In [V1, Theorem 7.23], the random variable M_{P,D_1} appears, where D_1 is independent of $P = (P_i)_{i \geq 1}$. We let $M_{P,D_1}^{(1)}$ and $M_{P,D_2}^{(2)}$ be two such random variables that are conditionally independent given $P = (P_i)_{i \geq 1}$ (but share the *same* $P = (P_i)_{i \geq 1}$ sequence). In terms of this notation, the main result on distances in $\mathrm{CM}_n(\boldsymbol{d})$ when the degrees have infinite mean is the following:

Theorem 7.23 (Distances in $\mathrm{CM}_n(\boldsymbol{d})$ with iid infinite mean degrees) *Consider* $\mathrm{CM}_n(\boldsymbol{d})$ *where the degrees* $\boldsymbol{d} = (d_v)_{v \in [n]}$ *are a sequence of iid copies of D satisfying (7.6.1) for some* $\tau \in (1,2)$. *Then, with o_1, o_2 chosen independently and uar from $[n]$,*

$$\lim_{n \to \infty} \mathbb{P}(\mathrm{dist}_{\mathrm{CM}_n(d)}(o_1, o_2) = 2) = 1 - \lim_{n \to \infty} \mathbb{P}(\mathrm{dist}_{\mathrm{CM}_n(d)}(o_1, o_2) = 3) = p_F \in (0,1).$$

The probability p_F can be identified as the probability that an outcome occurs both in $M_{P,D_1}^{(1)}$ and $M_{P,D_2}^{(2)}$, where D_1 and D_2 are two iid copies of D.

Sketch of proof of Theorem 7.23. By exchangeability, it suffices to consider $o_1 = 1$ and $o_2 = 2$. First, whp, we have both $d_1 \leq \log n$ and $d_2 \leq \log n$.

The event that $\mathrm{dist}_{\mathrm{CM}_n(d)}(o_1, o_2) = 1$ occurs precisely when one of the d_1 half-edges of vertex 1 is attached to one of the d_2 half-edges of vertex 2. With high probability, $\ell_n \geq n^{1/(\tau-1)-\varepsilon}$. On the event that $\ell_n \geq n^{1/(\tau-1)-\varepsilon}$, $d_1 \leq \log n$, and $d_2 \leq \log n$, the probability that $\mathrm{dist}_{\mathrm{CM}_n(d)}(o_1, o_2) = 1$ is bounded above by

$$\frac{(\log n)^2}{n^{1/(\tau-1)-\varepsilon}} = o(1), \tag{7.6.3}$$

so that $\mathrm{dist}_{\mathrm{CM}_n(d)}(o_1, o_2) \geq 2$ whp (see Exercise 7.35).

We note that the proof of [V1, Theorem 7.23] implies that $M_{P,d_1}^{(1)}$ equals the limit in distribution of the number of edges between vertex 1 and the vertices with the largest degrees. Indeed, $M_{P,d_1}^{(1)} = (B_1^{(1)}, B_2^{(1)}, \ldots)$, where $B_i^{(1)}$ is the number of edges between vertex 1 and the vertex with degree $d_{(n+1-i)}$. The same applies to vertex 2, where the limit is denoted by $M_{P,d_2}^{(2)}$ (again with the same P). As a result, the graph distance between vertices 1 and 2 equals 2 precisely when $M_{P,d_1}^{(1)}$ and $M_{P,d_2}^{(2)}$ have an identical outcome (meaning that $B_i^{(1)} > 0$ and $B_i^{(2)} > 0$ for some $i \geq 1$). We are left with proving that the graph distance between vertices 1 and 2 is whp bounded by 3.

By [V1, (2.6.17)], we have that $P_k k^{1/(\tau-1)} \xrightarrow{\mathbb{P}} 1/Z$ as $k \to \infty$. Thus, when K is large, the probability that vertex 1 is not connected to any of the vertices corresponding to $(d_{(n+1-i)})_{i \in [K]}$ converges to 0 when first $n \to \infty$ followed by $K \to \infty$.

Let $\mathbb{P}_n$ denote the conditional probability given the degrees $(d_v)_{v \in [n]}$. For $i \in [n]$, we let v_i be the vertex corresponding to the ith largest degree $d_{(n+1-i)}$. By Lemma 7.12,

$$\mathbb{P}_n(v_i \text{ not directly connected to } v_j) \leq \exp\left\{-d_{(n+1-i)}d_{(n+1-j)}/2\ell_n\right\}. \tag{7.6.4}$$

Moreover, $d_{(n+1-i)}, d_{(n+1-j)} \geq n^{1/(\tau-1)-\varepsilon}$ whp for n sufficiently large and any $\varepsilon > 0$, while whp $\ell_n \leq n^{1/(\tau-1)+\varepsilon}$. As a result, whp,

$$\mathbb{P}_n(v_i \text{ not directly connected to } v_j) \leq \exp\left\{-n^{1/(\tau-1)-3\varepsilon}\right\} = o_{\mathbb{P}}(1), \tag{7.6.5}$$

when $\varepsilon > 0$ is sufficiently small. Therefore, for fixed K and for every $i, j \in [K]$, the vertices v_i and v_j are whp neighbors. This implies that the vertices corresponding to the highest degrees whp form a *complete graph*.

We have already concluded that 1 is whp connected to v_i for some $i \leq K$. In the same way, we conclude that vertex 2 is whp connected to v_j for some $j \leq K$. Since v_i is whp directly connected to v_j, we conclude that

$$\mathbb{P}_n\left(\mathrm{dist}_{\mathrm{CM}_n(d)}(o_1, o_2) \leq 3\right) = 1 - o(1). \tag{7.6.6}$$

This completes the sketch of the proof of Theorem 7.23.

Fluctuation of Distances for Finite-Variance Degrees

We continue to study the *fluctuations* of the distances in the configuration model, starting with the case where the degrees have finite variance. We need a limit result from branching-process theory before we can identify the limiting random variables $(R_a)_{a \in (-1,0]}$.

Recall that $(Z_k)_{k \geq 0}$ denotes the unimodular branching process with root-offspring distribution $(p_k)_{k \geq 1}$, with $p_k = \mathbb{P}(D = k)$. The process $(Z_k/\mathbb{E}[D]\nu^{k-1})_{k \geq 1}$ is a martingale with expectation equal to 1, and consequently converges almost surely to a limit (see, e.g., [V1, Theorem 2.24]):

$$\lim_{n\to\infty} \frac{Z_n}{\mathbb{E}[D]\nu^{n-1}} = Y \quad \text{almost surely.} \tag{7.6.7}$$

In the theorem below we need two independent copies $Y^{(1)}$ and $Y^{(2)}$ of Y:

Theorem 7.24 (Limit law for typical distance in $\mathrm{CM}_n(\boldsymbol{d})$) *Consider $\mathrm{CM}_n(\boldsymbol{d})$ where the degrees $\boldsymbol{d} = (d_v)_{v\in[n]}$ are a sequence of iid copies of D satisfying that there exist $\tau > 3$ and $c < \infty$ such that, for all $x \geq 1$,*

$$[1 - F](x) \leq cx^{-(\tau-1)}. \tag{7.6.8}$$

Let $\nu = \mathbb{E}[D(D-1)]/\mathbb{E}[D] > 1$. For $k \geq 1$, let $a_k = \lfloor \log_\nu k \rfloor - \log_\nu k \in (-1, 0]$. Then, there exist random variables $(R_a)_{a\in(-1,0]}$ such that, as $n \to \infty$ and for all $k \in \mathbb{Z}$, with o_1, o_2 chosen independently and uar from $[n]$,

$$\mathbb{P}\big(\mathrm{dist}_{\mathrm{CM}_n(\boldsymbol{d})}(o_1, o_2) - \lfloor \log_\nu n \rfloor = k \mid \mathrm{dist}_{\mathrm{CM}_n(\boldsymbol{d})}(o_1, o_2) < \infty\big) \tag{7.6.9}$$
$$= \mathbb{P}(R_{a_n} = k) + o(1).$$

The random variables $(R_a)_{a\in(-1,0]}$ can be identified by

$$\mathbb{P}(R_a > k) = \mathbb{E}\big[\exp\{-\kappa\nu^{a+k}Y^{(1)}Y^{(2)}\}\big|Y^{(1)}Y^{(2)} > 0\big], \tag{7.6.10}$$

where $Y^{(1)}$ and $Y^{(2)}$ are independent limit copies of Y in (7.6.7) and $\kappa = \mathbb{E}[D]/(\nu - 1)$.

In words, Theorem 7.24 states that, for $\tau > 3$, the graph distance $\mathrm{dist}_{\mathrm{CM}_n(\boldsymbol{d})}(o_1, o_2)$ between two randomly chosen connected vertices grows as $\log_\nu n$, where n is the size of the graph, and that the fluctuations around this leading asymptotics *remain uniformly bounded* in n. Exercise 7.37 shows that $\mathrm{dist}_{\mathrm{CM}_n(\boldsymbol{d})}(o_1, o_2) - \lfloor \log_\nu n \rfloor$ converges in distribution along appropriately chosen subsequences.

The law of R_a is involved, and in most cases cannot be computed exactly. The reason for this is the fact that the random variables $Y^{(1)}$ and $Y^{(2)}$ that appear in its statement are hard to compute explicitly (see also [V1, Chapter 3]).

Let us give two examples where the law of Y is known. The first example is the r-regular random graphs, for which all degrees in the graph are equal to some $r \geq 3$. In this case, $\mathbb{E}[D] = r, \nu = r - 1$, and $Y = 1$ almost surely. In particular, $\mathbb{P}(\mathrm{dist}_{\mathrm{CM}_n(\boldsymbol{d})}(o_1, o_2) < \infty) = 1 + o(1)$. Therefore,

$$\mathbb{P}(R_a > k) = \exp\Big\{ -\frac{r}{r-2}(r-1)^{a+k}\Big\}, \tag{7.6.11}$$

and $\mathrm{dist}_{\mathrm{CM}_n(\boldsymbol{d})}(o_1, o_2)$ is asymptotically equal to $\log_{r-1} n$. Note that the distribution R_a depends explicitly on a, so that $\mathrm{dist}_{\mathrm{CM}_n(\boldsymbol{d})}(o_1, o_2) - \lfloor \log_\nu n \rfloor$ does *not* converge in distribution (see also Exercise 7.38).

The second example for which Y can be explicitly computed is where $p^\star$ is the probability mass function of a geometric random variable, in which case the branching-process generation sizes with offspring $p^\star$, conditioned to be positive, converge to an exponential random variable with parameter 1. This example corresponds to

$$p_j^\star = p(1-p)^{j-1}, \quad \text{so that} \quad p_j = \frac{1}{jc_p}p(1-p)^{j-2}, \quad \forall j \geq 1, \tag{7.6.12}$$

and c_p is a normalization constant. For $p > \frac{1}{2}$, Y has the same law as the sum of D_1 copies of a random variable that is equal to 0 with probability $(1 - p)/p$ and an exponential random variable with parameter 1 with probability $(2p - 1)/p$. Even in this simple case the computation of the exact law of R_a is non-trivial.

Fluctuation of Distances for Infinite-Variance Degrees

Next, we study the fluctuations of typical distances in $CM_n(d)$ in the setting where the degrees are iid and satisfy that there exist $\tau \in (2, 3)$, $\gamma \in [0, 1)$, and $C < \infty$ such that

$$x^{-\tau+1-C(\log x)^{\gamma-1}} \leq 1 - F(x) \leq x^{-\tau+1+C(\log x)^{\gamma-1}}, \qquad \text{for large } x. \qquad (7.6.13)$$

The condition in (7.6.13) is such that the results in Theorem 7.17 apply. Then we can identify the fluctuations of the typical distance in $CM_n(d)$ as follows:

Theorem 7.25 (Fluctuations of the graph distance $CM_n(d)$ for infinite-variance degrees) *Consider $CM_n(d)$ where the degrees $d = (d_v)_{v \in [n]}$ are a sequence of iid copies of D satisfying (7.6.13) for some $\tau \in (2, 3)$. Then there exist random variables $(R_a)_{a \in (-1,0]}$ such that, as $n \to \infty$ and for all $l \in \mathbb{Z}$, with o_1, o_2 chosen independently and uar from $[n]$,*

$$\mathbb{P}\left(\mathrm{dist}_{CM_n(d)}(o_1, o_2) = 2\left\lfloor \frac{\log \log n}{|\log(\tau - 2)|} \right\rfloor + l \,\Big|\, \mathrm{dist}_{CM_n(d)}(o_1, o_2) < \infty\right)$$

$$= \mathbb{P}(R_{a_n} = l) + o(1), \qquad (7.6.14)$$

where

$$a_n = \left\lfloor \frac{\log \log n}{|\log(\tau - 2)|} \right\rfloor - \frac{\log \log n}{|\log(\tau - 2)|} \in (-1, 0]. \qquad (7.6.15)$$

Here, the random variables $(R_a)_{a \in (-1,0]}$ are given by

$$\mathbb{P}(R_a > l) = \mathbb{P}\left(\min_{s \in \mathbb{Z}} \left[(\tau - 2)^{-s} Y^{(1)} + (\tau - 2)^{s - c_l} Y^{(2)} \right] \leq (\tau - 2)^{\lceil l/2 \rceil + a} \,\Big|\, Y_1 Y_2 > 0 \right),$$

where $c_l = 1$ if l is even, and zero otherwise, and $Y^{(1)}, Y^{(2)}$ are two independent copies of the limit random variable Y in Theorem 7.17.

In words, Theorem 7.25 states that for $\tau \in (2, 3)$ the graph distance $\mathrm{dist}_{CM_n(d)}(o_1, o_2)$ between two randomly chosen connected vertices grows proportionally to the log log of the size of the graph and that the fluctuations around this mean remain uniformly bounded in n.

We next discuss an extension, obtained by possibly truncating the degree distribution. In order to state the result, we make the following assumption that makes (7.6.13) more precise:

Condition 7.26 (Truncated infinite-variance degrees) *Fix $\varepsilon > 0$. There exists a $\beta_n \in (0, 1/(\tau - 1)]$ such that $F_n(x) = 1$ for $x \geq n^{\beta_n(1+\varepsilon)}$, while, for all $x \leq n^{\beta_n(1-\varepsilon)}$,*

$$1 - F_n(x) = \frac{L_n(x)}{x^{\tau-1}}, \qquad (7.6.16)$$

with $\tau \in (2, 3)$ and a function $L_n(x)$ that satisfies, for some constant $C > 0$ and $\gamma \in (0, 1)$, that, for all $x \leq n^{\beta_n(1-\varepsilon)}$,

$$x^{-C(\log x)^{\gamma-1}} \leq L_n(x) \leq x^{C(\log x)^{\gamma-1}}. \qquad (7.6.17)$$

Theorem 7.27 (Fluctuations of the distances $\mathrm{CM}_n(d)$ for truncated infinite-variance degrees) *Consider* $\mathrm{CM}_n(d)$ *where the degrees* $d = (d_v)_{v \in [n]}$ *satisfy Condition 7.26 for some* $\tau \in (2,3)$. *Assume that* $d_{\min} \geq 2$, *and that there exists* $\kappa > 0$ *such that*

$$\max\{d_{\mathrm{TV}}(F_n, F), d_{\mathrm{TV}}(F_n^\star, F^\star)\} \leq n^{-\kappa \beta_n}. \tag{7.6.18}$$

When $\beta_n \to 1/(\tau - 1)$, *we further require that the limit random variable* Y *in Theorem 7.17 has no point mass on* $(0, \infty)$. *Then, with* o_1, o_2 *chosen independently and uar from* $[n]$, *and conditional on* $o_1 \longleftrightarrow o_2$,

$$\mathrm{dist}_{\mathrm{CM}_n(d)}(o_1, o_2) - 2\frac{\log\log(n^{\beta_n})}{|\log(\tau - 2)|} - \frac{1}{\beta_n(\tau - 3)} \tag{7.6.19}$$

is a tight sequence of random variables.

Which of the two terms in (7.6.19) dominates depends sensitively on the choice of β_n. When $\beta_n \to \beta \in (0, 1/(\tau - 1)]$, the first term dominates. When $\beta_n = (\log n)^{-\gamma}$ for some $\gamma \in (0, 1)$, the second term dominates. Both terms are of the same order of magnitude when $\beta_n = \Theta(1/\log\log n)$.

For supercritical graphs, typical distances are at most of order $\log n$. The boundary point in (7.6.19) corresponds to $\beta_n = \Theta(1/\log n)$, in which case $n^{\beta_n} = \Theta(1)$ and Theorem 7.1 applies. Thus, even after truncation of the degrees, in the infinite-variance case, typical distances are always ultra-small.

7.7 NOTES AND DISCUSSION FOR CHAPTER 7

Notes on Section 7.2

Distances in the configuration model were first obtained in a non-rigorous way in Newman et al. (2000b, 2002); see also Cohen and Havlin (2003) for results on ultra-small distances. Theorem 7.1 is proved in van der Hofstad et al. (2005). Theorem 7.2 is proved by van der Hofstad et al. (2007a).

Notes on Section 7.3

Proposition 7.5 was adapted from (Janson, 2010b, Lemma 5.1). The path-counting techniques used in Section 7.3 are adaptations of those in Section 6.5.1. Comparisons with branching processes appear in many papers on the configuration model. We have strived for a construction that is the most transparent and complete. The proof of Theorem 7.13 is a slightly simplified version of the analysis in Reittu and Norros (2004).

Notes on Section 7.4

Theorem 7.17 is proved in Davies (1978), whose proof we have adapted. The proof of Davies (1978) relies on Laplace transforms. Darling (1970) proved a related result, under stronger conditions. Branching processes with infinite mean have attracted considerable attention; see, e.g., Schuh and Barbour (1977); Seneta (1973) and the references therein. There is a balance between the generality of the results and the conditions on the offspring distribution, and in our opinion Theorem 7.17 strikes this balance nicely, in that the result is relatively simple and the conditions fairly general.

Properties of the limit in Theorem 7.17 are hard to prove. For example, it is in general unknown whether Y has a density. This fact is proved, under stronger assumptions, by Seneta (1973, 1974).

Notes on Section 7.5

Theorem 7.19 is (Fernholz and Ramachandran, 2007, Theorem 5.1) when $p_1 > 0$, (Fernholz and Ramachandran, 2007, Theorem 5.14) when $n_1 = 0$, and (Fernholz and Ramachandran, 2007, Theorem 5.15) when $n_1 = n_2 = 0$. The proof by Fernholz and Ramachandran (2007) involves a precise analysis of the trees that are augmented to the 2-core and their maximal height, as well as an analysis showing that the pairs of vertices determined above really are at distances close to the diameter.

Theorem 7.20 is proved in Caravenna et al. (2019). A logarithmic lower bound on the diameter when $\tau \in (2, 3)$ was also proved by van der Hofstad et al. (2007b), but that result is substantially weaker than Theorem 7.19. A sharper result on the diameter of random regular graphs was obtained by Bollobás and Fernandez de la Vega (1982). For a nice discussion and results about the existence of a large k-core in the configuration model, we refer to Janson and Luczak (2007).

Notes on Section 7.6

Theorem 7.23 is proved in van den Esker et al. (2006). The explicit identification of the limit of $\mathbb{P}(\mathrm{dist}_{\mathrm{CM}_n(d)}(o_1, o_2) = 2)$ is novel. One might argue that including degrees larger than $n - 1$ is artificial in a network with n vertices. In fact, in many real-world networks, the degree is bounded by a physical constant. Therefore, van den Esker et al. (2006) also considered the case where the degrees are conditioned to be smaller than n^α, where α is an arbitrary positive number. In this setting, it turns out that the typical distance is equal to $k+3$ with high probability, whenever $\alpha \in (1/(\tau+k), 1/(\tau+k-1))$. It can be expected that a much more detailed picture can be derived when instead one conditions on the degrees being at most n^{β_n}, as in van der Hofstad and Komjáthy (2017b) for the $\tau \in (2, 3)$ case.

Theorem 7.24 is proved in van der Hofstad et al. (2005). Theorem 7.25 is proved in van der Hofstad et al. (2007a). The assumption that the limit Y in Theorem 7.17, conditional on $Y > 0$, has no point masses was implicitly made by van der Hofstad et al. (2007a). However, this assumption is not necessary for Theorem 7.25. Instead, it suffices to have that $\mathbb{P}(0 < Y \leq \varepsilon)$ vanishes as $\varepsilon \searrow 0$ (see, e.g., (van der Hofstad et al., 2007a, Lemma 4.6)), which is generally true.

Theorem 7.27 is proved in van der Hofstad and Komjáthy (2017b). There, the assumption on the random variable Y not having point mass in $(0, \infty)$ was also explicitly made and discussed in detail. Several extensions were proved in van der Hofstad and Komjáthy (2017b), the most precise result being (van der Hofstad and Komjáthy (2017b), Corollary 1.9).

7.8 Exercises for Chapter 7

Exercise 7.1 (Degree example)   *Let the degree sequence $d = (d_v)_{v \in [n]}$ be given by $d_v = 1 + (v \mod 3)$ as in Exercise 1.12. Let o_1, o_2 be two independent vertices chosen uar from $[n]$. Identify a number a such that, conditional on $\mathrm{dist}_{\mathrm{CM}_n(d)}(o_1, o_2) < \infty$, it holds that*

$$\mathrm{dist}_{\mathrm{CM}_n(d)}(o_1, o_2)/\log n \xrightarrow{\mathbb{P}} a. \tag{7.8.1}$$

Exercise 7.2 (Poisson degree example)   *Consider the degree sequence $d = (d_v)_{v \in [n]}$ satisfying the Poisson degree limit as formulated in (1.7.4) and (1.7.5) with $\lambda > 1$. Let o_1, o_2 be two independent vertices chosen uar from $[n]$. Identify a number a such that, conditional on $\mathrm{dist}_{\mathrm{CM}_n(d)}(o_1, o_2) < \infty$,*

$$\mathrm{dist}_{\mathrm{CM}_n(d)}(o_1, o_2)/\log n \xrightarrow{\mathbb{P}} a. \tag{7.8.2}$$

Exercise 7.3 (Power-law degree example)   *Consider the degree sequence $d = (d_v)_{v \in [n]}$ with $d_v = [1 - F]^{-1}(v/n)$, where F is the distribution of a random variable D having generating function, for $\alpha \in (0, 1)$,*

$$G_X(s) = s - (1 - s)^{\alpha + 1}/(\alpha + 1) \tag{7.8.3}$$

as in Exercise 1.14. Identify a number a such that, conditional on $\mathrm{dist}_{\mathrm{CM}_n(d)}(o_1, o_2) < \infty$,

$$\mathrm{dist}_{\mathrm{CM}_n(d)}(o_1, o_2)/\log \log n \xrightarrow{\mathbb{P}} a. \tag{7.8.4}$$

Exercise 7.4 (Branching-process approximation in Corollary 7.3)   *Use Corollary 2.19 to prove the branching-process approximation for $\mathrm{CM}_n(d)$ in (7.3.2) in Corollary 7.3. Hint: Use that, with $G_n = \mathrm{CM}_n(d)$, $Z_l^{(n;i)} = |\partial B_l^{(G_n)}(o_i)|$ for all $l \in [r]$ when $B_{r+1}^{(G_n)}(o_i)$ is a tree, and then use Theorem 4.1.*

Exercise 7.5 (Proof of logarithmic lower bound distances in Theorem 7.6)   *Let o_1, o_2 be two independent vertices chosen uar from $[n]$. Use Proposition 7.5 with $a = o_1, b = o_2, \mathcal{I} = [n]$ to prove the logarithmic lower bound on typical distances in Theorem 7.6.*

Exercise 7.6 (Proof of distances for $\tau = 3$ in Corollary 7.7) *Let o_1, o_2 be two independent vertices chosen uar from $[n]$. Use Theorem 7.6 to prove the logarithmic and logarithmic divided by log log lower bounds on typical distances in Corollary 7.7 when Conditions 1.7(a),(b) and (7.3.15) hold.*

Exercise 7.7 (Proof of logarithmic upper bound distances without degree truncation) *Check that the "giant is almost local" analysis in Section 4.3.1 can be performed without the degree-truncation argument of Theorem 1.11 when $\limsup_{n\to\infty} \mathbb{E}[D_n^3] < \infty$. Hint: Note that $\limsup_{n\to\infty} \mathbb{E}[D_n^3] < \infty$ implies that the Chebychev inequality can be used without degree truncation.*

Exercise 7.8 (Proof of logarithmic upper bound distances without degree truncation (cont.)) *Extend the proof in Exercise 7.7 to the case where $\limsup_{n\to\infty} \mathbb{E}[D_n^p] < \infty$ for some $p > 2$. Hint: Instead of the Chebychev inequality, use the Marcinkiewicz–Zygmund inequality (Gut, 2005, Corollary 8.2), a form of which states that, for iid random variables $(X_i)_{i=1}^m$ with $\mathbb{E}[X_i] = 0$ and all $q \in (1, 2]$,*

$$\mathbb{E}\Big[\Big|\sum_{i=1}^m X_i\Big|^q\Big] \leq n\mathbb{E}[|X_1|^q]. \tag{7.8.5}$$

Exercise 7.9 (Proof of logarithmic upper bound on typical distances in Theorem 7.9) *Use Proposition 7.10 to prove the logarithmic upper bound on typical distances in Theorem 7.9 by adapting the proof of the related result for $\mathrm{NR}_n(\boldsymbol{w})$ in Theorem 6.19.*

Exercise 7.10 (Alternative proof of log log typical distances in Theorem 7.11) *Give an alternative proof of the doubly logarithmic upper bound on typical distances in Theorem 7.11 by adapting the proof of the related result for $\mathrm{NR}_n(\boldsymbol{w})$ in Theorem 6.11.*

Exercise 7.11 (The hubs Γ_1 form whp a complete graph) *Use Lemma 7.12 and $\beta > \frac{1}{2}$ to show that, whp, the set of hubs in Γ_1 in (7.3.52) forms a complete graph, i.e., whp, every pair $i, j \in \Gamma_1$ are neighbors in $\mathrm{CM}_n(\boldsymbol{d})$.*

Exercise 7.12 (Second alternative proof of log log typical distances in Theorem 7.11 using the core) *Give an alternative proof of the doubly logarithmic upper bound on typical distances in Theorem 7.11 by using the diameter of the core in Theorem 7.13, and an application of the second-moment method for the existence of paths in Proposition 7.10.*

Exercise 7.13 (Typical distances when $\tau = 2$ in Theorem 7.16) *Recall the definition of $k_n^\star$ for the critical case $\tau = 2$ studied in Theorem 7.16. Show that $k_n^\star = o(\log^{\star p}(n))$ for every $p \geq 1$, where $\log^{\star p}(n)$ is obtained by taking the logarithm of n p times.*

Exercise 7.14 (Typical distances when $\tau = 2$ in Theorem 7.16) *Recall $k_n^\star$ from Exercise 7.13. Investigate heuristically the size of $k_n^\star$.*

Exercise 7.15 (Another example of typical distances when $\tau = 2$) *Adapt the upper bound on typical distances for $\tau = 2$ in Theorem 7.16 to degree sequences for which $\tau = 2$, in which (7.3.67) is replaced by $[1 - F_n](x) \geq c_1 e^{-c(\log x)^\gamma}/x$ for some $c, c_1, \gamma \in (0, 1)$, and all $x \leq n^\beta$ for some $\beta > \frac{1}{2}$.*

Exercise 7.16 (Infinite mean under conditions in Theorem 7.17) *Prove that $\mathbb{E}[X] = \infty$ when the conditions in Theorem 7.17 are satisfied. Extend this to show that $\mathbb{E}[X^s] = \infty$ for every $s > \alpha \in (0, 1)$.*

Exercise 7.17 (Example of infinite-mean branching process) *Prove that $\gamma(x) = (\log x)^{\gamma - 1}$, for some $\gamma \in [0, 1)$, satisfies the assumptions in Theorem 7.17.*

Exercise 7.18 (Telescoping sum identity for generation sizes in infinite-mean branching processes) *Consider an infinite-mean branching process as in Theorem 7.17. Prove the telescoping-sum identity (7.4.3) for $\alpha^k \log(Z_k \vee 1)$.*

Exercise 7.19 (Conditions in Theorem 7.17 for individuals with infinite line of descent) *Prove that $p^{(\infty)}$ in [V1, (3.4.2)] satisfies the conditions in Theorem 7.17 with the function $x \mapsto \gamma^\star(x)$ given by $\gamma^\star(x) = \gamma(x) + c/\log x$.*

Exercise 7.20 (Convergence for $Z_k + 1$) *Show that, under the conditions of Theorem 7.17, it also holds that $\alpha^k \log(Z_k + 1)$ converges to Y almost surely.*

Exercise 7.21 (Branching processes with infinite mean: case $X \geq 0$)   *Use the branching process of the number of vertices with infinite line of descent in [V1, Theorem 3.12] to extend the proof of Theorem 7.17 to the case where $X \geq 0$.*

Exercise 7.22 (Distributional identity of limit in Theorem 7.17)   *Let Y be the limit of $k \mapsto \alpha^k \log(Z_k \vee 1)$ in Theorem 7.17. Prove (7.4.34) by showing that*

$$Y \stackrel{d}{=} \alpha \max_{i=1}^{X} Y_i,$$

where X denotes the offspring variable of the infinite-mean branching process and $(Y_i)_{i \geq 1}$ is a sequence of iid copies of Y.

Exercise 7.23 (Exponential limit in Theorem 7.17)   *Let the offspring X have generating function*

$$G_X(s) = 1 - (1-s)^\alpha \tag{7.8.6}$$

with $\alpha \in (0,1)$ as in Exercise 1.5. Use (7.4.34) as well as Theorem 7.18 to show that the limit Y of $k \mapsto \alpha^k \log(Z_k \vee 1)$ in Theorem 7.17 has an exact exponential distribution.

Exercise 7.24 (Maximum process for infinite-mean branching processes)   *Recall the maximum process for infinite-mean branching processes, for which we let $Q_0 = 1$, and, given $Q_{k-1} = q_{k-1}$, let Q_k denote the maximal offspring of the q_{k-1} individuals in the $(k-1)$th generation. Show that $(Q_k)_{k \geq 0}$ is a Markov chain, for which the transition probabilities can be derived from (recall (7.4.39))*

$$\mathbb{P}(Q_k > q \mid Q_{k-1} = q_{k-1}) = 1 - F_X(q)^{q_{k-1}}. \tag{7.8.7}$$

Exercise 7.25 (Telescoping sum identity for maximum process, for infinite-mean branching processes)   *Recall the maximum process $(Q_k)_{k \geq 0}$ for infinite-mean branching processes from Exercise 7.24. Prove the telescoping-sum identity for $\alpha^k \log Q_k$ in (7.4.40).*

Exercise 7.26 (Convergence of the maximum process for infinite-mean branching processes)   *Recall the maximum process $(Q_k)_{k \geq 0}$ for infinite-mean branching processes from Exercise 7.24. Show $\alpha^k \log Q_k \xrightarrow{a.s.} Q_\infty$ under the conditions of Theorem 7.17, by adapting the proof of double-exponential growth of the generation sizes in Theorem 7.17. You may for simplicity assume that the offspring distribution satisfies $\mathbb{P}(X > k) = c/k^\alpha$ for all $k \geq k_{\min}$ and $c = k_{\min}^\alpha$.*

Exercise 7.27 (Diameter of "soup" of cycles)   *Prove that in a graph consisting solely of cycles, the diameter is equal to the longest cycle divided by 2.*

Exercise 7.28 (Longest cycle in a 2-regular graph)   *Let M_n denote the size of the longest cycle in a 2-regular graph. Prove that $M_n/n \xrightarrow{d} M$ for some M. What can you say about the distribution of M?*

Exercise 7.29 (Diameter result for $\mathrm{ER}_n(\lambda/n)$)   *Fix $\lambda > 1$, and recall the constants in the limit of $\mathrm{diam}(\mathrm{ER}_n(\lambda/n))/\log n$ in (7.5.4), as a consequence of Theorem 7.19. Prove that ν in Theorem 7.19 equals $\nu = \lambda$ and that μ in Theorem 7.19 equals $\mu = \mu_\lambda$, where $\mu_\lambda \in [0,1)$ is the dual parameter, i.e., the unique $\mu < 1$ satisfying*

$$\mu e^{-\mu} = \lambda e^{-\lambda}. \tag{7.8.8}$$

Exercise 7.30 (Diameter of $\mathrm{GRG}_n(\boldsymbol{w})$)   *Consider $\mathrm{GRG}_n(\boldsymbol{w})$ where the weights $\boldsymbol{w} = (w_v)_{v \in [n]}$ satisfy Conditions 1.1(a)–(c). Identify the limit in probability of $\mathrm{diam}(\mathrm{GRG}_n(\boldsymbol{w}))/\log n$. Can this limit be zero?*

Exercise 7.31 (Expectation of the number of minimally k-connected vertices)   *Recall the definition of minimally k-connected vertices in Definition 7.21. Prove that, for all $k \geq 1$,*

$$\mathbb{E}[M_k] = n_{d_{\min}} \prod_{i=1}^{d_{\min}(d_{\min}-1)^{k-1}} \frac{d_{\min}(n_{d_{\min}} - (i-1))}{\ell_n - 2i + 1}. \tag{7.8.9}$$

Exercise 7.32 (Second moment of the number of minimally k-connected vertices) *Recall the definition of minimally k-connected vertices in Definition 7.21. Prove that, for all k such that $d_{\min}(d_{\min} - 1)^{k-1} \leq \ell_n/8$,*

$$\mathbb{E}[M_k^2] \leq \mathbb{E}[M_k]^2 + \mathbb{E}[M_k]\left[\frac{d_{\min}}{d_{\min} - 2}(d_{\min} - 1)^k + \frac{2n_{d_{\min}}d_{\min}^2(d_{\min} - 1)^{2k}}{(d_{\min} - 2)\ell_n}\right]. \tag{7.8.10}$$

Exercise 7.33 (Concentration of the number of minimally k-connected vertices: proof of Lemma 7.22) *Recall the definition of minimally k-connected vertices in Definition 7.21. Use Exercises 7.31 and 7.32 to complete the proof of Lemma 7.22.*

Exercise 7.34 (Sum of Gamma variables is finite almost surely) *Fix $\tau \in (1,2)$. Let $(E_i)_{i\geq 1}$ be iid exponentials, and $\Gamma_i = \sum_{j=1}^i E_j$ be (dependent) Gamma variables. Show that $Z = \sum_{i\geq 1} \Gamma_i^{-1/(\tau-1)}$ is almost surely finite.*

Exercise 7.35 (Typical distance is at least 2 whp for $\tau \in (1,2)$) *Complete the argument that*

$$\mathbb{P}(\text{dist}_{\text{CM}_n(d)}(o_1, o_2) = 1) = o(1)$$

in the proof of the typical distance for $\tau \in (1,2)$ in Theorem 7.23.

Exercise 7.36 (Typical distance equals 2 whp for $\tau = 1$) *Let $(d_v)_{v\in[n]}$ be a sequence of iid copies of D with distribution function F satisfying that $x \mapsto [1 - F](x)$ is slowly varying at ∞. Prove that $\text{CM}_n(d)$ satisfies that $\text{dist}_{\text{CM}_n(d)}(o_1, o_2) \xrightarrow{\mathbb{P}} 2$. You may use without proof that $M_n/S_n \xrightarrow{\mathbb{P}} 1$, where $S_n = \sum_{v\in[n]} D_v$ and $M_n = \max_{v\in[n]} D_v$.*

Exercise 7.37 (Convergence along subsequences (van der Hofstad et al. (2005))) *Fix an integer n_1. Prove that, under the assumptions in Theorem 7.24, and conditional on $\text{dist}_{\text{CM}_n(d)}(o_1, o_2) < \infty$, along the subsequence $n_k = \lfloor n_1 \nu^{k-1} \rfloor$ the sequence of random variables $\text{dist}_{\text{CM}_n(d)}(o_1, o_2) - \lfloor \log_\nu n_k \rfloor$ converges in distribution to $R_{a_{n_1}}$ as $k \to \infty$.*

Exercise 7.38 (Non-convergence of graph distances for random regular graph) *Let $d_v = r$ for every $v \in [n]$ and let nr be even. Recall (7.6.11). Show that Theorem 7.24 implies that $\text{dist}_{\text{CM}_n(d)}(o_1, o_2) - \lfloor \log_\nu n \rfloor$ does not converge in distribution.*

Exercise 7.39 (Tightness of the hopcount (van der Hofstad et al. (2005))) *Prove that, under the assumptions in Theorem 7.24:*

(a) conditional on $\text{dist}_{\text{CM}_n(d)}(o_1, o_2) < \infty$ and whp, the random variable $\text{dist}_{\text{CM}_n(d)}(o_1, o_2)$ is in between $(1 \pm \varepsilon)\log_\nu n$ for any $\varepsilon > 0$;

(b) conditional on $\text{dist}_{\text{CM}_n(d)}(o_1, o_2) < \infty$, the random variables $\text{dist}_{\text{CM}_n(d)}(o_1, o_2) - \log_\nu n$ form a tight sequence, i.e.,

$$\lim_{K\to\infty} \limsup_{n\to\infty} \mathbb{P}\big(|\text{dist}_{\text{CM}_n(d)}(o_1, o_2) - \log_\nu n| \leq K \mid \text{dist}_{\text{CM}_n(d)}(o_1, o_2) < \infty\big) = 1. \tag{7.8.11}$$

As a consequence, prove that the same result applies to a uniform random graph with degrees $(d_v)_{v\in[n]}$. Hint: Make use of [V1, Theorem 7.21].

SMALL-WORLD PHENOMENA IN PREFERENTIAL ATTACHMENT MODELS

Abstract

In this chapter we investigate graph distances in preferential attachment models. We focus on typical distances as well as on the diameter of preferential attachment models. We again rely on path-counting techniques, as well as local limit results. Since the local limit is a rather involved quantity, some parts of our analysis are considerably harder than those in Chapters 6 and 7.

8.1 MOTIVATION: LOCAL STRUCTURE VERSUS SMALL-WORLD PROPERTIES

In Chapters 6 and 7, we saw that generalized random graphs and configuration models with finite-variance degrees are small worlds, whereas random graphs with infinite-variance degrees are ultra-small worlds. In the small-world setting, distances are roughly $\log_\nu n$, where ν describes the exponential growth of the branching-process approximation of local neighborhoods in the random graphs in question. For preferential attachment models with $\delta > 0$, for which the degree power-law exponent equals $\tau = 3 + \delta/m > 3$, it is highly unclear whether the neighborhoods grow exponentially, since the Pólya point tree that arises as the local limit in Theorem 5.8 is a rather intricate object (recall also Theorem 5.21). This local limit also makes path-counting estimates, the central method for bounding typical distances, much more involved.

The ultra-small behavior in generalized random graphs and configuration models, on the other hand, can be understood informally in terms of two effects. First, we note that such random graph models contain super-hubs, whose degrees are much larger than $n^{1/2}$ and which form a *complete* graph of connections. Second, vertices of large degree $d \gg 1$ are typically connected to vertices of much larger degree, more precisely of degree roughly $d^{1/(\tau-2)}$, by the power-iteration method. When combined, these two effects mean that it takes roughly $\log \log n/|\log(\tau - 2)|$ steps from a typical vertex to reach one of the super-hubs, and thus roughly $2 \log \log n/|\log(\tau-2)|$ steps to connect two typical vertices to each other. Of course the proofs are more technical, but this is the bottom line.

For preferential attachment models with $\delta \in (-m, 0)$ and $m \geq 2$, however, vertices of large degree d tend to be the old vertices, but old vertices are not necessarily connected to *much* older vertices, which would be necessary to increase their degree from d to $d^{1/(\tau-2)}$. However, vertices of degree $d \gg 1$ *do* tend to be connected to vertices that are *in turn* connected to a vertex of degree roughly $d^{1/(\tau-2)}$. This gives rise to a *two-step power-iteration* property. We conclude that distances seem about twice as large in preferential attachment models with infinite-variance degrees than in the corresponding generalized random graphs or configuration models, and that this can be explained by differences in the *local* connectivity structure. Unfortunately, owing to their dynamic nature, the results for preferential attachment models are harder to prove, and they are less complete.

The above explanations depend crucially on the *local structure* of the generalized random graph as well as the configuration model. In this chapter we will see that, for the preferential attachment model, such considerations need to be subtly adapted.

Organization of this Chapter

This chapter is organized as follows. We start in Section 8.2 by showing that preferential attachment trees (where $m = 1$, so that every vertex enters the random graph with one edge to earlier vertices) have logarithmic height and diameter. In Section 8.3 we investigate typical distances in $\mathrm{PA}_n^{(m,\delta)}$ for $m \geq 2$ and formulate our main results. In Section 8.4 we investigate path-counting techniques in preferential attachment models, which we use in Section 8.5 to prove logarithmic lower bounds on distances for $\delta > 0$ and in Section 8.6 to prove doubly logarithmic lower bounds for $\delta < 0$. In Section 8.7 we prove the matching doubly logarithmic upper bound for $\delta < 0$. In Section 8.8 we discuss the diameter of preferential attachment models, and complete the upper bound on typical distances, for $m \geq 2$. In Section 8.9 we discuss further results about distances in preferential attachment models. We close this chapter with notes and discussion in Section 8.10 and with exercises in Section 8.11.

Connectivity Notation for Preferential Attachment Models

Throughout this chapter, we work with the preferential attachment models defined in Section 1.3.5, given by $(\mathrm{PA}_n^{(m,\delta)}(a))_{n \geq 1}$, $(\mathrm{PA}_n^{(m,\delta)}(b))_{n \geq 1}$, and $(\mathrm{PA}_n^{(m,\delta)}(d))_{n \geq 1}$. We will write $(\mathrm{PA}_n^{(m,\delta)})_{n \geq 1}$ when we talk about all of these models at the same time. We write $u \overset{j}{\rightsquigarrow} v$ to denote that the jth edge from u is connected to v, where $u \geq v$. We further write $u \longleftrightarrow v$ when there exists a path of edges connecting u and v, so that v is in the connected component of u.

8.2 LOGARITHMIC DISTANCES IN PREFERENTIAL ATTACHMENT TREES

In this section, we investigate distances in scale-free trees, arising for $m = 1$. We explicitly treat $\mathrm{PA}_n^{(1,\delta)}(b)$, and we discuss the minor adaptations necessary for $\mathrm{PA}_n^{(1,\delta)}(a)$ and $\mathrm{PA}_n^{(1,\delta)}(d)$. Such results are interesting in their own right and at the same time provide natural upper bounds on distances for $m \geq 2$ owing to the fact that $\mathrm{PA}_n^{(m,\delta)}(a)$ and $\mathrm{PA}_n^{(m,\delta)}(b)$ can be obtained by collapsing blocks of m vertices in $(\mathrm{PA}_n^{(1,\delta/m)}(a))_{n \geq 1}$ and $(\mathrm{PA}_n^{(1,\delta/m)}(b))_{n \geq 1}$.

Let the height of a tree T on the vertex set $[n]$ be defined as

$$\mathrm{height}(T) = \max_{v \in [n]} \mathrm{dist}_T(1, v), \tag{8.2.1}$$

where $\mathrm{dist}_T(u, v)$ denotes the graph distance between vertices u and v in the tree T, and 1 denotes the root of the tree. We start by studying various distances in the tree $\mathrm{PA}_n^{(1,\delta)}$:

Theorem 8.1 (Distances in scale-free trees) *Consider* $\mathrm{PA}_n^{(1,\delta)}$ *with* $m = 1$ *and* $\delta > -1$, *and let* $\theta \in (0, 1)$ *be the non-negative solution of*

$$\theta + (1 + \delta)(1 + \log \theta) = 0. \tag{8.2.2}$$

Then, as $n \to \infty$, for $o \in [n]$ chosen uar,

$$\frac{\mathrm{dist}_{\mathrm{PA}_n^{(1,\delta)}(b)}(1, o)}{\log n} \xrightarrow{\mathbb{P}} \frac{1+\delta}{2+\delta}, \qquad \frac{\mathrm{height}(\mathrm{PA}_n^{(1,\delta)}(b))}{\log n} \xrightarrow{a.s.} \frac{1+\delta}{(2+\delta)\theta}, \qquad (8.2.3)$$

and, as $n \to \infty$, for $o_1, o_2 \in [n]$ chosen independently and uar,

$$\frac{\mathrm{dist}_{\mathrm{PA}_n^{(1,\delta)}(b)}(o_1, o_2)}{\log n} \xrightarrow{\mathbb{P}} \frac{2(1+\delta)}{2+\delta}, \qquad \frac{\mathrm{diam}(\mathrm{PA}_n^{(1,\delta)}(b))}{\log n} \xrightarrow{\mathbb{P}} \frac{2(1+\delta)}{(2+\delta)\theta}. \qquad (8.2.4)$$

The same results apply to $\mathrm{PA}_n^{(1,\delta)}(d)$ and $\mathrm{PA}_n^{(1,\delta)}(a)$, where in the latter we condition on 1 being connected to o in (8.2.3), and on o_1 being connected to o_2 in (8.2.4).

We will prove all these bounds except for the lower bound on the height in (8.2.3), which we discuss in Section 8.10. We start by proving the upper bounds in Theorem 8.1 for $\mathrm{PA}_n^{(1,\delta)}(b)$. As we will see, the proof indicates that the almost sure limit of the height in Theorem 8.1 does not depend on the precise starting configuration of the graph $\mathrm{PA}_2^{(1,\delta)}(b)$, which is useful in extending the results in Theorem 8.1 to $\mathrm{PA}_n^{(1,\delta)}(d)$ and $\mathrm{PA}_n^{(1,\delta)}(a)$. Exercises 8.1 and 8.2 explore properties of the constant θ.

In the proof of Theorem 8.1 we make use of the following result, which computes the probability mass function of the distance between vertex v and the root 1. Before stating the result, we need some more notation. Recall that we write $u \rightsquigarrow v$, which is the same as $u \overset{1}{\rightsquigarrow} v$, when in $(\mathrm{PA}_n^{(1,\delta)}(b))_{n \geq 1}$ the (single) edge from vertex u is connected to vertex v.

For $u = \pi_0 > \pi_1 > \cdots > \pi_k = 1$, and denoting $\vec{\pi} = (\pi_0, \pi_1, \ldots, \pi_k)$, we write the event that $\vec{\pi}$ is present in the tree $\mathrm{PA}_n^{(1,\delta)}(b)$ as

$$\{\vec{\pi} \subseteq \mathrm{PA}_n^{(1,\delta)}(b)\} = \bigcap_{i=0}^{k-1} \{\pi_i \rightsquigarrow \pi_{i+1}\}. \qquad (8.2.5)$$

The probability mass function of $\mathrm{dist}_{\mathrm{PA}_n^{(1,\delta)}(b)}(u, v)$ can be explicitly identified as follows:

Proposition 8.2 (Distribution of tree distance) *Fix $m = 1$ and $\delta > -1$. Then, for all $\pi_0 = u > \pi_1 > \cdots > \pi_{k-1} > \pi_k = v$,*

$$\mathbb{P}(\vec{\pi} \subseteq \mathrm{PA}_n^{(1,\delta)}(b)) = \left(\frac{1+\delta}{2+\delta}\right)^k \frac{\Gamma(u - \frac{1+\delta}{2+\delta})\Gamma(v)}{\Gamma(v + \frac{1}{2+\delta})\Gamma(u)} \prod_{i=1}^{k-1} \frac{1}{\pi_i - \frac{1+\delta}{2+\delta}}. \qquad (8.2.6)$$

Consequently, for all $u > v$,

$$\mathbb{P}(\mathrm{dist}_{\mathrm{PA}_n^{(1,\delta)}(b)}(u, v) = k) = \left(\frac{1+\delta}{2+\delta}\right)^k \frac{\Gamma(u - \frac{1+\delta}{2+\delta})\Gamma(v)}{\Gamma(v + \frac{1}{2+\delta})\Gamma(u)} \sum_{\vec{\pi}} \prod_{i=1}^{k-1} \frac{1}{\pi_i - \frac{1+\delta}{2+\delta}}, \qquad (8.2.7)$$

where the sum is over ordered vectors $\vec{\pi} = (\pi_0, \ldots, \pi_k)$ with $\pi_0 = u$ and $\pi_k = v$.

Remark 8.3 (Extension to $\mathrm{PA}_n^{(1,\delta)}(a)$) For $\mathrm{PA}_n^{(1,\delta)}(a)$, (8.2.6) is replaced by

$$\mathbb{P}(\vec{\pi} \subseteq \mathrm{PA}_n^{(1,\delta)}(a)) = \left(\frac{1+\delta}{2+\delta}\right)^k \frac{\Gamma(v - \frac{1}{2+\delta})\Gamma(u)}{\Gamma(u + \frac{1+\delta}{2+\delta})\Gamma(v)} \prod_{i=1}^{k-1} \frac{1}{\pi_i - \frac{1}{2+\delta}}. \qquad (8.2.8)$$

◀

Proof of Proposition 8.2. We claim that the events $\{\pi_i \rightsquigarrow \pi_{i+1}\}$ are *independent*, i.e., it holds that, for every $\vec{\pi} = (\pi_0, \ldots, \pi_k)$,

$$\mathbb{P}\Big(\bigcap_{i=0}^{k-1}\{\pi_i \rightsquigarrow \pi_{i+1}\}\Big) = \prod_{i=0}^{k-1} \mathbb{P}(\pi_i \rightsquigarrow \pi_{i+1}). \qquad (8.2.9)$$

We prove the independence in (8.2.9) by induction on $k \geq 1$. For $k = 1$, there is nothing to prove, and this initializes the induction hypothesis. To advance the induction hypothesis in (8.2.9), we condition on $\mathrm{PA}_{\pi_0-1}^{(1,\delta)}(b)$ to obtain

$$\mathbb{P}(\vec{\pi} \subseteq \mathrm{PA}_n^{(1,\delta)}(b)) = \mathbb{E}\Big[\mathbb{P}\Big(\bigcap_{i=0}^{k-1}\{\pi_i \rightsquigarrow \pi_{i+1}\}\Big|\mathrm{PA}_{\pi_1-1}^{(1,\delta)}(b)\Big)\Big]$$

$$= \mathbb{E}\Big[\mathbb{1}_{\bigcap_{i=1}^{k-1}\{\pi_i \rightsquigarrow \pi_{i+1}\}}\mathbb{P}\big(\pi_0 \rightsquigarrow \pi_1 \mid \mathrm{PA}_{\pi_0-1}^{(1,\delta)}(b)\big)\Big], \qquad (8.2.10)$$

since the event $\bigcap_{i=1}^{k-1}\{\pi_i \rightsquigarrow \pi_{i+1}\}$ is measurable with respect to $\mathrm{PA}_{\pi_0-1}^{(1,\delta)}(b)$ because $\pi_0 - 1 \geq \pi_i$ for all $i \in [k-1]$. Furthermore, from [V1, (8.2.2)],

$$\mathbb{P}\big(\pi_0 \rightsquigarrow \pi_1 \mid \mathrm{PA}_{\pi_0-1}^{(1,\delta)}(b)\big) = \frac{D_{\pi_1}(\pi_0 - 1) + \delta}{(2 + \delta)(\pi_0 - 1)}. \qquad (8.2.11)$$

In particular,

$$\mathbb{P}\big(\pi_0 \rightsquigarrow \pi_1\big) = \mathbb{E}\Big[\frac{D_{\pi_1}(\pi_0 - 1) + \delta}{(2 + \delta)(\pi_0 - 1)}\Big]. \qquad (8.2.12)$$

Therefore,

$$\mathbb{P}(\vec{\pi} \subseteq \mathrm{PA}_n^{(1,\delta)}(b)) = \mathbb{E}\Big[\mathbb{1}_{\bigcap_{i=1}^{k-1}\{\pi_i \rightsquigarrow \pi_{i+1}\}}\frac{D_{\pi_1}(\pi_0 - 1) + \delta}{(2 + \delta)(\pi_0 - 1)}\Big]$$

$$= \mathbb{P}\Big(\bigcap_{i=1}^{k-1}\{\pi_i \rightsquigarrow \pi_{i+1}\}\Big)\mathbb{E}\Big[\frac{D_{\pi_1}(\pi_0 - 1) + \delta}{(2 + \delta)(\pi_0 - 1)}\Big], \qquad (8.2.13)$$

since the random variable $D_{\pi_1}(\pi_0 - 1)$ depends only on how many edges are connected to π_1 after time π_1, and is thus independent of the event $\bigcap_{i=1}^{k-1}\{\pi_i \rightsquigarrow \pi_{i+1}\}$, which only depends on the attachment of the edges *up to and including* time π_1. We conclude that

$$\mathbb{P}(\vec{\pi} \subseteq \mathrm{PA}_n^{(1,\delta)}(b)) = \mathbb{P}(\pi_0 \rightsquigarrow \pi_1)\mathbb{P}\Big(\bigcap_{i=1}^{k-1}\{\pi_i \rightsquigarrow \pi_{i+1}\}\Big). \qquad (8.2.14)$$

The claim in (8.2.9) for k follows from the induction hypothesis.

Combining (8.2.19) with (8.2.9) and (8.2.12), we obtain that

$$\mathbb{P}(\vec{\pi} \subseteq \mathrm{PA}_n^{(1,\delta)}(b)) = \prod_{i=0}^{k-1} \mathbb{E}\Big[\frac{D_{\pi_{i+1}}(\pi_i - 1) + \delta}{(2 + \delta)(\pi_i - 1)}\Big]. \qquad (8.2.15)$$

By [V1, (8.11.3)], for all $n \geq s$ and for $\mathrm{PA}_n^{(1,\delta)}(b)$,

$$\mathbb{E}[D_s(n) + \delta] = (1 + \delta)\frac{\Gamma(n + \frac{1}{2+\delta})\Gamma(s)}{\Gamma(n)\Gamma(s + \frac{1}{2+\delta})}. \qquad (8.2.16)$$

As a result,

$$
\mathbb{P}(\pi_i \rightsquigarrow \pi_{i+1}) = \frac{1+\delta}{(2+\delta)(\pi_i - 1)} \frac{\Gamma(\pi_i - 1 + \frac{1}{2+\delta})\Gamma(\pi_{i+1})}{\Gamma(\pi_i - 1)\Gamma(\pi_{i+1} + \frac{1}{2+\delta})}
$$

$$
= \frac{1+\delta}{2+\delta} \frac{\Gamma(\pi_i - 1 + \frac{1}{2+\delta})\Gamma(\pi_{i+1})}{\Gamma(\pi_i)\Gamma(\pi_{i+1} + \frac{1}{2+\delta})}, \tag{8.2.17}
$$

so that

$$
\mathbb{P}(\vec{\pi} \subseteq \mathrm{PA}_n^{(1,\delta)}(b)) = \left(\frac{1+\delta}{2+\delta}\right)^k \prod_{i=0}^{k-1} \frac{\Gamma(\pi_i - 1 + \frac{1}{2+\delta})\Gamma(\pi_{i+1})}{\Gamma(\pi_i)\Gamma(\pi_{i+1} + \frac{1}{2+\delta})}
$$

$$
= \left(\frac{1+\delta}{2+\delta}\right)^k \frac{\Gamma(\pi_0 - \frac{1+\delta}{2+\delta})\Gamma(\pi_k)}{\Gamma(\pi_0)\Gamma(\pi_k + \frac{1}{2+\delta})} \prod_{i=0}^{k-1} \frac{\Gamma(\pi_i - 1 + \frac{1}{2+\delta})}{\Gamma(\pi_i + \frac{1}{2+\delta})}
$$

$$
= \left(\frac{1+\delta}{2+\delta}\right)^k \frac{\Gamma(u - \frac{1+\delta}{2+\delta})\Gamma(v)}{\Gamma(u)\Gamma(v + \frac{1}{2+\delta})} \prod_{i=1}^{k-1} \frac{1}{\pi_i - \frac{1+\delta}{2+\delta}}, \tag{8.2.18}
$$

which proves (8.2.6). Since the path between vertex u and v in $\mathrm{PA}_n^{(1,\delta)}(b)$ is *unique*,

$$
\mathbb{P}(\mathrm{dist}_{\mathrm{PA}_n^{(1,\delta)}(b)}(u, v) = k) = \sum_{\vec{\pi}} \mathbb{P}\left(\bigcap_{i=0}^{k-1} \{\pi_i \rightsquigarrow \pi_{i+1}\}\right), \tag{8.2.19}
$$

where again the sum is over all ordered vectors $\vec{\pi} = (\pi_0, \ldots, \pi_k)$ with $\pi_0 = u$ and $\pi_k = v$. Thus, (8.2.7) follows immediately from (8.2.6). This completes the proof of Proposition 8.2. $\qquad\square$

For the proof of Remark 8.3, (8.2.15) is replaced with

$$
\mathbb{P}(\vec{\pi} \subseteq \mathrm{PA}_n^{(1,\delta)}(a)) = \prod_{i=0}^{k-1} \mathbb{E}\left[\frac{D_{\pi_{i+1}}(\pi_i - 1) + \delta}{(2+\delta)(\pi_i - 1) + 1 + \delta}\right], \tag{8.2.20}
$$

and (8.2.16) is replaced with, for $n \geq s$,

$$
\mathbb{E}[D_s(n) + \delta] = (1+\delta)\frac{\Gamma(n+1)\Gamma(s - \frac{1}{2+\delta})}{\Gamma(n + \frac{1+\delta}{2+\delta})\Gamma(s)}. \tag{8.2.21}
$$

After these changes, the proof follows the same steps (see Exercise 8.3). $\qquad\square$

Proof of the upper bounds in Theorem 8.1 for $\mathrm{PA}_n^{(1,\delta)}(b)$. We first use Proposition 8.2 to prove that, for o chosen uar from $[n]$,

$$
\frac{\mathrm{dist}_{\mathrm{PA}_n^{(1,\delta)}(b)}(1, o)}{\log n} \leq (1 + o_{\mathbb{P}}(1))\frac{1+\delta}{2+\delta}, \tag{8.2.22}
$$

and, almost surely for large enough n,

$$
\frac{\mathrm{dist}_{\mathrm{PA}_n^{(1,\delta)}(b)}(1, n)}{\log n} \leq (1 + \varepsilon)\frac{(1+\delta)}{(2+\delta)\theta}, \tag{8.2.23}
$$

where θ is the non-negative solution of (8.2.2). We start by noting that (8.2.22) and (8.2.23) immediately prove the first upper bound in (8.2.3). Further, (8.2.23) implies that, almost surely for large n,

$$\frac{\text{height}(\text{PA}_n^{(1,\delta)}(b))}{\log n} \leq (1+\varepsilon)\frac{(1+\delta)}{(2+\delta)\theta}. \tag{8.2.24}$$

Indeed, $\text{height}(\text{PA}_n^{(1,\delta)}(b)) > \log n(1+\varepsilon)(1+\delta)/(2+\delta)\theta$ for n large precisely when there exists an m large (m must at least satisfy $m \geq \log n(1+\varepsilon)(1+\delta)/(2+\delta)\theta$) such that $\text{dist}_{\text{PA}_n^{(1,\delta)}(b)}(1,m) > \log m(1+\varepsilon)(1+\delta)/(2+\delta)\theta$. Since the latter almost surely does not happen for m large, by (8.2.23), it follows that (8.2.24) does not hold for n large. Thus, (8.2.22) and (8.2.23) also prove the second upper bound in (8.2.3).

By the triangle inequality,

$$\text{dist}_{\text{PA}_n^{(1,\delta)}(b)}(o_1, o_2) \leq \text{dist}_{\text{PA}_n^{(1,\delta)}(b)}(1, o_1) + \text{dist}_{\text{PA}_n^{(1,\delta)}(b)}(1, o_2), \tag{8.2.25}$$

$$\text{diam}(\text{PA}_n^{(1,\delta)}(b)) \leq 2\,\text{height}(\text{PA}_n^{(1,\delta)}(b)), \tag{8.2.26}$$

so that (8.2.22) and (8.2.23) imply the upper bounds in (8.2.4), and thus all those in Theorem 8.1, for $\text{PA}_n^{(1,\delta)}(b)$.

We proceed to prove (8.2.22) and (8.2.23) for $\text{PA}_n^{(1,\delta)}(b)$, and start with some preparations. We use (8.2.7) and symmetry to obtain

$$\mathbb{P}(\text{dist}_{\text{PA}_n^{(1,\delta)}(b)}(1, n) = k) = \left(\frac{1+\delta}{2+\delta}\right)^k \frac{\Gamma(n - \frac{1+\delta}{2+\delta})\Gamma(1)}{\Gamma(1 + \frac{1}{2+\delta})\Gamma(n)} \sum_{\vec{t}_{k-1}}^{*} \frac{1}{(k-1)!} \prod_{i=1}^{k-1} \frac{1}{t_i - \frac{1+\delta}{2+\delta}}, \tag{8.2.27}$$

where the sum now is over *all* vectors $\vec{t}_{k-1} = (t_1, \ldots, t_{k-1})$, with $1 < t_i < n$, having *distinct* coordinates. We can upper bound this sum by leaving out the restriction that the coordinates of $\vec{t}_{k-1}$ are distinct, so that

$$\mathbb{P}(\text{dist}_{\text{PA}_n^{(1,\delta)}(b)}(1, n) = k) \leq \left(\frac{1+\delta}{2+\delta}\right)^k \frac{\Gamma(n - \frac{1+\delta}{2+\delta})}{\Gamma(1 + \frac{1}{2+\delta})\Gamma(n)} \frac{1}{(k-1)!} \left(\sum_{s=2}^{n-1} \frac{1}{s - \frac{1+\delta}{2+\delta}}\right)^{k-1}. \tag{8.2.28}$$

Since $x \mapsto 1/x$ is monotonically decreasing and $(1+\delta)/(2+\delta) \in (0, 1)$,

$$\sum_{s=2}^{n-1} \frac{1}{s - \frac{1+\delta}{2+\delta}} \leq \sum_{s=2}^{n-1} \frac{1}{s-1} \leq 1 + \int_1^n \frac{1}{x} dx \leq \log(en). \tag{8.2.29}$$

Also, we use [V1, (8.3.9)] to bound $\Gamma(n - \frac{1+\delta}{2+\delta})/\Gamma(n) \leq C_\delta n^{-(1+\delta)/(2+\delta)}$, for some constant $C_\delta > 0$, so that

$$\mathbb{P}(\text{dist}_{\text{PA}_n^{(1,\delta)}(b)}(1, n) = k) \leq C_\delta n^{-\frac{1+\delta}{2+\delta}} \frac{\left(\frac{1+\delta}{2+\delta} \log(en)\right)^{k-1}}{(k-1)!}$$

$$= C_\delta \mathbb{P}\left(\text{Poi}\left(\frac{1+\delta}{2+\delta} \log(en)\right) = k - 1\right). \tag{8.2.30}$$

Now we are ready to prove (8.2.22) for $\mathrm{PA}_n^{(1,\delta)}(b)$. We note that o is chosen uar from $[n]$, so that, with C denoting a generic constant that may change from line to line,

$$
\begin{aligned}
\mathbb{P}(\mathrm{dist}_{\mathrm{PA}_n^{(1,\delta)}(b)}(1,o)=k) &= \frac{1}{n}\sum_{s=1}^{n}\mathbb{P}(\mathrm{dist}_{\mathrm{PA}_n^{(1,\delta)}(b)}(s,1)=k) \\
&\leq \frac{1}{n}\sum_{s=1}^{n} C_\delta s^{-\frac{1+\delta}{2+\delta}}\frac{\left(\frac{1+\delta}{2+\delta}\log(es)\right)^{k-1}}{(k-1)!} \\
&\leq \frac{\left(\frac{1+\delta}{2+\delta}\log(en)\right)^{k-1}}{n(k-1)!}\sum_{s=1}^{n}C_\delta s^{-\frac{1+\delta}{2+\delta}} \\
&\leq C\frac{\left(\frac{1+\delta}{2+\delta}\log(en)\right)^{k-1}}{(k-1)!}n^{-\frac{1+\delta}{2+\delta}} \\
&= C\mathbb{P}\left(\mathrm{Poi}\left(\frac{1+\delta}{2+\delta}\log(en)\right)=k-1\right).
\end{aligned}
\tag{8.2.31}
$$

Therefore,

$$
\mathbb{P}(\mathrm{dist}_{\mathrm{PA}_n^{(1,\delta)}(b)}(1,o)>k)\leq C\mathbb{P}\left(\mathrm{Poi}\left(\frac{1+\delta}{2+\delta}\log(es)\right)\geq k\right).
\tag{8.2.32}
$$

Fix $\varepsilon>0$ and take $k=k_n=\lceil\frac{(1+\varepsilon)(1+\delta)}{(2+\delta)}\log(en)\rceil$, to arrive at

$$
\begin{aligned}
&\mathbb{P}(\mathrm{dist}_{\mathrm{PA}_n^{(1,\delta)}(b)}(1,o)>k_n) \\
&\leq C\mathbb{P}\left(\mathrm{Poi}\left(\frac{1+\delta}{2+\delta}\log(en)\right)\geq\frac{(1+\varepsilon)(1+\delta)}{(2+\delta)}\log(en)\right)=o(1),
\end{aligned}
\tag{8.2.33}
$$

by the law of large numbers and for any $\varepsilon>0$, as required.

We continue by proving (8.2.23) for $\mathrm{PA}_n^{(1,\delta)}(b)$. By (8.2.30),

$$
\mathbb{P}(\mathrm{dist}_{\mathrm{PA}_n^{(1,\delta)}(b)}(1,n)>k)\leq K_\delta\mathbb{P}\left(\mathrm{Poi}\left(\frac{1+\delta}{2+\delta}\log(en)\right)\geq k\right).
\tag{8.2.34}
$$

Take $k_n=\lceil a\log n\rceil$ with $a>(1+\delta)/(2+\delta)$. We use the Borel–Cantelli lemma to see that $\mathrm{dist}(1,n)>k_n$ will occur only finitely often when (8.2.34) is summable. We then use the large-deviation bounds for Poisson random variables in [V1, Exercise 2.20] with $\lambda=(1+\delta)/(2+\delta)$ to obtain that

$$
\mathbb{P}(\mathrm{dist}_{\mathrm{PA}_n^{(1,\delta)}(b)}(1,n)>a\log n)\leq K_\delta n^{-p},
\tag{8.2.35}
$$

with

$$
p=a\log\left(\frac{a(2+\delta)}{1+\delta}\right)-a+\frac{1+\delta}{2+\delta}.
\tag{8.2.36}
$$

Let x be the solution of

$$
x(\log(x(2+\delta)/(1+\delta))-1)+\frac{1+\delta}{2+\delta}=1;
\tag{8.2.37}
$$

so that $x = (1 + \delta)/[(2 + \delta)\theta]$. Then, for every $a > x$,

$$\mathbb{P}(\text{dist}_{\text{PA}_n^{(1,\delta)}(b)}(1, n) > a \log n) = O(n^{-p}), \tag{8.2.38}$$

where p in (8.2.36) satisfies $p > 1$ for any $a > (1 + \delta)/[(2 + \delta)\theta]$. As a result, by the Borel–Cantelli lemma, the event $\{\text{dist}_{\text{PA}_n^{(1,\delta)}(b)}(1, n) > k_n\}$, with $k_n = \lceil a \log n \rceil$ and $a > (1 + \delta)/[(2 + \delta)\theta]$, occurs only finitely often, so that (8.2.23) holds. $\square$

Proof of the lower bound on $\text{dist}_{\text{PA}_n^{(1,\delta)}(b)}(1, o)$ *in Theorem 8.1 for* $\text{PA}_n^{(1,\delta)}(b)$. By (8.2.31),

$$\mathbb{P}(\text{dist}_{\text{PA}_n^{(1,\delta)}(b)}(1, o) \leq k) \leq C\mathbb{P}\left(\text{Poi}\left(\frac{1 + \delta}{2 + \delta}\log(en)\right) \leq k\right). \tag{8.2.39}$$

Fix $k_n = \lceil [(1 - \varepsilon)(1 + \delta)/(2 + \delta)] \log(2 + \delta)n \rceil$, and note that $\mathbb{P}(\text{dist}_{\text{PA}_n^{(1,\delta)}(b)}(1, o) \leq k_n) = o(1)$ by the law of large numbers. $\square$

To complete the proof that $\text{height}(\text{PA}_n^{(1,\delta)}(b))/\log n \xrightarrow{a.s.} \frac{(1+\delta)}{(2+\delta)\theta}$ in Theorem 8.1, we use the second-moment method to prove that $\text{height}(\text{PA}_n^{(1,\delta)}(b)) \leq (1 - \varepsilon)\frac{(1+\delta)}{(2+\delta)\theta}\log n$ has vanishing probability. Together with (8.2.24), this certainly proves that

$$\frac{\text{height}(\text{PA}_n^{(1,\delta)}(b))}{\log n} \xrightarrow{\mathbb{P}} \frac{(1+\delta)}{(2+\delta)\theta}.$$

However, since $\text{height}(\text{PA}_n^{(1,\delta)}(b))$ is a non-decreasing sequence of random variables, this also implies convergence almost surely, as we argue in more detail below Proposition 8.4.

We formalize the statement that $\text{height}(\text{PA}_n^{(1,\delta)}(b)) \leq \frac{(1-\varepsilon)(1+\delta)}{(2+\delta)\theta}\log n$ as follows:

Proposition 8.4 (Height of $\text{PA}_n^{(1,\delta)}(b)$ converges in probability) *For every* $\varepsilon > 0$ *there exists an* $\eta = \eta(\varepsilon) > 0$ *such that*

$$\mathbb{P}\left(\text{height}(\text{PA}_n^{(1,\delta)}(b)) \leq \frac{(1 - \varepsilon)(1 + \delta)}{(2 + \delta)\theta}\log n\right) \leq O(n^{-\eta}). \tag{8.2.40}$$

Proof of lower bound on $\text{height}(\text{PA}_n^{(1,\delta)}(b))$ *in Theorem 8.1 subject to Proposition 8.4.* Fix $\alpha > 0$, and take $n_k = n_k(\alpha) = e^{\alpha k}$. For any $\alpha > 0$, by Proposition 8.4 and the fact that $n_k^{-\eta}$ is summable in k, almost surely, $\text{height}(\text{PA}_{n_k}^{(1,\delta)}(b)) \geq \frac{(1-\varepsilon)(1+\delta)}{(2+\delta)\theta}\log n_k$. This proves the almost sure lower bound on $\text{height}(\text{PA}_n^{(1,\delta)}(b))$ along the subsequence $(n_k)_{k \geq 0}$. To extend this to an almost sure lower bound when $n \to \infty$, we use that $n \mapsto \text{height}(\text{PA}_n^{(1,\delta)}(b))$ is non-decreasing, so that, for every $n \in [n_{k-1}, n_k]$,

$$\text{height}(\text{PA}_n^{(1,\delta)}(b)) \geq \text{height}(\text{PA}_{n_{k-1}}^{(1,\delta)}(b))$$

$$\geq \frac{(1 - \varepsilon)(1 + \delta)}{(2 + \delta)\theta}\log n_{k-1}$$

$$\geq (1 - \varepsilon)(1 - \alpha)\frac{(1 + \delta)}{(2 + \delta)\theta}\log n, \tag{8.2.41}$$

where the third inequality follows from the almost sure lower bound on $\text{height}(\text{PA}_{n_{k-1}}^{(1,\delta)}(b))$. The above bound holds for all $\varepsilon, \alpha > 0$, so that letting $\varepsilon, \alpha \searrow 0$ proves our claim. $\square$

We omit the proof of Proposition 8.4, and refer to the notes and discussion in Section 8.10 for details. The proof relies on a continuous-time embedding of preferential attachment models, first invented by Pittel (1994), and the height of such trees, found using a beautiful argument by Kingman (1975) (see Exercises 8.4 and 8.5). In the remainder of the proofs in this chapter, we rely only on the upper bound in (8.2.24).

We continue by proving the lower bound on $\text{diam}\big(\text{PA}_n^{(1,\delta)}(b)\big)$ in Theorem 8.1:

Proof of the lower bound on $\text{diam}\big(\text{PA}_n^{(1,\delta)}(b)\big)$ *in Theorem 8.1.* We use the lower bound on $\text{height}(\text{PA}_n^{(1,\delta)}(b))$ in Theorem 8.1 and the decomposition of scale-free trees in Theorem 5.4. Theorem 5.4 states that $\text{PA}_n^{(1,\delta)}(b)$ can be decomposed into two scale-free trees having similar distributions as copies $\text{PA}_{S_1(n)}^{(1,\delta)}(b_1)$ and $\text{PA}_{n-S_1(n)}^{(1,\delta)}(b_2)$, where $(\text{PA}_n^{(1,\delta)}(b_1))_{n\geq 1}$ and $(\text{PA}_n^{(1,\delta)}(b_2))_{n\geq 1}$ are *independent* scale-free tree processes, and the law of $S_1(n)$ is described in (5.2.30). By this tree decomposition,

$$\text{diam}\big(\text{PA}_n^{(1,\delta)}(b)\big) \geq \text{height}\big(\text{PA}_{S_1(n)}^{(1,\delta)}(b_1)\big) + \text{height}\big(\text{PA}_{n-S_1(n)}^{(1,\delta)}(b_2)\big). \qquad (8.2.42)$$

The two trees $(\text{PA}_n^{(1,\delta)}(b_1))_{n\geq 1}$ and $(\text{PA}_n^{(1,\delta)}(b_2))_{n\geq 1}$ are not *exactly* equal in distribution to $(\text{PA}_n^{(1,\delta)}(b))_{n\geq 1}$, because the initial degrees of the starting vertices at time $n = 2$ are different. However, the precise almost sure scaling in Theorem 5.4 does not depend in a sensitive way on the degrees d_1 and d_2 of the first two vertices, and furthermore the height of the scale-free tree in Theorem 8.1 does not depend on the starting graphs $\text{PA}_2^{(1,\delta)}(b_1)$ and $\text{PA}_2^{(1,\delta)}(b_2)$ (see the remark below Theorem 8.1). Since $S_1(n)/n \xrightarrow{a.s.} U$, where U has a Beta distribution with parameters $a = (3+\delta)/(2+\delta)$ and $b = (1+\delta)/(2+\delta)$, we obtain that $\text{height}\big(\text{PA}_{S_1(n)}^{(1,\delta)}(b_1)\big)/\log n \xrightarrow{a.s.} \frac{(1+\delta)}{(2+\delta)\theta}$ and $\text{height}\big(\text{PA}_{n-S_1(n)}^{(1,\delta)}(b_2)\big)/\log n \xrightarrow{a.s.} \frac{(1+\delta)}{(2+\delta)\theta}$. Thus, we conclude that, almost surely for all large n,

$$\frac{\text{diam}\big(\text{PA}_n^{(1,\delta)}(b)\big)}{\log n} \geq (1-\varepsilon)\frac{2(1+\delta)}{(2+\delta)\theta}. \qquad (8.2.43)$$

We proceed by proving the convergence of $\text{dist}_{\text{PA}_n^{(1,\delta)}(b)}(o_1, o_2)/\log n$ in Theorem 8.1. We write

$$\text{dist}_{\text{PA}_n^{(1,\delta)}(b)}(o_1, o_2) = \text{dist}_{\text{PA}_n^{(1,\delta)}(b)}(1, o_1) + \text{dist}_{\text{PA}_n^{(1,\delta)}(b)}(1, o_2)$$
$$- \text{dist}_{\text{PA}_n^{(1,\delta)}(b)}(1, V), \qquad (8.2.44)$$

where V is the last vertex that is on both paths, i.e., from $1 \to o_1$ as well as from $1 \to o_2$. The asymptotics of the first two terms are identified in (8.2.3) in Theorem 8.1, so that it suffices to show that $\text{dist}_{\text{PA}_n^{(1,\delta)}(b)}(1, V) = o_{\mathbb{P}}(\log n)$, as stated in Lemma 8.5. $\qquad\square$

Lemma 8.5 (Distance to most recent common ancestor) *Consider* $\text{PA}_n^{(1,\delta)}(b)$. *Let* o_1, o_2 *be two vertices in* $[n]$ *chosen independently and uar from* $[n]$. *Let* V *be the oldest vertex that the paths from 1 to* o_1 *and 1 to* o_2 *have in common in* $\text{PA}_n^{(1,\delta)}(b)$. *Then,*

$$\frac{\text{dist}_{\text{PA}_n^{(1,\delta)}(b)}(1, V)}{\log n} \xrightarrow{\mathbb{P}} 0. \qquad (8.2.45)$$

We leave the proof of Lemma 8.5 as Exercise 8.16, which we postpone until we have discussed *path-counting techniques* for preferential attachment models.

Proof of Theorem 8.1 for $\mathrm{PA}_n^{(1,\delta)}(d)$ *and* $\mathrm{PA}_n^{(1,\delta)}(a)$. The proof of Theorem 8.1 for $\mathrm{PA}_n^{(1,\delta)}(d)$ follows the same line of argument as for $\mathrm{PA}_n^{(1,\delta)}(b)$, where we note that the only difference in $\mathrm{PA}_n^{(1,\delta)}(d)$ and $\mathrm{PA}_n^{(1,\delta)}(b)$ is in the graph for $n = 2$. We omit further details.

To prove Theorem 8.1 for $\mathrm{PA}_n^{(1,\delta)}(a)$, we note that the connected components of $\mathrm{PA}_n^{(1,\delta)}(a)$ are similar in distribution to single scale-free tree $\mathrm{PA}_{t_1}^{(1,\delta)}(b_1), \ldots, \mathrm{PA}_{t_{N_n}}^{(1,\delta)}(b_{N_n})$, apart from the initial degree of the root. Here t_i denotes the size of the ith tree at time n, and we recall that N_n denotes the total number of trees at time n. Since $N_n/\log n \overset{\mathbb{P}}{\longrightarrow} (1+\delta)/(2+\delta)$ by Exercise 5.26, whp the largest connected component has size at least $\varepsilon n/\log n$. Since $\log(\varepsilon n/\log n) = (1+o(1))\log n$, the distances in these trees are closely related to those in $\mathrm{PA}_n^{(1,\delta)}(b)$. Theorem 8.1 for $\mathrm{PA}_n^{(1,\delta)}(a)$ then follows similarly to the proof for $\mathrm{PA}_n^{(1,\delta)}(b)$. $\quad\square$

8.3 SMALL-WORLD PHENOMENA IN PREFERENTIAL ATTACHMENT MODELS

In the following sections we investigate typical distances in preferential attachment models for $m \geq 2$. These results are not as complete as those for the inhomogeneous random graphs or configuration models discussed in Chapters 6 and 7, respectively. This is partly due to the fact that the dynamics in preferential attachment models make them substantially harder to analyze than those static models.

By Theorem 5.27, $\mathrm{PA}_n^{(m,\delta)}(a)$ is whp connected when $m \geq 2$, and the same is trivially true for $\mathrm{PA}_n^{(m,\delta)}(b)$ and $\mathrm{PA}_n^{(m,\delta)}(d)$. In a connected graph, the typical distance is the graph distance between two vertices chosen uar from $[n]$. Recall further that the power-law degree exponent for $\mathrm{PA}_n^{(m,\delta)}(a)$ is equal to $\tau = 3 + \delta/m$. Therefore, $\tau > 3$ precisely when $\delta > 0$. For the generalized random graph and the configuration model, we have seen that distances are logarithmic in the size of the graph when $\tau > 3$ and doubly logarithmic when $\tau \in (2, 3)$. We will see similar results for preferential attachment models.

Logarithmic Distances in PA Models with $m \geq 2$ and $\delta > 0$

We start by investigating the case where $\delta > 0$, so that the power-law degree exponent $\tau = 3 + \delta/m$ satisfies $\tau > 3$. Here, the typical distance is logarithmic in the graph size:

Theorem 8.6 (Logarithmic bounds for typical distances for $\delta > 0$) *Consider* $\mathrm{PA}_n^{(m,\delta)}(a)$ *with $m \geq 2$ and $\delta > 0$. Let o_1, o_2 be chosen independently and uar from $[n]$. There exist $0 < c_1 \leq c_2 < \infty$ such that, as $n \to \infty$ and whp,*

$$c_1 \log n \leq \mathrm{dist}_{\mathrm{PA}_n^{(m,\delta)}(a)}(o_1, o_2) \leq c_2 \log n. \tag{8.3.1}$$

These results also apply to $\mathrm{PA}_n^{(m,\delta)}(b)$ *and* $\mathrm{PA}_n^{(m,\delta)}(d)$ *under identical conditions.*

It turns out that any $c_1 < 1/\log\nu$ suffices, where $\nu = \|\boldsymbol{T}_\kappa\| > 1$ is the operator norm of the offspring operator $\boldsymbol{T}_\kappa$ of the Pólya point tree defined in (8.5.61) below. It turns out that ν has an interpretation not only as the operator norm $\|\boldsymbol{T}_\kappa\|$ of $\boldsymbol{T}_\kappa$ but also as the *spectral radius* of $\boldsymbol{T}_\kappa$. As a result, we believe that $\mathrm{dist}_{\mathrm{PA}_n^{(m,\delta)}(a)}(o_1, o_2)/\log_\nu n \overset{\mathbb{P}}{\longrightarrow} 1$, but we lack a proof for the upper bound.

See Figure 8.1 for a simulation of the typical distances in $\mathrm{PA}_n^{(m,\delta)}(a)$ with $m = 2, \delta = -1$ (so that $\tau = 2.5$) and $m = 2, \delta = 1$ (so that $\tau = 3.5$), respectively.

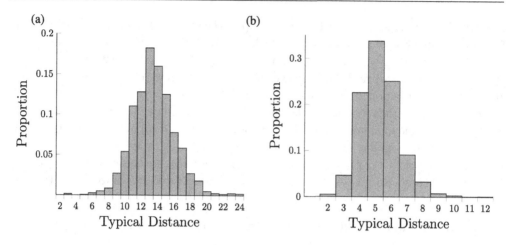

Figure 8.1 Typical distances between 2,000 pairs of vertices in the preferential attachment model with $n = 100,000$ and (a) $\tau = 2.5$; (b) $\tau = 3.5$.

Doubly Logarithmic Distances in PA Models with $m \geq 2$ and $\delta < 0$

We continue by discussing the case where $\delta \in (-m, 0)$, so that $\tau = 3 + \delta/m \in (2, 3)$. In this case, it turns out that distances again grow doubly logarithmically in the graph size:

Theorem 8.7 (Ultra-small typical distances for $\delta < 0$) *Consider* $\mathrm{PA}_n^{(m,\delta)}(a)$ *with* $m \geq 2$ *and* $\delta \in (-m, 0)$. *Let* o_1, o_2 *be chosen independently and uar from* $[n]$. *As* $n \to \infty$,

$$\frac{\mathrm{dist}_{\mathrm{PA}_n^{(m,\delta)}(a)}(o_1, o_2)}{\log\log n} \xrightarrow{\mathbb{P}} \frac{4}{|\log(\tau - 2)|}. \tag{8.3.2}$$

These results also apply to $\mathrm{PA}_n^{(m,\delta)}(b)$ *and* $\mathrm{PA}_n^{(m,\delta)}(d)$ *under identical conditions.*

Exercise 8.6 investigates an example of the above result. Interestingly, the limiting constant $4/|\log(\tau - 2)|$ appearing in Theorem 8.7 replaces the limit $2/|\log(\tau - 2)|$ in Theorem 6.3 for the Norros–Reittu model $\mathrm{NR}_n(\boldsymbol{w})$ and in Theorem 7.2 for the configuration model $\mathrm{CM}_n(\boldsymbol{d})$ when the power-law exponent τ satisfies $\tau \in (2, 3)$. Thus, typical distances are *twice as large* for $\mathrm{PA}_n^{(m,\delta)}$ compared with $\mathrm{CM}_n(\boldsymbol{d})$ with the same power-law exponent. This can be intuitively explained as follows. For the configuration model $\mathrm{CM}_n(\boldsymbol{d})$, vertices with degree $d \gg 1$ are likely to be directly connected to vertices of degree $\approx d^{1/(\tau-2)}$ (see, e.g., Lemma 7.12), which is the whole idea behind the power-iteration methodology.

For $\mathrm{PA}_n^{(m,\delta)}$, this is *not* the case. However, pairs of high-degree vertices are likely to be at distance 2, as whp there is a young vertex that connects to both older vertices. This makes distances in $\mathrm{PA}_n^{(m,\delta)}$ effectively twice as big as those for $\mathrm{CM}_n(\boldsymbol{d})$ with the same degree sequence. This effect is special for $\delta < 0$ and is studied in more detail in Exercises 8.7 and 8.8, while Exercise 8.9 shows that this effect is absent when $\delta > 0$.

Distances in Preferential Attachment Models with $m \geq 2$ and $\delta = 0$

We close this section by discussing the case $\delta = 0$, where the power-law exponent τ satisfies $\tau = 3$. For $\mathrm{NR}_n(\boldsymbol{w})$ and $\mathrm{CM}_n(\boldsymbol{d})$, distances grow as $\log n/\log\log n$ in this case (recall Theorem 6.22 and Corollary 7.7). The same turns out to be true for $\mathrm{PA}_n^{(m,0)}(a)$:

Theorem 8.8 (Typical distances for $\delta = 0$)   *Consider* $\mathrm{PA}_n^{(m,0)}(a)$ *with* $m \geq 2$. *Then, with* o_1, o_2 *chosen independently and uar from* $[n]$,

$$\mathrm{dist}_{\mathrm{PA}_n^{(m,0)}(a)}(o_1, o_2) \frac{\log \log n}{\log n} \xrightarrow{\mathbb{P}} 1. \tag{8.3.3}$$

The lower bounds in Theorem 8.8 also apply to $\mathrm{PA}_n^{(m,0)}(b)$ and $\mathrm{PA}_n^{(m,0)}(d)$.

Universality in Distances for Scale-Free Graphs

The available results are all consistent with the prediction that distances in preferential attachment models have asymptotics similar to the distances in generalized random graphs or configuration models with the same degree sequence. Such results suggest a strong form of *universality*, which is interesting in its own right. However, certain *local* effects of the graph may change graph distances somewhat, as exemplified by the fact that the distances in Theorem 8.7 are asymptotically twice as large as those for the Norros–Reittu model $\mathrm{NR}_n(w)$ in Theorem 6.3 and for the configuration model $\mathrm{CM}_n(d)$ in Theorem 7.2. This shows that the details of the model *do* matter.

Organization of the Proof of Small-World Properties of the PA Model

We prove small-world and ultra-small-world results in the following four sections. These proofs are organized as follows. We start in Section 8.4 by discussing *path-counting techniques* for preferential attachment models. While preferential attachment models lack the independence or weak dependence between the edge statuses of the graph that are present in inhomogeneous random graphs and configuration models, it turns out that the probability of the existence of paths can still be bounded from above by products of probabilities, owing to the inherent *negative correlations* between edge statuses. Therefore, the lower bounds on typical distances can be obtained in a very similar way to those for rank-1 inhomogeneous random graphs in Theorems 6.4 and 6.7 or to those for configuration models in Theorems 7.6 and 7.8. These bounds also apply to $\delta = 0$. The resulting lower bounds are obtained in Sections 8.5 and 8.6 for $\delta \geq 0$ and $\delta < 0$, respectively.

For the upper bounds, we use those in Section 8.2 for $m = 1$ and $\delta > 0$, which are obviously not sharp. This explains why the upper bound in Theorem 8.6 is suboptimal. For $\delta < 0$, instead, in Section 8.7 we again define the notion of a *core*, as for configuration models in Theorem 7.13. However, owing to the preferential attachment growth dynamics, the precise definition is more involved. The idea of the core is also used to analyze the diameter for $m \geq 2$ and $\delta \in (-m, 0)$ in Section 8.8.

These proofs together complete the proofs of the typical distances in $\mathrm{PA}_n^{(m,\delta)}$. Sub-results are stated explicitly, as these are sometimes sharper than the results stated here, and are thus interesting in their own right.

8.4 Path Counting in Preferential Attachment Models

In this section we study the probability that a certain path is present in $\mathrm{PA}_n^{(m,\delta)}$. Recall from Definition 6.5 that we call a k-step path $\vec{\pi} = (\pi_0, \pi_1, \ldots, \pi_k)$ *self-avoiding* when $\pi_i \neq \pi_j$ for all $1 \leq i < j \leq k$. The following proposition studies the probability that a path $\vec{\pi}$ is present in $\mathrm{PA}_n^{(m,\delta)}$:

Proposition 8.9 (Path counting in preferential attachment models)   *Consider* $\mathrm{PA}_n^{(m,\delta)}(a)$ *with* $m \geq 2$. *Denote* $\gamma = m/(2m+\delta)$. *Fix* $k \geq 0$ *and let* $\vec{\pi} = (\pi_0, \pi_1, \ldots, \pi_k)$ *be a k-step self-avoiding path. Then, there exists a constant $C > 0$ such that, for all $k \geq 1$,*

$$\mathbb{P}(\vec{\pi} \subseteq \mathrm{PA}_n^{(m,\delta)}(a)) \leq (Cm)^k \prod_{i=0}^{k-1} \frac{1}{(\pi_i \wedge \pi_{i+1})^\gamma (\pi_i \vee \pi_{i+1})^{1-\gamma}}. \tag{8.4.1}$$

This result also applies to $\mathrm{PA}_n^{(m,\delta)}(b)$ *and* $\mathrm{PA}_n^{(m,\delta)}(d)$.

Paths are formed by repeatedly forming edges. When $m = 1$, paths always go from younger to older vertices. When $m \geq 2$ this monotonicity property is lost, which makes proofs harder. We start by investigating intersections of events that specify which edges are present in $\mathrm{PA}_n^{(m,\delta)}$. More precise results appear in Propositions 8.14 and 8.15 below.

8.4.1 Negative Correlation of Connection Events

We recall that $\{u \overset{j}{\rightsquigarrow} v\}$, for $j \in [m]$, denotes the event that the jth edge of vertex u is attached to the earlier vertex v, where $u, v \in [n]$. It is a direct consequence of the definition of preferential attachment models that the event $\{u \overset{j}{\rightsquigarrow} v\}$ increases the preference for vertex v, and hence decreases (in a relative way) the preference for the other vertices in $[n] \setminus \{v\}$. Intuitively, another way of expressing this effect is to say that, for different $v_1 \neq v_2$, the events $\{u_1 \overset{j_1}{\rightsquigarrow} v_1\}$ and $\{u_2 \overset{j_2}{\rightsquigarrow} v_2\}$ are negatively correlated. We now formalize this result.

For an integer $n_v \geq 1$ and $i \in [n_v]$, we denote the event that the $j_i^{(v)}$th edge of vertex $u_i^{(v)}$ is attached to the earlier vertex v by

$$\mathcal{E}_{n_v,v} = \mathcal{E}_{n_v,v}\big((u_i^{(v)}, j_i^{(v)})_{i\in[n_v]}\big) = \bigcap_{i\in[n_v]} \{u_i^{(v)} \overset{j_i^{(v)}}{\rightsquigarrow} v\}. \tag{8.4.2}$$

The following lemma shows that the events $\mathcal{E}_{n_v,v}$, for *different* v, are *negatively correlated*:

Lemma 8.10 (Negative correlation for edge connections in preferential attachment models) *Consider* $\mathrm{PA}_n^{(m,\delta)}(a)$ *with* $m \geq 2$. *Fix* $k \geq 1$. *For distinct* $v_1, v_2, \ldots, v_k \in [n]$ *and all* $n_{v_1}, \ldots, n_{v_k} \geq 1$,

$$\mathbb{P}\Big(\bigcap_{t\in[k]} \mathcal{E}_{n_{v_t},v_t}\Big) \leq \prod_{t\in[k]} \mathbb{P}(\mathcal{E}_{n_{v_t},v_t}). \tag{8.4.3}$$

These results also apply to $\mathrm{PA}_n^{(m,\delta)}(b)$ *and* $\mathrm{PA}_n^{(m,\delta)}(d)$ *under identical conditions.*

Proof   We prove only the statement for $\mathrm{PA}_n^{(m,\delta)}(a)$; the proofs for $\mathrm{PA}_n^{(m,\delta)}(b)$ and $\mathrm{PA}_n^{(m,\delta)}(d)$ are very similar and are left as Exercises 8.10–8.13.

We define the edge number of the event $\{u \overset{j}{\rightsquigarrow} v\}$ to be $m(u-1)+j$, which is the order of the edge when we consider the edges as being attached in sequence in $\mathrm{PA}_{mn}^{(1,\delta/m)}(a)$.

We use induction on the largest edge number present in the events $\mathcal{E}_{n_{v_1},v_1}, \ldots, \mathcal{E}_{n_{v_k},v_k}$. The induction hypothesis is that (8.4.3) holds for all k, all distinct $v_1, v_2, \ldots, v_k \in [n]$, all $n_{v_1}, \ldots, n_{v_k} \geq 1$, and all choices of $u_i^{(v_s)}, j_i^{(v_s)}$ such that $\max_{i,s} m(u_i^{(v_s)} - 1) + j_i^{(v_s)} \leq e$, where induction is performed with respect to e.

To initialize the induction, we note that for $e = 1$ the induction hypothesis holds trivially, since $\bigcap_{s=1}^{k} \mathcal{E}_{n_{v_s}, v_s}$ can be empty or can consist of exactly one event, and in the latter case there is nothing to prove. This initializes the induction.

To advance the induction, we assume that (8.4.3) holds for all k, all distinct vertices $v_1, v_2, \ldots, v_k \in [n]$, all $n_{v_1}, \ldots, n_{v_k} \geq 1$, and all choices of $u_i^{(v_s)}, j_i^{(v_s)}$ such that we have $\max_{i,s} m(u_i^{(v_s)} - 1) + j_i^{(v_s)} \leq e - 1$, and we extend it to all k, all distinct vertices $v_1, v_2, \ldots, v_k \in [n]$, all $n_{v_1}, \ldots, n_{v_k} \geq 1$, and all choices of $u_i^{(v_s)}, j_i^{(v_s)}$ such that $\max_{i,s} m(u_i^{(v_s)} - 1) + j_i^{(v_s)} \leq e$. Clearly, by induction, we may restrict attention to the case for which $\max_{i,s} m(u_i^{(v_s)} - 1) + j_i^{(v_s)} = e$.

We note that there is a unique choice of u, j such that $m(u - 1) + j = e$. There are two possibilities: (1) either there is exactly *one* choice of s and $u_i^{(v_s)}, j_i^{(v_s)}$ such that $u_i^{(v_s)} = u, j_i^{(v_s)} = j$, or (2) there are at least *two* such choices. In the latter case, $\bigcap_{t \in [k]} \mathcal{E}_{n_{v_t}, v_t} = \varnothing$, since the eth edge is connected to a *unique* vertex. Hence, there is nothing to prove.

We are left with investigating the case where there exists a unique s and $u_i^{(v_s)}, j_i^{(v_s)}$ such that $u_i^{(v_s)} = u, j_i^{(v_s)} = j$. Denote the restriction of $\mathcal{E}_{n_{v_s}, v_s}$ to all *other* edges by

$$\mathcal{E}'_{n_{v_s}, v_s} = \bigcap_{i \in [n_v]: \, (u_i^{(v_s)}, j_i^{(v_s)}) \neq (u,j)} \{ u_i^{(v_s)} \overset{j_i^{(v_s)}}{\rightsquigarrow} v_s \}. \tag{8.4.4}$$

Then we can write

$$\bigcap_{t \in [k]} \mathcal{E}_{n_{v_t}, v_t} = \{ u \overset{j}{\rightsquigarrow} v_s \} \cap \mathcal{E}'_{n_{v_s}, v_s} \cap \bigcap_{t \in [k]: \, v_t \neq v_s} \mathcal{E}_{n_{v_t}, v_t}. \tag{8.4.5}$$

By construction, all edge numbers of events in $\mathcal{E}'_{n_v, v} \cap \bigcap_{i=1: \, s_i \neq s}^{k} \mathcal{E}_{n_{v_i}, v_i}$ are at most $e - 1$.

By conditioning, we obtain

$$\mathbb{P}\Big(\bigcap_{t \in [k]} \mathcal{E}_{n_{v_t}, v_t} \Big) \leq \mathbb{E}\Big[\mathbb{1}_{\mathcal{E}'_{n_{v_s}, v_s} \cap \bigcap_{t \in [k]: \, v_t \neq v_s} \mathcal{E}_{n_{v_t}, v_t}} \mathbb{P}(u \overset{j}{\rightsquigarrow} v_s \mid \mathrm{PA}_{e-1}^{(1, \delta/m)}(a)) \Big], \tag{8.4.6}$$

where we have used that the event $\mathcal{E}'_{n_v, v} \cap \bigcap_{t \in [k]: \, v_t \neq v_s} \mathcal{E}_{n_{v_t}, v_t}$ is measurable with respect to $\mathrm{PA}_{e-1}^{(1, \delta/m)}(a)$. We compute

$$\mathbb{P}(u \overset{j}{\rightsquigarrow} v_s \mid \mathrm{PA}_{e-1}^{(1, \delta/m)}(a)) = \frac{D_{v_s}(u - 1, j - 1) + \delta}{z_{u,j}}, \tag{8.4.7}$$

where we recall that $D_{v_s}(u - 1, j - 1)$ is the degree of vertex v_s after $j - 1$ edges of vertex u have been attached, and we write the normalization constant in (8.4.7) as

$$z_{u,j} = z_{u,j}(\delta, m) = (2m + \delta)(u - 1) + (j - 1)(2 + \delta/m) + 1 + \delta. \tag{8.4.8}$$

We wish to use the induction hypothesis. For this, we note that

$$D_{v_s}(u - 1, j - 1) = m + \sum_{(u',j'): \, mu' + j' \leq e - 1} \mathbb{1}_{\{u' \overset{j'}{\rightsquigarrow} v_s\}}, \tag{8.4.9}$$

where we recall that $e - 1 = m(u - 1) + j - 1$.

Each of the events $\{ u' \overset{j'}{\rightsquigarrow} v_s \}$ in (8.4.9) has edge number strictly smaller than e and occurs with a non-negative multiplicative constant. As a result, we may use the induction hypothesis for each of these terms. Thus, we obtain, using also that $m + \delta \geq 0$,

$$\mathbb{P}\Big(\bigcap_{t\in[k]}\mathcal{E}_{n_{v_t},v_t}\Big) \le \frac{m+\delta}{z_{u,j}}\mathbb{P}(\mathcal{E}'_{n_{v_s},v_s})\prod_{t\in[k]:\,v_t\ne v_s}\mathbb{P}(\mathcal{E}_{n_{v_t},v_t}) \tag{8.4.10}$$

$$+\sum_{(u',j'):\,mu'+j'\le e-1}\frac{\mathbb{P}(\mathcal{E}'_{n_v,v}\cap\{u'\xrightarrow{j'}v_s\})}{z_{u,j}}\prod_{t\in[k]:\,v_t\ne v_s}\mathbb{P}(\mathcal{E}_{n_{v_t},v_t}).$$

We use (8.4.9) to recombine the above as follows:

$$\mathbb{P}\Big(\bigcap_{t\in[k]}\mathcal{E}_{n_{v_t},v_t}\Big) \le \mathbb{E}\Big[\mathbb{1}_{\mathcal{E}'_{n_{v_s},v_s}}\frac{D_{v_s}(u-1,j-1)+\delta}{z_{u,j}}\Big]\prod_{t\in[k]:\,v_t\ne v_s}\mathbb{P}(\mathcal{E}_{n_{v_t},v_t}), \tag{8.4.11}$$

and the advancement of the induction hypothesis is complete when we note that

$$\mathbb{E}\Big[\mathbb{1}_{\mathcal{E}'_{n_{v_s},v_s}}\frac{D_{v_s}(u-1,j-1)+\delta}{z_{u,j}}\Big]=\mathbb{P}(\mathcal{E}_{n_{v_s},v_s}). \tag{8.4.12}$$

The claim in Lemma 8.10 follows by induction.   □

8.4.2 PROBABILITIES OF PATH CONNECTION EVENTS

We next study the probabilities of the events $\mathcal{E}_{n_v,v}$ when $n_v\le 2$:

Lemma 8.11 (Edge connection events for at most two edges)   *Consider* $\mathrm{PA}_n^{(m,\delta)}(a)$ *and denote* $\gamma=m/(2m+\delta)$. *There exist absolute constants* $M_1=M_1(\delta,m)$, $M_2=M_2(\delta,m)$, *such that the following statements hold:*

(a) For $m=1$ and any $u>v$,

$$\mathbb{P}(u\xrightarrow{1}v)=\frac{1+\delta}{2+\delta}\frac{\Gamma(u)\Gamma(v-\frac{1}{2+\delta})}{\Gamma(u+\frac{1+\delta}{2+\delta})\Gamma(v)}. \tag{8.4.13}$$

Consequently, for $m\ge 2$, any $j\in[m]$ and $u>v$,

$$\mathbb{P}(u\xrightarrow{j}v)=\frac{m+\delta}{2m+\delta}\frac{1}{u^{1-\gamma}v^\gamma}(1+o(1))\le\frac{M_1}{u^{1-\gamma}v^\gamma}; \tag{8.4.14}$$

the asymptotics in the first equality in (8.4.14) refer to the limit when v grows large.

(b) For $m=1$ and any $u_2>u_1>v$,

$$\mathbb{P}(u_1\xrightarrow{1}v,u_2\xrightarrow{1}v)=\frac{1+\delta}{2+\delta}\frac{\Gamma(u_2)\Gamma(u_1+\frac{1}{2+\delta})\Gamma(v+\frac{1+\delta}{2+\delta})}{\Gamma(u_2+\frac{1+\delta}{2+\delta})\Gamma(u_1+1)\Gamma(v+\frac{3+\delta}{2+\delta})}. \tag{8.4.15}$$

Consequently, for $m\ge 2$, any $j_1,j_2\in[m]$ and $u_2>u_1>v$,

$$\mathbb{P}\Big(u_1\xrightarrow{j_1}v,u_2\xrightarrow{j_2}v\Big)\le\frac{m+\delta}{2m+\delta}\frac{m+1+\delta}{2m+\delta}\frac{1}{(u_1u_2)^{1-\gamma}v^{2\gamma}}(1+o(1)) \tag{8.4.16}$$

$$\le\frac{M_2}{(u_1u_2)^{1-\gamma}v^{2\gamma}};$$

the asymptotics in the first equality in (8.4.16) refer to the limit when v grows large.

(c) The asymptotics in (8.4.14) and (8.4.16) also apply to $\mathrm{PA}_n^{(m,\delta)}(b)$ and $\mathrm{PA}_n^{(m,\delta)}(d)$.

Proof We prove only the results for $\mathrm{PA}_n^{(m,\delta)}(a)$ in parts (a) and (b); the proofs for $\mathrm{PA}_n^{(m,\delta)}(b)$ and $\mathrm{PA}_n^{(m,\delta)}(d)$ in part (c) are similar. Further, we will prove only (8.4.13) and (8.4.15); the bounds in (8.4.14) and (8.4.16) follow immediately from the Stirling-type formula in [V1, (8.3.9)].

For $m = 1$, part (a) follows from Remark 8.3. The proof of (8.4.14) for $m \geq 2$ follows by recalling that $u \overset{j}{\rightsquigarrow} v$ occurs when $(m-1)u + j \overset{1}{\rightsquigarrow} [mv] \setminus [m(v-1)]$, since $j \in [m]$ and $m \geq 2$ are fixed.

We proceed with the proof of part (b) and start with (8.4.15) for $m = 1$. Take $u_2 > u_1$. We compute

$$\mathbb{P}\Big(u_1 \overset{1}{\rightsquigarrow} v, u_2 \overset{1}{\rightsquigarrow} v\Big) = \mathbb{E}\Big[\mathbb{P}\big(u_1 \overset{1}{\rightsquigarrow} v, u_2 \overset{1}{\rightsquigarrow} v \,\big|\, \mathrm{PA}_{u_2-1}^{(1,\delta)}(a)\big)\Big]$$

$$= \mathbb{E}\Big[\mathbbm{1}_{\{u_1 \overset{1}{\rightsquigarrow} v\}} \frac{D_v(u_2-1)+\delta}{(u_2-1)(2+\delta)+1+\delta}\Big]. \tag{8.4.17}$$

We use the following iteration, for $u > u_1$:

$$\mathbb{E}\Big[\mathbbm{1}_{\{u_1 \overset{1}{\rightsquigarrow} v\}}\big(D_v(u)+\delta\big)\Big]$$

$$= \Big(1 + \frac{1}{(2+\delta)(u-1)+1+\delta}\Big)\mathbb{E}\Big[\mathbbm{1}_{\{u_1 \overset{1}{\rightsquigarrow} v\}}\big(D_v(u-1)+\delta\big)\Big]$$

$$= \frac{u}{u-1+\frac{1+\delta}{2+\delta}}\mathbb{E}\Big[\mathbbm{1}_{\{u_1 \overset{1}{\rightsquigarrow} v\}}\big(D_v(u-1)+\delta\big)\Big]$$

$$= \frac{\Gamma(u+1)\Gamma(u_1+\frac{1+\delta}{2+\delta})}{\Gamma(u+\frac{1+\delta}{2+\delta})\Gamma(u_1+1)}\mathbb{E}\Big[\mathbbm{1}_{\{u_1 \overset{1}{\rightsquigarrow} v\}}\big(D_v(u_1)+\delta\big)\Big]. \tag{8.4.18}$$

Therefore,

$$\mathbb{P}\Big(u_1 \overset{1}{\rightsquigarrow} v, u_2 \overset{1}{\rightsquigarrow} v\Big)$$

$$= \frac{1}{(u_2-1)(2+\delta)+1+\delta} \frac{\Gamma(u_2)\Gamma(u_1+\frac{1+\delta}{2+\delta})}{\Gamma(u_2-\frac{1}{2+\delta})\Gamma(u_1+1)}\mathbb{E}\Big[\mathbbm{1}_{\{u_1 \overset{1}{\rightsquigarrow} v\}}\big(D_v(u_1)+\delta\big)\Big]$$

$$= \frac{1}{2+\delta} \frac{\Gamma(u_2)\Gamma(u_1+\frac{1+\delta}{2+\delta})}{\Gamma(u_2+\frac{1+\delta}{2+\delta})\Gamma(u_1+1)}\mathbb{E}\Big[\mathbbm{1}_{\{u_1 \overset{1}{\rightsquigarrow} v\}}\big(D_v(u_1)+\delta\big)\Big]. \tag{8.4.19}$$

We thus need to compute $\mathbb{E}\Big[\mathbbm{1}_{\{u_1 \overset{1}{\rightsquigarrow} v\}}\big(D_v(u_1)+\delta\big)\Big]$. We use recursion to obtain

$$\mathbb{E}\Big[\mathbbm{1}_{\{u_1 \overset{1}{\rightsquigarrow} v\}}\big(D_v(u_1)+\delta\big)\,\big|\,\mathrm{PA}_{u_1-1}^{(m,\delta)}(a)\Big]$$

$$= (D_v(u_1-1)+1+\delta)\mathbb{P}\big(u_1 \overset{1}{\rightsquigarrow} v \,\big|\, \mathrm{PA}_{u_1-1}^{(m,\delta)}(a)\big)$$

$$= \frac{(D_v(u_1-1)+\delta)(D_v(u_1-1)+1+\delta)}{(u_1-1)(2+\delta)+1+\delta}, \tag{8.4.20}$$

since $D_v(u_1) = D_v(u_1-1)+1$ on the event $\{u_1 \overset{1}{\rightsquigarrow} v\}$. By [V1, Proposition 8.15],

$$\mathbb{E}[(D_v(u)+\delta)(D_v(u)+1+\delta)] = \frac{\Gamma(u+\frac{3+\delta}{2+\delta})\Gamma(v+\frac{1+\delta}{2+\delta})}{\Gamma(u+\frac{1+\delta}{2+\delta})\Gamma(v+\frac{3+\delta}{2+\delta})}(2+\delta)(1+\delta). \tag{8.4.21}$$

Consequently,

$$
\begin{aligned}
&\mathbb{E}\left[\mathbb{1}_{\{u_1 \overset{1}{\rightsquigarrow} v\}}\left(D_v(u_1 - 1) + \delta\right)\right] \\
&= \frac{\Gamma(u_1 + \frac{1}{2+\delta})\Gamma(v + \frac{1+\delta}{2+\delta})}{[(u_1 - 1)(2 + \delta) + 1 + \delta]\Gamma(u_1 - \frac{1}{2+\delta})\Gamma(v + \frac{3+\delta}{2+\delta})}(2 + \delta)(1 + \delta) \\
&= (1 + \delta)\frac{\Gamma(u_1 + \frac{1}{2+\delta})\Gamma(v + \frac{1+\delta}{2+\delta})}{\Gamma(u_1 + \frac{1+\delta}{2+\delta})\Gamma(v + \frac{3+\delta}{2+\delta})}.
\end{aligned}
\tag{8.4.22}
$$

Combining (8.4.19)–(8.4.22), we arrive at

$$
\begin{aligned}
&\mathbb{P}\left(u_1 \overset{1}{\rightsquigarrow} v, u_2 \overset{1}{\rightsquigarrow} v\right) \\
&= \frac{1 + \delta}{2 + \delta}\frac{\Gamma(u_2)\Gamma(u_1 + \frac{1+\delta}{2+\delta})}{\Gamma(u_2 + \frac{1+\delta}{2+\delta})\Gamma(u_1 + 1)} \times \frac{\Gamma(u_1 + \frac{1}{2+\delta})\Gamma(v + \frac{1+\delta}{2+\delta})}{\Gamma(u_1 + \frac{1+\delta}{2+\delta})\Gamma(v + \frac{3+\delta}{2+\delta})} \\
&= \frac{1 + \delta}{2 + \delta}\frac{\Gamma(u_2)\Gamma(u_1 + \frac{1}{2+\delta})\Gamma(v + \frac{1+\delta}{2+\delta})}{\Gamma(u_2 + \frac{1+\delta}{2+\delta})\Gamma(u_1 + 1)\Gamma(v + \frac{3+\delta}{2+\delta})},
\end{aligned}
\tag{8.4.23}
$$

as claimed in (8.4.15).

The proof of (8.4.16) for $m \geq 2$ follows by again recalling that $u \overset{j}{\rightsquigarrow} v$ occurs when $(m-1)u+j \overset{1}{\rightsquigarrow} [mv]\setminus[m(v-1)]$ and we replace δ by δ/m. Now there are two possibilities, depending on whether $m(u_1 - 1) + j_1 \overset{1}{\rightsquigarrow} v_1$ and $m(u_2 - 1) + j_2 \overset{1}{\rightsquigarrow} v_2$ hold for the *same* $v_1 = v_2 \in [mv] \setminus [m(v - 1)]$ or for two *different* $v_1, v_2 \in [mv] \setminus [m(v - 1)]$.

For $v_1 = v_2$, we use (8.4.15) to obtain a contribution that is asymptotically equal to

$$
\frac{m}{m^2}\frac{m + \delta}{2m + \delta}\frac{1}{(u_1 u_2)^{1-\gamma}v^{2\gamma}}(1 + o(1)),
\tag{8.4.24}
$$

where the factor m comes from the m distinct choices for $v_1 = v_2$ and the factor $1/m^2$ originates since we need to multiply u_1, u_2 and v in (8.4.15) by m.

For $v_1 \neq v_2$, we use the negative correlation in Lemma 8.10 to bound this contribution from above by the product of the probabilities in (8.4.14), so that the contribution is asymptotically bounded by

$$
\frac{m(m - 1)}{m^2}\left(\frac{m + \delta}{2m + \delta}\right)^2 \frac{1}{(u_1 u_2)^{1-\gamma}v^{2\gamma}}(1 + o(1)).
\tag{8.4.25}
$$

Summing (8.4.24) and (8.4.25) completes the proof of (8.4.16). $\qquad\square$

8.4.3 RESULTING PATH PROBABILITIES

With Lemmas 8.10 and 8.11 in hand, we are ready to prove Proposition 8.9:

Proof of Proposition 8.9. Before starting the proof, it is convenient to explain how to define paths in preferential attachment models:

Definition 8.12 (Edge-labeled paths in preferential attachment models with $m \geq 2$)   An *edge-labeled path* $\vec{\pi}^e$ is written as follows:

$$
\vec{\pi}^e = \{(\pi_0, j_0), \dots, (\pi_{k-1}, j_{k-1}), \pi_k\}.
\tag{8.4.26}
$$

Here j_i is such that $\pi_i \overset{j_i}{\leadsto} \pi_{i+1}$ or $\pi_{i+1} \overset{j_i}{\leadsto} \pi_i$, depending on whether $\pi_i > \pi_{i+1}$ or $\pi_i < \pi_{i+1}$. We call an edge-labeled path $\vec{\pi}^e$ *self-avoiding* when $\vec{\pi} = (\pi_0, \dots, \pi_k)$ is self-avoiding, where $\vec{\pi} = (\pi_0, \dots, \pi_k)$ denotes a path without edge labels. ◀

We have the bound

$$\mathbb{P}(\vec{\pi} \subseteq \mathrm{PA}_n^{(m,\delta)}) \leq \sum_{j_0, \dots, j_{k-1}} \mathbb{P}(\vec{\pi}^e \subseteq \mathrm{PA}_n^{(m,\delta)}). \qquad (8.4.27)$$

Since $\vec{\pi}$ is self-avoiding, we can write

$$\{\vec{\pi}^e \subseteq \mathrm{PA}_n^{(m,\delta)}\} = \bigcap_{s \in [k]} \mathcal{E}_{n_{v_s}, v_s}, \qquad (8.4.28)$$

with either

$$\mathcal{E}_{n_{v_s}, v_s} = \{u \overset{j}{\leadsto} v_s\}, \qquad (8.4.29)$$

where $u > v_s$ satisfy $\{u, v_s\} = \{\pi_i, \pi_{i+1}\}$ for some $i \in \{0, \dots, k-1\}$ and $j = j_i \in [m]$, or

$$\mathcal{E}_{n_{v_s}, v_s} = \{u_1 \overset{s_1}{\leadsto} v_s, u_2 \overset{s_2}{\leadsto} v_s\}, \qquad (8.4.30)$$

where $u_1, u_2 > v_s$ satisfy $(u_1, v_s, u_2) = (\pi_i, \pi_{i+1}, \pi_{i+2})$ for some $i \in \{0, \dots, k-1\}$ and $(s_1, s_2) = (j_i, j_{i+1}) \in [m]^2$.

In the first case, by (8.4.13),

$$\mathbb{P}(\mathcal{E}_{n_{v_s}, v_s}) = \mathbb{P}(u \overset{j}{\leadsto} v_s) \leq \frac{M_1}{u^{1-\gamma} v_s^{\gamma}}, \qquad (8.4.31)$$

whereas in the second case, according to (8.4.15),

$$\mathbb{P}(\mathcal{E}_{n_{v_s}, v_s}) = \mathbb{P}(u_1 \overset{j_1}{\leadsto} v_s, u_2 \overset{j_2}{\leadsto} v_s) \leq \frac{M_2}{(u_1 u_2)^{1-\gamma} v^{2\gamma}} = \frac{M_2}{u_1^{1-\gamma} v^{\gamma} u_2^{1-\gamma} v^{\gamma}}. \qquad (8.4.32)$$

In both cases M_i, $i = 1, 2$, is an absolute constant. Lemma 8.10 then yields (8.4.1) with $C = M_1 \vee M_2 \vee 1$, where the factor m^k originates from the number of possible choices of $j_i \in [m]$ for $i \in \{0, \dots, k-1\}$ (recall (8.4.27)). □

At this point, it would be appropriate to prove Lemma 8.5, but we leave this as Exercise 8.16.

8.5 LOGARITHMIC LOWER BOUNDS ON THE DISTANCE

In this section we prove lower bounds on distances in $\mathrm{PA}_n^{(m,\delta)}$ with $m \geq 2$ and $\delta > 0$, using the bounds, derived in Section 8.4, on the probability that a path exists in $\mathrm{PA}_n^{(m,\delta)}$; this is our main tool in this section. The following theorem is our main result:

Theorem 8.13 (Logarithmic lower distances for $\delta > 0$ and $m \geq 2$) *Consider* $\mathrm{PA}_n^{(m,\delta)}(a)$ *with $\delta > 0$ and $m \geq 2$. Then, as $n \to \infty$,*

$$\mathbb{P}(\mathrm{dist}_{\mathrm{PA}_n^{(m,\delta)}(a)}(o_1, o_2) \leq (1 - \varepsilon) \log_\nu n) \to 0, \qquad (8.5.1)$$

where $\nu > 1$ is the spectral radius of the offspring operator $\boldsymbol{T}_\kappa$ of the Pólya point tree defined in (8.5.61). These results also apply to $\mathrm{PA}_n^{(m,\delta)}(b)$ and $\mathrm{PA}_n^{(m,\delta)}(d)$.

The proof of Theorem 8.13 is organized as follows. We prove path-counting bounds for $\mathrm{PA}_n^{(m,\delta)}(a)$ and $\mathrm{PA}_n^{(m,\delta)}(b)$ with $\delta > 0$ in Section 8.5.1. Those for $\mathrm{PA}_n^{(m,\delta)}(d)$ are deferred to Section 8.5.2 and are based on the Pólya finite graph description of $\mathrm{PA}_n^{(m,\delta)}(d)$. We prove Theorem 8.13 in Section 8.5.3. In Section 8.5.4 we prove the resulting lower bounds for $\delta = 0$ and $m \geq 2$.

8.5.1 PATH COUNTING FOR $\delta > 0$

In this section we perform path-counting techniques for $m \geq 2$. Recall the offspring operator κ identified in Lemma 5.25, and let $\boldsymbol{T}_\kappa$ be the integral operator defined by

$$(\boldsymbol{T}_\kappa f)(x,s) = \sum_{t \in \{\mathrm{O},\mathrm{Y}\}} \int_0^1 \kappa((x,s)),(y,t))f(y,t)\mathrm{d}y, \qquad (8.5.2)$$

where $f \colon [0,1] \times \{\mathrm{O},\mathrm{Y}\} \to \mathbb{R}$. Introduce the function f by

$$f(y,s) = \int_0^1 \kappa_\varnothing(x,(y,s))\mathrm{d}x, \qquad s \in \{\mathrm{O},\mathrm{Y}\}, \quad y \in [0,1], \qquad (8.5.3)$$

where

$$\kappa_\varnothing(x,(y,s)) = \frac{c_s^{(\varnothing)}(\mathbb{1}_{\{x>y,s=\mathrm{O}\}} + \mathbb{1}_{\{x<y,s=\mathrm{Y}\}})}{(x \vee y)^\chi (x \wedge y)^{1-\chi}}, \qquad (8.5.4)$$

with $c_\mathrm{Y}^{(\varnothing)} = m, c_\mathrm{O}^{(\varnothing)} = m + \delta$. Further, let

$$f^\star(y,s) = f(y,s^\star), \qquad \text{where} \qquad \mathrm{O}^\star = \mathrm{Y}, \mathrm{Y}^\star = \mathrm{O}. \qquad (8.5.5)$$

Our main path-counting result is the following proposition, which will be crucial in obtaining the lower bound on typical distances:

Proposition 8.14 (Path-counting and multi-type branching processes) *Consider* $\mathrm{PA}_n^{(m,\delta)}(a)$ *for* $m \geq 2$. *For every* $\varepsilon > 0$, *there exists a* $K = K_\varepsilon$ *such that, with* o_1, o_2 *independently and uar chosen from* $[n]$,

$$\mathbb{P}(\mathrm{dist}_{\mathrm{PA}_n^{(m,\delta)}(a)}(o_1,o_2) = k) \leq \frac{K}{n}(1+\varepsilon)^k \langle f, \boldsymbol{T}_\kappa^{k-2} f^\star \rangle. \qquad (8.5.6)$$

Equivalently, with Z_k *the generation size of the multi-type branching process that started from an individual of uniform age and with type* $\varnothing$,

$$\mathbb{P}(\mathrm{dist}_{\mathrm{PA}_n^{(m,\delta)}(a)}(o_1,o_2) = k) \leq \frac{K}{n}(1+\varepsilon)^k \mathbb{E}[Z_k]. \qquad (8.5.7)$$

These results also apply to $\mathrm{PA}_n^{(m,\delta)}(b)$ *and* $\mathrm{PA}_n^{(m,\delta)}(d)$.

Below, we give a proof based on the negative correlations (8.4.3) in Lemma 8.10, as well as Lemma 8.11. In Section 8.5.2, we redo the analysis for $\mathrm{PA}_n^{(m,\delta)}(d)$ using its finite-graph Pólya version in Theorem 5.10.

Proof   We start by analyzing the consequences of (8.4.3) in Lemma 8.10 for the existence of paths. Note that

$$\mathbb{P}(\text{dist}_{\text{PA}_n^{(m,\delta)}}(o_1, o_2) = k) \leq \frac{1}{n^2} \sum_{\vec{\pi}^e} \mathbb{P}(\vec{\pi}^e \subseteq \text{PA}_n^{(m,\delta)}), \qquad (8.5.8)$$

where we sum over all self-avoiding edge-labeled paths $\vec{\pi}^e$, as in Definition 8.12.

Sum over the Edge Labels

We first perform the sum over the edge labels. Given the vertices $\vec{\pi} = (\pi_0, \ldots, \pi_k)$ in the edge-labeled path $\vec{\pi}^e$, the number of choices for the edge labels is fixed. Since $\vec{\pi}$ is self-avoiding, every vertex in it receives zero, one, or two edges from younger vertices. The number of ways in which we can choose the edges equals (recall (8.4.27))

$$m^{k-b}(m-1)^b, \qquad (8.5.9)$$

where $b = b(\vec{\pi})$ equals the number of i for which there are two edges from π_i: one to π_{i-1} and one to π_{i+1}. This occurs precisely when $\pi_i > \max\{\pi_{i-1}, \pi_{i+1}\}$.

We next give a convenient way to compute the powers of m and of $m - 1$. We associate a *label* with each of the vertices in $\vec{\pi}^e$, denoted by $\text{label}(\pi_i) \in \{\text{o}, \text{y}\}$, as follows:

$$\text{label}(\pi_i) = \begin{cases} \text{Y} & \text{when } \pi_i > \pi_{i-1}, \\ \text{o} & \text{when } \pi_i < \pi_{i-1}. \end{cases} \qquad (8.5.10)$$

Then, for a given set of vertices in $\vec{\pi}$, the number of ways in which we can choose the edge labels equals $m^{k-b}(m-1)^b$, with b equal to the number of i for which $\text{label}(\pi_i) = \text{Y}$ and $\text{label}(\pi_{i+1}) = \text{o}$ since this is equivalent to $\pi_i > \max\{\pi_{i-1}, \pi_{i+1}\}$, i.e., $b(\vec{\pi})$ equals the number of YO reversals in the vertices of $\vec{\pi}$:

$$b(\vec{\pi}) = \sum_{i=0}^{k-1} \mathbb{1}_{\{\text{label}(\pi_i)=\text{Y}, \text{label}(\pi_{i+1})=\text{o}\}}. \qquad (8.5.11)$$

This improves upon the obvious upper bound m^k, which was used in Proposition 8.9.

Connection Probabilities

We rely on the asymptotic equalities in (8.4.14) and (8.4.16) in Lemma 8.11. Fix $\varepsilon > 0$. We note that the bounds with an extra factor $1 + \varepsilon/2$ in (8.4.14) and $(1 + \varepsilon/2)^2$ in (8.4.16) can be used only when the inequality $\pi_i > M$ holds for some $M = M_\varepsilon$ for the relevant π_i.

For the contributions where $\pi_i \leq M$, we use the uniform bounds in (8.4.14) and (8.4.16). Note that the number of i for which $\pi_i \leq M$ can be at most M since $\varepsilon \pi$ is self-avoiding. Thus, in total, this gives rise to an additional factor $(M_1 \vee M_2)^M \equiv K$. We next look at the asymptotic equalities in (8.4.14) and (8.4.16), and conclude that the product over the constant equals

$$\left(\frac{m+\delta}{2m+\delta}\right)^{k-c} \left(\frac{m+1+\delta}{2m+\delta}\right)^c, \qquad (8.5.12)$$

where now $c = c(\vec{\pi})$ is the number of OY reversals in the vertices in $\vec{\pi}$ and is thus defined by

$$c(\vec{\pi}) = \sum_{i=0}^{k-1} \mathbb{1}_{\{\text{label}(\pi_i)=\text{o}, \text{label}(\pi_{i+1})=\text{Y}\}}. \qquad (8.5.13)$$

We can combine these factors effectively as

$$m^{k-b}(m-1)^b \left(\frac{m+\delta}{2m+\delta}\right)^{k-c} \left(\frac{m+1+\delta}{2m+\delta}\right)^c$$

$$= c^{(\varnothing)}_{\text{label}(\pi_1)} c^{(\varnothing)}_{\text{label}(\pi_k)*} \prod_{i=2}^{k-1} c_{\text{label}(\pi_{i-1}),\text{label}(\pi_i)} \equiv A(\vec{\pi}), \qquad (8.5.14)$$

where we recall (5.4.89) in Lemma 5.25 and $c^{(\varnothing)}_{\text{O}} = m$ and $c^{(\varnothing)}_{\text{Y}} = m + \delta$ below (8.5.4).

We are left with computing the powers of π_i for every i. Again, this depends on the precise order of the vertices in the path $\vec{\pi}$. Let p_i denote the number of edges pointing *towards* π_i, so that $p_i \in \{0, 1, 2\}$ and we have

$$p_i = p_i(\vec{\pi}) = \mathbb{1}_{\{\text{label}(\pi_i)=\text{O}\}} + \mathbb{1}_{\{\text{label}(\pi_{i+1})=\text{Y}\}}; \qquad (8.5.15)$$

here $\pi_0 = \mathbb{1}_{\{\text{label}(\pi_{i+1})=\text{Y}\}}$, and $p_k = \mathbb{1}_{\{\text{label}(\pi_k)=\text{O}\}}$.

Recall (8.4.14) and (8.4.16). We note that a factor $\pi_i^{-a_i}$ occurs, where, for $i \in [k-1]$,

$$a_i = a(p_i) = 2(1-\gamma) + (2\gamma-1)p_i = \begin{cases} 2(1-\gamma) & \text{when } p_i = 0, \\ 1 & \text{when } p_i = 1, \\ 2\gamma & \text{when } p_i = 2, \end{cases} \qquad (8.5.16)$$

whereas $a_0 = a_0(p_0) = (1-\gamma) + (2\gamma-1)p_0 = 1-\gamma$ when $p_0 = 0$ and $a_0 = a_0(p_0) = \gamma$ when $p_0 = 1$, and the same for $a_k(p_k)$. We conclude that

$$\mathbb{P}(\text{dist}_{\text{PA}_n^{(m,\delta)}}(o_1, o_2) = k) \leq \frac{K}{n^2}(1+\varepsilon)^k \sum_{\vec{\pi}} A(\vec{\pi}) \prod_{i=1}^k \pi_i^{-a(p_i)}$$

$$= \frac{K}{n^{k+2}}(1+\varepsilon)^k \sum_{\vec{\pi}} A(\vec{\pi}) \prod_{i=0}^k \left(\frac{n}{\pi_i}\right)^{a(p_i)}, \qquad (8.5.17)$$

where $b = b(\vec{\pi})$ and $p_i = p_i(\vec{\pi})$, and we have used that $\sum_{i=0}^k a(p_i) = k$. We rescale the sum and use that, uniformly for $a \geq M/n$, $b - a \geq 1/n$, and all $p \geq 0$,

$$\frac{1}{n} \sum_{l=\lceil na \rceil}^{\lceil nb \rceil} l^{-p} \leq (1+\varepsilon/4) \int_a^b x^{-p} dx, \qquad (8.5.18)$$

since $x \mapsto x^{-p}$ is decreasing. Also using that $(1 + \varepsilon/2)(1 + \varepsilon/4) \leq 1 + \varepsilon$ for $\varepsilon > 0$ sufficiently small, we can rewrite this as

$$\mathbb{P}(\text{dist}_{\text{PA}_n^{(m,\delta)}}(o_1, o_2) = k) \leq \frac{K}{n}(1+\varepsilon)^k \int_0^1 \cdots \int_0^1 A(\vec{x}) \prod_{i=0}^k x_i^{-a(p_i)} dx_0 \cdots dx_k.$$

$$(8.5.19)$$

Comparison with T_κ

We next relate the $(k+1)$-fold integral in (8.5.19) to the operator T_κ defined in (8.5.2). For this, we note that, by Lemma 5.25 and (8.5.3)–(8.5.5), $\langle f, T_\kappa^{k-2} f^\star \rangle$ equals

$$\int_0^1 \cdots \int_0^1 \sum_{t_1,\ldots,t_k} c_{\text{label}(x_1)}^{(\varnothing)} c_{\text{label}(x_k)^\star}^{(\varnothing)} \prod_{i=1}^{k-1} \frac{c_{t_{i-1},t_i}}{(x_{i-1} \vee x_i)^\chi (x_{i-1} \wedge x_i)^{1-\chi}} \mathrm{d}x_0 \cdots \mathrm{d}x_k.$$

(8.5.20)

Again, we note that $c_{\text{label}(x_1)}^{(\varnothing)}, c_{\text{label}(x_k)^\star}^{(\varnothing)}$ and $(c_{t_{i-1},t_i})_{i \in [k-1]}$ are determined by $(x_i)_{i=0}^{k-1}$ alone. It is not hard to see that, since $\chi = 1 - \gamma$,

$$\prod_{i=1}^k \frac{1}{(x_{i-1} \vee x_i)^\chi (x_{i-1} \wedge x_i)^{1-\chi}} = \prod_{i=0}^k x_i^{-a(p_i)},$$

(8.5.21)

as required, so that the powers of $x_1, \ldots, x_k$ in (8.5.19) and in (8.5.20) agree. Indeed, note that $a_i = a(p_i) = 2(1-\gamma) + p_i(2\gamma - 1)$ and $p_i = \mathbb{1}_{\{x_i < x_{i-1}\}} + \mathbb{1}_{\{x_i < x_{i+1}\}}$, so that, since $1 - \gamma = \chi$, $-a(p_i)$ equals the power of x_i in

$$(x_i \vee x_{i-1})^\chi (x_i \wedge x_{i-1})^{1-\chi} (x_i \vee x_{i+1})^\chi (x_i \wedge x_{i+1})^{1-\chi}.$$

(8.5.22)

We conclude that

$$\int_0^1 \cdots \int_0^1 A(\vec{x}) \prod_{i=0}^k x_i^{-a(p_i)} \mathrm{d}x_0 \cdots \mathrm{d}x_k = \langle f, T_\kappa^{k-2} f^\star \rangle,$$

(8.5.23)

as required. This completes the proof of Proposition 8.14. $\square$

8.5.2 Lower Bounds on Distances for $\text{PA}_n^{(m,\delta)}(d)$

In this subsection we prove a stronger form of Proposition 8.14 for $\text{PA}_n^{(m,\delta)}(d)$:

Proposition 8.15 (Path counting in $\text{PA}_n^{(m,\delta)}(d)$: upper bound) *Consider* $\text{PA}_n^{(m,\delta)}(d)$ *for* $m \geq 2$ *with* $\delta > -m$. *There exist* $C, p \geq 0$, *such that, for every* $k \geq 0$,

$$\mathbb{P}(\text{dist}_{\text{PA}_n^{(m,\delta)}(d)}(o_1, o_2) = k) \leq \frac{Ck^p}{n} \langle f, T_\kappa^{k-2} f^\star \rangle.$$

(8.5.24)

Proposition 8.15 is a substantial improvement compared with Proposition 8.14, since the correction factor is *polynomial* in k rather than exponential in k as in Proposition 8.14. Further, the proof is much more direct. Indeed, instead of using the negative correlations in Lemma 8.10, to prove Proposition 8.15, we rely on the explicit formulas for the edge probabilities in Lemma 5.12.

We start by exploring this, and give the proof of Proposition 8.15 at the end of this section. In the finite-graph Pólya urn version of $\text{PA}_n^{(m,\delta)}(d)$ in Theorem 5.10, different edges are *conditionally independent*, so that we obtain the following corollary, in which we recall that $\mathbb{P}_n$ denotes the conditional probability given $(\psi_1, \ldots, \psi_n)$:

Corollary 8.16 (Path probabilities in $\mathrm{PA}_n^{(m,\delta)}(d)$ conditioned on Beta variables) *Consider* $\mathrm{PA}_n^{(m,\delta)}(d)$ *with* $m \geq 2$ *and* $\delta > -m$. *For any edge-labeled self-avoiding path* $\vec{\pi}^e$ *as in Definition 8.12,*

$$\mathbb{P}_n(\vec{\pi}^e \subseteq \mathrm{PA}_n^{(m,\delta)}(d)) = \prod_{s=1}^{n} \psi_s^{p_s} \prod_{s=1}^{n} (1 - \psi_s)^{q_s}, \tag{8.5.25}$$

where $p_s = p_s^\pi$ *and* $q_s = q_s^\pi$ *are given by*

$$p_s^\pi = \sum_{i=0}^{k} \mathbb{1}_{\{\pi_i = s\}} [\mathbb{1}_{\{\pi_{i-1} > s\}} + \mathbb{1}_{\{\pi_{i+1} > s\}}], \quad q_s^\pi = \sum_{i=0}^{k-1} \mathbb{1}_{\{s \in (\pi_i, \pi_{i+1}) \cup (\pi_{i+1}, \pi_i)\}} \tag{8.5.26}$$

(with $\pi_{-1} = \pi_{k+1} = 0$ *by convention).*

We note that p_s equals the number of edges in the edge-labeled path $\vec{\pi}^e$ that point towards s. This is relevant since, in (5.3.29) in Lemma 5.12, every older vertex v in an edge receives a factor ψ_v. Further, again by (5.3.29) in Lemma 5.12, there are factors $1 - \psi_s$ for every $s \in (v, u)$, so q_s counts how many factors $1 - \psi_s$ occur for each $s \in [n]$.

Proof Recall Lemma 5.12. Multiply the factors $\mathbb{P}_n(\pi_i \overset{j_i}{\leadsto} \pi_{i+1})$ when $\pi_i > \pi_{i+1}$, or $\mathbb{P}_n(\pi_{i+1} \overset{j_i}{\leadsto} \pi_i)$ when $\pi_i < \pi_{i+1}$, and collect the powers of ψ_s and $1 - \psi_s$. $\qquad\square$

We see in Corollary 8.16 that we obtain expectations of the form $\mathbb{E}[\psi^a(1 - \psi)^b]$, where $\psi \sim \mathrm{Beta}(\alpha, \beta)$ and $a, b \geq 0$, which we can analyze using Lemma 5.14. Exercises 8.20 and 8.21 investigate the consequences of Corollary 8.16 and Lemma 5.14 for connection events as in Lemma 8.11. The above computation, when applied to Corollary 8.16, leads to the following expression for the probability of the existence of paths like that in Proposition 8.14:

Corollary 8.17 (Path probabilities in $\mathrm{PA}_n^{(m,\delta)}(d)$) *Consider* $\mathrm{PA}_n^{(m,\delta)}(d)$ *with* $m \geq 2$ *and* $\delta > -m$. *For any edge-labeled self-avoiding path* $\vec{\pi}^e$ *as in Definition 8.12,*

$$\mathbb{P}(\vec{\pi}^e \subseteq \mathrm{PA}_n^{(m,\delta)}(d)) = \prod_{s=1}^{n} \frac{(\alpha + p_s - 1)_{p_s} (\beta_s + q_s - 1)_{q_s}}{(\alpha + \beta_s + p_s + q_s - 1)_{p_s + q_s}}, \tag{8.5.27}$$

where $\alpha = m + \delta$, $\beta_s = (2s - 3)m + \delta(s - 1)$, *and* $p_s = p_s^\pi$ *and* $q_s = q_s^\pi$ *are defined in* (8.5.26).

Proof of Proposition 8.15. We start from (8.5.27) in Corollary 8.17. We note that $p_s^\pi \in \{0, 1, 2\}$ since $\vec{\pi}^e$ is an edge-labeled path, and so we have

$$(\alpha - 1)_0 = 1, \qquad (\alpha + 1 - 1)_1 = \alpha = m + \delta, \tag{8.5.28}$$

and

$$(\alpha + 2 - 1)_1 = \alpha(\alpha + 1) = (m + \delta)(m + 1 + \delta). \tag{8.5.29}$$

Thus, as in (8.5.12),

$$\prod_{s=1}^{n} (\alpha + p_s - 1)_{p_s} = \left(\frac{m + \delta}{2m + \delta}\right)^{k-c} \left(\frac{m + 1 + \delta}{2m + \delta}\right)^c, \tag{8.5.30}$$

where $c = c(\vec{\pi})$ is the number of OY reversals in the vertices in $\vec{\pi}$ defined in (8.5.13). The terms in (8.5.28) and (8.5.29) correspond to the prefactors in (8.4.14) and (8.4.16) in Lemma 8.11. Further, the factors m and $m-1$ arise owing to the sum over $(j_i)_{i=0}^{k-1}$. Combining with the sums over $(\pi_i)_{i=0}^{k}$ and $(j_i)_{i=0}^{k-1}$ gives rise to the factor $\langle f, T_\kappa^{k-2} f^\star \rangle$ as in Proposition 8.14, as we prove next.

We compute

$$\prod_{s=2}^{n} \frac{(\beta_s + q_s - 1)_{q_s}}{(\alpha + \beta_s + p_s + q_s - 1)_{p_s+q_s}}$$

$$= \prod_{s=2}^{n} \frac{1}{(\alpha + \beta_s + p_s + q_s - 1)_{p_s}} \frac{(\beta_s + q_s - 1)_{q_s}}{(\alpha + \beta_s + q_s - 1)_{q_s}}$$

$$= \prod_{s=2}^{n} \frac{1}{(\alpha + \beta_s + p_s + q_s - 1)_{p_s}} \prod_{i=0}^{q_s-1} \frac{\beta_s + i}{\alpha + \beta_s + i}$$

$$= \prod_{s=2}^{n} \frac{1}{(\alpha + \beta_s + p_s + q_s - 1)_{p_s}} \prod_{i=0}^{q_s-1} \left(1 - \frac{\alpha}{\alpha + \beta_s + i}\right). \tag{8.5.31}$$

Recall that $p_s = \sum_{i=0}^{k-1} \mathbb{1}_{\{\pi_i=s\}} [\mathbb{1}_{\{\pi_{i-1}>s\}} + \mathbb{1}_{\{\pi_{i+1}>s\}}]$, so that

$$\prod_{s=1}^{n} s^{-p_s} = \prod_{i=1}^{k} \pi_i^{-(\mathbb{1}_{\{\pi_{i-1}>\pi_i\}} + \mathbb{1}_{\{\pi_{i+1}>\pi_i\}})} = \prod_{i=1}^{k} \frac{1}{(\pi_{i-1} \wedge \pi_i)}. \tag{8.5.32}$$

Further, recall that $\alpha = m + \delta$ and $\beta_s = (2s-3)m + \delta(s-1)$. Therefore,

$$\prod_{s=1}^{n} \frac{1}{(\alpha + \beta_s + p_s + q_s - 1)_{p_s}} = (2m+\delta)^{-k} \prod_{s=1}^{n} s^{-p_s} \left(1 + O\left(\frac{p_s}{s}\right)\right) \tag{8.5.33}$$

$$= (2m+\delta)^{-k} \prod_{i=1}^{k} \frac{1}{(\pi_{i-1} \wedge \pi_i)} \prod_{s=1}^{n} \left(1 + O\left(\frac{p_s}{s}\right)\right).$$

As a result, we obtain the bound

$$\prod_{s=1}^{n} \left(1 + O(p_s/s)\right) \le \exp\left(O(1) \sum_{s=1}^{n} p_s/s\right)$$

$$\le \exp\left(O(1) \sum_{l=0}^{k} 1/(\pi_{\min} + l)\right) \le (1 + k/\pi_{\min})^{O(1)}, \tag{8.5.34}$$

where $\pi_{\min} = \min_{i=0}^{k} \pi_i$ and we have used that $p_s = 0$ for $s < \pi_{\min}$. This bounds the first factor in (8.5.31).

We continue by analyzing the second factor in (8.5.31). By a Taylor expansion,

$$\log\left(1 - \frac{\alpha}{\alpha + \beta_s + i}\right) = -\frac{\alpha}{\alpha + \beta_s + i} + O\left((\alpha + \beta_s + i)^{-2}\right), \tag{8.5.35}$$

so that

$$
\prod_{s=1}^{n} \prod_{i=0}^{q_s-1} \left(1 - \frac{\alpha}{\alpha + \beta_s + i} \right)
$$

$$
= \exp \left(\sum_{s=1}^{n} \sum_{i=0}^{q_s-1} -\frac{\alpha}{\alpha + \beta_s + i} + O((\alpha + \beta_s + i)^{-2}) \right)
$$

$$
= \exp \left(O(1) \sum_{s=1}^{n} q_s s^{-2} - \sum_{s=1}^{n} \sum_{i=0}^{q_s-1} \frac{\alpha}{\alpha + \beta_s + i} \right). \tag{8.5.36}
$$

Recall that $\gamma = (m + \delta)/(2m + \delta)$ to compute

$$
\sum_{s=1}^{n} \sum_{i=0}^{q_s-1} -\frac{\alpha}{\alpha + \beta_s + i} = \sum_{s=1}^{n} \sum_{i=0}^{q_s-1} \left(-\frac{m+\delta}{(2m+\delta)s} + O(s^{-2}) \right)
$$

$$
= O(1) \sum_{s=1}^{n} \frac{q_s}{s^2} - \gamma \sum_{s=1}^{n} \frac{q_s}{s}. \tag{8.5.37}
$$

Further, since $q_s = \sum_{i=0}^{k-1} \mathbb{1}_{\{s \in (\pi_i, \pi_{i+1}) \cup (\pi_{i+1}, \pi_i)\}}$ by (8.5.26), we have

$$
-\gamma \sum_{s=1}^{n} \frac{q_s}{s} = -\gamma \sum_{i=0}^{k-1} \sum_{s=\pi_i \wedge \pi_{i+1}+1}^{\pi_{i+1} \vee \pi_i - 1} \frac{1}{s}. \tag{8.5.38}
$$

We now use that

$$
\sum_{s=a+1}^{b-1} \frac{1}{s} = \log \left(\frac{b}{a} \right) + O\left(\frac{1}{a} \right), \tag{8.5.39}
$$

where the constant in $O(1/a)$ is uniform in a and b, to arrive at

$$
-\gamma \sum_{s=1}^{n} \frac{q_s}{s} = -\gamma \sum_{i=0}^{k-1} \log \left(\frac{\pi_{i+1} \vee \pi_i}{\pi_{i+1} \wedge \pi_i} \right) + O(1) \sum_{i=0}^{k-1} \frac{1}{(\pi_{i+1} \wedge \pi_i)}.
$$

Further,

$$
\sum_{s=1}^{n} \frac{q_s}{s^2} \le \sum_{i=0}^{k-1} \sum_{s=1}^{n} \frac{\mathbb{1}_{\{s > (\pi_i \wedge \pi_{i+1})\}}}{s^2} \le \sum_{i=0}^{k} \frac{c}{(\pi_i \wedge \pi_{i+1})}
$$

$$
\le 2c \sum_{l=0}^{k} \frac{1}{\pi_{\min} + l} \le 2c \log \left(1 + \frac{k}{\pi_{\min}} \right), \tag{8.5.40}
$$

since $\vec{\pi}^e$ is self-avoiding. Using this with $\pi_{\min} = 1$ yields the bound $2c \log (1 + k)$. We conclude that

$$
\prod_{s=1}^{n} \prod_{i=0}^{q_s-1} \left(1 - \frac{\alpha}{\alpha + \beta_s + i} \right) = \left(1 + \frac{k}{\pi_{\min}} \right)^{O(1)} \prod_{i=1}^{k} \frac{(\pi_{i-1} \wedge \pi_i)^\gamma}{(\pi_{i-1} \vee \pi_i)^\gamma}. \tag{8.5.41}
$$

We next rewrite the above in terms of $\langle f, T_\kappa^{k-2} f^* \rangle$, as in the proof of Proposition 8.14. By (8.5.33) and (8.5.41), we conclude that

$$\mathbb{P}(\vec{\pi}^e \subseteq \mathrm{PA}_n^{(m,\delta)}(d)) = \left(1 + \frac{k}{\pi_{\min}}\right)^{O(1)} \prod_{s=1}^n \frac{(\alpha + p_s - 1)_{p_s}}{(\pi_{i-1} \wedge \pi_i)^\gamma (\pi_{i-1} \vee \pi_i)^\gamma}. \tag{8.5.42}$$

By (8.5.9), there are $m^{k-b}(m-1)^b$ choices for the edge labels. By (8.5.30) and (8.5.14),

$$m^{k-b}(m-1)^b \prod_{s=1}^n (\alpha + p_s - 1)_{p_s} = A(\vec{\pi}). \tag{8.5.43}$$

Recall that $\{\vec{\pi} \subseteq \mathrm{PA}_n^{(m,\delta)}(d)\}$ denotes the event that there exists a labeling for which $\vec{\pi}^e \subseteq \mathrm{PA}_n^{(m,\delta)}(d)$ occurs. Then,

$$\mathbb{P}(\vec{\pi} \subseteq \mathrm{PA}_n^{(m,\delta)}(d)) \le \left(1 + \frac{k}{\pi_{\min}}\right)^{O(1)} \frac{1}{n^k} A(\vec{\pi}) \prod_{i=1}^k \frac{1}{(\pi_{i-1} \wedge \pi_i)^\gamma (\pi_{i-1} \vee \pi_i)^\gamma},$$

where $q_i = \mathrm{label}(\pi_i)$. Apart from the factor $(1 + k/\pi_{\min})^{O(1)}$, this agrees exactly with the summand in (8.5.17). Thus, we may follow the analysis of (8.5.17). Bounding $\pi_{\min} \ge 1$ and summing this over the vertices in $\vec{\pi}$, combined with an approximation of the discrete sum by an integral, leads to the claim in (8.5.24). $\qquad\square$

8.5.3 LOGARITHMIC LOWER BOUNDS ON DISTANCES: PROOF OF THEOREM 8.13

We now use Propositions 8.14 and 8.15 to prove Theorem 8.13:

Proof of lower bound on typical distances in Theorem 8.13. By Propositions 8.14 and 8.15, it obviously suffices to study $\langle f, T_\kappa^{k-2} f^* \rangle$. We write

$$\langle f, T_\kappa^{k-2} f^* \rangle = \sum_{s,t} \int_{S^2} f(x,s) \kappa^{*k-2}((x,s),(y,t)) f^*(y,t) \mathrm{d}x \mathrm{d}y, \tag{8.5.44}$$

where $\kappa^{*1}((x,s),(y,t)) = \kappa((x,s),(y,t))$, and we define recursively

$$\kappa^{*k}((x,s),(y,t)) = \sum_{r \in \{\circ, \mathrm{Y}\}} \int_0^1 \kappa^{*(k-1)}((x,s),(z,r)) \kappa((z,r),(y,t)) \mathrm{d}x. \tag{8.5.45}$$

We obtain the following bound, using (8.5.3) and (8.5.4):

$$f(y,s) = \int_0^1 \frac{c_s^{(\varnothing)}(\mathbb{1}_{\{x>y,s=\circ\}} + \mathbb{1}_{\{x<y,s=\mathrm{Y}\}})}{(x \vee y)^\chi (x \wedge y)^{1-\chi}} \mathrm{d}x \le \frac{c}{y^{1-\chi}}, \tag{8.5.46}$$

for some $c > 0$. Thus, also,

$$f^*(y,s) = f(y,s^*) \le \frac{c}{y^{1-\chi}}. \tag{8.5.47}$$

Note that $1 - \chi \in (0, \frac{1}{2})$, since $\delta > 0$. We use that $(x/y)^{1/2-\chi} + (y/x)^{1/2-\chi} \ge 2$, and

$$\frac{c}{x^{1-\chi}} \frac{c}{y^{1-\chi}} [(x/y)^{1/2-\chi} + (y/x)^{1/2-\chi}] = \frac{c^2 x^{2\chi-3/2}}{\sqrt{y}} + \frac{c^2 y^{2\chi-3/2}}{\sqrt{x}}. \tag{8.5.48}$$

This implies that

$$\langle f, T_{\kappa}^{k-2} f^{\star} \rangle \leq \frac{c^2}{2} \sum_{s,t} \int_{S^2} \kappa^{\star k-2}((x,s),(y,t)) \left(\frac{x^{2\chi-3/2}}{\sqrt{y}} + \frac{y^{2\chi-3/2}}{\sqrt{x}} \right) dx dy. \quad (8.5.49)$$

Let $q(y,t) = x_t / \sqrt{y}$ for some vector $x = (x_{\text{Y}}, x_{\text{O}})$. By Lemma 5.25,

$$\int_S \kappa(x,y) q(y) dy = \int_S \frac{c_{st}(\mathbb{1}_{\{x>y,t=\text{O}\}} + \mathbb{1}_{\{x<y,t=\text{Y}\}})}{(x \vee y)^\chi (x \wedge y)^{1-\chi}} \frac{x_t}{\sqrt{y}} dy$$

$$= c_{s\text{O}} x_{\text{O}} \int_0^x \frac{1}{x^\chi y^{1-\chi}} \frac{1}{\sqrt{y}} dy + c_{s\text{Y}} x_{\text{Y}} \int_x^1 \frac{1}{y^\chi x^{1-\chi}} \frac{1}{\sqrt{y}} dy. \quad (8.5.50)$$

We next compute, since $\chi > \frac{1}{2}$,

$$\int_0^x \frac{1}{x^\chi y^{1-\chi}} \frac{1}{\sqrt{y}} dy = \frac{1}{x^\chi} \int_0^x \frac{1}{y^{3/2-\chi}} dy = \frac{1}{(\chi - \frac{1}{2}) x^\chi} \left[\frac{1}{y^{1/2-\chi}} \right]_{y=0}^x \quad (8.5.51)$$

$$= \frac{1}{(\chi - \frac{1}{2}) x^\chi} \frac{1}{x^{1/2-\chi}} = \frac{2}{(2\chi - 1)\sqrt{x}},$$

and

$$\int_x^1 \frac{1}{y^\chi x^{1-\chi}} \frac{1}{\sqrt{y}} dy = \frac{1}{x^{1-\chi}} \int_x^1 \frac{1}{y^{1/2+\chi}} dy = \frac{1}{(\frac{1}{2} - \chi) x^{1-\chi}} [y^{1/2-\chi}]_{y=x}^1$$

$$= \frac{1}{(\frac{1}{2} - \chi) x^{1-\chi}} [1 - x^{1/2-\chi}]$$

$$= \frac{2}{(2\chi - 1)\sqrt{x}} - \frac{2}{(2\chi - 1) x^{1-\chi}} \leq \frac{2}{(2\chi - 1)\sqrt{x}}. \quad (8.5.52)$$

Thus, for $q(y,t) = x_t / \sqrt{y}$,

$$\sum_t \int_0^1 \kappa((x,s),(y,t)) q(y,t) dy = \frac{2(Mx)_s}{(2\chi - 1)} \left[\frac{1}{\sqrt{x}} - \frac{1}{x^{1-\chi}} \right] \leq \frac{2(Mx)_s}{(2\chi - 1)\sqrt{x}}, \quad (8.5.53)$$

where the matrix $M = (M_{s,t})_{s,t \in \{\text{O,Y}\}}$ is defined as follows:

$$M_{s,t} = \begin{pmatrix} c_{\text{OO}} & c_{\text{OY}} \\ c_{\text{YO}} & c_{\text{YY}} \end{pmatrix}. \quad (8.5.54)$$

By repeatedly using (8.5.53), the contribution of the $x^{2\chi-3/2}/\sqrt{y}$ term in (8.5.44) is bounded by

$$\left(\frac{2}{2\chi - 1} \right)^{k-2} \sum_{s,t} M_{s,t}^{k-2} \int_0^1 \frac{x^{2\chi-3/2}}{\sqrt{x}} dx = \int_0^1 x^{2(\chi-1)} dx \langle 1, M^{k-2} 1 \rangle \quad (8.5.55)$$

$$= \frac{1}{2} \left(\frac{2}{2\chi - 1} \right)^{k-1} \langle 1, M^{k-2} 1 \rangle,$$

where $1 = (1,1)^T$ is the constant vector. A similar computation, now computing the integrals from left to right instead, shows that the contribution due to $y^{2\chi-3/2}/\sqrt{x}$ is bounded by

$$\frac{1}{2}\left(\frac{2}{2\chi-1}\right)^{k-1}\langle(\mathbf{M}^*)^{k-2}\mathbf{1},\mathbf{1}\rangle=\frac{1}{2}\left(\frac{2}{2\chi-1}\right)^{k-1}\langle\mathbf{1},\mathbf{M}^{k-2}\mathbf{1}\rangle,\qquad(8.5.56)$$

so that we end up with

$$\langle f,\boldsymbol{T}_\kappa^{k-2}f^\star\rangle\leq\frac{c^2}{2}\left(\frac{2}{2\chi-1}\right)^{k-1}\langle\mathbf{1},\mathbf{M}^{k-2}\mathbf{1}\rangle.\qquad(8.5.57)$$

The matrix $\mathbf{M}$ has largest eigenvalue

$$\lambda_{\mathbf{M}}=\frac{m(m+\delta)+\sqrt{m(m-1)(m+\delta)(m+1+\delta)}}{2m+\delta},\qquad(8.5.58)$$

and smallest eigenvalue

$$\mu_{\mathbf{M}}=\frac{m(m+\delta)-\sqrt{m(m-1)(m+\delta)(m+1+\delta)}}{2m+\delta}<\lambda_{\mathbf{M}}.\qquad(8.5.59)$$

A simple computation using that $m>-\delta$ shows that $\mu_{\mathbf{M}}>0$. Thus,

$$\langle\mathbf{1},\mathbf{M}^{k-2}\mathbf{1}\rangle=\lambda_{\mathbf{M}}^{k-2}\langle\mathbf{1},\boldsymbol{u}\rangle^2+\mu_{\mathbf{M}}^{k-2}\langle\mathbf{1},\boldsymbol{v}\rangle^2\leq\lambda_{\mathbf{M}}^{k-2}[\langle\mathbf{1},\boldsymbol{u}\rangle^2+\langle\mathbf{1},\boldsymbol{v}\rangle^2]$$
$$=\lambda_{\mathbf{M}}^{k-2}\|\mathbf{1}\|=2\lambda_{\mathbf{M}}^{k-2},\qquad(8.5.60)$$

where $\boldsymbol{u}$ and $\boldsymbol{v}$ are the two right-eigenvectors of $\mathbf{M}$ corresponding to $\lambda_{\mathbf{M}}$ and $\mu_{\mathbf{M}}$, respectively. Denote

$$\nu\equiv\frac{2\lambda_{\mathbf{M}}}{2\chi-1}=2\frac{m(m+\delta)+\sqrt{m(m-1)(m+\delta)(m+1+\delta)}}{\delta},\qquad(8.5.61)$$

where the equality follows since $\chi=(m+\delta)/(2m+\delta)$. Then this proves that

$$\langle f,\boldsymbol{T}_\kappa^{k-2}f^\star\rangle\leq c^2\nu^{k-2}.\qquad(8.5.62)$$

Write

$$k_n^\star=\lceil(1-2\varepsilon)\log_\nu n\rceil.\qquad(8.5.63)$$

Then, by Proposition 8.14, with $K'=2Kc^2$ and for all n sufficiently large,

$$\mathbb{P}(\mathrm{dist}_{\mathrm{PA}_n^{(m,\delta)}(a)}(o_1,o_2)\leq k_n^\star)\leq\sum_{k=0}^{k_n^\star}\frac{K'}{n}[(1+\varepsilon)\nu]^{k-2}\leq\frac{K'k_n}{n}[(1+\varepsilon)\nu]^{k_n}$$
$$\leq\frac{K'k_n^\star}{n}n^{\log(1+\varepsilon)+1-2\varepsilon}=o(1),\qquad(8.5.64)$$

since $\log(1+\varepsilon)\leq 1+\varepsilon$, as required. The proof for $\mathrm{PA}_n^{(m,\delta)}(d)$ proceeds identically, now using Proposition 8.15 instead. $\qquad\square$

8.5.4 Extension to Lower Bounds on Distances for $\delta = 0$ and $m \geq 2$

In this subsection we investigate lower bounds on the distances in $\mathrm{PA}_n^{(m,\delta)}$ when $m \geq 2$ and $\delta = 0$ and prove the lower bound in Theorem 8.8. We again perform the proof for $\mathrm{PA}_n^{(m,0)} = \mathrm{PA}_n^{(m,0)}(a)$; the proof for $\mathrm{PA}_n^{(m,0)}(b)$ and $\mathrm{PA}_n^{(m,0)}(d)$ is identical.

We again start from Proposition 8.9, which as we will show implies that, for $\delta = 0$,

$$k_n^\star = \left\lceil \frac{\log n}{\log(2Cm \log n)} \right\rceil \tag{8.5.65}$$

is whp a lower bound for typical distances in $\mathrm{PA}_n^{(m,0)}$. Consider a path $\vec{\pi} = (\pi_0, \pi_1, \dots, \pi_k)$ (this path is not edge-labeled); then (8.4.1) in Proposition 8.9 implies that

$$\mathbb{P}(\vec{\pi} \subseteq \mathrm{PA}_n^{(m,0)}) \leq (Cm^2)^k \prod_{j=0}^{k-1} \frac{1}{\sqrt{\pi_j \pi_{j+1}}} = \frac{(Cm^2)^k}{\sqrt{\pi_0 \pi_k}} \prod_{j=1}^{k-1} \frac{1}{\pi_j}, \tag{8.5.66}$$

since $\gamma = 1/2$ for $\delta = 0$.

We use $\sum_{a=1}^n 1/\sqrt{a} \leq \int_0^n 1/\sqrt{x}\,dx = 2\sqrt{n}$ to compute that

$$\mathbb{P}(\mathrm{dist}_{\mathrm{PA}_n^{(m,0)}}(o_1, o_2) = k) \leq \frac{4(Cm)^k}{n^2} \sum_{\vec{\pi}} \frac{1}{\sqrt{\pi_0 \pi_k}} \prod_{j=1}^{k-1} \frac{1}{\pi_j}$$

$$\leq \frac{4(Cm)^k}{n} \sum_{1 \leq \pi_1, \dots, \pi_{k-1} \leq n} \prod_{j=1}^{k-1} \frac{1}{\pi_j}. \tag{8.5.67}$$

Thus,

$$\mathbb{P}(\mathrm{dist}_{\mathrm{PA}_n^{(m,0)}}(o_1, o_2) = k) \leq \frac{4(Cm)^k}{n} \left(\sum_{s=1}^n \frac{1}{s} \right)^{k-1}$$

$$\leq \frac{4(Cm)^k}{n} (\log n)^{k-1}. \tag{8.5.68}$$

As a result,

$$\mathbb{P}(\mathrm{dist}_{\mathrm{PA}_n^{(m,0)}}(o_1, o_2) \leq k_n^\star) \leq 4 \sum_{k \leq k_n^\star} \frac{4(Cm)^k}{n} (\log n)^{k-1}$$

$$\leq 4 \sum_{k \leq k_n^\star} \frac{1}{2^k} (\log n)^{-1} \to 0, \tag{8.5.69}$$

since $(2Cm \log n)^{k_n^\star} \leq n$, which follows from (8.5.65). This implies that the typical distances are whp at least $k_n^\star$ in (8.5.65). Since $k_n^\star \leq (1 - \varepsilon) \log n / \log \log n$, this completes the proof of the lower bound on the graph distances for $\delta = 0$ in Theorem 8.8. $\qquad\square$

In Exercise 8.18, the reader is asked to prove that the distance between vertices $n - 1$ and n is also whp at least $k_n^\star$ in (8.5.65). Exercise 8.19 considers whether the above proof implies that the distance between vertices 1 and 2 is whp at least $k_n^\star$ in (8.5.65) as well.

8.6 LOG LOG DISTANCE LOWER BOUND FOR INFINITE-VARIANCE DEGREES

In this section we prove the lower bound in Theorem 8.7 for $\delta < 0$. We do so in a more general setting, by assuming an upper bound on the existence of paths in a model that is inspired by Proposition 8.9:

Assumption 8.18 (Path probabilities) *There exist constants $\kappa > 0$ and $\gamma > 0$ such that, for all n and self-avoiding paths $\vec{\pi} = (\pi_0, \ldots, \pi_k) \in [n]^l$,*

$$\mathbb{P}(\vec{\pi} \subseteq \mathrm{PA}_n) \leq \prod_{i=1}^{k} \kappa(\pi_{i-1} \wedge \pi_i)^{-\gamma}(\pi_i \vee \pi_{i-1})^{\gamma-1}. \tag{8.6.1}$$

By Proposition 8.9, Assumption 8.18 is satisfied for $\mathrm{PA}_n^{(m,\delta)}$ with $\gamma = m/(2m + \delta)$ and $\kappa = Cm$. We expect doubly logarithmic distances in such networks if and only if $\delta \in (-m, 0)$, so that $\frac{1}{2} < \gamma < 1$. Theorem 8.19, which is the main result in this section, gives a lower bound on the typical distance in this case:

Theorem 8.19 (Doubly logarithmic lower bound on typical distances) *Let PA_n be a random graph that satisfies Assumption 8.18 for some $\gamma \in (\frac{1}{2}, 1)$. Let o_1 and o_2 be chosen independently and uar from $[n]$, and define*

$$\tau = \frac{2 - \gamma}{1 - \gamma} \in (2, 3). \tag{8.6.2}$$

Then, for every $\varepsilon > 0$, there exists $K = K_\varepsilon$ such that

$$\mathbb{P}\left(\mathrm{dist}_{\mathrm{PA}_n}(o_1, o_2) \geq \frac{4 \log \log n}{|\log(\tau - 2)|} - K\right) \geq 1 - \varepsilon. \tag{8.6.3}$$

We will prove Theorem 8.19 in the form where $|\log(\tau - 2)|$ is replaced with $\log(\gamma/(1 - \gamma))$. For $\mathrm{PA}_n^{(m,\delta)}$, $\gamma = m/(2m + \delta)$, so that

$$\frac{\gamma}{1 - \gamma} = \frac{m}{m + \delta} = \frac{1}{\tau - 2}, \tag{8.6.4}$$

where we recall from (1.3.63) that $\tau = 3 + \delta/m$. Therefore, Theorem 8.19 proves the lower bound in Theorem 8.7, and even extends this to lower tightness for the distances. However, Theorem 8.19 also applies to *other* random graph models that satisfy Assumption 8.18, such as inhomogeneous random graph models with appropriate kernels such as $p_{u,v} \leq \kappa(u \wedge v)^{-\gamma}(u \vee v)^{\gamma-1}$.

In fact, we will prove a slightly different version of Theorem 8.19, namely, that, uniformly in $u, v \geq \varepsilon n$, we can choose $K = K_\varepsilon > 0$ sufficiently large, so that, uniformly in $n \geq 1$,

$$\mathbb{P}\left(\mathrm{dist}_{\mathrm{PA}_n}(u, v) \leq \frac{4 \log \log n}{|\log(\tau - 2)|} - K\right) \leq \varepsilon. \tag{8.6.5}$$

Since $\mathbb{P}(o_1 \leq \varepsilon n) \leq \varepsilon$, this clearly implies Theorem 8.19 with ε in (8.6.3) replaced by 3ε.

The proof of Theorem 8.19 is based on a *constrained* or *truncated first-moment method*, similar to the method used for $\mathrm{NR}_n(\boldsymbol{w})$ in Theorem 6.7 and for $\mathrm{CM}_n(\boldsymbol{d})$ in Theorem 7.8. Owing to the fact that the path probabilities in Assumption 8.18 satisfy a rather different bound compared with those for $\mathrm{NR}_n(\boldsymbol{w})$ and $\mathrm{CM}_n(\boldsymbol{d})$, this truncated first-moment method looks rather different compared with those presented in the proof of Theorems 6.7 and 7.8. This difference also explains why distances are *twice as large* for PA_n in Theorem 8.19 compared with Theorems 6.7 and 7.8.

Let us now briefly explain the truncated first-moment method. We start with an explanation of the (unconstrained) first-moment bound and its shortcomings. Let $u, v \geq \varepsilon n$ be distinct vertices of PA_n. Then, by Assumption 8.18, for $k_n \in \mathbb{N}$,

$$\mathbb{P}(\mathrm{dist}_{\mathrm{PA}_n}(u, v) \leq 2k_n) = \mathbb{P}\left(\bigcup_{k=1}^{2k_n}\bigcup_{\vec{\pi}}\{\vec{\pi} \subseteq \mathrm{PA}_n\}\right) \leq \sum_{k=1}^{2k_n}\sum_{\vec{\pi}} p(\vec{\pi}), \qquad (8.6.6)$$

where $\vec{\pi} = (\pi_0, \ldots, \pi_k)$ is a self-avoiding path in PA_n with $\pi_0 = u$ and $\pi_k = v$, and we define

$$p(\vec{\pi}) = \prod_{j=1}^{k} p(\pi_{j-1}, \pi_j) \qquad \text{where} \qquad p(a, b) = \kappa(a \wedge b)^{-\gamma}(a \vee b)^{\gamma-1}. \qquad (8.6.7)$$

The shortcoming of the above bound is that the paths that contribute most to the total weight are those that connect u or v quickly to very old vertices. However, such paths are quite unlikely to be present. This explains why the very old vertices have to be removed in order to get a reasonable estimate, and why this leads to small errors when we do so. For this, and similarly to Section 6.3.2 for $\mathrm{NR}_n(\boldsymbol{w})$, we split the paths into *good* and *bad* paths:

Definition 8.20 (Good and bad paths for PA_n)   For a decreasing sequence $g = (g_l)_{l=0,\ldots,k}$ of positive integers, we consider a path $\vec{\pi} = (\pi_0, \ldots, \pi_k)$ to be *good* when $\pi_l \wedge \pi_{k-l} \geq g_l$ for all $l \in \{0, \ldots, k\}$. We denote the event that there exists a good path of length k between u and v by $\mathcal{E}_k(u, v)$. We further let $\mathcal{F}_l(v)$ denote the event that there exists a bad path of length l in PA_n starting at v. This means that there exists a path $\vec{\pi} \subseteq \mathrm{PA}_n$, with $v = \pi_0$, such that $\pi_0 \geq g_0, \ldots, \pi_{l-1} \geq g_{l-1}$, but $\pi_l < g_l$, i.e., a path that exceeds the threshold after exactly l steps.   ◀

For fixed vertices $u, v \geq g_0$, we thus obtain

$$\mathbb{P}(\mathrm{dist}_{\mathrm{PA}_n}(u, v) \leq 2k) \leq \sum_{l=1}^{k} \mathbb{P}(\mathcal{F}_l(u)) + \sum_{l=1}^{k} \mathbb{P}(\mathcal{F}_l(v)) + \sum_{l=1}^{2k} \mathbb{P}(\mathcal{E}_l(u, v)). \qquad (8.6.8)$$

The truncated first-moment estimate arises when one is bounding the events of the existence of certain good or bad paths by their expected numbers. Owing to the split into good and bad paths, these sums will now behave better than without this split. Inequality (8.6.8) is identical to the inequality (6.3.30) used in the proof of Theorem 6.7. However, the notion of *good* has changed owing to the fact that the vertices no longer have a *weight*, but rather an *age*.

By Assumption 8.18,

$$\mathbb{P}(\vec{\pi} \subseteq \mathrm{PA}_n) \leq p(\vec{\pi}). \qquad (8.6.9)$$

Thus, for $v \geq g_0$ and $l \in [k]$, and with $\vec{\pi} = (\pi_0, \ldots, \pi_l)$ with $\pi_0 = u$,

$$\mathbb{P}(\mathcal{F}_l(u)) \leq \sum_{\pi_1=g_1}^{n} \cdots \sum_{\pi_{l-1}=g_{l-1}}^{n} \sum_{\pi_l=1}^{g_l-1} p(\vec{\pi}) = \sum_{w=1}^{g_l-1} f_{l-1,n}(u, w), \tag{8.6.10}$$

where, for $l \in [k]$ and $u, v \in [n]$, we set

$$f_{l,n}(u, v) := \mathbb{1}_{\{v \geq g_0\}} \sum_{\pi_1=g_1}^{n} \cdots \sum_{\pi_{l-1}=g_{l-1}}^{n} p(u, \pi_1, \ldots, \pi_{l-1}, v), \tag{8.6.11}$$

and $f_{0,n}(u, v) = \mathbb{1}_{\{v=u\}} \mathbb{1}_{\{u \leq n\}}$.

We next investigate $\mathbb{P}(\mathcal{E}_l(u, v))$. Since p is symmetric, for all $l \leq 2k$, we have

$$\mathbb{P}(\mathcal{E}_l(u, v)) \leq \sum_{\pi_1=g_1}^{n} \cdots \sum_{\pi_{\lfloor l/2 \rfloor}=g_{\lfloor l/2 \rfloor}}^{n} \cdots \sum_{\pi_{l-1}=g_1}^{n} p(u, \pi_1, \ldots, \pi_{\lfloor l/2 \rfloor}) p(\pi_{\lfloor l/2 \rfloor}, \ldots, \pi_{l-1}, v)$$

$$= \sum_{\pi_{\lfloor l/2 \rfloor}=g_{\lfloor l/2 \rfloor}}^{n} f_{\lfloor l/2 \rfloor, n}(u, \pi_{\lfloor l/2 \rfloor}) f_{\lceil l/2 \rceil, n}(v, \pi_{\lfloor l/2 \rfloor}). \tag{8.6.12}$$

We conclude that all terms on the rhs of (8.6.8) are now bounded in terms of $f_{l,n}(u, v)$, as

$$\mathbb{P}(\text{dist}_{\text{PA}_n}(u, v) \leq 2k_n) \leq \sum_{k=1}^{k_n} \sum_{w=1}^{g_k-1} f_{k,n}(u, w) + \sum_{k=1}^{k_n} \sum_{w=1}^{g_k-1} f_{k,n}(v, w) \tag{8.6.13}$$

$$+ \sum_{k=1}^{2k_n} \sum_{\pi_{\lfloor k/2 \rfloor}=g_{\lfloor k/2 \rfloor}}^{n} f_{\lfloor k/2 \rfloor, n}(u, \pi_{\lfloor k/2 \rfloor}) f_{\lceil k/2 \rceil, n}(v, \pi_{\lfloor k/2 \rfloor}).$$

The remaining task in the proof is to choose $k_n \in \mathbb{N}$, as well as a decreasing sequence $(g_k)_{k=0}^{k_n}$ such that $2 \leq g_{k_n} \leq \cdots \leq g_0 \leq n$, that allow us to bound the rhs of (8.6.13).

We study $f_{l,n}(u, v)$ in detail in the remainder of this section. This analysis is quite involved. Using the recursive representation

$$f_{k+1,n}(u, v) = \sum_{w=g_k}^{n} f_{k,n}(u, w) p(w, v), \tag{8.6.14}$$

we establish upper bounds on $f_{k,n}(u, v)$ and use these to show that the rightmost term in (8.6.8) remains small when $k = k_n$ is chosen appropriately.

Our aim is to provide an upper bound of the form

$$f_{k,n}(u, v) \leq \alpha_k v^{-\gamma} + \beta_k v^{\gamma-1} \quad \text{for all } u, v \in [n], \tag{8.6.15}$$

for suitably chosen parameters $\alpha_k, \beta_k \geq 0$. Key to this choice is the following lemma:

Lemma 8.21 (Recursive bounds) *Let $\gamma \in (\frac{1}{2}, 1)$ and suppose that $2 \leq \ell \leq n$, $\alpha, \beta \geq 0$. Assume that $q_\ell \colon [n] \to [0, \infty)$ satisfies*

$$q_\ell(w) \leq \mathbb{1}_{\{w > \ell\}} (\alpha w^{-\gamma} + \beta w^{\gamma-1}) \quad \text{for all } w \in [n]. \tag{8.6.16}$$

Then there exists a constant $c = c(\gamma, \kappa) > 1$ such that, for all $u \in [n]$,

$$\sum_{w=1}^{n} q_\ell(w)p(w,u) \le c \left(\alpha \log(n/\ell) + \beta n^{2\gamma-1}\right) u^{-\gamma}$$

$$+ c\mathbb{1}_{\{u>\ell\}} \left(\alpha \ell^{1-2\gamma} + \beta \log(n/\ell)\right) u^{\gamma-1}. \tag{8.6.17}$$

Proof Throughout the remainder of this section, we frequently use the following bound:

$$\sum_{w=a}^{b} w^{-p} \le \begin{cases} \frac{1}{1-p} b^{1-p} & \text{for } p \in [0,1), \\ \log(b/(a-1)) & \text{for } p = 1, \\ \frac{1}{p-1}(a-1)^{1-p} & \text{for } p > 1. \end{cases} \tag{8.6.18}$$

We use (8.6.7) and (8.6.16) to rewrite

$$\sum_{w=1}^{n} q_\ell(w)p(w,u) = \sum_{w=u\vee\ell}^{n} q_\ell(w)p(w,u) + \mathbb{1}_{\{u>\ell\}} \sum_{w=\ell}^{u-1} q_\ell(w)p(w,u)$$

$$= \sum_{w=u\vee\ell}^{n} \kappa(\alpha w^{-\gamma} + \beta w^{\gamma-1}) w^{\gamma-1} u^{-\gamma}$$

$$+ \mathbb{1}_{\{u>\ell\}} \sum_{w=\ell}^{u-1} \kappa(\alpha w^{-\gamma} + \beta w^{\gamma-1}) w^{-\gamma} u^{\gamma-1}. \tag{8.6.19}$$

Simplifying the sums leads to, using $\gamma \in (\frac{1}{2}, 1)$ and (8.6.18),

$$\sum_{w=1}^{n} q_\ell(w)p(w,u) \le \kappa \left(\alpha \sum_{w=u\vee\ell}^{n} w^{-1} + \beta \sum_{w=u\vee\ell}^{n} w^{2\gamma-2}\right) u^{-\gamma}$$

$$+ \kappa\mathbb{1}_{\{u>\ell\}} \left(\alpha \sum_{w=\ell}^{u-1} w^{-2\gamma} + \beta \sum_{w=\ell}^{u-1} w^{-1}\right) u^{\gamma-1}$$

$$\le \kappa \left(\alpha \log\left(\frac{n}{\ell-1}\right) + \frac{\beta}{2\gamma-1} n^{2\gamma-1}\right) u^{-\gamma}$$

$$+ \kappa\mathbb{1}_{\{u>\ell\}} \left(\frac{\alpha}{1-2\gamma}(\ell-1)^{1-2\gamma} + \beta \log\left(\frac{u}{\ell-1}\right)\right) u^{\gamma-1}. \tag{8.6.20}$$

This immediately implies the assertion since $\ell \ge 2$ and $u \in [n]$ by assumption. $\square$

We aim to apply Lemma 8.21 iteratively. We use induction in k to prove that there exist $(g_k)_{k\ge0}$, $(\alpha_k)_{k\ge1}$, and $(\beta_k)_{k\ge1}$ such that (8.6.15) holds. The sequences $(g_k)_{k\ge0}$, $(\alpha_k)_{k\ge1}$ and $(\beta_k)_{k\ge1}$ are chosen as follows:

Definition 8.22 (Choices of parameters $(g_k)_{k\ge0}$, $(\alpha_k)_{k\ge1}$, and $(\beta_k)_{k\ge1}$) Fix $\kappa > 1$ and $\varepsilon \in (0,1)$. We define

$$g_0 = \lceil \varepsilon n \rceil, \qquad \alpha_1 = \kappa g_0^{\gamma-1}, \qquad \beta_1 = \kappa g_0^{-\gamma}, \tag{8.6.21}$$

and, recursively, for $k \geq 1$, we make the following choices:

(1) g_k is the smallest integer such that

$$\frac{1}{1-\gamma}\alpha_k g_k^{1-\gamma} \geq \frac{\varepsilon}{\pi^2 k^2};\tag{8.6.22}$$

(2) α_{k+1} is chosen as

$$\alpha_{k+1} = c\big(\alpha_k \log(n/g_k) + \beta_k n^{2\gamma-1}\big);\tag{8.6.23}$$

(3) β_{k+1} is chosen as

$$\beta_{k+1} = c\big(\alpha_k g_k^{1-2\gamma} + \beta_k \log(n/g_k)\big),\tag{8.6.24}$$

where $c = c(\kappa, \gamma) > 1$ is the constant appearing in Lemma 8.21.   ◄

One can check that $k \mapsto g_k$ is non-increasing, while $k \mapsto \alpha_k, \beta_k$ are non-decreasing. Further, since g_k is the *smallest* integer such that (8.6.22) holds, we have

$$\frac{1}{1-\gamma}\alpha_k(g_k-1)^{1-\gamma} < \frac{\varepsilon}{\pi^2 k^2},\tag{8.6.25}$$

which, since $g_k \geq 2$, in turn implies that

$$\frac{1}{1-\gamma}\alpha_k g_k^{1-\gamma} < 2^{1-\gamma}\frac{\varepsilon}{\pi^2 k^2},\tag{8.6.26}$$

which we crucially use below.

We recall $f_{k,n}(u,v)$ in (8.6.11), with $p(a,b)$ defined in (8.6.7), and that $f_{k,n}$ satisfies the recursion in (8.6.14). The following lemma derives recursive bounds on $f_{k,n}$:

Lemma 8.23 (Recursive bound on $f_{k,n}$)   *For the sequences in Definition 8.22, for every* $u, v \in [n]$ *and* $k \in \mathbb{N}$,

$$f_{k,n}(u,v) \leq \alpha_k v^{-\gamma} + \mathbb{1}_{\{v > g_{k-1}\}}\beta_k v^{\gamma-1}.\tag{8.6.27}$$

Proof   We prove (8.6.27) by induction on k. For $k = 1$, using $\alpha_1 = \kappa g_0^{\gamma-1}, \beta_1 = \kappa g_0^{-\gamma}$,

$$\begin{aligned}
f_{1,n}(u,v) = p(u,v)\mathbb{1}_{\{u \geq g_0\}} &\leq \kappa g_0^{\gamma-1}v^{-\gamma} + \mathbb{1}_{\{v > g_0\}}\kappa g_0^{-\gamma}v^{\gamma-1}\\
&= \alpha_1 v^{-\gamma} + \mathbb{1}_{\{v > g_0\}}\beta_1 v^{\gamma-1},
\end{aligned}\tag{8.6.28}$$

as required. This initializes the induction hypothesis.

We now proceed to advance the induction: suppose that g_{k-1}, α_k and β_k are such that (8.6.27) holds. We use the recursive property of $f_{k,n}$ in (8.6.14) and apply Lemma 8.21, with $q_{g_k}(w) = f_{k,n}(u,w)\mathbb{1}_{\{w > g_k\}}$, so that, by Definition 8.22,

$$\begin{aligned}
f_{k+1,n}(u,v) &\leq c\big[\alpha_k \log(n/g_k) + \beta_k n^{2\gamma-1}\big]v^{-\gamma}\\
&\quad + c\mathbb{1}_{\{v > g_k\}}\big[\alpha_k g_k^{1-2\gamma} + \beta_k \log(n/g_k)\big]v^{\gamma-1}\\
&= \alpha_{k+1}v^{-\gamma} + \mathbb{1}_{\{v > g_k\}}\beta_{k+1}v^{\gamma-1},
\end{aligned}\tag{8.6.29}$$

as required. This advances the induction hypothesis, and thus completes the proof.   □

We next use (8.6.15) to prove Theorem 8.19. We start with the contributions due to bad paths. Summing over (8.6.27) in Lemma 8.23, and using (8.6.18) and (8.6.26), we obtain

$$\sum_{w=1}^{g_k-1} f_{k,n}(v,w) \le \frac{1}{1-\gamma} \alpha_k g_k^{1-\gamma} 2^{1-\gamma} \le \frac{2\varepsilon}{\pi^2 k^2}, \tag{8.6.30}$$

which, when summed over all $k \ge 1$, is bounded by $\varepsilon/3$. Hence, together the first two summands on the rhs in (8.6.13) are smaller than $2\varepsilon/3$. This shows that the probability that there exists a bad path from either u or v is small, uniformly in $u, v \ge \varepsilon n$.

We continue with the contributions due to good paths, which is the most delicate part of the argument. For this, it remains for us to choose k_n as large as possible while ensuring that $g_{k_n} \ge 2$ and

$$\sum_{k=1}^{2k_n} \sum_{\pi_{\lfloor k/2 \rfloor} = g_{\lfloor k/2 \rfloor}}^{n} f_{\lfloor k/2 \rfloor, n}(u, \pi_{\lfloor k/2 \rfloor}) f_{\lceil k/2 \rceil, n}(v, \pi_{\lfloor k/2 \rfloor}) \le \varepsilon/3. \tag{8.6.31}$$

Proving (8.6.31) for the appropriate k_n is the main content of the remainder of this section.

Recall from Definition 8.22 that g_k is the largest integer satisfying (8.6.22) and that the parameters α_k, β_k are defined via equalities in (8.6.23) and (8.6.24). To establish lower bounds for the decay of g_k, we instead investigate the growth of $\eta_k = n/g_k > 0$ for k large:

Proposition 8.24 (Inductive bound on η_k) *Recall Definition 8.22, and let $\eta_k = n/g_k$. Let $\varepsilon \in (0,1)$. Then, there exists a constant $B = B_\varepsilon$ such that, for any $k = O(\log \log n)$,*

$$\eta_k \le e^{B(\tau-2)^{-k/2}}, \tag{8.6.32}$$

where we recall the degree power-law exponent $\tau = (2-\gamma)/(1-\gamma)$ from (8.6.2).

Exercise 8.22 asks the reader to relate the above bound to the growth of the Pólya point tree.

Before turning to the proof of Proposition 8.24, we comment on it. Recall that we are summing over $\pi_k \ge g_k$, which is equivalent to $n/\pi_k \le \eta_k$. The sums in (8.6.13) are such that the summands obey this bound for appropriate values of k.

Compare this with (6.3.30), where, instead, the weights obey $w_{\pi_k} \le b_k$. We see that η_k plays a similar role to b_k. Recall that w_{π_k} is indeed close to the degree of π_k in $\mathrm{GRG}_n(\boldsymbol{w})$, while the degree of vertex π_k in $\mathrm{PA}_n^{(m,\delta)}$ is close to $(n/\pi_k)^{1/(\tau-1)}$ by [V1, (8.3.12)]. Thus, the truncation $n/\pi_k \le \eta_k$ can be interpreted as a bound of order $e^{B(\tau-2)^{-k/2}}$ on the degree of π_k. Note, however, that $b_k \approx e^{(\tau-2)^{-k}}$ by (6.3.42), which grows roughly twice as quickly as η_k. This is again a sign that distances in PA_n are twice as large as those in $\mathrm{GRG}_n(\boldsymbol{w})$.

Before proving Proposition 8.24, we first derive a recursive bound on η_k:

Lemma 8.25 (Recursive bound on η_k) *Recall Definition 8.22, and let $\eta_k = n/g_k$. Then there exists a constant $C > 0$ independent of $\eta_0 = \varepsilon > 0$ such that*

$$\eta_{k+2}^{1-\gamma} \le C \left[\eta_k^\gamma + \eta_{k+1}^{1-\gamma} \log \eta_{k+1} \right]. \tag{8.6.33}$$

Proof By the definition of g_k in (8.6.22) and the fact that $\gamma - 1 < 0$,

$$\eta_{k+2}^{1-\gamma} = n^{1-\gamma} g_{k+2}^{\gamma-1} \le n^{1-\gamma} \frac{1}{1-\gamma} \frac{\pi^2}{\varepsilon} (k+2)^2 \alpha_{k+2}. \tag{8.6.34}$$

By the definition of α_k in (8.6.23),

$$n^{1-\gamma}\frac{1}{1-\gamma}\frac{\pi^2}{\varepsilon}(k+2)^2\alpha_{k+2}$$

$$= n^{1-\gamma}\frac{c}{1-\gamma}\frac{\pi^2}{\varepsilon}(k+2)^2\left[\alpha_{k+1}\log\eta_{k+1}+\beta_{k+1}n^{2\gamma-1}\right]. \tag{8.6.35}$$

We bound each of the two terms in (8.6.35) separately. By (8.6.26),

$$\alpha_{k+1}\le\frac{(1-\gamma)\varepsilon}{\pi^2(k+1)^2}2^{1-\gamma}g_{k+1}^{\gamma-1}. \tag{8.6.36}$$

We conclude that the first term in (8.6.35) is bounded by

$$n^{1-\gamma}\frac{c}{1-\gamma}\frac{\pi^2}{\varepsilon}(k+2)^2\alpha_{k+1}\log\eta_{k+1}$$

$$\le n^{1-\gamma}\frac{c2^{1-\gamma}}{1-\gamma}\frac{\pi^2}{\varepsilon}(k+2)^2\frac{(1-\gamma)\varepsilon}{\pi^2(k+1)^2}g_{k+1}^{\gamma-1}\log\eta_{k+1}$$

$$= c2^{1-\gamma}\frac{(k+2)^2}{(k+1)^2}\eta_{k+1}^{1-\gamma}\log\eta_{k+1}, \tag{8.6.37}$$

which is part of the second term in (8.6.33).

We now have to bound the second term in (8.6.35) by the rhs of (8.6.33). This term equals

$$n^{1-\gamma}\frac{c}{1-\gamma}\frac{\pi^2}{\varepsilon}(k+2)^2\beta_{k+1}n^{2\gamma-1}=\frac{c}{1-\gamma}\frac{\pi^2}{6\varepsilon}(k+2)^2\beta_{k+1}n^\gamma. \tag{8.6.38}$$

We use the definition of β_k in (8.6.24) to write

$$\frac{c}{1-\gamma}\frac{\pi^2}{\varepsilon}(k+2)^2\beta_{k+1}n^\gamma=\frac{c}{1-\gamma}\frac{\pi^2}{\varepsilon}(k+2)^2n^\gamma c\left[\alpha_k g_k^{1-2\gamma}+\beta_k\log\eta_k\right], \tag{8.6.39}$$

which again leads to two terms that we bound separately. For the first term in (8.6.39), we again use the fact that

$$\alpha_k\le 2^{1-\gamma}\frac{(1-\gamma)\varepsilon}{\pi^2 k^2}g_k^{\gamma-1},$$

to arrive at

$$\frac{c}{1-\gamma}\frac{\pi^2}{\varepsilon}(k+2)^2n^\gamma c\alpha_k g_k^{1-2\gamma}$$

$$\le\frac{c2^{1-\gamma}}{1-\gamma}\frac{\pi^2}{\varepsilon}(k+2)^2n^\gamma c\frac{(1-\gamma)\varepsilon}{\pi^2 k^2}g_k^{\gamma-1}g_k^{1-2\gamma}=c^22^{1-\gamma}\frac{(k+2)^2}{k^2}\eta_k^\gamma, \tag{8.6.40}$$

which contributes to the first term on the rhs of (8.6.33).

By Definition 8.22 we have $c\beta_k n^{2\gamma-1} \leq \alpha_{k+1}$, so that, using (8.6.36), the second term in (8.6.39) is bounded by

$$\frac{c}{1-\gamma}\frac{\pi^2}{\varepsilon}(k+2)^2 n^\gamma c\beta_k \log \eta_k$$

$$\leq \frac{c}{1-\gamma}\frac{\pi^2}{\varepsilon}(k+2)^2 \alpha_{k+1} n^{1-\gamma} \log \eta_k$$

$$\leq c2^{1-\gamma}\frac{(k+2)^2}{(k+1)^2}g_{k+1}^{\gamma-1}n^{1-\gamma} \log \eta_k$$

$$= c2^{1-\gamma}\frac{(k+2)^2}{(k+1)^2}\eta_{k+1}^{1-\gamma} \log \eta_k. \tag{8.6.41}$$

Since $k \mapsto g_k$ is decreasing, it follows that $k \mapsto \eta_k$ is increasing, so that

$$c2^{1-\gamma}\frac{(k+2)^2}{(k+1)^2}\eta_{k+1}^{1-\gamma} \log \eta_k \leq c2^{1-\gamma}\frac{(k+2)^2}{(k+1)^2}\eta_{k+1}^{1-\gamma} \log \eta_{k+1}, \tag{8.6.42}$$

which again contributes to the second term in (8.6.33). This proves that both terms in (8.6.39), and thus the second term in (8.6.39), are bounded by the rhs of (8.6.33).

Putting together all the bounds and taking a sufficiently large constant $C = C(\gamma)$, we obtain (8.6.33). $\qquad\square$

Proof of Proposition 8.24. We prove the proposition by induction on k, and start by initializing the induction. For $k = 0$,

$$\eta_0 = \frac{n}{g_0} = \frac{n}{\lceil \varepsilon n \rceil} \leq \varepsilon^{-1} \leq e^B, \tag{8.6.43}$$

when $B \geq \log(1/\varepsilon)$, which initializes the induction.

We next advance the induction hypothesis. Suppose that the statement is true for all $l \in [k-1]$; we will advance it to k. We use that, for all $z, w \geq 0$,

$$(z+w)^{1/(1-\gamma)} \leq 2^{1/(1-\gamma)}\left(z^{1/(1-\gamma)} + w^{1/(1-\gamma)}\right). \tag{8.6.44}$$

By Lemma 8.25, for a different constant C,

$$\eta_k \leq C\left[\eta_{k-2}^{\gamma/(1-\gamma)} + \eta_{k-1}(\log \eta_{k-1})^{1/(1-\gamma)}\right]$$
$$= C\left[\eta_{k-2}^{1/(\tau-2)} + \eta_{k-1}(\log \eta_{k-1})^{1/(1-\gamma)}\right]. \tag{8.6.45}$$

Using this inequality, we can write

$$\eta_{k-2} \leq C\left[\eta_{k-4}^{1/(\tau-2)} + \eta_{k-3}(\log \eta_{k-3})^{1/(1-\gamma)}\right], \tag{8.6.46}$$

so that, by (8.6.44),

$$\eta_k \leq C(2C)^{1/(\tau-2)}\left[\eta_{k-4}^{1/(\tau-2)^2} + \eta_{k-3}^{1/(\tau-2)}(\log \eta_{k-3})^{1/[(1-\gamma)(\tau-2)]}\right]$$
$$+ C\eta_{k-1}(\log \eta_{k-1})^{1/(1-\gamma)}. \tag{8.6.47}$$

Renaming $2C$ as C for simplicity, and iterating these bounds, we obtain

$$\eta_k \leq C^{\sum_{l=0}^{k/2}(\tau-2)^{-l}} \eta_0^{(\tau-2)^{-k/2}}$$

$$+ \sum_{i=1}^{k/2} C^{\sum_{l=0}^{i-1}(\tau-2)^{-l}} \eta_{k-2i+1}^{(\tau-2)^{-(i-1)}} \left(\log \eta_{k-2i+1}\right)^{(\tau-2)^{-(i-1)}/(1-\gamma)}. \tag{8.6.48}$$

For the first term in (8.6.48), we use the upper bound $\eta_0 \leq 1/\varepsilon$ to obtain

$$C^{\sum_{l=0}^{k/2}(\tau-2)^{-l}} \eta_0^{(\tau-2)^{-k/2}} \leq C^{\sum_{l=0}^{k/2}(\tau-2)^{-l}} e^{(B/2)(\tau-2)^{-k/2}} \tag{8.6.49}$$

$$\leq \tfrac{1}{2} e^{B(\tau-2)^{-k/2}},$$

when $B \geq 2\log(1/\varepsilon)$, and we use that $C^{\sum_{l=0}^{k/2}(\tau-2)^{-l}} \leq \tfrac{1}{2} e^{(B/2)(\tau-2)^{-k/2}}$ for B large, since C is independent of ε.

For the second term in (8.6.48), we use the induction hypothesis to obtain

$$\sum_{i=1}^{k/2} C^{\sum_{l=0}^{i-1}(\tau-2)^{-l}} \eta_{k-2i+1}^{(\tau-2)^{-(i-1)}} \left(\log \eta_{k-2i+1}\right)^{(\tau-2)^{-(i-1)}/(1-\gamma)} \tag{8.6.50}$$

$$\leq \sum_{i=1}^{k/2} C^{\sum_{l=0}^{i-1}(\tau-2)^{-l}} e^{B(\tau-2)^{-(k-1)/2}} \left[B(\tau-2)^{-(k-2i+1)/2}\right]^{(\tau-2)^{-(i-1)}/(1-\gamma)}.$$

We can write

$$e^{B(\tau-2)^{-(k-1)/2}} = e^{B(\tau-2)^{-k/2}} e^{B(\tau-2)^{-k/2}(\sqrt{\tau-2}-1)}. \tag{8.6.51}$$

Since $\sqrt{\tau-2}-1 < 0$, for $k = O(\log\log n)$ we can take B large enough that, uniformly in $k \geq 1$,

$$\sum_{i=1}^{k/2} C^{\sum_{l=0}^{i-1}(\tau-2)^{-l}} e^{B(\tau-2)^{-k/2}(\sqrt{\tau-2}-1)} \left[B(\tau-2)^{-(k-2i+1)/2}\right]^{(\tau-2)^{-(i-1)}/(1-\gamma)} < \tfrac{1}{2}. \tag{8.6.52}$$

We can now sum the bounds in (8.6.49) and (8.6.50)–(8.6.52) to obtain

$$\eta_k \leq \left(\tfrac{1}{2} + \tfrac{1}{2}\right) e^{B(\tau-2)^{-k/2}} = e^{B(\tau-2)^{-k/2}}, \tag{8.6.53}$$

as required. This advances the induction hypothesis, and thus completes the proof of Proposition 8.24 for $B \geq 2\log(1/\varepsilon)$. $\qquad\square$

We are now ready to complete the proof of Theorem 8.19:

Completion of the proof of Theorem 8.19. Recall that we were left with proving (8.6.31), i.e., that uniformly in $u, v \geq \varepsilon n$,

$$\sum_{k=1}^{2k_n} \sum_{\pi_{\lfloor k/2 \rfloor}=g_{\lfloor k/2 \rfloor}}^{n} f_{\lfloor k/2 \rfloor,n}(u, \pi_{\lfloor k/2 \rfloor}) f_{\lceil k/2 \rceil,n}(v, \pi_{\lfloor k/2 \rfloor}) \leq \varepsilon. \tag{8.6.54}$$

A crucial part of the proof will be the optimal choice of k_n. By Lemma 8.24,

$$g_k \geq n/\eta_k \geq n e^{B(\tau-2)^{-k/2}}. \tag{8.6.55}$$

We use Lemma 8.23 to obtain the estimate

$$
\begin{aligned}
f_{\lfloor k/2 \rfloor, n}(u, w) &\leq \left(\alpha_{\lfloor k/2 \rfloor} w^{-\gamma} + \mathbb{1}_{\{w > g_{\lfloor k/2 \rfloor - 1}\}} \beta_{\lfloor k/2 \rfloor} w^{\gamma-1} \right) \\
&\leq \left(\alpha_{\lceil k/2 \rceil} w^{-\gamma} + \mathbb{1}_{\{w > g_{\lceil k/2 \rceil - 1}\}} \beta_{\lceil k/2 \rceil} w^{\gamma-1} \right),
\end{aligned}
\tag{8.6.56}
$$

since $k \mapsto \alpha_k$ and $k \mapsto \beta_k$ are non-decreasing, while $k \mapsto g_k$ is non-increasing. Thus, again using Lemma 8.23,

$$
\begin{aligned}
\sum_{k=1}^{2k_n} \sum_{\pi_{\lfloor k/2 \rfloor} = g_{\lfloor k/2 \rfloor}}^{n} & f_{\lfloor k/2 \rfloor, n}(u, \pi_{\lfloor k/2 \rfloor}) f_{\lceil k/2 \rceil, n}(v, \pi_{\lfloor k/2 \rfloor}) \\
&\leq \sum_{k=1}^{2k_n} \sum_{w = g_{\lfloor k/2 \rfloor}}^{n} \left(\alpha_{\lceil k/2 \rceil} w^{-\gamma} + \mathbb{1}_{\{w > g_{\lceil k/2 \rceil - 1}\}} \beta_{\lceil k/2 \rceil} w^{\gamma-1} \right)^2 \\
&\leq 2 \sum_{k=1}^{2k_n} \sum_{w = g_{\lfloor k/2 \rfloor}}^{n} \left(\alpha_{\lceil k/2 \rceil}^2 w^{-2\gamma} + \mathbb{1}_{\{w > g_{\lceil k/2 \rceil - 1}\}} \beta_{\lceil k/2 \rceil}^2 w^{2(\gamma-1)} \right).
\end{aligned}
\tag{8.6.57}
$$

This gives two terms, which we estimate one at a time. For the first term, using that $\gamma > \frac{1}{2}$ and that $k \mapsto \alpha_k$ is non-decreasing, while $k \mapsto g_k$ is non-increasing, by (8.6.18), we find that

$$
\begin{aligned}
2 \sum_{k=1}^{2k_n} \sum_{w = g_{\lfloor k/2 \rfloor}}^{n} \alpha_{\lceil k/2 \rceil}^2 w^{-2\gamma} &\leq \frac{2}{2\gamma - 1} \sum_{k=1}^{2k_n} \alpha_{\lceil k/2 \rceil}^2 (g_{\lfloor k/2 \rfloor} - 1)^{1-2\gamma} \\
&\leq \frac{2}{2\gamma - 1} \sum_{k=1}^{2k_n} \alpha_{\lceil k/2 \rceil}^2 g_{\lceil k/2 \rceil}^{1-2\gamma} = \frac{4 \times 2^{2\gamma-1}}{2\gamma - 1} \frac{\eta_{k_n}}{n} \sum_{k=1}^{k_n} \alpha_k^2 g_k^{2-2\gamma}.
\end{aligned}
\tag{8.6.58}
$$

Using (8.6.26), we obtain

$$
\alpha_k^2 g_k^{2-2\gamma} \leq 2^{2(1-\gamma)} \left(k^{-2} \frac{\varepsilon}{\pi^2} (1-\gamma) \right)^2,
\tag{8.6.59}
$$

so that

$$
2 \sum_{k=1}^{2k_n} \sum_{w = g_{\lfloor k/2 \rfloor}}^{n} \alpha_{\lceil k/2 \rceil}^2 w^{-2\gamma} \leq C \frac{\eta_{k_n+1}}{n} \sum_{k \geq 1} \varepsilon^2 k^{-4} \leq \frac{C \eta_{k_n+1} \varepsilon^2}{n}.
\tag{8.6.60}
$$

This bounds the first term on the rhs of (8.6.57).

For the second term on the rhs of (8.6.57), we again use that $k \mapsto g_k$ is non-increasing, and that $2(\gamma - 1) > -1$ since $\gamma \in (\frac{1}{2}, 1)$, to obtain, using (8.6.18),

$$
2 \sum_{k=1}^{2k_n} \sum_{w = g_{\lceil k/2 \rceil - 1}}^{n} \beta_{\lceil k/2 \rceil}^2 w^{2(\gamma-1)} \leq \frac{4}{2\gamma - 1} \sum_{k=1}^{k_n} \beta_k^2 n^{2\gamma-1}.
\tag{8.6.61}
$$

By the definition of α_k in (8.6.23), we get $\beta_k n^{2\gamma-1} \leq \alpha_{k+1}$. Thus,

$$\sum_{k=1}^{k_n} \beta_k^2 n^{2\gamma-1} \leq \sum_{k=1}^{k_n} \alpha_{k+1}^2 n^{1-2\gamma} = \frac{1}{n}\eta_{k_n+1}^{2-2\gamma} \sum_{k=1}^{k_n} \alpha_{k+1}^2 g_{k+1}^{2-2\gamma}$$

$$\leq \frac{C\eta_{k_n+1}^{2-2\gamma}\varepsilon^2}{n} \leq \frac{C\eta_{k_n+1}\varepsilon^2}{n}, \tag{8.6.62}$$

as in (8.6.58)–(8.6.60), since $\gamma \in (\frac{1}{2}, 1)$.

We conclude that, using (8.6.32) in Proposition 8.24,

$$\sum_{k=1}^{2k_n} \sum_{\pi_{\lfloor k/2\rfloor}=g_{\lfloor k/2\rfloor}}^{n} f_{\lfloor k/2\rfloor,n}(u, \pi_{\lfloor k/2\rfloor}) f_{\lceil k/2\rceil,n}(v, \pi_{\lfloor k/2\rfloor})$$

$$\leq C\frac{\varepsilon^2\eta_{k_n+1}}{n} \leq \frac{C\varepsilon^2}{n}e^{B(\tau-2)^{-k_n/2}} \leq \frac{\varepsilon}{3}, \tag{8.6.63}$$

when finally choosing

$$k_n \leq \frac{2\log\log n}{|\log(\tau-2)|} - K, \tag{8.6.64}$$

for $K = K_\varepsilon$ sufficiently large. Thus, for $\varepsilon > 0$,

$$\mathbb{P}(\text{dist}_{\text{PA}_n}(u, v) \leq 2k_n) \leq 2\varepsilon/3 + \varepsilon/3 = \varepsilon, \tag{8.6.65}$$

whenever $u, v \geq g_0 = \lceil \varepsilon n \rceil$. Note that from (8.6.32) in Proposition 8.24 this choice also ensures that $g_{k_n} = n/\eta_{k_n} \geq 2$ for K sufficiently large. This implies Theorem 8.19. □

8.7 LOG LOG UPPER BOUNDS FOR PAMs WITH INFINITE-VARIANCE DEGREES

In this section we investigate $\text{PA}_n^{(m,\delta)}$ with $m \geq 2$ and $\delta \in (-m, 0)$ and prove a first step towards the upper bound in Theorem 8.7. This proof is divided into two key steps. In the first, in Theorem 8.26 in Section 8.7.1, we bound the diameter of the *core*, which consists of the vertices with degree at least a certain power of $\log n$. This argument is close in spirit to the argument used to bound the diameter of the core in the configuration model in Theorem 7.13, but substantial adaptations are necessary. After this, in Theorem 8.32, we derive a bound on the distance between a typical vertex having a small degree and the core.

8.7.1 DIAMETER OF THE CORE

In what follows, it will be convenient to prove Theorem 8.7 for $2n$ rather than for n. Clearly, this does not make any difference to the results. We will adapt the proof of Theorem 7.13 to $\text{PA}_{2n}^{(m,\delta)}$. We recall that $\tau = 3 + \delta/m$ from (1.3.63), so that $\delta \in (-m, 0)$ corresponds to $\tau \in (2, 3)$. Throughout this section, we fix $m \geq 2$.

We take $\sigma > 1/(3-\tau) = -m/\delta > 1$ and define the *core* Core_n of $\text{PA}_{2n}^{(m,\delta)}$ by

$$\text{Core}_n = \{v \in [n]: D_v(n) \geq (\log n)^\sigma\}, \tag{8.7.1}$$

i.e., all vertices in $[n]$ that at time n have degree at least $(\log n)^\sigma$. We emphasize that we require the degree to be large at time n, rather than at the final time $2n$ for $\text{PA}_{2n}^{(m,\delta)}$.

Let us explain the philosophy of the proof. Note that Core_n requires only information about $\text{PA}_n^{(m,\delta)}$, while we are going to study its diameter in $\text{PA}_{2n}^{(m,\delta)}$. This allows us to use the edges originating from vertices in $[2n] \setminus [n]$ as a *sprinkling* of the graph that will create *shortcuts* in $\text{PA}_n^{(m,\delta)}$. Such shortcuts shorten graph distances tremendously. We call the vertices that create such shortcuts *n-connectors*.

Basically, this argument shows that a vertex $v \in [n]$ of large degree $D_v(n) \gg 1$ will likely have an n-connector to a vertex $u \in [n]$ satisfying $D_u(n) \geq D_v(n)^{1/(\tau-2)}$. This is related to the *power-iteration* argument for the configuration model discussed below Proposition 7.15. However, for preferential attachment models, we emphasize that it takes *two* steps to link a vertex of large degree to another vertex of even larger degree. In the proof for the configuration model in Theorem 7.13, this happened in *only one* step. Therefore, distances in $\text{PA}_n^{(m,\delta)}$ for $\delta < 0$ are (at least in terms of upper bounds) *twice as large* as the corresponding distances for a configuration model with similar degree structure. Let us now state our main result, for which we require some notation.

For $A \subseteq [2n]$, we write

$$\text{diam}_{2n}(A) = \max_{i,j \in A} \text{dist}_{\text{PA}_{2n}^{(m,\delta)}}(i,j). \tag{8.7.2}$$

Then, the diameter of the core in the graph $\text{PA}_{2n}^{(m,\delta)}$ is bounded as follows:

Theorem 8.26 (Diameter of the core) *Consider $\text{PA}_{2n}^{(m,\delta)}$ with $m \geq 2$ and $\delta \in (-m, 0)$. For every $\sigma > 1/(3 - \tau)$, whp there exists a $K = K_\delta > 0$ such that*

$$\text{diam}_{2n}(\text{Core}_n) \leq \frac{4 \log \log n}{|\log (\tau - 2)|} + K. \tag{8.7.3}$$

These results apply to $\text{PA}_{2n}^{(m,\delta)}(a)$, as well as to $\text{PA}_{2n}^{(m,\delta)}(b)$ and $\text{PA}_{2n}^{(m,\delta)}(d)$.

The proof of Theorem 8.26 is divided into several smaller steps. We start by proving that the diameter of the *inner core* Inner_n, which is defined by

$$\text{Inner}_n = \{v \in [n] \colon D_v(n) \geq n^{1/[2(\tau-1)]}(\log n)^{-1/2}\}, \tag{8.7.4}$$

is whp bounded by some finite constant $K_\delta < \infty$. After this, we show that the distance from the *outer core*, given by $\text{Outer}_n = \text{Core}_n \setminus \text{Inner}_n$, to the inner core can be bounded by $2 \log \log n / |\log (\tau - 2)|$. This shows that the diameter of the outer core is bounded by $4 \log \log n / |\log (\tau - 2)| + K_\delta$, as required. We now give the details, starting with the diameter of the inner core:

Proposition 8.27 (Diameter of the inner core) *Consider $\text{PA}_{2n}^{(m,\delta)}$ with $m \geq 2$ and $\delta \in (-m, 0)$. Then, whp, there exists a constant $K = K_\delta$ such that*

$$\text{diam}_{2n}(\text{Inner}_n) < K_\delta. \tag{8.7.5}$$

These results apply to $\text{PA}_{2n}^{(m,\delta)}(a)$, as well as to $\text{PA}_{2n}^{(m,\delta)}(b)$ and $\text{PA}_{2n}^{(m,\delta)}(d)$.

Before proving Proposition 8.27, we first introduce the important notion of an n-connector between two sets of vertices $A, B \subseteq [n]$, which plays a crucial role throughout the proof:

Definition 8.28 (n-connector) Fix two sets of vertices A and B. We say that the vertex $j \in [2n] \setminus [n]$ is an n-*connector* between A and B if one of the edges incident to j connects to a vertex in A, while another edge incident to j connects to a vertex in B. Thus, when there exists an n-connector between A and B, the distance between A and B in $\mathrm{PA}_{2n}^{(m,\delta)}$ is at most 2. ◀

The next lemma gives bounds on the probability that an n-connector does not exist:

Lemma 8.29 (Connectivity sets in infinite-variance degree preferential attachment models) *Consider* $\mathrm{PA}_{2n}^{(m,\delta)}(a)$ *with* $m \geq 2$ *and* $\delta \in (-m, 0)$. *For any two sets of vertices* $A, B \subseteq [n]$, *there exists* $\eta = \eta(m, \delta) > 0$ *such that*

$$\mathbb{P}(\text{no } n\text{-connector for } A \text{ and } B \mid \mathrm{PA}_n^{(m,\delta)}(a)) \leq e^{-\eta D_A(n) D_B(n)/n}, \qquad (8.7.6)$$

where, for any $A \subseteq [n]$,

$$D_A(n) = \sum_{a \in A} D_a(n), \qquad (8.7.7)$$

denotes the total degree of vertices in A at time n. These results also apply to $\mathrm{PA}_{2n}^{(m,\delta)}(b)$ *and* $\mathrm{PA}_{2n}^{(m,\delta)}(d)$ *under identical conditions.*

Lemma 8.29 plays the same role for preferential attachment models as Lemma 7.12 for configuration models.

Proof We give only the proof for $\mathrm{PA}_{2n}^{(m,\delta)}(a)$; the proofs for $\mathrm{PA}_{2n}^{(m,\delta)}(b)$ and $\mathrm{PA}_{2n}^{(m,\delta)}(d)$ are identical. We note that for two sets of vertices A and B, conditional on $\mathrm{PA}_n^{(m,\delta)}(a)$ the probability that $j \in [2n] \setminus [n]$ is an n-connector for A and B is at least

$$\frac{(D_A(n) + \delta|A|)(D_B(n) + \delta|B|)}{[2n(2m + \delta)]^2}, \qquad (8.7.8)$$

independently of whether the other vertices are n-connectors.

Since $D_i(n) + \delta \geq m + \delta > 0$ for every $i \leq n$, and $\delta < 0$, for every $i \in B$, we have

$$D_i(n) + \delta = D_i(n)\left(1 + \frac{\delta}{D_i(n)}\right) \geq D_i(n)\left(1 + \frac{\delta}{m}\right) = D_i(n)\frac{m + \delta}{m}, \qquad (8.7.9)$$

and thus $D_A(n) + \delta|A| \geq D_A(n)(m + \delta)/m$. As a result, for $\eta = (m + \delta)^2/(2m(2m + \delta))^2 > 0$, the probability that $j \in [2n] \setminus [n]$ is an n-connector for A and B is at least $\eta D_A(n) D_B(n)/n^2$, independently of whether the other vertices are n-connectors. Therefore, the probability that there is no n-connector for A and B is, conditional on $\mathrm{PA}_n^{(m,\delta)}(a)$, bounded above by

$$\left(1 - \frac{\eta D_A(n) D_B(n)}{n^2}\right)^n \leq e^{-\eta D_A(n) D_B(n)/n}, \qquad (8.7.10)$$

as required. □

We now give the proof of Proposition 8.27:

Proof of Proposition 8.27. From [V1, Theorem 8.3 and Exercise 8.20] whp, Inner_n contains at least $\sqrt{n}$ vertices. Denote the first $\sqrt{n}$ vertices of Inner_n by I. We are relying on Lemma

8.29. Recall that $D_i(n) \geq n^{1/[2(\tau-1)]}(\log n)^{-1/2}$ for all $i \in I$. Observe that $n^{1/(\tau-1)-1} = o(1)$ for $\tau > 2$, so that, for any $i, j \in I$, the probability that there exists an n-connector for i and j is bounded below by

$$1 - \exp\{-\eta n^{1/(\tau-1)-1}(\log n)^{-1}\} \geq p_n \equiv n^{-(\tau-2)/(\tau-1)}(\log n)^{-2}, \qquad (8.7.11)$$

for n sufficiently large.

We wish to couple Inner_n to an Erdős–Rényi random graph with $N_n = \sqrt{n}$ vertices and edge probability p_n, which we denote by $\mathrm{ER}_{N_n}(p_n)$. For this, for $i, j \in [N_n]$, we say that an edge between i and j is present when there exists an n-connector connecting the ith and jth vertices in I.

We now prove that this graph is stochastically bounded below by $\mathrm{ER}_{N_n}(p_n)$. Note that (8.7.11) does not guarantee this coupling; instead we need to prove that the lower bound holds uniformly when i and j belong to I, independently of the previous edges. For this, we order the $N_n(N_n - 1)/2$ edges in an arbitrary way and bound the conditional probability that the lth edge is present, conditioning on all previous edges, from below by p_n for every l. This proves the claimed stochastic lower bound.

Indeed, the lth edge is present precisely when there exists an n-connector connecting the corresponding vertices, which we call i and j in I. Moreover, we shall not make use of the first vertices that were used to n-connect the previous edges. This removes at most $N_n(N_n - 1)/2 \leq n/2$ possible n-connectors, after which at least another $n/2$ remain. The probability that one of them is an n-connector for the ith and jth vertex in I is, for n sufficiently large, bounded below by

$$1 - \exp\{-\eta n^{1/(\tau-1)-2}(\log n)^{-1}n/2\} \geq p_n \equiv n^{-(\tau-2)/(\tau-1)}(\log n)^{-2},$$

using $1 - \mathrm{e}^{-x} \geq x/2$ for $x \in [0, 1]$ and $\eta/2 \geq 1/\log n$ for n sufficiently large. This proves the claimed stochastic domination of the random graph on I and $\mathrm{ER}(N_n, p_n)$. Next, we show that $\mathrm{diam}(\mathrm{ER}_{N_n}(p_n))$ is, whp, uniformly bounded by a constant.

For this we use the result in (Bollobás, 2001, Corollary 10.12), which gives sharp bounds on the diameter of an Erdős–Rényi random graph. Indeed, this result implies that if $p^d N^{d-1} - 2 \log N \to \infty$, while $p^{d-1}N^{d-2} - 2\log N \to -\infty$, then $\mathrm{diam}(\mathrm{ER}_N(p)) = d$, whp. In our case, $N = N_n = n^{1/2}$ and

$$p = p_n = n^{-(\tau-2)/(\tau-1)}(\log n)^{-2} = N^{-2(\tau-2)/(\tau-1)}(2\log N)^{-2},$$

which implies that, whp, $\frac{\tau-1}{3-\tau} < d \leq \frac{\tau-1}{3-\tau} + 1$. Thus, we obtain that the diameter of I in $\mathrm{PA}_{2n}^{(m,\delta)}$ is whp bounded by $2d \leq 2(\frac{\tau-1}{3-\tau} + 1)$. In Exercise 8.23, the reader is asked to prove an upper bound on the diameter of $\mathrm{ER}_n(p)$.

We finally show that for any $i \in \mathrm{Inner}_n \setminus I$, the probability that no n-connector exists between i and I is small. We again rely on Lemma 8.29. Since $D_I(n) \geq \sqrt{n}n^{1/[2(\tau-1)]}$ $(\log n)^{-1/2}$, and $D_i(n) \geq n^{1/[2(\tau-1)]}(\log n)^{-1/2}$, the probability that no n-connector exists between i and I is bounded above by $\exp\{-\eta n^{1/(\tau-1)-1/2}(\log n)^{-1}\}$, which is tiny since $\tau < 3$. This proves that the distance between any vertex $i \in \mathrm{Inner}_n \setminus I$ and I is whp bounded by 2, and, together with the fact that $\mathrm{diam}_{2n}(I) \leq 2(\frac{\tau-1}{3-\tau} + 1)$, this implies that diam_{2n} $(\mathrm{Inner}_n) \leq 2(\frac{\tau-1}{3-\tau} + 2) \equiv K_\delta$. $\qquad\qquad\square$

We proceed by studying the distances between the outer core $\text{Core}_n \setminus \text{Inner}_n$ and the inner core Inner_n:

Proposition 8.30 (Distance between outer and inner core) *Consider* $\text{PA}_{2n}^{(m,\delta)}(a)$ *with* $m \geq 2$ *and* $\delta \in (-m, 0)$. *The inner core* Inner_n *can whp be reached from any vertex in the outer core* Outer_n *using no more than* $\frac{2 \log \log n}{|\log (\tau - 2)|}$ *edges in* $\text{PA}_{2n}^{(m,\delta)}(a)$, *i.e., whp,*

$$\max_{i \in \text{Outer}_n} \min_{j \in \text{Inner}_n} \text{dist}_{\text{PA}_{2n}^{(m,\delta)}(a)}(i, j) \leq \frac{2 \log \log n}{|\log (\tau - 2)|}. \tag{8.7.12}$$

These results also apply to $\text{PA}_{2n}^{(m,\delta)}(b)$ *and* $\text{PA}_{2n}^{(m,\delta)}(d)$.

Proof Again, we give the proof only for $\text{PA}_{2n}^{(m,\delta)}(a)$. Recall that

$$\text{Outer}_n = \text{Core}_n \setminus \text{Inner}_n, \tag{8.7.13}$$

and define

$$\Gamma_1 = \text{Inner}_n = \{i : D_i(n) \geq u_1\}, \tag{8.7.14}$$

where

$$u_1 = n^{1/[2(\tau-1)]}(\log n)^{-1/2}. \tag{8.7.15}$$

We now recursively define a sequence u_k, for $k \geq 2$, so that, for any vertex $i \in [n]$ with degree at least u_k, the probability that there is no n-connector for the vertex i and the set

$$\Gamma_{k-1} = \{j : D_j(n) \geq u_{k-1}\}, \tag{8.7.16}$$

conditional on $\text{PA}_n^{(m,\delta)}$, is tiny; see Lemma 8.31 below. According to Lemma 8.29 and [V1, Exercise 8.20], this probability is at most

$$\exp\left\{ -\frac{\eta B n [u_{k-1}]^{2-\tau} u_k}{n} \right\}. \tag{8.7.17}$$

To make this sufficiently small, we define

$$u_k = D \log n (u_{k-1})^{\tau-2}, \tag{8.7.18}$$

with the constant D exceeding $(\eta B)^{-1}$. Note that the recursion in (8.7.18) is identical to that in (7.3.53). Therefore, by Lemma 7.14,

$$u_k = (D \log n)^{a_k} n^{b_k}, \tag{8.7.19}$$

where

$$b_k = \frac{(\tau - 2)^{k-1}}{2(\tau - 1)}, \quad a_k = \frac{1 - (\tau - 2)^{k-1}}{3 - \tau}. \tag{8.7.20}$$

The key step in the proof of Proposition 8.30 is the following lemma:

Lemma 8.31 (Connectivity between Γ_{k-1} and Γ_k) *Fix* $m, k \geq 2$ *and* $\delta \in (-m, 0)$. *Then the probability that there exists an* $i \in \Gamma_k$ *that is not at distance 2 from* Γ_{k-1} *in* $\text{PA}_{2n}^{(m,\delta)}$ *is bounded by* $n^{-\zeta}$ *for some* $\zeta > 0$.

Proof   We note that, by [V1, Exercise 8.20], with probability exceeding $1 - o(n^{-1})$, for all k, we have

$$\sum_{i \in \Gamma_{k-1}} D_i(n) \geq Bn[u_{k-1}]^{2-\tau}. \tag{8.7.21}$$

On the event that the bounds in (8.7.21) hold, by Lemma 8.29 we obtain that the conditional probability, given $\mathrm{PA}_n^{(m,\delta)}$, that there exists an $i \in \Gamma_k$ such that there is no n-connector between i and Γ_{k-1} is bounded, using Boole's inequality, by

$$n \exp\left\{ -\eta B[u_{k-1}]^{2-\tau} u_k \right\} = n e^{-\eta B D \log n} \leq n^{-(1+\zeta)}, \tag{8.7.22}$$

where we have used (8.7.18) and taken $D = 2(1 + \zeta)/(\eta B)$. $\square$

We now complete the proof of Proposition 8.30. Fix

$$k_n^\star = \left\lfloor \frac{\log \log n}{|\log(\tau - 2)|} \right\rfloor. \tag{8.7.23}$$

By Lemma 8.31, and since $k_n^\star n^{-\zeta} = o(1)$ for all $\zeta > 0$, the distance between $\Gamma_{k_n^\star}$ and Inner_n is at most $2k_n^\star$. Therefore, we are done if we can show that

$$\mathrm{Outer}_n \subseteq \{i \colon D_i(n) \geq (\log n)^\sigma\} \subseteq \Gamma_{k_n^\star} = \{i \colon D_i(n) \geq u_{k_n^\star}\}, \tag{8.7.24}$$

so that it suffices to prove that $(\log n)^\sigma \geq u_{k_n^\star}$ for any $\sigma > 1/(3 - \tau)$. This follows by (7.3.66), which implies that

$$u_{k_n^\star} = (\log n)^{1/(3-\tau)+o(1)}; \tag{8.7.25}$$

by picking n sufficiently large, we see that this is smaller than $(\log n)^\sigma$ for any $\sigma > 1/(3 - \tau)$. This completes the proof of Proposition 8.30. $\square$

Proof of Theorem 8.26. We note that whp $\mathrm{diam}_{2n}(\mathrm{Core}_n) \leq K_\delta + 2k_n^\star$, where $k_n^\star$ in (8.7.23) is the upper bound on $\max_{i \in \mathrm{Outer}_n} \min_{j \in \mathrm{Inner}_n} \mathrm{dist}_{\mathrm{PA}_{2n}^{(m,\delta)}}(i, j)$ in Proposition 8.30, and we have made use of Proposition 8.27. This proves Theorem 8.26. $\square$

8.7.2 CONNECTING THE PERIPHERY TO THE CORE

In this subsection we extend the results of the previous subsection and, in particular, study the distance between the vertices not in Core_n and the core. The main result is the following theorem:

Theorem 8.32 (Connecting the periphery to the core)   *Consider* $\mathrm{PA}_{2n}^{(m,\delta)}$ *with* $m \geq 2$ *and* $\delta \in (-m, 0)$. *For every* $\sigma > 1/(3 - \tau)$, *whp the distance between a uniformly chosen vertex* $o_1 \in [2n]$ *and* Core_n *in* $\mathrm{PA}_{2n}^{(m,\delta)}(a)$ *is bounded by* $C \log \log \log n$ *for some* $C > 0$. *These results also apply to* $\mathrm{PA}_{2n}^{(m,\delta)}(b)$ *and* $\mathrm{PA}_{2n}^{(m,\delta)}(d)$.

Together with Theorem 8.26, Theorem 8.32 proves the upper bound in Theorem 8.7:

Proof of the upper bound in Theorem 8.7.   Choose $o_1, o_2 \in [2n]$ independently and uar. Using the triangle inequality, we obtain the bound

$$\mathrm{dist}_{\mathrm{PA}_{2n}^{(m,\delta)}}(o_1, o_2)$$
$$\leq \mathrm{dist}_{\mathrm{PA}_{2n}^{(m,\delta)}}(o_1, \mathrm{Core}_n) + \mathrm{dist}_{\mathrm{PA}_{2n}^{(m,\delta)}}(o_2, \mathrm{Core}_n) + \mathrm{diam}_{2n}(\mathrm{Core}_n). \tag{8.7.26}$$

By Theorem 8.32, the first two terms are each whp bounded by $C \log \log \log n$. Further, by Theorem 8.26, the third term is bounded by $(1 + o_{\mathbb{P}}(1)) \frac{4 \log \log n}{|\log (\tau - 2)|}$. This completes the proof of the upper bound in Theorem 8.7. $\qquad\square$

Exercise 8.24 shows that $\mathrm{dist}_{\mathrm{PA}_{2n}^{(m,\delta)}}(o_1, o_2) - 2 \log \log n / |\log(\tau - 2)|$ is upper tight when $\mathrm{dist}_{\mathrm{PA}_{2n}^{(m,\delta)}}(o_1, \mathrm{Core}_n)$ is tight.

Proof of Theorem 8.32. We use the same ideas as in the proof of Theorem 8.26, but now start from a vertex of large degree at time n instead. We need to show that, for fixed $\varepsilon > 0$, a uniformly chosen vertex $o \in [(2 - \varepsilon)n]$ can whp be connected to Core_n using no more than $C \log \log \log n$ edges in $\mathrm{PA}_{2n}^{(m,\delta)}$. This is done in two steps.

In the **first step**, we explore the neighborhood of o in $\mathrm{PA}_{2n}^{(m,\delta)}$ until we find a vertex v_0 with degree $D_{v_0}(n) \geq u_0$, where u_0 will be determined below. Denote the set of all vertices in $\mathrm{PA}_{2n}^{(m,\delta)}$ that can be reached from o using exactly k different edges from $\mathrm{PA}_{2n}^{(m,\delta)}$ by S_k. Denote the first k for which there is a vertex in S_k whose degree at time n is at least u by

$$T_u^{(o)} = \inf \{ k \colon S_k \cap \{ v \colon D_v(n) \geq u \} \neq \varnothing \}. \tag{8.7.27}$$

Recall the local convergence in Theorem 5.26, as well as the fact that each vertex v has m older neighbors $v_1, \ldots, v_m$, whose ages $a_{v_1}, \ldots, a_{v_m}$ are distributed as $U_{v_i}^{(\tau-2)/(\tau-1)} a_v$, where a_v is the age of v. Therefore, whp, there is a vertex in S_k with arbitrarily small age, and thus also with arbitrarily large degree, at time n. As a result, there exists a $C = C_{u,\varepsilon}$ such that, for sufficiently large n,

$$\mathbb{P}(T_u^{(o)} \geq C_{u,\varepsilon}) \leq \varepsilon. \tag{8.7.28}$$

The **second step** is to show that a vertex v_0 satisfying $D_{v_0}(n) \geq u_0$ for sufficiently large u_0 can be joined to the core by using $O(\log \log \log n)$ edges. To this end, we apply Lemma 8.29 to obtain, for any vertex a with $D_a(n) \geq w_a$, the probability that there does not exist a vertex b with $D_b(n) \geq w_b$ that is connected to a by an n-connector, conditional on $\mathrm{PA}_n^{(m,\delta)}$, is at most

$$\exp \{ -\eta D_a(n) D_B(n) / n \}, \tag{8.7.29}$$

where $B = \{ b \colon D_b(n) \geq w_b \}$. Since, as in (8.7.21),

$$D_B(n) \geq w_b \# \{ b \colon D_b(n) \geq u_b \} \equiv w_b N_{\geq w_b}(n) \geq cn w_b^{2-\tau}, \tag{8.7.30}$$

we thus obtain that the probability that such a b does not exist is at most

$$\exp \{ -\eta' w_a w_b^{2-\tau} \}, \tag{8.7.31}$$

where $\eta' = \eta c$. Fix $\varepsilon > 0$ such that $(1 - \varepsilon)/(\tau - 2) > 1$. We then iteratively take $u_k = u_{k-1}^{(1-\varepsilon)/(\tau-2)}$, to see that the probability that there exists a k for which there does not exist a v_k with $D_{v_k}(n) \geq u_k$ is at most

$$\sum_{l=1}^{k} \exp \{ -\eta' u_{l-1}^{\varepsilon} \}. \tag{8.7.32}$$

Since, for $\varepsilon > 0$,

$$u_k = u_0^{\kappa^k}, \qquad \text{where} \qquad \kappa = (1 - \varepsilon)/(\tau - 2) > 1, \tag{8.7.33}$$

we obtain that the probability that there exists an l with $l \leq k$ for which there does not exist a v_l with $D_{v_l}(n) \geq u_l$ is at most

$$\sum_{l=1}^{k} \exp\left\{-\eta' u_0^{\varepsilon\kappa^{l-1}}\right\}. \tag{8.7.34}$$

Now fix $k = k_n = \lceil C \log\log\log n \rceil$ and choose u_0 sufficiently large that

$$\sum_{l=1}^{k_n} \exp\left\{-\eta' u_0^{\varepsilon\kappa^{l-1}}\right\} \leq \varepsilon/2. \tag{8.7.35}$$

Then we obtain that, with probability at least $1 - \varepsilon/2$, v_0 is connected in k_n steps to a vertex v_{k_n} with $D_{v_{k_n}}(n) \geq u_0^{\kappa^{k_n}}$. Since, for $C \geq 1/\log\kappa$, we have

$$u_0^{\kappa^{k_n}} \geq u_0^{\log\log n} \geq (\log n)^\sigma \tag{8.7.36}$$

when $\log u_0 \geq \sigma$, we obtain that $v_{k_n} \in \mathrm{Core}_n$ whp. $\qquad\square$

8.8 Diameters in Preferential Attachment Models

In this section we investigate the diameter in preferential attachment models. We start by discussing logarithmic bounds for $\delta > 0$, continue with the doubly logarithmic bounds for $\delta < 0$, and finally discuss the case where $\delta = 0$.

Logarithmic Bounds on the Diameter

We start by proving a logarithmic bound on the diameter for $(\mathrm{PA}_n^{(m,\delta)})_{n\geq 1}$, which obviously also implies the logarithmic *upper bound* on typical distances in Theorem 8.6:

Theorem 8.33 (Diameter of preferential attachment models)   *Consider* $\mathrm{PA}_n^{(m,\delta)}(a)$ *with* $m \geq 2$ *and* $\delta > -m$. *There exists a constant* c_2 *with* $0 < c_2 < \infty$ *such that, as* $n \to \infty$ *and whp,*

$$\mathrm{diam}(\mathrm{PA}_n^{(m,\delta)}(a)) \leq c_2 \log n. \tag{8.8.1}$$

Consequently, there exists a constant c_2 *with* $0 < c_2 < \infty$ *such that* $\mathrm{dist}_{\mathrm{PA}_n^{(m,\delta)}}(o_1, o_2) \leq c_2 \log n$ *whp as* $n \to \infty$. *These results also apply to* $\mathrm{PA}_n^{(m,\delta)}(b)$ *and* $\mathrm{PA}_n^{(m,\delta)}(d)$.

Proof   We start by proving the logarithmic bound on the diameter of $\mathrm{PA}_n^{(m,\delta)}(b)$, which is easier, since $\mathrm{PA}_n^{(m,\delta)}(b)$ is whp connected. Since $(\mathrm{PA}_n^{(m,\delta)}(b))_{n\geq 1}$ is obtained by collapsing m successive vertices in $(\mathrm{PA}_{mn}^{(1,\delta/m)}(b))_{n\geq 1}$, we have

$$\mathrm{diam}(\mathrm{PA}_n^{(m,\delta)}(b)) \preceq \mathrm{diam}(\mathrm{PA}_{mn}^{(1,\delta/m)}(b)), \tag{8.8.2}$$

and the result follows from Theorem 8.1.

    We next extend this result to $(\mathrm{PA}_n^{(m,\delta)}(a))_{n\geq 1}$. We know that $\mathrm{PA}_n^{(m,\delta)}(a)$ is whp connected for all n sufficiently large (recall Theorem 5.27). Fix a sufficiently large T, to be determined below. Then we can bound, for all $u, v \in [n]$,

$$\text{dist}_{\text{PA}_n^{(m,\delta)}(a)}(u, v) \leq \text{dist}_{\text{PA}_n^{(m,\delta)}(a)}(u, [T]) + \text{dist}_{\text{PA}_n^{(m,\delta)}(a)}(v, [T])$$
$$+ \text{diam}_{\text{PA}_n^{(m,\delta)}(a)}([T]). \tag{8.8.3}$$

Since $\text{PA}_T^{(m,\delta)}(a)$ is connected whp,

$$\text{diam}_{\text{PA}_n^{(m,\delta)}(a)}([T]) \leq \text{diam}(\text{PA}_T^{(m,\delta)}(a)), \tag{8.8.4}$$

which is a tight random variable. Then, similarly to (8.8.2),

$$\text{dist}_{\text{PA}_n^{(m,\delta)}(a)}(u, [T]) \leq \text{dist}_{\text{PA}_{mn}^{(1,\delta/m)}(a)}(u, [mT]). \tag{8.8.5}$$

As in the proof for $\text{PA}_n^{(m,\delta)}(b)$, whp, when T is sufficiently large,

$$\max_{u \in [mn]} \text{dist}_{\text{PA}_{mn}^{(1,\delta/m)}(a)}(u, [mT]) \leq c \log n. \tag{8.8.6}$$

A similar proof can be used for $\text{PA}_n^{(m,\delta)}(d)$; see Exercise 8.25. $\qquad\square$

Doubly Logarithmic Diameter

When $\delta < 0$, typical distances in $\text{PA}_n^{(m,\delta)}$ grow doubly logarithmically. The following theorem shows that this extends to the diameter of $\text{PA}_n^{(m,\delta)}$, and it identifies the exact constant:

Theorem 8.34 (Diameter of $\text{PA}_n^{(m,\delta)}$ for $\delta < 0$) *Consider $\text{PA}_n^{(m,\delta)}(a)$ with $m \geq 2$ and $\delta \in (-m, 0)$. As $n \to \infty$,*

$$\frac{\text{diam}(\text{PA}_n^{(m,\delta)}(a))}{\log \log n} \overset{\mathbb{P}}{\longrightarrow} \frac{4}{|\log(\tau - 2)|} + \frac{2}{\log m}. \tag{8.8.7}$$

These results also apply to $\text{PA}_n^{(m,\delta)}(b)$ and $\text{PA}_n^{(m,\delta)}(d)$.

Theorem 8.34 is an adaptation of Theorem 7.20 from $\text{CM}_n(\boldsymbol{d})$ to $\text{PA}_n^{(m,\delta)}$. Indeed, the first term on the rhs of (8.8.7) corresponds to the typical distances in Theorem 8.19, just as the first term on the rhs of (7.5.5) in Theorem 7.20 corresponds to the typical distances in Theorem 7.2. Further, the additional term $2/\log m$ can be interpreted as $2/\log(d_{\min} - 1)$, where $d_{\min}$ is the minimal degree of an *internal* vertex in the neighborhood of a vertex in $\text{PA}_n^{(m,\delta)}$, as in Theorem 7.20. In turn, this can be interpreted as twice $1/\log m$, where $\log \log n / \log m$ can be viewed as the depth of the worst trap.

We will be brief about the proof of the upper bound. We will explore the r-neighborhoods of vertices, where we take $r = (1+\varepsilon) \log \log n / \log m$. Then, these boundaries are so large, that, with probability of order $1 - o(1/n^2)$, pairs of such boundaries are quickly connected to the core Core_n. An application of Theorem 8.26 then completes the upper bound.

We now give some more details on the proof of the lower bound, which proceeds by defining the notion of a vertex being minimally k-connected. It means that the k-neighborhood of the vertex is as small as possible, so that its boundary has size m^k (rather than $d_{\min}(d_{\min} - 1)^{k-1}$ as it is in $\text{CM}_n(\boldsymbol{d})$). We then show that there are plenty of such minimally k-connected vertices as long as $k \leq (1 - \varepsilon) \log \log n / \log m$, similarly to the analysis in Lemma 7.22 for $\text{CM}_n(\boldsymbol{d})$. However, the analysis itself is quite a bit harder, owing to the dynamic nature of $\text{PA}_n^{(m,\delta)}$.

Definition 8.35 (Minimally k-connected vertices) Fix $G_n = \mathrm{PA}_n^{(m,\delta)}$, and define the set of minimally k-connected vertices by

$$\mathcal{M}_k = \{v \in [n]: D_v(n) = m, D_u(n) = m+1 \,\forall u \in B_{k-1}^{(G_n)}(v) \setminus \{v\},$$
$$B_k^{(G_n)}(v) \subseteq [n] \setminus [n/2]\}, \tag{8.8.8}$$

and let $M_k = |\mathcal{M}_k|$ denote its size. ◀

In words, $\mathcal{M}_k$ consists of those vertices in $[n] \setminus [n/2]$ whose k-neighborhood is minimally k-connected at time n. The following lemma shows that $M_k \xrightarrow{\mathbb{P}} \infty$ when $k \leq (1-\varepsilon)\log\log n / \log m$:

Lemma 8.36 (Many minimally k-connected vertices in $\mathrm{PA}_n^{(m,\delta)}$) *Consider* $\mathrm{PA}_n^{(m,\delta)}(a)$, $\mathrm{PA}_n^{(m,\delta)}(b)$ *and* $\mathrm{PA}_n^{(m,\delta)}(d)$ *with* $m \geq 2$ *and* $\delta \in (-m, 0)$. *For* $k \leq (1-\varepsilon)\log\log n / \log m$,

$$M_k \xrightarrow{\mathbb{P}} \infty. \tag{8.8.9}$$

Exercises 8.26–8.28 prove Lemma 8.36 for $\mathrm{PA}_n^{(m,\delta)}(d)$; they rely on the arguments in the proof of Proposition 5.22. The proofs for $\mathrm{PA}_n^{(m,\delta)}(a)$ and $\mathrm{PA}_n^{(m,\delta)}(b)$ are quite similar.

To complete the lower bound on $\mathrm{diam}(\mathrm{PA}_n^{(m,\delta)})$ in Theorem 8.34, we take two vertices $u, v \in \mathcal{M}_k$ with $k = (1-\varepsilon)\log\log n / \log m$. By definition, $\partial B_k^{(G_n)}(u), \partial B_k^{(G_n)}(v) \subset [n] \setminus [n/2]$. We can then adapt the proof of Theorem 8.19 to show that the distance between $\partial B_k^{(G_n)}(u)$ and $\partial B_k^{(G_n)}(v)$ is whp still bounded from below by $4\log\log n / |\log(\tau - 2)|$. Therefore, whp,

$$\mathrm{diam}(\mathrm{PA}_n^{(m,\delta)}) \geq \mathrm{dist}_{\mathrm{PA}_n^{(m,\delta)}}(u, v) = 2k + \mathrm{dist}_{\mathrm{PA}_n^{(m,\delta)}}(\partial B_k^{(G_n)}(u), \partial B_k^{(G_n)}(v))$$
$$\geq \frac{2(1-\varepsilon)\log\log n}{\log m} + \frac{4\log\log n}{|\log(\tau - 2)|}. \tag{8.8.10}$$

This gives an informal proof of the lower bound.

Critical Case $\delta = 0$

We close this section by discussing the diameter for $\delta = 0$:

Theorem 8.37 (Diameter of $\mathrm{PA}_n^{(m,0)}(a)$ for $\delta = 0$) *Fix* $m \geq 2$. *As* $n \to \infty$,

$$\mathrm{diam}(\mathrm{PA}_n^{(m,0)}(a)) \frac{\log\log n}{\log n} \xrightarrow{\mathbb{P}} 1. \tag{8.8.11}$$

Theorem 8.37 is one of the few results that have not been extended to $\mathrm{PA}_n^{(m,0)}(b)$ and $\mathrm{PA}_n^{(m,0)}(b)$. As we see, the diameter in Theorem 8.37 and the typical distances in Theorem 8.8 behave similarly. This is unique to $\delta = 0$.

8.9 RELATED RESULTS ON DISTANCES IN PREFERENTIAL ATTACHMENT MODELS

Distance Evolution for $\delta < 0$

We now study the *evolution* of the graph distances $n \mapsto \mathrm{dist}_{\mathrm{PA}_n^{(m,\delta)}}(i, j)$. In Exercise 8.29, the reader is asked to show that $n \mapsto \mathrm{dist}_{\mathrm{PA}_n^{(m,\delta)}}(i, j)$ is non-increasing for $n \geq i \vee j$. Exercises 8.30–8.34 study various aspects of the evolution of distances. The main result below identifies precisely how the distances between uniform vertices in $[n]$ decrease as time progresses. We indicate the dependence on the initial time explicitly and investigate how $t \mapsto \mathrm{dist}_{\mathrm{PA}_t^{(m,\delta)}}(o_1^{(n)}, o_2^{(n)})$ evolves for all $t \geq n$:

Theorem 8.38 (Evolution of distances) *Consider* $(\mathrm{PA}_n^{(m,\delta)}(b))_{n \geq 1}$ *with* $m \geq 2$ *and* $\delta \in (-m, 0)$. *Choose* $o_1^{(n)}, o_2^{(n)}$ *independently and uar from* $[n]$. *Then, for all* $t \geq n$,

$$\sup_{t \geq n} \left| \mathrm{dist}_{\mathrm{PA}_t^{(m,\delta)}(b)}(o_1^{(n)}, o_2^{(n)}) - 4 \left\lfloor \frac{\log \log n - \log\left(1 \vee \log(t/n)\right)}{|\log(\tau - 2)|} \right\rfloor \vee 4 \right| \qquad (8.9.1)$$

is a tight sequence of random variables.

Theorem 8.38 is very strong, as it describes rather precisely how the distances decrease as the graph $\mathrm{PA}_t^{(m,\delta)}$ grows. Further, Theorem 8.38 also proves that $\mathrm{dist}_{\mathrm{PA}_n^{(m,\delta)}}(o_1^{(n)}, o_2^{(n)}) - 4 \log \log n / |\log(\tau - 2)|$ is a tight sequence of random variables. While the lower tightness (i.e., the tightness of $[\mathrm{dist}_{\mathrm{PA}_n^{(m,\delta)}}(o_1^{(n)}, o_2^{(n)}) - 4 \log \log n / |\log(\tau - 2)|]_-$) follows from Theorem 8.19, the upper tightness (i.e., the tightness of $[\mathrm{dist}_{\mathrm{PA}_n^{(m,\delta)}}(o_1^{(n)}, o_2^{(n)}) - 4 \log \log n / |\log(\tau - 2)|]_+$) was not proved in Theorem 8.7 (recall (8.7.26)). Exercises 8.35 and 8.36 investigate what happens when $t = t_n = n \exp\{(\log n)^\alpha\}$ for $\alpha \in (0, 1)$ and $\alpha > 1$, respectively. This leaves open the interesting case where $\alpha = 1$.

The proof of Theorem 8.38 uses ideas similar to those used earlier in the present chapter, but significantly extends them in order to prove the uniformity in $t \geq n$. We do not present the full proof here. We do explain, however, why $\mathrm{dist}_{\mathrm{PA}_t^{(m,\delta)}}(o_1^{(n)}, o_2^{(n)}) = 2$ whp when $t = t_n \gg n^{-2m/\delta}$, which is an interesting case and explains why the supremum in (8.9.1) can basically be restricted to $t \in [n, n^{-2m/\delta + o(1)}]$. This sheds light on precisely what happens when $t = t_n = n \exp\{(\log n)^\alpha\}$ for $\alpha = 1$, a case that is left open above.

The probability that one of the m edges of vertex $n + t + 1$ connects to u, and another one to v (which certainly makes the distance between u and v equal to 2), is close to

$$m(m-1)\mathbb{E}\left[\frac{D_v(n+t) + \delta}{2m(n+t) + (n+t)\delta} \frac{D_u(n+t) + \delta}{2m(n+t) + (n+t)\delta} \,\Big|\, \mathrm{PA}_n^{(m,\delta)} \right]$$

$$= (1 + o_\mathbb{P}(1)) \frac{m(m-1)}{(2m+\delta)^2 t^2} \mathbb{E}\left[(D_v(n+t) + \delta)(D_u(n+t) + \delta) \,\Big|\, \mathrm{PA}_n^{(m,\delta)} \right]$$

$$= (1 + o_\mathbb{P}(1)) \frac{m(m-1)}{(2m+\delta)^2 t^2} (D_v(n) + \delta)(D_u(n) + \delta) \left(\frac{t}{n}\right)^{2/(2+\delta/m)}$$

$$= (1 + o_\mathbb{P}(1)) \frac{m(m-1)}{(2m+\delta)^2} t^{-2(m+\delta)/(2m+\delta)} n^{-2/(2+\delta/m)} (D_v(n) + \delta)(D_u(n) + \delta).$$

$$(8.9.2)$$

If we take $u = o_1^{(n)}, v = o_2^{(n)}$, we have that $D_v(n) \xrightarrow{d} D_1$, $D_u(n) \xrightarrow{d} D_2$, where (D_1, D_2) are two iid copies of the random variable with asymptotic degree distribution $\mathbb{P}(D = k) = p_k$ in (1.3.60). Thus, the conditional expectation of the total number of double attachments to both $o_1^{(n)}$ and $o_2^{(n)}$ up to time $n + t$ is close to

$$\sum_{s=1}^{t} \frac{m(m-1)(D_1+\delta)(D_2+\delta)}{(2m+\delta)^2} s^{-2(m+\delta)/(2m+\delta)} n^{-2/(2+\delta/m)}$$

$$\approx \frac{m(m-1)(D_1+\delta)(D_2+\delta)}{(2m+\delta)(-\delta)} n^{-2m/(2m+\delta)} t^{-\delta/(2m+\delta)}, \qquad (8.9.3)$$

which becomes $\Theta_{\mathbb{P}}(1)$ when $t = Kn^{-2m/\delta}$. The above events, for different t, are close to being independent. This suggests that the process of attaching to both $o_1^{(n)}$ and $o_2^{(n)}$ is, conditioning on their degrees (D_1, D_2), Poisson with some random intensity.

Distances in the Bernoulli PA Model

Recall the Bernoulli preferential attachment model $(\mathrm{BPA}_n^{(f)})_{n\geq 1}$ from Section 1.3.5. Its degree structure is understood much more generally than for preferential attachment models with a fixed number of edges (recall (1.3.67)). This allows one to zoom in on particular instances of the preferential attachment function. While $(\mathrm{BPA}_n^{(f)})_{n\geq 1}$ is a quite different model, in particular since whp the graph is not connected unlike $\mathrm{PA}_n^{(m,\delta)}$ for $m \geq 2$ and n large, in terms of distances it behaves similarly to the fixed out-degree models. This is exemplified by the following theorem, which applies in the infinite-variance degree setting:

Theorem 8.39 (Evolution of distances for scale-free $\mathrm{BPA}_n^{(f)}$)   *Consider* $\mathrm{BPA}_n^{(f)}$ *where the concave attachment rule f satisfies that there exists $\gamma \in (\frac{1}{2}, 1)$ such that*

$$f(k) = \gamma k + \beta. \qquad (8.9.4)$$

Then Theorem 8.38 extends to this setting when we restrict to $t \geq n$ such that $o_1^{(n)}$ and $o_2^{(n)}$ are connected in $\mathrm{BPA}_n^{(f)}$. In particular, conditional on this event,

$$\mathrm{dist}_{\mathrm{BPA}_n^{(f)}}(o_1^{(n)}, o_2^{(n)}) - 4\frac{\log\log n}{|\log(\tau-2)|} \qquad (8.9.5)$$

is a tight sequence of random variables.

The situation of affine preferential attachment functions f in (8.9.4) where $\gamma \in (0, \frac{1}{2})$, for which the degree power-law exponent satisfies $\tau = 1 + 1/\gamma > 3$, is not so well understood, but one can conjecture that again, the distance between $o_1^{(n)}$ and $o_2^{(n)}$ is whp logarithmic at some base related to the multi-type branching process that describes its local limit.

The following theorem, for which $\gamma = \frac{1}{2}$ so that $\tau = 3$, describes nicely how the addition of an extra power of a logarithm in the degree distribution affects the distances:

Theorem 8.40 (Critical case: interpolation)   *Consider* $\mathrm{BPA}_n^{(f)}$ *where the concave attachment rule f satisfies that there exists $\alpha > 0$ such that*

$$f(k) = \frac{k}{2} + \frac{\alpha}{2}\frac{k}{\log k} + o\left(\frac{k}{\log k}\right), \qquad (8.9.6)$$

Choose o_1, o_2 independently and uar from $[n]$. Then, conditional on $o_1 \longleftrightarrow o_2$,

$$\mathrm{dist}_{\mathrm{BPA}_n^{(f)}}(o_1, o_2) = (1 + o_{\mathbb{P}}(1))\frac{1}{1+\alpha}\frac{\log n}{\log\log n}. \qquad (8.9.7)$$

Exercise 8.37 shows that the degree distribution for f in (8.9.6) satisfies $\sum_{l>k} p_l \approx k^{-2}(\log k)^{-2\alpha}$, as for $\mathrm{GRG}_n(\boldsymbol{w})$ in Theorem 6.28. Comparing Theorems 8.40 and 6.28, we see that, for large α, the typical distances in $\mathrm{BPA}_n^{(f)}$ are about twice as large as those in $\mathrm{GRG}_n(\boldsymbol{w})$ with similar degrees. This gives an explanation of the occurrence of the extra factor 2 in Theorem 8.7 compared with Theorem 6.3 for the Norros–Reittu model $\mathrm{NR}_n(\boldsymbol{w})$, and Theorem 7.2 for the configuration model $\mathrm{CM}_n(\boldsymbol{d})$, when the power-law exponent τ satisfies $\tau \in (2,3)$. Note that this extra factor is absent *precisely* when $\alpha = 0$.

8.10 Notes and Discussion for Chapter 8

Notes on Section 8.2

Scale-free trees have received substantial attention in the literature; we refer to Bollobás and Riordan (2004b); Pittel (1994), and the references therein. The almost sure limit of $\mathrm{height}(\mathrm{PA}_n^{(1,\delta)}(d))/\log n$ in Theorem 8.1 is (Pittel, 1994, Theorem 1). Our proof of this result in Section 8.2 is incomplete, particularly since we have omitted the proof of Proposition 8.4. Proposition 8.4 can be proved using the beautiful result on the asymptotic height of branching-process trees due to Kingman (1975), of which Pittel (1994) makes crucial use. This approach is based on a continuous-time embedding of $\mathrm{PA}_n^{(1,\delta)}(d)$ that leads to a continuous-time branching process, for which exponential martingales and the subadditive ergodic theorem can be used to give a relatively short proof of the almost sure limit of the tree height. The adaptation to $\mathrm{PA}_n^{(1,\delta)}(b)$ is rather straightforward. We do not give more details, since we have not relied on the lower bound but have used the upper bound instead. See Exercises 8.4 and 8.5 for some background on this proof.

There is a close analogy between $\mathrm{PA}_n^{(1,\delta)}(a)$ and so-called *uniform recursive trees*. In uniform recursive trees, we grow a tree such that at time 1, we have a unique vertex called the root, with label 1, and, at time n, we add a vertex and connect it to a uniformly chosen vertex in the tree. See Smythe and Mahmoud (1994) for a survey of recursive trees.

A variant of a uniform recursive tree arises when the probability that a newly added vertex is attached to a particular vertex is proportional to the degree of that vertex (and, for the root, its degree plus one). This process is called a *random plane-oriented recursive tree*. For a uniform recursive tree of size n, it was proved in Pittel (1994) that the maximal distance between the root and any other vertex is whp equal to $\frac{1}{2\gamma} \log n(1 + o(1))$, where γ satisfies (8.2.2) with $\delta = 0$. It is not hard to see that this implies that the diameter of the uniform recursive tree is equal to $\frac{1}{\gamma} \log n(1 + o_{\mathbb{P}}(1))$.

Notes on Section 8.3

Theorem 8.6 was proved in Dommers et al. (2010). A weaker version of Theorem 8.7 was proved in Dommers et al. (2010). The current theorem was inspired by Dereich et al. (2012).

Theorem 8.8 was proved in Bollobás and Riordan (2004a), who, more precisely, proved the result for the diameter of $\mathrm{PA}_n^{(m,0)}$ in Theorem 8.37. This proof can rather easily be extended to the typical distances.

Notes on Section 8.4

The bound in Proposition 8.9 for $\delta = 0$ was proved in (Bollobás and Riordan, 2004a, Lemma 3) in a rather different way. The current version for all δ is (Dommers et al., 2010, Corollary 2.3), and its proof was also adapted from there. Lemma 8.10 is (Dommers et al., 2010, Lemma 2.1).

Notes on Section 8.5

Theorem 8.13 is stronger than the corresponding result in Dommers et al. (2010), since we expect the lower bound to be sharp. We thank Joost Jorritsma and Júlia Komjáthy for preliminary discussions on Theorem 8.13. The proofs in this section for $\delta = 0$ first appeared in (Bollobás and Riordan, 2004a, Section 4).

Notes on Section 8.6

The proof of Theorem 8.19 was adapted from Dereich et al. (2012).

Notes on Section 8.7

Theorem 8.26 is somewhat stronger than (Dommers et al., 2010, Theorem 3.1), who also first proved a weaker upper bound than that in Theorem 8.7. We follow the proof of (Dommers et al., 2010, Theorem 3.1). Theorem 8.32 was proved by Dereich et al. (2012); see also (Dommers et al., 2010, Theorem 3.6).

Notes on Section 8.8

Theorem 8.33 was proved in Dommers et al. (2010). Theorem 8.34 was proved in Caravenna et al. (2019). Theorem 8.37 was proved by Bollobás and Riordan (2004a). Its proof relies on the *linear cord diagram* (LCD) description of the $\delta = 0$ model by Bollobás and Riordan (2004a). This explains why the proof applies only to $\mathrm{PA}_n^{(m,0)}(a)$ and not immediately to $\mathrm{PA}_n^{(m,0)}(b)$ and $\mathrm{PA}_n^{(m,0)}(d)$.

Notes on Section 8.9

Theorems 8.38 and 8.39 were proved by Jorritsma and Komjáthy (2022), who studied the more general problem of *first-passage percolation* on the preferential attachment model. In first-passage percolation, the edges are weighted. These weights can be interpreted as the traversal time of an edge in a rumor-spreading model. Then, Jorritsma and Komjáthy (2022) studied the time it takes for a rumor to go from a random source to a random destination. They obtained sharp results for the leading-order asymptotics of this traversal time, and also in the dynamical setting. Theorem 8.40 is proved in Dereich et al. (2017).

8.11 EXERCISES FOR CHAPTER 8

Exercise 8.1 (Bound on θ)   *Prove that the solution θ of (8.2.2) satisfies $\theta < 1$. What does this imply for the diameter and typical distances in scale-free trees?*

Exercise 8.2 (Bound on θ)   *Prove that the solution θ of (8.2.2) satisfies $\theta \in (0, e^{-1})$.*

Exercise 8.3 (Extension Remark 8.3)   *Prove the path-presence probability in (8.2.8) in Remark 8.3 by using (8.2.20) and (8.2.21).*

Exercise 8.4 (Height of preferential attachment trees (Pittel (1994)))   *Check Pittel (1994) to see how Pittel identifies the almost sure limit of $\mathrm{height}(\mathrm{PA}_n^{(1,\delta)})/\log n$, using a continuous-time embedding and the identification of the almost sure limit of the height of a continuous-time branching process by Kingman (1975).*

Exercise 8.5 (Height of continuous-time branching process trees (Kingman (1975)))   *Check the beautiful argument by Kingman (1975) that identifies the almost sure limit of $\mathrm{height}(T_n)/\log n$, where T_n is a continuous-time branching process tree of size n. Verify that the conditions used by Kingman (1975) hold for the branching process used by Pittel (1994) to study preferential attachment trees.*

Exercise 8.6 (Doubly logarithmic distances for $\mathrm{PA}_n^{(m,\delta)}$)   *Fix $m = 4$ and $\delta = -2$. Let o_1, o_2 be independent uniform vertices in $[n]$. Find the constant a such that*

$$\frac{\mathrm{dist}_{\mathrm{PA}_n^{(m,\delta)}}(o_1, o_2)}{\log\log n} \xrightarrow{\mathbb{P}} a. \tag{8.11.1}$$

Exercise 8.7 (Early vertices are whp at distance 2 for $\delta < 0$)   *Let $\delta \in (-m, 0)$ and $m \geq 2$. Show that, for i, j fixed,*

$$\lim_{n\to\infty} \mathbb{P}(\mathrm{dist}_{\mathrm{PA}_n^{(m,\delta)}}(i,j) \leq 2) = 1. \tag{8.11.2}$$

Exercise 8.8 (All early vertices are whp at distance 2 for $\delta < 0$)   *Let $\delta \in (-m, 0)$ and $m \geq 2$. Extend Exercise 8.7 to the statement that, for $K \geq 1$ fixed,*

$$\lim_{n\to\infty} \mathbb{P}(\mathrm{dist}_{\mathrm{PA}_n^{(m,\delta)}}(i,j) \leq 2 \, \forall i,j \in [K]) = 1. \tag{8.11.3}$$

Exercise 8.9 (Early vertices are not at distance 2 when $\delta > 0$)   *Let $\delta > 0$ and $m \geq 2$. Show that*

$$\lim_{K\to\infty} \lim_{n\to\infty} \mathbb{P}(\mathrm{dist}_{\mathrm{PA}_n^{(m,\delta)}}(i,j) = 2 \, \forall i,j \in [K]) = 0. \tag{8.11.4}$$

Exercise 8.10 (Extension of negative correlations to $\mathrm{PA}_n^{(m,\delta)}(b)$) *Verify that the proof of the negative correlations in Lemma 8.10 applies also to $\mathrm{PA}_n^{(m,\delta)}(b)$.*

Exercise 8.11 (Computing vertex-attachment probabilities for $\mathrm{PA}_n^{(m,\delta)}(d)$) *Consider $\mathrm{PA}_n^{(m,\delta)}(d)$, and fix $v \in [n]$ and $n_v \geq 1$. Use the Pólya graph representation in Theorem 5.10, as well as Lemma 5.12, to compute $\mathbb{P}(\mathcal{E}_{n_v,v})$ in (8.4.2).*

Exercise 8.12 (Computing the vertex-attachment probabilities for $\mathrm{PA}_n^{(m,\delta)}(d)$ (cont.)) *Consider $\mathrm{PA}_n^{(m,\delta)}(d)$, and fix $k \geq 1$, $v_1, \ldots, v_k \in [n]$ distinct, and $n_{v_i} \geq 1$ for every $i \in [k]$. Use the Pólya graph representation in Theorem 5.10, as well as Lemma 5.12, to compute $\mathbb{P}\left(\bigcap_{t\in[k]} \mathcal{E}_{n_{v_t},v_t}\right)$.*

Exercise 8.13 (Negative correlations for $\mathrm{PA}_n^{(m,\delta)}(d)$) *Prove the negative correlations in Lemma 8.10 for $\mathrm{PA}_n^{(m,\delta)}(d)$ by combining Exercises 8.11 and 8.12.*

Exercise 8.14 (Negative correlations for $m = 1$) *Show that, for $m = 1$, Lemma 8.10 implies that if $(\pi_0, \ldots, \pi_k)$ contains different coordinates as $(\rho_0, \ldots, \rho_k)$ then*

$$\mathbb{P}\left(\bigcap_{i=0}^{k-1}\{\pi_i \rightsquigarrow \pi_{i+1}\} \cap \bigcap_{i=0}^{k-1}\{\rho_i \rightsquigarrow \rho_{i+1}\}\right) \leq \mathbb{P}\left(\bigcap_{i=0}^{k-1}\{\pi_i \rightsquigarrow \pi_{i+1}\}\right)\mathbb{P}\left(\bigcap_{i=0}^{k-1}\{\rho_i \rightsquigarrow \rho_{i+1}\}\right). \quad (8.11.5)$$

Exercise 8.15 (Extension of (8.4.15) to $\mathrm{PA}_n^{(1,\delta)}(b)$) *Prove that, for $\mathrm{PA}_n^{(1,\delta)}(b)$, (8.4.15) is replaced with*

$$\mathbb{P}\left(u_1 \overset{1}{\rightsquigarrow} v, u_2 \overset{1}{\rightsquigarrow} v\right) = (1+\delta)\frac{\Gamma(u_1 - \delta/(2+\delta))\Gamma(u_2 - (1+\delta)/(2+\delta))\Gamma(v)}{\Gamma(u_1 + 1/(2+\delta))\Gamma(u_2)\Gamma(v + 2/(2+\delta))}. \quad (8.11.6)$$

Exercise 8.16 (Most recent common ancestor in $\mathrm{PA}_n^{(1,\delta)}$) *Fix o_1, o_2 to be two vertices in $[n]$ chosen uar, and let V be the oldest vertex that the paths from 1 to o_1 and from 1 to o_2 have in common in $\mathrm{PA}_n^{(1,\delta)}$. Prove that $\mathrm{dist}_{\mathrm{PA}_n^{(1,\delta)}(b)}(1, V)/\log n \overset{\mathbb{P}}{\longrightarrow} 0$, as stated in Lemma 8.5.*

Exercise 8.17 (Most-recent common ancestor in $\mathrm{PA}_n^{(1,\delta)}$ (cont.)) *Fix o_1, o_2 to be two vertices in $[n]$ chosen uar, and let V be the oldest vertex that the paths from 1 to o_1 and that from 1 to o_2 have in common in $\mathrm{PA}_n^{(1,\delta)}$. Extend Exercise 8.16 to show that $\mathrm{dist}_{\mathrm{PA}_n^{(1,\delta)}(b)}(1, V)$ is tight.*

Exercise 8.18 (Distance between $n-1$ and n in $\mathrm{PA}_n^{(m,0)}$) *Show that $\mathrm{dist}_{\mathrm{PA}_n^{(m,0)}}(n-1, n) \geq k_n^\star$ whp, where $k_n^\star$ is defined in (8.5.65).*

Exercise 8.19 (Distance between vertices 1 and 2 in $\mathrm{PA}_n^{(m,0)}$) *Check the implications of the analysis in Section 8.5.4 for $\mathrm{dist}_{\mathrm{PA}_n^{(m,0)}}(1, 2)$. In particular, where does the proof that $\mathrm{dist}_{\mathrm{PA}_n^{(m,0)}}(1, 2) \geq k_n^\star$ whp, with $k_n^\star$ as in (8.5.65), fail, and why does this happen?*

Exercise 8.20 (Connection probabilities in $\mathrm{PA}_n^{(m,\delta)}(d)$) *Extend (8.4.14) and (8.4.16) in Lemma 8.11 to $\mathrm{PA}_n^{(m,\delta)}(d)$ using Corollary 8.16 and Lemma 5.14.*

Exercise 8.21 (Connection probabilities in $\mathrm{PA}_n^{(m,\delta)}(d)$) *Consider $\mathrm{PA}_n^{(m,\delta)}(d)$ for $m \geq 2$. Fix $\ell \geq 1$ and $u_i > v$ for all $i \in [\ell]$. What do Corollary 8.16 and Lemma 5.14 imply for $\mathbb{P}(u_i \overset{ji}{\rightsquigarrow} v \,\forall i \in [\ell])$ for $\ell = 3, 4$?*

Exercise 8.22 (Growth of Pólya point tree for $\delta < 0$) *Argue that Proposition 8.24 suggests that the size of the kth generation in the Pólya point tree is at most $\exp\left\{B(\tau - 2)^{-k/2}\right\}$ for some large constant B. Here, you may use without proof that the degree of a vertex of age a is comparable with $(n/a)^{1/(\tau-1)}$.*

Exercise 8.23 (Diameter of $\mathrm{ER}_n(p)$) *Let $p = n^{-\alpha}$ with $\alpha \in (0, 1)$. Show that whp $\mathrm{diam}(\mathrm{ER}_n(p)) \leq 1/(1-\alpha)$ when $1/(1-\alpha)$ is not an integer, and $\mathrm{diam}(\mathrm{ER}_n(p)) \leq 1/(1-\alpha) + 1$ when $1/(1-\alpha)$ is an integer. Compare this to the claimed bound of $\mathrm{diam}(\mathrm{Inner}_n)$ in the proof of Proposition 8.27.*

Exercise 8.24 (Upper tightness criterion for centered $\mathrm{dist}_{\mathrm{PA}_{2n}^{(m,\delta)}}(o_1, o_2)$) *Fix $\delta \in (-m, 0)$. Let o_1 be chosen uar from $[2n]$. Use (8.7.26) and Theorem 8.26 to show that $\mathrm{dist}_{\mathrm{PA}_{2n}^{(m,\delta)}}(o_1, o_2) - 4\log\log n / |\log(\tau-2)|$ is upper tight when $\mathrm{dist}_{\mathrm{PA}_{2n}^{(m,\delta)}}(o_1, \mathrm{Core}_n)$ is tight. Is it plausible that $\mathrm{dist}_{\mathrm{PA}_{2n}^{(m,\delta)}}(o_1, \mathrm{Core}_n)$ is tight?*

Exercise 8.25 (Upper bound on the diameter of $(\mathrm{PA}_n^{(m,\delta)}(d))_{n\geq 1}$) *Consider* $(\mathrm{PA}_n^{(m,\delta)}(d))_{n\geq 1}$, *and define the subtree of* $\mathrm{PA}_n^{(m,\delta)}(d)$ *by considering only the first edge of each vertex. Use the Pólya urn representation in Theorem 5.10 and an adaptation of the proof of (8.2.24) to show that* $\mathrm{height}(\mathrm{PA}_n^{(m,\delta)}(d)) \leq c_2 \log n$ *for some* $c_2 < \infty$.

Exercise 8.26 (Expectation of number of minimally k-connected vertices) *Consider* $\mathrm{PA}_n^{(m,\delta)}(d)$ *with* $m \geq 2$ *and* $\delta \in (-m, 0)$. *Recall the definition of minimally k-connected vertices in Definition 8.35. Prove that, for all* $k \leq (1-\varepsilon) \log \log n / \log m$,

$$\mathbb{E}[M_k] \geq n^{1-\varepsilon}. \tag{8.11.7}$$

Exercise 8.27 (Second moment of number of minimally k-connected vertices) *Consider* $\mathrm{PA}_n^{(m,\delta)}(d)$ *with* $m \geq 2$ *and* $\delta \in (-m, 0)$ *as in Exercise 8.26. Prove that, for all* $k \leq (1-\varepsilon) \log \log n / \log m$,

$$\mathbb{E}[M_k^2] = \mathbb{E}[M_k]^2 (1 + o(1)). \tag{8.11.8}$$

Exercise 8.28 (Concentration of number of minimally k-connected vertices: proof of Lemma 8.36) *Consider* $\mathrm{PA}_n^{(m,\delta)}(d)$ *with* $m \geq 2$ *and* $\delta \in (-m, 0)$ *as in Exercise 8.26. Use Exercises 8.26 and 8.27 to prove that* $M_k/\mathbb{E}[M_k] \xrightarrow{\mathbb{P}} 1$ *for all* $k \leq (1-\varepsilon) \log \log n / \log m$, *as in Lemma 8.36.*

Exercise 8.29 (Monotonicity of distances in $\mathrm{PA}_n^{(m,\delta)}$) *Fix* $m \geq 1$ *and* $\delta > -m$. *Show that* $n \mapsto \mathrm{dist}_{\mathrm{PA}_n^{(m,\delta)}}(i,j)$ *is non-decreasing for* $n \geq i \vee j$.

Exercise 8.30 (Distance evolution in $\mathrm{PA}_n^{(1,\delta)}$) *Fix* $m = 1$ *and* $\delta > -1$. *Show that* $n \mapsto \mathrm{dist}_{\mathrm{PA}_n^{(1,\delta)}}(i,j)$ *is constant for* $n \geq i \wedge j$.

Exercise 8.31 (Distance structure on $\mathbb{N}$ due to $\mathrm{PA}_n^{(m,\delta)}$) *Fix* $m \geq 1$ *and* $\delta > -m$. *Use Exercise 8.29 to show that* $\mathrm{dist}_{\mathrm{PA}_n^{(m,\delta)}}(i,j) \xrightarrow{a.s.} \mathrm{dist}_\infty(i,j) < \infty$ *as* $n \to \infty$ *for all* $i,j \geq 1$. *Thus,* dist_∞ *is a distance function on* $\mathbb{N}$.

Exercise 8.32 (Nearest neighbors on $\mathbb{N}$ due to $\mathrm{PA}_n^{(m,\delta)}$) *Fix* $m \geq 2$ *and* $\delta > -m$. *Recall* dist_∞ *from Exercise 8.31. Compute the asymptotics of* $\mathbb{P}(\mathrm{dist}_\infty(i,j) = 1)$ *for* j *large and* i *fixed.*

Exercise 8.33 (Infinitely many neighbors) *Fix* $m \geq 2$ *and* $\delta > -m$. *Recall* dist_∞ *from Exercise 8.31. Show that* $\#\{j : \mathrm{dist}_\infty(i,j) = 1\} = \infty$ *almost surely for all* $i \geq 1$, *so there are infinitely many vertices at distance 1 for every vertex in* $\mathbb{N}$.

Exercise 8.34 (Eventually distances are at most 2 in $\mathrm{PA}_n^{(m,\delta)}$) *Fix* $m \geq 2$ *and* $\delta \in (-m, 0)$. *Recall* dist_∞ *from Exercise 8.31. Show that* $\mathrm{dist}_\infty(i,j) \leq 2$ *almost surely for all* i,j.

Exercise 8.35 (Evolution of distances in $\mathrm{PA}_t^{(m,\delta)}(b)$: critical parametric choice) *Fix* $m \geq 2$ *and* $\delta \in (-m, 0)$. *Choose* $o_1^{(n)}, o_2^{(n)}$ *uar from* $[n]$, *and take* $t = t_n = n \exp\{(\log n)^\alpha\}$ *for some* $\alpha \in (0, 1)$. *Use Theorem 8.38 to identify* θ_α *such that*

$$\frac{\mathrm{dist}_{\mathrm{PA}_t^{(m,\delta)}(b)}(o_1^{(n)}, o_2^{(n)})}{\log \log n} \xrightarrow{\mathbb{P}} \theta_\alpha. \tag{8.11.9}$$

Exercise 8.36 (Tight distances in $\mathrm{PA}_t^{(m,\delta)}$ for $\delta < 0$) *Fix* $m \geq 2$ *and* $\delta \in (-m, 0)$. *Choose* $o_1^{(n)}, o_2^{(n)}$ *uar from* $[n]$, *and take* $t = t_n = n \exp\{(\log n)^\alpha\}$ *for some* $\alpha > 1$. *Use Theorem 8.38 to show that* $\mathrm{dist}_{\mathrm{PA}_t^{(m,\delta)}(b)}(o_1^{(n)}, o_2^{(n)})$ *is a tight sequence of random variables.*

Exercise 8.37 (Degree distribution in $\mathrm{BPA}_n^{(f)}$ with logarithmic corrections) *Show that the degree distribution of* $\mathrm{BPA}_n^{(f)}$ *with* f *in (8.9.6) in Theorem 8.40 satisfies* $\sum_{l > k} p_l = (1 + o(1)) ck^{-2}(\log k)^{-2\alpha}$ *for some constant* $c > 0$.

Part IV

Related Models and Problems

Summary of Part III

In Part III we investigated the *small-world* behavior of random graphs, extending the results on the existence and uniqueness of the giant component as informally described in Meta Theorem A on page 243. It turns out that the results are all quite similar, even though the details of the description of the models are substantially different. We can summarize the results obtained in the following *meta theorem:*

Meta Theorem B (Small-world and ultra-small-world characteristics) *In a random graph model with power-law degrees having power-law exponent τ, the typical distances of the giant component in a graph of size n are of order $\log \log n$ when $\tau \in (2, 3)$, while they are of order $\log n$ when $\tau > 3$. Further, these typical distances are highly* concentrated.

Informally, these results quantify the "six degrees of separation" paradigm in random graphs, where we see that random graphs with very heavy-tailed degrees have ultra-small typical distances, as could perhaps be expected.

Often, even the lines of proof of these results are similar, relying on clever path-counting techniques. In particular, the results show that in both generalized random graphs and configuration models, in the $\tau \in (2, 3)$ regime vertices of high degrees, say k, are typically connected to vertices of even higher degree, of order $k^{1/(\tau-2)}$. In the preferential attachment model, on the other hand, this is not true, yet vertices of degree k tend to be connected to vertices of degree $k^{1/(\tau-2)}$ in *two* steps, making typical distances roughly twice as large.

Overview of Part IV

In Part IV we study several related random graph models that can be seen as extensions of the simple models studied so far. They incorporate novel features, such as directed edges, clustering, communities, and/or geometry. The important aspect in Part IV will be to verify to what extent the main results informally described in Meta Theorems A (see the start of Part III) and B (see above) remain valid, and otherwise, to what extent they need to be adapted. We will not give complete proofs but instead informally explain why results are similar to those in Meta Theorems A and B or, instead, why they are different.

RELATED MODELS

Abstract

In this chapter we discuss some related random graph models that have been studied in the literature. We explain their relevance, as well as some of the properties in them. We discuss *directed* random graphs, random graphs with local and global *community structures*, as well as *spatial* random graphs.

Organization of this Chapter

We start in Section 9.1 by extensively discussing the real-world network example of *citation networks*. The aim there is to show that the models, as discussed so far, tend not to resemble some, or even many, real-world systems. Citation networks are directed, have substantial clustering, have a hierarchical community structure, and possibly even have a spatial component to them. While we focus on citation networks, we also highlight the fact that we could have chosen other real-world examples as well.

We continue in Section 9.2 by considering directed versions of the random graphs studied in this book. In Sections 9.3 and 9.4, we introduce several random graph models that have a *community structure*, in order to model the communities occurring in real-world networks. Section 9.3 studies the setting where communities are *macroscopic*, in the sense that there is a bounded number of communities even when the network size tends to infinity. In Section 9.4, instead, we look at the setting where the communities have a bounded average size. We argue that both settings are relevant. In Section 9.5, we discuss random graph models that have a *spatial* component to them, and explain how this spatial structure gives rise to highly clustered random graphs. We close this chapter with notes and discussion in Section 9.6 and with exercises in Section 9.7. We do not give all the relevant proofs but refer to the notes and discussion in Section 9.6 for details of where these proofs can be found.

Throughout this chapter, $\mathscr{C}_{\max}$ denotes the largest (strongly) connected component of the model under consideration and $\mathscr{C}_{(2)}$ denotes its second largest cluster (breaking ties arbitrarily when needed).

9.1 MOTIVATION: REAL-WORLD NETWORK MODELING

Here, we discuss real-world network models. We start in Section 9.1.1 by discussing *citation networks* in detail. In Section 9.1.2 we draw conclusions about network modeling.

9.1.1 EXAMPLE: CITATION NETWORKS

In this subsection we discuss citation networks as a real-world network example. This example shows how real-world networks are different from the stylized network models that we have discussed so far. In fact, many real-world networks are *directed*, in that edges point from one vertex to another. Also, real-world networks often display a pronounced *community structure*, in that certain parts are more densely connected than the rest of the network, and these communities are relevant in practice.

In *citation networks*, vertices denote scientific papers and the directed edges correspond to citations of one paper to another. Obviously, such citations are *directed*, since it makes a difference whether your paper cites mine, or my paper cites yours.

Citation networks *grow* in time. Indeed, papers do not disappear, so a citation, once made in a published paper, does not disappear either. Further, their growth is enormous. Figure 9.1(a) shows that the number of papers in various fields grows *exponentially* in time, meaning that more and more papers are being written. If you ever wondered why scientists seem to be ever more busy, then this may be an obvious explanation.

In Figure 9.1(a) we display the number of papers in three different domains, namely, *Probability and Statistics* (PS), *Electrical Engineering* (EE), and *Biotechnology and Applied Microbiology* (BT). The data comes from the Web of Science data base. While exponential growth is quite prominent in the data, it is somewhat unclear how this exponential growth arises. It could be due either to the fact that the number of journals that are listed in Web of Science grows over time or to the fact that journals contain more and more papers. However, the exponential growth was observed as early as the 1980's; see the book by Derek de Solla Price (1986), appropriately called *Little science, big science*.

As you can see, we have already restricted to certain *subfields* in science, the reason being that the publication and citation cultures in different fields are vastly different. Thus, we have attempted to go to a situation in which the networks that we investigate are somewhat more *homogeneous*. For this, it is relevant to be able to distinguish such fields, and to decide which papers (or journals) contribute to which field. This is a fairly daunting task. However, it is also an ill-defined task, as no subdomain is truly homogeneous. Let me restrict myself to probability and statistics, as I happen to know this area best. In probability and statistics, there are subdomains that are very pure, as well as areas that are highly applied such as applied statistics. These areas do indeed have different publication and citation cultures. Thus, science as a whole is probably *hierarchical,* in that large scientific disciplines can be identified, that can, in turn, be subdivided into smaller subdomains, etc. However, one should

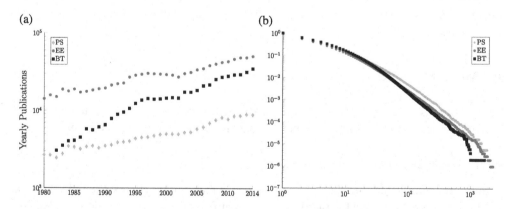

Figure 9.1 (a) Number of publications per year (logarithmic y axis). (b) Log–log plot for the in-degree distribution tail in citation networks.

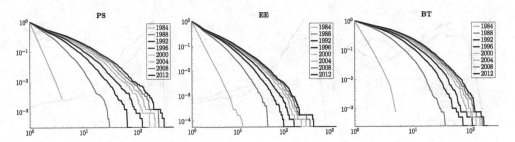

Figure 9.2 Degree distribution for papers from 1984 versus time.

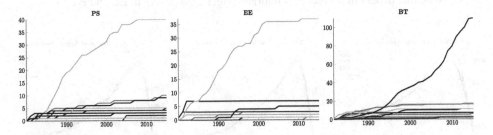

Figure 9.3 Time evolution for citations of 20 randomly chosen papers from 1980 for PS and EE, and from 1982 for BT.

stop somewhere, and the three scientific disciplines relating to Figure 9.1 are homogeneous enough to make our point.

Figure 9.1(b) shows the log–log plot for the in-degree distribution in these three citation networks. We notice that these data sets seem to have empirical power-law citation distributions. Thus, on average, papers attract few citations but the variability in the number of citations is rather substantial. We are also interested in the *dynamics* of the citation distribution of the papers published in a given year, as time proceeds. This can be observed in Figure 9.2. We see a *dynamical power law*, meaning that at any time the degree distribution of a cohort of papers from a given time period (in this case 1984) is close to a power law, but the exponent changes over time (and in fact decreases, which corresponds to heavier tails). When time grows quite large, the power law approaches a fixed value.

Interestingly, the existence of power-law in-degrees in citation networks also has a long history. Derek de Solla Price (1965) observed it and even proposed a model for it that relied on a preferential attachment mechanism, more than two decades before Barabási and Albert (1999) proposed the first preferential attachment model.

We wish to discuss two further properties of citation networks and their dynamics. In Figure 9.3 we see that the majority of papers stop receiving citations after some time, while a few others keep being cited for longer times. This inhomogeneity in the evolution of vertex in-degrees is not present in classical preferential attachment models, where the degree of *every* fixed vertex grows as a positive power of the graph size. Figure 9.3 shows that the number of citations of papers published in the same year can be rather different, and the majority of papers actually stop receiving citations quite soon. In particular, after a first increase the average increment of citations decreases over time (see Figure 9.4). We observe

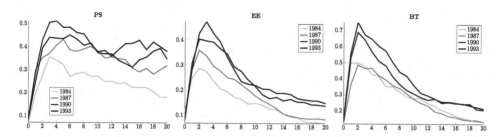

Figure 9.4 Average citation increment over a 20-year time window for papers published in different years. PS presents an aging effect different from EE and BT, showing that papers in PS receive citations longer than papers in EE and BT.

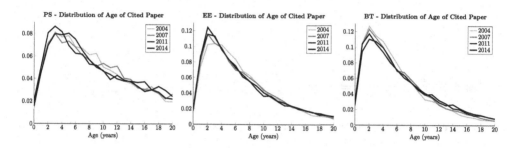

Figure 9.5 Distribution of the ages of cited papers, for PS, EE, and BT, for different citing years.

a difference in this aging effect between the PS data set and the other two data sets, due to the fact that in PS scientists tend to cite older papers than in EE or BT, again exemplifying the differences in citation and publication patterns in different fields. Nevertheless, the average increment of citations received by papers in different years tends to decrease over time for all three data sets.

A last characteriztic that we observe is the log-normal distribution of the age of cited papers. In Figure 9.5, we plot the distribution of cited papers, looking at references made by papers in different years. We have used a 20-year time window in order to compare different citing years. Notice that this log-normal distribution seems to be rather stable over time, and the shape of the curve is also similar for different fields.

Let us summarize the differences between citation networks and the random graph models that form the basis of network science. First, citation networks are *directed*, which is different from the typically *undirected* models that we have discussed so far. However, it is not hard to adapt our models to become directed, and we explain this in Section 9.2. Second, citation networks have a substantial *community structure*, in that parts of the network exist that are much more densely connected than the network as a whole. We can argue that both communities exist on a macroscopic scale, for example in terms of the various scientific disciplines of which science consists, as well as on a microscopic scale, where research networks of small groups of scientists create subnetworks that are more densely connected than the whole network. One could even argue that *geography* plays an important role in citation networks, since many collaborations between scientists are within their own university or country, even though we all work with various researchers around the globe.

Third, citation networks are *dynamic*, like preferential attachment models (PAMs), but their time evolution is quite different from PAMs, as the linear growth in PAMs is replaced by an exponential growth in citation networks. Further, papers in citation networks seem to *age*, as seen both in Figures 9.3 and 9.4, in that citation rates become smaller for large times, in such a way that typically paperscompletely stop receiving citations at some (random) point in time.

In conclusion, finding an appropriate *model* for citation networks is quite a challenge, and one should be quite humble in one's expectation that the standard models are in any way indicative of the complexity of real-world networks.

9.1.2 GENERAL REAL-WORLD NETWORK MODELS

We have concluded that real-world networks tend to have many properties that the models that we have discussed so far in this book do not have, in that they may be *directed*, may have *community structure*, and may have *spatial structure*. Further, real-world networks often evolve over time and should therefore be considered as *temporal networks*. In some cases, *several networks even work together*, and thus should be seen not as separate networks but as *multiplex networks*. For example, in transporting people, the railroad and road networks are both highly relevant. At larger distances, airline networks also become involved. Thus, to study how people move around the globe, we cannot study each of these networks in isolation. In science, the collaboration networks of authors and the citation networks of papers together give a much clearer picture of how science works than these networks in isolation, and in fact these more restricted views can offer useful insights as well.

One may become rather overwhelmed by the complexity that real-world networks provide. Indeed, high-level complex network science can be viewed to be part of *complexity theory*, the science of complex systems. The past decades have given rise to a bulk of insights, often based on relatively simple models such as those discussed so far. Indeed, often Box (1976) is quoted as saying that "All models are wrong but some are useful." Box (1979) gave a more elaborate version of this quote as follows:

> Now it would be very remarkable if any system existing in the real world could be exactly represented by any simple model. However, cunningly chosen parsimonious models often do provide remarkably useful approximations. For example, the law $PV = RT$ relating pressure P, volume V and temperature T of an "ideal" gas via a constant R is not exactly true for any real gas, but it frequently provides a useful approximation and furthermore its structure is informative since it springs from a physical view of the behavior of gas molecules. For such a model there is no need to ask the question "Is the model true?" If "truth" is to be the "whole truth" the answer must be "No." The only question of interest is "Is the model illuminating and useful?"

Thus, we should not feel discouraged at all! In particular, it is important to know when to include extra features into the model at hand, so that it becomes more "useful." For this, the first step is to come up with models that do incorporate these extra features. Many of the models discussed so far can rather straightforwardly be adapted to include features such as directedness, community structure, and geometry. Further, these properties can also be combined. The simpler models that do not have such features serve as a useful model for comparison, and can thus act as a "benchmark" for more complex situations. In this way, the understanding of simple models often helps one to understand more complex models since many properties, tools, and ideas can be extended to them. In some cases the extra features give rise to a richer behavior, which then merits being studied in full detail. In this way, network science has moved significantly forward compared with the models described so far. The aim of this chapter is to highlight some of the lessons learned.

We discuss directed random graphs in Section 9.2, random graphs with macroscopic or global communities in Section 9.3, random graphs with microscopic or local communities in Section 9.4, and spatial random graphs in Section 9.5.

9.2 DIRECTED RANDOM GRAPHS

Many real-world networks are *directed*, in the sense that edges are oriented. For example, in the World-Wide Web, the vertices are web pages, and the edges are the hyperlinks between them. One could forget about these directions, but that would discard a wealth of information. For example, in citation networks it makes a substantial difference whether my paper cites yours, or yours cites mine. See Figure 9.6 for the maximum out- and in-degrees in the KONECT data base.

This section is organized as follows. We start by defining directed graphs or digraphs. After this, we discuss various models invented for them. We discuss directed inhomogeneous random graphs in Section 9.2.1, directed configuration models in Section 9.2.2, and directed preferential attachment models in Section 9.2.3.

A *digraph* $G = (V(G), E(G))$ on the vertex set $V(G) = [n]$ has an edge set that is a subset of the set $E(G) \subseteq [n]^2 = \{(u, v) \colon u, v \in [n]\}$ of all *ordered* pairs of elements of $[n]$. Elements of G are called directed edges or *arcs*.

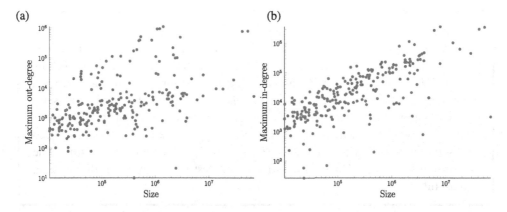

Figure 9.6 Maximum (a) out- and (b) in-degrees of the 229 networks of size larger than 10,000 from the KONECT data base.

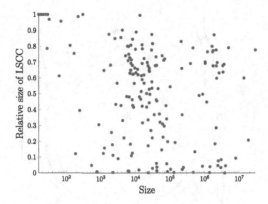

Figure 9.7 Proportion of vertices in the largest strongly connected component (LSSC) in the 229 networks of size larger than 10,000 from the KONECT data base.

Connectivity Structure of Digraphs

The edge orientations significantly impact the connectivity structure of digraphs. Indeed, a given vertex v has both a *forward* connected component consisting of all the vertices to which it is connected, as well as a *backward* connected component. Further, every vertex has a strongly connected component (SCC) consisting of those vertices to which there exist both a forward as well as a backward path. Exercise 9.1 shows that the SCCs in a graph are well defined. See Figure 9.7 for the proportion of vertices in the largest SCC in the KONECT data base.

The different notions of connectivity divide the graph up into several disjoint parts. Often, there is a unique largest SCC that contains a positive proportion of the graph. The IN-part of the graph consists of collections of vertices that are forward connected to the largest SCC, but not backward connected. Further, the OUT-parts of the graph are the collections of vertices that are backward connected to the largest SCC, but not forward connected. Finally, there are the parts of the graph that are in neither of these parts, and consist of their own SCC and IN- and OUT-parts. See Figure 9.8 for a description of these parts of the WWW, as well as an estimate of their relative sizes.

Local Convergence of Digraphs

It turns out that there are several ways to define local convergence for digraphs, owing to the fact that local convergence is defined in terms of *neighborhoods*, which in turn depend on the connectivity that one wishes to use. For the neighborhood $B_r^{(G)}(v)$, do we wish to explore all vertices that can be reached from v (which is relevant when v is the source of an infection), or all the vertices from which we can reach v (which is relevant when investigating whether v can be infected by others, and also in the case of PageRank)? One might also consider all edges at the same time, thus ignoring the directions of the edges in the local neighborhoods and possibly keeping these as *edge marks*.

Let $\mathrm{dist}_G(u, v)$ be the (directed) graph distance from u to v, i.e., the minimal number of directed edges needed to connect u to v. Note that, for digraphs, $\mathrm{dist}_G(u, v)$ and $\mathrm{dist}_G(v, u)$

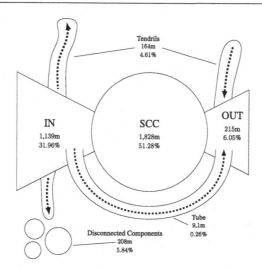

Figure 9.8 The WWW according to Broder et al. (2000), with updated numbers from Fujita et al. (2019).

may be different. We define $B_r^{(G)}(u) = (V(B_r^{(G)}(u)), E(B_r^{(G)}(u)), u)$, the *forward* exploration neighborhood, by

$$V(B_r^{(G)}(u)) = \{v \in V(G): \operatorname{dist}_G(u, v) \leq r\}, \tag{9.2.1}$$

$$E(B_r^{(G)}(u)) = \{(x, y) \in E(G): \operatorname{dist}_G(u, x), \operatorname{dist}_G(u, y) \leq r\}, \tag{9.2.2}$$

as in (2.3.1). When no confusion can arise wrt the graph that we are considering, we sometimes write $B_r^{(\mathrm{out})}(u)$ instead of $B_r^{(G)}(u)$, to indicate that we are following the out-edges in our exploration of the neighborhood.

For the *backward* neighborhood, $\operatorname{dist}_G(u, v) \leq r$ in the definition of $V(B_r^{(G)}(u))$ is replaced by $\operatorname{dist}_G(v, u) \leq r$, and $\operatorname{dist}_G(u, x), \operatorname{dist}_G(u, y)$ in the definition of $E(B_r^{(G)}(u))$ is replaced by $\operatorname{dist}_G(x, u), \operatorname{dist}_G(y, u)$, and we similarly sometimes write this as $B_r^{(\mathrm{in})}(u)$. One could also consider the forward–backward exploration neighborhood, which is the union of the forward and the backward exploration neighborhoods.

Finally, we also sometimes consider the *half-edge-marked directed* neighborhood, obtained by ignoring the directions of edges in the neighborhood (thus effectively replacing the graph by its undirected version), but edge-marking the edges to indicate their direction. Such a mark is indicated at each of the vertices in the edge, and thus can be thought of as a mark on the half-edges incident to the vertices, indicating the direction of the half-edge.

The notion of isomorphisms in Definition 2.3, and of the metric on directed rooted graphs in Definition 2.5, can straightforwardly be adapted. From this moment on, the definition of the forward local convergence of deterministic digraphs, and that of the forward local convergence of random digraphs in their various settings, can straightforwardly be adapted as well.

There is a minor catch, though. While the forward exploration neighborhood keeps track of the out-degrees $d_v^{(\mathrm{out})}$ for all $v \in B_{r-1}^{(\mathrm{out})}(u)$, it does *not* keep track of the *in-degrees* $d_v^{(\mathrm{in})}$ for $v \in B_{r-1}^{(\mathrm{out})}(u)$. Similarly, the backward exploration neighborhood keeps track of the in-degrees $d_v^{(\mathrm{in})}$ for all $v \in B_{r-1}^{(\mathrm{in})}(u)$, but not of the *out-degrees* $d_v^{(\mathrm{out})}$ for $v \in B_{r-1}^{(\mathrm{in})}(u)$. Below, we sometimes need these as well. For this, we add a *degree mark* to the vertices, which indicates the in-degrees in the forward exploration neighborhood, and the out-degrees of the vertices in the backward exploration neighborhood. Particularly in random graphs that are locally tree-like, the edges in the other direction than the one explored will often go to vertices that are far away, and thus are often not in the explored neighborhood themselves. Therefore, to use this information, we need to explicitly keep track of it.

Let us explain how this marking is defined. For a vertex v, let $m(v)$ denote its mark. We then define an isomorphism $\phi \colon V(G_1) \to V(G_2)$ between two labeled rooted graphs (G_1, o_1) and (G_2, o_2) to be an isomorphism between (G_1, o_1) and (G_2, o_2) that respects the marks, i.e., for which $m_1(v) = m_2(\phi(v))$ for every $v \in V(G_1)$, where m_1 and m_2 denote the degree-mark functions on G_1 and G_2 respectively. We then define R^* as in (2.2.2), and the metric on rooted degree-marked graphs as in (2.2.3). We call the resulting notion of local convergence (LC) *marked forward LC* and *marked backward LC*, respectively.

Even when we are considering forward–backward neighborhoods, the addition of marks is necessary. Indeed, while for the root of this graph we do know by construction its in- and out-degrees, for the other vertices in the forward and backward neighborhoods this information is still not available.

Exercises 9.3–9.5 investigate the various notions of local convergence for some directed random graphs that are naturally derived from undirected graphs.

While the above discussion may not be relevant for all questions that one may wish to investigate using local convergence techniques, it is useful for the discussion of PageRank, as we discuss next.

Convergence of PageRank

Recall the definition of PageRank from [V1, Section 1.5]. Let us first explain the solution in the absence of dangling ends, so that $d_v^{(\mathrm{out})} \geq 1$ for all $v \in [n]$. Let $G_n = (V(G_n), E(G_n))$ denote a digraph, where we let $n = |V(G_n)|$ denote the number of vertices. Fix the *damping factor* $\alpha \in (0, 1)$. Then we let the vector of PageRanks $(R_v^{(G_n)})_{v \in V(G_n)}$ be the unique solution to the equation

$$R_v^{(G_n)} = \alpha \sum_{u \to v} \frac{R_u^{(G_n)}}{d_u^{(\mathrm{out})}} + 1 - \alpha, \tag{9.2.3}$$

satisfying the normalization $\sum_{v \in V(G_n)} R_v^{(G_n)} = n$.

The damping parameter $\alpha \in (0, 1)$ guarantees that (9.2.3) has a *unique* solution. This solution can be understood in terms of the stationary distribution of a "bored surfer." Indeed, denote $\pi_v = R_v^{(G_n)}/n$, so that $(\pi_v)_{v \in V(G_n)}$ is a probability distribution that satisfies a similar relation to $(R_v^{(G_n)})_{v \in V(G_n)}$ in (9.2.3), namely

$$\pi_v = \alpha \sum_{u \to v} \frac{\pi_u}{d_u^{(\mathrm{out})}} + \frac{1 - \alpha}{n}. \tag{9.2.4}$$

Therefore, $(\pi_v)_{v \in V(G_n)}$ is the stationary distribution of a random walker, which, with probability α, jumps according to a simple random walk, i.e., it chooses any of the out-edges with equal probability, while, with probability $1 - \alpha$, the walker is bored, forgets about the search they were doing, and jumps to a uniform vertex in $V(G_n)$.

In the presence of dangling ends, we can just redistribute their mass equally over all vertices that are not dangling ends, so that (9.2.3) becomes

$$R_v^{(G_n)} = \alpha \sum_{u \to v} \frac{R_u^{(G_n)}}{d_u^{(\text{out})}} \mathbb{1}_{\{u \notin \mathcal{D}\}} + \frac{\alpha}{n} \sum_{u \in \mathcal{D}} R_u^{(G_n)} + 1 - \alpha, \qquad (9.2.5)$$

where $\mathcal{D} = \{v \colon d_v^{(\text{out})} = 0\} \subseteq V(G_n)$ denotes the collection of dangling vertices.

The damping factor α is quite crucial. When $\alpha = 0$, the stationary distribution is just $\pi_v = 1/n$ for every $v \in V(G_n)$, so that all vertices have PageRank 1. This is not very informative. On the other hand, PageRank possibly converges quite slowly when α is close to 1, and this is also not what we want. Experimentally, $\alpha = 0.85$ seems to work well and strikes a nice balance between these two extremes.

We next investigate the convergence of the PageRank distribution on a directed graph sequence $(G_n)_{n \geq 1}$ that converges locally:

Theorem 9.1 (Existence of asymptotic PageRank distribution) *Consider a sequence of directed random graphs $(G_n)_{n \in \mathbb{N}}$ and let $o_n \in V(G_n)$ be chosen uar. Then, the following hold:*

(a) *If G_n converges locally weakly in the marked backward sense to $(\bar{G}, \bar{o}) \sim \bar{\mu}$, then there exists a limiting distribution $R_{\varnothing}^{(\bar{G})}$, with $\mathbb{E}_\mu[R_{\varnothing}^{(\bar{G})}] \leq 1$, such that*

$$R_{o_n}^{(G_n)} \xrightarrow{d} R_{\varnothing}^{(\bar{G})}. \qquad (9.2.6)$$

(b) *If G_n converges locally in probability in the marked backward sense to $(G, o) \sim \mu$, then there exists a limiting distribution $R_{\varnothing}^{(G)}$, with $\mathbb{E}_\mu[R_{\varnothing}^{(G)}] \leq 1$, such that, for every $r > 0$ that is a continuity point of the distribution $R_{\varnothing}^{(G)}$,*

$$\frac{1}{n} \sum_{v \in V(G_n)} \mathbb{1}_{\{R_v^{(G_n)} > r\}} \xrightarrow{\mathbb{P}} \mu\left(R_{\varnothing}^{(G)} > r\right). \qquad (9.2.7)$$

Interestingly, the positivity of the damping factor also allows us to give a power-iteration formula for $R_v^{(G_n)}$, and thus for $R_{\varnothing}$. Indeed, let

$$A_{i,j}^{(G_n)} = \frac{\mathbb{1}_{\{j \to i\}}}{d_j^{(\text{out})}}, \qquad i, j \in V(G_n) \qquad (9.2.8)$$

denote the (normalized) adjacency matrix of the graph G_n. Then, $R_v^{(G_n)}$ can be computed as follows:

$$R_v^{(G_n)} = (1 - \alpha) \sum_{k=0}^{\infty} \alpha^k \sum_{i \in V(G_n)} (A^{(G_n)})_{v,i}^k. \qquad (9.2.9)$$

As a result, when G_n converges locally in probability in the marked backward sense with limit $(G, \varnothing)$, then $R_{\varnothing}^{(G)}$ can be computed:

$$R_{\varnothing}^{(G)} = (1 - \alpha) \sum_{k=0}^{\infty} \alpha^k \sum_{i \in V(G)} (A^{(G)})_{\varnothing,i}^k, \qquad (9.2.10)$$

where $A_{i,j}^{(G)}$ is the normalized adjacency matrix of the backwards local limit G. We refer to the notes in Section 9.6 for a discussion of Theorem 9.1, including consideration of an error in its original statement.

Power-Law Hypothesis for PageRank

Recall from [V1, Section 1.5] that the *PageRank power-law hypothesis* states that the PageRank distribution satisfies a power law with the *same* exponent as that of the in-degree. By Theorem 9.1, this can be rephrased by stating that $\mathbb{P}(R_{\varnothing} > r) \asymp r^{-(\tau_{\text{in}}-1)}$ when $\mathbb{P}(D_{\varnothing}^{(\text{in})} > r) \asymp r^{-(\tau_{\text{in}}-1)}$ (we are deliberately being vague about what $\asymp$ means in this context).

Exercises 9.6 and 9.7 investigate the implications of (9.2.10) for the power-law hypothesis for random graphs having *bounded* out-degrees.

9.2.1 DIRECTED INHOMOGENEOUS RANDOM GRAPHS

Let $(x_i)_{i \in [n]}$ be a sequence of variables, with values in a type space $\mathcal{S}$, such that the empirical distribution of $(x_i)_{i \in [n]}$ approximates a measure μ as $n \to \infty$. That is, we assume that, for each μ-continuous Borel set $\mathcal{A} \subseteq \mathcal{S}$, as $n \to \infty$, we have

$$\frac{1}{n}|\{i \in [n] \colon x_i \in \mathcal{A}\}| \to \mu(\mathcal{A}). \tag{9.2.11}$$

Here we say that a Borel set $\mathcal{A}$ is μ-continuous whenever its boundary $\partial \mathcal{A}$ has zero probability, i.e., $\mu(\partial \mathcal{A}) = 0$.

Given n, let G_n be the random digraph on the vertex set $(x_i)_{i \in [n]}$ with independent arcs having probabilities

$$p_{ij} = \mathbb{P}((x_i, x_j) \in E(G_n)) = 1 \wedge (\kappa(x_i, x_j)/n), \qquad i, j \in [n]. \tag{9.2.12}$$

We denote the resulting graph by $\text{DIRG}_n(\kappa)$. Combining $\mathcal{S}$, μ, and κ, we obtain a large class of inhomogeneous digraphs with independent arcs. Obviously, the model includes digraphs with power-law in-degree and/or out-degree distributions. This general undirected model was studied in detail in Chapter 3.

We will be a little less general here, since we assume that κ in (9.2.12) does not depend on n, which simplifies the exposition. Note that in the case of undirected graphs it is necessary to assume, in addition, that the kernel κ is symmetric. In the definition of digraphs G_n for $n \geq 2$, we do not require symmetry of the kernel.

Assumptions on the Kernel

We need to impose further conditions on the kernel κ, like those in Chapter 3. Namely, we need to assume that the kernel κ is irreducible $(\mu \times \mu)$-almost everywhere. That is, for any measurable $\mathcal{A} \subseteq \mathcal{S}$ with $\mu(\mathcal{A}) \notin \{0, 1\}$, the identity $(\mu \times \mu)(\{(s, t) \in \mathcal{A} \times (\mathcal{S} \setminus \mathcal{A}) \colon \kappa(s, t) = 0\}) = 0$ implies that either $\mu(\mathcal{A}) = 0$ or $\mu(\mathcal{S} \setminus \mathcal{A}) = 0$. In addition, we assume that κ is continuous almost everywhere on $(\mathcal{S} \times \mathcal{S}, \mu \times \mu)$ and that the number of arcs in $\text{DIRG}_n(\kappa)$, denoted by $|E(\text{DIRG}_n(\kappa))|$, satisfies that, as $n \to \infty$,

$$\frac{1}{n}|E(\text{DIRG}_n(\kappa))| \overset{\mathbb{P}}{\longrightarrow} \int_{\mathcal{S} \times \mathcal{S}} \kappa(s, t)\mu(\text{d}s)\mu(\text{d}t) < \infty. \tag{9.2.13}$$

Examples of Directed Inhomogeneous Random Graphs

We next discuss some examples of $\mathrm{DIRG}_n(\kappa)$.

Directed Erdős–Rényi random graph. The most basic example is the directed Erdős–Rényi random graph, for which $p_{uv} = p_{vu} = \lambda/n$. In this case, $\kappa(x_u, x_v) = \lambda$ and $\mathcal{S} = \{1\}$.

Finite-type directed inhomogeneous random graphs. Slightly more involved than the above are kernels of finite type, in which case $(s, r) \mapsto \kappa(s, r)$ takes on finitely many values and $\mathcal{S} = [t]$. Such kernels are also highly convenient to approximate more general models, as was exemplified in an undirected setting in Chapter 3.

Directed rank-1 inhomogeneous random graphs. We next generalize rank-1 inhomogeneous random graphs. For $v \in [n]$, let $w_v^{(\mathrm{in})}$ and $w_v^{(\mathrm{out})}$ be its respective in- and out-weights, which will have the interpretation of the asymptotic average in- and out-degrees, respectively, under a summation-symmetry condition on the weights. Then, the directed generalized random graph $\mathrm{DGRG}_n(\boldsymbol{w})$ has edge probabilities given by

$$p_{uv} = p_{uv}^{(\mathrm{DGRG})} = \frac{w_u^{(\mathrm{out})} w_v^{(\mathrm{in})}}{\ell_n + w_u^{(\mathrm{out})} w_v^{(\mathrm{in})}}, \tag{9.2.14}$$

where

$$\ell_n = \frac{1}{2} \sum_{v \in [n]} (w_v^{(\mathrm{out})} + w_v^{(\mathrm{in})}). \tag{9.2.15}$$

Let $(W_n^{(\mathrm{out})}, W_n^{(\mathrm{in})}) = (w_o^{(\mathrm{out})}, w_o^{(\mathrm{in})})$ denote the in- and out-weights of a uniformly chosen vertex $o \in [n]$. Similarly to Condition 1.1, we assume that

$$(W_n^{(\mathrm{out})}, W_n^{(\mathrm{in})}) \xrightarrow{d} (W^{(\mathrm{out})}, W^{(\mathrm{in})}) \tag{9.2.16}$$

and

$$\mathbb{E}[W_n^{(\mathrm{out})}] \to \mathbb{E}[W^{(\mathrm{out})}], \qquad \mathbb{E}[W_n^{(\mathrm{in})}] \to \mathbb{E}[W^{(\mathrm{in})}], \tag{9.2.17}$$

where $(W^{(\mathrm{out})}, W^{(\mathrm{in})})$ is the limiting in- and out-weight distribution. Exercises 9.8 investigates the expected number of edges in this setting.

When thinking of $w_v^{(\mathrm{out})}$ and $w_v^{(\mathrm{in})}$ as corresponding to the approximate out- and in-degrees of vertex $v \in [n]$, it is reasonable to assume that

$$\mathbb{E}[W^{(\mathrm{in})}] = \mathbb{E}[W^{(\mathrm{out})}]. \tag{9.2.18}$$

Indeed, if $D_v^{(\mathrm{out})}$ and $D_v^{(\mathrm{in})}$ denote the out- and in-degrees of vertex $v \in [n]$, we know that (recall Exercise 9.2)

$$\sum_{v \in [n]} D_v^{(\mathrm{out})} = \sum_{v \in [n]} D_v^{(\mathrm{in})}. \tag{9.2.19}$$

Thus, if indeed $w_v^{(\mathrm{out})}$ and $w_v^{(\mathrm{in})}$ are approximately equal to the out- and in-degrees of vertex $v \in [n]$ then also

$$\sum_{v \in [n]} w_v^{(\mathrm{out})} \approx \sum_{v \in [n]} w_v^{(\mathrm{in})}, \tag{9.2.20}$$

which, assuming (9.2.16), proves (9.2.18).

As discussed in more detail in Section 1.3.2 (see in particular (1.3.8) and (1.3.9)), many related versions of the rank-1 inhomogeneous random graph exist. We refrain from giving more details here.

Multi-Type Marked Branching Processes for $\mathrm{DIRG}_n(\kappa)$

For large n, the local convergence and phase transition in the digraph G_n can be described in terms of the survival probabilities of the related multi-type branching processes with type space $\mathcal{S}$. Let us now introduce the necessary mixed-Poisson branching processes. Given $s \in \mathcal{S}$, let $\mathcal{X}(s)$ and $\mathcal{Y}(s)$, respectively, denote the branching processes starting at an individual of type $s \in \mathcal{S}$ such that the number of children of types in a subset $\mathcal{A} \subseteq \mathcal{S}$ of an individual of type $r \in \mathcal{S}$ have Poisson distributions with means,

$$\int_{\mathcal{A}} \kappa(r, u) \mu(\mathrm{d}u), \qquad \text{and} \qquad \int_{\mathcal{A}} \kappa(u, r) \mu(\mathrm{d}u), \qquad (9.2.21)$$

respectively. These numbers are independent for disjoint subsets $\mathcal{A}$ and for different individuals. The two branching processes $\mathcal{X}(s)$ and $\mathcal{Y}(s)$ correspond to the forward and backward limits of $\mathrm{DIRG}_n(\kappa)$, respectively.

We now extend the discussion by defining the *marks* of these branching processes. With each individual of type r we associate independent marks having Poisson distributions with means $\int_{\mathcal{S}} \kappa(r, u) \mu(\mathrm{d}u)$ and $\int_{\mathcal{S}} \kappa(u, r) \mu(\mathrm{d}u)$, respectively. These random variables correspond to the "in-degrees" for the forward exploration process $\mathcal{X}(s)$ and the "out-degrees" for the backward exploration process $\mathcal{Y}(s)$. Finally, for the forward–backward setting, we let the marked branching processes $\mathcal{X}(s)$ and $\mathcal{Y}(s)$ be independent. We call these objects *Poisson marked branching processes with kernel* κ. As in Section 3.4.3, we let $\boldsymbol{T}_\kappa$ be defined as in (3.4.15), i.e., for $f: \mathcal{S} \to \mathbb{R}$, we let $(\boldsymbol{T}_\kappa f)(x) = \int_{\mathcal{S}} \kappa(x, y) f(y) \mu(\mathrm{d}y)$.

Local Convergence for $\mathrm{DIRG}_n(\kappa)$

The following theorem describes the local convergence of $\mathrm{DIRG}_n(\kappa)$:

Theorem 9.2 (Local convergence of $\mathrm{DIRG}_n(\kappa)$) *Suppose that κ is irreducible and continuous almost everywhere on $(\mathcal{S} \times \mathcal{S}, \mu \times \mu)$ and that (9.2.13) holds. Then, $\mathrm{DIRG}_n(\kappa)$ converges locally in probability in the marked forward, backward, and forward–backward sense to the above Poisson marked branching processes with kernel κ, where the law of the type of the root $\varnothing$ is μ.*

We do not give a proof of Theorem 9.2, but refer to Section 9.6 for its history. In Exercise 9.10 the reader is asked to determine the local limit of the directed Erdős–Rényi random graph. Exercise 9.11 proves Theorem 9.2 in the case of finite-type kernels, while Exercise 9.12 investigates the local convergence of the directed generalized random graph.

Giant Component in $\mathrm{DIRG}_n(\kappa)$

The critical point of the emergence of the giant strongly connected component is determined when the averaged joint survival probability

$$\zeta = \int_{\mathcal{S}} \zeta_{\mathcal{X}}(s) \zeta_{\mathcal{Y}}(s) \mu(\mathrm{d}s) \qquad (9.2.22)$$

becomes positive. Here $\zeta_{\mathcal{X}}(s)$ and $\zeta_{\mathcal{Y}}(s)$ denote the survival probabilities of $\mathcal{X}(s)$ and $\mathcal{Y}(s)$, respectively. The following theorem describes the phase transition in $\mathrm{DIRG}_n(\kappa)$:

Theorem 9.3 (Phase transition in $\mathrm{DIRG}_n(\kappa)$) *Suppose that κ is irreducible and continuous almost everywhere on $(\mathcal{S} \times \mathcal{S}, \mu \times \mu)$ and that (9.2.13) holds. Then*

$$|\mathscr{C}_{\max}|/n \xrightarrow{\mathbb{P}} \zeta, \qquad (9.2.23)$$

while $|\mathscr{C}_{(2)}|/n \xrightarrow{\mathbb{P}} 0$ and $|E(\mathscr{C}_{(2)})|/n \xrightarrow{\mathbb{P}} 0$.

Theorem 9.3 is the directed version of Theorem 3.19. Owing to the directed nature of the random graph involved, it is now more involved to identify when $\zeta > 0$. In the finite-type case, this is determined by the largest eigenvalue of the mean-offspring matrix exceeding 1; in the infinite-type case, this is less clear. Exercises 9.13 and 9.14 investigate the conditions for a giant component to exist for directed Erdős–Rényi and generalized random graphs.

9.2.2 DIRECTED CONFIGURATION MODELS

One way to obtain a directed version of $\mathrm{CM}_n(\boldsymbol{d})$ is to give each edge a direction, chosen with probability $\frac{1}{2}$, independently of the directions of all other edges. In such a model, however, the correlation coefficient between the in- and out-degree of vertices is close to 1, particularly when the degrees are large (see Exercise 9.15). In real-world applications, correlations between in- and out-degrees can be positive or negative, depending on the precise application, so we should aim to formulate a model that is more general, in that we prescribe *both* the in- and out-degrees of vertices.

Let $\boldsymbol{d}^{(\mathrm{in})} = (d_v^{(\mathrm{in})})_{v \in [n]}$ be a sequence of in-degrees, where $d_v^{(\mathrm{in})}$ denotes the in-degree of vertex v. Similarly, we let $\boldsymbol{d}^{(\mathrm{out})} = (d_v^{(\mathrm{out})})_{v \in [n]}$ be a sequence of out-degrees. Naturally, in order for a graph with in- and out-degree sequence $\boldsymbol{d} = (\boldsymbol{d}^{(\mathrm{in})}, \boldsymbol{d}^{(\mathrm{out})})$ to exist, we need that (recall Exercise 9.2)

$$\sum_{v \in [n]} d_v^{(\mathrm{in})} = \sum_{v \in [n]} d_v^{(\mathrm{out})}. \qquad (9.2.24)$$

We think of $d_v^{(\mathrm{in})}$ and $d_v^{(\mathrm{out})}$, respectively, as the number of in- and out-half-edges incident to vertex v. The directed configuration model $\mathrm{DCM}_n(\boldsymbol{d})$ is obtained by sequentially pairing each in-half-edge with a uniformly chosen out-half-edge, without replacement. The resulting graph is a random multi-graph, where each vertex v has in-degree $d_v^{(\mathrm{in})}$ and out-degree $d_v^{(\mathrm{out})}$. Similarly to $\mathrm{CM}_n(\boldsymbol{d})$, $\mathrm{DCM}_n(\boldsymbol{d})$ can have self-loops as well as multiple edges. A self-loop arises at vertex v when one of its in-half-edges pairs to one of its out-half-edges.

We now continue to investigate the strongly connected component of $\mathrm{DCM}_n(\boldsymbol{d})$. Let $(D_n^{(\mathrm{in})}, D_n^{(\mathrm{out})})$ denote the in- and out-degree of a vertex chosen uar from $[n]$. Assume, similarly to Condition 1.7(a),(b), that

$$(D_n^{(\mathrm{in})}, D_n^{(\mathrm{out})}) \xrightarrow{d} (D^{(\mathrm{in})}, D^{(\mathrm{out})}), \qquad (9.2.25)$$

and that

$$\mathbb{E}[D_n^{(\mathrm{in})}] \to \mathbb{E}[D^{(\mathrm{in})}], \qquad \text{and} \qquad \mathbb{E}[D_n^{(\mathrm{out})}] \to \mathbb{E}[D^{(\mathrm{out})}]. \qquad (9.2.26)$$

Naturally, by (9.2.24), this implies that (see Exercise 9.16)

$$\mathbb{E}[D^{(\text{out})}] = \mathbb{E}[D^{(\text{in})}]. \tag{9.2.27}$$

Exercise 9.17 investigates the convergence of the numbers of self-loops and multi-edges in $\text{DCM}_n(\boldsymbol{d})$.

Let

$$p_{k,l} = \mathbb{P}(D^{(\text{in})} = k, D^{(\text{out})} = l) \tag{9.2.28}$$

denote the asymptotic joint in- and out-degree distribution. We refer to $(p_{k,l})_{k,l \geq 0}$ simply as the asymptotic degree distribution of $\text{DCM}_n(\boldsymbol{d})$. The distribution $(p_{k,l})_{k,l \geq 0}$ plays a role for $\text{DCM}_n(\boldsymbol{d})$ similar to the role $(p_k)_{k \geq 0}$ plays for $\text{CM}_n(\boldsymbol{d})$. We further define

$$p_k^{*(\text{in})} = \sum_l l p_{k,l} / \mathbb{E}[D^{(\text{out})}], \qquad p_l^{*(\text{out})} = \sum_k k p_{k,l} / \mathbb{E}[D^{(\text{in})}]. \tag{9.2.29}$$

The distributions $(p_k^{*(\text{in})})_{k \geq 0}$ and $(p_k^{*(\text{out})})_{k \geq 0}$ correspond to the asymptotic forward in- and out-degrees of a uniformly chosen edge in $\text{DCM}_n(\boldsymbol{d})$.

Local Limit of the Directed Configuration Model

Let us now formulate the construction of the appropriate marked forward and backward branching processes that arises as the local limit of $\text{DCM}_n(\boldsymbol{d})$. For the forward branching process, we let the root have out-degree with distribution

$$p_l^{(\text{out})} = \mathbb{P}(D^{(\text{out})} = l) = \sum_{l \geq 0} p_{k,l}, \tag{9.2.30}$$

whereas every other vertex except the root has independent out-degree with law $(p_l^{*(\text{out})})_{l \geq 0}$. Further, for a vertex of out-degree l, we let the mark (corresponding to its asymptotic in-degree) be k with probability $p_{k,l} / p_l^{(\text{out})}$. For the marked backward branching process, we reverse the role of in- and out-. For the marked forward–backward branching process, we let the root have joint out- and in-degree distribution $(p_{k,l})_{k,l \geq 0}$, and define the forward and backward processes and marks as before. We call the above branching process the *marked unimodular branching process with degree distribution* $(p_{k,l})_{k,l \geq 0}$.

The following theorem describes the local convergence in $\text{DCM}_n(\boldsymbol{d})$:

Theorem 9.4 (Local convergence of $\text{DCM}_n(\boldsymbol{d})$) *Suppose that the out- and in-degrees in a directed configuration model $\text{DCM}_n(\boldsymbol{d})$ satisfy (9.2.25) and (9.2.26). Then $\text{DCM}_n(\boldsymbol{d})$ converges locally in probability in the marked forward, backward, and forward–backward sense to the above marked unimodular branching process with degree distribution $(p_{k,l})_{k,l \geq 0}$.*

It will not come as a surprise that Theorem 9.4 is the directed version of Theorem 4.1, and Exercise 9.18 asks the reader to prove Theorem 9.4 by adapting its proof.

Giant Component in the Directed Configuration Model

Let $\theta^{(\text{in})}$ and $\theta^{(\text{out})}$ be the survival probabilities of the branching processes with offspring distributions $(p_k^{*(\text{in})})_{k \geq 0}$ and $(p_k^{*(\text{out})})_{k \geq 0}$, respectively, and define

$$\zeta^{(\text{in})} = 1 - \sum_{k,l} p_{k,l} (1 - \theta^{(\text{in})})^l, \qquad \zeta^{(\text{out})} = 1 - \sum_{k,l} p_{k,l} (1 - \theta^{(\text{out})})^k. \tag{9.2.31}$$

Then, $\zeta^{(\text{out})}$ has the interpretation of the asymptotic probability that a uniform vertex has a large forward cluster, while $\zeta^{(\text{in})}$ has that of a uniform vertex having a large backward cluster. Further, let

$$\psi = \sum_{k,l} p_{k,l}(1 - \theta^{(\text{in})})^l(1 - \theta^{(\text{out})})^k, \tag{9.2.32}$$

so that ψ has the interpretation of the asymptotic probability that a uniform vertex has both a finite forward and a finite backward cluster. We conclude that $1 - \psi$ is the probability that a uniform vertex has either a large forward or a large backward cluster, and thus

$$\zeta = \zeta^{(\text{out})} + \zeta^{(\text{in})} - (1 - \psi) \tag{9.2.33}$$

has the interpretation of the asymptotic probability that a uniform vertex has *both* a large forward and a large backward cluster. Finally, we let

$$\nu = \sum_{k=0}^{\infty} k p_k^{\star(\text{in})} = \sum_{k,l} \frac{kl p_{k,l}}{\mathbb{E}[D^{(\text{out})}]} = \frac{\mathbb{E}[D^{(\text{in})} D^{(\text{out})}]}{\mathbb{E}[D^{(\text{out})}]}. \tag{9.2.34}$$

Alternatively, $\nu = \sum_{k=0}^{\infty} k p_k^{\star(\text{out})} = \mathbb{E}[D^{(\text{in})} D^{(\text{out})}]/\mathbb{E}[D^{(\text{in})}]$ by (9.2.27). The main result concerning the size of the giant is as follows:

Theorem 9.5 (Phase transition in $\text{DCM}_n(\boldsymbol{d})$) *Suppose that the out- and in-degrees in the directed configuration model* $\text{DCM}_n(\boldsymbol{d})$ *satisfy* (9.2.25) *and* (9.2.26).

(a) *When* $\nu > 1$, ζ *in* (9.2.33) *satisfies* $\zeta \in (0, 1]$ *and*

$$|\mathscr{C}_{\max}|/n \xrightarrow{\mathbb{P}} \zeta, \tag{9.2.35}$$

while $|\mathscr{C}_{(2)}|/n \xrightarrow{\mathbb{P}} 0$ *and* $|E(\mathscr{C}_{(2)})|/n \xrightarrow{\mathbb{P}} 0$.

(b) *When* $\nu \leq 1$, ζ *in* (9.2.33) *satisfies* $\zeta = 0$, *so that* $|\mathscr{C}_{\max}|/n \xrightarrow{\mathbb{P}} 0$ *and* $|E(\mathscr{C}_{\max})|/n \xrightarrow{\mathbb{P}} 0$.

Theorem 9.5 is the adaptation to $\text{DCM}_n(\boldsymbol{d})$ of the existence of the giant for $\text{CM}_n(\boldsymbol{d})$ in Theorem 4.9. In Exercise 9.19, the reader is asked to prove that the probability that the size of $|\mathscr{C}_{\max}|/n$ exceeds $\zeta + \varepsilon$ vanishes whenever a graph sequence converges locally in probability in the marked forward–backward sense. In Exercise 9.20, this is used to prove Theorem 9.5(b).

Logarithmic Typical Distances in the Directed Configuration Model

We continue by studying the small-world nature in the directed configuration model. Let $u, v \in [n]$, and let $\text{dist}_{\text{DCM}_n(d)}(u, v)$ denote the graph distance between u and v, i.e., the minimal number of *directed* edges needed to connect u to v, so that $\text{dist}_{\text{DCM}_n(d)}(u, v)$ is not necessarily equal to $\text{dist}_{\text{DCM}_n(d)}(v, u)$. The main result is as follows:

Theorem 9.6 (Logarithmic typical distances in $\text{DCM}_n(\boldsymbol{d})$) *Suppose that the out- and in-degrees in the directed configuration model* $\text{DCM}_n(\boldsymbol{d})$ *satisfy* (9.2.25) *and* (9.2.26), *and assume that*

$$\nu = \frac{\mathbb{E}[D^{(\text{in})} D^{(\text{out})}]}{\mathbb{E}[D^{(\text{out})}]} > 1. \tag{9.2.36}$$

Further, assume that

$$\mathbb{E}[(D_n^{(\mathrm{in})})^2] \to \mathbb{E}[(D^{(\mathrm{in})})^2] < \infty, \qquad \mathbb{E}[(D_n^{(\mathrm{out})})^2] \to \mathbb{E}[(D^{(\mathrm{out})})^2] < \infty. \qquad (9.2.37)$$

Then, conditional on $o_1 \to o_2$,

$$\frac{\mathrm{dist}_{\mathrm{DCM}_n(\boldsymbol{d})}(o_1, o_2)}{\log n} \xrightarrow{\mathbb{P}} \frac{1}{\log \nu}. \qquad (9.2.38)$$

Theorem 9.6 is the directed version of Theorem 7.1. The philosophy behind the proof is quite similar:.A breadth-first exploration process shows that $|\partial B_r^{(\mathrm{out})}(o_1)|$ grows roughly like ν^r, so in order to "catch" o_2, one would need $r \approx \log_\nu n = \log n / \log \nu$ (recall the discussion of the various directed neighborhoods below (9.2.1)). Of course, at this stage, the branching-process approximation starts to fail, which is why one needs to grow the neighborhoods from two sides and use that also $|\partial B_r^{(\mathrm{in})}(o_2)|$ grows roughly like ν^r.

It would be tempting to believe that (9.2.37) is more than what is needed; this is the content of Exercise 9.21. In Exercise 9.23, the reader is asked to prove that $\mathrm{dist}_{\mathrm{DCM}_n(\boldsymbol{d})}(o_1, o_2)$ $= o_{\mathbb{P}}(\log n)$ when $\nu = \infty$.

Doubly Logarithmic Typical Distances in the Directed Configuration Model

We continue by studying the ultra-small-world nature in the directed configuration model, for which we assume that there exist $\tau^{(\mathrm{in})}, \tau^{(\mathrm{out})} \in (2, 3)$ and that, for all $\delta > 0$, there exist $c_1 = c_1(\delta)$ and $c_2 = c_2(\delta)$ such that, uniformly in n,

$$c_1 x^{-(\tau^{(\mathrm{in})} - 2 + \delta)} \le \frac{1}{n} \sum_{v \in [n]} d_v^{(\mathrm{out})} \mathbb{1}_{\{d_v^{(\mathrm{in})} > v\}} \le c_2 x^{-(\tau^{(\mathrm{in})} - 2 - \delta)},$$

$$c_1 x^{-(\tau^{(\mathrm{out})} - 2 + \delta)} \le \frac{1}{n} \sum_{v \in [n]} d_v^{(\mathrm{in})} \mathbb{1}_{\{d_v^{(\mathrm{out})} > v\}} \le c_2 x^{-(\tau^{(\mathrm{out})} - 2 - \delta)}, \qquad (9.2.39)$$

where the upper bound holds for every $x \ge 1$ while the lower bound is required to hold only for $1 \le x \le n^\beta$ for some $\beta > \frac{1}{2}$. The main result is then as follows:

Theorem 9.7 (Doubly logarithmic typical distances in $\mathrm{DCM}_n(\boldsymbol{d})$) *Suppose that the out- and in-degrees in the directed configuration model* $\mathrm{DCM}_n(\boldsymbol{d})$ *satisfy* (9.2.25), (9.2.26), *and* (9.2.39). *Then, conditional on* $o_1 \to o_2$,

$$\frac{\mathrm{dist}_{\mathrm{DCM}_n(\boldsymbol{d})}(o_1, o_2)}{\log \log n} \xrightarrow{\mathbb{P}} \frac{1}{|\log(\tau^{(\mathrm{in})} - 2)|} + \frac{1}{|\log(\tau^{(\mathrm{out})} - 2)|}. \qquad (9.2.40)$$

Theorem 9.7 is the directed version of Theorem 7.2.

Logarithmic Diameter in the Directed Configuration Model

We next investigate the logarithmic asymptotics of the diameter of $\mathrm{DCM}_n(\boldsymbol{d})$. Here, we define the diameter of the digraph $\mathrm{DCM}_n(\boldsymbol{d})$ by

$$\mathrm{diam}(\mathrm{DCM}_n(\boldsymbol{d})) = \max_{u,v: \, u \to v} \mathrm{dist}_{\mathrm{DCM}_n(\boldsymbol{d})}(u, v). \qquad (9.2.41)$$

In order to state the result, we introduce some notation. Let

$$f(s, t) = \mathbb{E}[s^{D^{(\mathrm{in})}} t^{D^{(\mathrm{out})}}] \qquad (9.2.42)$$

be the bivariate generating function of $(D^{(\mathrm{in})}, D^{(\mathrm{out})})$. Recall that $\theta^{(\mathrm{in})}$ and $\theta^{(\mathrm{out})}$ are the survival probabilities of a branching process with offspring distributions $(p_k^{\star(\mathrm{in})})_{k \geq 0}$ and $(p_k^{\star(\mathrm{out})})_{k \geq 0}$, respectively. Write

$$\frac{1}{\nu^{(\mathrm{in})}} = \frac{1}{\mathbb{E}[D^{(\mathrm{in})}]} \frac{\partial^2}{\partial s \partial t} f(s,t)\Big|_{s=1-\theta^{(\mathrm{in})}, t=1}, \tag{9.2.43}$$

and

$$\frac{1}{\nu^{(\mathrm{out})}} = \frac{1}{\mathbb{E}[D^{(\mathrm{out})}]} \frac{\partial^2}{\partial s \partial t} f(s,t)\Big|_{s=1, t=1-\theta^{(\mathrm{out})}}. \tag{9.2.44}$$

Then, the diameter in the directed configuration model $\mathrm{DCM}_n(\boldsymbol{d})$ behaves as follows:

Theorem 9.8 (Logarithmic diameter in $\mathrm{DCM}_n(\boldsymbol{d})$)   *Suppose that the out- and in-degrees in the directed configuration model $\mathrm{DCM}_n(\boldsymbol{d})$ satisfy (9.2.25) and (9.2.26). Further, assume that (9.2.37) holds. Then, if $\nu = \mathbb{E}[D^{(\mathrm{in})}D^{(\mathrm{out})}]/\mathbb{E}[D^{(\mathrm{out})}] > 1$,*

$$\frac{\mathrm{diam}(\mathrm{DCM}_n(\boldsymbol{d}))}{\log n} \xrightarrow{\mathbb{P}} \frac{1}{\log \nu^{(\mathrm{in})}} + \frac{1}{\log \nu} + \frac{1}{\log \nu^{(\mathrm{out})}}. \tag{9.2.45}$$

Theorem 9.8 is the directed version of Theorem 7.19. The interpretation of the various terms is similar to that in Theorem 7.19: the terms involving $\nu^{(\mathrm{in})}$ and $\nu^{(\mathrm{out})}$ indicate the depths of the deepest traps, where a trap indicates that the neighborhood lives for a long time without gaining substantial mass, i.e., it is *thin*. The term involving $\nu^{(\mathrm{in})}$ is the depth of the largest in-trap, so that the in-neighborhood is thin, and that involving $\nu^{(\mathrm{out})}$ that of the largest out-trap, so that the out-neighborhood is thin. These numbers are determined by first taking r such that

$$\mathbb{P}(|\partial B_r^{(\mathrm{in/out})}(o)| \in [1, K]) \approx \Theta\left(\frac{1}{n}\right), \tag{9.2.46}$$

where $\partial B_r^{(\mathrm{in/out})}(o)$ corresponds to the ball of the backward r-neighborhood for $\nu^{(\mathrm{in})}$ and to the forward r-neighborhood for $\nu^{(\mathrm{out})}$, while K is arbitrary and large. Owing to large deviations for supercritical branching processes, one can expect that

$$\mathbb{P}(|\partial B_r^{(\mathrm{in/out})}(o)| \in [1, K]) \approx (\nu^{(\mathrm{in/out})})^r. \tag{9.2.47}$$

Then we can identify $r^{(\mathrm{in})} = \log_{\nu^{(\mathrm{in})}}(n)$ and $r^{(\mathrm{out})} = \log_{\nu^{(\mathrm{out})}}(n)$. The solutions to (9.2.47) are given by (9.2.43) and (9.2.44). For those special vertices u, v for which $|\partial B_{r^{(\mathrm{in})}}^{(\mathrm{in})}(u)| \in [1, K]$ and $|\partial B_{r^{(\mathrm{out})}}^{(\mathrm{out})}(v)| \in [1, K]$, it then takes around $\log_\nu(n)$ steps to connect $\partial B_{r^{(\mathrm{in})}}^{(\mathrm{in})}(u)$ to $\partial B_{r^{(\mathrm{out})}}^{(\mathrm{out})}(v)$, thus explaining the asymptotics in Theorem 9.8. Of course, proving that this heuristic is correct is quite a bit harder.

9.2.3 DIRECTED PREFERENTIAL ATTACHMENT MODELS

A directed preferential attachment model was introduced in [V1, Section 8.9]. It is known that the degrees obey a power law similar to that in [V1, Theorem 8.3]. For the definition of this model and the available results on degree structure, we refer the reader to [V1, Section 8.9]. Unfortunately, the type of properties investigated in the present volume have so far not been analyzed for this random graph model. In particular, there is no description of the strongly connected component, nor of the typical distances and diameters of this model.

One can also interpret normal preferential attachment models as directed graphs by orienting edges from young to old. This can be a useful perspective, for example when modeling temporal networks in which younger vertices can connect only to older vertices, such as in citation networks (recall Section 9.1.1). The connectivity structure of such directed versions is not particularly interesting. For example, the strongly connected component is always small (see Exercise 9.24). Below, we discuss the PageRank of this model.

PageRank of Directed Preferential Attachment Models

We close this section by discussing a very attractive result about PageRank on preferential attachment models. Recall the directed version of the preferential attachment model introduced above and the definition of the PageRank vector $(R_v^{(G_n)})_{v\in V(G_n)}$ in Section 9.2. Theorem 9.1 shows that the PageRank of a uniform vertex converges. The next theorem describes the power-law structure of the limiting PageRank:

Theorem 9.9 (Power-law PageRank distribution of directed PAM)   *Let $(R_v^{(G_n)})_{v\in V(G_n)}$ be the PageRank vector with damping factor α of the directed preferential attachment model G_n with $\delta \geq 0$ and $m \geq 1$, where edges in the normal preferential attachment model are directed from young to old. Let $R_\varnothing$ be the limiting distribution of the PageRank $R_{o_n}^{(G_n)}$ of a uniform vertex, as derived in Theorem 9.1. Then there exist constants $0 < c_1 \leq c_2 < \infty$ such that, for any $r \geq 1$,*

$$c_1 r^{-(2+\delta/m)/(1+(m+\delta)\alpha/m)} \leq \mu(R_\varnothing > r) \leq c_2 r^{-(2+\delta/m)/(1+(m+\delta)\alpha/m)}, \qquad (9.2.48)$$

where μ is the law of the local limit of this directed preferential attachment model.

Theorem 9.9 implies that the PageRank power-law hypothesis, as explained in [V1, Section 1.5] and restated in Section 9.2, is *false* in general. Indeed, the PageRank distribution obeys a power law, as formulated above, with exponent $\tau^{(\mathrm{PR})} = 1 + (2 + \delta/m)/(1 + (m + \delta)\alpha/m)$, while the in-degree obeys a power law with exponent $\tau = 3 + \delta/m$. Note that $\tau^{(\mathrm{PR})} \to 1/\alpha$ for $\delta \to \infty$, while $\tau^{(\mathrm{in})} = \tau = 3 + \delta/m \to \infty$ for $\delta \to \infty$. Thus, the power-law exponent of the directed preferential attachment PageRank remains *uniformly bounded* independently of δ, while that of the in-degree distribution grows infinitely large. This suggests that the PageRank distribution could have power-law tails even for random graphs with *thin-tailed* in-degree distributions.

Since the PageRank distribution obeys a power law, it is of interest to investigate the *maximal* PageRank in a network of size n. The theorem below gives a result for the very first vertex:

Theorem 9.10 (PageRank of first vertex in a directed preferential attachment tree)   *Let $(R_v^{(G_n)})_{v\in V(G_n)}$ be the PageRank vector with damping factor α of the directed preferential attachment tree G_n with $\delta \geq 0$ and $m = 1$, defined above. Then there exists a limiting random variable R such that*

$$n^{-(1+(1+\delta)\alpha)/(2+\delta)} R_1^{(G_n)} \xrightarrow{a.s.} R. \qquad (9.2.49)$$

Theorem 9.10 shows that the PageRank of vertex 1 has the same order of magnitude as the maximum of n random variables with power-law exponent $\tau^{(\mathrm{PR})} = 1 + (2 + \delta/m)/(1 + (m + \delta)\alpha/m)$ would have. It would be of interest to extend Theorem 9.10 to other values of m, as well as to the maximal PageRank $\max_{v\in[n]} R_v^{(G_n)}$.

9.3 RANDOM GRAPHS WITH COMMUNITY STRUCTURE: GLOBAL COMMUNITIES

Many real-world networks have communities that are global in size. For example, when dividing science into its core fields, citation networks have just such a global community structure, as discussed in Section 9.1.1. In Belgian telecommunication networks of who calls whom, the division into the French and the Flemish speaking parts is clearly visible, Blondel et al. (2008), while in US politics the division into Republicans and Democrats plays a pronounced effect on the network structure of social interactions between politicians, Mucha et al. (2010).

In this section we discuss random graph models for networks with a global community structure. The section is organized as follows. In Section 9.3.1 we discuss stochastic block models, which are the models of choice for networks with community structures. In Section 9.3.2 we consider degree-corrected stochastic block models, which are similar to stochastic block models but allow for more pronounced inhomogeneity in the degree structure. In Sections 9.3.3 and 9.3.4, we study configuration models and preferential attachment models with global communities, respectively. We introduce the models, state the most important results in them, and also discuss the topic of *community detection* in such models, a topic that has attracted considerable attention owing to its practical importance.

9.3.1 STOCHASTIC BLOCK MODEL

We have already encountered stochastic block models as inhomogeneous random graphs with finitely many types in Chapter 3. Here we repeat the definition, after which we discuss the extremely interesting and challenging community detection results.

Fix $t \geq 2$ and suppose we have a graph with t different types of vertices. Let $\mathcal{S} = [t]$. Let n_s denote the number of vertices of type s, and let $\mu_n(s) = n_s/n$. Let $\mathrm{IRG}_n(\kappa)$ be the random graph where two vertices of types s and r, respectively, are joined by an edge with probability $n^{-1}\kappa(s,r)$ (for $n \geq \max_{s,r \in [t]} \kappa(s,r)$). Then κ is equivalent to a $t \times t$ matrix, and the random graph $\mathrm{IRG}_n(\kappa)$ has vertices of t different types. We assume that the type distribution μ_n satisfies, for all $s \in [t]$,

$$\lim_{n \to \infty} \mu_n(s) = \lim_{n \to \infty} n_s/n = \mu(s). \tag{9.3.1}$$

Exercise 3.11 then shows that the resulting random graph is graphical as in Definition 3.3(a), so that the results in Chapters 3 and 6 apply. As a result, we will not spend much time on the degree distribution and the giant and graph distances in this model, as they were addressed there. Exercise 9.25 elaborates on the degree structure, while Exercise 9.26 investigates further the conditions for a giant to exist.

Let us mention that, for the stochastic block model to be a good model for networks with a global community structure, one would expect that the edge probabilities of the *internal* edges between vertices of the same type are larger than those of the *external* edges between vertices of different types. In terms of formulas, this means that $\kappa(s,s) > \kappa(s,r)$ for all $s, r \in \mathcal{S} = [t]$ with $s \neq r$. For example, the bipartite Erdős–Rényi random graph has a structure that is quite opposite to a random graph with global communities (as vertices only have neighbors of a different type).

Community Detection in Stochastic Block Models

We next discuss the topic of community detection in stochastic block models. Before we can say anything about when it is possible to detect communities, we must first define what this means. A *community detection algorithm* is an assignment $\hat{\sigma}: [n] \mapsto [t]$, where $\hat{\sigma}(v) = s$ means that the algorithm assigns type s to vertex v. In what follows, we assume that the communities have equal size. Then, in a random guess of group membership, a vertex is guessed to be of the correct type with probability $1/t$. As a result, we are impressed with the performance of a community detection algorithm only when it does far better than random guessing. This explains the following definition:

Definition 9.11 (Solvable community detection)   Consider a stochastic block model where there are the same number of vertices of each of the r types, and where $\sigma(v)$ denotes the *type* of vertex $v \in [n]$. We call a community detection problem *solvable* when there exists an algorithm $\hat{\sigma}: [n] \mapsto [t]$ and an $\varepsilon > 0$ such that, whp as $n \to \infty$,

$$\max_{p: \, [t] \to [t]} \frac{1}{n} \sum_{v \in [n]} \left[\mathbb{1}_{\{\hat{\sigma}(v) = (p \circ \sigma)(v)\}} - \frac{1}{t} \right] \geq \varepsilon, \qquad (9.3.2)$$

where the maximum is over all possible permutations p from $[t]$ to $[t]$. If such an algorithm does not exist then we call the problem *unsolvable*.   ◀

The maximum over permutations of the types in (9.3.2) is due to the fact that the type labels generally have no meaning in real-world networks, so that they can be permuted without changing anything. Exercise 9.27 shows that (9.3.2) is indeed false for random guessing.

Community detection is the most difficult when the degree distributions of vertices of all the different types are the same. This is not surprising, as otherwise one may aim to classify on the basis of the degrees of the graph. As a result, from now on, we assume that the expected degrees of all types of vertices are the same. Some ideas about how one can prove that the problem is solvable for unequal expected degrees can be obtained from Exercises 9.28 and 9.29.

We start by considering the case where there are just two types, so that we can take the edge probability p_{uv} to be a/n for vertices of the same type, and b/n for vertices of opposite types. Here we think of $a > b$. The question whether community detection is solvable is answered in the following theorem:

Theorem 9.12 (Stochastic block model threshold)   *Take n to be even. Consider a stochastic block model of two types, each having $n/2$ vertices, where the edge probability p_{uv} is a/n for vertices of the same type, and b/n for vertices of opposite types, where $a > b$. Then, the community detection problem is* solvable *as in Definition 9.11 when*

$$\frac{(a-b)^2}{2(a+b)} > 1, \qquad (9.3.3)$$

while it is unsolvable *when*

$$\frac{(a-b)^2}{2(a+b)} < 1. \qquad (9.3.4)$$

Theorem 9.12 is quite surprising. Indeed, it shows that not only should $a > b$ in order to have a chance to perform community detection but it also should be sufficiently large compared to $a + b$. Further, the transition in Theorem 9.12 is sharp, in the sense that (9.3.3) and (9.3.4) complement each other. It is unclear what happens in the critical case when $(a - b)^2 = 2(a + b)$. The solvable case in (9.3.3) is sometimes called an "achievability result," the unsolvable case in (9.3.4) an "impossibility result." We do not give the full proof of Theorem 9.12, as this is quite involved. The proof of the solvable case also shows that the proportion of pairs of vertices that are correctly classified to be of the same type converges to 1 when $(a - b)^2/[2(a + b)]$ grows large.

In Exercise 9.30, the reader is asked to show that (9.3.3) implies that $a - b > 2$ (and thus $a + b > 2$), and to conclude that a giant thus exists in this setting.

While the results for a general number of types r are less complete, there is an achievability result when $p_{uv} = a/n$ for vertices of the same type, and $p_{uv} = b/n$ for all vertices of different types, in which (9.3.3) is replaced by

$$\frac{(a - b)^2}{t(a + (t - 1)b)} > 1, \tag{9.3.5}$$

which indeed reduces to (9.3.3) for $t = 2$. Also, many results exist about whether efficient algorithms for community detection exist. In general, this means that not only should a detection algorithm exist that achieves (9.3.2), but it should also be computable in reasonable time (say $\Theta(n \log n)$ for fixed t). We refer to Section 9.6 for a more elaborate discussion on such results.

Let us continue this subsection by explaining how thresholds such as (9.3.3) and (9.3.5) can be interpreted. Interestingly, there is a close connection with multi-type branching processes. Consider a branching process with finitely many types. Kesten and Stigum (1966) asked in this context when it would be possible to estimate the type of the root while observing the types of the vertices in generation k for very large k. In this case, the expected offspring matrix equals $M_{s,r} = \kappa(s, r)\mu(r)$, which is a $t \times t$ matrix. Let $\lambda_1 > \lambda_2$ be the two largest eigenvalues of $\mathbf{M}$. Then, the Kesten–Stigum criterion is that estimation of the root type is possible with probability strictly larger than $1/t$ when

$$\frac{\lambda_2^2}{\lambda_1} > 1. \tag{9.3.6}$$

Next, consider a general finite-type inhomogeneous random graph, with limiting type distribution $\mu(s)$ and expected offspring matrix $M_{s,r} = \kappa(s, r)\mu(r)$. Obviously, the local limit of the stochastic block model is the above multi-type branching process, so a link between the two detection problems can indeed be expected. Under the condition in (9.3.6), it is believed that the community detection problem is solvable and even that communities can be detected in polynomial time. For $t = 2$, this is sharp, as we have seen above. For $t \geq 3$, the picture is much more involved. It is believed that, for $t \geq 4$, a double phase transition occurs: detection should be possible in polynomial time when $\lambda_2^2/\lambda_1 > 1$, much harder but still possible (i.e., the best algorithms take a time that is exponentially long in the size of the network) when $\lambda_2^2/\lambda_1 > c^\star$ for some $0 < c^\star < 1$, and information-theoretically impossible when $\lambda_2^2/\lambda_1 < c^\star$. However, this is not yet known in the general case.

The way to get from a condition like (9.3.6) to an algorithm for community detection is by using the two largest eigenvalues of the so-called *non-backtracking matrix* of the random graph defined below, and to obtain an estimate for the partition by using the eigenvectors corresponding to these eigenvectors. The leading eigenvalue converges to λ_1 in probability, while the second is bounded by $|\lambda_2|$. This, together with a good approximation to the corresponding eigenvectors, suggests a specific estimation procedure that we explain now.

Let $\mathbf{B}$ be the non-backtracking matrix of the graph G. This means that $\mathbf{B}$ is indexed by the oriented edges $\vec{E}(G) = \{(u,v) \colon \{u,v\} \in E(G)\}$, so that $\mathbf{B} = (B_{e,f})_{e,f \in \vec{E}(G)}$. For an edge $e \in \vec{E}(G)$, denote $e = (e_1, e_2)$, and write

$$B_{e,f} = \mathbb{1}_{\{e_2 = f_1, e_1 \neq f_2\}}, \qquad (9.3.7)$$

which indicates that e ends in the vertex in which f starts, but e is not the reversal of f. The latter property explains the name *non-backtracking* matrix.

Now we come to the eigenvalues. We restrict ourselves to the case where $t = 2$, even though some results extend with modifications to higher values of r. Let $\lambda_1(\mathbf{B})$ and $\lambda_2(\mathbf{B})$ denote the two leading eigenvalues of $\mathbf{B}$. Then, for the stochastic block model,

$$\lambda_1(\mathbf{B}) \xrightarrow{\mathbb{P}} \lambda_1, \qquad \lambda_2(\mathbf{B}) \xrightarrow{\mathbb{P}} \lambda_2, \qquad (9.3.8)$$

where we recall that $\lambda_1 > \lambda_2$ are the two largest eigenvalues of $\mathbf{M}$, where $M_{s,r} = \kappa(s,r)\mu(r)$. It turns out that for the Erdős–Rényi random graph with edge probability $(a+b)/(2n) \equiv \alpha/n$, the first eigenvalue $\lambda_1(\mathbf{B})$ satisfies $\lambda_1(\mathbf{B}) \xrightarrow{\mathbb{P}} \lambda_1 = \alpha$, while the second eigenvalue $\lambda_2(\mathbf{B})$ satisfies $\lambda_2(\mathbf{B}) \leq \sqrt{\alpha} + o_{\mathbb{P}}(1)$. Note that this does not follow from (9.3.8), since $\mathbf{M}$ is a 1×1 matrix. For the stochastic block model with $t = 2$ instead, $\lambda_2(\mathbf{B}) \xrightarrow{\mathbb{P}} \lambda_2 = (a-b)/2$. Thus, we can expect that the graph is a stochastic block model when

$$\frac{\lambda_2(\mathbf{B})^2}{\lambda_1(\mathbf{B})} > 1, \qquad (9.3.9)$$

while if the reverse inequality holds then we are not even sure whether the model is an Erdős–Rényi random graph or a stochastic block model. In the latter case the graph is so random and homogeneously distributed that we are not able to make a good estimate of the types of the vertices, which strongly suggests that this case is unsolvable. This at least informally explains (9.3.6).

Finally, we explain how the above analysis of eigenvalues can be used to estimate the types. Assume that $\lambda_2^2/\lambda_1 > 1$. Let $\xi_2(\mathbf{B}) \colon \vec{E}(G) \to \mathbb{R}$ denote the normalized eigenvector corresponding to $\lambda_2(\mathbf{B})$. We fix a constant $\theta > 0$. Then, we estimate that $\hat{\sigma}(v) = 1$ when

$$\sum_{e \colon e_2 = v} \xi_k(e) \geq \frac{\theta}{\sqrt{n}} \qquad (9.3.10)$$

and otherwise that $\hat{\sigma}(v) = 2$ for some deterministic threshold θ. This estimation can then be shown to achieve (9.3.2) owing to the sufficient separation of the eigenvalues.

9.3.2 Degree-Corrected Stochastic Block Model

While the stochastic block model is a nice model for networks with communities, it has degrees that have Poisson tails (recall Theorem 3.4). Thus, in order to account for the abundant inhomogeneities present in real-world networks, an adaptation of the stochastic block model has been proposed in which the degrees are more flexible. This is called the *degree-corrected stochastic block model* and is an inhomogeneous random graph that takes features of rank-1 inhomogeneous random graphs as well as of stochastic block models. Just like the rank-1 setting, there are various possible versions of the model. Here we stick to the version for which the strongest community detection results have been proved.

For each vertex v we sample a random vertex weight X_v, where we assume that $(X_v)_{v \in [n]}$ are iid. Conditional on $(X_v)_{v \in [n]}$, we then assume that the edge between vertices u and v is independently present with probability

$$p_{uv} = \left(\kappa(\sigma(u), \sigma(v)) \frac{X_u X_v}{n} \right) \wedge 1, \qquad (9.3.11)$$

with $\sigma(u) \in [t]$ the type of vertex u and $\kappa \colon [t] \times [t] \to [0, \infty)$ the type kernel.

Let us first discuss this setting in the simplest case where $t = 1$. When the weights $(x_v)_{v \in [n]}$ in (9.3.11) are fixed we could take them as $x_v = w_v \sqrt{n/\ell_n}$, where $\ell_n = \sum_{v \in [n]} w_v$, and obtain the Chung–Lu model $CL_n(\boldsymbol{w})$. This intuition is helpful in what follows. Unfortunately, if $(w_v)_{v \in [n]}$ are iid and $x_v = w_v \sqrt{n/\ell_n}$ then $(x_v)_{v \in [n]}$ are not exactly iid, so it is not obvious how to rigorously transfer results between the settings. As a result, we will stick to the setting in (9.3.11).

From now on, we assume that there are t types of vertices, each of which occurs roughly (or precisely, depending on the setting) equally often. We also assume that $\kappa(\sigma, \sigma')$ takes on two values: $\kappa(s, s) = a$, and $\kappa(s, r) = b$ when $s \neq r$. This leads us to a model very similar to the stochastic block model, except that the weight structure $(X_v)_{v \in [n]}$ adds some additional inhomogeneity to the vertex roles, according to which vertices with high weights generally have larger degrees than those with small weights.

Exercise 9.32 and 9.33 study the degree structure of the degree-corrected stochastic block model, while Exercise 9.34 investigates conditions for a giant to exist in this model.

Community Detection in Degree-Corrected Stochastic Block Models

We now come to the main result of this section, which involves the solvability of the estimation in degree-corrected stochastic block models:

Theorem 9.13 (Degree-corrected stochastic block model threshold) *Take n to be even. Consider a stochastic block model of two types, each having $n/2$ vertices, where the edge probabilities are given by (9.3.11), with $\kappa(s, s) = a$, and $\kappa(s, r) = b$ for $s \neq r$. Without loss of generality, assume that $\mathbb{E}[X] = 1$. The community detection problem is* unsolvable *as in Definition 9.11 when*

$$\frac{(a - b)^2 \mathbb{E}[X^2]}{2(a + b)} < 1. \qquad (9.3.12)$$

Assume further that there exists a $\beta > 8$ such that

$$\mathbb{P}(X > x) \leq \frac{1}{x^\beta}.$$ (9.3.13)

Then the community detection problem is solvable *as in Definition 9.11 if*

$$\frac{(a-b)^2\mathbb{E}[X^2]}{2(a+b)} > 1.$$ (9.3.14)

The impossibility result in (9.3.12) in Theorem 9.13 is extended to all $t \geq 2$ under the condition that $(a-b)^2\mathbb{E}[X^2] < t(a+b)$. The crux of the proof is to show that, for two vertices o_1, o_2 chosen uar, and with $G_n = ([n], E(G_n))$, the realization of the graph of the degree-corrected stochastic block model, we have

$$\mathbb{P}(\sigma(o_1) = s \mid \sigma(o_2), G_n) \xrightarrow{\mathbb{P}} \frac{1}{t}$$ (9.3.15)

for every $s \in [t]$. Thus, the type of o_2 gives us, asymptotically, no information about the type of o_1. This should make detection quite hard, and thus intuitively explains (9.3.12). The proof of the achievability result follows a spectral argument similar to that of the ordinary stochastic block model considered in Section 9.3.1, and we refrain from discussing it further here. We can expect the power-law bound in (9.3.13) to be too strict, in fact, and the results to extend to slightly milder assumptions.

9.3.3 CONFIGURATION MODELS WITH GLOBAL COMMUNITIES

The stochastic block model is an adaptation of the Erdős–Rényi random graph that incorporates global communities, and the degree-corrected stochastic block model is an adaptation of the Chung–Lu model. In a similar way, one can adapt the configuration model to incorporate global communities. Surprisingly, this has not attracted substantial attention in the literature, which is why this section is relatively short. We will discuss an obvious setting in which we let every vertex $v \in [n]$ have a type $\sigma(v) \in [t]$.

For a vertex $v \in [n]$ and a type $s \in [t]$, we let $d_v^{(s)}$ denote the number of half-edges to be connected from vertex v to vertices of type s. We first specify the structure of the graph between vertices of the same type. Let $\sum_{v: \sigma(v)=s} d_v^{(s)}$ be the total number of half-edges between vertices of type s, and assume that this number is even. We also assume that the graph on $\{v: \sigma(v) = s\}$ is a configuration model with $n_s = \#\{v \in [n]: \sigma(v) = s\}$ vertices and degrees $(d_v^{(s)})_{v: \sigma(v)=s}$.

For the structure of the edges between vertices of different types, we recall the *bipartite configuration model*. We suppose that we have vertices of two types, say types 1 and 2, and that there are n_1 and n_2 vertices of these two types, respectively. Let the type of vertex v again be given by $\sigma(v) \in \{1, 2\}$. Vertices of type 1 have degrees $(d_v)_{v: \sigma(v)=1}$ and vertices of type 2 have degrees $(d_v)_{v: \sigma(v)=2}$. Assuming that

$$\sum_{v: \sigma(v)=1} d_v = \sum_{v: \sigma(v)=2} d_v,$$ (9.3.16)

we pair the half-edges incident to vertices of type 1 uar to those incident to vertices of type 2, without replacement. As a result, there are edges only between vertices of types 1 and 2, and the total number of edges is given in (9.3.16).

Using the above definition, we let the edges between vertices of types s, r be given by a bipartite configuration model between the vertices in $\{v \colon \sigma(v) = s\}$ and $\{v \colon \sigma(v) = r\}$, where the former have degrees $(d_v^{(r)})_{v \colon \sigma(v)=s}$ and the latter have degrees $(d_v^{(s)})_{v \colon \sigma(v)=r}$. To make the construction feasible, we assume that

$$\sum_{v \colon \sigma(v)=s} d_v^{(r)} = \sum_{v \colon \sigma(v)=r} d_v^{(s)} \qquad \text{for all } s, r \in [t]. \tag{9.3.17}$$

Special cases of this model are the configuration model, for which $t = 1$, and the bipartite configuration model itself, for which $t = 2$ and $d_v^{(r)} = 0$ for every v with $\sigma(v) = r$.

Let $\mu_n(s)$ denote the proportion of vertices of type s. We again assume, as in (9.3.1) that the type distribution $\mu_n(s) = n_s/n$ satisfies that, for all $s \in [t]$,

$$\lim_{n \to \infty} \mu_n(s) = \lim_{n \to \infty} n_s/n = \mu(s). \tag{9.3.18}$$

Also, in order to describe the local and global properties of the configuration model with global communities, one should make assumptions similar to those for the original configuration model in Condition 1.7 but now for the *matrix* of degree distributions. For example, it is natural to assume that, for all $s \in [t]$, the joint distribution function of all the type degrees satisfies

$$F_n^{(s)}(x_1, \dots, x_t) = \frac{1}{n_s} \sum_{v \colon \sigma(v)=s} \mathbb{1}_{\{d_v^{(1)} \le x_1, \dots, d_v^{(t)} \le x_t\}} \to F^{(s)}(x_1, \dots, x_t), \tag{9.3.19}$$

for all $x_1, \dots, x_t \in \mathbb{R}$ and some limiting joint distribution $F^{(s)} \colon \mathbb{R}^t \to [0, 1]$. Further, it is natural to assume that an adaptation of Condition 1.7(b) holds for all these degrees, such that, for all $s, r \in [t]$, we have

$$\frac{1}{n_s} \sum_{v \colon \sigma(v)=s} d_v^{(r)} \to \mathbb{E}[D^{(s,r)}], \tag{9.3.20}$$

where $D^{(s,r)}$ is the rth coordinate of the random vector whose distribution function is given by $F^{(s)}$, i.e.,

$$\mathbb{P}(D^{(s,r)} \le x_r) = \lim_{x_1, \dots, x_{r-1}, x_{r+1}, \dots, x_t \to \infty} F^{(s)}(x_1, \dots, x_t). \tag{9.3.21}$$

While the configuration model, as well as its bipartite version, has attracted substantial attention, the above extension has not. Exercises 9.35–9.37 informally investigate some of properties of this extension.

9.3.4 PREFERENTIAL ATTACHMENT MODELS WITH GLOBAL COMMUNITIES

The preferential attachment model with global communities is naturally a dynamic random graph model, where now every vertex n has a type $\sigma(n) \in [t]$ with t the total number of communities. The graph is considered to be *directed*, with all edges directed from young to old. Each vertex comes in with out-degree equal to m, as in the normal preferential attachment model. We will see that the extension studied in the literature is most similar to

$(\mathrm{PA}_n^{(m,0)}(b))_{n\geq0}$, but later on an extension to $(\mathrm{PA}_n^{(m,\delta)}(b))_{n\geq0}$ will be discussed as well. In a similar way, extensions to $(\mathrm{PA}_n^{(m,\delta)}(a))_{n\geq0}$ can be formulated.

We start with an initial graph at time n_0 given by G_{n_0}, in which we assume that every vertex has out-degree m and a label $\sigma(v)$ for all $v \in [n_0]$. The graph then evolves as follows. Let the graph G_n at time n be given. At time $n+1$, let vertex $n+1$ have a type $\sigma(n+1)$ that is chosen in an iid way from $[t]$, where

$$\mu(s) = \mathbb{P}(\sigma(n+1) = s) \qquad (9.3.22)$$

is the type distribution. We consider the m edges incident to vertex $n+1$ to be *half-edges*, similarly to the construction of the configuration model. We give all the out-half-edges incident to a vertex v the label $\sigma(v)$. Further, let the matrix $\kappa\colon [t] \times [t] \to [0,\infty)$ be an "affinity" matrix. If $\sigma(n+1) = s$ then give each half-edge of label r a weight $\kappa(s,r)$. Choose a half-edge according to these weights, meaning that a half-edge x with label r has a probability proportional to $\kappa(s,r)$ of being chosen as the pair of any of the m half-edges incident to vertex $n+1$ when $\sigma(n) = s$. We do this for all m half-edges incident to vertex $n+1$ independently. This creates the graph G_{n+1} at time $n+1$. The dynamics is iterated indefinitely.

Degree Distribution of PAMs with Global Communities

Let $\eta_n(s)$ denote the fraction of half-edges with label s, and let $\eta_n = (\eta_n(s))_{s\in[t]}$ denote the empirical distribution of the labels of the half-edges. For a probability distribution η on $[t]$, and a type $s \in [t]$, let

$$h_s(\eta) = \mu(s) + \sum_{r\in[t]} \mu(r) \frac{\kappa(r,s)\eta(s)}{\sum_{r'\in[t]} \kappa(s,r')\eta(r')} - 2\eta(s). \qquad (9.3.23)$$

Then, the half-edge label distribution satisfies

$$\eta_n(s) \xrightarrow{a.s.} \eta^\star(s), \qquad \text{where} \qquad h_s(\eta^\star) = 0 \ \forall s \in [t]. \qquad (9.3.24)$$

The probability distribution $\eta^\star$ that solves $h_s(\eta^\star) = 0$ for all $s \in [t]$ can be shown to be *unique*. We next define the crucial parameters in the model.

For $s, r \in [t]$, let

$$\theta^\star(s, r) = \frac{\kappa(s,r)}{\sum_{r'\in[t]} \kappa(s,r')\eta^\star(r')}, \qquad (9.3.25)$$

and write

$$\theta^\star(s) = \sum_{r\in[t]} \mu(r)\theta^\star(s, r). \qquad (9.3.26)$$

We let $n_s = \#\{v\colon \sigma(v) = s\}$ denote the type count. Next, we study the degree distribution in the above preferential attachment model with global communities. For $s \in [r]$, define

$$P_k^{(s)} = \frac{1}{n_s} \sum_{v\in[n]} \mathbb{1}_{\{D_v(n)=k, \sigma(v)=s\}} \qquad (9.3.27)$$

to be the degree distribution of the types in the model, where $D_v(n)$ denotes the degree of vertex v at time n and n_s equals the number of vertices of type $s \in [t]$. The main result on the degree distribution is as follows:

Theorem 9.14 (Degrees in preferential attachment models with global communities) *In the above preferential attachment models with global communities, for every $s \in [r]$,*

$$P_k^{(s)}(n) \xrightarrow{a.s.} p_k(\theta^\star(s)), \tag{9.3.28}$$

where $\theta^\star(s)$ is defined in (9.3.26) and where, for every $\theta > 0$, we write

$$p_k(\theta) = \frac{\Gamma(m + 1/\theta)}{\theta \Gamma(m)} \frac{\Gamma(k)}{\Gamma(k + 1 + 1/\theta)}. \tag{9.3.29}$$

Exercise 9.38 shows that the limiting distribution in (9.3.29) is indeed a probability distribution. Theorem 9.14 shows that the degree distribution has a power-law tail, as can be expected from the fact that the preferential attachment mechanism is intrinsically present. Moreover, (9.3.28) also shows that the degrees of vertices of type s satisfy a power law with exponent that depends sensitively on the type through the key parameters $(\theta^\star(s))_{s \in [t]}$. Exercise 9.39 shows that the global degree distribution also converges, as can be expected by the convergence of the degree distribution of the types. Further, Exercise 9.40 shows that the global degree distribution has a power-law tail with exponent $\tau = 1 + 1/\max_{s \in [r]} \theta^\star(s)$, provided that $\mu(s^\star) > 0$ for at least one $s^\star \in [t]$ satisfying $\theta^\star(s^\star) = \max_{s \in [r]} \theta^\star(s)$.

Related properties of preferential attachment models with global communities seem not to have been investigated, so we will move to the community detection problem.

Community Detection in PAMs with Global Communities

It is important to discuss what we assume to be known in a community detection problem. We think of the graph as being *directed*, i.e., that the edges are directed from young to old. Further, we assume that m and the probability distribution $(\mu(s))_{s \in [t]}$ are known. Finally, and probably most importantly, we assume that the labels or ages of the vertices in the graph are known. Some of these parameters can be estimated (m being the easiest). While for stochastic block models it is clear that the case where the communities are all equally large and have the same inter- and intra-community edge probabilities is the most challenging, this is not obvious in the present case. However, to mimic the stochastic block model setting, one can keep in mind the case where $\mu(s) = 1/t$ as a key example. Also, the setting where $\kappa(s, s) = a$ for all $s \in [t]$ and $\kappa(s, r) = b$ for all $s, r \in [t]$ with $s \neq r$ is particularly interesting; here, to model community structure, to take $a > b$ is natural. By scaling invariance, we may assume that $b = 1$. In this case, $\eta^\star(s) = 1/t$ for all $s \in [t]$, $\theta^\star(s, r) = ar/[2(a+t-1)]$ for all $s, r \in [t]$ with $s \neq r$, while $\theta^\star(s, s) = a/[2(a+t-1)]$. Also, $\theta^\star(s) = \frac{1}{2}$ for all $s \in [t]$.

The aim is to estimate the type $\sigma(v)$ for all $v \in [n]$ from the above information. This can be done by several algorithms. One algorithm performs an estimation of the label of v on the basis of the neighbors of v. A second algorithm does it on the basis of the degree of v at time n. Let Err_n denote the fraction of errors in the above algorithms. It can be shown that

$$\mathrm{Err}_n \xrightarrow{\mathbb{P}} \mathrm{Err}, \tag{9.3.30}$$

for some limiting constant Err depending on the algorithm. The precise form of Err is known but it is difficult to obtain rigorously as it relies on a continuous-time branching-process approximation of the graph evolution. In particular, in the setting where $\mu(s) = 1/t$, $\kappa(s, s) = a > 1$ for all $s \in [t]$, and $\kappa(s, r) = 1$ for all $s, r \in [t]$ with $s \neq r$, it is unclear whether Err $< (t-1)/t$.

Many more detailed result can be proved, for example that the probability that a vertex label of vertex v is estimated wrongly converges uniformly for all $v \in [n] \setminus [\delta n]$ for any $\delta > 0$. Also, there exists an algorithm that estimates $\sigma(v)$ correctly whp provided that $v = o(n)$. We refrain from discussing such results further.

9.4 RANDOM GRAPHS WITH COMMUNITY STRUCTURE: LOCAL COMMUNITIES

In the previous section we investigated settings where the models have a *finite* number of communities, making the communities *global*. This setting is realiztic when we would like to partition a network of choice into a finite number of parts, for example corresponding to the main scientific fields in citation or collaboration networks, or the continents in the Internet. However, in many other settings this is not realiztic. Indeed, most communities of social networks correspond to smaller entities, such as school classes, families, sports teams, etc. In most real-world settings, it is not even clear what communities look like. As a result, *community detection* has become an art.

The topic is relevant, since most models (including the models with global community structure from Section 9.4) have rather low clustering. For example, consider a general inhomogeneous random graph $\text{IRG}_n(\kappa_n)$ with kernel κ_n. Assume that $\kappa_n(x, y) \leq n$. Then, the expected number of triangles in an $\text{IRG}_n(\kappa_n)$ is close to

$$\mathbb{E}[\# \text{ triangles in } \text{IRG}_n(\kappa_n)] = \frac{1}{6n^3} \sum_{i,j,k \in [n]} \kappa_n(x_i, x_j)\kappa_n(x_j, x_k)\kappa_n(x_k, x_i), \quad (9.4.1)$$

Under relatively weak conditions on the kernel κ_n, it follows that

$$\mathbb{E}[\# \text{ triangles in } \text{IRG}_n(\kappa_n)]$$
$$\to \frac{1}{6} \int_{\mathcal{S}^3} \kappa(x_1, x_2)\kappa(x_2, x_3)\kappa(x_3, x_1)\mu(dx_1)\mu(dx_2)\mu(dx_3). \quad (9.4.2)$$

Therefore, the clustering coefficient converges to zero at speed $1/n$. In many real-world networks, particularly in social networks, the clustering coefficient is strictly positive. See Figure 9.9 for the clustering coefficients in the KONECT data base.

In this section we discuss random graph models in which most communities are quite small, in that the average community size is bounded. However, since the community sizes can also be quite large, we might call them *mesoscopic* rather than *microscopic*. See Figures 9.10 and 9.11 for some empirical data on real-world networks. Figure 9.10 shows that there is enormous variability in the tail probabilities of the degree distributions, community-size distribution, and inter-community degrees in real-world networks. Figure 9.11 shows that the edge-densities of communities generally decrease with their sizes, where the edge density of

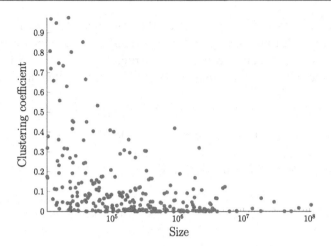

Figure 9.9 Clustering coefficients in the 727 networks of size larger than 10,000 from the KONECT data base.

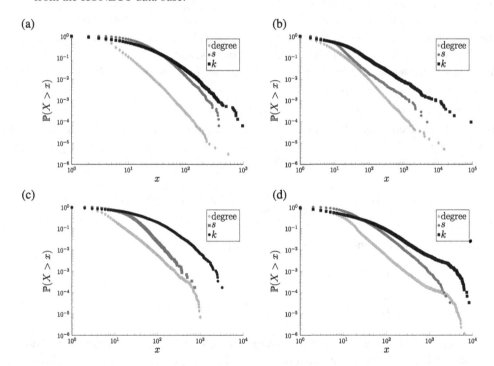

Figure 9.10 Tail probabilities of the degrees, community sizes s, and inter-community degrees k in real-world networks. (a) AMAZON co-purchasing network. (b) GOWALLA social network. (c) English word relations. (d) GOOGLE web graph. Figures taken from Stegehuis et al. (2016b).

a community of size s is given by $2e_{\text{in}}/[s(s-1)]$, with e_{in} the number of edges within the community. Here, the communities were extracted (or detected) using the so-called Louvain method.

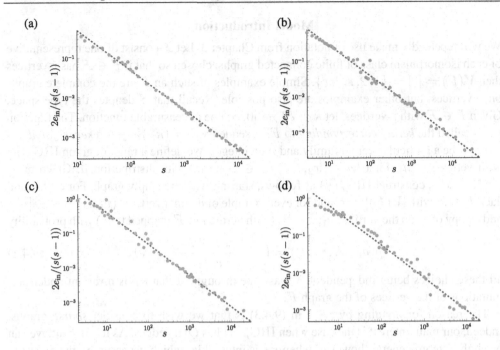

Figure 9.11 The relation between the community edge density $2e_{in}/(s^2 - s)$ and the community size s can be approximated by a power law. (a) AMAZON co-purchasing network, (b) GOWALLA social network, (c) English word relations, (d) GOOGLE web graph. Figures taken from Stegehuis et al. (2016b).

We conclude that we need to be flexible in our community structure to be able to real-iztically model the community structure in real-world networks. Below, we consider several models that attempt to do so. In Section 9.4.1 we start by discussing inhomogeneous random graphs with community structures. We continue in Section 9.4.2 by describing the hierarchical configuration model as well as some close cousins; this is followed by a discussion of random intersection graphs in Section 9.4.3 and exponential random graphs in Section 9.4.4.

9.4.1 INHOMOGENEOUS RANDOM GRAPHS WITH LOCAL COMMUNITIES

In this subsection we discuss a model, similar to the inhomogeneous random graph $IRG_n(\kappa_n)$, that incorporates clustering. The idea behind this model is that instead of only adding *edges* independently, we can also add other graphs on r vertices in an independent way. For example, we could study a graph where each pair of vertices is independently connected with probability λ/n, as for $ER_n(\lambda/n)$, but also each collection of triples forms a triangle with probability μ/n^2, independently for all triplets and independently of the status of the edges. Here the exponent $1/n^2$ is chosen so as to make the expected number of triangles containing a vertex bounded. See Exercise 9.41 for the clustering in such random graphs. In social networks, complete graphs of size 4, 5, etc., are present more often than in the usual random graph. Therefore, we also wish to add those independently.

Model Introduction

We will repeatedly make use of notation from Chapter 3. Let $\mathcal{F}$ consist of one representative of each isomorphism class of finite connected graphs, chosen so that if $F \in \mathcal{F}$ has r vertices then $V(F) = [r] = \{1, 2, \ldots, r\}$. Simple examples of such an F are the complete graphs on r vertices, but other examples are also possible. Recall that $\mathcal{S}$ denotes the type space. Given $F \in \mathcal{F}$ with r vertices, let $\kappa_F \colon \mathcal{S}^r \to [0, \infty)$ be a measurable function. The function κ_F is called the *kernel corresponding to* F. A sequence $\widetilde{\kappa} = (\kappa_F)_{F \in \mathcal{F}}$ is a *kernel family*.

Let $\widetilde{\kappa}$ be a particular kernel family and n an integer. We define a random graph $\mathrm{IRG}_n(\widetilde{\kappa})$ with vertex set $[n]$. First let $x_1, x_2, \ldots, x_n \in \mathcal{S}$ be iid with distribution μ. Given $\boldsymbol{x} = (x_1, \ldots, x_n)$, construct $\mathrm{IRG}_n(\widetilde{\kappa})$ as follows, starting with the empty graph. For each r and each $F \in \mathcal{F}$ with $|V(F)| = r$, and for every r-tuple of distinct vertices $(v_1, \ldots, v_r) \in [n]^r$, add a copy of F on the vertices $v_1, \ldots, v_r$ (with vertex i of F mapped to v_i) with probability

$$p(v_1, \ldots, v_r; F) = \left(\frac{\kappa_F(x_{v_1}, \ldots, x_{v_r})}{n^{r-1}} \right) \wedge 1, \qquad (9.4.3)$$

all these choices being independent. We assume throughout that κ_F is invariant under permutations of the vertices of the graph F.

The reason for dividing by n^{r-1} in (9.4.3) is that we wish to consider *sparse* graphs. Indeed, our main interest is the case when $\mathrm{IRG}_n(\widetilde{\kappa})$ has $O(n)$ edges. As it turns out, we can be slightly more general; however, when κ_F is integrable (which we assume), the expected number of added copies of each graph F is $O(n)$. Below, all incompletely specified integrals are taken with respect to the appropriate r-fold product measure μ^r on $\mathcal{S}^r$.

In the special case where all κ_F are zero apart from κ_{K_2}, the kernel corresponding to an edge, we recover (essentially) a special case of the inhomogeneous random graph model discussed in Chapter 3. In this case, given $\boldsymbol{x}$, two vertices i and j are joined with probability

$$\frac{\kappa_{K_2}(x_i, x_j) + \kappa_{K_2}(x_j, x_i)}{n} + O\left(\frac{(\kappa_{K_2}(x_i, x_j) + \kappa_{K_2}(x_j, x_i))^2}{n^2} \right). \qquad (9.4.4)$$

The correction term will never matter, so we may as well replace κ_{K_2} by its symmetrized version.

For any kernel family $\widetilde{\kappa}$, let κ_e be the corresponding *edge kernel*, defined by

$$\kappa_e(x, y) \qquad\qquad\qquad\qquad\qquad\qquad\qquad\qquad\qquad\qquad\qquad\qquad (9.4.5)$$

$$= \sum_F \sum_{\{i,j\} \in E(F)} \int_{\mathcal{S}^{|V(F) \setminus \{i,j\}|}} \kappa_F(x_1, \ldots, x_{i-1}, x, x_{i+1}, \ldots, x_{j-1}, y, x_{j+1}, \ldots, x_{|V(F)|}),$$

where the second sum runs over all $2|E(F)|$ ordered pairs (i, j) with $\{i, j\} \in E(F)$, and we integrate over all variables apart from x and y. Note that the sum need not always converge. Since every term is positive this causes no problems: we simply allow $\kappa_e(x, y) = \infty$ for some x, y. Given x_i and x_j, the probability that i and j are joined in $\mathrm{IRG}_n(\widetilde{\kappa})$ is at most $\kappa_e(x_i, x_j)/n$. In other words, κ_e captures the *edge probabilities* in $\mathrm{IRG}_n(\widetilde{\kappa})$, but not the correlations.

Number of Edges

Before proceeding to deeper properties, let us note that the expected number of added copies of F is $(1 + o(1))n \int_{\mathcal{S}^{|V(F)|}} \kappa_F$. Unsurprisingly, the actual number turns out to be concentrated about this mean. Let

$$\xi(\widetilde{\kappa}) = \sum_{F \in \mathcal{F}} |E(F)| \int_{S^{|V(F)|}} \kappa_F = \frac{1}{2} \int_{S^2} \kappa_e \leq \infty \qquad (9.4.6)$$

be the *asymptotic edge density* of $\widetilde{\kappa}$. Since every copy of F contributes $|E(F)|$ edges, the following theorem is almost obvious, provided that we can ignore overlapping edges:

Theorem 9.15 (Edge density in $\mathrm{IRG}_n(\widetilde{\kappa})$) *As $n \to \infty$,*

$$\mathbb{E}[|E(\mathrm{IRG}_n(\widetilde{\kappa}))|/n] \to \xi(\widetilde{\kappa}) \leq \infty. \qquad (9.4.7)$$

Moreover, if $\xi(\widetilde{\kappa}) < \infty$ then

$$|E(\mathrm{IRG}_n(\widetilde{\kappa}))|/n \xrightarrow{\mathbb{P}} \xi(\widetilde{\kappa}). \qquad (9.4.8)$$

In other words, Theorem 9.15 states that if $\xi(\widetilde{\kappa}) < \infty$ then $|E(\mathrm{IRG}_n(\widetilde{\kappa}))| = \xi(\widetilde{\kappa})n + o_{\mathbb{P}}(n)$, and if $\xi(\widetilde{\kappa}) = \infty$ then $|E(\mathrm{IRG}_n(\widetilde{\kappa}))| > Cn$ whp for every constant C. We conclude that the model is sparse when $\xi(\widetilde{\kappa}) < \infty$. This is certainly true when κ_F is uniformly bounded, but it may also be true more generally under some integrability assumptions.

Existence of a Giant

We next consider the emergence of the giant component. For this, the linear operator T_{κ_e}, defined by

$$(T_{\kappa_e} f)(x) = \int_S \kappa_e(x, y) f(y) \mu(dy), \qquad (9.4.9)$$

where κ_e is defined in (9.4.5), is crucial. We need to impose some integrability condition on our kernel family:

Definition 9.16 (Integrable and irreducible kernel families)

(a) A kernel family $\widetilde{\kappa} = (\kappa_F)_{F \in \mathcal{F}}$ is *integrable* if

$$\int \widetilde{\kappa} = \sum_{F \in \mathcal{F}} |V(F)| \int_{S^{|V(F)|}} \kappa_F < \infty. \qquad (9.4.10)$$

(b) A kernel family $\widetilde{\kappa} = (\kappa_F)_{F \in \mathcal{F}}$ is *edge integrable* if

$$\sum_{F \in \mathcal{F}} |E(F)| \int_{S^{|V(F)|}} \kappa_F < \infty; \qquad (9.4.11)$$

we write this as $\xi(\kappa) < \infty$. This means that the expected number of edges in $\mathrm{IRG}_n(\widetilde{\kappa})$ is $O(n)$, see Theorem 9.15, and thus the expected degree of a uniform vertex is bounded.

(c) We say that a symmetric edge kernel $\kappa_e \colon S^2 \to [0, \infty)$ is *reducible* if

$$\exists \mathcal{A} \subset S \text{ with } 0 < \mu(\mathcal{A}) < 1 \text{ such that } \kappa_e = 0 \text{ almost everywhere on } \mathcal{A} \times (S \setminus \mathcal{A});$$

otherwise κ_e is *irreducible*. ◀

We are now ready to formulate our main result concerning the phase transition in $\mathrm{IRG}_n(\widetilde{\kappa})$:

Theorem 9.17 (Giant in clustered inhomogeneous random graphs) *Let $\widetilde{\kappa} = (\kappa_F)_{F \in \mathcal{F}}$ be an irreducible, integrable kernel family. Then there exists a $\zeta(\widetilde{\kappa}) \in [0, 1)$ such that*

$$\frac{1}{n}|\mathscr{C}_{\max}| \xrightarrow{\;\mathbb{P}\;} \zeta(\widetilde{\kappa}), \qquad and \qquad \frac{1}{n}|\mathscr{C}_{(2)}| \xrightarrow{\;\mathbb{P}\;} 0. \tag{9.4.12}$$

Theorem 9.17 is proved by showing that the branching process that captures the "local structure" of $\mathrm{IRG}_n(\widetilde{\kappa})$ is a good approximation and indeed describes the existence of the giant. Along the way, it is also proved that $\mathrm{IRG}_n(\widetilde{\kappa})$ converges locally to this "branching process." This is however not a branching process in the usual sense, but instead a branching process that describes the connections *between* the subgraphs that act as communities.

For Theorem 9.17 to be useful, we would like to know something about $\zeta(\widetilde{\kappa})$, which can be calculated from $x \mapsto \zeta_{\widetilde{\kappa}}(x)$, and which, in turn, equals the largest solution to the functional equation

$$f(x) = 1 - e^{-(\mathbf{S}_{\widetilde{\kappa}}f)(x)}, \tag{9.4.13}$$

for some non-linear operator $\mathbf{S}_{\widetilde{\kappa}}$. We can think of $\zeta_{\widetilde{\kappa}}(x)$ as the probability that a vertex of type $x \in \mathcal{S}$ has a "large" connected component. The question of when $\zeta(\widetilde{\kappa}) > 0$ is settled in the following theorem:

Theorem 9.18 (Condition for existence of giant component) *Let $\widetilde{\kappa}$ be an integrable clique kernel. Then $\zeta(\widetilde{\kappa}) > 0$ if and only if $\|T_{\kappa_e}\| > 1$. Furthermore, if $\widetilde{\kappa}$ is irreducible and $\|T_{\kappa_e}\| > 1$, then $\zeta_{\widetilde{\kappa}}(x)$ is the unique non-zero solution to the functional equation (9.4.13), and $\zeta_{\widetilde{\kappa}}(x) > 0$ holds for almost every x.*

In general, $\|T_{\kappa_e}\|$ may be rather hard to calculate. If we suppose that the symmetrized version of $\sum_{F \in \mathcal{F}:\ |V(F)|=r} \kappa_F(x_1, \ldots, x_r)$ is constant for each $r \geq 2$, however, this can be done. Indeed, say that this symmetrized kernel equals c_r. Then $\kappa_e(x, y) = \sum_r r(r-1)c_r = 2\xi(\kappa)$ for all x and y, so that

$$\|T_{\kappa_e}\| = 2\xi(\kappa). \tag{9.4.14}$$

This is perhaps surprising: it tells us that, for such kernels, the critical point where a giant component emerges is determined only by the total number of edges added; the size of the cliques in which they live does not matter, even though, for example, the third edge in every triangle might be "wasted."

9.4.2 Configuration Models with Local Community Structure

In this subsection we investigate several models that are related to the configuration model and yet have a pronounced local community structure. We start by discussing the hierarchical configuration model and its properties. We then look at a particular version that goes under the name of the household model, and we close by describing a model in which triangles are explicitly added.

Hierarchical Configuration model: Model Introduction

Consider the configuration model $\mathrm{CM}_N(d)$ with a degree sequence $d = (d_i)_{i \in [N]}$ satisfying Condition 1.7(a),(b), and now using N rather than n. We replace each of the vertices by a small "local" graph to create a *community* or *household* structure. Thus, vertex i is replaced by a local graph $G_i = (V(G_i), E(G_i))$. We assign each of the d_i half-edges incident to vertex i to a vertex in G_i in an arbitrary way. Thus, vertex i is replaced by a pair consisting of the community graph G_i and the inter-community degrees $d^{(b)} = (d_u^{(b)})_{u \in V(G_i)}$ satisfying $\sum_{u \in V(G_i)} d_u^{(b)} = d_i$. Naturally, the size of the graph becomes $n = \sum_{v \in [N]} |V(G_v)|$.

As a result, we obtain a graph with two levels of hierarchy; its local structure is described by the local graphs $(G_i)_{i \in [N]}$ whereas its global structure is described by the configuration model $\mathrm{CM}_N(d)$. This model is called the *hierarchical configuration model*. A natural assumption is that the degree sequence $d = (d_i)_{i \in [N]}$ satisfies Condition 1.7(a),(b) with n replaced by N, while the empirical distribution of the graphs satisfies, as $N \to \infty$,

$$\mu_n(H, \vec{d}) = \frac{1}{N} \sum_{i \in [N]} \mathbb{1}_{\{G_i = H, (d_u^{(b)})_{u \in V(G_i)} = \vec{d}\}} \to \mu(H, \vec{d}), \tag{9.4.15}$$

for every connected graph H of size $|V(H)|$ and degree vector $\vec{d} = (d_h)_{h \in |V(H)|}$ and some probability distribution μ on graphs with integer marks associated with the vertices. We assume that $\mu_n(H, \vec{d}) = 0$ for all H that are disconnected. Indeed, we think of the graphs $(G_i)_{i \in [N]}$ as describing the local community structure of the graph, so it makes sense to assume that all $(G_i)_{i \in [N]}$ are connected. In particular, (9.4.15) shows that a typical community has bounded size.

We often also make assumptions on the average size of the community of a random vertex. For this, it is necessary to impose that, with $\mu_n(H) = \sum_{\vec{d}} \mu_n(H, \vec{d})$ and $\mu(H) = \sum_{\vec{d}} \mu(H, \vec{d})$ the community distribution,

$$\sum_H |V(H)| \mu_n(H) = \frac{1}{n} \sum_{i \in [N]} |V(G_i)| \to \sum_H |V(H)| \mu(H) < \infty. \tag{9.4.16}$$

Equation (9.4.16) indicates that the community of a random vertex has a tight size, since the community size of a random vertex has a size-biased community distribution (see Exercise 9.43). The degree structure of the hierarchical configuration model is determined by the model description. We next discuss the giant and the distances in the hierarchical configuration model.

Giant in the Hierarchical Configuration Model

The main result concerning the size of the giant is the following theorem:

Theorem 9.19 (Giant in hierarchical configuration model) *Assume that the inter-community degree sequence $d = (d_i)_{i \in [N]}$ satisfies Conditions 1.7(a),(b) with N replacing n and with limit D, while the communities satisfy (9.4.15) and (9.4.16). Then, there exists $\zeta \in [0, 1]$ such that*

$$\frac{1}{n}|\mathscr{C}_{\max}| \xrightarrow{\mathbb{P}} \zeta, \qquad \frac{1}{n}|\mathscr{C}_{(2)}| \xrightarrow{\mathbb{P}} 0. \tag{9.4.17}$$

Write $\nu = \mathbb{E}[D(D-1)]/\mathbb{E}[D]$. Then, $\zeta > 0$ precisely when $\nu > 1$.

Since the communities $(G_i)_{i\in[N]}$ are connected, the sizes of the clusters in the hierarchical configuration model are closely related to those in $\mathrm{CM}_N(\boldsymbol{d})$. Indeed, for $v \in [n]$, let i_v denote the vertex for which $v \in V(G_{i_v})$. Then

$$|\mathscr{C}(v)| = \sum_{i\in\mathscr{C}'(i_v)} |V(G_i)|, \tag{9.4.18}$$

where $\mathscr{C}'(i)$ denotes the connected component of i in $\mathrm{CM}_N(\boldsymbol{d})$. This allows one to move back and forth between the hierarchical configuration model and the corresponding configuration model $\mathrm{CM}_N(\boldsymbol{d})$ that describes the inter-community connections.

It also allows us to identify the limit ζ. Let $\xi \in [0,1]$ be the extinction probability of the local limit of $\mathrm{CM}_N(\boldsymbol{d})$ of a vertex of degree 1, so that a vertex of degree d survives with probability $1 - \xi^d$. Then,

$$\zeta = \sum_H \sum_{\vec{d}} |V(H)|\mu(H,\vec{d})[1 - \xi^d], \tag{9.4.19}$$

where $d = \sum_{v\in V(H)} d_v$. Further, $\xi = 1$ precisely when $\nu \le 1$; see, e.g., Theorem 4.9. This explains the result in Theorem 9.19. In Exercise 9.45, the reader is asked to fill in the details.

Before moving to graph distances, we discuss an example of the hierarchical configuration model that has attracted attention under the name *configuration model with household structure*.

Configuration Model with Household Structure

In the configuration model with household structure, each G_i is a complete graph of size $|V(G_i)|$, while each inter-community degree equals 1, i.e., $d_u^{(b)} = 1$ for all $u \in V(G_i)$. As a result, the degree of a vertex $v \in [n]$ is equal to $|V(G_{i_v})|$, where we recall that i_v is such that $v \in V(G_{i_v})$. This is a rather interesting example, where every vertex is in a "household," and every household member has precisely one connection to the outside. This connects it to another household. In this case, the degree distribution is equal to

$$F_n(x) = \frac{1}{n} \sum_{v\in[n]} \mathbb{1}_{\{|V(G_{i_v})|\le x\}} = \frac{1}{n} \sum_{i\in[N]} |V(G_i)|\mathbb{1}_{\{|V(G_i)|\le x\}}. \tag{9.4.20}$$

As a result, the degree distribution in the household model is the size-biased degree in the configuration model that describes its inter-community structure. In particular, this implies that if the limiting degree distribution D in $\mathrm{CM}_N(\boldsymbol{d})$ is a power law with exponent τ' then the limiting degree distribution in the household model is a power law with exponent $\tau = \tau' - 1$. This is sometimes called a *power-law shift* and is clearly visible in Figure 9.12.

Graph Distances in the Hierarchical Configuration Model

We next turn to the graph distances in the hierarchical configuration model. Again, a link can be made to the distances in the configuration model, particularly when the diameters of the community graphs $(G_i)_{i\in[N]}$ are uniformly bounded. Indeed, in this case, one can expect the graph distances in the hierarchical configuration model to be of the same order of magnitude as in the configuration model $\mathrm{CM}_N(\boldsymbol{d})$.

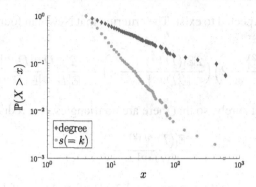

Figure 9.12 Degree distribution of household model follows a power law with a smaller exponent than the community size distribution and outside degree distribution

This link is easiest to formulate for the household model. Indeed, there the diameter of the community graphs is 1 since all community graphs are complete graphs. Denote the hierarchical configuration model by $\text{HCM}_n(\mathbf{G})$. Then, unless $u = v$ or the half-edge incident to u is being paired to that incident to v, $\text{dist}_{\text{HCM}_n(\mathbf{G})}(u, v) = 2 \, \text{dist}_{\text{CM}_N(d)}(i_u, i_v) - 1$, where we recall that i_v is such that $v \in V(G_{i_v})$. Thus, distances in $\text{HCM}_n(\mathbf{G})$ are asymptotically twice as large as those in $\text{CM}_N(d)$. The reason is that paths in the household model alternate between intra-community edges and inter-community edges, because the inter-community degrees are all equal to 1 so there is no way to jump to a vertex using an inter-community edge and leave through an inter-community edge again. This is different from general hierarchical configuration models.

Configuration Model with Clustering

We close this section on adaptations of the configuration model with local communities by introducing a model that has not yet attracted a lot of attention in the mathematical community.

The low clustering of $\text{CM}_n(d)$ can be resolved by introducing households as described above. Alternatively, and in the spirit of clustered inhomogeneous random graphs as described in Section 9.4.1, we can also introduce clustering directly. In the configuration model with clustering, we assign two types of "degrees" to a vertex $v \in [n]$. We let $d_v^{(\text{si})}$ denote the number of simple half-edges incident to vertex v, and we let $d_v^{(\text{tr})}$ denote the number of triangles of which vertex v is part. In this terminology, the degree d_v of a vertex is equal to $d_v = d_v^{(\text{si})} + 2d_v^{(\text{tr})}$. We can then say that there are $d_v^{(\text{si})}$ half-edges incident to vertex v and $d_v^{(\text{tr})}$ "third-triangles" of pairs of half-edges.

The graph is built by (a) recursively choosing two half-edges uar without replacement, and pairing them into edges (as for $\text{CM}_n(d)$), and (b) choosing triples of third-triangles uar and without replacement, and drawing edges between the three vertices incident to the third-triangles that are chosen.

Let $\left(D_n^{(\text{si})}, D_n^{(\text{tr})}\right)$ denote the number of simple edges and triangles incident to a uniform vertex in $[n]$, and assume that $\left(D_n^{(\text{si})}, D_n^{(\text{tr})}\right) \xrightarrow{d} \left(D^{(\text{si})}, D^{(\text{tr})}\right)$ for some limiting distribution $\left(D^{(\text{si})}, D^{(\text{tr})}\right)$. Newman (2009) performed a generating function analysis to investigate when a giant component is expected to exist. The criterion that Newman found is that a giant exists when

$$\left(\frac{\mathbb{E}[(D^{(\mathrm{si})})^2]}{\mathbb{E}[D^{(\mathrm{si})}]} - 2\right)\left(\frac{2\mathbb{E}[(D^{(\mathrm{tr})})^2]}{\mathbb{E}[D^{(\mathrm{tr})}]} - 3\right) < \frac{2\mathbb{E}[D^{(\mathrm{si})}D^{(\mathrm{tr})}]}{\mathbb{E}[D^{(\mathrm{si})}]\mathbb{E}[D^{(\mathrm{tr})}]}. \tag{9.4.21}$$

When $D^{(\mathrm{tr})} = 0$ almost surely, so that there are no triangles, this reduces to

$$\frac{\mathbb{E}[(D^{(\mathrm{si})})^2]}{\mathbb{E}[D^{(\mathrm{si})}]} - 2 > 0, \tag{9.4.22}$$

which is equivalent to $\nu = \mathbb{E}[D^{(\mathrm{si})}(D^{(\mathrm{si})} - 1)]/\mathbb{E}[D^{(\mathrm{si})}] > 1$ (recall Theorem 4.9).

It would be of interest to analyze this model mathematically. While the extra triangles do create extra clustering in the graph, in that the graph is no longer locally tree-like, the community structure of the graph is less clear. Of course, the above setting can be generalized to arbitrary cliques and possible other community structures, but this would make the mathematical analysis substantially more involved. Exercises 9.46 and 9.47 investigate the local and global clustering coefficients, respectively.

9.4.3 RANDOM INTERSECTION GRAPHS

In most of the above models the local communities to which vertices belong *partition* the vertex space. However, in most real-world applications, particularly in social networks, often the vertices are not just part of *one* community, but of *several*. We now present a model, the random intersection graph, where the group memberships are rather general. In a random intersection graph, vertices are members of *groups*. The group memberships arise in a random way. Once the group memberships are chosen, the random intersection graph is constructed by giving an edge to two vertices precisely when they are both members of the same group. Formally, let $V(G_n)$ denote the vertex set and $A(G_n)$ the collection of groups. Let $M(G_n) = \{(v, a) : v \text{ is in group } a\} \subseteq V(G_n) \times A(G_n)$ denote the group memberships. Then, the edge set of the random intersection graph with these group memberships is

$$E(G_n) = \{\{u, v\} : (u, a), (v, a) \in M(G_n) \text{ for some } a \in A(G_n)\}. \tag{9.4.23}$$

Thus the random intersection graph is a deterministic function of the group memberships. The operation in (9.4.23), going from a bipartite graph to a unipartite graph, is sometimes called *one-mode projection*. In turn, the groups give rise to a community structure in the resulting network. Since vertices can be in several groups, the groups no longer partition the graph. It is even possible that pairs of vertices are both in several groups, even though we will see that this is rare.

Group memberships occur through some random process. There are many possibilities for this. We will discuss several different versions of the model, as introduced in the literature, and will then focus on one particular type of model to state the main results. In general, we will require that our models are *sparse*. Suppose that the probability that vertex v is part of group a equals p_{va} and that group a has on average m_a group elements. Then the average degree of vertex v equals

$$\mathbb{E}[D_v] = \sum_{a \in A(G_n)} p_{va}(m_a - 1), \tag{9.4.24}$$

which we aim to keep bounded. We now discuss several different choices.

Random Intersection Graphs with Independent Group Membership

In the most studied model there are n vertices, $m = m_n$ groups, often with $m = \beta n^\alpha$ for some $\alpha > 0$, and each vertex is independently connected to each group with probability p_n. In this case, $p_{va} = p_n$ and $m_a = np_n$, so that (9.4.24) turns into

$$\mathbb{E}[D_v] \approx np_n^2 m = \beta n^{1+\alpha} p_n^2, \tag{9.4.25}$$

and choosing $p = \gamma n^{-(1+\alpha)/2}$ yields an expected degree $\beta\gamma^2$. A more flexible version is obtained by giving a weight w_v to each of the vertices, for example in an iid way, letting

$$p_{va} = (\gamma w_v p_n) \wedge 1, \tag{9.4.26}$$

and making all the edges between vertices and groups conditionally independent given the weights $(w_v)_{v \in [n]}$. In the theorem below, we assume that $(w_v)_{v \in [n]}$ is a sequence of iid random variables with finite mean:

Theorem 9.20 (Degrees in random intersection graph with iid vertex weights) *Consider the above random intersection graph, with $m = \beta n^\alpha$ groups, vertex weights $(w_v)_{v \in [n]}$ that are iid copies of $W \sim F$ and have finite mean, and group membership probabilities $p_{va} = (\gamma w_v n^{-(1+\alpha)/2}) \wedge 1$. Then, for any $v \in [n]$:*

(a) $D_v \xrightarrow{\mathbb{P}} 0$ *when* $\alpha < 1$;

(b) $D_v \xrightarrow{d} \sum_{i=1}^{X} Y_i$ *when* $\alpha = 1$, *where* $(Y_i)_{i \geq 1}$ *are iid* $\mathsf{Poi}(\gamma)$ *random variables and* $X \sim \mathsf{Poi}(\beta\gamma W)$;

(c) $D_v \xrightarrow{d} X$ *where* $X \sim \mathsf{Poi}(\beta\gamma^2 W)$ *when* $\alpha > 1$.

Theorem 9.20 can be understood as follows. The expected number of groups to which individual v belongs is roughly $(\beta n^\alpha) \times (\gamma w_v n^{-(1+\alpha)/2}) = \beta\gamma w_v n^{-(1-\alpha)/2}$. When $\alpha < 1$, this is close to zero, so that $D_v = 0$ whp. For $\alpha = 1$, it is close to Poisson, with parameter $\beta\gamma w_v$, and the number of other individuals in each of these groups is approximately $\mathsf{Poi}(\gamma)$ distributed. For $\alpha > 1$, individual v belongs to a number of groups that tends to infinity when $n \to \infty$, while each group has expected vanishing size $n^{(1-\alpha)/2}$. The latter means that group sizes are generally 0 or 1, asymptotically independently, giving rise to the Poisson distribution specified in part (c). Part (b) is the most interesting, and interpolates between the two extremes.

Random Intersection Graphs with Prescribed Groups

We next discuss a setting in which the number of groups per vertex and the group sizes are deterministic and are obtained by randomly pairing vertices to groups. As such, the random intersection graph is obtained by (9.4.23), where now the edges between vertices in $V(G_n)$ and groups in $A(G_n)$ are modeled as a *bipartite configuration model*. Again, the number of groups $m = |A(G_n)|$ grows asymptotically linearly in the number of vertices $n = |V(G_n)|$.

In more detail, vertex $v \in [n]$ belongs to $d_v^{(\mathrm{ve})}$ groups, while group $g \in [m]$ has size $d_g^{(\mathrm{gr})}$. Here n is the number of individuals while m is the number of groups. Naturally, in order for the model to be well defined we require

$$\sum_{v \in [n]} d_v^{(\mathrm{ve})} = \sum_{a \in [m]} d_a^{(\mathrm{gr})}. \tag{9.4.27}$$

The edges in this random intersection graph are given by (9.4.23).

We will now focus on this particular setting, for concreteness. We refer to the extensive discussion in Section 9.6 for more details and references to the literature.

Local Limit of Random Intersection Graphs with Prescribed Groups

The local limit of the random intersection graph with prescribed degrees and groups is described in the following theorem:

Theorem 9.21 (Local limit of random intersection graphs with prescribed groups) *Consider the random intersection graph with prescribed groups, where the number of groups satisfies $m_n = \beta n$. Assume that the group membership sequence $\boldsymbol{d}^{(\mathrm{ve})} = (d_v^{(\mathrm{ve})})_{v \in [n]}$ and the group size sequence $\boldsymbol{d}^{(\mathrm{gr})} = (d_a^{(\mathrm{gr})})_{v \in [m_n]}$ both satisfy Conditions 1.7(a),(b), where (9.4.27) also holds (and, for $\boldsymbol{d}^{(\mathrm{gr})}$, n in Conditions 1.7(a),(b) is replaced by m_n). Then the model converges locally in probability to a so-called clique tree, where*

▷ *the number of groups in which the root participates has law $D^{(\mathrm{ve})}$, which is the limiting law of $D_n^{(\mathrm{ve})} = d_o^{(\mathrm{ve})}$ with $o \in [n]$ uar;*
▷ *the number of groups in which every other vertex participates has law $Y^\star - 1$, where $Y^\star - 1$ is the size-biased version of $D^{(\mathrm{ve})}$;*
▷ *the numbers of vertices per group are iid random variables with law $X^\star$, where $X^\star$ is the size-biased version of the limiting law of $D_n^{(\mathrm{gr})} = d_V^{(\mathrm{gr})}$ where now $V \in [m]$ is a uniformly chosen group.*

By Theorem 9.21, the degree distribution of the random intersection graphs with prescribed groups is equal to

$$D = \sum_{i=1}^{D^{(\mathrm{ve})}} (X_i^\star - 1), \tag{9.4.28}$$

which can be compared with Theorem 9.20(b). The intuition behind Theorem 9.21 is that the random intersection graph can easily be obtained from the bipartite configuration model by making all group members direct neighbors. By construction, the local limit of the bipartite configuration model can be described by an alternating branching process of the size-biased vertex and group distributions. Note that the local limit in Theorem 9.21 is *not* a tree; vertices are in multiple groups. However, Theorem 9.21 does imply that the probability that a uniform vertex has a neighbor with which shares two group memberships vanishes; see Exercise 9.48. Thus, the *overlap* between groups is generally a single vertex.

Theorem 9.21 also has implications for the clustering coefficients. Indeed, by Theorem 2.23 the local clustering coefficient for the random intersection graph converges; by Theorem 2.22, the same holds for the global clustering coefficient under a finite-second moment-condition on the degrees. See Exercises 9.49 and 9.50 for more details.

Giant in Random Intersection Graphs with Prescribed Degrees

We now investigate the existence and size of the giant component in the random intersection graph with prescribed degrees and groups:

Theorem 9.22 (Giant in random intersection graphs with prescribed degrees) *Consider the random intersection graphs with prescribed degrees and groups, where the number of groups satisfies $m_n = \beta n$. Assume that the group membership sequence $\boldsymbol{d}^{(ve)} = (d_v^{(ve)})_{v \in [n]}$ and the group size sequence $\boldsymbol{d}^{(gr)} = (d_a^{(gr)})_{v \in [m_n]}$ both satisfy Conditions 1.7(a),(b) (and, for $\boldsymbol{d}^{(gr)}$, n in Conditions 1.7(a),(b) is replaced by m_n), where (9.4.27) also holds. Then there exists $\zeta \in [0, 1]$ such that*

$$\frac{1}{n} |\mathscr{C}_{\max}| \xrightarrow{\mathbb{P}} \zeta, \qquad \frac{1}{n} |\mathscr{C}_{(2)}| \xrightarrow{\mathbb{P}} 0, \qquad (9.4.29)$$

where ζ is the survival probability of the local limit in Theorem 9.21. Further, $\zeta > 0$ precisely when $\nu > 1$, where

$$\nu = \frac{\mathbb{E}[(D^{(ve)} - 1)D^{(ve)}]}{\mathbb{E}[D^{(ve)}]} \frac{\mathbb{E}[(D^{(gr)} - 1)D^{(gr)}]}{\mathbb{E}[D^{(gr)}]}. \qquad (9.4.30)$$

In terms of the intuitive argument just below Theorem 9.21, the survival probability of the random intersection graph is identical to that of the bipartite configuration model. Owing to its alternating nature in odd and even generations, the bipartite configuration model needs to be treated with care. However, the numbers of individuals in odd or in even generations are normal branching processes with offspring distribution $\sum_{i=1}^{Y^\star - 1}(X_i^\star - 1)$ for the even generations and $\sum_{i=1}^{X^\star - 1}(Y_i^\star - 1)$ for the odd generations. These branching processes have a positive survival probability when the expected value ν of their number of offspring satisfies $\nu > 1$. Obviously, the expected numbers of offspring in the branching processes describing even and odd generations agree, by Wald's identity. This leads us to ν in (9.4.30).

9.4.4 EXPONENTIAL RANDOM GRAPHS AND MAXIMAL ENTROPY

Suppose we have a real-world network for which we observe a large number of occurrences of certain subgraphs $F \in \mathcal{F}$. For example, though the model is sparse, we do see a linear number of triangles. How can we model such a network? This is particularly relevant in the area of social science, where early on it was observed that social networks have much more clustering, i.e., many more triangles, than one might expect on the basis of many classical random graph models. This raises the question how to devise models that have similar features.

One solution may be to take the subgraph counts as given, and use a random graph model with precisely these subgraph counts. For example, considering the subgraphs to be stars of any order, this would fix the degrees of all the vertices, which would lead us to a uniform graph with prescribed degrees. However, this model is notoriously difficult to work with, and even to simulate. It becomes more difficult only when more involved quantities such as the number of triangles or of triangles per vertex are considered. Therefore, this solution may be practically impossible. Also, it may be that the numbers we observe are merely the end product of a random process, so that we should see the realizations as an indication of their *mean*, that is, as constituting a *soft constraint* rather than a hard constraint.

The exponential random graph is a way to leverage the randomness and still obtain a model that one can write down. Indeed, let $\mathcal{F}$ be a collection of subgraphs and suppose we observe that, on our favorite real-world network, the number of occurrences of subgraph F equals α_F for every $F \in \mathcal{F}$. Let us now write down what this might mean. Let F be a graph on $|V(F)| = m$ vertices. For a graph G on n vertices and m vertices $v_1, \ldots, v_m$, let $G|_{(v_i)_{i \in [m]}}$ be the subgraph spanned by $(v_i)_{i \in [m]}$. This means that the vertex set of $G|_{(v_i)_{i \in [m]}}$ equals $[m]$, while its edge set equals $\{\{i, j\} : \{v_i, v_j\} \in E(G)\}$. The number of occurrences of F in G can then be written as

$$N_F(G) = \sum_{v_1, \ldots, v_m \in V(G)} \mathbb{1}_{\{G|_{(v_i)_{i \in [m]}} = F\}}. \tag{9.4.31}$$

Here, it is convenient to recall that we may equivalently write $G = ([n], (x_{uv})_{1 \le u < v \le n})$, where $x_{i,j} \in \{0, 1\}$ and $x_{i,j} = 1$ if and only if $\{i, j\} \in E(G)$. Then, we can write $N_F(G) = N_F(x)$.

In order to define a measure, we can take a so-called *exponential family* of the form

$$p_{\vec{\beta}}(x) = \frac{1}{Z_n(\vec{\beta})} e^{\sum_{F \in \mathcal{F}} \beta_F N_F(x)}, \tag{9.4.32}$$

where $Z_n(\vec{\beta})$ is the normalization constant:

$$Z_n(\vec{\beta}) = \sum_x e^{\sum_{F \in \mathcal{F}} \beta_F N_F(x)}, \tag{9.4.33}$$

and $\vec{\beta} = (\beta_F)_{F \in \mathcal{F}}$ is a collection of parameters. In order to ensure that $p_{\vec{\beta}}$ has the correct mean values for N_F, we choose $\vec{\beta} = (\beta_F)_{F \in \mathcal{F}}$ as the solution to

$$\sum_x N_F(x) p_{\vec{\beta}}(x) = \alpha_F \qquad \text{for all } F \in \mathcal{F}. \tag{9.4.34}$$

In this case, $\mathbb{E}[N_F(X)] = \alpha_F$ for all $F \in \mathcal{F}$ when $\mathbb{P}(X = x) = p_{\vec{\beta}}(x)$. Further, when *conditioning* on $N_F(x) = q_F$ for some parameters $(q_F)_{F \in \mathcal{F}}$, the conditional exponential random graph is *uniform* over the set of graphs with this property. This is a *conditioning property* of exponential random graphs.

We next discuss two examples that we know quite well, and that arise as exponential random graphs with certain specific subgraph counts:

Example 9.23 (Example: $\mathrm{ER}_n(\lambda/n)$ and edge subgraphs) Take $N_F(x) = N_{K_2}(x) = \sum_{u,v \in [n]} x_{uv} = 2|E(G_n)|$, so that we put a restriction on the expected number of edges or complete graphs of size 2 in the graph. Then we see that, with $G_n = ([n], (x_{uv})_{1 \le u < v \le n})$,

$$Z_n(\vec{\beta}) = \sum_x e^{2\beta|E(G_n)|} = (1 + e^{2\beta})^{\binom{n}{2}}, \tag{9.4.35}$$

and

$$p_{\vec{\beta}}(x) = \frac{1}{Z_n(\vec{\beta})} e^{2\beta|E(G_n)|} = \prod_{1 \le u < v \le n} \frac{e^{2\beta x_{uv}}}{1 + e^{2\beta}}. \tag{9.4.36}$$

Thus, the different edges are independent, and an edge is present with probability $e^{2\beta}/(1 + e^{2\beta})$, and absent with probability $1/(1 + e^{2\beta})$. In a sparse setting, we aim at

$$\mathbb{E}[|E(G_n)|] = \frac{\lambda}{2}(n - 1), \tag{9.4.37}$$

so that the average degree per vertex is precisely equal to λ. The constraint in (9.4.34) thus reduces to

$$\binom{n}{2}\frac{e^{2\beta}}{1 + e^{2\beta}} = \frac{\lambda}{2}(n - 1). \tag{9.4.38}$$

This leads to $\mathrm{ER}_n(\lambda/n)$, where

$$\frac{e^{2\beta}}{1 + e^{2\beta}} = \frac{\lambda}{n}, \tag{9.4.39}$$

that is, $e^{2\beta} = \lambda/(n - \lambda)$. This shows that the $\mathrm{ER}_n(\lambda/n)$ is an example of an exponential random graph with a constraint on the expected number of edges in the graph. Further, by the conditioning property of exponential random graphs, conditional on $\mathrm{ER}_n(\lambda/n) = m$, the distribution is uniform over all graphs with m edges. ◀

Example 9.24 (Example: $\mathrm{GRG}_n(\boldsymbol{w})$ and vertex degrees)   The second example arises when we fix the expected degrees of all the vertices. This occurs when we take $N_v(x) = \sum_{u\in[n]} x_{vu} = d_v^{(G_n)}$ for every $v \in [n]$. In this case, with $G_n = ([n], (x_{uv})_{1\leq u<v\leq n})$, we have

$$Z_n(\vec{\beta}) = \sum_x e^{\sum_{v\in[n]} \beta_v d_v^{(G_n)}} = \sum_x e^{\sum_{1\leq u<v\leq n}(\beta_u+\beta_v)x_{uv}} = \prod_{1\leq u<v\leq n}(1 + e^{\beta_u+\beta_v}) \tag{9.4.40}$$

and

$$p_{\vec{\beta}}(x) = \frac{1}{Z_n(\vec{\beta})}e^{\sum_{v\in[n]} \beta_v d_v^{(G_n)}} = \prod_{1\leq u<v\leq n}\frac{e^{(\beta_u+\beta_v)x_{uv}}}{1 + e^{\beta_i+\beta_j}}. \tag{9.4.41}$$

Thus the different edges are still independent, edge $\{u, v\}$ being present with probability $e^{\beta_u+\beta_v}/(1 + e^{\beta_u+\beta_j})$ and absent with probability $1/(1 + e^{\beta_u+\beta_v})$. In a sparse setting, we aim for

$$\mathbb{E}[d_v^{(G_n)}] = \alpha_v, \tag{9.4.42}$$

so that the average degree of vertex v is precisely equal to α_v. The constraint in (9.4.34) thus reduces to

$$\sum_{j\neq v}\frac{e^{\beta_v+\beta_j}}{1 + e^{\beta_v+\beta_j}} = \alpha_v. \tag{9.4.43}$$

This leads to $\mathrm{GRG}_n(\boldsymbol{w})$, where

$$\frac{w_v}{\sqrt{\sum_{u\in[n]} w_u}} = e^{\beta_v}. \tag{9.4.44}$$

Thus, $\mathrm{GRG}_n(\boldsymbol{w})$ is an example of an exponential random graph with a constraint on the expected number of edges in the graph. Further, by the conditioning property of exponential random graphs, conditional on $d_v^{(G_n)} = \alpha_v$ for all $v \in [n]$, the distribution is uniform over all graphs with these degrees. This gives an alternative proof of Theorem 1.4.

We note that this does not *exactly* fit the format as in (9.4.31) since we have fixed the expected vertex degrees rather than subgraph counts. However, the model where we use the number of k-stars for all k in (9.4.31) is closely related to the model where we fix the expected degrees of all the vertices. ◄

Now that we have discussed two quite nice examples of an exponential random graph, let us discuss its intricacies. The above choices, in Examples 9.23 and 9.24, are quite special in the sense that the exponent in (9.4.32) is *linear* in the edge occupation statuses $(x_{uv})_{1 \leq u < v \leq n}$. This gives rise to exponential random graphs that have *independent edges*. However, when investigating more intricate subgraph counts, such as *triangles*, this linearity no longer holds. Indeed, the number of triangles is a cubic function of $(x_{uv})_{1 \leq u < v \leq n}$. In such cases the edges will no longer be independent, making the exponential random graph very hard to study.

Indeed, the exponential form in (9.4.32) naturally leads to large deviations in random graphs, a topic that is much better understood in a *dense* setting where the number of edges grows proportionally to n^2. In the sparse setting such problems are hard, and sometimes ill-defined, for example since the model may have *phase transitions* (see, e.g., Häggström and Jonasson (1999)). Such phase transitions imply that the *estimation problem* of finding parameters $\vec{\beta} = (\beta_F)_{F \in \mathcal{F}}$ such that the expected subgraph counts are exactly as intended may be ill defined. We refer to the notes and discussion in Section 9.6 for more background and references.

9.5 SPATIAL RANDOM GRAPHS

The models described so far do not incorporate geometry at all. Yet, geometry may be relevant (see, e.g., Wong et al. (2006) and the references therein). In many networks the vertices are located somewhere in space and their locations may indeed be relevant. People who live closer to one another are more likely to know each other, even though we all know people who live far away from us. This is a very direct link to the geometric properties of networks. However, the geometry may also be much more indirect or *latent*. For example, people who have similar interests are also more likely to know one another. Thus, when we are associating a whole bunch of attributes with vertices in the network, vertices with similar attributes (age, interests, hobbies, profession, music preference, etc.) may be more likely to know each other. In any case, we are rather directly led to studying networks where the vertices are embedded in some general geometric space. These are what we refer to as *spatial networks*.

One further aspect of spatial random graphs deserves to be mentioned. Owing to the fact that nearby vertices are more likely to be neighbors, conversely it is also true that two neighbors of a vertex are more likely to be connected. Therefore, geometry rather naturally leads to clustering.

9.5.1 SMALL-WORLD MODEL

The small-world model was arguably the first spatial model to be proposed in the context of complex networks. We again refer to the notes and discussion in Section 9.6 for more background and references, including the history of the model. The aim was to describe how the small-world effect can arise in a simple and natural way through the addition of *long-range edges*. Here we call an edge *long range* when it connects a pair of vertices that are far away in an underlying (or extrinsic) geometry, in which the vertices are located in some geometric space with a natural distance on it. Such long-range edges can lead to substantial *shortcuts* and thus significantly decrease graph distances. Let us describe a first version of the small-world network.

We start with a finite torus, and add random long-range connections to it, independently for each pair of vertices. This gives rise to a graph that is a small perturbation of the original lattice, but has occasional long-range connections that are crucial in order to shrink graph distances. From a practical point of view, we can think of the original graph as being the local description of acquaintances in a social network, while the shortcuts describe the occasional long-distance acquaintances. The main idea is that, even though the shortcuts only form a small part of the connections in the graph, they are crucial in order to make it a small world.

Small-World Behavior in the Continuous Circle Model

The simplest version of the model is obtained by taking a circle of circumference n, and adding a Poisson number of shortcuts with parameter $n\rho/2$, where the starting and end-points of the shortcuts are chosen uar independently of each other. This model is called the *continuous circle model*.

Distance is measured as usual along the circle, and the shortcuts have, by convention, length zero. Thus, one can think of this model as a circle where the points along the random shortcut are identified, thus creating a puncture in the circle. Multiple shortcuts then lead to multiple punctures of the circle, and the distance is then the usual distance along the punctured graph. Bear in mind that this is somewhat different from the usual graph distance, which counts the number of edges along the shortest path between pairs of vertices. The following result describes this punctured graph distance:

Theorem 9.25 (Distance in continuous circle model) *Let D_n be the distance between two uniformly chosen points along the punctured circle in the continuous circle model. Then, for every $\rho > 0$, as $n \to \infty$,*

$$D_n(2\rho)/\log(\rho n) \xrightarrow{\mathbb{P}} 1. \tag{9.5.1}$$

More precisely,

$$\rho(D_n - \log(\rho n)/2) \xrightarrow{d} T, \tag{9.5.2}$$

where T is a random variable satisfying

$$\mathbb{P}(T > t) = \int_0^\infty \frac{e^{-y}dy}{1 + e^{2t}}. \tag{9.5.3}$$

The random variable T can be described by

$$\mathbb{P}(T > t) = \mathbb{E}[e^{-e^{2t}W^{(1)}W^{(2)}}], \tag{9.5.4}$$

where $W^{(1)}$, $W^{(2)}$ are two independent exponential random variables with parameter 1. Alternatively, it can be seen that $T = (G_1 + G_2 - G_3)/2$, where G_1, G_2, G_3 are three independent Gumbel distributions having density $f_G(x) = e^x e^{-e^x}$ on $\mathbb{R}$.

Interestingly, the method of proof of Theorem 9.25 is quite close to that of Theorem 7.24. Indeed, again the parts of the graph that can be reached in a distance at most t are analyzed. Let P_1 and P_2 be two uniform points along the circle, so that D_n has the same distribution as the distance between P_1 and P_2. Denote by $R^{(1)}(t)$ and $R^{(2)}(t)$ the parts of the graph that can be reached within a distance t from P_1 and P_2, respectively. Then $D_n = 2T_n$, where T_n is the first time that $R^{(1)}(t)$ and $R^{(2)}(t)$ have a non-zero intersection. The proof then consists of showing that, up to time T_n, the processes $R^{(1)}(t)$ and $R^{(2)}(t)$ are close to certain continuous-time branching processes, primarily owing to the fact that the probability that there are two intervals that overlap is quite small. The random variables $W^{(1)}$ and $W^{(2)}$ can be viewed as appropriate martingale limits of these branching processes.

Comparing Theorem 9.25 with Theorem 7.24, we see that the rescaled distance D_n, after subtraction of the correct multiple of $\log n$, converges in distribution, while in Theorem 7.24, convergence is at best along subsequences. This is due to the fact that D_n is a *continuous* random variable, while graph distances are integer-valued. Therefore, graph distances suffer from discretization effects. In the next paragraph, we will see that the graph distances in the small-world model suffer from similar issues.

Small-World Behavior in Discrete Small-World Model

We now extend the analysis to graph distances in the small-world model where the long-range edges are counted as having distance 1 rather than zero. Further, the vertices are positioned on a *discrete* torus. Thus, these are the distances in the small-world model when the latter is considered as a graph. We let the total number of vertices be $n = Lk$ and every vertex be connected to k of its closest neighbors by an (undirected) edge. The extra variable k is useful, as $k > 2$ allows the model to have significant *clustering*, which the nearest-neighbor version for $k = 2$ does not have. "Shortcuts" between all pairs of vertices are added independently with probability λ/n, and we call these *long-range* edges. Thus, the small-world model is realized by taking the union of the discrete torus, which gives rise to the short-range edges, with $\mathrm{ER}_n(\lambda/n)$ so that each possible long-range edge is present independently with probability λ/n. Exercise 9.51 investigates the degree structure of this graph.

Define

$$\nu = \tfrac{1}{2}\left[\lambda + 1 + \sqrt{(\lambda+1)^2 + 4\lambda(2k-1)}\right] > \lambda + 1. \tag{9.5.5}$$

The main result concerning small-world properties in the small-world model is as follows:

Theorem 9.26 (Distance in small-world model) *Let G_n be the above discrete small-world model. Assume that $n \to \infty$ and that $\rho = \lambda k$ remains bounded. Let $\mathrm{dist}_{G_n}(o_1, o_2)$ denote the graph distance between two uniformly chosen vertices $o_1, o_2 \in [n]$. Then, with ν as in* (9.5.5),

$$\mathrm{dist}_{G_n}(o_1, o_2)/\log n \xrightarrow{\mathbb{P}} \frac{1}{\log \nu}. \tag{9.5.6}$$

A more precise version of the fluctuations of $\mathrm{dist}_{G_n}(o_1, o_2) - \lceil \log_\nu n \rceil$ is also known. Further, the case where $\rho = \lambda_n k \to 0$ has been studied, and there the behavior is closely related to that in Theorem 9.25. See Section 9.6 for an extensive discussion. The parameter

ν arises as the largest eigenvalue of the offspring matrix of an appropriate two-type branching process that describes the local neighborhoods in the discrete small-world model. This branching process has two types in that there is a difference between an interval starting *immediately after* a shortcut, and intervals that have previously been found, owing to the "hesitation" arising from the fact that a long-range edge now has length 1 rather than 0.

9.5.2 HYPERBOLIC RANDOM GRAPHS

Here we consider the hyperbolic random graph, where vertices are in a disk of radius R and are connected if their hyperbolic distance is at most R. This is sometimes called the Poincaré disk model. These graphs are very different from general inhomogeneous random graphs, because the geometry creates random graphs that are more clustered.

The hyperbolic random graph has two key parameters, denoted by ν and α. The model samples n vertices on a disk of radius $R = 2\log(n/\nu)$, where the density of the radial coordinate r of a vertex $p = (r, \phi)$ is

$$\rho(r) = \alpha \frac{\sinh(\alpha r)}{\cosh(\alpha R) - 1}. \tag{9.5.7}$$

This measure is natural as it gives rise to the uniform measure on the hyperbolic disk. Here ν parametrizes the average degree of the generated networks and $-\alpha$ parametrizes the so-called negative curvature of the space. The angle ϕ of p is sampled uniformly from $[0, 2\pi]$, so that the points have a spherically symmetric distribution. Then, two vertices are connected when their hyperbolic distance is at most R. Here, the hyperbolic distance $x = \text{dist}_{\mathbb{H}}(u, v)$ between two points at polar coordinates $u = (r, \phi)$ and $v = (r', \phi')$ is given by the hyperbolic law of cosines,

$$\cosh(x) = \cosh(r)\cosh(r') - \sinh(r)\sinh(r')\cos(\|\phi - \phi'\|), \tag{9.5.8}$$

where $\|\phi - \phi'\| = \pi - |\pi - |\phi - \phi'||$ is the difference between the two angles (which is the Euclidean distance on the circle).

Degree Structure of Hyperbolic Random Graphs

Let $P_k^{(n)} = \frac{1}{n}\sum_{v\in[n]} \mathbb{1}_{\{D_v=k\}}$ denote the degree distribution in the hyperbolic random graph. As explained informally in the previous paragraph, we may expect that the degree distribution obeys a power law. This is the content of the following theorem:

Theorem 9.27 (Power-law degrees in hyperbolic random graphs) *As $n \to \infty$, there exists a probability distribution $(p_k)_{k\geq 0}$ such that*

$$P_k^{(n)} \xrightarrow{\mathbb{P}} p_k, \tag{9.5.9}$$

where $(p_k)_{k\geq 0}$ obeys an asymptotic power law, i.e., there exists a $c > 0$ such that

$$p_k = ck^{-\tau}(1 + o(1)), \tag{9.5.10}$$

where $\tau = 2\alpha + 1$.

(a)   (b)

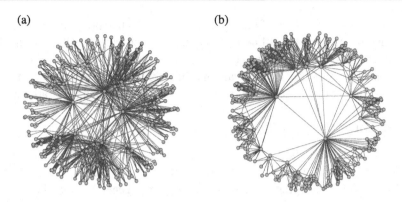

Figure 9.13 Examples of hyperbolic graphs for $n = 250$, with (a) $\tau = 2.5$ and (b) $\tau = 3.5$, and average degree approximately 5.

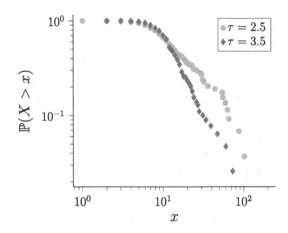

Figure 9.14 Degree distributions of the hyperbolic graphs in Figure 9.13.

We deduce that the model is scale-free, meaning that the asymptotic degree distribution has infinite variance, precisely when $\alpha \in (\frac{1}{2}, 1)$; otherwise the degree distribution obeys a power law with a larger degree exponent. Let us informally explain the power law in Theorem 9.27. For a vertex v with radial coordinate r_v, we define its *type* t_v by

$$t_v = e^{(R-r_v)/2}. \tag{9.5.11}$$

Then, the degree D_v of vertex v can be approximated by a Poisson random variable with mean t_v, so that D_v is of order t_v. Furthermore, the random variables $(t_v)_{v \geq 1}$ are distributed as a power law with exponent $\tau = 2\alpha + 1$, so that the degrees have a power-law distribution as well, with the same exponent.

The exact form of p_k involves several special functions. Its identification is quite impressive, the proof of which is rather involved. For the purpose of this book, however, the exact shape of p_k is not so relevant.

Much more is known about the local structure of the hyperbolic random graph, for example, its local limit has been identified. We postpone this discussion to the next subsection, in which we discuss the local limit in geometric inhomogeneous random graphs. It turns out that we can interpret the hyperbolic random graph as a special case which is interesting in its own right.

Giant in Hyperbolic Random Graphs

We next study the giant in hyperbolic random graphs, the result being somewhat surprising:

Theorem 9.28 (Giant in hyperbolic random graphs) *Consider the hyperbolic random graph with parameters $\alpha > \frac{1}{2}$ and $\nu > 0$.*

(a) *For $\alpha > 1$, $|\mathscr{C}_{\max}|/n \overset{\mathbb{P}}{\longrightarrow} 0$ for all $\nu > 0$.*

(b) *For $\alpha \in (\frac{1}{2}, 1)$, there exists a $\zeta > 0$ such that $|\mathscr{C}_{\max}|/n \overset{\mathbb{P}}{\longrightarrow} \zeta$ whp, and $|\mathscr{C}_{(2)}|/n \overset{\mathbb{P}}{\longrightarrow} 0$ for all $\nu > 0$.*

(c) *For $\alpha = 1$, there exist $\pi/8 \leq \nu_0 \leq \nu_1 \leq 20\pi$ and $\zeta > 0$ such that $|\mathscr{C}_{\max}|/n \overset{\mathbb{P}}{\longrightarrow} \zeta$ and $|\mathscr{C}_{(2)}|/n \overset{\mathbb{P}}{\longrightarrow} 0$ for all $\nu > \nu_1$, while $|\mathscr{C}_{\max}|/n \overset{\mathbb{P}}{\longrightarrow} 0$ for $\nu \leq \nu_0$.*

We see that a giant component exists only in the scale-free regime, which is quite surprising. In various other random graphs, a giant component exists in settings where the degrees have finite variance, particularly when there are sufficiently many edges. This turns out not to be the case for hyperbolic random graphs. The fact that no giant exists when $\alpha > 1$ and $\tau > 3$ is explained in more detail in Section 9.5.3; see below Theorem 9.34. There, it is explained that the hyperbolic random graph is a special case of a one-dimensional geometric inhomogeneous random graph. Giants in one dimension exist only when the degrees have infinite variance. We postpone a further discussion to that section.

More precise results include the fact that $|\mathscr{C}_{\max}| = \Theta_{\mathbb{P}}(R^2(\log\log R)^3 n^{1/\alpha})$ for $\alpha > 1$, while the second largest component is at most polylogarithmic (i.e., at most a power of $\log n$) for $\alpha \in (\frac{1}{2}, 1)$.

Ultra-Small Distances in Hyperbolic Random Graphs

Obviously, there is little point in studying typical distances in the hyperbolic random graph when there is no giant component, even though such results do shed light on the component structure of smaller components. Thus, we restrict to the setting for which $\alpha \in (\frac{1}{2}, 1)$, where typical distances are *ultra-small*:

Theorem 9.29 (Ultra-Small Distances in Hyperbolic Random Graphs) *Let G_n be the hyperbolic random graph with parameters $\alpha \in (\frac{1}{2}, 1)$ and $\nu > 0$. Then, with o_1, o_2 two independent vertices chosen uar from $[n]$, conditional on $o_1 \longleftrightarrow o_2$,*

$$\frac{\mathrm{dist}_{G_n}(o_1, o_2)}{\log\log n} \overset{\mathbb{P}}{\longrightarrow} \frac{2}{|\log(2\alpha - 1)|}. \tag{9.5.12}$$

Note that $2\alpha - 1 = \tau - 2$, so that Theorem 9.29 agrees with the ultra-small nature of the random graphs studied in this book.

Hyperbolic Embeddings of Complex Networks

Can hyperbolic geometries be used to map real-world complex networks efficiently? On the one hand, hyperbolic random graphs have very small distances, while on the other hand they

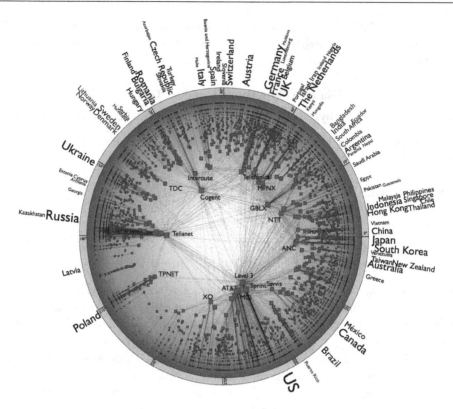

Figure 9.15 Hyperbolic embedding of the Internet at Autonomous Systems level, by Boguná et al. (2010) (see (Boguná et al., 2010, Figure 3)).

also have the necessary clustering to make them appropriate models. Of course, the question of *how* to embed them precisely is highly relevant and also quite difficult.

The example that has attracted the most attention is the Internet. In Figure 9.15, you can see a hyperbolic embedding of the Internet, as performed by Boguná et al. (2010). We see that the regions on the boundary of the outer circle can be grouped in a fairly natural way, where the countries in which the autonomous systems reside seem to be grouped according to their geography, with some exceptions (for example, it is not clear why Kenya is almost next to the Netherlands). This hyperbolic geometry is first of all quite interesting, but it could also be helpful in sustaining the ever growing Internet traffic.

9.5.3 GEOMETRIC INHOMOGENEOUS RANDOM GRAPHS

We continue by defining the geometric inhomogeneous random graph (GIRG). Here vertices are located in a general metric space $\mathcal{X} \subseteq \mathbb{R}^d$ for some dimension $d \geq 1$. We let $(X_v)_{v \in [n]}$ denote their locations, and assume that $(X_v)_{v \in [n]}$ are iid and have some law μ on $\mathcal{X}$. Often, we assume that $\mathcal{X} = [-\frac{1}{2}, \frac{1}{2}]^d$ is the cube of width 1 in d dimensions centered at the origin, with periodic boundary conditions, and μ the uniform measure. Further, we assume that each vertex v has a *weight* W_v associated with it, where $(W_v)_{v \in [n]}$ are assumed to be iid, for example as power-law random variables. We write $(x_v)_{v\in[n]}$ and $(w_v)_{v\in[n]}$ for the realizations of $(X_v)_{v\in[n]}$ and $(W_v)_{v\in[n]}$.

The edges are *conditionally independent* given $(x_v)_{v\in[n]}$ and $(w_v)_{v\in[n]}$, where the conditional probability that the edge between u and v is present is equal to

$$p_{u,v} = \kappa_n(\|x_u - x_v\|, w_u, w_v), \tag{9.5.13}$$

for some $\kappa\colon [0,\infty)^3 \to [0,\infty)$. A prime example of such a GIRG is the so-called product GIRG, for which

$$\kappa_n(t, w_u, w_v) = \lambda\Big(1 \wedge \Big(\frac{w_u w_v}{\sum_{i\in[n]} w_i}\Big)^{\max\{\alpha,1\}} t^{-d\alpha}\Big), \tag{9.5.14}$$

where $\alpha, \lambda > 0$ are appropriate parameters. Often, we assume that the vertex weights obey a power law, i.e.,

$$\mathbb{P}(W > w) = L(w)w^{-(\tau-1)} \tag{9.5.15}$$

for some slowly varying function $L\colon [0,\infty) \to (0,\infty)$. As is usual, the literature treats a variety of models and settings, and we refer to the notes and discussion in Section 9.6 for more details. To describe the local limit of GIRGs, we need to assume a more restrictive setting:

Assumption 9.30 (Limiting connection probabilities exist) *Assume the following:*

(a) *the vertex weights $(w_v)_{v\in[n]}$ satisfy Condition 1.1(a) for some limiting random variable W;*
(b) *the vertex locations $(x_v)_{v\in[n]}$ are a sequence of iid uniform locations on $[-\frac{1}{2}, \frac{1}{2}]^d$ that are independent of $(w_v)_{v\in[n]}$;*
(c) *there exists a function $\kappa\colon [0,\infty)^3 \to [0,\infty)$ such that $\kappa_n(n^{-1/d}t, x_n, y_n) \to \kappa(t, x, y)$ for all $x_n \to x$ and $y_n \to y$, where κ satisfies that there exists $\alpha > 0$ such that, for all t large enough,*

$$\mathbb{E}[\kappa(t, W_1, W_2)] \le t^{-\alpha}, \tag{9.5.16}$$

with W_1, W_2 two copies of the limiting random variable W in part (a).

Hyperbolic Random Graph as a One-Dimensional Product GIRG

We now explain that the hyperbolic random graph can be interpreted as a special case of the product GIRG. To this end, we embed the disk of the native hyperbolic model into our model with dimension 1. Hence, we reduce the geometry of the hyperbolic disk to the geometry of a circle but gain additional freedom, as we can choose the weights of vertices.

Notice that a single point on the hyperbolic disk has measure zero, so we can assume that no vertex has radius $r_v = 0$. Recall that $\mathrm{dist}_\mathbb{H}$ denotes the hyperbolic distance defined in (9.5.8). Here, we also treat the *soft* hyperbolic graph, for which the probability that there exists an edge between the vertices $u = (r_u, \phi_u)$ and $v = (r_v, \phi_v)$ equals

$$p_\mathbb{H}(\mathrm{dist}_\mathbb{H}(u, v)) = \Big(1 + e^{(d_\mathbb{H}(u,v) - R_n)/(2T)}\Big)^{-1}. \tag{9.5.17}$$

In the limit $T \searrow 0$, this gives rise to the (hard) hyperbolic graph studied in the previous subsection. The identification is given in the following theorem:

Theorem 9.31 (Hyperbolic random graphs as product GIRGs)

(a) *The soft hyperbolic random graph is a GIRG that satisfies Assumption 9.30 with limiting W satisfying (9.5.15) with parameters*

$$d = 1, \qquad \tau = 2\alpha + 1, \qquad \alpha = 1/T, \tag{9.5.18}$$

and limiting connection probabilities given by

$$\kappa(t, w_u, w_v) = \frac{1}{1 + \left(\frac{ct}{w_u w_v}\right)^{\alpha}}, \tag{9.5.19}$$

where $c = \sqrt{2\pi}/\nu$.

(b) *The hard hyperbolic random graph can be mapped to a one-dimensional threshold product GIRG, where vertices $u, v \in [n]$ are connected precisely when $\|x_u - x_v\| \leq \nu w_u w_v / \pi$.*

Sketch of proof of Theorem 9.31. We define the mapping

$$w_v := e^{(R - r_v)/2} \qquad \text{and} \qquad x_v := \phi_v / 2\pi, \tag{9.5.20}$$

where, for $v \in [n]$, its radial coordinate equals r_v and its angular coordinate equals ϕ_v. The map $(r, \phi) \mapsto (e^{(R-r)/2}, \phi/2\pi)$ is a bijection from the hyperbolic space to $[0, e^{R/2}] \times \mathbb{T}_{1,1}$, where $\mathbb{T}_{1,1} = [-\frac{1}{2}, \frac{1}{2}]$ with periodic boundary conditions. Therefore, it also has an inverse, which we denote by $g(w_v, x_v) = (r_v, \phi_v)$. Then, for $i, j \in [n]$, we let

$$p_{i,j} = p_{i,j}((w_i, x_i), (w_j, x_j)) = p_{\mathbb{H}}(\text{dist}_{\mathbb{H}}(g(w_i, x_i), g(w_j, x_j))). \tag{9.5.21}$$

This maps the soft hyperbolic random graph to an explicit GIRG.

We next verify the conditions on the edge probabilities that are required for GIRGs. We start by computing from (9.5.7) that

$$\mathbb{P}(r_v \leq r) = \int_0^r \rho(r) \mathrm{d}r = \int_0^r \alpha \frac{\sinh(\alpha r)}{\cosh(\alpha R) - 1} \mathrm{d}r = \frac{\cosh(\alpha r) - 1}{\cosh(\alpha R) - 1}. \tag{9.5.22}$$

Since $R_n = 2\log(n/\nu)$ is large,

$$\mathbb{P}(r_v \leq r) = \frac{e^{\alpha(r-R)} + e^{-\alpha(R+r)}}{1 + e^{-2R}} = e^{\alpha(r-R)}(1 + o(1)). \tag{9.5.23}$$

Therefore, the distribution of w_v satisfies

$$\begin{aligned} \mathbb{P}(w_v > w) = \mathbb{P}(r_v < R - 2\log w) &= e^{-2\alpha \log w}(1 + o(1)) \\ &= w^{-(\tau - 1)}(1 + o(1)), \end{aligned} \tag{9.5.24}$$

as required by (9.5.15) and since $\tau = 2\alpha + 1$. In particular, this implies that $r_v \in [R, R - A]$ for *most* vertices in $[n]$ when A is large. Thus, most vertices are close to the boundary of the hyperbolic disk.

We next investigate the role of the radial coordinate. For this, we assume that $r_u \in [0, R]$, which is true almost surely, and $r_u + r_v \geq R$, which occurs whp for R large. We use

$$\cosh(x \pm y) = \cosh(x)\cosh(y) \pm \sinh(x)\sinh(y) \tag{9.5.25}$$

to rewrite (9.5.8) for $u = (r_u, \phi_u)$ and $v = (r_v, \phi_v)$ as

$$\cosh(\mathrm{dist}_\mathbb{H}(u,v)) = \cosh(r_u - r_v) + \sinh(r_u)\sinh(r_v)[1 - \cos(\|\phi_u - \phi_v\|)]. \tag{9.5.26}$$

When $\|\phi_u - \phi_v\|$ is small, as in Assumption 9.30, we thus obtain that $\cosh(\mathrm{dist}_\mathbb{H}(u,v))$ equals

$$\cosh(r_u - r_v) + \tfrac{1}{2}\sinh(r_u)\sinh(r_v)\|\phi_u - \phi_v\|^2(1 + o(1)). \tag{9.5.27}$$

Denote $s_n = \mathrm{dist}_\mathbb{H}(u,v) - R_n$, and note that

$$e^{s_n} = 2\cosh(\mathrm{dist}_\mathbb{H}(u,v))e^{-R_n} - e^{-s_n - 2R_n}. \tag{9.5.28}$$

Further, multiplying (9.5.27) by e^{-R_n}, with $R_n = 2\log(n/\nu)$, leads to

$$
\begin{aligned}
&\cosh(\mathrm{dist}_\mathbb{H}(u,v))e^{-R_n} \\
&= \tfrac{1}{2}e^{r_u - R_n}e^{r_v - R_n}e^{R_n}\|\phi_u - \phi_v\|^2(1 + o(1)) + \cosh(r_u - r_v)e^{-R_n} \\
&= \frac{2\pi^2}{\nu^2}\frac{n^2(x_u - x_v)^2}{(w_u w_v)^2}(1 + o(1)) + \nu^2\frac{(w_u/w_v)^2 + (w_u/w_v)^2}{n^2}. \tag{9.5.29}
\end{aligned}
$$

Combining (9.5.28) and (9.5.29), and substituting this into (9.5.17) for $x_v = x + t/n$, leads to

$$
\begin{aligned}
p_\mathbb{H}(\mathrm{dist}_\mathbb{H}(u,v)) &= \left(1 + e^{s_n/(2T)}\right)^{-1} = \left(1 + \left(\frac{2\pi^2}{\nu^2}\frac{t^2}{(w_u w_v)^2}\right)^{1/(2T)}\right)^{-1}(1 + o(1)) \\
&= \left(1 + \left(c\frac{t}{w_u w_v}\right)^\alpha\right)^{-1}(1 + o(1)), \tag{9.5.30}
\end{aligned}
$$

where $c = \sqrt{2\pi}/\nu$. $\qquad\square$

Since hyperbolic random graphs are special cases of the GIRG, some results can simply be deduced from the analysis for the GIRG. Also, the fact that hyperbolic graphs arise as a *one-dimensional* GIRG helps us to explain why there is no giant for $\alpha > 1$.

Degree Structure of Product Geometric Inhomogeneous Random Graphs

We start by analyzing the degree structure of the graph. Let $P_k^{(n)}$ denote the proportion of vertices with degree k in the hyperbolic random graph, which obeys a power law:

Theorem 9.32 (Power-law degrees in product GIRGs) *Let the edge probabilities in the product geometric inhomogeneous random graph be given by*

$$p_{u,v} = \lambda\left(1 \wedge \left(\frac{W_u W_v}{n}\right)^\alpha \|X_u - X_v\|^{-d\alpha}\right), \tag{9.5.31}$$

where $\alpha > 1$, $(W_u)_{u\in[n]}$ are iid random variables satisfying $\mathbb{P}(W > w) = c_W w^{-(\tau-1)}(1 + o(1))$ for some $c_W > 0$ and $w \to \infty$, and $(X_u)_{u\in[n]}$ are iid uniform random variables on $[-\frac{1}{2}, \frac{1}{2}]^d$. As $n \to \infty$, there exists a probability distribution $(p_k)_{k\geq 0}$ such that

$$P_k^{(n)} \xrightarrow{\;\mathbb{P}\;} p_k, \tag{9.5.32}$$

where $(p_k)_{k \geq 0}$ obeys an asymptotic power law, i.e., there exists a constant $c > 0$ such that

$$\sum_{l > k} p_l = ck^{-(\tau-1)}(1 + o(1)). \tag{9.5.33}$$

Exercise 9.54 asks the reader to investigate the proof of Theorem 9.32.

Local Limit of Geometric Inhomogeneous Random Graphs

We next study the local limit of geometric inhomogeneous random graphs. Before stating the result, we introduce the local limit. We fix a Poisson process with intensity 1 on $\mathbb{R}^d$, with an additional point at the origin that will serve as our point of reference. With each point $x \in \mathbb{R}^d$ in this process, we associate an independent copy of the limiting weight W_x. Then we draw an edge between two vertices x, y with probability $\kappa(\|x - y\|, W_x, W_y)$, where these variables are *conditionally independent* given the points of the Poisson process and the weights of its points. Call the above process the *Poisson infinite GIRG with edge probabilities given by κ*. When $\kappa \in \{0, 1\}$, we are dealing with a threshold Poisson infinite GIRG with edge statuses given by κ. The main result is as follows:

Theorem 9.33 (Local limit of GIRG) *The GIRG, with $p_{u,v}$ as in* (9.5.13) *under Assumption 9.30, converges locally in probability to the Poisson infinite GIRG with edge probabilities given by κ.*

Together with Theorem 9.31, Theorem 9.33 identifies the local limit of both the soft as well as the hard hyperbolic random graph, as well as for the product GIRG in (9.5.14), for which, with W as in Assumption 9.30,

$$\kappa(t, w_u, w_v) = \lambda \left(\frac{w_u w_v}{\mathbb{E}[W]} \right)^{\max\{\alpha,1\}} t^{-d\alpha}. \tag{9.5.34}$$

Giant in Geometric Inhomogeneous Random Graphs

We next study the giant in GIRGs. First, we define the class of GIRGs to which these results apply. We denote the graph by $G_n = (V(G_n), E(G_n))$, and let the vertices $V(G_n)$ of G_n be the points of a Poisson point process on $\frac{1}{2}[-n^{1/d}, n^{1/d}] \times [1, \infty)$ with intensity measure

$$\mu(dx, dw) = dx \times (\tau - 1)w^{-\tau}dw. \tag{9.5.35}$$

For $p \geq 0$, $\alpha \in [0, \infty]$, $\sigma \in [0, \infty)$, and $\beta > 0$, we let

$$p((x, w), (x', w')) = \begin{cases} p \min \left\{ 1, \left(\frac{\beta(x \wedge x')(x \vee x')^{\sigma}}{\|x - x'\|} \right)^{\alpha} \right\} & \text{for } \alpha < \infty, \\ p \mathbb{1}_{\{\beta(x \wedge x')(x \vee x')^{\sigma} \geq \|x - x'\|\}} & \text{for } \alpha = \infty. \end{cases} \tag{9.5.36}$$

For $u, v \in V(G_n)$, we independently let $\{u, v\} \in E(G_n)$ with probability $p(u, v)$, and $\{u, v\} \notin E(G_n)$ otherwise.

Theorem 9.34 (Giant in GIRGs) *Let G_n be the GIRG defined in* (9.5.35) *and* (9.5.36) *with $p \geq 0$, $\alpha \in [0, \infty]$, $\sigma \in [0, \infty)$, and $\beta > 0$. Then $|\mathscr{C}_{\max}|/n \xrightarrow{\mathbb{P}} \zeta$ whp, where ζ is the survival probability of the Poisson infinite GIRG with edge probabilities given by* (9.5.35) *and* (9.5.36)*, while $|\mathscr{C}_{(2)}|/n \xrightarrow{\mathbb{P}} 0$ for all $\alpha > 0$.*

Special cases of the GIRG defined in (9.5.35) and (9.5.36) are the product GIRG, as well as certain cases of the GIRG in Assumption 9.30. By the construction of the inhomogeneous Poisson process in (9.5.35) and conditional on $|V(G_n)| = n'$, the variables $((x_i, w_i))_{i \in [n']}$ are iid with distribution that is uniform on $\frac{1}{2}[-n^{1/d}, n^{1/d}]$ for the x_i's, and w_i iid copies of a Pareto random variable W with $\mathbb{P}(W > w) = w^{-(\tau-1)}$ for $w \geq 1$.

Let us complete this discussion by considering the situation when $d = 1$ using the local convergence statement in Theorem 9.33. Here, there is no infinite component in the local limit for $\tau > 3$, so that no giant exists in the pre-limit either by Corollary 2.27.

We call a vertex u *strongly isolated* when there does not exist an occupied edge $\{v_1, v_2\} \in E(G_n)$ with $v_1 \leq u \leq v_2$ in the graph. In particular, this means that the connected component of u is finite. We prove that the expected number of occupied edges $\{v_1, v_2\}$ with $v_1 \leq u \leq v_2$ is bounded. Indeed, for the local limit of the hard hyperbolic graph in $d = 1$, this expected number is equal to

$$\sum_{v_1 \leq u \leq v_2} \mathbb{E}\left[\frac{1}{1 + \left(\frac{\|v_1 - v_2\|}{cW_{v_1}W_{v_2}}\right)^{\alpha}}\right] \leq C \sum_{v_1 \leq u \leq v_2} \mathbb{E}\left[\frac{(W_{v_1}W_{v_2})^{\alpha}}{(W_{v_1}W_{v_2})^{\alpha} + \|v_1 - v_2\|^{\alpha}}\right]$$

$$\leq C \sum_{k \geq 1} k\mathbb{E}\left[\frac{X^{\alpha}}{X^{\alpha} + k^{\alpha}}\right], \tag{9.5.37}$$

where $X = W_1 W_2$ is the product of two independent W variables. Now, when $\mathbb{P}(W > w) = w^{-(\tau-1)}$, it is not hard to see that

$$\mathbb{P}(X^{\alpha} > x) \leq C\frac{\log x}{x^{(\tau-1)/\alpha}}. \tag{9.5.38}$$

In turn, this implies that

$$\mathbb{E}\left[\frac{X^{\alpha}}{X^{\alpha} + k^{\alpha}}\right] \leq C\frac{\log k}{k^{\tau-1}}. \tag{9.5.39}$$

We conclude that, when multiplied by k, this is summable when $\tau > 3$ so that the expected number of occupied edges $\{v_1, v_2\}$ with $v_1 \leq u \leq v_2$ is bounded. In turn, this suggests that this number equals zero with strictly positive probability (beware, these numbers are not independent for different u), and this in turn suggests that there is a positive proportion of them. However, when there is a positive proportion of strongly isolated vertices, there cannot be an infinite component. This intuitively explains why the existence of the giant component is restricted to $\tau \in (2, 3)$. Thus, the absence of a giant in hyperbolic graphs with power-law exponent τ with $\tau > 3$ is intimately related to the fact that this model is inherently one-dimensional.

Ultra-Small Distances in Product Geometric Inhomogeneous Random Graphs

We continue by discussing ultra-small distances in product GIRGs:

Theorem 9.35 (Ultra-small distances in GIRGs) *For* $\tau \in (2, 3)$, *with* G_n *the product GIRG in (9.5.14) and (9.5.15), conditional on* o_1, o_2 *being connected,*

$$\frac{\text{dist}_{G_n}(o_1, o_2)}{\log \log n} \xrightarrow{\mathbb{P}} \frac{2}{|\log(\tau - 2)|}. \tag{9.5.40}$$

We return to the question of the distances in GIRGs for $\tau > 3$ in Section 9.5.5, where we discuss *scale-free percolation*, for which the strongest results have been shown. It is natural to believe that similar results may also be true for product GIRGs.

Clustering in Geometric Inhomogeneous Random Graphs

We now study the clustering properties of spatial inhomogeneous random graph.

Theorem 9.36 (Clustering in GIRGs) *The local clustering coefficient in the geometric inhomogeneous random graph, with p_{uv} as in (9.5.13), under Assumption 9.30, converges in probability to a positive constant.*

Theorem 9.36 follows directly from Theorem 9.33 combined with Theorem 2.23. One could say that Theorem 9.36 exemplifies why the spatial inhomogeneous random graphs are so relevant. Indeed, they allow us to build random graphs that have non-vanishing clustering, as well as all the other properties that real-world networks have, and for which the models discussed in this book have been invented. For $\tau \in (2, 3)$, it is not clear that the *global* clustering coefficient is also strictly positive, recall Theorem 2.22, since the square of the degree of a uniform vertex is not uniformly integrable.

9.5.4 Spatial Preferential Attachment Models

We discuss three spatial preferential attachment models (SPAMs), and recall that we have already encountered an example in Section 9.3.4 (Theorem 9.14). There, such models were considered to model *community structure*, for which it is most natural to let the vertices have types in a finite set. Here we substantially extend this notion, and instead consider the vertices to have a spatial location in some abstract measurable space.

Geometric Preferential Attachment Model with Uniform Locations

Let S be a bounded metric space and μ be the uniform measure on S. We require that $\mu(\mathcal{B}_r(u))$ is independent of $u \in S$, where $\mathcal{B}_r(u) = \{x \in S : |x - u| \leq r\}$ denotes the extrinsic or geometric ball of radius r around $u \in S$. Let $G_n = (V(G_n), E(G_n))$ denote the graph at time n. Then, we let $V(G_n)$ be a subset of S of size n.

The process $(G_n)_{n \geq 0}$ evolves as follows. At time $n = 0$, G_0 is the empty graph. At time $n + 1$, given G_n we obtain G_{n+1} in the following way. Let x_{n+1} be chosen uar from S, and denote $V(G_{n+1}) = V(G_n) \cup \{x_{n+1}\}$. We assign m edges to the vertex x_{n+1}, which we connect independently of each other to vertices in $V(G_n) \cap \mathcal{B}_r(x_{n+1})$. Denote

$$D_n(x_{n+1}) = \sum_{v \in V(G_n) \cap \mathcal{B}_r(x_{n+1})} D_v^{(n)}, \tag{9.5.41}$$

where $D_v^{(n)}$ denotes the degree of vertex $v \in V(G_n)$ in G_n. Thus, $D_n(x_{n+1})$ denotes the total degree of all vertices located in $\mathcal{B}_r(x_{n+1})$. The m edges are connected to vertices $(y_1, \ldots, y_m)$ conditionally independently given (G_n, x_{n+1}). Thus, for all $v \in V(G_n) \cap \mathcal{B}_r(x_{n+1})$,

$$\mathbb{P}(y_i = v \mid G_n) = \frac{D_v^{(n)}}{\max(D_n(x_{n+1}), \alpha m A_r n)}, \tag{9.5.42}$$

while

$$\mathbb{P}(y_i = x_{n+1} \mid G_n) = 1 - \frac{D_n(x_{n+1})}{\max(D_n(x_{n+1}), \alpha m A_r n)}, \tag{9.5.43}$$

where we denote $A_r = \mu(\mathcal{B}_r(u))$; $\alpha \geq 0$ is a parameter, and $r = r_n$ is a radius to be chosen appropriately. The parameter r may depend on the size of the graph. The degree sequence of the model that arises is characterized in the following theorem:

Theorem 9.37 (Degrees in preferential attachment models with uniform locations on the sphere) *Let S be the surface of a three-dimensional unit ball. Take $r_n = n^{\beta-1/2} \log n$, where $\beta \in (0, \frac{1}{2})$ is a constant. Finally, let $\alpha > 2$ and $m \geq 1$. In the above geometric preferential attachment model given by (9.5.41)–(9.5.43),*

$$P_k^{(n)} \xrightarrow{\mathbb{P}} p_k, \tag{9.5.44}$$

where $p_k = Ck^{-(\alpha+1)}(1 + o(1))$ for k large.

Theorem 9.37 allows for $r = r_n = o(1)$, so that vertices can make connections only to vertices that are close by.

We next discuss a setting where r_n remains fixed. Let us first introduce the model. We again assume that S is a metric space, and μ is the uniform measure on S. Further, let $\alpha \colon \mathbb{R}^+ \to \mathbb{R}^+$ be an attractiveness function. The graph process is denoted by $(G_n)_{n\geq0}$. Here G_0 is assumed to be a connected graph with n_0 vertices and e_0 edges. We let the spatial locations $(X_i)_{i\geq1}$ be iid draws from μ in S. Each vertex enters the graph with m edges to be connected to the vertices already in the graph. Denote the receiving vertices by $(V_i^{(n+1)})_{i\in[m]}$. Conditional on G_n and X_{n+1}, we let the vertices $(V_i^{(n+1)})_{i\in[m]}$ be conditionally iid with

$$\mathbb{P}(V_i^{(n+1)} = u \mid X_{n+1}, G_n) = \frac{(\deg_{G_n}(u) + \delta)\alpha(|u - X_{n+1}|)}{\sum_{i\in[n]}(\deg_{G_n}(X_i) + \delta)\alpha(|X_i - X_{n+1}|)}, \tag{9.5.45}$$

where $\delta > -m$ is a parameter in the model. Let

$$I_k = \int_S \alpha(|x - u|)^k \mu(\mathrm{d}x). \tag{9.5.46}$$

The main result about the degree distribution in the above model is as follows:

Theorem 9.38 (Degrees in preferential attachment models with uniform locations) *Let S be a general metric space, and let μ be the uniform measure on it. Let $m \geq 1$ and $\delta \geq 0$. For $m = 1$ assume that $I_1 < \infty$, and, for $m \geq 2$, that $I_2 < \infty$ in (9.5.46). Let α be continuous, and for $\delta = 0$ assume that $\alpha(r) \geq \alpha_0 > 0$. In the geometric preferential attachment model described in (9.5.45),*

$$P_k^{(n)} \xrightarrow{\mathbb{P}} p_k = (2 + \delta/m)\frac{\Gamma(k + \delta)\Gamma(m + 2 + \delta + \delta/m)}{\Gamma(m + \delta)\Gamma(k + 3 + \delta + \delta/m)}. \tag{9.5.47}$$

The asymptotic degree distribution in (9.5.47) is identical to that in the non-geometric preferential attachment model; see, e.g., (1.3.60). This is true since we are working on a fixed metric space, where vertices become more and more dense. Thus, locally, the behavior is very similar to a normal preferential attachment model, i.e., the spatial effects are "washed away." Remarkably, $\delta \in (-m, 0)$ is not allowed even though that model is perfectly well defined.

So far, we have discussed settings where $I_1 < \infty$, so that, in particular, the geometric component is not very pronounced. We continue to study a setting where $r \mapsto \alpha(r)$ is quite large for small r, so that the proximity of the vertices does become much more pronounced:

Theorem 9.39 (Stretched exponential degrees in geometric preferential attachment models) *Let S be a general metric space, and let μ be the uniform measure on it. Let $\alpha(r) = r^{-s}$ for $s > d$. In the geometric preferential attachment model described in (9.5.45), for any $\gamma \in (0, (s - d)/(2s - d))$, there exists a constant C such that, almost surely,*

$$P_k^{(n)} \leq Ce^{-k^\gamma}. \tag{9.5.48}$$

Thus, we see that if the geometric restriction is highly pronounced, then degree distributions are no longer power-law. We refer to the notes and discussion in Section 9.6 for a brief discussion of related results.

We next discuss a setting where the location of points is not uniform, and thus locally the power-law degree exponent can vary.

Geometric Preferential Attachment Model with Non-Uniform Locations

Again, we assume that S is a bounded metric space, and μ is a (not necessarily uniform) measure on S. This means that $\mu(\mathcal{B}_r(u))$ might depend on u. As before we let $\alpha \colon S \times S \to \mathbb{R}^+$ be an attractiveness function. Here, for simplicity, we restrict to the setting where S is finite, denote $|S| = t$, and for simplicity take $S = [t]$.

We let $(X_v)_{v \geq 1}$ be iid random variables with measure μ. For $s \in [t]$, let

$$P_{s,k}^{(n)} = \frac{1}{n} \sum_{v \in [n]} \mathbb{1}_{\{D_v(n) = k, X_v = s\}} \tag{9.5.49}$$

be the degree distribution of types in the model, where $D_v(n)$ denotes the degree of vertex v at time n. The main result on the degree distribution is as follows:

Theorem 9.40 (Degrees in preferential attachment models with uniform locations) *Let $S = [t]$ be a finite space, and let μ be a measure on it. Assume that $\alpha(x, y) \geq \alpha_0 > 0$. In the geometric preferential attachment model described in (9.5.45),*

$$P_{s,k}^{(n)} \xrightarrow{\mathbb{P}} p_{s,k} = \frac{2\mu_s}{\phi_s} \frac{\Gamma(k)\Gamma(m + \phi(s)^{-1})}{\Gamma(m)\Gamma(k + 1 + \phi(s)^{-1})}, \tag{9.5.50}$$

where $\phi(s)$ satisfies

$$\phi(s) = \sum_{r \in [t]} \frac{\alpha(s, r)}{\sum_{j \in [t]} \alpha(j, r)\nu(j)} \mu(r); \tag{9.5.51}$$

$\nu(r)$ is the limiting proportion of edges at spatial location r.

Theorem 9.40 implies that asymptotic limiting proportions of edges at the various spatial locations exist. Exercise 9.52 investigates the degree distribution of the model.

Spatial Preferential Attachment as Influence

We close the discussion of spatial PAMs by considering a spatial preferential attachment model with local influence regions, inspired by the Web graph. The model is *directed*, but it can easily be adapted to an undirected setting. In preferential attachment models, new vertices should be "aware" of the degrees of the vertices already present. In reality, it is quite hard to observe vertex degrees and, therefore, we let vertices instead have a *region of influence* in some metric space, for example the torus $[0,1]^d$ for some dimension d, for which the metric equals

$$d(x,y) = \min\{\|x - y + u\|_\infty : u \in \{0,1,-1\}^d\}. \tag{9.5.52}$$

When a new vertex arrives, it is uniformly located somewhere in the unit cube, and it connects to each of the older vertices in whose region of influence it lands independently and with fixed probability p. These regions of influence evolve as time proceeds, in such a way that the volume of the influence region of the vertex v at time n is equal to

$$R_v(n) = \frac{a_1 D_v(n) + a_2}{n + a_3}, \tag{9.5.53}$$

where now $D_v(n)$ is the in-degree of vertex v at time n and a_1, a_2, a_3 are parameters which are chosen such that $pa_1 \le 1$. One of the main results is that this model is a scale-free graph process with limiting degree distribution $(p_k)_{k \ge 0}$ satisfying (1.1.9) with $\tau = 1 + 1/(pa_1) \in [2, \infty)$:

Theorem 9.41 (Degrees in preferential attachment models with influence) *In the above preferential attachment models with influence where the volume of the influence region is given by* (9.5.53), *for* $k \le (n^{1/8}/\log n)^{4pa_1/(2pa_1+1)}$,

$$P_k^{(n)} = \frac{1}{n} \sum_{v \in [n]} \mathbb{1}_{\{D_v(n)=k\}} \xrightarrow{\mathbb{P}} p_k, \tag{9.5.54}$$

where

$$p_k = \frac{p^k}{1 + kpa_1 + pa_2} \prod_{j=0}^{k-1} \frac{ja_1 + a_2}{1 + jpa_1 + a_2}. \tag{9.5.55}$$

In Exercise 9.53, the reader is asked to prove that $p_k = ck^{-(1+1/(pa_1))}$ for the p_k in (9.5.55).

9.5.5 COMPLEX NETWORK MODELS ON THE HYPERCUBIC LATTICE

So far, we have studied *finite* random graphs having geometry. However, there is a large body of literature that studies *infinite* random graphs. Usually these models live on the hypercubic lattice and are related to *percolation*. See Grimmett (1999); Kesten (1982); Bollobás and Riordan (2006) for extensive introductions to percolation.

The main questions in percolation relate to the existence and structure of infinite connected components. Many of these models obey a *phase transition* in terms of the infinite connected component. Sometimes such models are also formulated in terms of Poisson processes on the d-dimensional reals, and then the topic touches upon stochastic geometry. See, for example, Meester and Roy (1996) for an introduction to continuum percolation.

(a)   (b)

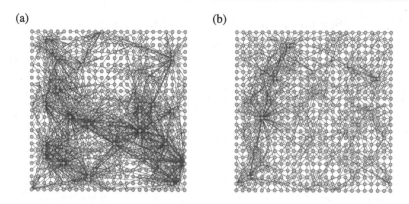

Figure 9.16 Examples of scale-free percolation graphs, with (a) $\tau = 2.5$ and (b) $\tau = 3.5$.

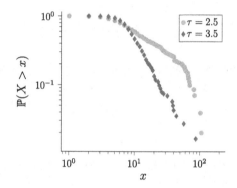

Figure 9.17 Degree distributions of the scale-free percolation graphs in Figure 9.16.

How are these models related to the types of random graphs studied in this section? First of all, infinite models can arise as local limits of finite models. Indeed, we have already seen in Theorem 9.33 that the local limit of a geometric inhomogeneous random graph is formulated in terms of a Poisson process on $\mathbb{R}^d$. However, models on the hypercubic lattice can be expected to arise as local limits of models where the spatial locations of the vertices are on a grid, such as $[n^{1/d}]^d$ (where, to simplify notation, we assume that $n^{1/d}$ is an integer).

For models on $\mathbb{Z}^d$, the definition of power-law degrees needs some adaptation. Indeed, we say that an infinite random graph has power-law degrees when

$$p_k = \mathbb{P}(D_o = k), \tag{9.5.56}$$

where D_x is the degree of vertex $x \in \mathbb{Z}^d$ and $o \in \mathbb{Z}^d$ is the origin, satisfies (1.4.4) for some $\tau > 1$. This is a reasonable definition. Indeed, let $\mathcal{B}_r(o) = [-r, r]^d \cap \mathbb{Z}^d$ be a cube of width $2r$ around the origin $o \in \mathbb{Z}^d$, denote $n = (2r + 1)^d$, and, for each $k \geq 0$, let

$$P_k^{(n)} = \frac{1}{n} \sum_{x \in \mathcal{B}_r(o)} \mathbb{1}_{\{D_x = k\}}, \tag{9.5.57}$$

which, assuming translation invariance and ergodicity, converges almost surely to p_k.

In this section, we discuss *scale-free percolation*, which is a hypercubic lattice model like the inhomogeneous random graph, and *spatial configuration models*. There are currently no results on preferential attachment models on hypercubic lattices.

Scale-Free Percolation

Let each vertex $x \in \mathbb{Z}^d$ be equipped with an iid weight W_x. Conditional on the weights $(W_x)_{x \in \mathbb{Z}^d}$, the edges in the graph are independent and the probability that there is an edge between x and y is

$$p_{xy} = 1 - e^{-\lambda(W_x W_y)^\alpha / |x-y|^{\alpha d}}, \tag{9.5.58}$$

for $\alpha, \lambda \in (0, \infty)$. Here, the parameter $\alpha > 0$ describes the *long-range nature* of the model, while we think of $\lambda > 0$ as the *percolation parameter*. In terms of the *weight distribution*, we are mainly interested in settings where the W_x have unbounded support in $[0, \infty)$ and particularly when they vary substantially, as in (9.5.15). The name *scale-free* percolation is justified by the following theorem:

Theorem 9.42 (Power-law degrees for power-law weights) *Fix $d \geq 1$, consider scale-free percolation as in (9.5.58), and assume that the vertex weights are iid random variables satisfying (9.5.15).*

(a) Let $\alpha \leq 1$ or $\tau \leq 2$. Then $\mathbb{P}(D_o = \infty \mid W_o > 0) = 1$.
(b) Let $\alpha > 1$ and $\tau > 2$. Then there exists $s \mapsto \ell(s)$ that is slowly varying at infinity such that

$$\mathbb{P}(D_o > s) = s^{-(\tau-1)}\ell(s). \tag{9.5.59}$$

When edges are present *independently* but not with the same probability, it is impossible to have infinite-variance degrees in the long-range setting (see Exercise 9.55). Thus, the vertex weights have a pronounced effect on the structure of the graph, particularly when $\tau \in (2,3)$ (see also Exercise 9.56).

We continue by studying the percolative properties of scale-free percolation. Denote the connected component or cluster of x by $\mathscr{C}(x) = \{y : x \longleftrightarrow y\}$, and the number of vertices in $\mathscr{C}(x)$ by $|\mathscr{C}(x)|$. Further, define the *percolation probability* as

$$\theta(\lambda) = \mathbb{P}(|\mathscr{C}(o)| = \infty), \tag{9.5.60}$$

and the *critical percolation threshold* λ_c as

$$\lambda_c = \inf\{\lambda : \theta(\lambda) > 0\}. \tag{9.5.61}$$

It is a priori unclear whether $\lambda_c < \infty$, but $\lambda_c < \infty$ holds in most cases. Indeed, if $\mathbb{P}(W = 0) < 1$, then $\lambda_c < \infty$ in all $d \geq 2$. As for GIRGs, $d = 1$ is special and the results are not optimal. Indeed, $\lambda_c < \infty$ for $\alpha \in (1, 2]$ and $\mathbb{P}(W = 0) < 1$, while $\lambda_c = \infty$ for $\alpha > 2$ and $\tau > 3$.

More interesting is whether $\lambda_c = 0$ or not. The following theorem shows that this depends on whether the degrees have infinite variance:

Theorem 9.43 (Positivity of the critical value) *Fix $d \geq 1$, and consider scale-free percolation as in (9.5.58), with iid weights $(W_x)_{x \in \mathbb{Z}^d}$ satisfying (9.5.15), $\alpha > 1$, and $\tau > 2$.*

(a)   (b)

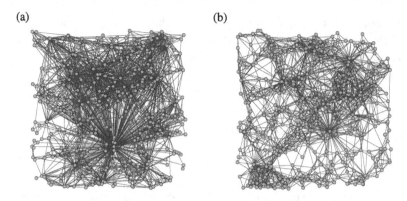

Figure 9.18 Examples of a scale-free percolation graph on the plane, with (a) $\tau = 2.5$ and (b) $\tau = 3.5$.

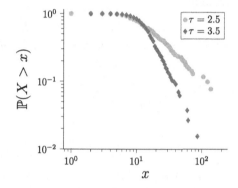

Figure 9.19 Degree distributions of the scale-free percolation on the planar graphs in Figure 9.18.

(a) Assume that $\tau > 3$. Then, $\theta(\lambda) = 0$ for small $\lambda > 0$, that is, $\lambda_c > 0$.
(b) Assume that $\tau \in (2, 3)$. Then, $\theta(\lambda) > 0$ for every $\lambda > 0$, that is, $\lambda_c = 0$.

   In percolation with independent edges, instantaneous percolation in the form $\lambda_c = 0$ can occur only when the degree of the graph is infinite. The *randomness* in the vertex weights facilitates instantaneous percolation, for which $\lambda_c = 0$, in scale-free percolation. We see a similar phenomenon for rank-1 inhomogeneous random graphs, such as the Norros–Reittu model and the configuration model, where the giant is robust (recall, e.g., Theorem 3.20).

   We close our discussion on scale-free percolation by studying the *graph distances*. In finite graphs, typical distances are obtained by choosing two vertices uar from the vertex set, and studying how the graph distance between them evolves when the network size $n \to \infty$. For infinite models, however, we replace this by studying the graph distances between far-away vertices, i.e., we study $\text{dist}_G(x, y)$ for $|x - y|$ large. By translation invariance, this is the same as studying $\text{dist}_G(0, x)$ for $|x|$ large. Below is the main result:

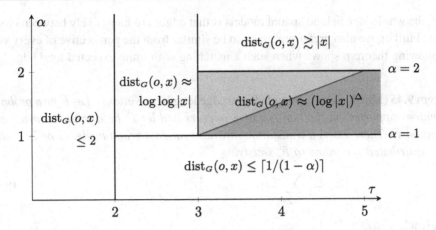

Figure 9.20 Distances in scale-free percolation. Figure adapted from Hao and Heydenreich (2023).

Theorem 9.44 (Distances in scale-free percolation) *Fix $d \geq 1$, consider scale-free perco-lation as in (9.5.58), with iid weights $(W_x)_{x \in \mathbb{Z}^d}$ satisfying (9.5.15), $\alpha > 1$, and $\tau > 2$, and let $\lambda > \lambda_c$. Then, conditional on $o \longleftrightarrow x$,*

(a) for $\tau \in (2, 3)$ and $\alpha > 1$,

$$\frac{\mathrm{dist}_G(o, x)}{\log \log |x|} \xrightarrow{\mathbb{P}} \frac{2}{|\log (\tau - 2)|}; \tag{9.5.62}$$

(b) for $\tau > 3$ and $\alpha \in (1, 2)$, whp for every $\varepsilon > 0$,

$$(\log |x|)^{\Delta' - \varepsilon} \leq \mathrm{dist}_G(o, x) \leq (\log |x|)^{\Delta + \varepsilon}, \tag{9.5.63}$$

for some $0 < \Delta' \leq \Delta < \infty$;

(c) for $\tau > 3$ and $\alpha > 2$, whp, for some $c > 0$,

$$\frac{\mathrm{dist}_G(o, x)}{|x|} \geq c. \tag{9.5.64}$$

Theorem 9.44 shows that the graph distances depend sensitively on all the parameters in the model. Explicit formulas are known for Δ, Δ' in Theorem 9.44(b). For example, $\Delta = \Delta' = \log 2 / \log (2/\alpha)$ when $\alpha(\tau - 1) > 2$, but $\Delta' < \Delta$ in other cases. See Figure 9.20 for a pictorial representation of Theorem 9.44, where the lighter shaded area signifies that $\Delta < \Delta'$, while the darker shaded area signifies that $\Delta' = \Delta$.

Spatial Configuration Models on the Lattice

Let $(D_x)_{x \in \mathbb{Z}^d}$ be iid integer-valued random variables. Can we construct a simple graph where vertex x has exactly degree D_x? This is a *matching problem*, where we can think of D_x as the number of half-edges incident to vertex x and we wish to pair the half-edges, as in a configuration model. However, it is also desirable that the edges are not too long.

Indeed, the whole idea behind spatial models is that edges are more likely between close-by vertices. Further, we also prefer our graph to be similar from the perspective of every vertex. The following theorem shows when such a matching with finite expected total edge length exists:

Theorem 9.45 (Matching with bounded total edge length per vertex)   *Let F be a probability distribution supported on the non-negative integers and let D be a random variable with distribution F. There exists a translation invariant random graph model G on $\mathbb{Z}^d$, with iid degrees distributed according to F, satisfying*

$$\mathbb{E}\Big[\sum_{x \in \mathbb{Z}^d} \mathbb{1}_{\{\{o,x\} \in E(G)\}} |x - o| \Big] < \infty, \tag{9.5.65}$$

precisely when $\mathbb{E}[D^{(d+1)/d}] < \infty$.

Since it is hard even to *construct* the spatial configuration model, it may not come as a surprise that few results are known for this particular model.

We close this section by discussing a particular matching for $d = 1$ that is rather natural. Give each half-edge a *direction* uar, meaning that it points to the left or to the right with equal probability and independently across the edges. The edges are then obtained by pairing half-edges pointing to each other, first exhausting all possible connections between nearest neighbors, then linking second-nearest neighbors, and so on.

Assuming that D has finite mean, it is known that this algorithm leads to a well-defined configuration, but that the expected length of the longest edge attached to a given vertex is infinite. Indeed, let N be the length of the edge to the furthest neighbor of o. Then, $N < \infty$ almost surely. If we assign directions with a probability that is not equal to $\frac{1}{2}$ then $N = \infty$ with positive probability. Further, let N_1 denote the length of the first edge incident to o. Then, $\mathbb{E}[N_1] = \infty$ when $\mathbb{E}[D] < \infty$. Thus, edges have finite (spatial) length, but their lengths have infinite mean.

9.6 NOTES AND DISCUSSION FOR CHAPTER 9

In this chapter we have given an extensive introduction to random graph models for networks that are directed, and have community structures and geometry. Obviously, these additional features can also be combined, but there is only a limited literature on that, and we refrain from discussing such models.

We have not been able to cover all the relevant models that have attracted attention in the literature. Particularly for *dynamic* random graphs, we have not discussed some of the relevant models. Examples are *copying* or *duplication models*; these are dynamic random graphs in which new vertices copy a portion of the neighbors of an older vertex (Kumar et al. (2000)).

Another class of dynamic models that has attracted considerable attention are models aimed at delaying or accelerating the birth of the giant component. Indeed, one can obtain the combinatorial Erdős–Rényi random graph by adding edges uniformly one by one, until the desired number of edges is added; then the distribution is the same as that for the Erdős–Rényi random graph with the same number of edges. For this process there is a giant when the number of added edges is $m = cn$ with $c > \frac{1}{2}$, while there is not giant for $m = cn$ with $c < \frac{1}{2}$.

This process can be modified using a "power of choice" by considering a *pair* of edges at each time. Achlioptas raised the question whether is it possible to select one of the two exposed edges at each stage in such a way as to either delay or accelerate the birth of a giant? Spencer and Wormald (2007) called this aptly *birth control for giants*. In general, one can select the edge for which the connected components on either side are the smallest in some sense. Bohman and Frieze (2001, 2002) studied settings where the first edge is taken when it connects two isolated vertices, but otherwise the second edge is chosen. They showed that there is no giant yet when the number of added edges is $m = cn$ with $c > 0.535$, so that this indeed delays the birth of the giant. Spencer and Wormald (2007) narrowed down the birth of the giant as being in between $m = 0.8293n$ and $m = 0.9823n$.

Intuitively, one may guess that the birth of the giant is delayed most when the chosen edge minimizes the *product* of the connected component sizes of the vertices in the edge. This is the so-called *product rule*, and is sometimes also called *explosive percolation* since the size of the giant grows very fast after the giant is first formed. In fact, this led Achlioptas et al. (2009) to the conjecture that the limiting size of the giant for $m = cn$ might be *discontinuous* around the critical value. This, however, turns out not to be true, as proved by Riordan and Warnke (2011). It is as yet unclear how the limiting proportion of vertices in the giant grows slightly beyond the critical value, though.

The notes to this chapter are substantially more extensive than those in other chapters, because many models are being discussed, and we only have limited space. As before, the notes can be used to learn more about the models and to get pointers to the literature.

Notes on Section 9.1

Citation networks. Our discussion of citation networks is inspired by, and follows, Garavaglia et al. (2017). For citation networks, there is a rich literature proposing models for them using preferential attachment schemes and adaptations of these, mainly in the complex-networks literature. Aging effects, i.e., taking account of the *age of a paper* in its likelihood of obtaining citations, have been extensively considered as the starting point to investigate the dynamics of citation networks; see Wang et al. (2009, 2008); Hajra and Sen (2005, 2006); Csárdi (2006). Here the idea is that old papers are less likely to be cited than new papers. Such aging has been observed in many citation network data sets and makes PAMs with weight functions depending only on the degree ill-suited for them. Indeed, PAMs could more aptly be called *the-old-get-richer* models, i.e., in general *old* vertices have the highest degrees. In citation networks, however, papers with many citations appear all the time. Wang et al. (2013) investigated a model that incorporates these effects; see also Wang et al. (2014) for a comment on the methods in that paper. On the basis of empirical data, they suggested a model where the aging function follows a log-normal distribution with paper-dependent parameters, and the preferential attachment function is the identity. Wang et al. (2013) *estimated* the fitness function rather than using the more classical approach where the latter is taken to be an iid sample of random variables.

Notes on Section 9.2

Local convergence for PageRank as stated in Theorem 9.1 was proved by Garavaglia et al. (2020), where the notion of the marked backward local limit was also introduced. We have extended the discussion on local convergence for directed graphs here, since these other notions are useful in other contexts as well, for example in studying the strongly connected component of directed random graphs.

We also take the opportunity here to correct the statement of (Garavaglia et al., 2020, Theorem 2.1). Indeed, in the proof of (Garavaglia et al., 2020, Theorem 2.1(1)), the more general result that $\mathbb{P}(R_{o_n}^{(G_n)} > r) \to \mu(R_\varnothing > r)$ for *every* $r \geq 0$ is stated. However, the proof of this result is incomplete, as noted by Francesco Caravenna and Federica Finazzi. The statement of convergence for all continuity points r of $r \mapsto \mu(R_\varnothing > r)$ does follow by using an appropriate Markov inequality. Similarly, (Garavaglia et al., 2020, Theorem 2.1(2)) states the convergence in (9.2.7) *for every* $r \geq 0$, while the proof can only be extended to show (9.2.7) for continuity points of $r \mapsto \mu(R_\varnothing \leq r)$. We thank Francesco and Federica for notifying us of this oversight.

In more detail, for $N \geq 1$, let $R_{o_n}^{(N,G_n)}$ denote the contribution due to $k \leq N$ on the rhs of (9.2.9) and, similarly, let $R_\varnothing^{(N)}$ be that of the rhs in (9.2.10). Then, (Garavaglia et al., 2020, Lemma 4.1) states that $\mathbb{E}[R_{o_n}^{(G_n)} - R_{o_n}^{(N,G_n)}] \leq \alpha^{N+1}$. Below (Garavaglia et al., 2020, (29)), it is claimed that this implies that also

$$|\mathbb{P}(R_{o_n}^{(G_n)} > r) - \mathbb{P}(R_{o_n}^{(N,G_n)} > r)| = \mathbb{P}(R_{o_n}^{(G_n)} > r, R_{o_n}^{(N,G_n)} \leq r) \leq \alpha^{N+1}, \qquad (9.6.1)$$

but this is incorrect. However, by the Markov inequality, we do get that, for every $\varepsilon > 0$,

$$\mathbb{P}(R_{o_n}^{(G_n)} > r, R_{o_n}^{(N,G_n)} \leq r - \varepsilon) \leq \mathbb{P}(R_{o_n}^{(G_n)} - R_{o_n}^{(N,G_n)} \geq \varepsilon)$$

$$\leq \frac{1}{\varepsilon}\mathbb{E}[R_{o_n}^{(G_n)} - R_{o_n}^{(N,G_n)}] \leq \alpha^{N+1}/\varepsilon. \qquad (9.6.2)$$

Continuing the argument below (Garavaglia et al., 2020, (29)) with the above bound instead, this shows that the stated convergence follows provided that $\mu(R_\varnothing > r)$ is continuous at the value r. See Exercises 9.57 and 9.58 for a more precise analysis.

Directed inhomogeneous random graphs. Bloznelis et al. (2012) studied the general directed inhomogeneous random graph considered in this section and proved Theorem 9.3. Cao and Olvera-Cravioto (2020) continued this analysis and generalized it substantially. While the local convergence result in Theorem 9.2 has not been proved anywhere explicitly, it is the leading idea in the identification of the phase transition as well as in the description of the limiting branching processes and joint degree distribution. Garavaglia et al. (2020) investigated the marked directed forward local convergence for directed rank-1 inhomogeneous random graphs.

Cao and Olvera-Cravioto (2020) specifically investigated general rank-1 inhomogeneous digraphs. Our choice of edge probabilities in (9.2.14) and (9.2.15) is slightly different from that of Cao and Olvera-Cravioto (2020), particularly since the factor $\frac{1}{2}$ in (9.2.15) is absent in Cao and Olvera-Cravioto (2020). We have added it so as to make $w_v^{(\mathrm{in})}$ approximately the average in-degree of vertex v. If this were to be true for every v, then we would also need that

$$\sum_{v \in [n]} w_v^{(\mathrm{in})} \approx \sum_{v \in [n]} w_v^{(\mathrm{out})},$$

which would imply the limiting statement in (9.2.18). Lee and Olvera-Cravioto (2020) used the results in Cao and Olvera-Cravioto (2020) to prove that the limiting PageRank of such directed generalized random graphs exists and that the solution obeys the same recurrence relation as on a branching-process tree. In particular, under certain independence assumptions, this implies that the PageRank power-law hypothesis is valid for such models.

Directed configuration models were investigated by Cooper and Frieze (2004), who proved Theorem 9.5. In fact, the results in Cooper and Frieze (2004) also prove detailed bounds on the strongly connected component in the subcritical regime, as well as precise bounds on the number of vertices whose forward and backward clusters are large and the asymptotic size of forward and backward clusters. A substantially simpler proof was given by Cai and Perarnau (2021).

Both Cooper and Frieze (2004) as well as Cai and Perarnau (2021) made additional assumptions on the degree distribution. In particular, they assumed that $\mathbb{E}[D_n^{(\mathrm{in})} D_n^{(\mathrm{out})}] \to \mathbb{E}[D^{(\mathrm{in})} D^{(\mathrm{out})}] < \infty$, which we do not. Further, Cooper and Frieze (2004) assumed that $\boldsymbol{d}$ is *proper*, which is a technical requirement on the degree sequence stating that (a) $\mathbb{E}[(D_n^{(\mathrm{in})})^2] = O(1)$, $\mathbb{E}[(D_n^{(\mathrm{out})})^2] = O(1)$; (b) $\mathbb{E}[D_n^{(\mathrm{in})}(D_n^{(\mathrm{out})})^2] = o(n^{1/12} \log n)$. In view of the fact that such conditions do not appear in Theorem 4.9, these conditions can be expected to be suboptimal for Theorem 9.5 to hold, and we next explain how they can be avoided by a suitable *degree-truncation argument*, which we now explain:

Assume that the out- and in-degrees in the directed configuration model $\mathrm{DCM}_n(\boldsymbol{d})$ satisfy (9.2.25) and (9.2.26). By Exercise 9.19 below, $|\mathscr{C}_{\max}| \leq n(\zeta + \varepsilon)$ whp for n large and any $\varepsilon > 0$. Exercise 9.19 is proved by an adaptation of the proof of Corollary 2.27 in the undirected setting. Thus, we need to show only that $|\mathscr{C}_{\max}| > n(\zeta - \varepsilon)$ whp for n large and any $\varepsilon > 0$

Fix $b > 1$ large. We now construct a lower bounding directed configuration model where all the degrees are bounded above by b. This is similar to the degree-truncation argument for the undirected configuration model discussed in Section 1.3.3 (recall Theorem 1.11). When v is such that $d_v = d_v^{(\mathrm{out})} + d_v^{(\mathrm{in})} \geq b$, we split v into $n_v = \lceil d_v/b \rceil$ vertices, and we deterministically redistribute all out- and in-half-edges over the n_v vertices in an arbitrary way such that the out- and in-degrees of all the vertices that used to correspond to v now have both out- and in-degree bounded by b. Denote the corresponding random graph by DCM'_n.

The resulting degree sequence again satisfies (9.2.25) and (9.2.26). Moreover, for $b > 1$ large and by (9.2.25) and (9.2.26), the limits in (9.2.25) and (9.2.26) for the new degree sequence are quite close to the original limits of the old degree sequence, while the degrees are now *bounded*. As a result, we can apply the original result in Cooper and Frieze (2004) or Cai and Perarnau (2021) to the new setting.

Denote the size of the largest SCC in DCM'_n by $|\mathscr{C}'_{\max}|$. Obviously, since we split vertices, $|\mathscr{C}'_{\max}| \leq |\mathscr{C}_{\max}| + \sum_{v \in [n]} (n_v - 1)$. Therefore, $|\mathscr{C}_{\max}| \geq |\mathscr{C}'_{\max}| - \sum_{v \in [n]} (n_v - 1)$. Take $b > 0$ sufficiently large that $\sum_{v \in [n]} (n_v - 1) \leq \varepsilon n/3$ and that $\zeta' \geq \zeta - \varepsilon/3$, where ζ' is the forward–backward survival probability of the limiting directed DCM'_n and ζ that of $\mathrm{DCM}_n(\boldsymbol{d})$. Finally, for every $\varepsilon > 0$, whp $|\mathscr{C}'_{\max}| \geq n(\zeta' - \varepsilon/3)n$. As a result, we obtain that, again whp and as required,

$$|\mathscr{C}_{\max}| \geq |\mathscr{C}'_{\max}| - n\varepsilon/3 \geq n(\zeta' - 2\varepsilon/3) \geq n(\zeta - \varepsilon). \tag{9.6.3}$$

Chen and Olvera-Cravioto (2013) studied a way to obtain nearly iid in- and out-degrees in the directed configuration model. Here, the problem is that if $((d_v^{(\text{in})}, d_v^{(\text{out})}))_{v\in[n]}$ is an iid bivariate distribution with equal means, then $\sum_{v\in[n]}(d_v^{(\text{in})} - d_v^{(\text{out})})$ has Gaussian fluctuations at best, so that it will likely not be zero. Chen and Olvera-Cravioto (2013) indicated how the excess in- or out-half-edges can be removed so as to keep the degrees close to iid. Further, they showed that the removal of self-loops and multiple directed edges does not significantly change the degree distribution (so that, in particular, one would assume that the local limits would be the same, but Chen and Olvera-Cravioto (2013) considered only the degree distribution).

Theorem 9.6 was first proved by van der van der Hoorn and Olvera-Cravioto (2018) under stronger assumptions, but then the claim proved is also much stronger. Indeed, van der van der Hoorn and Olvera-Cravioto (2018) not only identified the first-order asymptotics, as in Theorem 9.6, but also fluctuations like those stated for the undirected configuration model in Theorem 7.24. This proof is substantially harder than that of (Cai and Perarnau, 2023, Proposition 7.7). Theorem 9.7 was proved by Colenbrander (2022). Theorem 9.8 was proved by Cai and Perarnau (2023).

Directed preferential attachment models and its PageRank. The backward local limit of the preferential attachment model with random out-degrees was identified by Banerjee et al. (2023), using a collapsing procedure on a continuous-time branching process in the spirit of Garavaglia and van der Hofstad (2018). Theorem 9.9 is (Banerjee and Olvera-Cravioto, 2022, Theorem 3.1), Theorem 9.10 is (Banerjee and Olvera-Cravioto, 2022, Theorem 3.3).

Notes on Section 9.3

Stochastic block models (SBMs) were introduced by Holland et al. (1983) in the context of group structures in social networks. Their introduction was inspired by the work of Fienberg and Wasserman (1981). There, the problem of community detection was also discussed. Of course, we have already seen SBMs in the context of inhomogeneous random graphs, so our focus here is on the detection problem. The definition of "solvable" in Definition 9.11, in particular (9.3.2), appeared in Bordenave et al. (2018).

The precise threshold in Theorem 9.12 was conjectured by Decelle et al. (2011), on the basis of a non-rigorous belief propagation method. The proof of the impossibility result of the block model threshold conjecture was first given by Mossel et al. (2015). The proof of the solvable part of the block model threshold conjecture was given by Mossel et al. (2018) and independently by Massoulié (2014). Mossel et al. (2016) gave an algorithm that maximizes the fraction of vertices labeled correctly. The results were announced by Krzakala et al. (2013). The achievability result for a general number t of types in (9.3.5) was proved by Abbe and Sandon (2018) and Bordenave et al. (2018), whose presentation we follow. The convergence of the eigenvalues in (9.3.8) is (Bordenave et al., 2018, Theorem 4) and the estimation of the types in (9.3.10) is investigated in (Bordenave et al., 2018, Theorem 5). There, it is also shown that this choice satisfies the solvability condition (9.3.2) in Definition 9.11. We have simplified the presentation substantially by considering $t = 2$ and equal sizes of the groups of the two types.

Degree-corrected stochastic block models were first introduced by Karrer and Newman (2011). The consistent estimation of communities in degree-corrected SBM was investigated by Zhao et al. (2012), under the assumption that the average degree tends to infinity for weak consistency (meaning that *most* pairs of vertices that are in the same community are correctly estimated to be such), and grows faster than $\log n$ for strong consistency (meaning that *all* pairs of vertices that are in the same community are correctly estimated to be such). Sparse settings suffer from a similar threshold phenomenon as that for the original stochastic block model, as derived in Mossel et al. (2018) and Massoulié (2014). The impossibility result in Theorem 9.13 was proved by Gulikers et al. (2018). The solvable parts were proved by Gulikers et al. (2017a,b).

Preferential attachment models with community structure. Jordan (2013) investigated preferential attachment models in general metric spaces. When one takes these metric spaces as discrete sets, one can interpret the geometric location as a type or community label. This interpretation was proposed by Hajek and Sankagiri (2019), where our results on community detection are also proved. We return to the geometric interpretation of the model in Section 9.5.4. The result in (9.3.24) is stated in (Jordan, 2013, Theorem 2.1). The asymptotics of the degree distribution in Theorem 9.14 is (Jordan, 2013, Theorem 2.2). (Jordan, 2013, Theorem 2.3) further contains some estimates of the number of vertices in given regions and with given degrees. The convergence of the proportion of errors in (9.3.30) is stated in (Hajek and Sankagiri, 2019, Proposition 8), to which we refer for the formula for Err.

A preferential attachment model with community structure, phrased as a coexistence model, was introduced by Antunović et al. (2016). In their model, contrary to the setting of Jordan (2013), the vertices choose their type on the basis of the types of their neighbors, thus creating the possibility of denser connectivity between vertices of the same type and thus community structure. The focus of Antunović et al. (2016) is the *coexistence* of all the different types or rather a *winner-takes-it-all* phenomenon, depending on the precise probability of choosing a type. Even with two types, the behavior is quite involved and depends sensitively on the distribution of type choices. For example, for the majority rule (where the type of the new vertex is the majority-type of its older neighbors), the winner type takes all, while if this probability is *linear* in the number of older neighbors of a given type, there is always coexistence.

Notes on Section 9.4

Empirical properties of real-world network with community structure were studied by Stegehuis et al. (2016b).

Inhomogeneous random graphs with communities were introduced by Bollobás et al. (2011).

Configuration models with community structure. The hierarchical configuration model was introduced by van der Hofstad et al. (2017). The fit to real-world networks, particularly in the context of epidemics, was studied by Stegehuis et al. (2016a). The configuration model with household structure was investigated in Ball et al. (2009, 2010) in the context of epidemics on social networks. Particularly when studying epidemics on networks, clustering is highly relevant, as it can slow down the spread of infectious diseases. The configuration model with clustering was defined by Newman (2009), who studied it, though not rigorously.

Random intersection graphs were introduced by Singer (1996) and further studied in Fill et al. (2000); Karoński et al. (1999); Stark (2004). Theorem 9.20 is (Deijfen and Kets, 2009, Theorem 1.1), where the authors also proved that clustering can be tuned. The model was also investigated for more general distributions of groups per vertex by Godehardt and Jaworski (2003); Jaworski et al. (2006). Random intersection graphs with prescribed degrees and groups are studied in a non-rigorous way in Newman (2003); Newman and Park (2003). We refer to Bloznelis et al. (2015) for a survey of recent results.

Rybarczyk (2011) studied various properties of the random intersection graph when each vertex is in precisely d groups that are all chosen uar from the collection of groups. In particular, Rybarczyk (2011) proves results on the giant as in Theorem 9.22, as well as on the diameter of the graph, which is $\Theta_{\mathbb{P}}(\log n)$ when the model is sparse.

Bloznelis (2009, 2010a,b) studied a general random intersection model, where the sizes of groups are iid random variables, and the sets of the vertices in them are chosen uar from the vertex set. His results include distances (Bloznelis (2009)) and component sizes (Bloznelis (2010a,b)). Bloznelis (2013) studied the degree and clustering structure in this setting.

Theorem 9.21 is proved in Kurauskas (2022); see also van der Hofstad et al. (2021). Both papers investigate more general settings: Kurauskas (2022) also allows for settings with *independent* group memberships, while van der Hofstad et al. (2021) also allows for more *general group structures* than the complete graph. Theorem 9.22 is proved in van der Hofstad et al. (2022).

Random intersection graphs with communities. van der Hofstad et al. (2021) proposed a model that combines the random intersection graph with more general communities than complete graphs. van der Hofstad et al. (2021) identified the local limit, as well as the nature of the overlaps between different communities. van der Hofstad et al. (2022) identified the giant component, also when percolation is being performed on the model. See Vadon et al. (2019) for an informal description of the model, aimed at a broad audience.

Exponential random graphs. For a general introduction to exponential random graphs, we refer to Snijders et al. (2006) and Wasserman and Pattison (1996). Frank and Strauss (1986) discussed the notion of *Markov graphs*, for which the edges of the graph form a Markov field. The general exponential random graph is only a Markov field when the subgraphs are restricted to edges, stars of any kind, and triangles. This is exemplified by Example 9.24, where general degrees are used and give rise to a model with independent edges. Kass and Wasserman (1996) discussed relations to Bayesian statistics.

For a discussion on the relation between statistical mechanics and exponential models, we refer to Jaynes (1957). Let us now explain the relation between exponential random graphs and entropy maximization. Let $(p_x)_{x \in \mathcal{X}}$ be a probability measure on a general discrete set $\mathcal{X}$. We define its *entropy* by

$$H(p) = - \sum_{x \in \mathcal{X}} p_x \log p_x. \tag{9.6.4}$$

Entropy measures the amount of randomness in a system. Shannon (1948) proved that the entropy is the unique quantity that is positive, increases with increasing uncertainty, and is additive for independent sources of uncertainty, so it is a very natural quantity.

The relation to exponential random graphs is that they are the random graphs that, for fixed expected values of the subgraph counts $N_F(G_n)$, optimize the entropy. Indeed, recall that $X = (X_{i,j})_{i \leq i < j \leq n}$ are the edge statuses of the graph, so that G_n is uniquely characterized by X. Then maximize $H(p)$ over all the p such that $\sum_x N_F(x) = \alpha_F$ for some given α_F and all subgraphs $F \in \mathcal{F}$, where $\mathcal{F}$ is an appropriate set of subgraphs. Then, using Lagrange multipliers, the optimization problem reduces to

$$p_{\vec{\beta}}(x) = \frac{1}{Z} e^{\sum_{F \in \mathcal{F}} \beta_F N_F(x)}, \tag{9.6.5}$$

where $Z = Z_n(\vec{\beta})$ is the normalization constant given in (9.4.33), and $\vec{\beta} = (\beta_F)_{F \in \mathcal{F}}$ is chosen as the solution to (9.4.34). This implies that, indeed, the exponential random graph model optimizes the entropy under this subgraph constraint.

See also Kass and Wasserman (1996) for a discussion of maximum entropy and a reference to its long history as well as a critique of the method.

An important question is how one can find the appropriate $\vec{\beta} = (\beta_F)_{F \in \mathcal{F}}$ such that (9.4.34) holds. This is particularly difficult, since the computation of the normalization constant $Z_n(\vec{\beta})$ in (9.4.33) is quite hard. Often, Markov chain Monte Carlo (MCMC) techniques are used to sample efficiently from $p_{\vec{\beta}}$. In this case, such MCMC techniques perform a form of Glauber dynamics, for which (9.4.32) is the stationary distribution. One can then try to solve (9.4.34) by keeping track of the value of N_F in the simulation. However, these methods can be very slow, as well as daunting, since the behavior of $N_F(X)$ under $p_{\vec{\beta}}$ may undergo a *phase transition* in $\vec{\beta}$, making $\sum_x N_F(x) p_{\vec{\beta}}(x)$ very sensitive to small changes of $\vec{\beta}$. See in particular Chatterjee and Diaconis (2013) for a discussion on this topic.

Bhamidi et al. (2011) (see also Bhamidi et al. (2008)) studied the mixing time of the exponential random graph when edges are changed dynamically in a Glauber way. The results were somewhat disappointing, since either edges are close to being iid as in an Erdős–Rényi random graph or the mixing is very slow. These results, however, apply only to *dense* settings where the number of edges grows quadratically with the number of vertices. This problem is closely related to large deviations on dense Erdős–Rényi random graphs. See also Chatterjee (2017) for background on such large deviations and Chatterjee and Varadhan (2011) for the original paper.

Results on *sparse* exponential random graphs are limited. We refer to Chakraborty et al. (2021) for a discussion of how sparse exponential random graphs with a linear number of triangles can be obtained.

For more background on sufficient statistics and their relation to symmetries, we refer to Diaconis (1992). For a discussion of the relation between information theory and exponential models, we refer to Shore and Johnson (1980).

Notes on Section 9.5

The small-world model was first introduced by Watts and Strogatz (1998). Newman and Watts (1999) presented a minor adaptation called the Newman–Watts model, which is such that shortcuts are *added* rather than edges being *rewired*. This has the advantage that the model remains connected. Small-world models were first analyzed in Moore and Newman (2000); Newman et al. (2000a); Newman and Watts (1999), and a non-rigorous mean-field analysis of distances in small-world models was first performed in Newman et al. (2000a). There are various ways of adding long-range connections (for example by rewiring the existing edges), and we have focussed on the models by Barbour and Reinert (2001, 2004, 2006), for which the strongest mathematical results have been obtained. See Barbour and Reinert (2001) for a discussion of the differences between the exact and mean-field analyses. The definition of the great-circle model by Ball et al. (1997) actually precedes Moore and Newman (2000); Newman et al. (2000a); Newman and Watts (1999). See Ball and Neal (2002, 2004, 2008) for further results on this model.

Theorem 9.25 is (Barbour and Reinert, 2001, Theorem 3.9). The description of the limiting random variable in terms of Gumbel distributions can be found in (Barbour and Reinert, 2006, page 1242). The proof of Theorem 9.25 was extended by Barbour and Reinert (2006) to deal with discrete tori where the usual distances on the torus are considered (but the shortcuts still do not count as single edges).

Theorem 9.26 follows from (Barbour and Reinert, 2006, Theorem 4.1), which even allows λ, ρ, and k to depend on n, as long as $\lambda_n n \to \infty$ and $\rho_n k_n$ remains bounded. When $\rho_n \to 0$, the result in (9.5.2) in Theorem 9.25 is retrieved for the discrete small-world model.

A related small-world model was considered by Turova and Vallier (2010), who studied a union of subcritical percolation on a finite cube and the Erdős-Rényi random graph. Using the methodology of

Bollobás et al. (2007), they showed that the phase transition is similar to the one described in Theorem 3.19.

Hyperbolic random graphs were introduced by Krioukov et al. (2010), who considered hyperbolic random graphs in $d \geq 2$. The mathematical results restrict to $d = 1$. The first rigorous results were obtained by Gugelmann et al. (2012), who proved Theorem 9.27 and identified the exact asymptotic degree distribution (see (Gugelmann et al., 2012, Theorem 2.2)), as well as the positivity of the clustering in the graph (see (Gugelmann et al., 2012, Theorem 2.1)). The result about the maximal degree in the hyperbolic graph is stated in (Gugelmann et al., 2012, Theorem 2.4). The sharpest results on the degree distribution and clustering were proved by Fountoulakis et al. (2021), who identified the *exact* clustering coefficient in terms of various special functions in an involved paper.

Bode et al. (2015) proved the main parts of Theorem 9.28. More precisely, instead of the law of large numbers for the giant, Bode et al. (2015) proved a linear lower bound on $|\mathscr{C}_{\max}|$. Kiwi and Mitsche (2019) proved that the second largest component is at most polylogarithmic for $\alpha \in [\frac{1}{2}, 1)$ (i.e., at most a power of $\log n$), and at most n^γ for some $\gamma \in (0, 1)$ for $\alpha = 1$. The law of large numbers follows from these results by a "giant is almost local" proof, and using the local convergence proved in van der Hofstad et al. (2023). See van der Hofstad (2021) for further details. Bode et al. (2015) also showed that $|\mathscr{C}_{\max}| = \Theta_{\mathbb{P}}(R^2 (\log \log R)^3 n^{1/\alpha})$ for $\alpha > 1$. For $\alpha \in (\frac{1}{2}, 1)$, Friedrich and Krohmer (2018) and Kiwi and Mitsche (2015) (see also Friedrich and Krohmer (2015)) studied the diameter of these hyperbolic random graphs, and Abdullah et al. (2017) identified their ultra-small-world behavior in Theorem 9.29. Bläsius et al. (2018) studied the size of the largest cliques in hyperbolic graphs.

Geometric inhomogeneous random graphs (GIRG)s. Product GIRGs were introduced by Bringmann et al. (2019) and Bringmann et al. (2020). The relation between the hyperbolic random graph and the product GIRG as described in Theorem 9.31 can be found in (Bringmann et al., 2017, Section 7), where limits are derived up to constants. Theorem 9.31 was first proved in (Komjáthy and Lodewijks, 2020, Section 7) under conditions slightly different from Assumption 9.30. The current statement of Assumption 9.30 is Assumptions 1.5–1.7 in van der Hofstad et al. (2023). An adaptation of the proof of Theorem 9.31 can be found in Section 2.1.3 in van der Hofstad et al. (2023). Komjáthy and Lodewijks (2020) also studied its weighted distances focussing on the case where $\tau \in (2, 3)$. Theorem 9.34 was proved in Bringmann et al. (2020) for product GIRGs with $\tau \in (2, 3)$, except for the law of large numbers of the giant. This again follows by a "giant is almost local" proof combined with the bound on the second largest component, proved in Bringmann et al. (2020), and the identification of the local limit in van der Hofstad et al. (2023). The result for the model in (9.5.35) and (9.5.36) is (Jorritsma et al., 2023, Corollary 2.3). The GIRGs in (9.5.35) and (9.5.36) are called *interpolating kernel-based spatial random graphs* by Jorritsma et al. (2023). The main focus of Jorritsma et al. (2023) is the size of the second largest connected component $|\mathscr{C}_{(2)}|$, on which the authors proved sharp polylogarithmic bounds with the correct exponent.

The local convergence in probability in Theorem 9.33 is proved in van der Hofstad et al. (2023) using path-counting techniques. Local weak convergence for product GIRGs was proved under slightly different assumptions by Komjáthy and Lodewijks (2020) (see (Komjáthy and Lodewijks, 2020, Assumption 2.5)). In more detail, a coupling version of Theorem 9.33 is stated in (Komjáthy and Lodewijks, 2020, Claim 3.3), where a blown-up version of the product GIRG is bounded from below and above by the limiting model with slightly smaller and larger intensities, respectively. Take a vertex uar in the product GIRG. Then whp it is also present in the lower and upper bounding Poisson infinite GIRG. Similarly, whp none of the edges within a ball of intrinsic radius r will be different in the three models, which proves local weak convergence. Local convergence in probability would follow from a coupling of the neighborhoods of two uniformly chosen vertices in the GIRG to two independent limiting copies. Such independence is argued in (Komjáthy and Lodewijks, 2020, proof of Theorem 2.12), in particular in the text around (Komjáthy and Lodewijks, 2020, (3.16)). It can be expected that a hyperbolic random graph in d dimensions can be mapped to a product GIRG in $d-1$ dimensions. The one-dimensional nature of the model for $d = 2$ discussed below Theorem 9.34 should thus not arise when $d \geq 3$, and one can expect a giant to exist for $\tau > 3$ also.

Local limits as arising in Theorem 9.33, in turn, were studied by Hirsch (2017). Fountoulakis (2015) studied an early version of a geometric Chung–Lu model.

Spatial preferential attachment models. Our exposition follows Jordan (2010, 2013); Flaxman et al. (2006, 2007). Jordan (2010) treats the case of uniform locations of the vertices, a problem first suggested by Flaxman et al. (2006, 2007). Jordan (2013) studies preferential attachment models where the vertices are located in a general metric space with not-necessarily-uniform location of the vertices. This is more difficult, as then the power-law degree exponent depends on the location of the vertices. Theorem 9.37 follows from (Flaxman et al., 2006, Theorem 1(a)), which is quite a bit sharper, as it states detailed concentration results as well. Further results involve the proof of connectivity of the resulting graph and an upper bound on the diameter of order $O(\log{(n/r)})$ when $r \geq n^{-1/2} \log n$, $m \geq K \log n$ for some large enough K, and $\alpha \geq 0$. Flaxman et al. (2007) generalized these results to the setting where, instead of a unit ball, a smoother version is used, while the majority of points were still within a distance $r_n = o(1)$. Theorem 9.38 is (Jordan, 2010, Theorem 2.1). Theorem 9.39 is (Jordan and Wade, 2015, Theorem 2.4). (Jordan and Wade, 2015, Theorem 2.2) shows that the degree distribution for $\alpha(r) = \exp\{(\log(1/r))^\gamma\}$ for $\gamma > \frac{3}{2}$ is the same as the so-called *online nearest-neighbor graph*, for which (Jordan and Wade, 2015, Theorem 2.1) shows that the limiting degree distribution has exponential tails. Manna and Sen (2002) studied geometric preferential attachment models from a simulation perspective.

Theorem 9.40 is (Jordan, 2013, Theorems 2.1 and 2.2). (Jordan, 2013, Theorem 2.3) contains partial results for the setting where $\mathcal{S}$ is infinite. These results are slightly weaker, as they do not characterize the degree power-law exponent exactly.

Aiello et al. (2008) gave an interpretation of spatial preferential attachment models in terms of *influence regions* and proved Theorem 9.41 (see (Aiello et al., 2008, Theorem 1.1)). Further results involve the study of maximal in-degrees and the total number of edges. See also Janssen et al. (2016) for a version with non-uniform locations.

Jacob and Mörters (2015) studied the degree distribution and local clustering in a related geometric preferential attachment model. Jacob and Mörters (2017) studied the robustness of the giant component in that model, and also presented heuristics that distances are ultra-small in the case where the degrees have infinite variance.

For a relation between preferential attachment graphs with so-called fertility and aging, and a geometric competition-induced growth model for networks, we refer to Berger et al. (2004, 2005) and the references therein. Zuev et al. (2015) studied how geometric preferential attachment models give rise to soft communities.

Complex network models on the hypercubic lattice. Below we give references to the literature.

Scale-free percolation was introduced by Deijfen et al. (2013). We have adapted the parameter choices, so that the model is closer to the geometric inhomogeneous random graph. In particular, in Deijfen et al. (2013), (9.5.58) is replaced with

$$p_{xy} = 1 - e^{-\lambda W_x W_y / |x-y|^\alpha}, \tag{9.6.6}$$

and then the power-law exponent for the degrees is such that $\mathbb{P}(D_o > k) \approx k^{-\gamma}$, where $\gamma = \alpha(\tau - 1)/d$ and τ is the weight power-law exponent as in (9.5.15). The current set-up has the advantage that the degree power-law exponent agrees with that of the weight distribution.

The fact that $\lambda_c < \infty$ holds in most cases is (Deijfen et al., 2013, Theorem 3.1). Theorem 9.43(a) is (Deijfen et al., 2013, Theorem 4.2), Theorem 9.43(b) is (Deijfen et al., 2013, Theorem 4.4). Deprez et al. (2015) showed that the percolation function is continuous when $\alpha \in (d, 2d)$, i.e., $\theta(\lambda_c) = 0$. However, in full generality, continuity of the percolation function at $\lambda = \lambda_c$ when $\lambda_c > 0$ is unknown.

Theorem 9.44(a) was proved in Deijfen et al. (2013); van der Hofstad and Komjáthy (2017a), see in particular Corollary 1.4 in van der Hofstad and Komjáthy (2017a). Theorem 9.44(b) was proved in Heydenreich et al. (2017); Hao and Heydenreich (2023), following up on similar results for long-range percolation proved by Biskup (2004); Biskup and Lin (2019). In long-range percolation, edges are present independently, and the probability that the edge $\{x, y\}$ is present equals $|x - y|^{-\alpha d + o(1)}$ for some $\alpha > 0$. In this case, detailed results exist for the limiting behavior of $\text{dist}_G(o, x)$ depending on the value of α. For example, in Benjamini et al. (2004), it is shown that the diameter of this infinite percolation model is equal to $\lceil 1/(1 - \alpha) \rceil$ almost surely when $\alpha \in (0, 1)$. Theorem 9.44(c) is (Deprez et al., 2015, Theorem 8(b)). Deprez and Wüthrich (2019) investigated graph distances in the continuum scale-free percolation model; a related result was proved in the long-range percolation setting by Sönmez (2021), who also addressed bounds on graph distances for $\alpha \in \{1, 2\}$.

There is some follow-up work on scale-free percolation. Hirsch (2017) proposed a continuum model for scale-free percolation. Deprez et al. (2015) argued that scale-free percolation can be used to model real-life networks. Heydenreich et al. (2017) established recurrence and transience criteria for random walks on the infinite connected component. For long-range percolation this was proved by Berger (2002).

Spatial configuration models on the lattice were introduced by Deijfen and Jonasson (2006); see also Deijfen and Meester (2006). In our exposition, we follow Jonasson (2009), who studied more general underlying graphs, such as trees or other infinite transitive graphs. Theorem 9.45 is (Jonasson, 2009, Theorem 3.1). (Jonasson, 2009, Theorem 3.2) extended Theorem 9.45 to settings where the degrees are not iid, but rather translation invariant. In this case, it is still necessary for (9.5.65) that $\mathbb{E}[D^{(d+1)/d}] < \infty$, but this may not be enough. Sharper conditions are restricted to the setting where $d = 1$ (and where no condition of the form $\mathbb{E}[D^k] < \infty$ suffices) and $d = 2$ (for which it suffices that $\mathbb{E}[D^{(d+1)/(d-1)+\alpha}] < \infty$ for some $\alpha > 0$, but if $\mathbb{E}[D^{(d+1)/(d-1)-\alpha}] = \infty$ for some $\alpha > 0$, then there exist translation-invariant matchings for which (9.5.65) fails).

We next discuss the properties of the model in $d = 1$ where the directions of the half-edges are chosen independently. (Deijfen and Meester, 2006, Proposition 2.1) shows that, when the direction is chosen with probability $p \neq \frac{1}{2}$, the maximal edge length N is infinite with positive probability. (Deijfen and Meester, 2006, Theorem 2.1) shows that $N < \infty$ almost surely when $p = \frac{1}{2}$, while (Deijfen and Meester, 2006, Theorem 4.1) implies that $\mathbb{E}[N] = \infty$ when $p = \frac{1}{2}$.

Deijfen (2009) studied a related model where the vertices are a Poisson point process on $\mathbb{R}^d$. This model was further studied by Deijfen et al. (2012). In the latter paper, the surprising result is shown that for *any* sequence of iid degrees of the points of the Poisson process, there are translation-invariant matchings that percolate, as well as matchings that do not. Further, this matching can be a factor, where a translation-invariant matching is called a *factor* if it is a deterministic function of the Poisson process and of the degrees of the vertices in the Poisson process, that is, if it does not involve any additional randomness. See (Deijfen et al., 2012, Theorem 1.1) for more details.

A threshold scale-free percolation model. We finally discuss the results by Yukich (2006) on another infinite geometric random graph model. We start by taking an iid sequence $(W_x)_{x \in \mathbb{Z}^d}$ of random variables on $[1, \infty)$ satisfying (9.5.15) with a constant slowly varying function. Fix $\delta \in (0, 1]$. The edge $\{x, y\} \in \mathbb{Z}^d \times \mathbb{Z}^d$ appears in the random graph precisely when

$$|x - y| \leq \delta \min\{W_x^{\tau/d}, W_y^{\tau/d}\}. \tag{9.6.7}$$

We can think of the ball of radius $\delta W_x^{\tau/d}$ as being the *region of influence* of x, and two vertices are connected precisely when each of them lies in the influence region of the other. This motivates the choice in (9.6.7). The parameter δ can be interpreted as the probability that nearest neighbors are connected, and in what follows we restrict ourselves to $\delta = 1$, in which case the infinite connected component turns out to equal $\mathbb{Z}^d$. We denote the resulting (infinite) random graph by G. Yukich (2006) parametrized the model slightly differently, and replaced $W_x^{\tau/d}$ by U_x^{-q} for a uniform random variable U_x.

The threshold model in (9.6.7) is quite different from a threshold scale-free percolation model as in (9.5.58), where an edge would be present when $|x - y| \leq (W_x W_y)^{1/d}$. The product structure creates rather different asymptotics compared with the minimum in (9.6.7).

We next discuss the properties of this model, starting with its power-law nature. (Yukich, 2006, Theorem 1.1) shows that the limit

$$\lim_{k \to \infty} k^{\tau-1} \mathbb{P}(D_o > k) \tag{9.6.8}$$

exists, so that the model has a power-law degree sequence with power-law exponent τ (recall (1.4.3)). The intuitive explanation of (9.6.8) is as follows. Suppose we condition on the value of $W_o = w$. Then, the conditional distribution of D_o given that $W_o = w$ is equal to

$$D_o = \sum_{x \in \mathbb{Z}^d} \mathbb{1}_{\{|x| \leq \min\{W_o^{\tau/d}, W_x^{\tau/d}\}\}} = \sum_{x \,:\, |x| \leq w^{\tau/d}} \mathbb{1}_{\{|x| \leq W_x^{\tau/d}\}}. \tag{9.6.9}$$

Note that the random variables $(\mathbb{1}_{\{|x| \leq W_x^{\tau/d}\}})_{x \in \mathbb{Z}^d}$ are independent Bernoulli random variables with parameter equal to

$$\mathbb{P}(\mathbb{1}_{\{|x| \leq W_x^{\tau/d}\}} = 1) = \mathbb{P}(W \geq |x|^{d/\tau}) = |x|^{-d(\tau-1)/\tau}. \tag{9.6.10}$$

In order for $D_o \geq k$ to occur, for k large, we must have that $W_o = w$ is quite large, and, in this case, a central limit theorem should hold for D_o, with mean equal to

$$\mathbb{E}[D_o \mid W_o = w] = \sum_{x \,:\, |x| \leq w^{\tau/d}} |x|^{-d(\tau-1)/\tau} = cw(1 + o(1)), \tag{9.6.11}$$

for some explicit constant $c = c(\tau, d)$. Furthermore, the conditional variance of D_o given that $W_o = w$ is bounded above by its conditional expectation, so that the conditional distribution of D_o given that $W_o = w$ is highly concentrated. We omit further details, and merely note that this heuristic can be made precise by using standard concentration results. Assuming sufficient concentration, we obtain that the probability that $D_o \geq k$ is asymptotically equal to the probability that $W > w_k$, where w_k is determined by the equation

$$\mathbb{E}[D_o \mid W_o = w_k] = c w_k (1 + o(1)) = k, \tag{9.6.12}$$

so that $w_k = (1 + o(1))k/c$. This suggests that

$$\mathbb{P}(D_o > k) = \mathbb{P}(W > k/c)(1 + o(1)) = (k/c)^{-(\tau-1)}(1 + o(1)), \tag{9.6.13}$$

which explains (9.6.8).

We next turn to distances in this model. For $x, y \in \mathbb{Z}^d$, we denote by $\mathrm{dist}_G(x, y)$ the graph distance between the vertices x and y, i.e., the minimal number of edges in G connecting x and y. The main result in Yukich (2006) is the following theorem:

Theorem 9.46 (Ultra-small distances for model in (9.6.7))   *For all $d \geq 1$ and all $\tau > 1$, whp, as $|x| \to \infty$,*

$$\mathrm{dist}_G(o, x) \leq 8 + 4 \log \log |x|. \tag{9.6.14}$$

The result in Theorem 9.46 shows that distances in the model given by (9.6.7) are *much* smaller than those in normal percolation models. Recall Meta Theorem B at the start of Part III. While Theorem 9.46 resembles the results in Meta Theorem B, the differences reside in the fact that distances are ultra-small *independently* of the exact value of the degree power-law exponent.

Again, the result in Theorem 9.46 can be compared with similar results for *long-range percolation* (recall the discussion of scale-free percolation).

9.7 EXERCISES FOR CHAPTER 9

Exercise 9.1 (Topology of the strongly connected component for digraphs)   *Let G be a digraph. Prove that if u and v are such that u is connected to v and v is connected to u, then the strongly connected components of u and v are the same.*

Exercise 9.2 (The sums of the out- and of the in-degrees of a digraph agree)   *Let G be a digraph for which $d_v^{(\mathrm{out})}$ and $d_v^{(\mathrm{in})}$ denote the out- and in-degree of $v \in V(G)$. Show that*

$$\sum_{v \in V(G)} d_v^{(\mathrm{out})} = \sum_{v \in V(G)} d_v^{(\mathrm{in})}. \tag{9.7.1}$$

Exercise 9.3 (Local convergence for randomly directed graphs)   *Let $(G_n)_{n \geq 1}$ be a random graph sequence that converges locally in probability. Give each edge e a random orientation, by orienting $e = \{u, v\}$ as $e = (u, v)$ with probability $\frac{1}{2}$ and as $e = (v, u)$ with probability $\frac{1}{2}$, independently across edges. Show that the resulting digraph converges locally in probability in the marked forward, backward, and forward–backward senses.*

Exercise 9.4 (Local convergence for randomly directed graphs (cont.))   *In the setting of Exercise 9.3, assume that the convergence of (G_n) is locally weakly. Conclude that the resulting digraph converges locally weakly in the marked forward, backward, and forward–backward senses.*

Exercise 9.5 (Local convergence for directed version of $\mathrm{PA}_n^{(m,\delta)}(d)$)   *Consider the edges in $\mathrm{PA}_n^{(m,\delta)}(d)$ to be oriented from young to old, so that the resulting digraph has out-degree m and random in-degrees. Use Theorem 5.8 and 5.21 to show that this digraph converges locally in probability in the marked forward, backward, and forward–backward senses.*

Exercise 9.6 (Power-law lower bound for PageRank on directed version of $\mathrm{PA}_n^{(m,\delta)}(d)$)   *Recall the directed version of $\mathrm{PA}_n^{(m,\delta)}(d)$ in Exercise 9.5. Use (9.2.10) to show that there exists a constant $c = c(\alpha, \delta, m) > 0$ such that*

$$\mathbb{P}(R_\varnothing > r) \geq c r^{-\tau}, \qquad where \qquad \tau = 3 + \frac{\delta}{m} \tag{9.7.2}$$

is the power-law exponent of $\mathrm{PA}_n^{(m,\delta)}(d)$. What does this say about the PageRank power-law hypothesis for the directed version of $\mathrm{PA}_n^{(m,\delta)}(d)$?

Exercise 9.7 (Power-law lower bound for PageRank on digraphs with bounded out-degrees) *Let $(G_n)_{n\geq 1}$ be a random digraph sequence that converges locally in probability in the marked backward sense to $(D, \varnothing) \sim \mu$. Assume that there exist $0 < a, b < \infty$ such that $d_v^{(\mathrm{out})} \in [a, b]$ for all $v \in V(G_n)$. Assume that $\mu(D_\varnothing^{(\mathrm{in})} > r) \geq cr^{-\gamma}$ for some $\gamma > 0$. Use (9.2.10) to show that there exists a constant $c' > 0$ such that $\mu(R_\varnothing > r) \geq c'r^{-\gamma}$.*

Exercise 9.8 (Mean number of edges in $\mathrm{DGRG}_n(\boldsymbol{w})$) *Consider the directed generalized random graph, as formulated in (9.2.14) and (9.2.15). Assume that the weight-regularity condition in (9.2.16) holds. Let X_{ij} be the indicator that there is a directed edge from i to j (with $X_{ii} = 0$ for all $i \in [n]$ by convention). Show that*

$$\frac{1}{n}\mathbb{E}\Big[\sum_{i,j\in[n]} X_{ij}\Big] \to \frac{\mathbb{E}[W^{(\mathrm{in})}]\mathbb{E}[W^{(\mathrm{out})}]}{\mathbb{E}[W^{(\mathrm{in})} + W^{(\mathrm{out})}]}. \tag{9.7.3}$$

Conclude that the limit equals $\mathbb{E}[W^{(\mathrm{in})}] = \mathbb{E}[W^{(\mathrm{out})}]$ when the symmetry condition in (9.2.18) holds.

Exercise 9.9 (Number of edges in $\mathrm{DGRG}_n(\boldsymbol{w})$) *In the setting of (9.7.3) in Exercise 9.8, show that*

$$\frac{1}{n}\sum_{i,j\in[n]} X_{ij} \xrightarrow{\mathbb{P}} \frac{\mathbb{E}[W^{(\mathrm{in})}]\mathbb{E}[W^{(\mathrm{out})}]}{\mathbb{E}[W^{(\mathrm{in})} + W^{(\mathrm{out})}]}, \tag{9.7.4}$$

which equals $\frac{1}{2}\mathbb{E}[W^{(\mathrm{in})}] = \frac{1}{2}\mathbb{E}[W^{(\mathrm{out})}]$ when the symmetry condition in (9.2.18) holds.

Exercise 9.10 (Local limit of directed Erdős–Rényi random graph) *Use Theorem 9.2 to describe the local limit of the directed Erdős–Rényi random graph.*

Exercise 9.11 (Local convergence for finite-type directed inhomogeneous random graphs) *Adapt the proof of Theorem 3.14 to prove Theorem 9.2 in the case of finite-type kernels. Here, we recall that a kernel κ is called finite type when $(s, r) \mapsto \kappa(s, r)$ takes on finitely many values.*

Exercise 9.12 (Local convergence for $\mathrm{DGRG}_n(\boldsymbol{w})$) *Consider the directed generalized random graph as formulated in (9.2.14) and (9.2.15). Assume that the weight-regularity condition in (9.2.16) holds. Use Theorem 9.2 to determine the local limit in probability of $\mathrm{DGRG}_n(\boldsymbol{w})$. Is the local limit of the forward and backward neighborhoods a single- or a multi-type branching process?*

Exercise 9.13 (Phase transition for directed Erdős–Rényi random graph) *For the directed Erdős–Rényi random graph, show that ζ in Theorem 9.3 satisfies $\zeta > 0$ precisely when $\lambda > 1$.*

Exercise 9.14 (Phase transition for directed generalized random graph) *Consider the directed generalized random graph, as formulated in (9.2.14) and (9.2.15). Assume that the weight-regularity condition in (9.2.16) holds. What is the condition on the asymptotic weight distribution $(W^{(\mathrm{out})}, W^{(\mathrm{in})})$ in (9.2.16) that is equivalent to $\zeta > 0$ in Theorem 9.3?*

Exercise 9.15 (Correlations of out- and in-degrees of a randomly directed graph) *In an undirected graph G, randomly direct each edge by orienting $e = \{u, v\}$ as (u, v) with probability $\frac{1}{2}$ and as (v, u) with probability $\frac{1}{2}$, as in Exercise 9.3. Let $v \in V(G)$ be a vertex in G of degree d_v. What is the correlation coefficient between its out- and in-degrees in the randomly directed version of G? Note: The correlation coefficient $\rho(X, Y)$ between two random variables X and Y is equal to $\mathrm{Cov}(X, Y)/\sqrt{\mathrm{Var}(X)\mathrm{Var}(Y)}$.*

Exercise 9.16 (Equivalence of convergence of in- and out-degrees in $\mathrm{DCM}_n(\boldsymbol{d})$) *Show that (9.2.24) implies that $\mathbb{E}[D^{(\mathrm{out})}] = \mathbb{E}[D^{(\mathrm{in})}]$ when $(D_n^{(\mathrm{in})}, D_n^{(\mathrm{out})}) \xrightarrow{d} (D^{(\mathrm{in})}, D^{(\mathrm{out})})$, $\mathbb{E}[D_n^{(\mathrm{in})}] \to \mathbb{E}[D^{(\mathrm{in})}]$, and $\mathbb{E}[D_n^{(\mathrm{out})}] \to \mathbb{E}[D^{(\mathrm{in})}]$.*

Exercise 9.17 (Self-loops and multiple edges in $\mathrm{DCM}_n(\boldsymbol{d})$) *Argue that the proof of [V1, Proposition 7.13] can be adapted to show that the number of self-loops in $\mathrm{DCM}_n(\boldsymbol{d})$ converges to a Poisson random variable with parameter $\mathbb{E}[D^{(\mathrm{in})}D^{(\mathrm{out})}]$ when $(D_n^{(\mathrm{in})}, D_n^{(\mathrm{out})}) \xrightarrow{d} (D^{(\mathrm{in})}, D^{(\mathrm{out})})$ and*

$$\mathbb{E}[D_n^{(\mathrm{in})}D_n^{(\mathrm{out})}] \to \mathbb{E}[D^{(\mathrm{in})}D^{(\mathrm{out})}]. \tag{9.7.5}$$

What can you say about the number of multiple edges in $\mathrm{DCM}_n(\boldsymbol{d})$? Note: No proof is expected, a reasonable argument suffices.

Exercise 9.18 (Local convergence for $\mathrm{DCM}_n(\boldsymbol{d})$ in Theorem 9.4) *Give a proof of the local limit result in Theorem 9.4 by suitably adapting the proof of Theorem 4.1.*

Exercise 9.19 (One-sided law of large numbers for SSC) *Adapt the proof of Corollary 2.27 to show that when $G_n = ([n], E(G_n))$ converges locally in probability in the forward–backward sense to (G, o) having distribution μ, then the size of the largest strongly connected component $|\mathscr{C}_{\max}|$ satisfies that, for every $\varepsilon > 0$ fixed,*

$$\mathbb{P}(|\mathscr{C}_{\max}| \leq n(\zeta + \varepsilon)) \to 1, \tag{9.7.6}$$

where $\zeta = \mu(|\mathscr{C}(o)| = \infty)$ is the forward–backward survival probability of the limiting graph (G, o) (i.e., the probability that both the forward and the backward component of o have infinite size).

Exercise 9.20 (Subcritical directed configuration model) *Let $\mathrm{DCM}_n(\boldsymbol{d})$ be a directed configuration model that satisfies the degree-regularity conditions in (9.2.25) and (9.2.26). Let $\mathscr{C}_{\max}$ denote its largest strongly connected component. Use Exercise 9.19 to show that $|\mathscr{C}_{\max}|/n \overset{\mathbb{P}}{\longrightarrow} 0$ when $\zeta = 0$, where $\zeta = \mu(|\mathscr{C}(o)| = \infty)$ is the forward–backward survival probability of the limiting graph (G, o). This proves the subcritical result in Theorem 9.5(b).*

Exercise 9.21 (Logarithmic growth of typical distances in the directed configuration model) *Let $\mathrm{DCM}_n(\boldsymbol{d})$ be a directed configuration model that satisfies the degree-regularity conditions in (9.2.25) and (9.2.26). Argue heuristically why the logarithmic typical distance result in Theorem 9.6 remains valid when (9.2.37) is replaced by the weaker condition that $(D_n^{(\mathrm{in})} D_n^{(\mathrm{out})})_{n \geq 1}$ is uniformly integrable. Also, give an example where this uniform integrability is true, but (9.2.37) is not.*

Exercise 9.22 (Logarithmic growth of typical distances in the directed configuration model (cont.)) *Let $\mathrm{DCM}_n(\boldsymbol{d})$ be a directed configuration model that satisfies the degree-regularity conditions in (9.2.25) and (9.2.26). Give a formal result of the claim in Exercise 9.21 by a suitable degree-truncation argument, as explained above (9.6.3).*

Exercise 9.23 (Ultra-small distances in the directed configuration model) *Let $\mathrm{DCM}_n(\boldsymbol{d})$ be a directed configuration model that satisfies the degree-regularity conditions in (9.2.25) and (9.2.26). Use the degree-truncation argument, as explained above (9.6.3), to show that $\mathrm{dist}_{\mathrm{DCM}_n(\boldsymbol{d})}(o_1, o_2) = o_{\mathbb{P}}(\log n)$ when $\nu = \infty$.*

Exercise 9.24 (Strongly connected component in temporal networks) *Let G be a temporal network, in which vertices have a time label of their birth and edges are oriented from younger to older vertices. What do the strongly connected components of G look like?*

Exercise 9.25 (Degree structure in stochastic block models) *Recall the definition of the stochastic block model in Section 9.3.1, and assume that the type regularity condition in (9.3.1) holds. What is the asymptotic expected degree of this model? When do all vertices have the same asymptotic expected degree?*

Exercise 9.26 (Giant in stochastic block models) *Recall the definition of the stochastic block model in Section 9.3.1, and assume that the type regularity condition in (9.3.1) holds. When is there a giant component?*

Exercise 9.27 (Random guessing in stochastic block models) *Consider a stochastic block model with t types as introduced in Section 9.3.1 and assume that each of the types occurs equally often. Let $\hat{\sigma}(v)$ be a random guess, so that $(\hat{\sigma}(v))_{v \in [n]}$ is an iid vector, with $\hat{\sigma}(v) = s$ with probability $1/t$ for every $s \in [t]$. Show that (9.3.2) does not hold, i.e., show that the probability that*

$$\max_{p:\, [t] \to [t]} \frac{1}{n} \sum_{v \in [n]} \left[\mathbb{1}_{\{\hat{\sigma}(v) = (p \circ \sigma)(v)\}} - \frac{1}{t} \right] \geq \varepsilon$$

vanishes.

Exercise 9.28 (Degree structure in stochastic block models with unequal expected degrees) *Let n be even. Consider the stochastic block model with two types and $n/2$ vertices with each type. Let $p_{ij} = a_1/n$ when i, j have type 1, $p_{ij} = a_2/n$ when i, j have type 2, and $p_{ij} = b/n$ when i, j have different types. For $i \in \{1, 2\}$ and $k \in \mathbb{N}_0$, let $N_{i,k}(n)$ denote the number of vertices of type i and degree k. Show that*

$$\frac{N_{i,k}(n)}{n} \overset{\mathbb{P}}{\longrightarrow} \mathbb{P}(\mathrm{Poi}(\lambda_i) = k), \tag{9.7.7}$$

where $\lambda_i = (a_i + b)/2$.

Exercise 9.29 (Community detection in stochastic block models with unequal expected degrees)   *In the setting of Exercise 9.28, assume that $a_1 > a_2$. Consider the following greedy community detection algorithm: let $\hat{\sigma}(v) = 1$ for the $n/2$ vertices $v \in [n]$ of highest degree, and $\hat{\sigma}(v) = 2$ for the remaining vertices (breaking ties randomly when necessary). Argue that this algorithm achieves the solvability condition in (9.3.2).*

Exercise 9.30 (Parameter conditions for solvable stochastic block models)   *Consider the stochastic block model in the setting of Theorem 9.12, and assume that $(a - b)^2 > 2(a + b)$, so that the community detection problem is solvable. Show that $a - b > 2$ and thus also $a + b > 2$. Conclude that this model has a giant.*

Exercise 9.31 (Parameter conditions for solvable stochastic block models (cont.))   *In the setting of Exercise 9.30, show that also the vertices of type 1 only (resulting in an Erdős–Rényi random graph of size $n/2$ and edge probability a/n) have a giant. What are the conditions for the vertices of type 2 to have a giant?*

Exercise 9.32 (Degree structure in degree-corrected stochastic block models)   *Recall the definition of the degree-corrected stochastic block model in (9.3.11) in Section 9.3.2, and assume that the type regularity condition in (9.3.1) holds. Assume further that $\mathbb{E}[X_u^p] < \infty$ for some $p > 1$. What is the asymptotic expected degree of a vertex v of weight x_v in this model? What are the restrictions on $(\kappa(s,r))_{s,r \in S}$ such that the expected degree of vertex v with weight x_v is equal to $x_v(1 + o_{\mathbb{P}}(1))$?*

Exercise 9.33 (Equal average degrees in degree-corrected stochastic block models)   *Recall the definition of the degree-corrected stochastic block model in (9.3.11) in Section 9.3.2, and assume that (9.3.1) holds. Let the number of types $t \geq 2$ be arbitrary and assume that $\kappa(s,r) = b$ for all $s \neq r$, while $\kappa(s,s) = a$. Assume that $\mu(s) = 1/t$ for every $s \in [t]$. Compute the asymptotic average degree of a vertex of type s, and show that it is independent of s.*

Exercise 9.34 (Giant in the degree-corrected stochastic block models)   *Recall the definition of the degree-corrected stochastic block model in Section 9.3.2, and assume that the type regularity condition in (9.3.1) holds. When is there a giant component?*

Exercise 9.35 (Degrees in configuration models with global communities)   *Recall the definition of the configuration models with global communities in Section 9.3.3, and assume that the degree regularity conditions in (9.3.18), (9.3.19), and (9.3.20) hold. What is the asymptotic average degree of this model? When do all vertices have the same asymptotic expected degree?*

Exercise 9.36 (Local limit in configuration models with global communities)   *Recall the definition of the configuration models with global communities in Section 9.3.3, and assume that (9.3.18), (9.3.19), and (9.3.20) hold. What is the local limit of this model? Note: No proof is expected; a reasonable argument suffices.*

Exercise 9.37 (Giant in configuration models with global communities)   *Recall the definition of the configuration model with global communities in Section 9.3.3, and assume that (9.3.18), (9.3.19), and (9.3.20) hold. When is there a giant component? Note: No proof is expected; a reasonable argument suffices.*

Exercise 9.38 (Degree distribution in preferential attachment model with global communities)   *Show that $(p_k(\theta))_{k \geq m}$ in (9.3.29) is a probability distribution for all θ, i.e., show that $\sum_{k \geq m} p_k(\theta) = 1$ and $p_k(\theta) \geq 0$ for all $k \geq 1$.*

Exercise 9.39 (Degree distribution in preferential attachment model with global communities)   *In the preferential attachment model with global communities studied in Theorem 9.14, show that also the global degree distribution given by $P_k(n) = \frac{1}{n} \sum_{v \in [n]} \mathbb{1}_{\{D_v(n) = k\}}$ converges almost surely.*

Exercise 9.40 (Power-law degrees in preferential attachment model with global communities)   *In the preferential attachment models with global communities studied in Theorem 9.14, show that the global degree distribution has a power-law tail with exponent $\tau = 1 + 1/\max_{s \in [s]} \theta^*(s)$, provided that $\mu(s^*) > 0$ for at least one $s^* \in [r]$ satisfying $\theta^*(s^*) = \max_{s \in [s]} \theta^*(s)$.*

Exercise 9.41 (Clustering in model with edges and triangles)   *Show that the global clustering coefficient in the model where each pair of vertices is independently connected with probability λ/n, as for $\mathrm{ER}_n(\lambda/n)$, and each triple forms a triangle with probability μ/n^2, independently for all triplets and independently of the status of the edges, converges to $\mu/((\lambda + \mu)^2 + \mu)$.*

Exercise 9.42 (Local limit in inhomogeneous random graph with communities) *Recall the definition of the inhomogeneous random graph with communities in Section 9.4.1. What is the local limit of this model? Note: No proof is expected; a reasonable argument suffices.*

Exercise 9.43 (Size-biased community size distribution in HCM) *In the hierarchical configuration model introduced in Section 9.4.2, choose a vertex o uar from $[n]$. Let G_o be the community containing o. Show that (9.4.16) implies that $|V(G_o)|$ converges in distribution, and identify its limiting distribution.*

Exercise 9.44 (Local limit in hierarchical configuration model) *Recall the definition of the hierarchical configuration model in Theorem 9.19. What is the local limit of this model? Note: No proof is expected; a reasonable argument suffices.*

Exercise 9.45 (Law of large numbers for $|\mathscr{C}_{\max}|$ in hierarchical configuration model) *Use Theorem 4.9 to prove the law of large numbers for the giant in the hierarchical configuration model in Theorem 9.19, and prove that ζ is given by (9.4.19).*

Exercise 9.46 (Local clustering for configuration model with clustering) *Recall the configuration model with clustering defined in Section 9.4.2. Let $(D_n^{(\mathrm{si})}, D_n^{(\mathrm{tr})})$ denote the number of simple edges and triangles incident to a uniform vertex in $[n]$, and assume that $(D_n^{(\mathrm{si})}, D_n^{(\mathrm{tr})}) \xrightarrow{d} (D^{(\mathrm{si})}, D^{(\mathrm{tr})})$ for some limiting distribution $(D^{(\mathrm{si})}, D^{(\mathrm{tr})})$. Compute the local clustering coefficient of this model under the extra assumptions that $\mathbb{E}[D_n^{(\mathrm{si})}] \to \mathbb{E}[D^{(\mathrm{si})}] < \infty$ and $\mathbb{E}[D_n^{(\mathrm{tr})}] \to \mathbb{E}[D^{(\mathrm{tr})}] < \infty$.*

Exercise 9.47 (Local clustering for configuration model with clustering) *In the setting of Exercise 9.46, compute the global clustering coefficient of this model under the extra assumption that also $\mathbb{E}[(D_n^{(\mathrm{si})})^2] \to \mathbb{E}[(D^{(\mathrm{si})})^2] < \infty$ and $\mathbb{E}[(D_n^{(\mathrm{tr})})^2] \to \mathbb{E}[(D^{(\mathrm{tr})})^2] < \infty$.*

Exercise 9.48 (Single overlap in random intersection graph) *Consider the random intersection graph with prescribed communities as defined in Section 9.4.3, under the conditions of Theorem 9.21. Show that it is unlikely for a uniform vertex to have a neighbor with which it shares two groups.*

Exercise 9.49 (Local clustering in the random intersection graph) *Consider the random intersection graph with prescribed communities as defined in Section 9.4.3, under the conditions of Theorem 9.21. Show that the local clustering coefficient converges. When is this limit strictly positive?*

Exercise 9.50 (Global clustering in the random intersection graph) *Consider the random intersection graph with prescribed communities as defined in Section 9.4.3, under the conditions of Theorem 9.21. What are the conditions on the group membership and size distributions that imply that the convergence of the global clustering coefficient in Theorem 2.22 follows? When is the limit of the global clustering coefficient strictly positive?*

Exercise 9.51 (Degree distribution in the discrete small-world model) *Recall the discrete small-world model in Section 9.5.1 as studied in Theorem 9.26, but now with $\lambda, \rho > 0$ and k fixed. What is the limit of the probability that a uniform vertex has degree l for $l \geq 0$?*

Exercise 9.52 (Degree distribution in the geometric preferential attachment model with non-uniform locations) *Recall that (9.5.50) in Theorem 9.40 identifies the degree distribution of the geometric preferential attachment model at each of the elements $z_i \in S$. Conclude what the degree distribution is of the entire graph. Does it obey a degree power law, and, if so, what is the degree power-law exponent?*

Exercise 9.53 (Power-law degrees for the spatial preferential attachment model with influence) *Prove that, for p_k in (9.5.55) and for k large, we have*

$$p_k = ck^{-(1+1/(pa_1))}(1 + o(1)), \tag{9.7.8}$$

so that the spatial preferential attachment model with influence indeed has a power-law degree distribution.

Exercise 9.54 (Degree distribution in the GIRG) *Investigate the degree distribution in the GIRG in Theorem 9.32 using a second-moment method.*

Exercise 9.55 (Degree moments in scale-free percolation (Deijfen et al. (2013))) *Recall that o denotes the degree of the origin in the scale-free percolation model defined in (9.5.58). Show that $\mathbb{E}[D_o^p] < \infty$ when $p < \tau - 1$ and $\mathbb{E}[D_o^p] = \infty$ when $p > \tau - 1$. In particular, the variance of the degrees is finite precisely when $\tau > 3$.*

Exercise 9.56 (Positive correlation between edge statuses in scale-free percolation)   *Show that, for scale-free percolation, for all x, y, z distinct, and for $\lambda > 0$,*

$$\mathbb{P}(\{x, y\} \text{ and } \{x, z\} \text{ occupied}) \geq \mathbb{P}(\{x, y\} \text{ occupied}) \, \mathbb{P}(\{x, z\} \text{ occupied}), \qquad (9.7.9)$$

the inequality being strict when $\mathbb{P}(W_o = 0) < 1$. In other words, the edge statuses are positively correlated.

Exercise 9.57 (Local convergence of PageRank)   *Assume that G_n converges locally in probability in the marked backward sense to $(G, o) \sim \mu$. Use (9.6.2) to show that, whp, and for every $\eta, \varepsilon > 0$,*

$$\frac{1}{n} \sum_{v \in V(G_n)} \mathbb{1}_{\{R_v^{(G_n)} > r\}} \leq \mu(R_o^{(G)} > r - \varepsilon) + \eta, \qquad (9.7.10)$$

and

$$\frac{1}{n} \sum_{v \in V(G_n)} \mathbb{1}_{\{R_v^{(G_n)} > r\}} \geq \mu(R_o^{(G)} > r) - \eta. \qquad (9.7.11)$$

Exercise 9.58 (Local convergence of PageRank (cont.))   *Use Exercise 9.57 to complete the proof of Theorem 9.1(ii).*

METRIC SPACE STRUCTURE OF ROOTED GRAPHS

Abstract

In this appendix we highlight some properties of and results about metric spaces, including separable metric spaces and Borel measures on them, as used throughout this book. We also present some missing details in the proof that the space of rooted graphs is a separable metric space. Finally, we discuss what compact sets look like in this topology and relate this to tightness criteria.

A.1 METRIC SPACES

In this section we discuss metric spaces. We start by defining the terms metric and metric space:

Definition A.1 (Metric and metric space) Let $\mathcal{X}$ be a space. A *metric* on $\mathcal{X}$ is a function $d_{\mathcal{X}} : \mathcal{X}^2 \mapsto [0, \infty)$ such that

(a) $0 \leq d_{\mathcal{X}}(x, y) < \infty$ for all $x, y \in \mathcal{X}$;
(b) $d_{\mathcal{X}}(x, y) = d_{\mathcal{X}}(y, x)$ for all $x, y \in \mathcal{X}$;
(c) $d_{\mathcal{X}}(x, z) \leq d_{\mathcal{X}}(x, y) + d_{\mathcal{X}}(y, z)$ for all $x, y, z \in \mathcal{X}$.

Let $\mathcal{X}$ be a space and $d_{\mathcal{X}}$ a metric on it. Then $(\mathcal{X}, d_{\mathcal{X}})$ is called a *metric space*. ◀

We next discuss desirable properties of metric spaces:

Definition A.2 (Complete and separable metric spaces) Let $(\mathcal{X}, d_{\mathcal{X}})$ be a metric space.

(a) We say that $(\mathcal{X}, d_{\mathcal{X}})$ is *complete* when every Cauchy sequence has a limit. Here, a Cauchy sequence is a sequence $(x_n)_{n \geq 1}$ with $x_n \in \mathcal{X}$ such that, for every $\varepsilon > 0$ there exists an $N = N(\varepsilon)$ such that $d_{\mathcal{X}}(x_n, x_m) \leq \varepsilon$ for all $n, m \geq N$.
(b) We say that $(\mathcal{X}, d_{\mathcal{X}})$ is *separable* if it contains a countable subset that is dense in $(\mathcal{X}, d_{\mathcal{X}})$. This means that there exists a countable set $\mathcal{A} \subseteq \mathcal{X}$ such that, for every $x \in \mathcal{X}$, there exists a sequence $(a_n)_{n \geq 1}$ with $a_n \in \mathcal{A}$ such that $d_{\mathcal{X}}(a_n, x) \to 0$.
(c) The metric space $(\mathcal{X}, d_{\mathcal{X}})$ is called *Polish* when it is separable and complete. ◀

A.2 LOCAL TOPOLOGIES

In this section we discuss local topologies in general. The metric $d_{\mathcal{G}_\star}$ on rooted graphs turns out to be a special case of such a local topology.

As before, let $(\mathcal{X}, d_{\mathcal{X}})$ be a general metric space. Let $x \in \mathcal{X}$. Suppose that, for every $r \geq 1$, there is a notion of a *restriction of x to radius r*, which we denote by $[x]_r$ and which is such that $[x]_r \in \mathcal{X}$. We formalize this notion as follows:

Definition A.3 (Restriction) Let $(\mathcal{X}, d_{\mathcal{X}})$ be a metric space. For every $r \geq 1$, a *restriction* of $x \in \mathcal{X}$ to radius r is a continuous function $[\cdot]_r : \mathcal{X} \mapsto \mathcal{X}$ that satisfies the following properties:

(Closure) $[x]_r \in \mathcal{X}$ for all $x \in \mathcal{X}$;
(Compatibility) $[[x]_r]_s = [x]_s$ for all $x \in \mathcal{X}$ and $r \geq s$;
(Coherence) For any sequence $x_1, x_2, \ldots \in \mathcal{X}$ satisfying $[x_i]_j = x_j$ for all $j \in [i]$ there exists a unique (infinite) element $x \in \mathcal{X}$ such that $[x]_r = x_r$ for all $r \geq 0$. ◀

For example, for a rooted graph (G, o), we can let $[(G, o)]_r = B_r^{(G)}(o)$ denote the r-neighborhood of o, seen as a rooted graph. There are many more examples of restrictions, such as $[x]_r = (x_1, \ldots, x_r, 0, \ldots)$ for $x = (x_i)_{i \geq 1}$, so that this restriction is equal to x in its first r coordinates, and equal to zero otherwise.

We next define a *local topology*:

Definition A.4 (Local topology) Let $[\cdot]_r : \mathcal{X} \mapsto \mathcal{X}$ be a restriction. Define, for $x, y \in \mathcal{X}$,

$$d_{\mathrm{loc}}(x, y) = \sum_{r \geq 1} \frac{d_{\mathcal{X}}([x]_r, [y]_r) \wedge 1}{r(r+1)}. \tag{A.2.1}$$

The *local topology* is defined through the metric d_{loc}.   ◀

The metric $d_{\mathcal{X}}$ may depend strongly on "far away" changes, in the sense that $d_{\mathcal{X}}(x, y)$ could be large even when $[x]_r = [y]_r$ for most small r. The metric d_{loc} derived from $d_{\mathcal{X}}$ does not suffer from such effects, in that it is at most $1/(R + 1)$ when $[x]_r = [y]_r$ for all $r \leq R$. This explains the name *local topology*. Further, convergence for d_{loc} is equivalent to local convergence:

Remark A.5 (Local convergence)   The metric d_{loc} is *local* in the sense that $x_n \to x$ for d_{loc} *precisely* when $[x_n]_r \to [x]_r$ for $d_{\mathcal{X}}$ for every $r \geq 1$, as we will see in Theorem A.7 below.   ◀

When we define $d_{\mathcal{X}}([x]_r, [y]_r) = \mathbb{1}_{\{[x]_r \neq [y]_r\}}$, we obtain with $R^* = \sup\{r : [x]_r = [y]_r\}$ that

$$d_{loc}(x, y) = \sum_{r \geq 1} \frac{d_{\mathcal{X}}([x]_r, [y]_r) \wedge 1}{r(r + 1)} = \sum_{r \geq R^* + 1} \frac{1}{r(r + 1)} = \frac{1}{R^* + 1}, \tag{A.2.2}$$

as in (2.2.2) and (2.2.3) for rooted graphs. Thus, the topology on rooted graphs can be seen as a special case of a local topology. In a similar way, the metric on marked rooted graphs in (2.3.14) and (2.3.15) can be viewed as a local topology. In this section we discuss local topologies in the general setting.

We next show that local topologies form a Polish space:

Theorem A.6 (Local topologies form a Polish space)   *Assume that $\{[x]_r : x \in \mathcal{X}\}$ is Polish for every $r \geq 1$. Then the space $(\mathcal{X}, d_{loc})$ is a Polish space, that is, $(\mathcal{X}, d_{loc})$ is a metric, separable, and complete space. Furthermore, a subset $A \subset \mathcal{X}$ is pre-compact (meaning that its closure is compact) if and only if the sets $\{[x]_r : x \in A\}$ are pre-compact for every $r \geq 0$.*

Proof   Let us first show that d_{loc} is a distance. The symmetry and triangle inequality will then follow directly. The fact that $d_{loc}(x, y) = 0$ precisely when $x = y$ is also easy, since if $[x]_r = [y]_r$ for all $r > 0$ then $x = y$ by (A.2.1).

We next show the separability of $(\mathcal{X}, d_{loc})$. For any $x \in \mathcal{X}$, we have $d_{loc}(x, [x]_r) \leq 1/(r + 1)$ and we have assumed that the set $\{[x]_r : x \in \mathcal{X}\}$ of all restrictions of elements in $\mathcal{X}$ to radius r is separable. Thus, $(\mathcal{X}, d_{loc})$ arises as a union over r of dense countable sets in $\{[x]_r : x \in \mathcal{X}\}$, so that $(\mathcal{X}, d_{loc})$ itself is countable and dense for d_{loc}.

For the completeness of $(\mathcal{X}, d_{loc})$, we let $(x_n)_{n \geq 1}$ be a Cauchy sequence for d_{loc}. Then, for every r, the restriction $[x_n]_r$ is again Cauchy and, by the completeness of $\{[x]_r : x \in \mathcal{X}\}$, $[x_n]_r$ thus converges for $d_{\mathcal{X}}$ to a certain element $y_r \in \{[x]_r : x \in \mathcal{X}\}$. By the continuity of $x \mapsto [x]_r$ we deduce that $[y_s]_r = y_r$ for any $s \geq r$, and so by the coherence property (A.2.1), we can define a unique element $y \in \mathcal{X}$ such that $y_r = [y]_r$. It is then clear that $x_n \to y$ for d_{loc}.

We complete the proof by characterizing the compacts. The condition in the theorem is clearly necessary for A to be pre-compact, for otherwise there exists $r_0 \geq 0$ and a sequence $(x_n)_{n \geq 1}$ in A whose restrictions of radius r_0 are all at a distance at most ε from each other. Such a sequence cannot admit a convergent subsequence. Conversely, a subset A satisfying the condition of the theorem is easily seen to be pre-compact for d_{loc}: just cover it with restrictions of radius $1/(r + 1)$ centered on a $1/(r + 1)$-net for $d_{\mathcal{X}}$ of A to get a $1/(r + 1)$-net for d_{loc}.   □

We next proceed to discuss the convergence of random variables on $(\mathcal{X}, d_{loc})$. We first recall that a random variable X is a measurable function from the underlying probability space $(\Omega, \mathscr{F}, \mathbb{P})$ with values in the Polish space $(\mathcal{X}, d_{loc})$ endowed with the Borel σ-field denoted by $\mathcal{B}_{loc}$. Therefore, the natural notion of convergence in distribution states that the sequence of random variables $(X_n)_{n \geq 0}$ converges in distribution (for the local topology) towards a random variable X, which we denote as $X_n \xrightarrow{d} X$, if, for any bounded continuous function $h : \mathcal{X} \to \mathbb{R}$,

$$\mathbb{E}[h(X_n)] \to \mathbb{E}[h(X)]. \tag{A.2.3}$$

The main result of this section is the following theorem:

Theorem A.7 (Convergence of finite-dimensional distributions implies tightness)   *Assume that $\{[x]_r : x \in \mathcal{X}\}$ is Polish for every $r \geq 1$. The local topology satisfies the following properties:*

(a) *A family $(X_i)_{i \in I}$ of random variables with values in $\mathcal{X}$ is tight in the local topology if and only if the family $([X_i]_r)_{i \in I}$ is tight for every $r \geq 0$.*

(b) *Let X_1 and X_2 be two random variables with values in $(\mathcal{X}, \mathrm{d}_{\mathrm{loc}})$ such that $\mathbb{P}([X_1]_r \in \mathcal{A}) = \mathbb{P}([X_2]_r \in \mathcal{A})$ for any $\mathcal{A} \in \mathcal{B}_{\mathrm{loc}}$ and any $r \geq 0$. Then $X_1 \stackrel{d}{=} X_2$.*

(c) *$X_n \stackrel{d}{\longrightarrow} X$ in the local topology when, for every $r \geq 1$ and Borel sets $\mathcal{A} \in \mathcal{B}_{\mathrm{loc}}$,*

$$\mathbb{P}([X_n]_r \in \mathcal{A}) \to \mu([X]_r \in \mathcal{A}), \tag{A.2.4}$$

where μ is a probability measure on $\mathcal{X}$.

Theorem A.7 is remarkable, since the convergence of $\mathbb{P}([X_n]_r \in \mathcal{A})$ is equivalent to the *convergence of finite-dimensional distributions*. Normally, one would expect to need this convergence to be combined with *tightness* to obtain convergence in distribution. Owing to the special nature of the local topology introduced in this section (recall also Remark A.5), however, this convergence, combined with the fact that the limit is a probability measure, implies tightness.

Proof of Theorem A.7   Part (a) follows directly from the compactness statement in Theorem A.6.

For part (b), we consider the family of events

$$\mathcal{M} = \Big\{ \{ x \in \mathcal{X} : [x]_r \in \mathcal{A} \} : \mathcal{A} \in \mathcal{B}_{\mathrm{loc}}, r \geq 0 \Big\}. \tag{A.2.5}$$

It is easy to see that the family $\mathcal{M}$ generates the Borel σ-field on $\mathcal{X}$ and moreover that $\mathcal{M}$ is stable under finite intersections. It follows from the monotone class theorem that two random variables X_1 and X_2 agreeing on $\mathcal{M}$ have the same law.

For part (c), above we have already seen that the sets $\{ [x]_r \in \mathcal{A} \}$ are stable under finite intersections, and it is easy to see that any open sets of the local topology can be written as a countable union of those sets. The result then follows from (Billingsley, 1968, Theorem 2.2). In particular, we deduce that for a sequence of random variables $(X_n)_{n \geq 1}$ to converge in distribution it is necessary and sufficient that $(X_n)_{n \geq 1}$ be tight and that $\mathbb{P}([X_n]_r \in \mathcal{A})$ converges for every $r \geq 0$ and every $\mathcal{A} \in \mathcal{B}_{\mathrm{loc}}$. The two conditions are necessary, since if $\mathbb{P}([X_n]_r \in \mathcal{A})$ converges to a limit that does not have full mass, then $X_n \stackrel{d}{\longrightarrow} X$ does not hold. From this, we conclude that if $\mathbb{P}([X_n]_r \in \mathcal{A}) \to \mu([X]_r \in \mathcal{A})$ where μ has full mass then $X_n \stackrel{d}{\longrightarrow} X$ does indeed follow. $\qquad\square$

A.3 PROPERTIES OF THE METRIC $\mathrm{d}_{\mathscr{G}_\star}$ ON ROOTED GRAPHS

For $(G, o) \in \mathscr{G}_\star$, let

$$[G, o] = \{ (G', o') : (G', o') \simeq (G, o) \} \tag{A.3.1}$$

denote the equivalence class in $\mathscr{G}_\star$ corresponding to (G, o). We further let

$$[\mathscr{G}_\star] = \{ [G, o] : (G, o) \in \mathscr{G}_\star \} \tag{A.3.2}$$

denote the set of equivalence classes in $\mathscr{G}_\star$. This is the set on which the distance $\mathrm{d}_{\mathscr{G}_\star}$ acts.

In this section we prove that $([\mathscr{G}_\star], \mathrm{d}_{\mathscr{G}_\star})$ is a Polish space:

Theorem A.8 (Rooted graphs form a Polish space) $\mathrm{d}_{\mathscr{G}_\star}$ *is a well defined metric on $[\mathscr{G}_\star]$. Further, the metric space $([\mathscr{G}_\star], \mathrm{d}_{\mathscr{G}_\star})$ is Polish.*

We give an explicit proof of Theorem A.8, even though completeness and separability might also be concluded from Theorem A.6, together with the observation that $\{ [G, o]_r : (G, o) \in \mathscr{G}_\star \}$ is Polish for every $r \geq 1$. This must be the case since completeness is obvious while separability follows because $\mathrm{d}_{\mathscr{G}_\star}((G_n, o_n), (G_m, o_m)) \leq \varepsilon$ implies that $B_r^{(G_n)}(o_n) \simeq B_r^{(G_m)}(o_m)$ for all $r \leq 1/\varepsilon - 1$.

The proof of Theorem A.8 is divided into several steps. These proof steps are a little involved, since we need to deal with isomorphism classes of rooted graphs, rather than rooted graphs themselves. This requires us to show that statements hold *irrespective* of the representative rooted graph chosen. We start in Proposition A.10 below by showing that $\mathrm{d}_{\mathscr{G}_\star}$ is an ultrametric, which is a slightly stronger property than being a metric, and which also implies that $(\mathscr{G}_\star, \mathrm{d}_{\mathscr{G}_\star})$ is a metric space. In Proposition A.12, we show that the metric space $(\mathscr{G}_\star, \mathrm{d}_{\mathscr{G}_\star})$ is complete, and in Proposition A.14, we show that it is separable. After that, we can complete the proof of Theorem A.8.

In the remainder of this section, we often work with r-neighborhoods $B_r^{(G)}(o)$ of o in G. We emphasize that we consider $B_r^{(G)}(o)$ to be a *rooted* graph, with root o (recall (2.2.1)).

A.3.1 ULTRAMETRIC PROPERTY OF THE SPACE $([\mathscr{G}_\star], d_{\mathscr{G}_\star})$ OF ROOTED GRAPHS

In this subsection we prove that $d_{\mathscr{G}_\star} : \mathscr{G}_\star \times \mathscr{G}_\star \to [0,1]$ is an ultrametric. One of the problems that we have to resolve is that the space of rooted graphs is defined only up to isomorphisms, which means that we have to make sure that $d_{\mathscr{G}_\star}((G_1, o_1), (G_2, o_2))$ is independent of the exact representative we choose in the equivalence classes of (G_1, o_1) and (G_2, o_2). That is the content of the following proposition:

Proposition A.9 ($d_{\mathscr{G}_\star}$ is well defined on $[\mathscr{G}_\star]$)   *The equality*

$$d_{\mathscr{G}_\star}((\hat{G}_1, \hat{o}_1), (\hat{G}_2, \hat{o}_2)) = d_{\mathscr{G}_\star}((G_1, o_1), (G_2, o_2))$$

holds whenever $(\hat{G}_1, \hat{o}_1) \simeq (G_1, o_1)$ and $(\hat{G}_2, \hat{o}_2) \simeq (G_2, o_2)$. Consequently, $d_{\mathscr{G}_\star} : [\mathscr{G}_\star] \times [\mathscr{G}_\star] \to [0, \infty)$ is well defined.

We continue to study the metric structure of $d_{\mathscr{G}_\star}$ by showing that it is an *ultrametric*:

Proposition A.10 (Ultrametricity)   *The map $d_{\mathscr{G}_\star} : \mathscr{G}_\star \times \mathscr{G}_\star \to [0,1]$ is an ultrametric, meaning that:*

(a) $d_{\mathscr{G}_\star}((G_1, o_1), (G_2, o_2)) = 0$ *precisely when* $(G_1, o_1) \simeq (G_2, o_2)$;
(b) $d_{\mathscr{G}_\star}((G_1, o_1), (G_2, o_2)) = d_{\mathscr{G}_\star}((G_2, o_2), (G_1, o_1))$ *for all* $(G_1, o_1), (G_2, o_2) \in \mathscr{G}_\star$;
(c) $d_{\mathscr{G}_\star}((G_1, o_1), (G_3, o_3)) \leq \max\{d_{\mathscr{G}_\star}((G_1, o_1), (G_2, o_2)), d_{\mathscr{G}_\star}((G_2, o_2), (G_3, o_3))\}$ *for all* $(G_1, o_1), (G_2, o_2), (G_3, o_3) \in \mathscr{G}_\star$.

We prove will Propositions A.9 and A.10 below. Before giving their proofs, we state and prove an important ingredient in them:

Lemma A.11 (Local neighborhoods determine the graph)   *Let (G_1, o_1) and (G_2, o_2) be two connected locally finite rooted graphs such that $B_r^{(G_1)}(o_1) \simeq B_r^{(G_2)}(o_2)$ for all $r \geq 0$. Then $(G_1, o_1) \simeq (G_2, o_2)$.*

Proof   We use a subsequence argument. Fix $r \geq 0$, and consider the isomorphism $\phi_r : B_r^{(G_1)}(o_1) \to B_r^{(G_2)}(o_2)$, which exists by assumption. Extend ϕ_r to (G_1, o_1) by defining

$$\psi_r(v) = \begin{cases} \phi_r(v) & \text{for } v \in V(B_r^{(G_1)}(o_1)); \\ o_2 & \text{otherwise.} \end{cases} \tag{A.3.3}$$

Our aim is to use $(\psi_r)_{r \geq 0}$ to construct an isomorphism between (G_1, o_1) and (G_2, o_2).

Set $V_r^{(G_1)} = V(B_r^{(G_1)}(o_1))$. Let $\psi_r|_{V_0^{(G_1)}}$ be the restriction of ψ_r to $V_0^{(G_1)} = \{o_1\}$. Then we know that $\psi_r(v) = o_2$ for every $v \in V_0^{(G_1)}$ and $r \geq 0$. We next let $\psi_r|_{V_1^{(G_1)}}$ be the restriction of ψ_r to $V_1^{(G_1)}$. Then, $\psi_r|_{V_1^{(G_1)}}$ is an isomorphism between $B_1^{(G_1)}(o_1)$ and $B_1^{(G_2)}(o_2)$ for every r. Since there are only *finitely* many such isomorphisms, the same isomorphism, say ϕ_1', needs to be repeated infinitely many times in the sequence $(\psi_r|_{V_1^{(G_1)}})_{r \geq 1}$. Let $\mathbb{N}_1$ denote the values of r for which

$$\psi_r|_{V_1^{(G_1)}} = \phi_1' \qquad \forall r \in \mathbb{N}_1. \tag{A.3.4}$$

Now we extend this argument to $k = 2$. Let $\psi_r|_{V_2^{(G_1)}}$ be the restriction of ψ_r to $V_2^{(G_1)}$. Again, $\psi_r|_{V_2^{(G_1)}}$ is an isomorphism between $B_2^{(G_1)}(o_1)$ and $B_2^{(G_2)}(o_2)$ for every r. Since there are again only finitely many such isomorphisms, the same isomorphism, say ϕ_2', needs to be repeated infinitely many times in the sequence $(\psi_r|_{V_2^{(G_1)}})_{r \in \mathbb{N}_1}$. Let $\mathbb{N}_2$ denote the values of $r \in \mathbb{N}_1$ for which

$$\psi_r|_{V_2^{(G_1)}} = \phi_2' \qquad \forall r \in \mathbb{N}_2. \tag{A.3.5}$$

We next generalize this argument to general $k \geq 2$. Let $\psi_r|_{V_k^{(G_1)}}$ be the restriction of ψ_r to $V_k^{(G_1)}$. Again, $\psi_r|_{V_k^{(G_1)}}$ is an isomorphism between $B_k^{(G_1)}(o_1)$ and $B_k^{(G_2)}(o_2)$ for every r. Since there are again only finitely many such isomorphisms, the same isomorphism, say ϕ_k', needs to be repeated infinitely many times in the sequence $(\psi_r|_{V_k^{(G_1)}})_{r \in \mathbb{N}_{k-1}}$. Let $\mathbb{N}_k$ denote the values of $r \in \mathbb{N}_{k-1}$ for which

$$\psi_r|_{V_k^{(G_1)}} = \phi_k' \qquad \forall r \in \mathbb{N}_k. \tag{A.3.6}$$

Then, we see that $\mathbb{N}_k$ is a sequence of decreasing infinite sets.

Let us define ψ'_ℓ to be the first element of the sequence $(\psi_r)_{r \in \mathbb{N}_\ell}$. Then, it follows that $\psi'_\ell(v) = \phi'_k(v)$ for all $\ell \geq k$ and all $v \in V_k^{(G_1)}$.

Denote $U_0 = \{o_1\}$ and $U_k = V_k^{(G_1)} \setminus V_{k-1}^{(G_1)}$. Since we are assuming that $V(G_1)$ is connected, we have that $\cup_{k \geq 0} U_k = V(G_1)$, and this union is disjoint.

It follows that the functions $(\psi'_\ell)_{\ell \geq 1}$ converge pointwise to

$$\psi(v) = \psi'_\infty(v) = \sum_{k \geq 1} \phi'_k(v) \mathbb{1}_{\{v \in U_k\}}. \tag{A.3.7}$$

We claim that ψ is the desired isomorphism between (G_1, o_1) and (G_2, o_2). The map ψ is clearly bijective, since $\phi'_k \colon U_k \to \phi'_k(U_k)$ is bijective. Further, let $u, v \in V(G_1)$. Denote

$$k = \max\{\text{dist}_{G_1}(o_1, u), \text{dist}_{G_1}(o_1, v)\}.$$

Then $u, v \in V_k^{(G_1)}$. Because ϕ'_k is an isomorphism between $B_k^{(G_1)}(o_1)$ and $B_k^{(G_2)}(o_2)$, it follows that $\phi'_k(u), \phi'_k(v) \in V(B_k^{(G_2)}(o_2))$, and further that $\{\phi'_k(u), \phi'_k(v)\} \in E(B_k^{(G_2)}(o_2))$ precisely when $\{u, v\} \in E(B_k^{(G_1)}(o_1))$. Since $\psi = \phi'_k$ on $V_k^{(G_1)}$, it then also follows that $\{\psi(u), \psi(v)\} \in E(B_k^{(G_2)}(o_2))$ precisely when $\{u, v\} \in E(B_k^{(G_1)}(o_1))$, as required. Finally, $\psi(o_1) = \phi_k(o_1)$ and $\phi_k(o_1) = o_2$ for every $k \geq 0$. This completes the proof. $\qquad\square$

Proof of Proposition A.9. We note that if $(\hat{G}_1, \hat{o}_1) \simeq (G_1, o_1)$ and $(\hat{G}_2, \hat{o}_2) \simeq (G_2, o_2)$ then we have that $B_r^{(G_1)}(o_1) \simeq B_r^{(G_2)}(o_2)$ if and only if $B_r^{(\hat{G}_1)}(\hat{o}_1) \simeq B_r^{(\hat{G}_2)}(\hat{o}_2)$. Therefore $d_{\mathscr{G}_\star}((G_1, o_1), (G_2, o_2))$ is independent of the exact choice of representative in the equivalence class of (G_1, o_1) and (G_2, o_2). In particular, $d_{\mathscr{G}_\star}((G_1, o_1), (G_2, o_2))$ is constant on such equivalence classes. This makes $d_{\mathscr{G}_\star}([G_1, o_1], [G_2, o_2])$ well defined for $[G_1, o_1], [G_2, o_2] \in [\mathscr{G}_\star]$. $\qquad\square$

Proof of Proposition A.10. (a) Assume that $d_{\mathscr{G}_\star}((G_1, o_1), (G_2, o_2)) = 0$. Then we have $B_r^{(G_1)}(o_1) \simeq B_r^{(G_2)}(o_2)$ for all $r \geq 0$, so that, by Lemma A.11, we also have that $(G_1, o_1) \simeq (G_2, o_2)$ as required.

The proof of (b) is trivial and omitted.

For (c) and $i, j \in [3]$, let

$$r_{ij} = \sup\{r \colon B_r^{(G_i)}(o_i) \simeq B_r^{(G_j)}(o_j)\}. \tag{A.3.8}$$

Then $B_r^{(G_1)}(o_1) \simeq B_r^{(G_3)}(o_3)$ for all $r \leq r_{13}$ and $B_r^{(G_2)}(o_2) \simeq B_r^{(G_3)}(o_3)$ for all $r \leq r_{23}$. We conclude that $B_r^{(G_1)}(o_1) \simeq B_r^{(G_2)}(o_2)$ for all $r \leq \min\{r_{13}, r_{23}\}$, so that $r_{12} \geq \min\{r_{13}, r_{23}\}$. This implies that

$$1/(r_{12} + 1) \leq \max\{1/(r_{13} + 1), 1/(r_{23} + 1)\},$$

which in turn implies the claim (recall (2.2.2)). $\qquad\square$

A.3.2 COMPLETENESS OF THE SPACE $([\mathscr{G}_\star], d_{\mathscr{G}_\star})$ OF ROOTED GRAPHS

In this subsection we prove that $([\mathscr{G}_\star], d_{\mathscr{G}_\star})$ is complete:

Proposition A.12 (Completeness) *The metric space $([\mathscr{G}_\star], d_{\mathscr{G}_\star})$ is complete.*

Before giving the proof of Proposition A.12, we state and prove an important ingredient in it:

Lemma A.13 (Coherence: compatible rooted graph sequences have a limit) *Let $((G_r, o_r))_{r \geq 0}$ be connected locally finite rooted graphs that are compatible, meaning that $B_r^{(G_s)}(o_s) \simeq (G_r, o_r)$ for all $r \leq s$. Then there exists a connected locally finite rooted graph (G, o) such that $(G_r(o_r), o_r) \simeq B_r^{(G)}(o)$. Moreover, (G, o) is unique up to isomorphisms.*

It is implicit in Lemma A.13 that $(G_r(o_r), o_r) = B_r^{(G_r)}(o_r)$.

Proof The crucial point is that (G_r, o_r) might not have the same vertex sets for different r. Therefore, we first create a version $(G'_r, o'_r) \simeq (G_r, o_r)$ that does.

For this, we first construct isomorphic copies of (G_r, o_r) on a common node set, in a compatible way. To do this, denote $V_r = [N_r]$, where $N_r = |V(B_r^{(G_r)}(o_r))|$. We define a sequence of bijections

$\phi_r \colon V(B_r^{(G_r)}(o_r)) \to V_r$ recursively as follows. Let ϕ_0 be the unique isomorphism from $V(B_0^{(G_r)}(o_0)) = \{o_0\}$ to $V_0 = \{1\}$.

Let ψ_r be an isomorphism between (G_{r-1}, o_{r-1}) and $B_{r-1}^{(G_r)}(o_r)$, and η_r an arbitrary bijection between $V(G_r) \setminus V(B_{r-1}^{(G_r)}(o_r))$ to $V_r \setminus V_{r-1}$. Define

$$\phi_r(v) = \begin{cases} \phi_{r-1}(\psi_r^{-1}(v)) & \text{for } v \in V(B_{r-1}^{(G_r)}(o_r)); \\ \eta_r(v) & \text{for } v \in V(G_r) \setminus V(B_{r-1}^{(G_r)}(o_r)), \end{cases} \tag{A.3.9}$$

so that ϕ_r is a bijection from $V(G_r)$ to V_r.

Then we define $(G_r', o_r') = (\phi_r(G_r), \phi_r(o_r))$, where $\phi_r(G_r)$ is the graph consisting of the vertex set $\{\phi_r(v) \colon v \in V(G_r)\} = V_r$ and edge set $\{\{\phi_r(u), \phi_r(v)\} \colon \{u, v\} \in E(G_r)\}$.

Let us derive some properties of (G_r', o_r'). First of all, we see that $o_r' = 1$ for every $r \geq 0$, by construction. Further, ϕ_r is a bijection, so that $B_r^{(G_r')}(o_r') \simeq B_r^{(G_r)}(o_r) = (G_r, o_r)$.

Now we are ready to define (G, o). We define the root as $o = 1$, the vertex set of G by $V(G) = \bigcup_{r \geq 1} V(G_r') = \bigcup_{r \geq 1} V_r$, and the edge set $E(G) = \bigcup_{r \geq 1} E(G_r')$. Then it follows that $B_r^{(G)}(o) = B_r^{(G_r')}(o_r') \simeq B_r^{(G_r)}(o_r) = (G_r, o_r)$, as required. Further, (G, o) is locally finite and connected, since (G_r, o_r) is so for every $r \geq 0$. Finally, to verify uniqueness apart from isomosphisms, note that if (G', o') also satisfies that $B_r^{(G')}(o') \simeq B_r^{(G_r)}(o_r) = (G_r, o_r)$ for every $r \geq 0$, then $B_r^{(G')}(o') \simeq B_r^{(G)}(o)$ for every $r \geq 0$, so that $(G, o) \simeq (G', o')$ by Lemma A.11 as required. □

Proof of Proposition A.12. To verify the completeness of the metric space $([\mathscr{G}_\star], d_{\mathscr{G}_\star})$, fix a Cauchy sequence $([G_n, o_n])_{n \geq 1}$ with representative rooted graphs $((G_n, o_n))_{n \geq 1}$. Then, for every $\varepsilon > 0$, there exists an $N = N(\varepsilon) \geq 0$ such that, for every $n, m \geq N$,

$$d_{\mathscr{G}_\star}([G_n, o_n], [G_m, o_m]) \leq \varepsilon. \tag{A.3.10}$$

By Proposition A.9,

$$d_{\mathscr{G}_\star}([G_n, o_n], [G_m, o_m]) = d_{\mathscr{G}_\star}((G_n, o_n), (G_m, o_m)), \tag{A.3.11}$$

so that from now on we can work with the representatives instead. Since $d_{\mathscr{G}_\star}((G_n, o_n), (G_m, o_m)) \leq \varepsilon$, we obtain that $B_r^{(G_n)}(o_n) \simeq B_r^{(G_m)}(o_m)$ for all $r \leq 1/\varepsilon - 1$ and $n, m \geq N$.

Equivalently, the fact that $([G_n, o_n])_{n \geq 1}$ is a Cauchy sequence implies that, for every $r \geq 1$, there exists an n_r such that, for all $n \geq n_r$,

$$B_r^{(G_n)}(o_n) \simeq B_r^{(G_{n_r})}(o_{n_r}). \tag{A.3.12}$$

Clearly, we may select n_r such that $r \mapsto n_r$ is strictly increasing. Define $(G_r', o_r') = B_r^{(G_{n_r})}(o_{n_r})$. Then $((G_r', o_r'))_{r \geq 0}$ forms a compatible sequence as in Lemma A.13. By Lemma A.13, there exists a locally finite rooted graph (G, o) such that $B_r^{(G)}(o) \simeq (G_r', o_r')$. But then also

$$B_r^{(G_n)}(o_n) \simeq B_r^{(G_{n_r})}(o_{n_r}) = (G_r', o_r'). \tag{A.3.13}$$

This, in turn, implies that, for all $n \geq n_r$,

$$d_{\mathscr{G}_\star}([G, o], [G_n, o_n]) = d_{\mathscr{G}_\star}((G, o), (G_n, o_n)) \leq 1/(r+1). \tag{A.3.14}$$

Since $r \geq 1$ is arbitrary, we conclude that $[G_n, o_n]$ converges to $[G, o]$, which is in $[\mathscr{G}_\star]$, so that $([\mathscr{G}_\star], d_{\mathscr{G}_\star})$ is complete. □

A.3.3 SEPARABILITY OF THE SPACE $([\mathscr{G}_\star], d_{\mathscr{G}_\star})$ OF ROOTED GRAPHS

In this subsection we prove that $([\mathscr{G}_\star], d_{\mathscr{G}_\star})$ is separable:

Proposition A.14 (Separability) *The metric space $([\mathscr{G}_\star], d_{\mathscr{G}_\star})$ is separable.*

Proof We need to show that there exists a countable dense subset in $([\mathscr{G}_\star], d_{\mathscr{G}_\star})$. Consider the set of all *finite* rooted graphs, which is certainly countable. Fix $[G, o] \in [\mathscr{G}_\star]$ with representative (G, o). Then $B_r^{(G)}(o)$ is a finite rooted graph for all $r \geq 0$. Finally, $d_{\mathscr{G}_\star}(B_r^{(G)}(o), (G, o)) \leq 1/(r+1)$, so that $B_r^{(G)}(o)$ converges to (G, o) when $r \to \infty$. Thus, the space of isomorphism classes of finite rooted graphs is dense and countable. This completes the proof that $(\mathscr{G}_\star, d_{\mathscr{G}_\star})$ is separable. □

A.3.4 $([\mathscr{G}_\star], \mathrm{d}_{\mathscr{G}_\star})$ IS POLISH: PROOF OF THEOREM A.8

Here we use the above results to complete the proof of Theorem A.8:

Proof of Theorem A.8. The function $\mathrm{d}_{\mathscr{G}_\star}$ is well defined on $[\mathscr{G}_\star] \times [\mathscr{G}_\star]$ by Proposition A.9. Proposition A.10 implies that $\mathrm{d}_{\mathscr{G}_\star}$ is an (ultra)metric on $[\mathscr{G}_\star]$. Finally, Proposition A.12 proves that $([\mathscr{G}_\star], \mathrm{d}_{\mathscr{G}_\star})$ is complete, while Proposition A.14 proves that $([\mathscr{G}_\star], \mathrm{d}_{\mathscr{G}_\star})$ is separable. Thus, $([\mathscr{G}_\star], \mathrm{d}_{\mathscr{G}_\star})$ is a Polish space. □

A.3.5 THE LAWS OF NEIGHBORHOODS DETERMINE DISTRIBUTIONS ON $\mathscr{G}_\star$

In this subsection we show that the laws of neighborhoods determine distributions on $\mathscr{G}_\star$, as was crucially used in the proof of Theorem 2.7:

Proposition A.15 (Laws of neighborhoods determine distributions) *Let μ and μ' be two distributions on $\mathscr{G}_\star$ such that $\mu(B_r^{(G)}(o) \simeq H_\star) = \mu'(B_r^{(G)}(o) \simeq H_\star)$ for all $r \geq 1$. Then $\mu = \mu'$.*

Proof This is Theorem A.7(ii), and here we also give a direct proof. The measures μ and μ' satisfy $\mu = \mu'$ precisely when $\mu(\mathscr{H}_\star) = \mu'(\mathscr{H}_\star)$ for every measurable $\mathscr{H}_\star \subseteq \mathscr{G}_\star$. Fix $\mathscr{H}_\star \subseteq \mathscr{G}_\star$. For $r \geq 0$, denote

$$\mathscr{H}_\star(r) = \{(G,o) \colon \exists (G',o') \in \mathscr{H}_\star \text{ such that } B_r^{(G)}(o) \simeq B_r^{(G')}(o')\}. \tag{A.3.15}$$

Thus, $\mathscr{H}_\star(r)$ contains those rooted graphs whose r-neighborhood is the same as that of a rooted graph in $\mathscr{H}_\star$. Clearly, $\mathscr{H}_\star(r) \searrow \mathscr{H}_\star$ as $r \to \infty$. Therefore, also $\mu(\mathscr{H}_\star(r)) \searrow \mu(\mathscr{H}_\star)$ and $\mu'(\mathscr{H}_\star(r)) \searrow \mu'(\mathscr{H}_\star)$.
Finally, note that $(G,o) \in \mathscr{H}_\star(r)$ if and only if $B_r^{(G)}(o) \in \mathscr{H}_\star(r)$. Thus,

$$\mu(\mathscr{H}_\star(r)) = \sum_{H_\star \in \mathscr{H}_\star(r)} \mu(B_r^{(G)}(o) \simeq H_\star) \tag{A.3.16}$$

(where we realize that the fact that the sum is over equivalence classes makes the events $\{B_r^{(G)}(o) \simeq H_\star\}$ disjoint). Since $\mu(B_r^{(G)}(o) \simeq H_\star) = \mu'(B_r^{(G)}(o) \simeq H_\star)$, we conclude that

$$\mu(\mathscr{H}_\star(r)) = \sum_{H_\star \in \mathscr{H}_\star(r)} \mu(B_r^{(G)}(o) \simeq H_\star) = \sum_{H_\star \in \mathscr{H}_\star(r)} \mu'(B_r^{(G)}(o) \simeq H_\star) = \mu'(\mathscr{H}_\star(r)), \tag{A.3.17}$$

so that $\mu(\mathscr{H}_\star) = \mu'(\mathscr{H}_\star)$, as required. □

A.3.6 COMPACT SETS IN $([\mathscr{G}_\star], \mathrm{d}_{\mathscr{G}_\star})$ AND TIGHTNESS

In this subsection we investigate compact sets in the metric space $([\mathscr{G}_\star], \mathrm{d}_{\mathscr{G}_\star})$, after which we formulate a convenient tightness condition. First, we recall the definition of compactness:

Definition A.16 (Compact sets and tightness on general metric spaces) Let $\mathcal{X}$ be a general metric space. A set $\mathcal{K}$ is *compact* when every collection of open sets covering $\mathcal{K}$ has a finite subset. A sequence of random variables $(X_n)_{n \geq 1}$ living on a general metric space $\mathcal{X}$ is *tight* when, for every $\varepsilon > 0$, there exists a compact set $\mathcal{K} = \mathcal{K}_\varepsilon$ such that $\limsup_{n \to \infty} \mathbb{P}(X_n \in \mathcal{K}_\varepsilon^c) \leq \varepsilon$. ◄

The notion of tightness is thus intimately related to compact sets. For real-valued random variables, $\mathcal{K} = [-K, K]$ are convenient compact sets. For $([\mathscr{G}_\star], \mathrm{d}_{\mathscr{G}_\star})$, we first investigate what compact sets look like:

Theorem A.17 (Compact sets in $([\mathscr{G}_\star], \mathrm{d}_{\mathscr{G}_\star})$) *For $(G,o) \in \mathscr{G}_\star$ and $r \geq 1$, define*

$$\Delta_r(G,o) = \max\{d_v^{(G)} \colon v \in V(B_r^{(G)}(o))\}, \tag{A.3.18}$$

where $d_v^{(G)}$ denotes the degree of $v \in V(G)$. Then, a closed family of equivalence classes of rooted graphs $[\mathcal{K}] \subseteq [\mathscr{G}_\star]$ is compact if and only if

$$\sup_{(G,o) \in \mathcal{K}} \Delta_r(G,o) < \infty \qquad \text{for all } r \geq 1. \tag{A.3.19}$$

Proof Recall from (Rudin, 1991, Theorem A.4) that a closed set $\mathcal{K}$ is compact when it is totally bounded, meaning that, for every $\varepsilon > 0$, the set $\mathcal{K}$ can be covered by finitely many balls of radius ε. As a result, for every $r \geq 1$, there must be graphs $(F_1, o_1), \ldots, (F_\ell, o_\ell)$ such that $\mathcal{K}$ is covered by the finitely many open sets

$$\{(G, o) \colon B_r^{(G)}(o) \simeq B_r^{(F_i)}(o_i)\}. \tag{A.3.20}$$

Equivalently, every $(G, o) \in \mathcal{K}$ satisfies $B_r^{(G)}(o) \simeq B_r^{(F_i)}(o_i)$ for some $i \in [\ell]$. In turn, this is equivalent to the statement that the set

$$\mathcal{A}_r = \{B_r^{(G)}(o) \colon (G, o) \in \mathcal{K}\} \tag{A.3.21}$$

is finite for every $r \geq 1$.

We finally prove that $\mathcal{A}_r$ is finite for every $r \geq 1$ precisely when (A.3.19) holds. Denote $\Delta_r = \sup_{(G,o) \in \mathcal{K}} \Delta_r(G, o)$. If Δ_r is finite for every $r \geq 1$, then, because every $(G, o) \in \mathcal{K}$ is connected, the graphs $B_r^{(G)}(o)$ can have at most

$$|V(B_r^{(G)}(o))| \leq 1 + \Delta_r + \cdots + \Delta_r^r$$

many vertices, so that $\mathcal{A}_r$ is finite. On the other hand, when $\Delta_r = \infty$, $\mathcal{K}$ contains a sequence of rooted graphs (G_i, o_i) such that $\Delta_r(G_i, o_i) \to \infty$, so that we also have $|V(B_r^{(G_i)}(o_i))| \to \infty$. Since rooted graphs with different numbers of vertices are non-isomorphic (recall Exercise 2.1), this shows that $\mathcal{A}_r$ is infinite. $\qquad\square$

We continue by giving a more convenient *tightness criterium* for local weak convergence:

Theorem A.18 (Tightness criterion for local weak convergence)   *Let $(G_n)_{n \geq 1}$ be a sequence of (possibly random) graphs with $|V(G_n)| \to \infty$. Let $d_{o_n}^{(G_n)}$ denote the degree of o_n in G_n, where o_n is chosen uar from the vertex set $V(G_n)$ of G_n. Then $((G_n, o_n))_{n \geq 1}$ is tight when $(d_{o_n}^{(G_n)})_{n \geq 1}$ forms a uniformly integrable sequence of random variables.*

The needed uniform integrability in Theorem A.18 is quite suggestive. Indeed, in many random graph models, such as the configuration model, the degree of a random *neighbor* of a vertex has the *size-biased* degree distribution (recall (1.4.18)). When $(d_{o_n}^{(G_n)})_{n \geq 1}$ forms a uniformly integrable sequence of random variables, there exists a subsequence along which $D_n^\star$, the size-biased version of $D_n = d_{o_n}^{(G_n)}$, converges in distribution (see Exercise 2.11).

For the configuration model, Conditions 1.7(a),(b) imply that $(d_{o_n}^{(G_n)})_{n \geq 1}$ is a tight sequence of random variables (see Exercise 2.12). Further, [V1, Theorem 7.25] discusses how Conditions 1.7(a),(b) imply convergence of the degrees of *neighbors* of the uniform vertex o_n, a distribution that is given by $D_n^\star$. Of course, for local weak convergence to hold, one certainly needs that the degrees of neighbors converge in distribution. Thus, at least for the configuration model, we can fully understand why the uniform integrability of $(d_{o_n}^{(G_n)})_{n \geq 1}$ is needed.

In general, however, local weak convergence does *not* imply that $(d_{o_n}^{(G_n)})_{n \geq 1}$ is uniformly integrable (see Exercises 2.13 and 2.14). This is due to the fact that a small proportion of vertices may have degrees that are very large.

Proof of Theorem A.18.   Let $\mathcal{A}$ be a family of finite graphs. For a graph G, let o denote a random vertex drawn uar from $V(G)$, let $U(G) = (G, o)$ be the rooted graph obtained by rooting G at o, and let μ_G be its law. We need to show that if $\{d_o^{(G)} \colon G \in \mathcal{A}\}$ is a uniformly integrable sequence of random variables, then the family $\mathcal{A}$ is tight. Let

$$f(d) = \sup_{G \in \mathcal{A}} \mathbb{E}\big[d_o^{(G)} \mathbb{1}_{\{d_o^{(G)} > d\}}\big]. \tag{A.3.22}$$

By assumption, $\lim_{d \to \infty} f(d) = 0$. Write $m(G) = \mathbb{E}\big[d_o^{(G)}\big]$. Thus, $1 \leq m(G) \leq f(0) < \infty$. Write $\mu_G^\star$ for the degree-biased probability measure on $\{(G, v) \colon v \in V(G)\}$, that is,

$$\mu_G^\star[(G, v)] = \frac{d_v^{(G)}}{m(G)} \times \mu_G[(G, v)], \tag{A.3.23}$$

and o_G for the corresponding root. Since $\mu_G \leq m(G)\mu_G^\star \leq f(0)\mu_G^\star$, it suffices to show that $\{\mu_G^\star \colon G \in \mathcal{A}\}$ is tight. Note that $\{d_{o_G}^{(G)} \colon G \in \mathcal{A}\}$ is tight by assumption.

For $r \in \mathbb{N}$, let $F_r^M(v)$ be the event such that there is some vertex at distance at most r from v whose degree is larger than M. Let X be a uniform random neighbor of o_G. Because $\mu_G^\star$ is a stationary measure for a simple random walk, $F_r^M(o_G)$ and $F_r^M(X)$ have the same probability. Also,

$$\mathbb{P}\left(F_{r+1}^M(o_G) \,\middle|\, d_{o_G}^{(G)}\right) \leq d_{o_G}^{(G)} \mathbb{P}\left(F_r^M(X) \,\middle|\, d_{o_G}^{(G)}\right). \tag{A.3.24}$$

We claim that, for all $r \in \mathbb{N}$ and $\varepsilon > 0$, there exists $M < \infty$ such that $\mathbb{P}\left(F_r^M(X)\right) \leq \varepsilon$ for all $G \in \mathcal{A}$. This clearly implies that $\{\mu_G^\star : G \in \mathcal{A}\}$ is tight. We prove the claim by induction on r.

The statement for $r = 0$ is trivial. Given that the property holds for r, let us now show it for $r + 1$. Given $\varepsilon > 0$, choose d sufficiently large that $\mathbb{P}(d_{o_G}^{(G)} > d) \leq \varepsilon/2$ for all $G \in \mathcal{A}$. Also, choose M sufficiently large that $\mathbb{P}(F_r^M(o_G)) \leq \varepsilon/(2d)$ for all $G \in \mathcal{A}$. Write F for the event that $d_{o_G}^{(G)} > d$. Then, by conditioning on $d_{o_G}^{(G)}$, we see that

$$\begin{aligned}
\mathbb{P}\left(F_{r+1}^M(o_G)\right) &\leq \mathbb{P}(F) + \mathbb{E}\left[\mathbf{1}_{F^c}\mathbb{P}\left(F_{r+1}^M(o_G) \,\middle|\, d_{o_G}^{(G)}\right)\right] \\
&\leq \varepsilon/2 + \mathbb{E}\left[\mathbf{1}_{F^c} d_{o_G}^{(G)} \mathbb{P}\left(F_r^M(o_G) \,\middle|\, d_{o_G}^{(G)}\right)\right] \\
&\leq \varepsilon/2 + \mathbb{E}\left[\mathbf{1}_{F^c} d\, \mathbb{P}\left(F_r^M(o_G) \,\middle|\, d_{o_G}^{(G)}\right)\right] \\
&\leq \varepsilon/2 + d\,\mathbb{P}\left(F_r^M(o_G)\right) \leq \varepsilon/2 + d\varepsilon/(2d) = \varepsilon, \tag{A.3.25}
\end{aligned}$$

for all $G \in \mathcal{A}$, which proves the claim. $\qquad\square$

A.4 Notes and Discussion

In Section A.1 we drew inspiration from Howes (1995) and Rudin (1987, 1991). Section A.2 is, to a large extent, based on (Curien, 2018, Section 1.2). I am grateful to Nicolas Curien for sharing his material, and allowing me to use it in this book. Definition A.3 is inspired by discussions with Simon Irons. Section A.3 is based to a large extent on (Leskelä, 2019, Appendix B); some parts of the presented material are copied almost verbatim from there. I am grateful to Lasse Leskelä for making me aware of the subtleties of the proof, as well as sharing his preliminary version of these notes.

Random variables on general metric graphs can be hard to fathom. For example, single probability measures may not be tight. Since the space of equivalence classes of rooted graphs is Polish (see Theorem A.8), (Parthasarathy, 1967, Theorem 3.2 in Chapter 2) implies that single probability measures are tight.

The tightness statement in Theorem A.18 is (Benjamini et al., 2015, Theorem 3.1). Benjamini et al. (2015) used the term *network* instead of a marked graph. We avoid the term networks here, as it may cause confusion with the complex networks in the real world that form the inspiration for this book. A related proof can be found in Angel and Schramm (2003).

Aldous (1991) investigated local weak convergence in the context of *finite trees* and called trees rooted at a uniform vertex *fringe trees*. For fringe trees, the uniform integrability of the degree of a random vertex *is* equivalent to tightness of the resulting tree in the local weak sense (see (Aldous, 1991, Lemma 4(ii))).

GLOSSARY

$\mathrm{Exp}(\lambda)$	Exponential random variable with parameter λ and expected value $1/\lambda$	41
$\mathrm{Gam}(r, \lambda)$	Gamma random variable with parameters λ and r, and expected value r/λ	41
whp	A sequence of events $(\mathcal{E}_n)_{n\geq 1}$ occurs with high probability (whp) when $\lim_{n\to\infty} \mathbb{P}(\mathcal{E}_n) = 1$	41
$\xrightarrow{d}$	Convergence in distribution	41
$\xrightarrow{\mathbb{P}}$	Convergence in probability	41
$\xrightarrow{a.s.}$	Convergence almost surely	41
$(\mathscr{G}_\star, \mathrm{d}_{\mathscr{G}_\star})$	Metric space of (equivalence classes of) rooted graphs.	50
$\mathscr{C}(v)$	Connected component of $v \in V(G)$	51
$\mathscr{C}_{\max}$	Connected component of maximal size	75
$Z_{\geq k}$	Number of vertices in connected component of size at least k	75

REFERENCES

Abbe, E., and Sandon, C. 2018. Proof of the achievability conjectures for the general stochastic block model. *Comm. Pure Appl. Math.*, **71**(7), 1334–1406.

Abdullah, M. A., Bode, M., and Fountoulakis, N. 2017. Typical distances in a geometric model for complex networks. *Internet Math.*, pp. 38.

Achlioptas, D., D'Souza, R., and Spencer, J. 2009. Explosive percolation in random networks. *Science*, **323**(5920), 1453–1455.

Aiello, W., Bonato, A., Cooper, C., Janssen, J., and Pralat, P. 2008. A spatial web graph model with local influence regions. *Internet Math.*, **5**(1-2), 175–196.

Aldous, D. 1985. Exchangeability and related topics. Pages 1–198 of: *École d'été de probabilités de Saint-Flour, XIII–1983*. Lecture Notes in Math., vol. **1117**. Springer.

Aldous, D. 1991. Asymptotic fringe distributions for general families of random trees. *Ann. Appl. Probab.*, **1**(2), 228–266.

Aldous, D., and Lyons, R. 2007. Processes on unimodular random networks. *Electron. J. Probab.*, **12**(54), 1454–1508.

Aldous, D., and Steele, J.M. 2004. The objective method: probabilistic combinatorial optimization and local weak convergence. Pages 1–72 of: *Probability on discrete structures*. Encyclopaedia Math. Sci., vol. **110**. Springer.

Anantharam, V., and Salez, J. 2016. The densest subgraph problem in sparse random graphs. *Ann. Appl. Probab.*, **26**(1), 305–327.

Andreis, L., König, W., and Patterson, R. 2021. A large-deviations principle for all the cluster sizes of a sparse Erdős–Rényi graph. *Random Structures Algorithms*, **59**(4), 522–553.

Andreis, L., König, W., and Patterson, R. 2023. A large-deviations principle for all the components in a sparse inhomogeneous random graph. *Probab. Theory Rel. Fields*, **186**(1-2), 521–620.

Angel, O., and Schramm, O. 2003. Uniform infinite planar triangulations. *Comm. Math. Phys.*, **241**(2-3), 191–213.

Antunović, T., Mossel, E., and Rácz, M. 2016. Coexistence in preferential attachment networks. *Combin. Probab. Comput.*, **25**(6), 797–822.

Arratia, R., Barbour, AD., and Tavaré, S. 2003. *Logarithmic combinatorial structures: a probabilistic approach*. EMS Monographs in Mathematics. European Mathematical Society.

Artico, I., Smolyarenko, I., Vinciotti, V., and Wit, EC. 2020. How rare are power-law networks really? *Proc. Roy. Soc. A*, **476**(2241), 20190742.

Athreya, K., and Ney, P. 1972. *Branching processes*. Springer-Verlag. Die Grundlehren der mathematischen Wissenschaften, Band 196.

Backhausz, Á., and Szegedy, B. 2022. Action convergence of operators and graphs. *Canad. J. Math.*, **74**(1), 72–121.

Ball, F., Mollison, D., and Scalia-Tomba, G. 1997. Epidemics with two levels of mixing. *Ann. Appl. Probab.*, **7**(1), 46–89.

Ball, F., and Neal, P. 2002. A general model for stochastic SIR epidemics with two levels of mixing. *Math. Biosci.*, **180**, 73–102. John A. Jacquez memorial volume.

Ball, F., and Neal, P. 2004. Poisson approximations for epidemics with two levels of mixing. *Ann. Probab.*, **32**(1B), 1168–1200.

Ball, F., and Neal, P. 2008. Network epidemic models with two levels of mixing. *Math. Biosci.*, **212**(1), 69–87.

Ball, F., and Neal, P. 2017. The asymptotic variance of the giant component of configuration model random graphs. *Ann. Appl. Probab.*, **27**(2), 1057–1092.

Ball, F., Sirl, D., and Trapman, P. 2009. Threshold behaviour and final outcome of an epidemic on a random network with household structure. *Adv. Appl. Probab.*, **41**(3), 765–796.

Ball, F., Sirl, D., and Trapman, P. 2010. Analysis of a stochastic SIR epidemic on a random network incorporating household structure. *Math. Biosci.*, **224**(2), 53–73.

Banerjee, S., Deka, P., and Olvera-Cravioto, M. 2023. *Local weak limits for collapsed branching processes with random out-degrees.* arXiv:2302.00562 [math.PR].

Banerjee, S., and Olvera-Cravioto, M. 2022. PageRank asymptotics on directed preferential attachment networks. *Ann. Appl. Probab.*, **32**(4), 3060–3084.

Barabási, A.-L. 2002. *Linked: The new science of networks.* Perseus Publishing.

Barabási, A.-L. 2016. *Network science.* Cambridge University Press.

Barabási, A.-L. 2018. Love is all you need: Clauset's fruitless search for scale-free networks. Blog post available at www.barabasilab.com/post/love-is-all-you-need.

Barabási, A.-L., and Albert, R. 1999. Emergence of scaling in random networks. *Science*, **286**(5439), 509–512.

Barbour, A. D., and Reinert, G. 2001. Small worlds. *Random Structures Algorithms*, **19**(1), 54–74.

Barbour, A. D., and Reinert, G. 2004. Correction: "Small worlds" [Random Structures Algorithms **19**(1) (2001) 54–74; MR1848027]. *Random Structures Algorithms*, **25**(1), 115.

Barbour, A. D., and Reinert, G. 2006. Discrete small world networks. *Electron. J. Probab.*, **11**(47), 1234–1283 (electronic).

Barbour, A. D., and Röllin, A. 2019. Central limit theorems in the configuration model. *Ann. Appl. Probab.*, **29**(2), 1046–1069.

Beirlant, J., Goegebeur, Y., Segers, J., and Teugels, J. 2006. *Statistics of extremes: theory and applications.* John Wiley and Sons.

Bender, E. A., and Canfield, E. R. 1978. The asymptotic number of labelled graphs with given degree sequences. *J. Combin. Theory (A)*, **24**, 296–307.

Benjamini, I., Kesten, H., Peres, Y., and Schramm, O. 2004. Geometry of the uniform spanning forest: transitions in dimensions $4, 8, 12, \ldots$ *Ann. Math. (2)*, **160**(2), 465–491.

Benjamini, I., Lyons, R., and Schramm, O. 2015. Unimodular random trees. *Ergodic Theory Dynam. Systems*, **35**(2), 359–373.

Benjamini, I., and Schramm, O. 2001. Recurrence of distributional limits of finite planar graphs. *Electron. J. Probab.*, **6**(23), 13 pp. (electronic).

Berger, N. 2002. Transience, recurrence and critical behavior for long-range percolation. *Comm. Math. Phys.*, **226**(3), 531–558.

Berger, N., Borgs, C., Chayes, J., and Saberi, A. 2014. Asymptotic behavior and distributional limits of preferential attachment graphs. *Ann. Probab.*, **42**(1), 1–40.

Berger, N., Borgs, C., Chayes, J. T., D'Souza, R. M., and Kleinberg, R. D. 2004. Competition-induced preferential attachment. Pages 208–221 of: *Automata, languages and programming.* Lecture Notes in Comput. Sci., vol. **3142**. Springer.

Berger, N., Borgs, C., Chayes, J. T., D'Souza, R. M., and Kleinberg, R. D. 2005. Degree distribution of competition-induced preferential attachment graphs. *Combin. Probab. Comput.*, **14**(5-6), 697–721.

Bhamidi, S., Bresler, G., and Sly, A. 2008. Mixing time of exponential random graphs. Pages 803–812 of: *FOCS '08: Proceedings of the 2008 49th Annual IEEE Symposium on Foundations of Computer Science.* IEEE Computer Society.

Bhamidi, S., Bresler, G., and Sly, A. 2011. Mixing time of exponential random graphs. *Ann. Appl. Probab.*, **21**(6), 2146–2170.

Bhamidi, S., Evans, S., and Sen, A. 2012. Spectra of large random trees. *J. Theoret. Probab.*, **25**(3), 613–654.

Bhattacharya, A., Chen, B., van der Hofstad, R., and Zwart, B. 2020. *Consistency of the PLFit estimator for power-law data.* arXiv:2002.06870 [math.PR].

Billingsley, P. 1968. *Convergence of probability measures.* John Wiley and Sons.

Bingham, N. H., Goldie, C. M., and Teugels, J. L. 1989. *Regular variation.* Encyclopedia of Mathematics and its Applications, vol. **27**. Cambridge University Press.

Biskup, M. 2004. On the scaling of the chemical distance in long-range percolation models. *Ann. Probab.*, **32**(4), 2938–2977.

Biskup, M., and Lin, J. 2019. Sharp asymptotic for the chemical distance in long-range percolation. *Random Structures Algorithms*, **55**(3), 560–583.

Bläsius, T., Friedrich, T., and Krohmer, A. 2018. Cliques in hyperbolic random graphs. *Algorithmica*, **80**(8), 2324–2344.

Blondel, V. D., Guillaume, J.-L., Lambiotte, R., and Lefebvre, E. 2008. Fast unfolding of communities in large networks. *J. Statist. Mech.: Theory and Experiment*, **2008**(10).

Bloznelis, M. 2009. *A note on log log distances in a power law random intersection graph*. arXiv:0911.5127 [math.PR].

Bloznelis, M. 2010a. Component evolution in general random intersection graphs. *SIAM J. Discrete Math.*, **24**(2), 639–654.

Bloznelis, M. 2010b. The largest component in an inhomogeneous random intersection graph with clustering. *Electron. J. Combin.*, **17**(1), Research Paper 110, 17.

Bloznelis, M. 2013. Degree and clustering coefficient in sparse random intersection graphs. *Ann. Appl. Probab.*, **23**(3), 1254–1289.

Bloznelis, M., Godehardt, E., Jaworski, J., Kurauskas, V., and Rybarczyk, K. 2015. Recent progress in complex network analysis: models of random intersection graphs. Pages 69–78 of: *Data science, learning by latent structures, and knowledge discovery*. Springer.

Bloznelis, M., Götze, F., and Jaworski, J. 2012. Birth of a strongly connected giant in an inhomogeneous random digraph. *J. Appl. Probab.*, **49**(3), 601–611.

Bode, M., Fountoulakis, N., and Müller, T. 2015. On the largest component of a hyperbolic model of complex networks. *Electron. J. Combin.*, **22**(3), Paper 3.24, 46.

Boguná, M., Papadopoulos, F., and Krioukov, D. 2010. Sustaining the internet with hyperbolic mapping. *Nature Commun.*, **1**(1), 1–8.

Bohman, T., and Frieze, A. 2001. Avoiding a giant component. *Random Structures Algorithms*, **19**(1), 75–85.

Bohman, T., and Frieze, A. 2002. Addendum to "Avoiding a giant component" [*Random Structures Algorithms* **19**(1) (2001), 75–85; MR1848028]. *Random Structures Algorithms*, **20**(1), 126–130.

Boldi, P., Rosa, M., Santini, M., and Vigna, S. 2011. Layered label propagation: a multiresolution coordinate-free ordering for compressing social networks. Pages 587–596 of: *Proceedings of the 20th International Conference on the World Wide Web*. ACM Press.

Boldi, P., and Vigna, S. 2004. The WebGraph Framework I: compression techniques. Pages 595–601 of: *Proc. 13th International World Wide Web Conference (WWW 2004)*. ACM Press.

Bollobás, B. 1980. A probabilistic proof of an asymptotic formula for the number of labelled regular graphs. *European J. Combin.*, **1**(4), 311–316.

Bollobás, B. 2001. *Random graphs*. Second edn. Cambridge Studies in Advanced Mathematics, vol. **73**. Cambridge University Press.

Bollobás, B., and Fernandez de la Vega, W. 1982. The diameter of random regular graphs. *Combinatorica*, **2**(2), 125–134.

Bollobás, B., Janson, S., and Riordan, O. 2005. The phase transition in the uniformly grown random graph has infinite order. *Random Structures Algorithms*, **26**(1-2), 1–36.

Bollobás, B., Janson, S., and Riordan, O. 2007. The phase transition in inhomogeneous random graphs. *Random Structures Algorithms*, **31**(1), 3–122.

Bollobás, B., Janson, S., and Riordan, O. 2011. Sparse random graphs with clustering. *Random Structures Algorithms*, **38**(3), 269–323.

Bollobás, B., and Riordan, O. 2004a. The diameter of a scale-free random graph. *Combinatorica*, **24**(1), 5–34.

Bollobás, B., and Riordan, O. 2004b. Shortest paths and load scaling in scale-free trees. *Phys. Rev. E*, **69**, 036114.

Bollobás, B., and Riordan, O. 2006. *Percolation*. Cambridge University Press.

Bollobás, B., and Riordan, O. 2015. An old approach to the giant component problem. *J. Combin. Theory Ser. B*, **113**, 236–260.

Bollobás, B., Riordan, O., Spencer, J., and Tusnády, G. 2001. The degree sequence of a scale-free random graph process. *Random Structures Algorithms*, **18**(3), 279–290.

Bordenave, C. 2016. *Lecture notes on random graphs and probabilistic combinatorial optimization*. Version April 8, 2016. Available at www.math.univ-toulouse.fr/~bordenave/coursRG.pdf.

Bordenave, C., and Caputo, P. 2015. Large deviations of empirical neighborhood distribution in sparse random graphs. *Probab. Theory Rel. Fields*, **163**(1-2), 149–222.

Bordenave, C., and Lelarge, M. 2010. Resolvent of large random graphs. *Random Structures Algorithms*, **37**(3), 332–352.

Bordenave, C., Lelarge, M., and Massoulié, L. 2018. Nonbacktracking spectrum of random graphs: community detection and nonregular Ramanujan graphs. *Ann. Probab.*, **46**(1), 1–71.

Bordenave, C., Lelarge, M., and Salez, J. 2011. The rank of diluted random graphs. *Ann. Probab.*, **39**(3), 1097–1121.

Bordenave, C., Lelarge, M., and Salez, J. 2013. Matchings on infinite graphs. *Probab. Theory Rel. Fields*, **157**(1-2), 183–208.

Box, G. E. P. 1976. Science and statistics. *J. Amer. Statist. Assoc.*, **71**(356), 791–799.

Box, G. E. P. 1979. Robustness in the strategy of scientific model building. Pages 201–236 of: *Robustness in statistics*. Elsevier.

Bringmann, K., Keusch, R., and Lengler, J. 2017. Sampling geometric inhomogeneous random graphs in linear time. In: *Proceeding of the 25th Annual European Symposium on Algorithms (ESA 2017)*. Schloss Dagstuhl-Leibniz-Zentrum fuer Informatik.

Bringmann, K., Keusch, R., and Lengler, J. 2019. Geometric inhomogeneous random graphs. *Theoret. Comput. Sci.*, **760**, 35–54.

Bringmann, K., Keusch, R., and Lengler, J. 2020. *Average distance in a general class of scale-free networks with underlying geometry*. arXiv: 1602.05712 [cs.DM].

Britton, T., Deijfen, M., and Martin-Löf, A. 2006. Generating simple random graphs with prescribed degree distribution. *J. Statist. Phys.*, **124**(6), 1377–1397.

Broder, A., Kumar, R., Maghoul, F., Raghavan, P., Rajagopalan, S., Stata, R., Tomkins, A., and Wiener, J. 2000. Graph structure in the Web. *Computer Networks*, **33**, 309–320.

Broido, A., and Clauset, A. 2019. Scale-free networks are rare. *Nature Commun.*, **10**(1), 1017.

Cai, X. S., and Perarnau, G. 2021. The giant component of the directed configuration model revisited. *ALEA Lat. Am. J. Probab. Math. Statist.*, **18**(2), 1517–1528.

Cai, X. S., and Perarnau, G. 2023. The diameter of the directed configuration model. *Ann. Inst. Henri Poincaré Probab. Stat.*, **59**(1), 244–270.

Callaway, D. S., Hopcroft, J. E., Kleinberg, J. M., Newman, M. E. J., and Strogatz, S. H. 2001. Are randomly grown graphs really random? *Phys. Rev. E*, **64**, 041902.

Cao, J., and Olvera-Cravioto, M. 2020. Connectivity of a general class of inhomogeneous random digraphs. *Random Structures Algorithms*, **56**(3), 722–774.

Caravenna, F., Garavaglia, A., and van der Hofstad, R. 2019. Diameter in ultra-small scale-free random graphs. *Random Structures Algorithms*, **54**(3), 444–498.

Chakraborty, S., van der Hofstad, R., and den Hollander, F. 2021. *Sparse random graphs with many triangles*. arXiv:2112.06526 [math.PR].

Chatterjee, S. 2017. *Large deviations for random graphs*. Lecture Notes in Mathematics, vol. **2197**. Springer. Lecture notes from the 45th Probability Summer School held in Saint-Flour, June 2015.

Chatterjee, S., and Diaconis, P. 2013. Estimating and understanding exponential random graph models. *Ann. Statist.*, **41**(5), 2428–2461.

Chatterjee, S., and Durrett, R. 2009. Contact processes on random graphs with power law degree distributions have critical value 0. *Ann. Probab.*, **37**(6), 2332–2356.

Chatterjee, S., and Varadhan, S. R. S. 2011. The large deviation principle for the Erdős–Rényi random graph. *European J. Combin.*, **32**(7), 1000–1017.

Chen, N., and Olvera-Cravioto, M. 2013. Directed random graphs with given degree distributions. *Stoch. Syst.*, **3**(1), 147–186.

Chung, F., and Lu, L. 2001. The diameter of sparse random graphs. *Adv. Appl. Math.*, **26**(4), 257–279.

Chung, F., and Lu, L. 2002a. The average distances in random graphs with given expected degrees. *Proc. Natl. Acad. Sci. USA*, **99**(25), 15879–15882 (electronic).

Chung, F., and Lu, L. 2002b. Connected components in random graphs with given expected degree sequences. *Ann. Comb.*, **6**(2), 125–145.

Chung, F., and Lu, L. 2003. The average distance in a random graph with given expected degrees. *Internet Math.*, **1**(1), 91–113.

Chung, F., and Lu, L. 2004. Coupling online and offline analyses for random power law graphs. *Internet Math.*, **1**(4), 409–461.

Chung, F., and Lu, L. 2006a. *Complex graphs and networks.* CBMS Regional Conference Series in Mathematics, vol. **107**.

Chung, F., and Lu, L. 2006b. The volume of the giant component of a random graph with given expected degrees. *SIAM J. Discrete Math.*, **20**, 395–411.

Clauset, A., Shalizi, C., and Newman, M. E. J. 2009. Power-law distributions in empirical data. *SIAM Review*, **51**(4), 661–703.

Cohen, R., and Havlin, S. 2003. Scale-free networks are ultrasmall. *Phys. Rev. Lett.*, **90**, 058701, 1–4.

Colenbrander, D. 2022. *Ultra-small world phenomenon in the directed configuration model.* M.Phil. thesis, Eindhoven University of Technology.

Collevecchio, A., Cotar, C., and LiCalzi, M. 2013. On a preferential attachment and generalized Pólya's urn model. *Ann. Appl. Probab.*, **23**(3), 1219–1253.

Cooper, C., and Frieze, A. 2004. The size of the largest strongly connected component of a random digraph with a given degree sequence. *Combin. Probab. Comput.*, **13**(3), 319–337.

Corten, R. 2012. Composition and structure of a large online social network in the Netherlands. *PLOS ONE*, **7**(4), 1–8.

Coscia, M. 2021. *The atlas for the aspiring network scientist.* arXiv:2101.00863 [cs.CY].

Csárdi, G. 2006. Dynamics of citation networks. Pages 698–709 of: *Proceedings of the International Conference on Artificial Neural Networks 2006.* Lecture Notes in Computer Science, vol. **4131**. Springer.

Curien, N. 2018. *Random graphs: the local convergence perspective.* Version October 17, 2018. Available at www.imo.universite-paris-saclay.fr/~curien/enseignement.html.

Danielsson, J., de Haan, L., Peng, L., and de Vries, C. G. 2001. Using a bootstrap method to choose the sample fraction in tail index estimation. *J. Multivariate Anal.*, **76**(2), 226–248.

Darling, D. A. 1970. The Galton–Watson process with infinite mean. *J. Appl. Probab.*, **7**, 455–456.

Davies, P. L. 1978. The simple branching process: a note on convergence when the mean is infinite. *J. Appl. Probab.*, **15**(3), 466–480.

Decelle, A., Krzakala, F., Moore, C., and Zdeborová, L. 2011. Asymptotic analysis of the stochastic block model for modular networks and its algorithmic applications. *Phys. Rev. E*, **84**(6), 066106.

Deijfen, M. 2009. Stationary random graphs with prescribed iid degrees on a spatial Poisson process. *Electron. Commun. Probab.*, **14**, 81–89.

Deijfen, M., van den Esker, H., van der Hofstad, R., and Hooghiemstra, G. 2009. A preferential attachment model with random initial degrees. *Ark. Mat.*, **47**(1), 41–72.

Deijfen, M., Häggström, O., and Holroyd, A. 2012. Percolation in invariant Poisson graphs with i.i.d. degrees. *Ark. Mat.*, **50**(1), 41–58.

Deijfen, M., van der Hofstad, R., and Hooghiemstra, G. 2013. Scale-free percolation. *Ann. Inst. Henri Poincaré (B) Prob. Statist.*, **49**(3), 817–838.

Deijfen, M., and Jonasson, J. 2006. Stationary random graphs on $\mathbb{Z}$ with prescribed iid degrees and finite mean connections. *Electron. Commun. Probab.*, **11**, 336–346 (electronic).

Deijfen, M., and Kets, W. 2009. Random intersection graphs with tunable degree distribution and clustering. *Probab. Engrg. Inform. Sci.*, **23**(4), 661–674.

Deijfen, M., and Meester, R. 2006. Generating stationary random graphs on $\mathbb{Z}$ with prescribed independent, identically distributed degrees. *Adv. Appl. Probab.*, **38**(2), 287–298.

Dembo, A., and Montanari, A. 2010a. Gibbs measures and phase transitions on sparse random graphs. *Braz. J. Probab. Statist.*, **24**(2), 137–211.

Dembo, A., and Montanari, A. 2010b. Ising models on locally tree-like graphs. *Ann. Appl. Probab.*, **20**(2), 565–592.

Deprez, P., Hazra, R., and Wüthrich, M. 2015. Inhomogeneous long-range percolation for real-life network modeling. *Risks*, **3**(1), 1–23.

Deprez, P., and Wüthrich, M. 2019. Scale-free percolation in continuum space. *Commun. Math. Statist.*, **7**(3), 269–308.

Dereich, S., Mönch, C., and Mörters, P. 2012. Typical distances in ultrasmall random networks. *Adv. Appl. Probab.*, **44**(2), 583–601.

Dereich, S., Mönch, C., and Mörters, P. 2017. Distances in scale free networks at criticality. *Electron. J. Probab.*, **22**, Paper No. 77, 38.

Dereich, S., and Mörters, P. 2009. Random networks with sublinear preferential attachment: degree evolutions. *Electron. J. Probab.*, **14**, 1222–1267.

Dereich, S., and Mörters, P. 2011. Random networks with concave preferential attachment rule. *Jahresber. Dtsch. Math.-Ver.*, **113**(1), 21–40.

Dereich, S., and Mörters, P. 2013. Random networks with sublinear preferential attachment: the giant component. *Ann. Probab.*, **41**(1), 329–384.

Diaconis, P. 1992. Sufficiency as statistical symmetry. Pages 15–26 of: *American Mathematical Society centennial publications, Vol. II*.

Ding, J., Kim, J. H., Lubetzky, E., and Peres, Y. 2010. Diameters in supercritical random graphs via first passage percolation. *Combin. Probab. Comput.*, **19**(5-6), 729–751.

Ding, J., Kim, J. H., Lubetzky, E., and Peres, Y. 2011. Anatomy of a young giant component in the random graph. *Random Structures Algorithms*, **39**(2), 139–178.

Dommers, S., van der Hofstad, R., and Hooghiemstra, G. 2010. Diameters in preferential attachment graphs. *J. Statist. Phys.*, **139**, 72–107.

Dorogovtsev, S. N., Mendes, J. F. F., and Samukhin, A. N. 2000. Structure of growing networks with preferential linking. *Phys. Rev. Lett.*, **85**(21), 4633–4636.

Dort, L., and Jacob, E. 2023. *Local weak limit of dynamical inhomogeneous random graphs*. arXiv: 2303.17437 [math.PR].

Draisma, G., de Haan, L., Peng, L., and Pereira, T. 1999. A bootstrap-based method to achieve optimality in estimating the extreme-value index. *Extremes*, **2**(4), 367–404.

Drees, H., Janßen, A., Resnick, S., and Wang, T. 2020. On a minimum distance procedure for threshold selection in tail analysis. *SIAM J. Math. Data Sci.*, **2**(1), 75–102.

Drmota, M. 2009. *Random trees: an interplay between combinatorics and probability*. Springer.

Durrett, R. 2003. Rigorous result for the CHKNS random graph model. Pages 95–104 of: *Discrete random walks (Paris, 2003)*. Association of Discrete Mathematics and Theoretical Computer Sciience.

Durrett, R. 2007. *Random graph dynamics*. Cambridge Series in Statistical and Probabilistic Mathematics. Cambridge University Press.

Eckhoff, M., Goodman, J., van der Hofstad, R., and Nardi, F. R. 2013. Short paths for first passage percolation on the complete graph. *J. Statist. Phys.*, **151**(6), 1056–1088.

Elek, G. 2007. On limits of finite graphs. *Combinatorica*, **27**(4), 503–507.

Erdős, P., Greenhill, C., Mezei, T., Miklós, I., Soltész, D., and Soukup, L. 2022. The mixing time of switch Markov chains: a unified approach. *European J. Combin.*, **99**, Paper No. 103421, 46.

Erdős, P., and Rényi, A. 1959. On random graphs. I. *Publ. Math. Debrecen*, **6**, 290–297.

Erdős, P., and Rényi, A. 1960. On the evolution of random graphs. *Magyar Tud. Akad. Mat. Kutató Int. Közl.*, **5**, 17–61.

Erdős, P., and Rényi, A. 1961a. On the evolution of random graphs. *Bull. Inst. Internat. Statist.*, **38**, 343–347.

Erdős, P., and Rényi, A. 1961b. On the strength of connectedness of a random graph. *Acta Math. Acad. Sci. Hungar.*, **12**, 261–267.

van den Esker, H., van der Hofstad, R., and Hooghiemstra, G. 2008. Universality for the distance in finite variance random graphs. *J. Statist. Phys.*, **133**(1), 169–202.

van den Esker, H., van der Hofstad, R., Hooghiemstra, G., and Znamenski, D. 2006. Distances in random graphs with infinite mean degrees. *Extremes*, **8**, 111–140.

Faloutsos, C., Faloutsos, P., and Faloutsos, M. 1999. On power-law relationships of the internet topology. *Computer Commun. Rev.*, **29**, 251–262.

Federico, L. 2023. Almost-2-regular random graphs. *Australas. J. Combin.*, **86**, 76–96.

Federico, L., and van der Hofstad, R. 2017. Critical window for connectivity in the configuration model. *Combin. Probab. Comput.*, **26**(5), 660–680.

Fernholz, D., and Ramachandran, V. 2007. The diameter of sparse random graphs. *Random Structures Algorithms*, **31**(4), 482–516.

Fienberg, S., and Wasserman, S. 1981. Categorical data analysis of single sociometric relations. *Sociological Methodology*, **12**, 156–192.

Fill, J., Scheinerman, E., and Singer-Cohen, K. 2000. Random intersection graphs when $m = \omega(n)$: an equivalence theorem relating the evolution of the $G(n, m, p)$ and $G(n, p)$ models. *Random Structures Algorithms*, **16**(2), 156–176.

Flaxman, A., Frieze, A., and Vera, J. 2006. A geometric preferential attachment model of networks. *Internet Math.*, **3**(2), 187–205.

Flaxman, A., Frieze, A., and Vera, J. 2007. A geometric preferential attachment model of networks II. In: *Proceedings of Workshop on Algorithms and Models for the Web Graph 2007.*

Fountoulakis, N. 2015. On a geometrization of the Chung-Lu model for complex networks. *J. Complex Netw.*, **3**(3), 361–387.

Fountoulakis, N., van der Hoorn, P., Müller, T., and Schepers, M. 2021. Clustering in a hyperbolic model of complex networks. *Electronic J. Probab.*, **26**, 1–132.

Frank, O., and Strauss, D. 1986. Markov graphs. *J. Amer. Statist. Assoc.*, **81**(395), 832–842.

Friedrich, T., and Krohmer, A. 2015. On the diameter of hyperbolic random graphs. Pages 614–625 of: *Automata, languages, and programming. Part II.* Lecture Notes in Computer Science, vol. **9135**. Springer.

Friedrich, T., and Krohmer, A. 2018. On the diameter of hyperbolic random graphs. *SIAM J. Discrete Math.*, **32**(2), 1314–1334.

Fujita, Y., Kichikawa, Y., Fujiwara, Y., Souma, W., and Iyetomi, H. 2019. Local bow-tie structure of the web. *Applied Netw. Sci.*, **4**(1), 1–15.

Gamarnik, D., Nowicki, T., and Swirszcz, G. 2006. Maximum weight independent sets and matchings in sparse random graphs. Exact results using the local weak convergence method. *Random Structures Algorithms*, **28**(1), 76–106.

Gao, P., and Greenhill, C. 2021. Mixing time of the switch Markov chain and stable degree sequences. *Discrete Appl. Math.*, **291**, 143–162.

Gao, P., van der Hofstad, R., Southwell, A., and Stegehuis, C. 2020. Counting triangles in power-law uniform random graphs. *Electron. J. Combin.*, **27**(3), Paper No. 3.19, 28.

Gao, P., and Wormald, N. 2016. Enumeration of graphs with a heavy-tailed degree sequence. *Adv. Math.*, **287**, 412–450.

Garavaglia, A., Hazra, R., van der Hofstad, R., and Ray, R. 2022. *Universality of the local limit in preferential attachment models.* arXiv:2212.05551 [math.PR].

Garavaglia, A., and van der Hofstad, R. 2018. From trees to graphs: collapsing continuous-time branching processes. *J. Appl. Probab.*, **55**(3), 900–919.

Garavaglia, A., van der Hofstad, R., and Litvak, N. 2020. Local weak convergence for PageRank. *Ann. Appl. Probab.*, **30**(1), 40–79.

Garavaglia, A., van der Hofstad, R., and Woeginger, G. 2017. The dynamics of power laws: fitness and aging in preferential attachment trees. *J. Statist. Phys.*, **168**(6), 1137–1179.

Gilbert, E. N. 1959. Random graphs. *Ann. Math. Statist.*, **30**, 1141–1144.

Gleiser, P., and Danon, L. 2003. Community structure in jazz. *Adv. Complex Systems*, **06**(04), 565–573.

Godehardt, E., and Jaworski, J. 2003. Two models of random intersection graphs for classification. Pages 67–81 of: *Exploratory data analysis in empirical research.* Stud. Classification Data Anal. Knowledge Organ. Springer.

Goñi, J., Esteban, F., de Mendizábal, N., Sepulcre, J., Ardanza-Trevijano, S., Agirrezabal, I., and Villoslada, P. 2008. A computational analysis of protein–protein interaction networks in neurodegenerative diseases. *BMC Systems Biology*, **2**(1), 52.

Grimmett, G. 1999. *Percolation.* 2nd edn. Springer.

Gugelmann, L., Panagiotou, K., and Peter, U. 2012. Random hyperbolic graphs: degree sequence and clustering. Pages 573–585 of: *Proceedings of the International Colloquium on Automata, Languages, and Programming.* Springer.

Gulikers, L., Lelarge, M., and Massoulié, L. 2017a. Non-backtracking spectrum of degree-corrected stochastic block models. Pages 1–27 of: *Proceedings of the 8th Innovations in Theoretical Computer Science Conference*. LIPIcs. Leibniz Int. Proc. Inform., vol. **67**. Schloss Dagstuhl–Leibniz-Zentrum für Informatik. Art. No. 44.

Gulikers, L., Lelarge, M., and Massoulié, L. 2017b. A spectral method for community detection in moderately sparse degree-corrected stochastic block models. *Adv. Appl. Probab.*, **49**(3), 686–721.

Gulikers, L., Lelarge, M., and Massoulié, L. 2018. An impossibility result for reconstruction in the degree-corrected stochastic block model. *Ann. Appl. Probab.*, **28**(5), 3002–3027.

Gut, A. 2005. *Probability: a graduate course*. Springer Texts in Statistics. Springer.

Häggström, O., and Jonasson, J. 1999. Phase transition in the random triangle model. *J. Appl. Probab.*, **36**(4), 1101–1115.

Hajek, B. 1990. Performance of global load balancing by local adjustment. *IEEE Trans. Inform. Theory*, **36**(6), 1398–1414.

Hajek, B. 1996. Balanced loads in infinite networks. *Ann. Appl. Probab.*, **6**(1), 48–75.

Hajek, B., and Sankagiri, S. 2019. Community recovery in a preferential attachment graph. *IEEE Trans. Inform. Theory*, **65**(11), 6853–6874.

Hajra, K.B., and Sen, P. 2005. Aging in citation networks. *Physica A. Statist. Mech. Applic.*, **346**(1-2), 44–48.

Hajra, K.B., and Sen, P. 2006. Modelling aging characteristics in citation networks. *Physica A: Statist. Mech. Applic.*, **368**(2), 575–582.

Hall, P. 1981. Order of magnitude of moments of sums of random variables. *J. London Math. Soc.*, **24**(2), 562–568.

Hall, P., and Welsh, A. 1984. Best attainable rates of convergence for estimates of parameters of regular variation. *Ann. Statist.*, **12**(3), 1079–1084.

Halmos, P. 1950. *Measure theory*. Van Nostrand.

Hao, N., and Heydenreich, M. 2023. Graph distances in scale-free percolation: the logarithmic case. *J. Appl. Probab.*, **60**(1), 295–313.

Hardy, G. H., Littlewood, J. E., and Pólya, G. 1988. *Inequalities*. Cambridge Mathematical Library. Cambridge University Press. Reprint of the 1952 edition.

Harris, T. 1963. *The theory of branching processes*. Die Grundlehren der Mathematischen Wissenschaften, Band 119. Springer-Verlag.

Hatami, H., Lovász, L., and Szegedy, B. 2014. Limits of locally-globally convergent graph sequences. *Geom. Funct. Anal.*, **24**(1), 269–296.

Heydenreich, M., Hulshof, T., and Jorritsma, J. 2017. Structures in supercritical scale-free percolation. *Ann. Appl. Probab.*, **27**(4), 2569–2604.

Hill, B. M. 1975. A simple general approach to inference about the tail of a distribution. *Ann. Statist.*, **3**(5), 1163–1174.

Hirsch, C. 2017. From heavy-tailed Boolean models to scale-free Gilbert graphs. *Braz. J. Probab. Statist.*, **31**(1), 111–143.

van der Hofstad, R. 2017. *Random graphs and complex networks. Volume 1*. Cambridge Series in Statistical and Probabilistic Mathematics. Cambridge University Press.

van der Hofstad, R. 2021. *The giant in random graphs is almost local*. arXiv:2103.11733 [math.PR].

van der Hofstad, R., and Komjáthy, J. 2017a. *Explosion and distances in scale-free percolation*. arXiv:1706.02597 [math.PR].

van der Hofstad, R., and Komjáthy, J. 2017b. When is a scale-free graph ultra-small? *J. Statist. Phys.*, **169**(2), 223–264.

van der Hofstad, R., and Litvak, N. 2014. Degree–degree dependencies in random graphs with heavy-tailed degrees. *Internet Math.*, **10**(3-4), 287–334.

van der Hofstad, R., Hooghiemstra, G., and Van Mieghem, P. 2005. Distances in random graphs with finite variance degrees. *Random Structures Algorithms*, **27**(1), 76–123.

van der Hofstad, R., Hooghiemstra, G., and Znamenski, D. 2007a. Distances in random graphs with finite mean and infinite variance degrees. *Electron. J. Probab.*, **12**(25), 703–766 (electronic).

van der Hofstad, R., Hooghiemstra, G., and Znamenski, D. 2007b. A phase transition for the diameter of the configuration model. *Internet Math.*, **4**(1), 113–128.

van der Hofstad, R., van der Hoorn, P., and Maitra, N. 2023. Local limits of spatial inhomogeneous random graphs. *Adv. Appl. Probab.*, 1–48.

van der Hofstad, R., Komjáthy, J., and Vadon, V. 2021. Random intersection graphs with communities. *Adv. Appl. Probab.*, **53**(4), 1061–1089.

van der Hofstad, R., Komjáthy, J., and Vadon, V. 2022. Phase transition in random intersection graphs with communities. *Random Structures Algorithms*, **60**(3), 406–461.

van der Hofstad, R., van Leeuwaarden, J. S. H., and Stegehuis, C. 2017. Hierarchical configuration model. *Internet Math.* arXiv:1512.08397 [math.PR].

Holland, P., Laskey, K., and Leinhardt, S. 1983. Stochastic blockmodels: first steps. *Social Netw.*, **5**(2), 109–137.

Holme, P. 2019. Rare and everywhere: perspectives on scale-free networks. *Nature Commun.*, **10**(1), 1016.

van der Hoorn, P., and Olvera-Cravioto, M. 2018. Typical distances in the directed configuration model. *Ann. Appl. Probab.*, **28**(3), 1739–1792.

Howes, N. 1995. *Modern analysis and topology*. Universitext. Springer-Verlag.

Jacob, E., and Mörters, P. 2015. Spatial preferential attachment networks: power laws and clustering coefficients. *Ann. Appl. Probab.*, **25**(2), 632–662.

Jacob, E., and Mörters, P. 2017. Robustness of scale-free spatial networks. *Ann. Probab.*, **45**(3), 1680–1722.

Janson, S. 2004. Functional limit theorems for multitype branching processes and generalized Pólya urns. *Stochastic Process. Appl.*, **110**(2), 177–245.

Janson, S. 2008. The largest component in a subcritical random graph with a power law degree distribution. *Ann. Appl. Probab.*, **18**(4), 1651–1668.

Janson, S. 2009. Standard representation of multivariate functions on a general probability space. *Electron. Commun. Probab.*, **14**, 343–346.

Janson, S. 2010a. Asymptotic equivalence and contiguity of some random graphs. *Random Structures Algorithms*, **36**(1), 26–45.

Janson, S. 2010b. Susceptibility of random graphs with given vertex degrees. *J. Combin.*, **1**(3-4), 357–387.

Janson, S. 2011. *Probability asymptotics: notes on notation.* arXiv:1108.3924 [math.PR].

Janson, S. 2020a. Asymptotic normality in random graphs with given vertex degrees. *Random Structures Algorithms*, **56**(4), 1070–1116.

Janson, S. 2020b. Random graphs with given vertex degrees and switchings. *Random Structures Algorithms*, **57**(1), 3–31.

Janson, S., and Luczak, M. 2007. A simple solution to the k-core problem. *Random Structures Algorithms*, **30**(1-2), 50–62.

Janson, S., and Luczak, M. 2008. Asymptotic normality of the k-core in random graphs. *Ann. Appl. Probab.*, **18**(3), 1085–1137.

Janson, S., and Luczak, M. 2009. A new approach to the giant component problem. *Random Structures Algorithms*, **34**(2), 197–216.

Janson, S., Łuczak, T., and Rucinski, A. 2000. *Random graphs.* Wiley-Interscience Series in Discrete Mathematics and Optimization. Wiley-Interscience.

Janssen, J., Prałat, P., and Wilson, R. 2016. Nonuniform distribution of nodes in the spatial preferential attachment model. *Internet Math.*, **12**(1-2), 121–144.

Jaworski, J., Karoński, M., and Stark, D. 2006. The degree of a typical vertex in generalized random intersection graph models. *Discrete Math.*, **306**(18), 2152–2165.

Jaynes, E. T. 1957. Information theory and statistical mechanics. *Phys. Rev.*, **106**(2), 620–630.

Jonasson, J. 2009. Invariant random graphs with iid degrees in a general geography. *Probab. Theory Rel. Fields*, **143**(3-4), 643–656.

Jordan, J. 2010. Degree sequences of geometric preferential attachment graphs. *Adv. Appl. Probab.*, **42**(2), 319–330.

Jordan, J. 2013. Geometric preferential attachment in non-uniform metric spaces. *Electron. J. Probab.*, **18**, no. 8, 15.

Jordan, J., and Wade, A. 2015. Phase transitions for random geometric preferential attachment graphs. *Adv. Appl. Probab.*, **47**(2), 565–588.

Jorritsma, J., and Komjáthy, J. 2022. Distance evolutions in growing preferential attachment graphs. *Ann. Appl. Probab.*, **32**(6), 4356–4397.

Jorritsma, J., Komjáthy, J., and Mitsche, D. 2023. *Cluster-size decay in supercritical kernel-based spatial random graphs*. arXiv:2303.00724 [math.PR].

Kallenberg, O. 2002. *Foundations of modern probability*. Second edn. Springer.

Kallenberg, O. 2017. *Random measures, theory and applications*. Probability Theory and Stochastic Modelling, vol. **77**. Springer.

Karoński, M., Scheinerman, E., and Singer-Cohen, K. 1999. On random intersection graphs: the subgraph problem. *Combin. Probab. Comput.*, **8**(1-2), 131–159.

Karp, R.M. 1990. The transitive closure of a random digraph. *Random Structures Algorithms*, **1**(1), 73–93.

Karrer, B., and Newman, M. E. J. 2011. Stochastic blockmodels and community structure in networks. *Phys. Rev. E*, **83**(1), 016107.

Kass, R.E., and Wasserman, L. 1996. The selection of prior distributions by formal rules. *J. Amer. Statist. Assoc.*, **91**(435), 1343–1370.

Kesten, H. 1982. *Percolation theory for mathematicians*. Progress in Probability and Statistics, vol. **2**. Birkhäuser.

Kesten, H., and Stigum, B. P. 1966. A limit theorem for multidimensional Galton-Watson processes. *Ann. Math. Statist.*, **37**, 1211–1223.

Kingman, J. F. C. 1975. The first birth problem for an age-dependent branching process. *Ann. Probab.*, **3**(5), 790–801.

Kiwi, M., and Mitsche, D. 2015. A bound for the diameter of random hyperbolic graphs. Pages 26–39 of: *2015 Proceedings of the 12th Workshop on Analytic Algorithmics and Combinatorics (ANALCO)*. SIAM.

Kiwi, M., and Mitsche, D. 2019. On the second largest component of random hyperbolic graphs. *SIAM J. Discrete Math.*, **33**(4), 2200–2217.

Komjáthy, J, and Lodewijks, B. 2020. Explosion in weighted hyperbolic random graphs and geometric inhomogeneous random graphs. *Stochastic Process. Appl.*, **130**(3), 1309–1367.

Krioukov, D., Kitsak, M., Sinkovits, R., Rideout, D., Meyer, D., and Boguñá, M. 2012. Network cosmology. *Sci. Rep.*, **2**.

Krioukov, D., Papadopoulos, F., Kitsak, M., Vahdat, A., and Boguñá, M. 2010. Hyperbolic geometry of complex networks. *Phys. Rev. E*, **82**(3), 036106, 18.

Krzakala, F., Moore, C., Mossel, E., Neeman, J., Sly, A., Zdeborová, L., and Zhang, P. 2013. Spectral redemption in clustering sparse networks. *Proc. National Acad. Sci.*, **110**(52), 20935–20940.

Kumar, R., Raghavan, P., Rajagopalan, S., Sivakumar, D., Tomkins, A., and Upfal, E. 2000. Stochastic models for the Web graph. Pages 57–65 of: *Proceedings of the 42nd Annual IEEE Symposium on Foundations of Computer Science*.

Kunegis, J. 2013. KONECT: the Koblenz network collection. Pages 1343–1350 of: *Proceedings of the 22nd International Conference on World Wide Web*.

Kunegis, J. 2017. *The Koblenz network collection*. arXiv:1402.5500 [cs.SI].

Kurauskas, V. 2022. On local weak limit and subgraph counts for sparse random graphs. *J. Appl. Probab.*, **59**(3), 755–776.

Last, G., and Penrose, M. 2018. *Lectures on the Poisson process*. Institute of Mathematical Statistics Textbooks, vol. **7**. Cambridge University Press.

Lee, J., and Olvera-Cravioto, M. 2020. PageRank on inhomogeneous random digraphs. *Stochastic Process. Appl.*, **130**(4), 2312–2348.

Leskelä, L. 2019. *Random graphs and network statistics*. Available at `http://math.aalto.fi/~lleskela/LectureNotes004.html`.

Leskovec, J., Kleinberg, J., and Faloutsos, C. 2007. Graph evolution: densification and shrinking diameters. *ACM Trans. Knowledge Discovery from Data (TKDD)*, **1**(1), 2.

Leskovec, J., and Krevl, A. 2014 (Jun). *SNAP Datasets: Stanford large network dataset collection*. `http://snap.stanford.edu/data`.

Leskovec, J., Lang, K., Dasgupta, A., and Mahoney, M. 2009. Community structure in large networks: natural cluster sizes and the absence of large well-defined clusters. *Internet Math.*, **6**(1), 29–123.

Litvak, N., and van der Hofstad, R. 2013. Uncovering disassortativity in large scale-free networks. *Phys. Rev. E*, **87**(2), 022801.

Lo, T. Y. Y. 2021. *Weak local limit of preferential attachment random trees with additive fitness.* arXiv:2103.00900 [math.PR].

Lovász, L. 2012. *Large networks and graph limits.* American Mathematical Society Colloquium Publications, vol. **60**. American Mathematical Society, Providence, RI.

Łuczak, T. 1992. Sparse random graphs with a given degree sequence. Pages 165–182 of: *Random graphs, Vol. 2* (Poznań, 1989). Wiley.

Lyons, R. 2005. Asymptotic enumeration of spanning trees. *Combin. Probab. Comput.*, **14**(4), 491–522.

Manna, S., and Sen, P. 2002. Modulated scale-free network in Euclidean space. *Phys. Rev. E*, **66**(6), 066114.

Massoulié, L. 2014. Community detection thresholds and the weak Ramanujan property. Pages 694–703 of: *Proceedings of the 2014 ACM Symposium on Theory of Computing.* ACM.

McKay, B. D. 1981. Subgraphs of random graphs with specified degrees. *Congressus Numerantium*, **33**, 213–223.

McKay, B. D. 2011. Subgraphs of random graphs with specified degrees. In: *Proceedings of the International Congress of Mathematicians 2010.* Hindustan Book Agency.

McKay, B. D., and Wormald, N. 1990. Asymptotic enumeration by degree sequence of graphs of high degree. *European J. Combin.*, **11**(6), 565–580.

Meester, R., and Roy, R. 1996. *Continuum percolation.* Cambridge Tracts in Mathematics, vol. **119**. Cambridge University Press.

Milewska, M., van der Hofstad, R., and Zwart, B. 2023. *Dynamic random intersection graph: Dynamic local convergence and giant structure.* arXiv: 2308.15629 [math.PR].

Molloy, M., and Reed, B. 1995. A critical point for random graphs with a given degree sequence. *Random Structures Algorithms*, **6**(2-3), 161–179.

Molloy, M., and Reed, B. 1998. The size of the giant component of a random graph with a given degree sequence. *Combin. Probab. Comput.*, **7**(3), 295–305.

Molloy, M., Surya, E., and Warnke, L. 2022. *The degree-restricted random process is far from uniform.* arXiv: 2211.00835v1 [math.CO].

Moore, C., and Newman, M. E. J. 2000. Epidemics and percolation in small-world networks. *Phys. Rev. E*, **61**, 5678–5682.

Mossel, E., Neeman, J., and Sly, A. 2015. Reconstruction and estimation in the planted partition model. *Probab. Theory Rel. Fields*, **162**(3-4), 431–461.

Mossel, E., Neeman, J., and Sly, A. 2016. Belief propagation, robust reconstruction and optimal recovery of block models. *Ann. Appl. Probab.*, **26**(4), 2211–2256.

Mossel, E., Neeman, J., and Sly, A. 2018. A proof of the block model threshold conjecture. *Combinatorica*, **38**(3), 665–708.

Mucha, P. J., Richardson, T., Macon, K., Porter, M. A., and Onnela, J.-P. 2010. Community structure in time-dependent, multiscale, and multiplex networks. *Science*, **328**(5980), 876–878.

Nair, J., Wierman, A., and Zwart, B. 2022. *The fundamentals of heavy tails: properties, emergence, and estimation.* Cambridge Series in Statistical and Probabilistic Mathematics. Cambridge University Press.

Newman, M. E. J. 2003. Properties of highly clustered networks. *Phys. Rev. E*, **68**(2), 026121.

Newman, M. E. J. 2009. Random graphs with clustering. *Phys. Rev. Lett.*, **103**(Jul), 058701.

Newman, M. E. J. 2010. *Networks: an introduction.* Oxford University Press.

Newman, M. E. J., Moore, C., and Watts, D. J. 2000a. Mean-field solution of the small-world network model. *Phys. Rev. Lett.*, **84**, 3201–3204.

Newman, M. E. J., and Park, J. 2003. Why social networks are different from other types of networks. *Phys. Rev. E*, **68**(3), 036122.

Newman, M. E. J., Strogatz, S., and Watts, D. 2000b. Random graphs with arbitrary degree distribution and their application. *Phys. Rev. E*, **64**, 026118, 1–17.

Newman, M. E. J., Strogatz, S., and Watts, D. 2002. Random graph models of social networks. *Proc. National Acad. Sci.*, **99**, 2566–2572.

Newman, M. E. J., Watts, D. J., and Barabási, A.-L. 2006. *The structure and dynamics of networks*. Princeton Studies in Complexity. Princeton University Press.

Newman, M. E. J., and Watts, D.J. 1999. Scaling and percolation in the small-world network model. *Phys. Rev. E*, **60**, 7332–7344.

Norros, I., and Reittu, H. 2006. On a conditionally Poissonian graph process. *Adv. Appl. Probab.*, **38**(1), 59–75.

Noutsos, D. 2006. On Perron–Frobenius property of matrices having some negative entries. *Linear Algebra Appl.*, **412**(2-3), 132–153.

O'Connell, N. 1998. Some large deviation results for sparse random graphs. *Probab. Theory Rel. Fields*, **110**(3), 277–285.

Parthasarathy, K. R. 1967. *Probability measures on metric spaces*. Probability and Mathematical Statistics, No. 3. Academic Press.

Pemantle, R. 2007. A survey of random processes with reinforcement. *Probab. Surv.*, **4**, 1–79 (electronic).

Pickands III, J. 1968. Moment convergence of sample extremes. *Ann. Math. Statistics*, **39**, 881–889.

Pittel, B. 1994. Note on the heights of random recursive trees and random m-ary search trees. *Random Structures Algorithms*, **5**(2), 337–347.

Price, D. J. de Solla. 1965. Networks of scientific papers. *Science*, **149**, 510–515.

Price, D. J. de Solla. 1986. *Little science, big science... and beyond*. Columbia University Press.

Reittu, H., and Norros, I. 2004. On the power law random graph model of massive data networks. *Performance Evaluation*, **55**(1-2), 3–23.

Resnick, S. 2007. *Heavy-tail phenomena*. Springer Series in Operations Research and Financial Engineering. Springer. (Probabilistic and statistical modeling).

Riordan, O., and Warnke, L. 2011. Explosive percolation is continuous. *Science*, **333**(6040), 322–324.

Riordan, O., and Wormald, N. 2010. The diameter of sparse random graphs. *Combin., Probab. & Comput.*, **19**(5-6), 835–926.

Ross, S. M. 1996. *Stochastic processes*. Second edn. Wiley Series in Probability and Statisticss. John Wiley and Sons.

Ruciński, A., and Wormald, N. 2002. Connectedness of graphs generated by a random d-process. *J. Aust. Math. Soc.*, **72**(1), 67–85.

Rudas, A., Tóth, B., and Valkó, B. 2007. Random trees and general branching processes. *Random Structures Algorithms*, **31**(2), 186–202.

Rudin, W. 1987. *Real and complex analysis*. McGraw-Hill.

Rudin, W. 1991. *Functional analysis*. International Series in Pure and Applied Mathematics. McGraw-Hill.

Rybarczyk, K. 2011. Diameter, connectivity, and phase transition of the uniform random intersection graph. *Discrete Math.*, **311**(17), 1998–2019.

Salez, J. 2013. Weighted enumeration of spanning subgraphs in locally tree-like graphs. *Random Structures Algorithms*, **43**(3), 377–397.

Schuh, H.-J., and Barbour, A. D. 1977. On the asymptotic behaviour of branching processes with infinite mean. *Adv. Appl. Probab.*, **9**(4), 681–723.

Seneta, E. 1973. The simple branching process with infinite mean. I. *J. Appl. Probab.*, **10**, 206–212.

Seneta, E. 1974. Regularly varying functions in the theory of simple branching processes. *Adv. Appl. Probab.*, **6**, 408–420.

Shannon, C. E. 1948. A mathematical theory of communication. *Bell System Tech. J.*, **27**, 379–423, 623–656.

Shepp, L. A. 1989. Connectedness of certain random graphs. *Israel J. Math.*, **67**(1), 23–33.

Shore, J. E., and Johnson, R. W. 1980. Axiomatic derivation of the principle of maximum entropy and the principle of minimum cross-entropy. *IEEE Trans. Inform. Theory*, **26**(1), 26–37.

Simon, H. A. 1955. On a class of skew distribution functions. *Biometrika*, **42**, 425–440.

Singer, K. 1996. *Random intersection graphs*. ProQuest LLC, Ann Arbor, MI. PhD Thesis, The Johns Hopkins University.

Smythe, R., and Mahmoud, H. 1994. A survey of recursive trees. *Teor. Ĭmovĭr. Mat. Statist.*, 1–29.

Snijders, T. A., Pattison, P., Robbins, G., and Handcock, M. 2006. New specifications for exponential random graph models. *Sociological Methodology*, **36**(1), 99–153.

Söderberg, B. 2002. General formalism for inhomogeneous random graphs. *Phys. Rev. E*, **66**(6), 066121, 6.

Söderberg, B. 2003a. Properties of random graphs with hidden color. *Phys. Rev. E*, **68**(2), 026107, 12.

Söderberg, B. 2003b. Random graph models with hidden color. *Acta Phys. Polonica B*, **34**, 5085–5102.

Söderberg, B. 2003c. Random graphs with hidden color. *Phys. Rev. E*, **68**(1), 015102, 4.

Sönmez, E. 2021. Graph distances of continuum long-range percolation. *Braz. J. Probab. Statist.*, **35**(3), 609–624.

Spencer, J., and Wormald, N. 2007. Birth control for giants. *Combinatorica*, **27**(5), 587–628.

Stark, D. 2004. The vertex degree distribution of random intersection graphs. *Random Structures Algorithms*, **24**(3), 249–258.

Stegehuis, C., van der Hofstad, R., and van Leeuwaarden, J. S. H. 2016a. Epidemic spreading on complex networks with community structures. *Sci. Rep.*, **6**, 29748.

Stegehuis, C., van der Hofstad, R., and van Leeuwaarden, J. S. H. 2016b. Power-law relations in random networks with communities. *Phys. Rev. E*, **94**, 012302.

Sundaresan, S., Fischhoff, I., Dushoff, J., and Rubenstein, D. 2007. Network metrics reveal differences in social organization between two fission–fusion species, Grevy's zebra and onager. *Oecologia*, **151**(1), 140–149.

Turova, T. S. 2011. The largest component in subcritical inhomogeneous random graphs. *Combin., Probab. Comput.*, **20**(01), 131–154.

Turova, T. S., and Vallier, T. 2010. Merging percolation on $\mathbf{Z}^d$ and classical random graphs: phase transition. *Random Structures Algorithms*, **36**(2), 185–217.

Ugander, J., Karrer, B., Backstrom, L., and Marlow, C. 2011. *The anatomy of the Facebook social graph*. arXiv:1111.4503 [cs.SI].

Vadon, V., Komjáthy, J., and van der Hofstad, R. 2019. A new model for overlapping communities with arbitrary internal structure. *Applied Network Science*, **4**(1), 42.

Voitalov, I., van der Hoorn, P., van der Hofstad, R., and Krioukov, D. 2019. Scale-free networks well done. *Phys. Rev. Res.*, **1**(3), 033034.

Wang, D., Song, C., and Barabási, A. L. 2013. Quantifying long-term scientific impact. *Science*, **342**(6154), 127–132.

Wang, J., Mei, Y., and Hicks, D. 2014. Comment on "Quantifying long-term scientific impact". *Science*, **345**(6193), 149.

Wang, M., Yu, G., and Yu, D. 2008. Measuring the preferential attachment mechanism in citation networks. *Physica A: Statist. Mech. Appli.*, **387**(18), 4692 – 4698.

Wang, M., Yu, G., and Yu, D. 2009. Effect of the age of papers on the preferential attachment in citation networks. *Physica A: Statist. Mech. Applic.*, **388**(19), 4273–4276.

Wasserman, S., and Pattison, P. 1996. Logit models and logistic regressions for social networks. *Psychometrika*, **61**(3), 401–425.

Watts, D. J. 1999. *Small worlds. The dynamics of networks between order and randomness*. Princeton Studies in Complexity. Princeton University Press.

Watts, D. J. 2003. *Six degrees. The science of a connected age*. W. W. Norton & Co.

Watts, D. J., and Strogatz, S. H. 1998. Collective dynamics of 'small-world' networks. *Nature*, **393**, 440–442.

Wong, L.H., Pattison, P., and Robins, G. 2006. A spatial model for social networks. *Physica A: Statist. Mech. Applic.*, **360**(1), 99–120.

Wormald, N. 1981. The asymptotic connectivity of labelled regular graphs. *J. Combin. Theory Ser. B*, **31**(2), 156–167.

Wormald, N. 1999. Models of random regular graphs. Pages 239–298 of: *Surveys in combinatorics, 1999 (Canterbury)*. London Math. Soc. Lecture Note Series, vol. **267**. Cambridge University Press.

Yukich, J. E. 2006. Ultra-small scale-free geometric networks. *J. Appl. Probab.*, **43**(3), 665–677.

Yule, G. U. 1925. A mathematical theory of evolution, based on the conclusions of Dr. J. C. Willis, F.R.S. *Phil. Trans. Roy. Soc. London, B*, **213**, 21–87.

Zhao, Y., Levina, E., and Zhu, J. 2012. Consistency of community detection in networks under degree-corrected stochastic block models. *Ann. Statist.*, **40**(4), 2266–2292.

Zuev, K., Boguná, M., Bianconi, G., and Krioukov, D. 2015. Emergence of soft communities from geometric preferential attachment. *Sci. Rep.*, **5**, 9421.

INDEX

Printed in the United States
by Baker & Taylor Publisher Services